Heat Shock Proteins

Volume 8

Series editors
Alexzander A.A. Asea, Ph.D., Professor and VD for Research Innovations, Deanship for Scientific Research, University of Dammam, Dammam, Saudi Arabia
Stuart K. Calderwood, Department of Radiation Oncology, Molecular & Cellular Radiation Oncology, Harvard Medical School, Beth Israel Deaconess Medical Center, Boston, Massachusetts, USA

Heat Shock Proteins: key mediators of Health and Disease. Heat shock proteins (HSP) are essential molecules conserved through cellular evolution required for cells to survive the stresses encountered in the environment and in the tissues of the developing and aging organism. These proteins play the essential roles in stress of preventing the initiation of programmed cell death and repairing damage to the proteome permitting resumption of normal metabolism. Loss of the HSP is lethal either in the short-term in cases of acute stress or in the long-term when exposure to stress is chronic. Cells appear to walk a fine line in terms of HSP expression. If expression falls below a certain level, cells become sensitive to oxidative damage that influences aging and protein aggregation disease. If HSP levels rise above the normal range, inflammatory and oncogenic changes occur. It is becoming clear that HSP are emerging as remarkably versatile mediators of health and disease. The aim of this series of volumes is to examine how HSP regulation and expression become altered in pathological states and how this may be remedied by pharmacological and other interventions.

More information about this series at http://www.springer.com/series/7515

Robert M. Tanguay • Lawrence E. Hightower
Editors

The Big Book on Small Heat Shock Proteins

Springer

Editors
Robert M. Tanguay
Department of Molecular Biology, Medical Biochemistry & Pathology Medical School Lab Cell & Developmental Genetics IBIS
Université Laval
Québec, QC, Canada

Lawrence E. Hightower
Department of Molecular and Cell Biology
University of Connecticut
Storrs, CT, USA

ISSN 1877-1246 ISSN 1877-1254 (electronic)
Heat Shock Proteins
ISBN 978-3-319-16076-4 ISBN 978-3-319-16077-1 (eBook)
DOI 10.1007/978-3-319-16077-1

Library of Congress Control Number: 2015940537

Springer Cham Heidelberg New York Dordrecht London

Printed on acid-free paper

Springer International Publishing AG Switzerland is part of Springer Science+Business Media (www.springer.com)

Preface

This book is based on a workshop entitled "The Small HSP World" and had the mission to bring together investigators studying small heat shock proteins (sHSPs). It was held at Le Bonne Entente in Quebec City (Quebec, Canada) from October 2 to October 5, 2014. Forty-four scientists from 14 different countries participated in the workshop sponsored by the Cell Stress Society International (CSSI). The small number of participants stimulated interesting discussions and the resulting informal atmosphere was appreciated by everybody. A brief review of this meeting appeared in Volume 20.2 of *Cell Stress and Chaperones.*

There are twenty-five chapters in this book. The chapters are from the best researchers working in this field. These include AP Arrigo (Lyon), JLP Benesch (Oxford), IJ Benjamin (Wisconsin), J Buchner-M Haslbeck- S Weinkauf (TUM, Garching), R Benndorf (Ann Arbor), WC Boelens (Nijmegen), S Carra , Z Chang (Beijing), RW Currie (Halifax), H Ecroyd (Wollongong), C Emanuelsson (Lund), X Fu (Beijing), C Garrido (Dijon), N Golenhofen (Ulm), NB Gusev (Moscow), LE Hightower (Storrs), HH Kampinga (Groningen), JN Lavoie (Québec), TH MacRae (Halifax), RA Quinlan (Durham), RM Tanguay (Québec), E Vierling (Amherst), M Toth-L Vigh (Szeged), SD Weeks (Leuven), T Wu (Wuhan), X Fu (Beijing) and their collaborators and colleagues. Briefly, the book starts with the structure and dynamics of small heat shock proteins, moving to their molecular and cellular functions as chaperones and regulators of cellular processes and finishing with their involvement in diseases. Although this is quite broad, the structural aspects of sHSP complexes are the unifying theme of the book.

Québec, QC, Canada — Robert M. Tanguay
Storrs, CT, USA — Lawrence E. Hightower

Contents

Part I
General Introduction

Chapter 1
The Multicolored World of the Human HSPB Family

Harm H. Kampinga, Romy de Boer, and Nico Beerstra

Abstract Small heat shock proteins (sHSPs) have been studied for 40 years now, initially as proteins up-regulated upon heat stress in which they can help cells to better withstand or recover from heat damage. In humans, we now know that there is a 'colorful' family of ten different sHSP members (referred to as HSPB1-HSPB10) that together seem to fulfill many more, only partially overlapping, functions than just protecting cells from heat damage. Here, we summarize how the ten human HSPB members are thought to contribute to the protein quality control in cells and how they may do so via different mechanisms. Also, we will summarize what is known to date about the role of individual HSPB members in neurodegenerative diseases and in cardiomyopathies.

Keywords HSPB oligomerization • HSPB chaperone activity • HSPB cytoskeletal protection • Acute stress • Chronic diseases • Neurodegeneration • Cardiac diseases

1.1 Introduction

Small heat shock proteins (sHSPs) were first discovered as proteins that were up-regulated after a heat shock treatment together with several other HSPs, including Hsp70s and Hsp40s (Tissières et al. 1974) . Whereas not all bacteria have sHsps, most eukaryotic genomes do contain multiple sHSP genes ranging from 2 in yeast to over 19 in *Arabidopsis thaliana* (Kappé et al. 2003, 2010; Kriehuber et al. 2010; Kampinga et al. 2009). There has been some dispute about the number of human small HSPs (Kampinga et al. 2009), but given that the presence of an alpha-crystallin domain documents the evolutionary relatedness of the sHSPs (Kappé et al. 2010) it seems justified to conclude there are only ten, here further referred to as HSPB1-HSPB10.

H.H. Kampinga (✉) • R. de Boer • N. Beerstra
Department of Cell Biology, University Medical Center Groningen, University of Groningen, Groningen, The Netherlands
e-mail: h.h.kampinga@umcg.nl

R.M. Tanguay, L.E. Hightower (eds.), *The Big Book on Small Heat Shock Proteins*, Heat Shock Proteins 8, DOI 10.1007/978-3-319-16077-1_1

Information on the human HSPB structure and function has mainly been based on cell free in vitro data and on cell biological data with the stress-inducible human HSPB1 and HSPB5 or HSP members from other organisms. These data have revealed that these members can act as ATP-independent molecular chaperones (Jakob et al. 1993; Kampinga et al. 1994; Richter et al. 2010) that block the aggregation of un- or misfolded proteins, and that they can protect cytoskeletal integrity or assist in cytoskeletal recovery upon stress, both functions potentially contributing to the increased survival of cells when exposed to stress-conditions that impede protein homeostasis and/or disrupt the cytoskeleton (Kampinga and Garrido 2012). Although all human HSPB family share the – above mentioned- α-crystallin domain of about 80 amino acid (Kampinga and Garrido 2012; Caspers et al. 1995), which likely represents the domain that is most important for some structural and functional characteristics of these proteins, there is little homology outside this domain, suggesting the individual members may also be functionally distinct.

In this chapter, we will summarize what is known about the similarities and differences between the ten human HSPB members and highlight how different members may contribute to the protein quality control in cells via different mechanisms. Also, we will summarize what is known to date about the ability of individual HSPB members to suppress toxicity and aggregation in neurodegenerative diseases and how they may play a role in cardiomyopathies. As several of the items will be dealt with in high detail elsewhere in this book, here we have tried to provide a comparative overview on main features only. This thus implies that we had to generalize and simplify at several places, thus evidently sometimes leading to omission of important details. More specific information and citations to original findings can be found in the several excellent reviews we refer to throughout this book chapter.

1.2 Structural Features and Chaperone Activity of Purified HSPBs In Vitro

1.2.1 HSPB Oligomerization

Dynamic oligomerization has been suggested as an important feature for the functionality of canonical members like HSPB1 and HSPB5. In general, small HSPs are very polydispersed, meaning that they can exist as dimers (considered the "building block") and heterologous homo- or hetero-oligomers of up to 50 subunits. Oligomerization is a very dynamic process and subunit exchange can be accelerated by stresses such as heat or low pH. Within living cells, posttranslational modifications (in particular phosphorylations) are thought to further stimulate oligomeric changes and this is believed to enable the sHSP to respond very specifically different stress stimuli. Several HSPB members contain phosphorylation sites and phosphorylation usually leads to de-oligomerization, whereas de-phosphorylation enhances oligomerization (Arrigo and Gibert 2012; Lanneau et al. 2008; Parcellier et al. 2005; Rogalla et al. 1999). The formation of different structures could provide a mechanism for regulating the sHSP binding activity and intracellular (re)distribution (van Montfort et al. 2001; Garrido et al. 2012; Boncoraglio et al. 2012; Acunzo et al. 2012).

However, not all HSPBs may form oligomers inside cells. The C-terminal region of most human HSPB members contains a conserved triad motif, IXI/V (Ile-Xaa-Ile/Val), in which Xaa is typically a proline residue that participates in the stabilization of the high molecular weight oligomers (Garrido et al. 2012; de Jong et al. 1993). HSPB6 and HSPB8 do not contain this conserved IXI/V motif and are mainly found in low molecular weight oligomers or complexed as dimers to other chaperones (Carra et al. 2012). HSPB7 does contain this motif and forms oligomers in vitro (Yang et al. 2011). Also other data show that purified HSPB7 can oligomerize and even form hetero-oligomers with HSPB8 (Sun et al. 2004). When expressed in cells, however, both HSPB7 (Vos et al. 2010) and HSPB8 (Carra et al. 2008) mainly seem to exist as dimers, suggesting that in cells other partners may exist that prevent these HSPB members from oligomerizing. For HSPB2 and HSPB3, it has been found that they form well-defined hetero-oligomers in a unique 3 to 1 subunit ratio (den Engelsman et al. 2009). Finally, no information exists regarding the oligomeric structure of HSPB9 and HSPB10 so far.

1.2.2 HSPB Chaperone-Like Activity

Using purified proteins, it has been shown that several canonical HSPB members are able to bind to (partially) unfolded proteins. As a result, these clients are maintained in a semi-soluble state from which they can be recovered. As all HSPBs lack ATPase activity, client release from HSPBs generally requires the help of ATP-dependent chaperones (HSP70) or proteases (HSP104) that further process the client (Haslbeck 2015). Whereas in vitro this supports refolding of the client, in cells HSPB1-bound substrates might also undergo ubiquitinylation and (via HSP70 or by other means) be released to be degraded by the proteasome or via autophagy (Carra 2015).

Extensive comparison on the various HSPB members for their ability to act in this chaperone like manner is yet lacking. In Table 1.1, we have summarized what is known for four often-used model substrates and the ability of the HSPB members to maintain them soluble upon thermal or chemically-induced unfolding. For nearly all substrates, most members do seem to be capable of suppressing aggregation, suggesting that HSPBs all have the intrinsic potential to bind un- or misfolded clients in a rather promiscuous manner. These data with purified proteins also imply that the post-translational modifications as many of them undergo in living cells and the effect of these on the oligomeric dynamics of some of the HSPB members are not a pre-requisite for the chaperone-like activities of HSPBs per se. This all,

Table 1.1 Some characteristics of purified HSPBs

Name	Alt. name	Number of amino acid residues	Oligomerization		Ability to preventing thermal aggregation of citrate synthase, luciferase or alcohol dehydrogenase or of DTT-induced aggregation of insulin[c]			
			Homo	Hetero	Citrate synthase	Luciferase	Alcohol dehydrogenase	Insulin
HSPB1	HSP27	205	Yes	Yes	+	+	+	+
HSPB2	MKBP	182	Yes	With B3	+	NA	+	+
HSPB3	HSPL27	150	Yes	With B2	±	NA	+	−
HSPB4	CRYAA	173	Yes	Yes	+	+	+	+
HSPB5	CRYAB	175	Yes	Yes	+	+	+	+
HSPB6	HSP20	160	?	?[a]	NA	NA	+	±
HSPB7	cvHSP	170	?	?[b]	NA	NA	NA	+[d]
HSPB8	HSP22	196	?	?[b]	+	NA	+	+
HSPB9	CT51	159	NA	NA	NA	NA	NA	NA
HSPB10	ODF1	250	NA	NA	NA	NA	NA	NA

[a]Still debated
[b]In vitro, they can but in cells they may normally not do so
[c]See references (Jakob et al. 1993; Ehrnsperger et al. 1997; Prabhu et al. 2012; Asthana et al. 2012; Derham and Harding 1999; Mymrikov et al. 2011; Chowdary et al. 2004; Lin et al. 2014)
[d]Rat HSPB7

however, not necessarily implies that the different HSPBs may not have substrate specificity or at least show differential affinity for different substrates. Up to now, no systematic comparisons have been done where all members were compared at equal stoichiometry using the same substrates. Also, it is not clear what the impact of (equal) binding to client proteins will be for the subsequent fate of that protein and whether or not the HSPB-partners (e.g. ATP dependent chaperones for client release) may differ. At least our comparison in cells suggests that such differences do exist (see Sect. 1.4).

1.3 Human HSPB: Regulation and Expression in Cells and Tissues

Unlike their name suggests, most members of the HSPB family are not induced by heat shock, except for the canonical members HSPB1, HSPB4 and HSPB5; HSPB8 was found to be only moderately heat-inducible. In fact, the different members are expressed constitutively in several cells and tissues, albeit some more ubiquitously than others (Table 1.2) (Vos et al. 2009). In tissue culture cell lines, most HSPBs are present in the cytosolic compartment, sometimes associated with cytoskeletal elements. Upon heat stress, reallocation to stress fibers can be observed of HSPB1 and HSPB5. The same proteins are also found in SC35 speckles, RNA-containing granules in nuclei, where HSPB7 also is also present under normal growth conditions.

Table 1.2 HSPB expression in cells and tissues

Name	Heat shock inducible	Main tissue distribution	Intracellular localization under normal growth conditions	Intracellular localization after heat stress
HSPB1	Yes	Ubiquitous	Cytosol/cytoskeletal elements	Cytoskeleton/nuclear speckles
HSPB2	No	Muscle	Cytosolic granules	Mitochondria-associated
HSPB3	No	Muscle	Unknown	Unknown
HSPB4	Yes	Eye lens	Cytosol/cytoskeletal elements	Unknown
HSPB5	Yes	Ubiquitous	Cytosol/cytoskeletal elements	Cytoskeleton/nuclear speckles
HSPB6	No	Muscle, brain	Cytosol/cytoskeletal elements	Unknown
HSPB7	No	Muscle	Cytosol/nuclear speckles	Cytosol/nuclear speckles
HSPB8	Yes[b]	Muscle, brain	Cytosol	Cytosol/cytosolic stress granules
HSPB9	NA	Testis	Cytosol/nucleus	Unknown
HSPB10	NA	Testis	Sperm cell tails	Unknown

See references (Mymrikov et al. 2011; Vos et al. 2009; Nakagawa et al. 2001; den Engelsman et al. 2013) for more details and original references

[a]Moderately and maybe restricted to certain cell lines or under certain conditions

HSPB8 is often found associated with cytosolic, RNA stress granules upon stress and for HSPB2 it was suggested that it becomes associated with mitochondria upon heat shock (Table 1.2). The association of several members with the cytoskeletal elements (before or after stress) may be consistent with their role in protecting or repairing the cytoskeleton upon stress and as a care-taker function at sarcomeric structures in skeletal and heart muscles, in which they are so highly expressed (Table 1.1; see Sect. 4.2.2). Whether or not HSPB2 has a role in mitochondrial stabilization is not known and neither is the role of the association of some HSPB members with RNA-containing compartments like the SC35 speckles in the nucleus (HSPB1, HSPB5 or HSPB7) or with cytoplasmatic stress granules (HSPB8).

1.4 Human HSPB: Functional Diversity In Vivo (Cell Lines, Animal Models)

Investigations on comparing the different functions of the various HSP members are complicated by the fact that their constitutive expression can vary dramatically between different cell types (Table 1.2) (Vos et al. 2009) and that several members can also hetero-oligomerise (Table 1.1); this implies that any modulation of one HSPB member expression may thus have differential impact depending on the cell type used. Yet, below we have tried to provide an overview of phenotypes induced by altered expression of single HSPB members as reported in the literature in relation to their role in resistance to acute stress resistance and their roles in disease. In fact, only our own lab has systematically compared the effects of overexpression of single HSPB members on the resistance towards several forms of stresses within one and the same cells line. This, of course, also has several limitations as these effects may not translate to other cell types and especially findings of the lack of an effect by the modulation of a single HSPB member are largely inconclusive. However, we believe such systematic comparison does provide some insight in possible differences in client specificity between the various HSPB members and differences in their mechanism of action. In the summarizing Table 1.3, we therefore distinguished the data from our own studies within one cell line from those observed in the literature, often done with different cell lines.

1.4.1 Protection of HSPB8 Members Against Acute Stress Conditions

HSPBs have been implied in cellular resistance against many forms of acute stress. Besides heat shock, these stressors include various forms of oxidative stress and several anti-cancer agents. The role of HSPBs in the oxidative stress has been linked to the ability to modulate the redox status in cells, which may have particular relevance for cardiac diseases (Christians et al. 2012). More details on this can be found elsewhere in this book, but since it is yet unclear how to link this functionality of the HSPBs to their chaperone or cytoskeletal related functions, it will not be further discussed here. The same applies to resistance to anti-cancer drugs, for which in several cases the mechanism of the HSPB-mediated protection is yet poorly understood. Here, we will focus on heat shock as the prototype for an acute proteotoxic form of stress.

1.4.1.1 Cytoskeletal Protection

The first demonstration that the overexpression of a single HSPB member (HSPB1) could protect cells from the lethal effects of a heat shock was made by the group of Jacques Landry (Landry et al. 1989). Next, this cell line was found to recover more rapidly from heat-induced cytoskeletal damage (Lavoie et al. 1993), consistent with the above mentioned association of HSPB1 with the cytoskeleton upon stress (Mounier and Arrigo 2002). Other members that have been shown to exert (protective) effects on the cytoskeletal stability under acute forms of stress or e.g. during differentiation include HSPB2, HSPB3, HPSB4, HSPB5, HSPB6, HSPB9, and likely also HSPB7 (Table 1.3). Here, HSPB1 seems rather promiscuous, whilst some HSPB members may reveal some specificity with regards to the type of cytoskeletal elements they can stabilize (Table 1.3) (Boncoraglio et al. 2012; Carra et al. 2012; Mymrikov et al. 2011; Mounier and Arrigo 2002; Sun and MacRae 2005; Vos et al. 2008; Sugiyama et al. 2000; Suzuki et al. 1998; Wettstein et al. 2012).

Table 1.3 Protective effects of HSPB members in against acute and chronis, disease related stresses (see text for references)

	Protection against acute stress			Protective role in chronic diseases																	
		Chaperone-like activity		Neurodegenerative diseases (anti aggregations)															Non-neurodegenerative diseases		
		Other	Luc refolding	CAG			ALS TDP43	ALS SOD1		PD α-synuclein		PD Parkin	AD extracellular Aβ			AD tau			DCM HSPB5mt	AF	Disease related to
Name	Cytoskeletal stability	C	C*	C	C*	V	C*	C	V	C	V	C*	CF	C	V	C	V		C/C*	C*	HSPB mutations
HSPB1	Actin, intermediate filaments, microtubules	Yes	++	O	0	0 M	0	+/–	+/–	+	NA	+++	+	+	+	+	+		++	+++	Neuropathy
HSPB2	Myofibrils	MDPK	0	NA	+/–	NA	NA	NA	NA	NA	NA	+++	0	NA	NA	NA	NA		0	0	ND
HSPB3	Myofibrils	MDPK	0	NA	0	NA	NA	NA	NA	NA	NA	0	0	NA	NA	NA	NA		0	0	Neuropathy
HSPB4	Intermediate filaments, microtubules	Yes	+	NA	+	NA	NA	NA	NA	NA	NA	++	NA	NA	NA	NA	NA		++	0	Cataract
HSPB5	Intermediate filaments, microtubules	Yes	+	0	0	0	0	+/–	0	+/–	+ Dm	0	+	–	NA	NA	NA		++	0	Cataract myopathy cardiomyopathy
HSPB6	Actin	NA	+/–	NA	++	NA	+	NA	NA	NA	NA	0	NA	NA	NA	NA	NA		+	++	Cardiomyopathy?
HSPB7	Sarcomeres?	NA	0	NA	+++	++ Dm	+	NA	NA	NA	NA	+++	NA	NA	NA	NA	NA		0	++	Cardiomyopathy?
HSPB8	NA	NA	0	+	+	+ Dm	++	++	NA	NA	NA	0	?	?	NA	NA	NA		+	0	Neuropathy
HSPB9	NA	NA	0	NA	++	+ Dm	+	NA	NA	NA	NA	+	NA	NA	NA	NA	NA		0	0	ND
HSPB10	Flagellar axoneme in sperm	NA	0	NA	0	NA	NA	NA	NA	NA	NA	0	NA	NA	NA	NA	NA		0	0	ND

C = cellular models; C*: comparison done within one cell type in our laboratory; CF = cell free experiments, using purified proteins; V = in vivo, using either mice (*M*) or Drosophila (*Dm*)

0 = no effect; +/– = minor effect or contradictory results; + mild protection; ++ clear protection; +++ = strong protection; – = enhanced effect; NA = not analysed; ND = so far, no mutants detected

1.4.1.2 Cellular Chaperone Activities

In parallel to being protected against heat-induced cytoskeletal damage, the HSPB1 overexpressing lines developed in the Landry group, were found to show a more rapid recovery from heat-induced protein aggregation (Kampinga et al. 1994). Subsequently, HSPB1 overexpression was found to increase the cellular capacity to refold the ectopically expressed model protein firefly luciferase, an action that was dependent on a functional Hsp70 machine (Bryantsev et al. 2007). These data actually formed the basis for and were consistent with the above-mentioned model of chaperone activity of sHSPs (Jakob et al. 1993; Kampinga et al. 1994; Haslbeck 2015; Cohen and Dillin 2008). However, as a note of caution, until today there is no full evidence that these effects can be entirely uncoupled from possible indirect effects related to the cytoskeletal stabilizing actions of the HSPBs. It can be envisioned that by virtue of cytoskeletal stabilization, the other chaperone machines have less substrates to deal with and thus can better handle more or gain access to heat-unfolded proteins.

1.4.2 Protective Effects of HSPB Members in Chronic Disease Conditions

1.4.2.1 Neurodegenerative Diseases

Many different neurodegenerative diseases share features of protein aggregates in the neurons of patients and several mechanistic links exist between the aging process and toxic protein aggregation (Cohen and Dillin 2008). Improving protein quality control has therefore been a goal of many investigations and here we will summarize those that have been aimed at doing so via manipulating the expression or activity of HSPB members. Whereas the primary goal of such research usually is to identify potential disease modifiers, we will here also use it to illustrate potential differences in substrate preferences and modes of actions of the HSPBs. What these data also will illustrate is that not all aggregation-prone proteins that cause disease give rise to similar (types of) aggregates and that they may require different handling by the protein quality control system. However, the latter will not be addressed in detail.

CAG Repeat Diseases

CAG repeat diseases (or polyglutamine (polyQ) diseases) are caused by expansions of CAG trinucleotide repeats within the coding region of several genes, resulting in the production of proteins with an expanded polyglutamine stretch (Bauer and Nukina 2009). Until now, nine polyQ diseases have been described: spinocerebellar ataxias (SCA) types 1, 2, 6, 7, 17, Machado-Joseph disease (MJD/SCA3),

Huntington's disease (HD), dentatorubral pallidoluysian atrophy (DRPLA), and spinal bulbar muscular atrophy X-linked type 1 (SMAX1/SBMA) (Matos et al. 2011; Takahashi et al. 2010). Except for SMAX1/SBMA, all of these diseases are autosomal dominantly inherited. The expansion in different disease-related genes results in an aggregation-prone expanded polyglutamine (Q) tract and aggregation propensity is strongly related to the CAG repeat length. Likewise, the age at which these diseases become clinically manifest (age of onset: (AO)) is inversely correlated to the CAG repeat length (Gusella and MacDonald 2000). This, together with many other lines of evidence including findings that several polyQ knockout mice do not develop polyQ diseases (Bauer and Nukina 2009; Gatchel and Zoghbi 2005; Morfini et al. 2005), implies that the disease is due to a gain of toxic function driven or initiated by protein aggregation, although both the types of aggregates that lead to toxicity as well as the mechanisms leading to neuronal dysfunction are still heavily disputed. In this chapter, the emphasis will be on HD and SCA3, which have the highest prevalence of polyQ diseases worldwide (Fan et al. 2014), but given the generic mode of action by which CAG-repeat diseases are initiated, the concepts shown likely apply to all polyQ diseases.

There are many cellular and animal models for CAG repeat diseases. The models typically are based on the expression of (a fragment of) the repeat-containing gene in cells or animals followed by the (semi) quantitative analysis of aggregation and toxicity readouts. Using these models, most emphasis with regards to putative chaperone-mediated rescues has been on the ATP-dependent chaperones (Kakkar et al. 2014). For this chapter, we will summarize what has been done so far with HSPB members (Table 1.3). Also here, the initial focus was on the classical HSPB1 and HSPB5 and the results were disappointing as overexpression of neither one of them could decrease aggregation in cell models, albeit that HSPB1 seemed to provide some level of protection against neuronal cell death (Wyttenbach et al. 2002; Carra et al. 2005), suggesting it may either compensate for or protect against some of the consequences of the aggregates. Also, in vivo neither one of these two members seemed to be effective in reducing aggregation or disease symptoms. Whereas, some data in Drosophila polyQ models suggested some suppression of aggregation by HSPB5 (Tue et al. 2012), other data with another fly model were negative (Robertson et al. 2010). Also in a polyQ mouse model, these HSPB members were found to be ineffective (Zourlidou et al. 2007).

In a comparative study done in our lab, using the same HEK293 cells as for the luciferase refolding assay, the ineffectiveness of HSPB1 and HSPB5 to prevent polyQ aggregation was confirmed (Vos et al. 2010). Using a series of experiments with polyQ proteins with different expansions, we discovered that some other members actually could effectively suppress polyQ aggregation, the order of effectiveness being: HSPB7>HSPB9=HSPB6>HSPB8>HSPB4>HSPB2. Furthermore, the effectiveness of HSPB7 and HSPB9 (Vos et al. 2010; Zijlstra and Kampinga unpublished data) as well as that of HSPB8 (Carra et al. 2010) to prevent polyQ aggregation was confirmed in other cell lines and also extended to Drosophila in vivo models, where they could ameliorate polyQ mediated degeneration (Table 1.3).

Intriguingly, the anti-aggregation activities of the most effective polyQ suppressing HSPB members were associated with either enhanced proteolytic degradation of the soluble poly Q protein (HSPB2, HSPB4, HSPB6, HSPB9) (Zijlstra and Kampinga unpublished data; Yang et al. in preparation) or with autophagosomal clearance of the polyQ aggregates (HSPB7, HSPB8) (Vos et al. 2010; Carra et al. 2008, 2009). For HSPB8, this function was highly dependent on its association with BAG3, a HSP70 co-chaperone (Carra et al. 2008; see Carra 2015 for detailed information). The effect of HSPB7 was clearly distinct from that of HSPB8 and would be consistent with a model whereby HSPB7 prevents polyQ fibrils to grow beyond a size that can be handled by the autophagic machinery (Vos et al. 2011). However, this needs further elucidation. Yet, it is striking to see that all the polyQ suppressing HSPB members were ineffective in assisting the cells to refold heat-denatured luciferase (Table 1.3), which could be a matter of differential substrate specificity or differential substrate handling, e.g. related to partnerships that these HSPB members may have with other protein quality control components. Also this speculation, however, needs further testing.

Amyotrophic Lateral Sclerosis (ALS)

Amyotrophic lateral sclerosis (ALS) is a lethal neurodegenerative disorder and also known as a conformational disease in which misfolding and aggregation of proteins such as SOD1 and TDP-43 are central features (Johnson et al. 2009; Kerman et al. 2010). Most cases of ALS are sporadic (SALS), but about 10 % of the ALS cases is inherited, usually dominantly (familial ALS (FALS)) (Sreedharan and Brown 2013). Multiple genetic mutations have been associated with FALS, including genes encoding for TDP-43, SOD1, SQSTM1, C9ORF72, VCP, OPTN, FUS, PFN1, and UBQLN2. For HSPB related protection, so far only data exist on some SOD1 and TDP-43 mutants and these will be discussed below (Table 1.3) (see also (Carra et al. 2013) for a more detailed overview).

HSPB *Effects on SOD1 Mutants*

Population-based studies show that mutations in the SOD1 gene account for ~12 % of familial cases and ~1 % of sporadic cases (Renton et al. 2014). More than 160 mutations in SOD1 are exclusively associated with ALS. Although ALS was initially thought to be due to a (partial) loss of normal SOD1 function, which is the detoxification of superoxide (Sreedharan and Brown 2013), several lines of evidence, including knockout mice that do not show an ALS phenotype (Mulligan and Chakrabartty 2013; Sreedharan et al. 2008), now also demonstrate that it is due to a gain of toxic function, associated with the unusual propensity of mutant SOD1 to misfold and aggregate (Mulligan and Chakrabartty 2013; Talbot 2014).

Only a limited amount of data is available regarding HSPB mediated protection against mutant SOD1 aggregation. In cell free systems, both purified HSPB1 and HSPB5 were found to be able to bind to and suppress the aggregation of SOD mutants (Patel et al. 2005; Yerbury et al. 2013). In cells, however, HSPB1 overexpression did result in some protective effects against aggregate related toxicity, but it only mildly reduced (An et al. 2009) or did not reduce (Patel et al. 2005) mutant SOD1 aggregation. In line with these cellular data, overexpression of HSPB1 showed no (Krishnan et al. 2008) or only a very mild (Sharp et al. 2008) protective effect during early stages of disease in SOD-mice. Inversely, HSPB5 knockout mice show no enhanced degeneration when crossed with SOD1 mouse models (Yerbury et al. 2013). The only other member investigated in mutant SOD1 models is HSPB8. In cell models, HSPB8 overexpression strongly reduced mutant SOD1 aggregation, an effect that was associated with enhanced degradation of the mutant but not wildtype SOD, implying that HSPB8 selectively recognized and degraded the misfolded conformation (Crippa et al. 2010). Again, functional interaction with BAG3 was important for the actions of HSPB8 on mutant SOD (Crippa et al. 2010) (Table 1.3). So, far no data of HSPB8 in in vivo SOD-1 models have been reported.

HSPB *Effects on TDP-43 Mutants*

TDP-43 is an ubiquitously expressed DNA/RNA binding protein with several roles, including gene transcription, RNA splicing, RNA shuttling and translation, and microRNA biogenesis (Sreedharan and Brown 2013). TDP-43 mutations are linked to sporadic and non-SOD1 familial ALS (Johnson et al. 2009) and up to 40 autosomal dominant missense mutations in the TDP-43 gene exist that are related to FALS. All of these mutations, except one, reside in the C-terminal glycine-rich domain, which is involved in ribonucleoprotein binding and splicing (Renton et al. 2014).

So far only two studies have investigated effects of HSPB members on TDP-43 aggregation. Crippa et al. showed that HSPB8 reduced aggregation of TDP-43 in the same cell model in which they also found effects on mutant SOD aggregation (Crippa et al. 2010). We confirmed the effectiveness of HSPB8 on reducing mutant TDP-43 aggregation in the HEK293 cells that were also used for all our other studies on HSPBs; in this study, we also found that HSPB1, HSPB5, HSPB7 were rather ineffective, whilst HSPB9 had only had a minor effect (Carra et al. 2013) (Table 1.3).

Together, the data on both mutant ALS and mutant TDP-43 suggest that HSPB8 is the most potent protector, although this requires substantiation, especially using in vivo models. This is clearly distinct from what is seen for polyQ aggregation (Table 1.3), which suggests that, albeit that both diseases are caused by toxic aggregates, the disease-causing mutants require different handling by the protein quality control system. Inversely, these data further underline the functional difference between the HSPB members.

Parkinson's Disease (PD)

Parkinson's Disease (PD) is the second most common neurodegenerative disease with a wide range of motoric and non-motoric symptoms. PD is mostly hallmarked by the presence of Lewy bodies, aggregates consisting of the 14 kDa α-synuclein protein, inside neuronal cells (Wakabayashi et al. 2013). Although thought to be mostly sporadic, both PD-causative dominant mutations and recessive mutations have been found to cause PD (Trinh and Farrer 2013). The dominant mutants include mutations in the SNCA gene that encodes α-synuclein and the Leucine-rich repeat kinase 2 (LRRK2). Interestingly, wildtype α-synuclein is an intrinsically disordered protein (Solanki et al. 2014) and the SNCA mutations linked to PD either lead to abnormally high expression or further destabilisation of the structure of α-synuclein, both increasing its risk to initiate aggregation. So, comparable to CAG repeat diseases and ALS, also in PD a toxic gain of function mechanisms seems to apply in which α-synuclein aggregation seems to be a central feature.

HSPB *Effects on α-Synuclein*

Using purified wildtype α-synuclein that, due to its disordered structure, already has a high propensity to aggregate, it was found that all HSPB members tested so far (HSPB1, HSPB2, HSPB3, HSPB5, HSPB6 and HSPB8) could reduce its aggregation, with HSPB6 having only weak activity (Prabhu et al. 2012; Bruinsma et al. 2011). Although several cell and in vivo models of PD are readily available, surprisingly little is known about the effect of sHSPs in such models. In a cell model for α-synuclein, HSPB1 overexpression reduced aggregation and toxicity, whereas HSPB5 overexpression had no significant effect on aggregation and only marginally protected against α-synuclein-induced cytotoxicity (Outeiro et al. 2006). However, whether this indeed suggests that HSPB1 is more potent in case of preventing α-synuclein aggregation is unclear and may also be due to differential transgene expression. Zourlidou et al. (2004) also suggested that HSPB1 is protective against alpha synuclein mediated cell death, but here no data on aggregation were provided, meaning that no conclusion can be made regarding the mode of protection (chaperoning α-synuclein versus reducing consequences of its aggregation). The only member tested in vivo, using a PD model in drosophila, is HSPB5 (Tue et al. 2012). Here, HSPB5 was found to suppress eye degeneration induced by α-synuclein, but again no information on aggregation of α-synuclein was reported, not allowing distinguishing between direct or indirect effects. So, although little can be stated up to date about possible α-synuclein chaperoning effects by the various HSPB members (Table 1.3) yet several indirect lines of evidence suggest that HSPBs may play a role in α-synucleinopathies (reviewed in Cox et al. 1842).

HSPB *Effects on the PARK2 C289G Mutant*

Dysfunction of degradation pathways is another hallmark in PD. An increasing body of evidence suggests that the ubiquitin-proteasome system and (especially) the autophagy-lysosome system are indeed impaired in PD. In line, there are a number of early-onset recessive parkinsonisms linked to mutations that lead to impaired quality control (Trinh and Farrer 2013), including several mutations in the E3 ligase parkin (also known as PARK2). Whereas the recessive inheritance of these parkinsonisms suggests loss of function, one of the PARK2 mutations, the C289G mutation that is located within the RING1 domain of PARK2, leads to PARK2 destabilization and aggregation (Gu et al. 2003). This PARK2 C289C mutant therefore may also exert dominant effects by a gained toxic (aggregation-related) function. In support of this, some evidence is accumulating that heterozygous carriers of this mutation have a higher risk for late onset PD (Khan et al. 2002, 2003; Hilker et al. 2001). The aggregates caused by this mutant in cells are amorphic, non-amyloidogenic and thus presumably quite distinct from those generated by α-synuclein and certainly polyQ proteins. This led us to test whether prevention of aggregation of this PARK2 C289G substrate would require different HSPB members than e.g. prevention of polyQ aggregation (Minoia et al. 2014). Except for the overlap of HSPB7, which protects against aggregation of both polyQ proteins and PARK2 C289G, this was indeed the case (Table 1.3) as overexpression of HSPB1, HSPB2, HSPB4 and HSPB7 protected against PARK2 C289G aggregation (Minoia et al. 2014), whereas in the same cell lines and using similar conditions, these were not effective in preventing polyQ aggregation (Carra et al. 2008) (Table 1.3). On one hand, this again could point to some differences in substrate specificity of the various HSPB members. On the other hand, a striking finding was that the protective actions of these HSPBs were not impaired upon inactivation of the ATP-dependent HSP70 chaperone machines (Minoia et al. 2014). A similar finding was also already made previously for the HSPB7-mediated protection against polyQ aggregation (Vos et al. 2010), but was in particular surprising for the canonical HSPB1 as it did require a functional HSP70 machine for its effects on facilitating refolding of heat-denatured luciferase (Sect. 1.4.1.2) (Bryantsev et al. 2007). In case of prevention of PARK2 C289G aggregation, all effective HSPB members depended on protein degradation pathways. This challenges the canonical model for HSPB-related chaperone actions and suggests that (1) they do act as chaperones also for clients like PARK2 C289G and interact with other client release factors or (2) that their protection on PARK2 C289G aggregation are mediated by indirect (cytoskeletal stabilizing) effects that allow other chaperones to deal with the misfolded PARK2 C289G. Anyhow, these data further illustrate that there are differences in the functionality of HSPB members and the manner by which they contribute to protein homeostasis.

Alzheimer's Disease (AD)

Alzheimer's disease (AD) is the most common form of dementia, affecting more than 35 million people worldwide. The pathological mechanisms of AD are very broad and complex and many factors are involved. However, the main pathological hallmarks of AD are the formation of extracellular amyloid plaques (Aβ) and of intracellular neurofibrillary tangles (hyperphosphorylated tau) that somehow, and maybe in a combined manner, are thought to ultimately lead to neurodegeneration (Ittner and Götz 2011; Giacobini and Gold 2013).

The mechanisms of Aβ toxicity are still unclear, but a hypothetic pathway of Aβ neurotoxicity is the induction of physical damage to cell structures by Aβ oligomers (not fibrils or monomers) leading to, membrane permeabilization, mitochondrial failure, causing oxidative stress and ultimately leading to synaptic failure (Querfurth and LaFerla 2010). The classical hypotheses assume that toxicity and degeneration is mediated by extracellular Aβ related to the extracellular plaques in post-mortem brain. However, there are data that suggest Aβ42 also accumulates intracellularly (Walsh et al. 2000; LaFerla et al. 2007; Zheng et al. 2013) and that the extracellular and intracellular pools are connected. Upon compromised degradation of Aβ, in proteasomes as well as lysosomes, this would next lead to accumulation of intracellular Aβ (Domert et al. 2014; Shimura et al. 2004). In such a scenario, elevated intracellular expression of HSPBs may thus also have therapeutic potential in AD.

HSPB *Effects on Aβ*

In cell free systems, HSPB1, HSPB5 and especially HSPB8 could suppress amyloid formation by Aβ, whereas a combination of HSPB2 and HSPB3 was ineffective. Prevention of aggregation was paralleled by reduced toxicity of the Aβ peptide when these were added to cells extracellularly (Wilhelmus et al. 2006a). However, in another study by the same group, co-incubation of HspB8 with Aβ (1–40) did, but with the more disease-associated Aβ (1–42) did not affect beta-sheet formation or Aβ-mediated cell death of cultured cerebrovascular cells (Wilhelmus et al. 2006b). An earlier study on HSPB5 showed that HSPB5 indeed also reduced aggregation, but increased (rather that decreased) toxicity when added together with Aβ to cells, suggesting that prevention of fibril formation by HSPB5 may lead to an increase of a more toxic non-fibrillar Aβ oligomer (Stege et al. 1999) as was indeed later found to be the case (Raman et al. 2005). For HSPB1, on the other hand, it has been found that substoichiometric amounts could sequester toxic Aβ oligomers and convert them into large nontoxic aggregate, thus abolishing the toxicity of Aβ oligomers on mouse neuroblastoma cells (Ojha et al. 2011). The same authors also showed that cultured neurons from HspB1-deficient mice were more sensitive to oligomer-mediated toxicity than those from wild-type mice; it is however, obscure how this intracellular depletion relates to the detoxification of toxic extracellular Aβ oligomers, and would suggest that indirect effects of HSPB1 depletion have rendered these neurons more sensitive to the extracellular Aβ oligomers. Yet, a study using an Aβ x HSPB1 double transgenic mouse model confirmed the suggestion that increased HSPB1 expression not only delayed several AD-related disease features, but also significantly reduced amyloid plaques formation in the brain (Tóth et al. 2013). Yet again, these data provide no information on whether this is due to direct HSPB1 mediated chaperoning of Aβ and how this may detoxify Aβ or due to indirect effects of HSPB1 overexpression.

So, whilst these data suggest that different HSPB members differently affect the formation of Aβ oligomers and fibrils (Table 1.3), outcomes in terms of neurotoxicity have so far been confusing and difficult to relate to these in vitro chaperoning effects. Good cellular models for intracellular Aβ (eventually combined with extracellular Aβ treatments) that could shed more light on putative effects of intracellular chaperoning by HSPBs are yet lacking, maybe due to the fact that free Aβ-peptides are rapidly degraded (E. A. Reits, personal communication). We are currently testing

Aβ-peptides-GFP fusion constructs to do so, but preliminary data suggest the GFP tagging interferes with Aβ-amyloid formation.

HSBP Effects on Tau

The major component of the neurofibrillary tangles in post-mortem brains of AD patients is hyperphosphorylated tau (Iqbal et al. 2010). Tau is a microtubule-associated protein that under normal circumstances promotes the assembly of microtubules and helps to stabilize them, both important for vesicle transport. The function of tau is regulated by phosphorylation status and in AD, all six isoforms of tau are hyperphosphorylated, which is associated with its aggregation that leads to both the loss of its function and even counteracts the assembly of microtubules (toxic gain of function). Hyperphosphorylation of tau makes it more resistant to proteolysis, which explains its accumulation in AD (Giacobini and Gold 2013).

Despite the fact that several HSPB members are, like tau, associated with microtubules (Table 1.3) and the finding that several HSPB members are up-regulated in relation to tau (and Aβ) aggregation (see Carra et al. 2013 for more details) surprisingly few studies have actually directly tested whether or not HSPB member up-regulation may reduce tau aggregation and pathology. In a study by Shimura et al. (2004) HSPB1 was found to bind preferentially to the pathological hyperphosphorylated tau, which was associated with reduced tau expression and tau-mediated cell death. Also in other studies, increased HSPB1 and HSPB5 expressions were inversely correlated with tau phosphorylation and expression levels (Shimura et al. 2004; Sahara et al. 2007), suggesting (but not demonstrating) that these two HSPB members indeed may directly chaperone phosphorylated tau towards degradation. Yet, in line with these in vitro studies, the group of Dickey (Abisambra et al. 2010, 2011) showed that delivery of HSPB1 expressing adeno-associated viral particles to tau transgenic mice resulted in reduced neuronal tau levels. In parallel, this delayed the normally rapid decay in hippocampal long-term potentiation (LTP) intrinsically seen in these tau transgenic mice. This effect, like both the cytoskeletal stabilization and chaperone effects of HSPB1 (Table 1.3), was dependent on the phosphorylation dynamics of HSPB1.

Although the limited data set for HSPB actions on tau allows no conclusions in terms of functional differences between HSPB members, the data do again suggest that HSPB1 is not only associated with client refolding, such as after acute heat stress was seen for refolding (Table 1.3), but can also assist in client degradation (phosphorylated tau and mutant PARKIN C289G). What determines fate destination when chaperoned by HSPB1 remains to be elucidated.

1.4.2.2 Other Non-neurodegenerative Diseases

Various HSPB members have been implied in many other diseases, including cancer (where they may be particularly important in metastasis (Zoubeidi and Gleave 2012), immunology (van Noort et al. 2012), cataract (Clark et al. 2012), and cardiac diseases (Table 1.3, see also below). Here, we would like to shortly discuss two cardiac diseases, dilated cardiomyopathies and atrial fibrillation, in which HSPB members were shown to be protective.

Dilated Cardiomyopathies (DCM)

The first cardiac disease where HSPBs are implied concerns a group of familial dilated cardiomyopathies (DCM) that (are) caused by several rare mutations in more than 30 different genes, most of which encode sarcomeric or sarcomeric-associated proteins, and that result in structural changes (hypertrophy) and contractile dysfunction (Hershberger et al. 2010). Recently, also mutations in various molecular chaperones have been found to lead to dominantly heritable forms of DCM, including mutations in HSPB5 (Hershberger et al. 2010, 2013; Rodríguez and Willis 2010).

Whereas loss of function (haploinsufficiency) may be one explanation by which the HSPB mutants may cause DCM, at least three different DCM-causing mutations (R120G, 450delA and 464delCT) lead to their aggregation when expressed in cells (Zhang et al. 2010), suggesting that also toxic gain-of-function (via aggregation) may play a role. Interestingly, HSPB1 was found to be able to counteract this aggregation, likely by enhancing the proteasomal degradation of these clients (Zhang et al. 2010). Ongoing work in our laboratory (Hussein et al. in preparation) confirms that HSPB1 can rescue the aggregation of these mutants and furthermore show that, besides HSPB1, also overexpression of HSPB4 and HSPB5 itself effectively prevents the aggregation of these three HSPB5 mutants. HSPB6 was also active but less, whilst the other five HSPB members were ineffective (Table 1.3).

Atrial Fibrillation (AF)

Atrial fibrillation (AF) is the most common clinical arrhythmia associated with significant morbidity and mortality. AF is a progressive disease initiated by (reversible) electrophysiological changes followed by irreversible structural damage to the contractile apparatus of cardiomyocytes, which in a perpetuating cycle finally results in myolysis (Hoogstra-Berends et al. 2012). Given that HSPBs were suggested as playing an important role in maintenance of cytoskeletal elements under conditions of acute stress (Sect. 1.4.1; Table 1.3), we decided to test whether and which HSPB members could potentially protect against AF-related myolysis, using pacing of HL-1 atrial myocytes as a model for AF (Brundel et al. 2006a). Initial experiments showed activating the heat shock response attenuated pacing induced myolysis and that HSPB1 upregulation was both required and sufficient for protection (Brundel et al. 2006a, b). Next, we tested all ten HSPB members for AF protection and found that in addition to HSPB1, also overexpression of HSPB6, HSPB7 and HSPB8 protected against tachypacing-induced damage, including the formation of F-actin stress fibers (Ke et al. 2011). Yet, our data further suggested that their mode of action was different, with HSPB1, HSPB6 and HSPB7 protection being due to direct prevention of F-actin formation, whilst HSPB8-protection was indirect.

So, also in the heart, different HSPB members seem to be required to handle different types of proteotoxic challenges and through different mode of actions.

1.4.3 HSPB Mutation Causing Disease

Last but not least, several diseases are directly caused by mutations in HSPBs. This has been recently reviewed extensively elsewhere (Boncoraglio et al. 2012; Carra et al. 2012, 2013; Mymrikov et al. 2011; Kakkar et al. 2014). In Table 1.3, we therefore only summarize the main picture that emerges from these so-called chaperonopathies. This reveals that depending on the HSPB member, mutations primarily affect (motor)neurons, skeletal or cardiac muscle or lead to cataract. Although likely also related to tissue specific expression, these data further support the possibility of functional diversification of the different HSPB members.

1.5 Summary and Perspective

The overall data and analysis of Table 1.3 clearly suggest that many things about functions and modes of actions of HSPBs are not (yet) 'crystal clear' and neither seem 'black or white'. Yet, a number of main points, illustrating the multicolored faces of the small HSPs, seem to emerge:

1. Different HSPB members may have partially overlapping abilities to handle different client, but do also show clear distinct differences with respect to client-preference.
2. Different HSPB members have different modes of client handling. HSPB1 and HSPB5 (and HSPB9) are mostly associated with HSP70 dependent refolding of acutely unfolded clients or proteasomal degradation of terminally misfolded clients. HSPB8 and HSPB7, albeit in mechanistically distinct manners, are most associated with dealing with terminally misfolded clients via interactions with the autophagosomal lysosomal pathway. More molecular insight in each of these actions is, however, needed.
3. The absence of an effect of a certain HSPB on a certain client in a particular cell line has to be interpreted with some caution since not all the required partners (including the correct HSPB partners for the formation of the correct hetero-oligomeric complexes) may be expressed in each cell line for a certain member to be effective. Clearly, more comparative data on the functional activity of all HSPB members on different clients and in different cell lines are needed to really get an insight in their differential modes of chaperone actions.

4. Certain actions of the HSPBs, including those of the canonical HSPB1 and HSPB5, may not be dependent on the HSP70 machine. So, experiments should be directed to other partners of HSPB members that under certain conditions or for certain clients might be able to cause HSPB client release.
5. Yet, since all HSPB members can interact with cytoskeletal elements, we still also need more data to conclusively demonstrate that their observed action on the diverse un- or misfolded clients are really due to direct chaperone-like actions (that all of them also seem to have in vitro) or whether they are mediated via indirectly facilitating the actions of other protein quality control machines in the cell.
6. Even more so, although not extensively discussed here, the world in which HSPB members may function may be even more colourful as they are associated with many other intracellular responses, including effects on the cellular redox status and mitochondrial functioning which, in turn, may have multiple consequences for the cellular robustness and impede on cellular ageing.

References

Abisambra JF, Blair LJ, Hill SE, Jones JR, Kraft C, Rogers J, Koren J 3rd, Jinwal UK, Banko J, Dickey CA et al (2010) Phosphorylation dynamics regulate Hsp27-mediated rescue of neuronal plasticity deficits in tau transgenic mice. J Neurosci 30:15374–15382

Abisambra JF, Jinwal UK, Jones JR, Blair LJ, Koren J 3rd, Dickey CA (2011) Exploiting the diversity of the heat-shock protein family for primary and secondary tauopathy therapeutics. Curr Neuropharmacol 9:623–631

Acunzo J, Katsogiannou M, Rocchi P (2012) Small heat shock proteins HSP27 (HspB1), αB-crystallin (HspB5) and HSP22 (HspB8) as regulators of cell death. Int J Biochem Cell Biol 44:1622–1631

An JJ, Lee YP, Kim DW, Sohn EJ, Jeong HJ, Kang HW, Shin MJ, Kim MJ, Ahn EH, Jang SH, Kang JH, Kang TC, Won MH, Kwon OS, Cho SW, Lee KS, Park J, Eum WS, Choi SY (2009) Transduced HSP27 protein protects neuronal cell death by enhancing FALS-associated SOD1 mutant activity. BMB Rep 42:136–141

Arrigo AP, Gibert B (2012) HspB1 dynamic phospho-oligomeric structure dependent interactome as cancer therapeutic target. Curr Mol Med 12:1151–1163

Asthana A, Raman B, Ramakrishna T, Rao CM (2012) Structural aspects and chaperone activity of human HspB3: role of the "C-terminal extension". Cell Biochem Biophys 64:61–72

Bauer PO, Nukina N (2009) The pathogenic mechanisms of polyglutamine diseases and current therapeutic strategies. J Neurochem 110:1737–1765

Boncoraglio A, Minoia M, Carra S (2012) The family of mammalian small heat shock proteins (HSPBs): implications in protein deposit diseases and motor neuropathies. Int J Biochem Cell Biol 44:1657–1669

Bruinsma IB, Bruggink KA, Kinast K, Versleijen AA, Segers-Nolten IM, Subramaniam V, Verbeek MM (2011) Inhibition of alpha-synuclein aggregation by small heat shock proteins. Proteins 79:2956–2967

Brundel BJ, Henning RH, Ke L, van Gelder IC, Crijns HJ, Kampinga HH (2006a) Heat shock protein upregulation protects against pacing-induced myolysis in HL-1 atrial myocytes and in human atrial fibrillation. J Mol Cell Cardiol 41:555–562

Brundel BJ, Shiroshita-Takeshita A, Qi X, Yeh YH, Chartier D, van Gelder IC, Henning RH, Kampinga HH, Nattel S (2006b) Induction of heat shock response protects the heart against atrial fibrillation. Circ Res 99:1394–1402

Bryantsev AL, Kurchashova SY, Golyshev SA, Polyakov VY, Wunderink HF, Kanon B, Budagova KR, Kabakov AE, Kampinga HH (2007) Regulation of stress-induced intracellular sorting and chaperone function of Hsp27 (HspB1) in mammalian cells. Biochem J 407:407–417
Carra S (2015) Role of HSPB8 in the proteostasis network: from protein synthesis to protein degradation and beyond. In: Tanguay RM, Hightower LE (eds) The big book on small heat shock proteins. Springer, Dordrecht, pp XX–XX
Carra S, Sivilotti M, Chávez Zobel AT, Lambert H, Landry J (2005) HspB8, a small heat shock protein mutated in human neuromuscular disorders, has in vivo chaperone activity in cultured cells. Hum Mol Genet 14:1659–1669
Carra S, Seguin SJ, Lambert H, Landry J (2008) HspB8 chaperone activity toward poly(Q)-containing proteins depends on its association with Bag3, a stimulator of macroautophagy. J Biol Chem 283:1437–1444
Carra S, Brunsting JF, Lambert H, Landry J, Kampinga HH (2009) HspB8 participates in protein quality control by a non-chaperone-like mechanism that requires eIF2{alpha} phosphorylation. J Biol Chem 284:5523–5532
Carra S, Boncoraglio A, Kanon B, Brunsting JF, Minoia M, Rana A, Vos MJ, Seidel K, Sibon OC, Kampinga HH (2010) Identification of the Drosophila ortholog of HSPB8: implication of HSPB8 loss of function in protein folding diseases. J Biol Chem 285:37811–37822
Carra S, Crippa V, Rusmini P, Boncoraglio A, Minoia M, Giorgetti E, Kampinga HH, Poletti A (2012) Alteration of protein folding and degradation in motor neuron diseases: implications and protective functions of small heat shock proteins. Prog Neurobiol 97:83–100
Carra S, Rusmini P, Crippa V, Giorgetti E, Boncoraglio A, Cristofani R, Naujock M, Meister M, Minoia M, Kampinga HH, Poletti A (2013) Different anti-aggregation and pro-degradative functions of the members of the mammalian sHSP family in neurological disorders. Philos Trans R Soc Lond B Biol Sci 368:20110409
Caspers GJ, Leunissen JA, de Jong WW (1995) The expanding small heat-shock protein family, and structure predictions of the conserved "alpha-crystallin domain". J Mol Evol 40:238–248
Chowdary TK, Braman B, Ramakrishna T, Rao CM (2004) Mammalian Hsp22 is a heat-inducible small heat-shock protein with chaperone-like activity. Biochem J 381:379–387
Christians ES, Ishiwata T, Benjamin IJ (2012) Small heat shock proteins in redox metabolism: implications for cardiovascular diseases. Int J Biochem Cell Biol 44:1632–1645
Clark AR, Lubsen NH, Slingsby C (2012) sHSP in the eye lens: crystallin mutations, cataract and proteostasis. Int J Biochem Cell Biol 44:1687–1697
Cohen E, Dillin A (2008) The insulin paradox: aging, proteotoxicity and neurodegeneration. Nat Rev Neurosci 9:759–767
Cox D, Carver JA, Ecroyd H (1842) Preventing α-synuclein aggregation: the role of the small heat-shock molecular chaperone proteins. Biochim Biophys Acta 2014:1830–1843
Crippa V, Sau D, Rusmini P, Boncoraglio A, Onesto E, Bolzoni E, Galbiati M, Fontana E, Marino M, Carra S, Bendotti C, De Biasi S, Poletti A (2010) The small heat shock protein B8 (HspB8) promotes autophagic removal of misfolded proteins involved in amyotrophic lateral sclerosis (ALS). Hum Mol Genet 19:3440–3456
de Jong WW, Leunissen JA, Voorter CE (1993) Evolution of the alpha-crystallin/small heat-shock protein family. Mol Biol Evol 10:103–126
den Engelsman J, Boros S, Dankers PY, Kamps B, Vree Egberts WT, Böde CS, Lane LA, Aquilina JA, Benesch JL, Robinson CV, de Jong WW, Boelens WC (2009) The small heat-shock proteins HSPB2 and HSPB3 form well-defined heterooligomers in a unique 3 to 1 subunit ratio. J Mol Biol 393:1022–1032
den Engelsman J, van de Schootbrugge C, Yong J, Pruijn GJ, Boelens WC (2013) Pseudophosphorylated αB-crystallin is a nuclear chaperone imported into the nucleus with help of the SMN complex. PLoS One 8(9):e73489
Derham BK, Harding JJ (1999) Alpha-crystallin as a molecular chaperone. Prog Retin Eye Res 18:463–509
Domert J, Rao SB, Agholme L, Brorsson AC, Marcusson J, Hallbeck M, Nath S (2014) Spreading of amyloid-β peptides via neuritic cell-to-cell transfer is dependent on insufficient cellular clearance. Neurobiol Dis 65:82–92

Ehrnsperger M, Gräber S, Gaestel M, Buchner J (1997) Binding of non-native protein to Hsp25 during heat shock creates a reservoir of folding intermediates for reactivation. EMBO J 16:221–229
Fan HC, Ho LI, Chi CS, Chen SJ, Peng GS, Chan TM, Lin SZ, Harn HJ (2014) Polyglutamine (PolyQ) diseases: genetics to treatments. Cell Transplant 23:441–458
Garrido C, Paul C, Seigneuric R, Kampinga HH (2012) The small heat shock proteins family: the long forgotten chaperones. Int J Biochem Cell Biol 44:1588–1592
Gatchel JR, Zoghbi HY (2005) Diseases of unstable repeat expansion: mechanisms and common principles. Nat Rev Genet 6:743–755
Giacobini E, Gold G (2013) Alzheimer disease therapy–moving from amyloid-β to tau. Nat Rev Neurol 9:677–686
Gu W-J, Corti O, Araujo F, Hampe C, Jacquier S, Lücking CB, Abbas N, Duyckaerts C, Rooney T, Pradier L, Ruberg M, Brice A (2003) The C289G and C418R missense mutations cause rapid sequestration of human Parkin into insoluble aggregates. Neurobiol Dis 14:357–364
Gusella JF, MacDonald ME (2000) Molecular genetics: unmasking polyglutamine triggers in neurodegenerative disease. Nat Rev Neurosci 1:109–115
Haslbeck M (2015) Regulation of the chaperone function of small Hsps. In: Tanguay RM, Hightower LE (eds) The big book on small heat shock proteins. Springer, Dordrecht, pp XX–XX
Hershberger RE, Morales A, Siegfried JD (2010) Clinical and genetic issues in dilated cardiomyopathy: a review for genetics professionals. Genet Med 12:655–667
Hershberger RE, Hedges DJ, Morales A (2013) Dilated cardiomyopathy: the complexity of a diverse genetic architecture. Nat Rev Cardiol 10:531–547
Hilker R, Klein C, Ghaemi M, Kis B, Strotmann T, Ozelius LJ, Lenz O, Vieregge P, Herholz K, Heiss WD, Pramstaller PP (2001) Positron emission tomographic analysis of the nigrostriatal dopaminergic system in familial parkinsonism associated with mutations in the parkin gene. Ann Neurol 49:367–376
Hoogstra-Berends F, Meijering RA, Zhang D, Heeres A, Loen L, Seerden JP, Kuipers I, Kampinga HH, Henning RH, Brundel BJ (2012) Heat shock protein-inducing compounds as therapeutics to restore proteostasis in atrial fibrillation. Trends Cardiovasc Med 22:62–68
Hussein R et al (in preparation)
Iqbal K, Liu F, Gong CX, Grundke-Iqbal I (2010) Tau in Alzheimer disease and related tauopathies. Curr Alzheimer Res 7:656–664
Ittner LM, Götz J (2011) Amyloid-β and tau- a toxic pas de deux in Alzheimer's disease. Nat Rev Neurosci 12:65–72
Jakob U, Gaestel M, Engel K, Buchner J (1993) Small heat shock proteins are molecular chaperones. J Biol Chem 268:1517–1520
Johnson BS, Snead D, Lee JJ, McCaffery JM, Shorter J, Gitler AD (2009) TDP-43 is intrinsically aggregation-prone, and amyotrophic lateral sclerosis-linked mutations accelerate aggregation and increase toxicity. J Biol Chem 284:20329–20339
Kakkar V, Meister-Broekema M, Minoia M, Carra S, Kampinga HH (2014) Barcoding heat shock proteins to human diseases: looking beyond the heat shock response. Dis Model Mech 7:421–434
Kampinga HH, Garrido C (2012) HSPBs: small proteins with big implications in human disease. Int J Biochem Cell Biol 44:1706–1710
Kampinga HH, Brunsting JF, Stege GJ, Konings AW, Landry J (1994) Cells overexpressing Hsp27 show accelerated recovery from heat-induced nuclear protein aggregation. Biochem Biophys Res Commun 204:1170–1177
Kampinga HH, Hageman J, Vos MJ, Kubota H, Tanguay RM, Bruford EA, Cheetham ME, Chen B, Hightower LE (2009) Guidelines for the nomenclature of the human heat shock proteins. Cell Stress Chaperones 14:105–111
Kappé G, Franck E, Verschuure P, Boelens WC, Leunissen JA, de Jong WW (2003) The human genome encodes 10 alpha-crystallin-related small heat shock proteins: HspB1–10. Cell Stress Chaperones 8:53–61
Kappé G, Boelens WC, de Jong WW (2010) Why proteins without an alpha-crystallin domain should not be included in the human small heat shock protein family HSPB. Cell Stress Chaperones 15:457–461

Ke L, Meijering RA, Hoogstra-Berends F, Mackovicova K, Vos MJ, Van Gelder IC, Henning RH, Kampinga HH, Brundel BJ (2011) HSPB1, HSPB6, HSPB7 and HSPB8 protect against RhoA GTPase-induced remodeling in tachypaced atrial myocytes. PLoS One 6(6):e20395
Kerman A, Liu HN, Croul S, Bilbao J, Rogaeva E, Zinman L, Robertson J, Chakrabartty A (2010) Amyotrophic lateral sclerosis is a non-amyloid disease in which extensive misfolding of SOD1 is unique to the familial form. Acta Neuropathol 119:335–344
Khan NL, Brooks DJ, Pavese N, Sweeney MG, Wood NW, Lees AJ, Piccini P (2002) Progression of nigrostriatal dysfunction in a parkin kindred: an [18F]dopa PET and clinical study. Brain J Neurol 125:2248–2256
Khan NL, Graham E, Critchley P, Schrag AE, Wood NW, Lees AJ, Bhatia KP, Quinn N (2003) Parkin disease: a phenotypic study of a large case series. Brain J Neurol 126:1279–1292
Kriehuber T, Rattei T, Weinmaier T, Bepperling A, Haslbeck M, Buchner J (2010) Independent evolution of the core domain and its flanking sequences in small heat shock proteins. FASEB J 24:3633–3642
Krishnan J, Vannuvel K, Andries M, Waelkens E, Robberecht W, Van Den Bosch L (2008) Overexpression of Hsp27 does not influence disease in the mutant SOD1(G93A) mouse model of amyotrophic lateral sclerosis. J Neurochem 106:2170–2183
LaFerla FM, Green KN, Oddo S (2007) Intracellular amyloid-beta in Alzheimer's disease. Nat Rev Neurosci 8:499–509
Landry J, Chrétien P, Lambert H, Hickey E, Weber LA (1989) Heat shock resistance conferred by expression of the human HSP27 gene in rodent cells. J Cell Biol 109:7–15
Lanneau D, Brunet M, Frisan E, Solary E, Fontenay M, Garrido C (2008) Heat shock proteins: essential proteins for apoptosis regulation. J Cell Mol Med 12:743–761
Lavoie JN, Gingras-Breton G, Tanguay RM, Landry J (1993) Induction of Chinese hamster HSP27 gene expression in mouse cells confers resistance to heat shock. HSP27 stabilization of the microfilament organization. J Biol Chem 268:3420–3429
Lin W, Yang Z, Lu Y, Zhao X (2014) Refined purification of large amounts of rat cvHsp/HspB7 and partial biological characterization in vitro. Protein Pept Lett 21:503–510
Matos CA, de Macedo-Ribeiro S, Carvalho AL (2011) Polyglutamine diseases: the special case of ataxin-3 and Machado-Joseph disease. Prog Neurobiol 95:26–48
Minoia M, Grit C, Kampinga HH (2014) HSPA1A-independent suppression of PARK2 C289G protein aggregation by human small heat shock proteins. Mol Cell Biol 34:3570–3578
Morfini G, Pigino G, Brady ST (2005) Polyglutamine expansion diseases: failing to deliver. Trends Mol Med 11:64–70
Mounier N, Arrigo AP (2002) Actin cytoskeleton and small heat shock proteins: how do they interact? Cell Stress Chaperones 7:167–176
Mulligan VK, Chakrabartty A (2013) Protein misfolding in the late-onset neurodegenerative diseases: common themes and the unique case of amyotrophic lateral sclerosis. Proteins 81:1285–1303
Mymrikov EV, Seit-Nebi AS, Gusev NB (2011) Large potentials of small heat shock proteins. Physiol Rev 91:1123–1159
Nakagawa M, Tsujimoto N, Nakagawa H, Iwaki T, Fukumaki Y, Iwaki A (2001) Association of HSPB2, a member of the small heat shock protein family, with mitochondria. Exp Cell Res 271:161–168
Ojha J, Masilamoni G, Dunlap D, Udoff RA, Cashikar AG (2011) Sequestration of toxic oligomers by HspB1 as a cytoprotective mechanism. Mol Cell Biol 31:3146–3157
Outeiro TF, Klucken J, Strathearn KE, Liu F, Nguyen P, Rochet JC, Hyman BT, McLean PJ (2006) Small heat shock proteins protect against alpha-synuclein-induced toxicity and aggregation. Biochem Biophys Res Commun 351(3):631–638
Parcellier A, Schmitt E, Brunet M, Hammann A, Solary E, Garrido C (2005) Small heat shock proteins HSP27 and alphaB-crystallin: cytoprotective and oncogenic functions. Antioxid Redox Signal 7:404–413
Patel YJ, Payne Smith MD, de Belleroche J, Latchman DS (2005) Hsp27 and Hsp70 administered in combination have a potent protective effect against FALS-associated SOD1-mutant-induced cell death in mammalian neuronal cells. Brain Res Mol Brain Res 134:256–274

Prabhu S, Rama B, Ramakrishn T, Rao C (2012) HspB2/myotonic dystrophy protein kinase binding protein (MKBP) as a novel molecular chaperone: structural and functional aspects. PLoS One 7(1):e29810

Querfurth HW, LaFerla FM (2010) Alzheimer's disease. N Engl J Med 4:329–344

Raman B, Ban T, Sakai M, Pasta SY, Ramakrishna T, Naiki H, Goto Y, Rao CM (2005) AlphaB-crystallin, a small heat-shock protein, prevents the amyloid fibril growth of an amyloid beta-peptide and beta2-microglobulin. Biochem J 392:573–581

Renton AE, Chiò A, Traynor BJ (2014) State of play in amyotrophic lateral sclerosis genetics. Nat Neurosci 17:17–23

Richter K, Haslbeck M, Buchner J (2010) The heat shock response: life on the verge of death. Mol Cell 40:253–266

Robertson AL, Headey SJ, Saunders HM, Ecroyd H, Scanlon MJ, Carver JA, Bottomley SP (2010) Small heat-shock proteins interact with a flanking domain to suppress polyglutamine aggregation. Proc Natl Acad Sci U S A 107:10424–10429

Rodríguez JE, Willis MS (2010) The therapeutic potential of heat shock proteins in cardiomyopathies due to mutations in cardiac structural proteins. J Mol Cell Cardiol 49:904–907

Rogalla T, Ehrnsperger M, Preville X, Kotlyarov A, Lutsch G, Ducasse C, Paul C, Wieske M, Arrigo AP, Buchner J, Gaestel M (1999) Regulation of Hsp27 oligomerization, chaperone function, and protective activity against oxidative stress/tumor necrosis factor alpha by phosphorylation. J Biol Chem 274:18947–18956

Sahara N, Maeda S, Yoshiike Y, Saito Y, Murayama S, Takashima A et al (2007) Molecular chaperone-mediated tau protein metabolism counteracts the formation of granular tau oligomers in human brain. J Neurosci Res 85:3098–3108

Sharp PS, Akbar MT, Bouri S, Senda A, Joshi K, Chen HJ, Latchman DS, Wells DJ, de Belleroche J (2008) Protective effects of heat shock protein 27 in a model of ALS occur in the early stages of disease progression. Neurobiol Dis 30:42–55

Shimura H, Miura-Shimura Y, Kosik KS (2004) Binding of tau to heat shock protein 27 leads to decreased concentration of hyperphosphorylated tau and enhanced cell survival. J Biol Chem 279:17957–17962

Solanki A, Neupane K, Woodside MT (2014) Single-molecule force spectroscopy of rapidly fluctuating, marginally stable structures in the intrinsically disordered protein α-synuclein. Phys Rev Lett 112:158103

Sreedharan J, Brown RH Jr (2013) Amyotrophic lateral sclerosis: problems and prospects. Ann Neurol 74:309–316

Sreedharan J, Blair IP, Tripathi VB, Hu X, Vance C, Rogelj B, Ackerley S, Durnall JC, Williams KL, Buratti E, Baralle F, de Belleroche J, Mitchell JD, Leigh PN, Al-Chalabi A, Miller CC, Nicholson G, Shaw CE (2008) TDP-43 mutations in familial and sporadic amyotrophic lateral sclerosis. Science 319:1668–1672

Stege GJ, Renkawek K, Overkamp PS, Verschuure P, van Rijk AF, Reijnen-Aalbers A, Boelens WC, Bosman GJ, de Jong WW (1999) The molecular chaperone alphaB-crystallin enhances amyloid beta neurotoxicity. Biochem Biophys Res Commun 262:152–156

Sugiyama Y, Suzuki A, Kishikawa M, Akutsu R, Hirose T, Waye MM, Tsui SK, Yoshida S, Ohno S (2000) Muscle develops a specific form of small heat shock protein complex composed of MKBP/HSPB2 and HSPB3 during myogenic differentiation. J Biol Chem 275:1095–1104

Sun Y, MacRae TH (2005) The small heat shock proteins and their role in human disease. FEBS J 272:2613–2627

Sun X, Fontaine JM, Rest JS, Shelden EA, Welsh MJ, Benndorf R (2004) Interaction of human HSP22 (HSPB8) with other small heat shock proteins. J Biol Chem 279:2394–2402

Suzuki A, Sugiyama Y, Hayashi Y, Nyu-i N, Yoshida M, Nonaka I, Ishiura S, Arahata K, Ohno S (1998) MKBP, a novel member of the small heat shock protein family, binds and activates the myotonic dystrophy protein kinase. J Cell Biol 140:1113–1124

Takahashi T, Katada S, Onodera O (2010) Polyglutamine diseases: where does toxicity come from? What is toxicity? Where are we going? J Mol Cell Biol 2:180–191

Talbot K (2014) Amyotrophic lateral sclerosis: cell vulnerability or system vulnerability? J Anat 224:45–51
Tissières A, Mitchell HK, Tracy UM (1974) Protein synthesis in salivary glands of Drosophila melanogaster: relation to chromosome puffs. J Mol Biol 84:389–398
Tóth ME, Szegedi V, Varga E, Juhász G, Horváth J, Borbély E, Csibrány B, Alföldi R, Lénárt N, Penke B, Sántha M (2013) Overexpression of Hsp27 ameliorates symptoms of Alzheimer's disease in APP/PS1 mice. Cell Stress Chaperones 18:759–771
Trinh J, Farrer M (2013) Advances in the genetics of Parkinson disease. Nat Rev Neurol 9:445–454
Tue NT, Shimaji K, Tanaka N, Yamaguchi M (2012) Effect of αB-crystallin on protein aggregation in Drosophila. J Biomed Biotechnol 2012:252049
van Montfort RL, Basha E, Friedrich KL, Slingsby C, Vierling E (2001) Crystal structure and assembly of a eukaryotic small heat shock protein. Nat Struct Biol 8:1025–1030
van Noort JM, Bsibsi M, Nacken P, Gerritsen WH, Amor S (2012) The link between small heat shock proteins and the immune system. Int J Biochem Cell Biol 44:1670–1679
Vos MJ, Hageman J, Carra S, Kampinga HH (2008) Structural and functional diversities between members of the human HSPB, HSPH, HSPA, and DNAJ chaperone families. Biochemistry 47:7001–7011
Vos MJ, Kanon B, Kampinga HH (2009) HSPB7 is a SC35 speckle resident small heat shock protein. Biochim Biophys Acta 1793:1343–1353
Vos MJ, Zijlstra MP, Kanon B, van Waarde-Verhagen MA, Brunt ER, Oosterveld-Hut HM, Kampinga HH et al (2010) HSPB7 is the most potent polyQ aggregation suppressor within the HSPB family of molecular chaperones. Hum Mol Genet 19:4677–4693
Vos MJ, Zijlstra MP, Carra S, Sibon OC, Kampinga HH (2011) Small heat shock proteins, protein degradation and protein aggregation diseases. Autophagy 7:101–103
Wakabayashi K, Tanji K, Odagiri S, Miki Y, Mori F, Takahashi H (2013) The Lewy body in Parkinson's disease and related neurodegenerative disorders. Mol Neurobiol 47:495–508
Walsh DM, Tseng BP, Rydel RE, Podlisny MB, Selkoe DJ (2000) The oligomerization of amyloid beta-protein begins intracellularly in cells derived from human brain. Biochemistry 39:10831–10839
Wettstein G, Bellaye PS, Micheau O, Bonniaud P (2012) Small heat shock proteins and the cytoskeleton: an essential interplay for cell integrity? Int J Biochem Cell Biol 44:1680–1686
Wilhelmus MM, Boelens WC, Otte-Höller I, Kamps B, de Waal RM, Verbeek MM (2006a) Small heat shock proteins inhibit amyloid-beta protein aggregation and cerebrovascular amyloid-beta protein toxicity. Brain Res 1089:67–78
Wilhelmus MM, Boelens WC, Otte-Höller I, Kamps B, Kusters B, Maat-Schieman ML, de Waal RM, Verbeek MM (2006b) Small heat shock protein HspB8: its distribution in Alzheimer's disease brains and its inhibition of amyloid-beta protein aggregation and cerebrovascular amyloid-beta toxicity. Acta Neuropathol 111:139–149
Wyttenbach A, Sauvageot O, Carmichael J, Diaz-Latoud C, Arrigo AP, Rubinsztein DC (2002) Heat shock protein 27 prevents cellular polyglutamine toxicity and suppresses the increase of reactive oxygen species caused by huntingtin. Hum Mol Genet 11:1137–1151
Yang Z, Wang Y, Lu Y, Zhao X (2011) Molecular characterization of rat cvHsp/HspB7 in vitro and its dynamic molecular architecture. Mol Med Rep 4:105–111
Yerbury JJ, Gower D, Vanags L, Roberts K, Lee JA, Ecroyd H (2013) The small heat shock proteins αB-crystallin and Hsp27 suppress SOD1 aggregation in vitro. Cell Stress Chaperones 18:251–257
Zhang H, Rajasekaran NS, Orosz A, Xiao X, Rechsteiner M, Benjamin IJ (2010) Selective degradation of aggregate-prone CryAB mutants by HSPB1 is mediated by ubiquitin-proteasome pathways. J Mol Cell Cardiol 49:918–930
Zheng L, Calvo-Garrido J, Hallbeck M, Hultenby K, Marcusson J, Cedazo-Minguez A, Terman A (2013) Intracellular localization of amyloid-β peptide in SH-SY5Y neuroblastoma cells. J Alzheimers Dis 37:713–733

Zoubeidi A, Gleave M (2012) Small heat shock proteins in cancer therapy and prognosis. Int J Biochem Cell Biol 44:1646–1656

Zourlidou A, Payne Smith MD, Latchman DS (2004) HSP27 but not HSP70 has a potent protective effect against alpha-synuclein-induced cell death in mammalian neuronal cells. J Neurochem 88:1439–1448

Zourlidou A, Gidalevitz T, Kristiansen M, Landles C, Woodman B, Wells DJ, Latchman DS, de Belleroche J, Tabrizi SJ, Morimoto RI, Bates GP (2007) Hsp27 overexpression in the R6/2 mouse model of Huntington's disease: chronic neurodegeneration does not induce Hsp27 activation. Hum Mol Genet 16:1078–1090

Chapter 2
Immense Cellular Implications Associated to Small Stress Proteins Expression: Impacts on Human Pathologies

André-Patrick Arrigo, Benjamin Ducarouge, Fabrice Lavial, and Benjamin Gibert

Abstract In addition to being potent chaperones that protect cells against the accumulation of unfolded proteins under stress conditions, mammalian small heat shock proteins (small Hsps) regulate many vital cellular processes in normal and pathological cells. Indeed, these Hsps are constitutively expressed in many tissues and show dramatic changes in their levels of expression in most human pathologies. They are characterized by a large spectrum of activities and are particularly active in protein conformational and inflammatory diseases as well as in cancer pathologies. It is now believed that the immense cellular implications of small Hsps results from their ability to interact, through particular structural changes, with many different client proteins that are subsequently modulated in their activities or half-lifes. Here, we have integrated functionally and structurally the recent data in the literature concerning the interactions of mammalian small Hsps with specific clients. Further analysis with geneMANIA software and database confirmed the incredibly large number of functions associated with these Hsps. The consequences for human pathologies as well as putative therapeutic strategies are discussed, particularly when the expression of small Hsps is harmful (as in some cancer pathologies) or when it appears beneficial for patients.

Keywords Mammalian small Hsps • Oligomeric complexes • Clients • Protein interactomes • Protein predictomes • Cellular implications • Pathologies

---

A.-P. Arrigo (✉)
Apoptosis, Cancer and Development Laboratory, Lyon Cancer Research Center, INSERM U1052-CNRS UMR5286, Claude Bernard University Lyon 1, 28 rue Laennec, 69008 Lyon, France

Apoptosis, Cancer and Development Laboratory, Lyon Cancer Research Center, U1052-CNRS UMR5286, Centre Léon Bérard, 28 rue Laennec, 69008 Lyon, France
e-mail: parrigo@me.com; arrigo@univ-lyon1.fr

B. Ducarouge • F. Lavial • B. Gibert
Apoptosis, Cancer and Development Laboratory, Lyon Cancer Research Center, INSERM U1052-CNRS UMR5286, Claude Bernard University Lyon 1, 28 rue Laennec, 69008 Lyon, France

R.M. Tanguay, L.E. Hightower (eds.), *The Big Book on Small Heat Shock Proteins*, Heat Shock Proteins 8, DOI 10.1007/978-3-319-16077-1_2

2.1 Many Functions Associated with Small Hsps in Addition to Their Protective Role in Stress Condition

The last decade has been characterized by an incredible jump in the interest in the ten mammalian small Hsps. Indeed, until the turn of the century, these stress proteins were considered as exotic chaperones that did not use ATP for their activity. These "forgotten chaperones", as they were called in 2002 (Solari and Garrido 2002), are now stars among Hsps to judge by the large number of scientific and medical publications dealing with their particular behaviors and functions that fill the current literature. This renewed interest is probably linked to their constitutive expression in normal and pathological conditions as well as to the large number of unrelated functions associated with their over- or under-expression in many different cell types. Interest has also been generated by the growing number of pathological mutations in their genes that induce degenerative or myopathic diseases and by their newly described ability to be secreted.

2.1.1 Stress Conditions, Chaperone Activity and Anti-aggregation Properties

Early studies dealing with HspB1 and HspB5 revealed their enhanced expression under heat shock conditions as well as their ATP-independent chaperone property (Jakob et al. 1993; Rogalla et al. 1999). It was shown that large oligomeric structures formed by small Hsps store stress-altered polypeptides in a refolding competent state that can interfere with their propensity to aggregate (Bellyei et al. 2007; Carra et al. 2005; Ehrnsperger et al. 1997, 2000; Ganea 2001; Haslbeck et al. 2005; Horwitz et al. 1992; Jakob et al. 1993; Lee et al. 1997; Markossian et al. 2009). These altered polypeptides can subsequently be refolded by the ATP-dependent Hsp70, Hsp90 and co-chaperones "foldase" machines (Buchner 1999; Bukau and Horwich 1998; Freeman and Morimoto 1996; Lee and Vierling 2000) or degraded by the CHIP-ubiquin-26S proteasome machine (McDonough and Patterson 2003). The dynamic oligomerization/phosphorylation status of small Hsps, and particularly HspB1, is an essential factor of this process (Arrigo et al. 1988; Lelj-Garolla and Mauk 2005, 2006; Paul et al. 2010; Preville et al. 1998b; Rogalla et al. 1999; Simon et al. 2013). The cytoskeleton is one of the primary targets protected by HspB1 and HspB5 in response to stress (Bellomo and Mirabelli 1992; Welch and Feramisco 1985) as well as in normal growth conditions. This property probably relies, at least in the case of HspB1, on the fact that phosphorylated small HspB1 oligomers modulate F-actin fiber growth and, indirectly, extracellular matrix organization (Dalle-Donne et al. 2001; Mounier and Arrigo 2002; Perng et al. 1999).

Under stress conditions, HspB1 and HspB5 stabilize microtubules (Hino et al. 2000; Preville et al. 1996; Xi et al. 2006). HspB5 is also very active in maintaining intermediate filaments homeostasis, particularly in muscle cells where it associates with desmin (Bennardini et al. 1992; Djabali et al. 1999). Moreover, HspB1 and HspB5 share an intriguing anti-oxidant property which appears linked to the chaperoning of several anti-oxidant enzymes, particularly G6PDH (glucose 6-phosphate dehydrogenase) (Arrigo 2001, 2007b, 2013; Arrigo et al. 2005; Firdaus et al. 2006a; Mehlen et al. 1996a; Paul and Arrigo 2000; Preville et al. 1998a, 1999; Rogalla et al. 1999; Yan et al. 2002). Consequently, damage such as protein and nucleic acid oxidation as well as lipid peroxidation is reduced and the positive effect of these Hsps towards mitochondrial $\Delta\Phi m$ increases ATP levels, which favors the activity of ATP-dependent chaperones (Mehlen et al. 1996a; Preville et al. 1999).

Only HspB1, HspB5 and HspB8 molecular chaperones are induced under stress conditions. Interestingly, constitutively expressed small Hsps, such as HspB2, HspB3, HspB4, HspB6 and HspB7, also display chaperone activities or at least anti-aggregation and pro-degradative functions (Carra et al. 2013). The anti-aggregation and anti-fibrillation properties of mammalian small Hsps are summarized in Table 2.1. Depending on the substrate, some Hsps perform these tasks better than others, suggesting that they do not all have the same chaperone-like activity. For example, HspB4 can chaperone HspB5 once in the alpha-crystallin complex (Andley 2007), while HspB3 (Asthana et al. 2012) and HspB2 exhibit significant chaperone-like activity towards specific target proteins and can attenuate the ordered amyloid fibril formation of α-synuclein (Prabhu et al. 2012). The major substrates recognized by small Hsps can be mutated polypeptides that cause degenerative or myopathic diseases (i.e. desmin, polyQ proteins, SOD, α-synuclein) or proteins that are prone to aggregate. It is also important to mention that small Hsp mutants can induce the aggregation of their substrates, such as the R120G missense mutation in HspB5 which is genetically linked to a desmin-related myopathy consequently of the aggregation of desmin (Bova et al. 1999; Vicart et al. 1998). Similarly, the P182L mutant of HspB1 leads to motor neuronopathies as a result of the formation of aggregates that sequestrate Neurofilament middle chain subunit (NF-M) and p150 Dynactin (Ackerley et al. 2005). Equally, proteins that interact with mutant small Hsps can counteract aggregation, as for example the chaperone-like effect of Bag3 towards aggregated HspB8 mutant (Hishiya et al. 2011). As a result of its interaction with Bag3, HspB8 also has the ability to trigger macroautophagy (Carra 2009; Carra et al. 2008b). This favors the elimination of aggregated polypeptides generated by heat (Nivon et al. 2009) or oxidative stress (Keller et al. 2004; Kiffin et al. 2006). Interestingly, HspB6 also appears to play a role in the Bag-3/HspB8 complex that triggers macroautophagy (Fuchs et al. 2010). Less information is available concerning HspB9 and HspB10 in spite of their ability to interact with particular polypeptides (see Table 2.2).

Table 2.1 Small Hsps activity against aggregation or fibrillation of proteins

Client	HspB	Resulting effects	Interacting structure/sequence	Observations made			References
				In cells	In vitro	By adding HspB to cells	
α-synuclein (wt, mutant)	HspB1	Inhibition of fibrillation	nd	X	X	–	Nemes et al. (2004) and Outeiro et al. (2006)
	HspB5	Inhibition of fibrillation	Known HspB5 sequence	X	X	–	Bruinsma et al. (2011); Outeiro et al. (2006) and Ghosh et al. (2008)
	HspB8	Inhibition of fibrillation	nd	–	X	–	Bruinsma et al. (2011)
	HspB3/B2	Inhibition of fibrillation	nd	–	X	–	Bruinsma et al. (2011) and Prabhu et al. (2012)
	HspB6	Inhibition of fibrillation	nd	–	X	–	Bruinsma et al. (2011)
Aβ-amyloid (wt, mutant)	HspB8	Inhibition of aggregation	nd	X	X	–	Wilhelmus et al. (2006a)
	HspB5	Inhibition of aggregation	Known HspB5 sequence	X	X	–	Ghosh et al. (2008) and Wilhelmus et al. (2006b)
	HspB5	Inhibition of fibrils elongation	nd	–	X	–	Shammas et al. (2011)
	HspB1	Inhibition of aggregation	nd	X	X	–	Wilhelmus et al. (2006b)
	HspB6	Inhibition of aggregation	nd	X	X	–	Wilhelmus et al. (2006b)
PolyQ proteins	HspB1	Inhibition of aggregation	nd	X	X	–	Robertson et al. (2010) and Vos et al. (2010)
	HspB5	Inhibition of aggregation	nd	X	X	–	Robertson et al. (2010) and Vos et al. (2010)
	HspB8	Inhibition of aggregation	nd	X	X	–	Carra et al. (2008a) and Vos et al. (2010)
	HspB7	Inhibition of aggregation	nd	X	X	–	Vos et al. (2010)
Catalase	HspB5	Protection against inactivation	HspB5/catalase ratio: 1:2	–	X	–	Hook and Harding (1996)

SOD1	HspB1	Inhibition of aggregation	Interaction with SOD1 aggregates	–	X	–	Yerbury et al. (2012)
	HspB5	Inhibition of aggregation	Increased mutant SOD1 solubility	X	X	–	Shinder et al. (2001) and Yerbury et al. (2012)
	HspB8	Inhibition of aggregation	Increased mutant SOD1 solubility	X	–	–	Crippa et al. (2010)
TDP-43	HspB8	Inhibition of aggregation	nd	X	–	–	Crippa et al. (2010)
Parkin	HspB1	Neuronal aggregates stabilization	nd	X	–	–	Nemes et al. (2004)
Phosphorylated Tau	HspB1	P-Tau degradation	nd	X	–	–	Shimura et al. (2004)
Desmin	HspB5	Inhibition of aggregation	Known HspB5 sequence	–	X	–	Ghosh et al. (2007b)
Vimentin	HspB1	Inhibition of aggregation	nd	X	X	–	Perng et al. (1999)
	HspB5	Inhibition of aggregation	Known HspB5 sequence	X	X	–	Djabali et al. (1997, 1999) and Perng et al. (1999)
Tubulin	HspB5	Inhibition of aggregation	Known HspB5 sequence	–	X	–	Ghosh et al. (2007a) and Ohto-Fujita et al. (2007)
Serpin	HspB5	Inhibition of aggregation	nd	–	X	–	Devlin et al. (2003)
PrPc	HspB5	Inhibition of aggregation	nd	X	X	–	Sun et al. (2005)
Apolipoprotein-CII	HspB5	Inhibition of aggregation	nd	–	X	–	Hatters et al. (2001)
β2-microglobulin	HspB5	Inhibition of fibrillation	Known HspB5 sequence	–	X	–	Ghosh et al. (2008)
Transthyretin	HspB5	Inhibition of fibrillation	Known HspB5 sequence	–	X	–	Ghosh et al. (2008)
p150 dynactin	**HspB1**	Formation of aggregates	Sequestration by HspB1 P182L mutant	X	–	–	Ackerley et al. (2005)
NF-M	**HspB1**	Formation of aggregates	Sequestration by HspB1 P182L mutant	X	–	–	Ackerley et al. (2005)
Desmin	**HspB5**	Formation of aggregates	Sequestration by HspB5 R120G mutant	X	–	–	Bova et al. (1999)
Bag3	**HspB5**	Inhibits aggregation of mut.HspB5	Ile-Pro-Val regions (Bag3)	X	X	–	Hishiya et al. (2011)

List of abbreviations: **HspB** (bold) mutant HspB, *nd* not determined, *SOD1* Copper-Zinc superoxide dismutase, *TDP-43* major disease protein in ubiquitin-positive, tau-, and alpha-synuclein-negative frontotemporal dementia, *PrPc* bovine prion protein, *NF-M* Neurofilament middle chain subunit; a protein kinase of the MLK family

Table 2.2 Some of the polypeptides interacting with mammalian small Hsps

Client	HspB	Consequences	Interacting structure/sequence	Observations made			References
				In cells	In vitro	by adding HspB to cells	
Cytoskeleton, cell adhesion, tissue integrity, epithelial to mesenchymal transition (MET)							
F-actin	HspB1	Protects F-actin integrity	Small P-oligomers (HspB1)	X	–	–	Ke et al. (2011) and Mounier and Arrigo (2002)
	HspB5	Protection of F-actin integrity	P-HspB5	X	–	–	Singh et al. (2007) and Wang and Spector (1996)
	α-crystallin	Protection of F-actin integrity	α-crystallin/actin	X	–	–	Del Vecchio et al. (1984) and Wang and Spector (1996)
	HspB6	Inhibition of stress fibers formation	nd	X	X	–	Ke et al. (2011)
	HspB7	Inhibition of stress fibers formation	nd	X	X	–	Ke et al. (2011)
	HspB8	Inhibition Rho GTPase/stress fibers	nd	X	X	–	Ke et al. (2011)
	HspB1	Translocation to F-actin by proteasome inhibition	nd	X	–	–	Verschuure et al. (2002)
	HspB2	Translocation to F-actin by proteasome inhibition	nd	X	–	–	Verschuure et al. (2002)
	HspB3	Translocation to F-actin by proteasome inhibition	nd	X	–	–	Verschuure et al. (2002)
	HspB5	Translocation to F-actin by proteasome inhibition	nd	X	–	–	Verschuure et al. (2002)
Actin	HspB5	Chaperoning and polymerization of actin	Known HspB5 sequence	–	X	–	Ghosh et al. (2007b)
Tubulin	HspB1	Chaperoning of α/β tubulin, protects microtubules	nd	X	–	–	Hino et al. (2000) and Preville et al. (1996)

	HspB5	Microtubule assembly; inhibits tubulin aggregation	Known HspB5 sequence	–	X	–	Ghosh et al. (2007a) and Ohto-Fujita et al. (2007)
MAPs	HspB5	Inhibition of microtubules aggregation	nd	X	–	–	Xi et al. (2006)
Vimentin IF	HspB5	Favors and protects VIF interaction	nd	X	X	–	Perng et al. (1999)
Vimentin IF	HspB1	Favors and protects VIF interaction	nd	X	X	–	Perng et al. (1999)
Vimentin	HspB1	Chaperoning of vimentin	nd	X	X	–	Perng et al. (1999)
	HspB5	Chaperoning of vimentin, protects IF	Known HspB5 sequence	X	X	–	Djabali et al. (1997, 1999) and Perng et al. (1999)
Peripherin	HspB5	Chaperoning of peripherin, protects IF	Known HspB5 sequence	X	X	–	Djabali et al. (1997, 1999)
Desmin	HspB5	Chaperoning of desmin	Known HspB5 sequence	–	X	–	Ghosh et al. (2007b)
GFAP IF	HspB1	Favors GFAP filaments interaction	nd	X	X	–	Perng et al. (1999)
GFAP IF	HspB5	Favors GFAP filaments interaction	nd	X	X	–	Perng et al. (1999)
GFAP	HspB5	Stabilization/degradation GFAP	nd	X	–	–	Perng et al. (1999) and Tang et al. (2010)
Keratin	HspB1	Chaperoning of keratin, keratin networks disassembly	nd	X	–	–	Duverger et al. (2004) and Perng et al. (1999)
Titin	HspB5	Protection of integrity	i-band (titin)	X	–	–	Bullard et al. (2004)
Neurofilaments	HspB1	Protection of integrity	nd	X	–	–	Bjorkdahl et al. (2008)
	HspB5	Protection of integrity	nd	X	–	–	Bjorkdahl et al. (2008)
β-catenin	HspB1	Cell adhesion	nd	X	–	–	Fanelli et al. (2008)
	HspB5	Cell adhesion	Known HspB5 sequence	–	X	–	Ghosh et al. (2007c)

(continued)

Table 2.2 (continued)

Client	HspB	Consequences	Interacting structure/sequence	Observations made			References
				In cells	In vitro	by adding HspB to cells	
E-cadherin	HspB5	Cell adhesion	nd	X	–	–	Huang et al. (2012)
Cadherin-16	HspB5	Cadherin-16-cytoskeleton connection	nd	X	–	–	Thedieck et al. (2008)
Snail	HspB1	Promotes MET	nd	X	–	–	Wettstein et al. (2013)
DSTN	HspB8	Destrin, actin depolymerization	nd	X	–	–	Havugimana et al. (2012)
Filensin	HspB5	Chaperoning of filensin	nd	X	–	–	Muchowski et al. (1999)
Phakinin	HspB5	Chaperoning of phakinin	nd	X	–	–	(Muchowski et al. 1999)
GRIFIN	HspB5	?	nd	X	–	–	(Barton et al. 2009)
HDAC6	HspB1	Stabilization of HDAC6 half-life	HspB1 500-700 kDa oligomers	X	–	–	Gibert et al. (2012a)
		Stimulates microtubules deacetylation					
p66Shc	HspB1	F-actin disruption	p66Shc Ser36 phosphorylation, abrogates HspB1 protective function	X	–	–	Arany et al. (2012)
14-3-3gamma	HspB6	Displaces 14-3-3 binding partners	P-HspB6	–	X	–	Chernik et al. (2007) and Seit-Nebi and Gusev (2010)
TCTEL1	HspB9	Testis specific	nd	X	–	–	de Wit et al. (2004)
Sperm dense fiber	HspB10	Linkage of sperm head to tail	nd	X	–	–	Gastmann et al. (1993) and Yang et al. (2012)

Transduction pathways, cell activation							
Membrane signaling proteins							
CD10	HspB1	?	nd	X	–	–	Dall'Era et al. (2007)
β2-microglobulin	HspB5	Inhibition of fibrillation known	HspB5 sequence – F, D beta strands of β2-microglobulin	X	–	–	Ghosh et al. (2008)
TLR4 ligand	HspB8	TLR4 dependent dendritic cells activation	nd	–	–	X	Roelofs et al. (2006)
	HspB4	TLR4 dependent dendritic cells activation	nd	–	–	X	Roelofs et al. (2006)
TLR3	HspB1	Angiogenesis activation	nd	–	–	X	Thuringer et al. (2013)
TLR2	HspB5	Macrophage activation	nd	–	–	X	van Noort et al. (2013)
TLR1	HspB5	Macrophage activation	nd	–	–	X	van Noort et al. (2013)
CD14	HspB5	Macrophage activation	nd	–	–	X	van Noort et al. (2013)
Plasminogen	HspB1	Modulates plasminogen binding to ECs	nd	–	–	X	Dudani et al. (2007)
Angiostatin	HspB1	Modulates angiostatin binding to ECs	nd	–	–	X	Dudani et al. (2007)
Growth factors, receptors, transduction pathway factors							
NGF-beta	HspB5	Chaperoning of NGF-beta	Known HspB5 sequence	–	X	–	Ghosh et al. (2007c)
FGF-2	HspB5	Chaperoning of FGF-2	Known HspB5 sequence	–	X	–	Ghosh et al. (2007c)
VEGF	HspB5	Chaperoning of VEGF	Known HspB5 sequence	–	X	–	Ghosh et al. (2007c) and Kerr and Byzova (2010)
	HspB5	Intracrine VEGF signaling	nd	X	–	–	Ruan et al. (2011)
	HspB1	Angiogenesis activation	nd	X	–	X	Choi et al. (2014)

(continued)

Table 2.2 (continued)

Client	HspB	Consequences	Interacting structure/sequence	Observations made			References
				In cells	In vitro	by adding HspB to cells	
Insulin	HspB5	Insulin stabilization	Known HspB5 sequence	–	X	–	Ghosh et al. (2007c)
AR	HspB1	AR stabilization	nd	X	–	–	Zoubeidi et al. (2007)
Her2	HspB1	Her2 stabilization	nd	X	–	–	Kang et al. (2008)
ERβ	HspB1	Estrogen signaling	P-HspB1	X	–	–	Al-Madhoun et al. (2007)
14-3-3zeta	HspB6	Modulates 14-3-3 interactome affinity	P-HspB6	–	X	–	Sluchanko et al. (2011)
14-3-3gamma	HspB6	Displaces 14-3-3 binding partners	P-HspB6	–	X	–	Chernik et al. (2007) and Seit-Nebi and Gusev (2010)
Beta-arrestin2	HSPB1	Essential for TRAIL-triggered Src-Akt/ERK survival signaling	P-HspB1 (ser78/82)	X	–	–	Qi et al. (2014)
Src	HspB1	Essential for TRAIL-triggered Src-Akt/ERK survival signaling	P-HspB1 (ser78/82)	X	–	–	Qi et al. (2014)
DAXX	HspB1	Blocks DAXX-mediated apoptosis	nd	X	–	–	Charette and Landry (2000)
TRAF6	HspB1	TRAF6 ubiquitination, IKKβ activation	P-HspB1 (ser78/82)	X	–	–	Wu et al. (2009)
Protein kinases, phosphatases							
PI3K	HspB6	Inhibits PI3K activity and Akt pathway	nd	X	–	–	Matsushima-Nishiwaki et al. (2013)
PTEN	HspB1	Stabilization of PTEN levels	nd	X	–	–	Cayado-Gutierrez et al. (2012)
Akt, P38, MK2	HspB1	Akt activation, scaffold MK2 to Akt	aa117-128 (Akt)	X	–	–	Rane et al. (2001) and Wu et al. (2007)

PKCΔ	HspB1	Inhibits HspB1 activity	PKCΔ V5 region (7 a.a.)	X	–	–	Kim et al. (2007) and Lee and Lee (2010)
RhoA, PKCα	HspB1	Muscle contraction	P-HspB1	X	–	–	Patil et al. (2004)
Phk	HspB1	?	Small HspB1 oligomers	X	–	–	Chebotareva et al. (2010)
p90Rsk	HspB1	HspB1 phosphorylation	nd	X	X	–	Zoubeidi et al. (2010)
PRKD1	HspB1	HspB1 phosphorylation	HspB1 P-serine 82	X	–	–	Doppler et al. (2005)
PPM1A	HspB1	?	nd	X	–	–	Wang et al. (2011)
Calponin, PKC	HspB1	Contraction of smooth muscles	nd	X	–	–	Patil et al. (2004)
IKKβ	HspB5	Stimulates IKKβ kinase activity	nd	X	–	–	Adhikari et al. (2011)
IKKα	HspB1	Suppresses NF-κB activation (SAP)	P-HspB1	X	–	–	Kammanadiminti and Chadee (2006)
IKKβ	HspB1	Suppresses NF-κB activation (SAP)	P-HspB1	X	–	–	Kammanadiminti and Chadee (2006)
DMPK	HspB2	Activates DMPK	Oligomers of HspB2	–	X	–	Prabhu et al. (2012)
ASK1	HspB1	Inhibited by HspB1-DAXX interaction	nd	X	–	–	Charette and Landry (2000)
Apoptotic, autophagic and redox factors, aging							
Caspase-3	HspB1	Pro-caspase-3 stabilization	150-200 kDa HspB1 oligomers	X	–	–	Gibert et al. (2012a) and Pandey et al. (2000)
	HspB5	Negative regulation of activity	nd	X	–	–	Hu et al. (2012)
	HspB4	Negative regulation of activity	nd	X	–	–	Hu et al. (2012)
Cytochrome c	HspB1	Inhibits cytochrome c apoptotic activity	a.a. 51-88 of HspB1	X	–	–	Bruey et al. (2000a)
Bcl-xs	HspB5	Inhibits translocation mitochondria	nd	X	X	–	Mao et al. (2004)
	HspB4	Inhibits translocation mitochondria	nd	X	X	–	Mao et al. (2004)

(continued)

Table 2.2 (continued)

Client	HspB	Consequences	Interacting structure/sequence	Observations made			References
				In cells	In vitro	by adding HspB to cells	
Bax	HspB5	Inhibits translocation mitochondria	nd	X	X	–	Hu et al. (2012) and Mao et al. (2004)
	HspB4	Inhibits translocation mitochondria	nd	X	X	–	Hu et al. (2012) and Mao et al. (2004)
p53	HspB5	Inhibits translocation mitochondria	nd	X	–	–	Liu et al. (2007)
	HspB5	Inhibits p53 mediated apoptosis	DNA-binding domain of p53	X	–	–	Watanabe et al. (2009)
GranzymeA	HspB1	GranzymeA stimulation	Mono/dimers of HspB1	X	–	–	Beresford et al. (1998)
DAXX	HspB1	Inhibits Fas apoptosis	Small HspB1oligomers	X	–	–	Charette and Landry (2000)
PEA-15	HspB1	Inhibits Fas apoptosis	nd	X	–	–	Hayashi et al. (2012)
CIAPIN1	HspB8	?	nd	X	–	–	Havugimana et al. (2012)
XIAP	HspB5	Overrides XIAP activity	nd	X	–	–	Lee et al. (2012)
HDAC1	HspB5	Enhances apoptosis of RPE cells	nd	X	–	–	Noh et al. (2008)
14-3-3gamma	HspB6	Displaces 14-3-3 binding partners	P-HspB6	–	X	–	Chernik et al. (2007) and Seit-Nebi and Gusev (2010)
14-3-3zeta	HspB6	Modulates 14-3-3 interactome affinity	P-HspB6	–	X	–	Sluchanko et al. (2011)
P66Shc	HspB1	F-actin disruption, apoptosis	nd	X	–	–	Arany et al. (2012)
G6PDH	HspB1	Stimulation of activity	Small HspB1 P-oligomers	X	–	–	Arrigo (2013), Cosentino et al. (2011) and Preville et al. (1999)

Bag3	HspB8	Co-chaperone	β4, β8 hydrophobic grooves (HspB8) IPV (Ile-Pro-Val) motifs (Bag3)	–	X	–	Carra et al. (2008b) and Fuchs et al. (2010)
	HspB6	Co-chaperone	β4, β8 hydrophobic grooves (HspB6) IPV (Ile-Pro-Val) motifs (Bag3)	–	X	–	Fuchs et al. (2010)
Transcription/translation, gene expression							
Transcription factors							
STAT-2	HspB1	STAT-2 stabilization	200–600 kDa oligomers of HspB1	X	–	–	Gibert et al. (2012a)
STAT-3	HspB1	STAT-3 stabilization	nd	X	–	–	Rocchi et al. (2005)
P-Stat-3	HspB1	P-STAT-3 stabilization	nd	X	–	–	Gibert et al. (2012a)
HSF-1	HspB1	HSF sumoylation	Large HspB1oligomers	X	–	–	Brunet Simioni et al. (2009)
GATA-1	HspB1	GATA-1 degradation	P-HspB1	X	–	–	de Thonel et al. (2010)
Snail	HspB1	Snail stabilization	nd	X	–	–	Wettstein et al. (2013)
p53	HspB1	?	nd	X	–	–	Sun et al. (2007)
Activators of transcription factors							
IKKα	HspB1	Suppression of NF-κB activation (SAP)	P-HspB1	X	–	–	Kammanadiminti and Chadee (2006)
IKKβ	HspB1	Suppression of NF-κB activation (SAP)	P-HspB1	X	–	–	Kammanadiminti and Chadee (2006)
IKKβ	HspB5	Activation of NF-κB, cell adhesion	nd	X	–	–	Adhikari et al. (2011) and Dieterich et al. (2013)

(continued)

Table 2.2 (continued)

Client	HspB	Consequences	Interacting structure/sequence	Observations made			References
				In cells	In vitro	by adding HspB to cells	
Spliceosome assembly, pre-mRNA processing							
SAM68	HspB8	Inhibits SAM68 activity	a.a. 62-133 of HspB8	X	X	–	Badri et al. (2006)
Ddx20	HspB8	Ribonucleoprotein processing	nd	X	–	–	Sun et al. (2010)
EFTUD2	HspB1	Spliceosome modulation	nd	X	–	–	Hegele et al. (2012)
SC35	HspB7	Localized to SC35 splicing speckles	nd	X	–	–	Vos et al. (2009)
	HspB5	HspB5-HDAC1 localized to SC35	nd	X	–	–	Noh et al. (2008)
Translation initiation factors							
eIF4G	HspB1	Inhibition translation during HS	nd	X	–	–	Cuesta et al. (2000)
eIF4E	HspB1	Tumor cell survival	nd	X	–	–	Andrieu et al. (2010)
mRNA half-life							
AUF1	HspB1	AUF1 degradation	P-HspB1	X	–	–	Knapinska et al. (2011) and Sinsimer et al. (2008)
Ribosomes							
p90Rsk	HspB1	HspB1 phosphorylation	nd	X	–	–	Zoubeidi et al. (2010)
Regulators of protein degradation – ubiquitination by sHsps-E3							
Smad/Smurf2	HspB1	HspB1 ubiquitination/ degradation	nd	X	–	–	Sun et al. (2011
TIF1gamma	HspB5	Disruption of Smad4 ubiquitination	nd	X	–	–	Bellaye et al. (2014)
p27kip1	HspB1	Favors p27kip1 ubiquitination/ degradation	nd	X	–	–	Parcellier et al. (2006)

Ubiquitin	HspB1	Protein degradation	nd	X	–	–	Parcellier et al. (2003)
HDM2	HspB1	HDM2 stabilization	nd	X	–	–	O'Callaghan-Sunol et al. (2007)
Fbx4	HspB5	Cyclin D1 ubiquitination/ degradation	nd	X	X	–	Lin et al. (2006)
TRAF6	HspB1	TRAF6 ubiquitination	P-HspB1	X	–	–	Wu et al. (2009)
C8/α7 proteasome subunits	HspB5	Proteasome assembly/ degradation of HspB5 bound proteins	nd	X	–	–	Boelens et al. (2001)
Protein sumoylation							
Ubc9 (UBE2I)	HspB1	Modulation of HSF-1 activity by sumoylation	nd	X	–	–	Ahner et al. (2012) and Brunet Simioni et al. (2009)
F508del CFTR	HspB1	Stimulates the degradation of Ubc9-HspB1 sumoylated F508del CFTR	nd	X	–	–	Ahner et al. (2012)
Protein deacetylation							
KDAC8	HspB1	Acetylation of HspB1	nd	X	–	–	Chen et al. (2013)
KDAC8	HspB6	Acetylation of HspB6	nd	X	–	–	Chen et al. (2013)
HDAC6	HspB1	Microtubules deacetylation	HspB1 500–700 kDa oligomers	X	–	–	Gibert et al. (2012a)
HDAC1	HspB5	Stimulates apoptosis of RPE cells	nd	X	–	–	Noh et al. (2008)
Proteins of the complement, acute phase proteins and coagulation factors							
Coagulation factors V, X; Complement C1qA, 1qB, 1qC, C1s, C1r, C5, C3, C2, C6, C7, C8, C9; Phosphatidylinositol-glycan-specific phospholipase D; vitamin K-dependent protein S cartilage acidic protein 1; mannosyl-oligosaccharide 1,2-alpha-mannosidase 1A; serpin A10 protein Z-dependent protease inhibitor; Insulin-like growth factor-binding protein; phenylcysteine oxidase 1; carboxypeptidase B2 and N subunit 2; thrombosporin; ficolin-3; platelet factor 4; glutathione peroxidase							
	HspB5	Modulation of inflammation	nd	–	–	X	Rothbard et al. (2012)

(continued)

Table 2.2 (continued)

Client	HspB	Consequences	Interacting structure/sequence	Observations made: In cells	Observations made: In vitro	Observations made: by adding HspB to cells	References
Plasma proteins							
Factor XIII	HspB1	Platelet FXIII regulation	P-HspB1 a.a. 484–503 of Factor XIII	X	–	–	Zhu et al. (1994)
Protein transport							
XPORT	HspB1	Transport of TRP and Rh1	nd	X	–	–	Rosenbaum et al. (2011)
SMN	HspB5	SMN nuclear import and assembly	P-HspB5	X	–	–	den Engelsman et al. (2005)
Cell cycle, tumor suppressors							
Cyclin D1	HspB5	Ubiquitination by HspB5- FBX4	nd	X	X	–	Lin et al. (2006)
Myc	HspB1	?	nd	X	–	–	Agrawal et al. (2010) and Koch et al. (2007)
p53	HspB1	?	nd	X	–	–	Sun et al. (2007)
	HspB5	Inhibition translocation mitochondria	nd	X	–	–	Liu et al. (2007)
	HspB5	Inhibition of p53 mediated apoptosis	DNA-binding domain of p53	X	–	–	Watanabe et al. (2009)
Base excision repair enzymes							
UDG	HspB1	?	nd	X	X	–	Mendez et al. (2000)
HAP1	HspB1	?	nd	X	X	–	Mendez et al. (2000)
Senescence							
HDM2	HspB1	Inhibition of p53 induced senescence *via* HDM2 stabilization	nd	X	–	–	O'Callaghan-Sunol et al. (2007)

Nucleases							
Rpp20 (POP7)	HspB1	Enhances RNase P activity	nd	X	–	–	Jiang and Altman (2001)
Golgi proteins							
GM130	HspB5	Golgi vesicles	nd	X	–	–	Gangalum and Bhat (2009)
Lens proteins							
HspB4	HspB5	Chaperoning	Hetero-oligomeric complex	X	X	–	Groenen et al. (1994) and Skouri-Panet et al. (2012)
HspB5	HspB4	Chaperoning	Hetero-oligomeric complex	–	X	–	Skouri-Panet et al. (2012)
	HspB1	Chaperoning	Hetero-oligomeric complex	–	X	–	Skouri-Panet et al. (2012)
betaB2-crystallin	HspB5	Chaperoning	Hetero-oligomeric complex	X	–	–	Fu and Liang (2002)
gammaC-crystallin	HspB5	Chaperoning	Hetero-oligomeric complex	X	–	–	Fu and Liang (2002)
MIP26/AQP0	HspB4	Modulation of refractive index	nd	X	–	–	Liu and Liang (2008)
sHsps co-chaperones, negative regulators							
p66Shc	HspB1	Negative regulator of HspB1	Requires p66Shc Ser36 phosphorylation	X	–	–	Arany et al. (2012)
PASS1	HspB1	Negative regulator of HspB1	a.a. 108–208 of PASS1	X	X	–	Liu et al. (2000)
Hic-5 (ARA55)	HspB1	Negative regulator of HspB1	C-terminal domain of HspB1 hic-5 LIM domains	X	–	–	Jia et al. (2001)

(continued)

Table 2.2 (continued)

Client	HspB	Consequences	Interacting structure/sequence	Observations made: In cells	Observations made: In vitro	Observations made: by adding HspB to cells	References
Bag3	HspB8	Co-chaperone	β4, β8 hydrophobic grooves (HspB8)	–	X	–	Carra et al. (2008b) and Fuchs et al. (2010)
			IPV (Ile-Pro-Val) motifs (Bag3)				
	HspB6	Co-chaperone	β4, β8 hydrophobic grooves (HspB6) IPV (Ile-Pro-Val) motifs (Bag3)	–	X	–	Fuchs et al. (2010)
Viral proteins							
NS5A (Hepatitis C)	HspB1	?	a.a. 1-122 of HspB1	X	–	–	Choi et al. (2004)
			a.a. 1-181 of NS5A				
EBNA5 (EBV)	HspB1	?	?	X	–	–	Forsman et al. (2008)
Some signaling pathways modulated by small Hsps but the targeted proteins are unknown							
?	HspB1	(de)phosphorylation cascade controlling the activity of the splicing regulator SRp38					Marin-Vinader et al. (2006)
?	HspB1	Pro-inflammatory cell signalling and gene expression, may act upstream of TAK1					Alford et al. (2007)

?	HspB4	Modulation of PKCalpha, RAF/MEK/ERK and AKT signaling pathways preventing UVA-induced apoptosis	Liu et al. (2004)
?	HspB5	Modulation of PKCalpha, RAF/MEK/ERK and AKT signaling pathways preventing UVA-induced apoptosis	Liu et al. (2004)
?	HspB5	Inhibition of RAS activation inducing p53 dependent apoptosis	Li et al. (2005)
?	HspB8	TAK1-dependent death pathways	Smith et al. (2012)

The polypeptides listed here have been demonstrated to interact in vivo and/or in vitro with the different small Hsps. The interactions detected using yeast 2-hybrid approaches that were not confirmed by immunoprecipitation or pull-down experiments were not considered. List of abbreviations: nd: not determined; P-: phosphorylated; a.a.: amino acids; 200–400 kDa: oligomers with native size of 200–400 kDa. HspB1: also known as Hsp27, Hsp28 and Hsp25 in murine; HspB2: also known as MKBP; HspB4: alphaA-crystallin; HspB5: alphaB-crystalline; α-crystallin: HspB4:HspB5 complex (3:1); HspB6: also known as Hsp20; HspB7: also known as cvHsp; HspB8: also known as Hsp22; VIF: Vimentin intermediate filaments; GFAP: Glial fibrillary acidic protein; F508del CFT: deletion F508 of CFT responsible for most cystic fibrosis pathologies; CD10: 100 kDa transmembrane metallo-endopeptidase; VEGF: vascular endothelial growth factor; FGF-2: Fibroblast growth factor 2; NGF-beta: Nerve growth factor beta; Her2: Human Epidermal Growth Factor Receptor-2; AR: androgen receptor; ERβ: estrogen receptor β; Hic-5 (ARA55): androgen receptor associated protein 55, a focal adhesion protein and steroid receptor co-activator; TRP: transient receptor potential channels; XPORT: exit protein of TRP and Rh1; Rh1: rhodopsin; HDAC6: histone deacetylase 6; KDAC8: lysine deacetylase 8; TRAF6: tumor necrosis factor receptor-associated factor 6; DAXX: death domain-associated protein 6; p90rsk: p90 ribosomal S6 kinase; IF: intermediate filaments;

(continued)

Table 2.2 (continued)

Factor XIII: transglutaminase, platelet Factor XIII; PhK: rabbit skeletal muscle phosphorylase kinase; MK2: MAPK-activated protein kinase-2; P38: P38 MAP Kinase; PKCΔ: protein kinase CΔ; Akt: also known as protein kinase B (PKB); TAK1: TGF-β activated kinase 1; p27kip1: cyclin-dependent kinase inhibitor p27kip1; DMPK: myotonic dystrophy protein kinase; PTEN: phosphatase and TENsin homolog; PEA-15: astrocytic phosphoprotein PEA-15; HDM2: human double minute 2; Bax: Bcl-2-associated X protein; Bag3: Bcl-2 associated athanogene 3; CIAPIN1: Anamorsin, a cytokine-induced inhibitor of apoptosis; ASK1: Apoptosis signaling kinase 1; Ubc6: ubiquitin conjugating enzyme E2 6; Ubc9: SUMO conjugating enzyme E2 9; GFAP: glial fibrillary acidic protein; GATA-1: globin transcription factor 1; HSF-1: heat shock factor 1; NF-κB: nuclear factor kappaB; IKKα and IKKβ: I kappa B kinase alpha and beta; SAP: soluble amebic proteins that suppress NF-κB activation through the binding of HspB1 to IKKα and IKKβ; STAT-2 and -3: signal transducer and activator of transcription 2 and 3; Fbx4: Fbox only protein 4; eIF4E, eukaryotic translation initiation factor 4E; eIF4G: eukaryotic translation initiation factor 4G; Smad-Smurf2: Smad ubiquitination regulatory factor 2, E3 ubiquitin protein ligase 2; TIF1gamma: ubiquitious nuclear protein Transcriptional Intermediary Factor 1gamma; Snail: zinc finger protein that binds and inhibits E-cadherin promoter to induce EMT (Epithelial-to-Mesenchymal Transition); SC35: serine/arginine-rich splicing factor 2; PASS1: protein associated with small stress proteins 1; Ddx20: DEAD box protein Ddx20 (gemin3, DP103); DSTN: Destrin or actin depolymerizing factor or ADF; G6PDH: glucose 6-phosphate dehydrogenase; p66Shc: 66 kDa isoform of ShcA (Src homology 2 domain containing transforming protein 2); SAM68: c-Src kinase during mitosis. MAPs: Microtubule-associated proteins; GRIFIN: galectin-related interfiber protein; SMN: survival motor neuron protein; SOD1: Cu/Zn-superoxide dismutase; UDG: uracil DNA glycosylase; HAP1: human AP endonuclease; TCTEL1: a light chain component of cytoplasmic and flagellar dynein; EFTUD2: U5-116KD, Snu114, Snrp116, elongation factor Tu GTP-binding domain-containing protein 2; PRKD1: Serine/threonine-protein kinase D1; PPM1A: protein phosphatase 1A (formerly 2C), magnesium-dependent, alpha isoform; ECs: endothelial cells; MIP26/AQP0: major lens fiber membrane protein; RPE: retinal pigment epithelial cells

2.1.2 *Enormous Cellular Implications Associated with Constitutively Expressed Small Hsps*

Mammalian small Hsps are expressed in the absence of apparent stress in specific tissues of developing and adult organisms as well as in pathological conditions (Arrigo 2012b; Bhat and Nagineni 1989; Gernold et al. 1993; Huang et al. 2007; Klemenz et al. 1993; Mymrikov et al. 2011; Quraishe et al. 2008; Srinivasan et al. 1992; Tanguay et al. 1993). For example, HspB1 and HspB6 are highly abundant in muscles. However, the overall tissue distribution of these two proteins is different since HspB6 is specific to muscles (Seit-Nebi and Gusev 2010) while HspB1 is expressed in almost all tissues. Similarly, HspB5, which forms with HspB4 the lens alpha-crystallin complex is also expressed in the heart, skeletal muscle fibers, brain and kidney while HspB4 is also present in pancreas. In contrast, HspB9 and HspB10 are restricted to testis expression (de Wit et al. 2004; Yang et al. 2012). Other important points concern the expression of these proteins in pathological conditions as well as the drastic effects (neuropathies, myopathies, cardiomyopathies, cataracts) induced by some of their mutations (i.e. mutations in HspB1, HspB3, HspB4, HspB5, HspB6 and HspB8) (Benndorf et al. 2014; Kwok et al. 2011; Mymrikov et al. 2011; Vicart et al. 1998). So, what is the function of these Hsps in specific tissues? (see Sect. 2.1.2.1).

2.1.2.1 Small Hsps Client Concept

The recent literature is quite abundant in descriptions of new functions associated with constitutively expressed small Hsps. Moreover, each small Hsp appears to have its own panel of activities (Fig. 2.1). An intriguing point is the unrelated nature of those activities distributed in almost all essential cellular pathways or activities, from cytoskeleton homeostasis to signal transduction pathways, gene expression and cell death (see Fig. 2.1). To understand why so many activities are associated with small Hsps, we must first explain their particular structural organization. Indeed, these proteins share, as a result of their crystallin homology, complex oligomeric structures that allow for the formation of dynamic homo and hetero-oligomeric structures (from 50 to >700 kDa, depending on the small Hsps) (Arrigo 2007a; 2011, Arrigo et al. 1988; Basha et al. 2011; Garrido 2002; Simon et al. 2013). Moreover, phosphorylation plays a key role in the case of HspB1, HspB5 and HspB4. These Hsps bear several serine sites phosphorylated by specific kinases, including stress and MAP kinases. Another key parameter is the cellular environment that modulates, in a dynamic and reversible way, the oligomeric organization and phosphorylation of some of these proteins, such as HspB1 (Arrigo et al. 1988; Arrigo 2000, 2007b, 2011; Arrigo and Gibert 2012; Bruey et al. 2000b; Mehlen and Arrigo 1994; Mehlen et al. 1997a; Paul et al. 2010). This suggests an intracellular sensor activity associated with small Hsps that can record changes in cellular environment. For example, HspB1 reorganizes differently its phosphorylation and oligomerization status in cells exposed to different apoptotic inducers (Paul et al. 2010). What could this mean? Since HspB1 is an anti-apoptotic protein its structural changes could instruct the cell

Fig. 2.1 Large spectrum of cellular activities associated with mammalian small Hsps. The functional activities of the different members of the small Hsps family are presented in a cartoon where each Hsp is characterized by a specific color

to choose the best strategy to counteract the effects of a particular apoptotic inducer. How can this be done? Do small Hsps have multiple enzymatic activities because of their complex oligomeric organization, and are they thus pleotropic polypeptides, or are they acting via chaperone-like activities towards other polypeptides? Recently published reports revealed that the novel activities of small Hsps often correlate with their ability to interact with different polypeptides. Hence, could the apparent pleotropic effects of small Hsps be indirect and, as previously described for Hsp90

(Georgakis and Younes 2005; Neckers et al. 1999), result from the modulation of the activity and/or half-life of many clients? (list of Hsp90 clients: http://www.picard.ch/downloads). To clarify this point, we analyzed three polypeptides pro-caspase-3, HDAC6 and STAT-2 interacting with HspB1 in HeLa cells and discovered that their half-life was greatly enhanced by interacting with HspB1 (Gibert et al. 2012a), which confirmed that, in the same cell, HspB1 can recognize different protein clients. The updated list of the major proteins interacting with mammalian small Hsps and the cellular consequences mediated by these interactions is presented in Table 2.2, see also (Arrigo 2013; Arrigo and Gibert 2012, 2013; Ciocca et al 2013). Clients are listed according to their activity in major cellular functions, such as transduction pathways, apoptosis, protein degradation, translation, transcription, cytoskeletal organization and homeostasis or cell adhesion. When available, information is given about the structural organization of small Hsps or their corresponding clients involved in the interactions. The little information already available confirms the important role played by the oligomerization and phosphorylation patterns of small Hsps. Several consequences can result from small Hsps/clients interactions, such as modulation of half-life, enzymatic activity, structural organization or modification of the client. For example, some clients interact with HspB1 to increase their half-life and thus avoid their rapid proteolytic degradation (Her2 oncogene, pro-caspase 3, HDM2, the histone deacetylase HDAC6, Androgen Receptor AR and the transcription factors STAT-2 and STAT-3) while the opposite effect occurs for the rapidly degraded PTEN polypeptide when it is bound to HspB1. The transcription factor HSF1 is sumoylated as a result of its interaction with HspB1 coupled to the Ubc-9 like sumoylating enzyme UBE21. Moreover, some cellular effects mediated by small Hsps are well known but the targeted proteins are still not defined. One striking example is the modulation of the TAK-1 inflammation pathway by HspB8 (see Table 2.2).

Two major questions arise from these observations: (i) what are the cellular consequences induced by the interaction of small Hsps to so many protein targets and (ii) how do small Hsps recognize client protein targets?

(i) Concerning the first question one can easily conclude by analyzing Table 2.2 that small Hsps modulate the maturation and activity of a wide range of client proteins including regulators of the life and death of the cell and signal transducer polypeptides, such as kinases and transcription factors. Therefore, by regulating a large repertoire of cellular functions small Hsps have a huge importance on normal biology, disease and evolutionary processes. Hence, as does Hsp90 (McClellan et al. 2007; Moulick et al. 2011; Taipale et al. 2010), these Hsps appear as global regulators of cell systems through their chaperone/client interactome systems. However, it is difficult to obtain a realistic view of the global cellular consequences generated by small Hsps interactomes. To meet this challenge we have performed protein interaction networks analysis using the geneMANIA software and database (Warde-Farley et al. 2010) (http://www.genemania.org/). This web interface shows the relationships between gene products and predicts their functional association in biological processes, pathways or diseases. Such data can help elucidate cellular pathways, create functional links between gene products and diseases, and can enable investigators to extract significantly more information about the cellular impact generated by the expression of small Hsps than by relying solely on primary literature (Table 2.2). However, care must be taken when using these data since some interactions are only predicted. An example presented in Fig. 2.2 illustrates the proteins interacting with HspB1, HspB5, HspB6 and HspB8. Only 100 proteins interacting with the four Hsps are analyzed, so some clients mentioned in Table 2.2 are not listed while new ones are mentioned. Nevertheless, this analysis further confirms that small Hsps interact with a wide spectrum of polypeptides and consequently modulate many different cellular pathways, as for example those dealing with protein kinases, gene expression, cell adhesion and migration, cell death, catabolic processes, responses to stimulation, confirming their broad implications in cell biology.

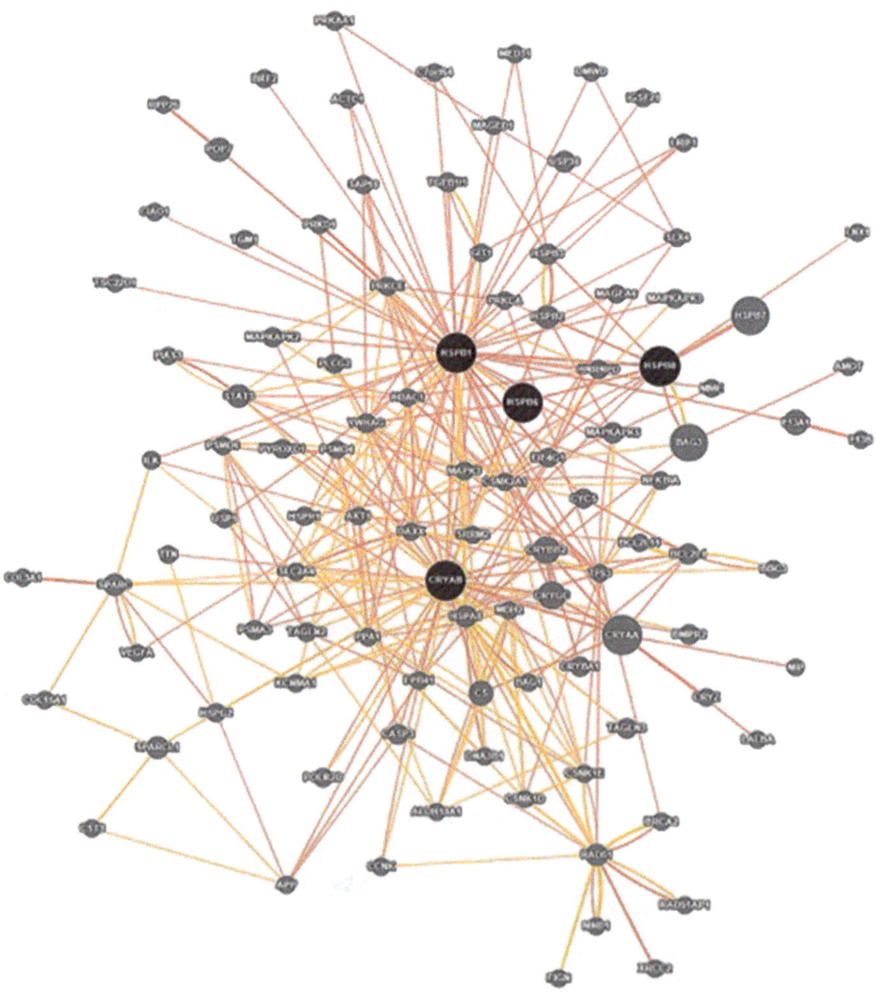

Fig. 2.2 Human HspB1, HspB5, HspB6 and HspB8 protein interactomes and predictomes as proposed by GeneMANIA software and database including BioGRID and PathwayCommons.

Fig. 2.2 (continued) Analyzed Hsps are indicated in *black* while interacting proteins are in *grey*. Physical interactions (*red lines*) and predicted (*orange lines*) ones were analyzed. The software was set to analyze up to hundred gene products and at most hundred related attributes. Automatically selected weighting method. Predicted interactions could be for instance, two proteins known to interact in another organism, such as S. cerevisiae. Abbreviations: *CRYAB* HspB5, *CRYAA* HspB4, *HSPA8* heat shock 70 kDa protein 8, *HSPH1* heat shock 105 kDa/110 kDa protein 1, *DNAJB1* DnaJ (Hsp40) homolog, subfamily B, member 1, *CRYGC* crystallin, gamma C, *CRYBB2* crystallin, beta B2, *CRYZ* crystallin, zeta (quinone reductase), *F13A1* coagulation factor XIII, A1 polypeptide, *BAG3* BCL2-associated athanogene 3, *CS* citrate synthase, *POP7* processing of precursor 7, ribonuclease P/MRP subunit (S. cerevisiae), *STAT-3* signal transducer and activator of transcription 3 (acute-phase response factor), *SPARCL1* SPARC-like 1 (hevin), *RAD51* RAD51 homolog (S. cerevisiae), *SPARC* secreted protein, acidic, cysteine-rich (osteonectin), *USP38* ubiquitin specific peptidase 38, *BCL2L1* BCL2-like 1, *MAPKAPK5* mitogen-activated protein kinase-activated protein kinase 5, *CRYBA1* crystallin, beta A1, *TAGLN3* transgelin 3, *CASP3* caspase 3, apoptosis-related cysteine peptidase, *BMPR2* bone morphogenetic protein receptor, type II (serine/threonine kinase), *CYCS* cytochrome c, somatic, *MAPKAPK2* mitogen-activated protein kinase-activated protein kinase 2, *TGFB1I1* transforming growth factor beta 1 induced transcript 1, *YWHAG* tyrosine 3-monooxygenase/tryptophan 5-monooxygenase activation protein, gamma polypeptide, *PSMA3* proteasome subunit, alpha type, 3, *MAPKAPK3* mitogen-activated protein kinase-activated protein kinase 3, *POLR2D* polymerase (RNA) II (DNA directed) polypeptide D, *TAGLN2* transgelin 2, *PLCG2* phospholipase C, gamma 2 (phosphatidylinositol-specific), *PYROXD1* pyridine nucleotide-disulphide oxidoreductase domain 1, *TGM1* transglutaminase 1 (K polypeptide epidermal type I, protein-glutamine-gamma-glutamyltransferase), *USP1* ubiquitin specific peptidase 1, *EIF4G1* eukaryotic translation initiation factor 4 gamma, 1, *HNRNPD* heterogeneous nuclear ribonucleoprotein D (AU-rich element RNA binding protein 1, 37 kDa), *PRKCE* protein kinase C, epsilon, *HSPG2* heparan sulfate proteoglycan 2, *PRKAA1* protein kinase, AMP-activated, alpha 1 catalytic subunit, *DMWD* dystrophia myotonica, WD repeat containing, *PRKD1* protein kinase D1, *ILK* integrin-linked kinase; *MAGED1* melanoma antigen family D, 1, *SAP18* Sin3A-associated protein, 18 kDa, *GIT1* G protein-coupled receptor kinase interacting ArfGAP 1, *MAPK3* mitogen-activated protein kinase 3, *MAGEA6* melanoma antigen family A, 6, *BRF2* BRF2, subunit of RNA polymerase III transcription initiation factor, BRF1-like, *CCNK* cyclin K, *IGSF21* immunoglobin superfamily, member 21, *MME* membrane metallo-endopeptidase, *PSMD4* proteasome 26S subunit, non-ATPase, 4, *PSMD6* proteasome 26S subunit, non-ATPase, 6, *TTN* titin, *CIAO1* cytosolic iron-sulfur protein assembly 1, *DAXX* death-domain associated protein, *EPB41* erythrocyte membrane protein band 4.1 (elliptocytosis 1, RH-linked), *PPA1* pyrophosphatase (inorganic) 1, *ACTC1* actin, alpha, cardiac muscle 1. *AKT1* v-akt murine thymoma viral oncogene homolog 1, *KCNMA1* potassium large conductance calcium-activated channel, subfamily M, alpha member 1, *LNX1* ligand of numb-protein X, *MED31* mediator complex subunit 31, *C7orf64* chromosome 7 open reading frame 64, *NFKBIA* nuclear factor of kappa light polypeptide gene enhancer in B-cells inhibitor, alpha, *SLC2A4* solute carrier family 2 (facilitated glucose transporter) member 4, *TP53* tumor protein p53, *TSC22D1* TSC22 domain family, member 1, *ALDH18A1* aldehyde dehydrogenase 18 family, member A1, *AMOT* angiomotin, *APP* amyloid beta (A4) precursor protein, *BAG1* BCL2-associated athanogene, *BBC3* BCL2 binding component 3, *BCL2L11* BCL2-like 11 (apoptosis facilitator), *BRCA2* breast cancer 2, early onset, *COL15A1* collagen, type XV, alpha 1, *COL3A1* collagen, type III, alpha 1, *CSNK1D* casein kinase 1, delta, *CSNK1E* casein kinase 1,epsilon, *CSNK2A1* casein kinase 2, alpha 1 polypeptide, *CST3* cystatin C, *F13B* coagulation factor XIII, B polypeptide, *FIGN* fidgetin, *HDAC1* histone deacetylase 1, *LALBA* alpha-lactalbumin, *LRIF1* ligand dependent nuclear receptor interacting factor 1, *MDH2* malate dehydrogenase 2, NAD (mitochondrial), *MIP* major intrinsic protein of lens fiber, *MND1* meiotic nuclear divisions 1 homolog (S. cerevisiae), *PIAS3* protein inhibitor of activated STAT-3, *PRKCA* protein kinase C, alpha RAD51AP1: RAD51 associated protein 1, *RPP25* ribonuclease P/MRP 25 kDa subunit, *SLX4* SLX4 structure-specific endonuclease subunit homolog (S. cerevisiae), *SRRM2* serine/arginine repetitive matrix 2, *VEGFA* vascular endothelial growth factor A

(ii) As for the second question, we believe that small Hsps act as does Hsp90 to recognize clients by taking advantage of a variety of conformational states to interact with co-chaperones and clients (Hessling et al. 2009; Mickler et al. 2009). Compared to other mammalian small Hsps, HspB1 has the most dynamic phospho-oligomeric organization, a property that could explain its ability to recognize a large number of protein clients probably through the rapid generation of interacting platforms (Arrigo and Gibert 2012, 2013; Ciocca et al. 2013; Gibert et al. 2011, 2012a; Paul et al. 2010). Consequently, HspB1 dynamic interactome may allow cells to respond quickly and mount the most effective response to a particular condition. However, an unanswered question is how small Hsps generate specific interacting platforms to act on client repertoire. At least in the case of HspB1, the phenomenon may depend on the complex patterns of MAPKAPK2,3-dependent phosphorylation of three serines sites located in the N-terminal domain of HspB1 (Arrigo and Gibert 2012, 2013; Paul et al. 2010; Rouse et al. 1994; Simon et al. 2013; Stokoe et al. 1992). Our recent observations favor this hypothesis since in growing HeLa cells pro-caspase-3 interacts mainly with the serine 15 phosphorylated small oligomers of HspB1 while HDAC6 is recovered at the level of the large serine 82 phosphorylated oligomers. In contrast, STAT-2 binds to the medium and large sized HspB1 oligomers (Arrigo and Gibert 2013; Gibert et al. 2012a). Thus, in growing HeLa cells, the specific phospho-oligomeric organization of HspB1 consists of signaling structures that recognize and bind at least three different polypeptides and subsequently modulate their half-life. This observation confirms the hypothesis that the dynamic structural plasticity of small Hsps structure can lead to at least 300 different stoichiometries that favor the recognition of many particular target proteins (Stengel et al. 2010).

An increased complexity arises by taking into account another fundamental property of small Hsps. Once they are expressed in the same cells, they have the ability to interact with each other and form multiple combinatorial oligomeric structures (Table 2.3, see also Arrigo 2013; Bukach et al. 2009; den Engelsman et al. 2009 #3479; Saha and Das 2004; Simon et al. 2007; Zantema et al. 1992). Since interaction between two small Hsps mutually affects the structure and chaperone activity of both partners (Aquilina et al. 2013; Bukach et al. 2009; den Engelsman et al. 2009; Gibert et al. 2013; Mymrikov et al. 2012; Simon et al. 2013; Skouri-Panet et al. 2012), it cannot be excluded that the chimeric oligomers can recognize novel client proteins and/or are unable to bind those interacting with parental small Hsps. Moreover, not all sHsps interact equally efficiently with each other in vitro (Mymrikov et al. 2012). In that respect, the molecular ratio between small Hsp partners is often different (e.g. 3:1 in the case of HspB4:HspB5 and HspB2:HspB3 complexes). In vivo, the phenomenon is probably even more complex since modifications, such as phosphorylation, which depend on the type of cell considered and its physiology are of prime importance (Paul et al. 2010; Simon et al. 2013). For example, in cells expressing an equimolar ratio of HspB1 and HspB5, only 90 % of

Table 2.3 Chimeric hetero-oligomeric HspB complexes

Client	HspB	Resulting effects	Interacting structure/sequence	Observations made		References
				In cells	In vitro	
HspB1	HspB5	HspB1 chaperoning	400–800 kDa hetero-oligomers	X	X	Fu and Liang (2003) and Simon et al. (2013)
	HspB6	?	Hetero-oligomers formation	–	X	Bukach et al. (2009) and Mymrikov et al. (2012)
	HspB8	?	Hetero-oligomers formation	X	X	Sun et al. (2004)
HspB2	HspB3	Role in myogenic differentiation	150 kDa oligomers (B2:B3 ratio 3:1)	X	X	den Engelsman et al. (2009) and Sugiyama et al. (2000)
HspB3	HspB2	Role in myogenic differentiation	150 kDa oligomers (B2:B3 ratio 3:1)	X	X	den Engelsman et al. (2009) and Sugiyama et al. (2000)
	HspB8	?	Hetero-oligomers formation	X	X	Fontaine et al. (2005)
HspB4	HspB5	HspB4 chaperoning	Alpha-crystallin (B4:B5 ratio: 3:1)	X	X	Groenen et al. (1994), Skouri-Panet et al. (2012), Sreelakshmi and Sharma (2006), and Srinivas et al. (2008)
HspB5	HspB1	HspB5 chaperoning	400–800 kDa hetero-oligomers	X	X	Fu and Liang (2003) and Simon et al. (2013)
	HspB6	?	Hetero-oligomers formation	–	X	Bukach et al. (2009) and Mymrikov et al. (2012)
HspB6	HspB1	?	Hetero-oligomers formation	–	X	Mymrikov et al. (2012)
	HspB8	?	Hetero-oligomers formation	X	X	Fontaine et al. (2005)
HspB7	HspB8	?	Hetero-oligomers formation	X	X	Sun et al. (2004)
HspB8	HspB1	?	Hetero-oligomers formation	X	X	Sun et al. (2004)
	HspB2	?	Hetero-oligomers formation	X	X	Sun et al. (2004)
	HspB3	?	Hetero-oligomers formation	X	X	Fontaine et al. (2005)
	HspB6	?	Hetero-oligomers formation	X	X	Fontaine et al. (2005)
	HspB7	?	Hetero-oligomers formation	X	X	Sun et al. (2004)
	HspB5	?	Hetero-oligomers formation	X	X	Fontaine et al. (2005) and Mymrikov et al. (2012)
HspB9	–	–	–	–	–	–
HspB10	–	–	–	–	–	–

Listed here are the interactions between small Hsps as demonstrated in vitro or in vivo using conventional biochemical approaches when these proteins are expressed in same cell

Table 2.4 HspB chimeric hetero-oligomeric complexes: biological consequences and targeted polypeptides

Client	HspB complex	Resulting effects	Interacting structure/sequence	Observations made		References
				In cells	In vitro	
?	HspB2/ HspB3	Muscle development	?	X	–	den Engelsman et al. (2009) and Sugiyama et al. (2000)
Bag3	HspB8/ HspB6	Chaperone Bag3 activity	β4, β8 hydrophobic grooves (HspB8/B6) IPV (Ile-Pro-Val) motifs (Bag3)	X	X	Carra et al. (2008b) and Fuchs et al. (2010)
G6PDH	HspB1/ HspB5	Stimulation of activity	Interaction with hyperphosphorylated HspB1, no interaction with HspB5	X	–	Arrigo (2013)

Listed here are the clients that interact with small Hsps consequently of the formation of hetero-oligomeric Hsps complexes

HspB1 forms chimeric molecules with HspB5. This enhances the phosphorylation of the remaining 10 % of non interacting HspB1 which can now recognize a new client, G6PDH, and can stimulate its detoxicant enzymatic activity (Table 2.4) (Arrigo 2013). Unfortunately, no clear data are yet available concerning the protein targets recognized by chimeric small Hsps (Table 2.4), as for example in the case of HspB2:HspB3 complex involved in the development of muscle cells. Similarly, it is not known whether Bag3, which interacts with HspB8 and HspB6, can bind to HspB8:HspB6 complex to modulate autophagy. Another important consequence of the above mentioned property of small Hsps is the dominant effect of a mutated small Hsp that can dramatically spread between other interacting members of the family (Diaz-Latoud et al. 2005; Fontaine et al. 2006; Simon et al. 2013). These pathological interactions can also lead to the accumulation of cytoplasmic protein aggregates linked to diseases.

2.2 Examples Illustrating the Broad Spectrum of Positive or Negative Roles of Small Hsps in Human Pathologies

Nowadays, the medical literature is filled with reports explaining that the level of expression of small Hsps is highly modulated, as they are often upregulated in pathological conditions such as protein conformational disorders (neurodegenerative diseases, myopathies, cataracts), inflammatory diseases and cancers. Many functions were attributed to HspB1 and HspB5 and, probably due to their more recent discovery, less frequently to the other small Hsps. As mentioned above, these proteins probably act by interacting with pathology specific clients. Based on earlier observations, we proposed that the upregulation of these proteins had a negative effect (for the patient) in cancer pathologies while it was positive in the case of degenerative diseases (Arrigo and Simon 2010; Arrigo 2005; Arrigo et al. 2007). The most recent studies have complicated this hypothesis since, as described below (Sect. 2.2.3), one small Hsp can be beneficial in one type of cancer and harmful in another. In fact, from a patient point of view, the major effects mediated by these interactions will depend on the friendly or hostile nature of the interacting clients. Thus, more work is needed to increase our knowledge of the pathology-dependent clients that interact with small Hsps, and future therapeutic interventions will have to be carefully planned to avoid dramatic off-target effects for patients.

2.2.1 Degenerative Diseases

2.2.1.1 Protective Role of Small Hsps

Elevated levels of Hsps, such as HspB1, HspB5 and high molecular weight Hsps, are observed in cells with altered protein folding homeostasis as a result of the expression of proteins prone to aggregate or fibrillate (see Table 2.1). Hence, high levels of these Hsps are observed in cortical Lewy bodies, Alzheimer's disease plaques containing β-amyloid peptide, granules of neurones expressing polyQ mutants of Huntingtin polypeptide, Rosenthal fibers of Alexander disease, Creutzfeldt-Jakob altered neurons, neurofibrillary tangles, α-synuclein deposit associated with Parkinson's disease, SOD1 aggregates in amyotrophic lateral sclerosis, myopathy-associated inclusion body such as muscle cells expressing mutated desmin as well as in neurones from cerebral ischemia or heart cells altered by myocardial infarction or atrial fibrillation (Bruinsma et al. 2011; Brundel et al. 2008; Goldfarb et al. 2004; Muchowski 2002; Muchowski and Wacker 2005; Renkawek et al. 1994; Wyttenbach 2004; Yerbury et al. 2012). In these cells, HspB1 and HspB5 trigger a beneficial protection by reducing the formation of pathological protein aggregates (Eaton et al. 2000; Efthymiou et al. 2004; Latchman 2005; Lewis et al. 1999). Protective activity has recently been reported for other small Hsps, such as HspB2, HspB3, HspB6, HspB7, HspB8 (Bruinsma et al. 2011; Brundel et al. 2008; Carra et al. 2005, 2008a; Ke et al. 2011; Vos et al. 2010). However, these Hsps are effective in their own way in counteracting protein aggregation or fibrillation. For example HspB7, which, unlike HspB1, does not improve the refolding of heat-denatured polypeptides, is nevertheless the most efficient small Hsp in suppressing polyQ aggregation and polyQ-induced cellular toxicity (Vos et al. 2010). Taken together these observations lead to the conclusion that small Hsps are beneficial proteins that interfere with pathological processes leading to neurodegenerative, myopathic, cardiomyopathic, cataract and retinal diseases (Andley 2007; Firdaus et al. 2006a; Lee et al. 2006; Outeiro et al. 2006; Perrin et al. 2007; Wilhelmus et al. 2006a, b; Wyttenbach et al. 2002). This conclusion was further supported by mutations which inhibit the chaperone activity of HspB1, HspB3, HspB4, HspB5, HspB6 and HspB8 and provoke pathological diseases, such as amyotrophic lateral sclerosis (ALS), axonal Charcot-Marie-Tooth disease, inherited peripheral and motor neuropathies, myofibrillar myopathies, cardiomyopathies and cataracts (Ackerley et al. 2005; Benndorf et al. 2014; Bova et al. 1999; Datskevich et al. 2012; Dierick et al. 2007; Elicker and Hutson 2007; Evgrafov et al. 2004; Kijima et al. 2005; Vicart et al. 1998). However, depending on the clients that are recognized by these Hsps, the consequences of their mutations will vary, with HspB1, HspB3, and HspB8 causing motor neuropathies, while HspB5 induces particular myopathies called αB-crystallinopathies (Benndorf et al. 2014).

2.2.1.2 Oxidative Stress Generated by Aggregated Polypeptides

In addition to their anti-aggregation and fibrillation properties the fact that at least HspB1 and HspB5 can act as anti-oxidant molecules (Arrigo 1998, 2013; Arrigo et al. 2005; Chen et al. 2006; Firdaus et al. 2006a, b; Mehlen et al. 1996a; Wyttenbach et al. 2002) is of prime importance as it can counteracts some of the harmful effects induced by aggregated polypeptides. Indeed, a disregulated intracellular redox leading to permanent oxidative conditions is a common feature observed in many degenerative diseases and in cells bearing aggregated polypeptides (Bharath et al. 2002; Browne et al. 1999; Choi et al. 2005; Firdaus et al. 2006b; Fox et al. 2007; Halliwell 2001; Jenner and Olanow 1996; Tabner et al. 2001; Turnbull et al. 2003). This phenomenon is a consequence of Huntingtin, β-amyloid and α-synuclein being metal homeostasis modulating or direct iron/copper binding polypeptides (Hilditch-Maguire et al. 2000; Huang et al. 2004). Hydroxyl radical over-production through the metal-mediated alteration of the hydroxyl radical generating Fenton reaction is thus a common feature of cells containing these aggregated polypeptides (Halliwell and Gutteridge 1984; Sayre et al. 2000 #1935; Shoham and Youdim 2000). Hydroxyl radicals stimulate protein aggregation and interfere with proteasome function (Firdaus et al. 2006a, b; Janue et al. 2007; Liu et al. 2006; Wyttenbach et al. 2002). These observations lead to the conclusion that some small Hsps, as HspB5 (Bjorkdahl et al. 2008; Ousman et al. 2007), could be considered as therapeutic agents to treat degenerative diseases.

2.2.2 Inflammation

HspB1 is essential for both IL-1 and TNF-induced pro-inflammatory signaling pathways leading to the expression of pro-inflammatory genes, such as cyclooxygenase-2, IL-6, and IL-8 (Alford et al. 2007). Increased cyclooxygenase-2 and IL-6 expression appears to occur through the stabilisation of their respective mRNAs as a result of the enhanced activation of the kinase downstream of p38 MAPK, MK2 by HspB1. The client(s) targeted by HspB1 to perform this task are still unknown, but may reside at the level or more upstream of the pivotal kinase TAK1. This study also shows that in this context many signaling events depend on HspB1, such as downstream signalling by p38 MAPK, JNK and their activators (MKK-3, -4, -6, -7) and IKKβ. In that respect, it is worth noting that HspB1 can interact with the activating kinases IKKα and IKKβ of the transcription factor NF-κB (Dodd et al. 2009). Another role has been proposed for HspB1 through its association with the AUF1- and signal transduction-regulated complex, ASTRC, that regulates mRNA degradation machinery. This could lead to a mechanism that combines proinflammatory cytokine induction with monocyte adhesion and motility (Sinsimer et al. 2008). HspB5 also plays several roles in inflammation. The first one describes HspB5 as a new regulator of leukocyte recruitment, through its ability to enhance NF-κB pro-inflammatory signaling pathways and the expression of endothelial adhesion molecule during endothelial activation (Dieterich et al. 2013). No putative client has yet been described to support this activity. The second activity concerns a role for HspB5 as an extracellular protein (see Sect. 2.2.4) and deals with its ability, when added to the plasma of patients suffering of multiple sclerosis, rheumatoid arthritis, and amyloidosis as well of mice with experimental allergic encephalomyelitis, to interact with some relative apparent selectivity with at least 70 different pro-inflammatory mediators (acute phase proteins, members of the complement cascade, and coagulation factors) (Rothbard et al. 2012) (see Table 2.2). Of great interest, the presence of exogenous HspB5 decreased inflammation as a result of a reduced concentration of these mediators. Using a similar approach, another study points to the activation of an immune-regulatory macrophage response and inhibition of lung inflammation using HspB5-loaded microparticles (van Noort et al. 2013). These observations, as well as that of Kurnellas et al. (2012), confirm that exogenous HspB5 could be used as an anti-inflammation therapeutic agent. HspB1 and HspB5 also have beneficial protective roles against inflammation since their anti-oxidant properties may favor their interference with tumor necrosis factor (TNFα) signaling pathways, as observed in the case of asthma (Alford et al. 2007; Kammanadiminti and Chadee 2006; Mehlen et al. 1995; Merendino et al. 2002). Taken together, these observations suggest crucial, but different, roles for HspB1 and HspB5 in inflammatory processes.

2.2.3 Cancers

Multiple molecular alterations are key characteristics of most cancer cells. However, an overall view of the major proteins involved in oncogenic signaling pathways is currently beyond reach. In that respect, small Hsps are among the proteins whose expression is altered in cancer cells. It is now well recognized that they have key roles in cancer biology as a result of their interaction with specific clients that modulate tumor development through their activity at the level of apoptosis, mitotic signaling pathways, angiogenesis, cell escape and survival, senescence, epithelial-to-mesenchymal transition (EMT) and metastasis (Arrigo and Gibert 2014).

In recent years, the major small Hsps reported to play important roles in cancer pathologies were HspB1 and HspB5 (Arrigo 2007a; Arrigo and Simon 2010; Arrigo and Gibert 2014; Arrigo et al. 2007; Calderwood et al. 2006; Ciocca and Calderwood 2005). Recent observations now include HspB4, HspB6 and HspB8 as well as the intriguing dual pro- and anti-tumorigenic properties of some small Hsps.

2.2.3.1 Pro-tumorigenic Effects of Small Hsps

Elevated levels of expression of HspB1 and HspB5 were the first indicators of the putative role of small Hsps in some cancer cells. It was first discovered that a high level of expression of these proteins protects against apoptotic death (Mehlen et al. 1996b) and is pro-tumorigenic (Garrido et al. 1998). Recent studies have analyzed their mode of action favoring tumor development.

Protection Against Cell Death, Apoptosis

Protection against apoptotic cell death by HspB1 was discovered in 1996 (Mehlen et al. 1996b, 1997b; Samali and Cotter 1996). This property suggested that the high level of expression of HspB1 observed in many cancer cells could promote carcinogenesis, tumor maintenance and dissemination, an assumption demonstrated two years later (Garrido et al. 1998). HspB1 anti-apoptotic property is a consequence of its interaction with many client proteins in the initiation and execution phases of apoptosis (Arrigo 2012a; Arrigo and Gibert 2014; Ciocca et al. 2013). In fact, based on the signal transduction-dependent dynamic reorganization of its phosphorylation and oligomerization status (Paul et al. 2010; Rogalla et al. 1999), HspB1 can interact with the more appropriate clients to counteract apoptotic processes. This leads to the hypothesis that HspB1 has multiple strategies to counteract inducer-specific intrinsic and extrinsic apoptosis (Arrigo 2011; Paul et al. 2010). For example, by acting towards F-actin and t-Bid translocation, HspB1 reduces cytochrome c (Paul et al. 2002) and Smac-diablo (Chauhan et al. 2003) release from mitochondria. In addition, it also decreases apoptosome and caspase-9 activation by a direct interaction with cytosolic cytochrome c (Bruey et al. 2000a; Garrido et al. 1999). A surprising effect occurs at the level of procaspase-3 whose activation is negatively regulated by phosphorylated small oligomers of HspB1 (Arrigo and Gibert 2013; Gibert et al. 2012a; Pandey et al. 2000). In the meantime, HspB1 increases procaspase-3 half-life by down-regulating its degradation by the ubiquitin-proteasome machinery (Gibert et al. 2012a). Among the death receptor pathways that are under the control of HspB1 are Fas, TNFα and TRAIL (Mehlen et al. 1995, 1996b; Zhuang et al. 2009). In the Fas signal transduction mechanism, phosphorylated dimers of HspB1 abolished the link between activated Fas receptor and apoptotic signaling kinase1 (Ask1) by interacting with DAXX (Charette et al. 2000). The protection against TNFα mediated transduction death signal is less well documented. Nevertheless, HspB1 may protect cells directly through the classical apoptotic machinery and/or its ability to interfere with the oxidative stress generated by this inflammatory cytokine (Mehlen et al. 1995, 1996a). In contrast (see below section "Stimulation of cell survival pathways, senescence"), the inhibitory effect of HspB1 against TRAIL induced death does not appear to occur at the level of the apoptotic machinery but rather through the stimulation a cell survival mechanism (Qi et al. 2014).

HspB5 and HspB4 have also been reported as anti-apoptotic proteins (Andley et al. 2000; Kamradt et al. 2005) and several reports mention their action towards tumorigenicity (Arrigo 2007a; Chen et al. 2012; Kase et al. 2009; Mahon et al. 1987; Rigas et al. 2009). Their anti-apoptotic modes of action differ from that of HspB1, however. Indeed, in addition to their action towards caspase-3, these Hsps negatively regulate members of the Bcl-2 family, Bcl-X_L, Bcl-XS and Bax, as well as cytoplasmic p53 by interfering with their redistribution into mitochondria in apoptotic conditions (Hu et al. 2012; Liu et al. 2007; Mao et al. 2004). HspB5 was also shown to modulate p53 level (Watanabe et al. 2009). Moreover, both HspB4 and HspB5 can prevent apoptosis through interactions with clients involved in regulating signaling Raf/MEK/ERK and PKCαlpha pathways (Liu et al. 2004). Moreover, HspB5 modulates the activity of XIAP, an endogenous inhibitor of caspases (Lee et al. 2012), and inhibits RAS activation responsive to the calcium-activated Raf/MEK/ERK signaling pathway mediated p53-dependent apoptosis (Li et al. 2005). HspB5 expression can also be correlated with pERK1/2 expression (van de Schootbrugge et al. 2013b). However, it is important to note that these particular properties are usually tissue specific; for example, in pancreatic cancer cells HspB4 has a surprising opposite effect and acts as a negative regulator of carcinogenesis (Deng et al. 2010) (see below section "Anti-tumorigenic Effects"). HspB5 also protects retinal pigment epithelial cells through its association with HDAC1 on SC35 speckles (Noh et al. 2008), which suggests that HspB5 knockout could be beneficial to vitreoretinopathy therapy.

It is also interesting to note that 14-3-3 polypeptide is a client of phosphorylated HspB6. Hence, this Hsp can compete with the large number of regulator proteins interacting with 14-3-3 and indirectly modulate many cellular processes, such as those involved in actin cytoskeleton reorganization or Bad mediated apoptosis (Chernik et al. 2007; Seit-Nebi and Gusev 2010; Sluchanko et al. 2011; Zha et al. 1997).

Stimulation of Cell Survival Pathways, Senescence

HspB1 still appears as being the major small Hsp involved in the stimulation of cell survival pathways through its interaction with specific clients. Among those pathways, the Akt signaling cascade is a major one which includes key factors such as Akt, PI3K, PTEN, mitogen-activated protein kinase kinase-3,6, BAD and Forkhead transcription factors. In cancer cells, high expression levels of HspB1 result in its interaction with Akt and PTEN. HspB1 action towards Akt kinase activity and the stimulation of the degradation of the phosphatase PTEN stimulate the PI3K/Akt signaling pathway and thus enhance the survival of these pathological cells (Cayado-Gutierrez et al. 2012; Rane et al. 2003; Wu et al. 2007). An interesting survival pathway also modulated by HspB1 is the PEA-15 molecular switch linking cell proliferation to Fas-induced apoptosis. In that regard, the interaction of HspB1 with PEA-15 inhibits

Fas-induced apoptosis and promotes cell survival and proliferation (Hayashi et al. 2012). Another example concerns the Src-Akt/ERK pro-survival signaling transduction triggered by TRAIL death receptor. Analysis of the molecular mechanism revealed that phosphorylated HspB1 activates the pathway by interacting with Src and by scaffolding protein beta-arrestin2 (Qi et al. 2014). The signaling complex made of phospho-HspB1/beta-arrestin2/Src appears therefore to be responsible for activating the TRAIL-triggered Src-Akt/ERK pro-survival pathway. HspB1 also appears to act in signaling pathways promoting survival of gliomas, but the molecular mechanism is not yet known (Golembieski et al. 2008; McClung et al. 2012).

In addition to improving cell survival, HspB1 has a p53 dependent negative action towards the oncogene-induced senescence (OIS) pathway which normally blocks cancer progression (O'Callaghan-Sunol et al. 2007). Indeed, HspB1 depletion usually induces a senescent-like phenotype in cancer cells. Among the morphological changes that were observed one can note a drastic reduction in the mitotic index through induction of p21waf expression (O'Callaghan-Sunol et al. 2007) and a particular cellular multi-nucleation which appears to be the result of the degradation of HDAC6 (Gibert et al. 2012a), an HspB1 client acting as a powerful contributor to oncogenic pathways activation (Lee et al. 2008). HDAC6 is proteolytically stabilized by HspB1 serine 82 phosphorylated oligomers (Arrigo and Gibert 2013; Gibert et al. 2012a). Among the other clients and/or pathways effective in supporting the negative effect of HspB1 towards senescence are the p53 stabilizator HDM2, an ubiquitin ligase (E3) that targets p53 for degradation (O'Callaghan-Sunol et al. 2007; Yang et al. 2005) and the PI3K/AKT induced OIS (Ghosh et al. 2013).

Cell Escape, Epithelial-to-Mesenchymal Transition (EMT), Metastasis

In addition to counteracting cell death and promoting cell survival pathways, HspB1 and HspB5 have been shown to bear tumorigenic (Garrido et al. 1998, 2006) and pro-metastatic (Bausero et al. 2006; Lemieux et al. 1997; Nagaraja et al. 2012b) properties. In that regard, several clients interacting with these proteins have been identified (Arrigo and Gibert 2014) that are particularly active at the level of the cytoskeleton and extracellular matrix (Arrigo and Gibert 2013; Gibert et al. 2012a; Lavoie et al. 1993; Mounier and Arrigo 2002; Perng et al. 1999; Wettstein et al. 2012; Xi et al. 2006). For example, in cancer cells, HspB1 is necessary for F-actin mediated cytokinesis and interferes with the accumulation of giant polynucleated cells (Gibert et al. 2012a). Another important client interacting with both HspB1 and HspB5 is β-catenin (Fanelli et al. 2008; Ghosh et al. 2007c) and the resulting effect is a modulation of cadherin-catenin cell adhesion proteins (Fanelli et al. 2008). At least in the case of HspB1, the interaction plays a crucial role in promoting tumor growth. Among the other clients of HspB1, one can cite several metalloproteinases (Bausero et al. 2006; Xu et al. 2006) as well as SPARC (secreted protein, acidic and rich in cysteine), a polypeptide that plays an important role in cell adhesion and migration (Golembieski et al. 2008; McClung et al. 2012; Schultz et al. 2012). In several cancer pathologies, HspB5 also promotes cell migration and invasion. For example, HspB5 induces the EGF- and anchorage-independent growth of human breast basal-like tumors through the constitutive activation of the MAPK kinase/ERK (MEK/ERK) pathway and transforms immortalized human mammary epithelial cells in invasive mammary carcinomas that have the same aspect as basal-like breast tumors (Gruvberger-Saal and Parsons 2006; Moyano et al. 2006). At least in the kidney, HspB5 can participate in maintaining tissue integrity by interacting with Ksp-cadherin-16 and promoting its connection to the cytoskeleton (Thedieck et al. 2008).

HspB1 is still the major small Hsp that stimulates metastasis (Bausero et al. 2004, 2006; Gibert et al. 2012b; Nagaraja et al. 2012a, b). Epithelial-to-mesenchymal transition (EMT) is the major parameter controlling metastasis that appears under the control of HspB1 (Shiota et al. 2013; Wei et al. 2011). Indeed, HspB1 modulates the expression of pro-metastatic genes (Nagaraja et al. 2012b), such as those dependent on STAT3/Twist signaling by enhancing the binding of the transcription factor STAT3 to the promoter of the Twist gene (Shiota et al. 2013). This transcriptional event generates two hallmarks of EMT: N-cadherin up-regulation and E-cadherin downregulation. It is therefore possible that the interaction of HspB1 with phosphorylated and activated STAT3 could be one of the key events regulating this phenomenon (Gibert et al. 2012a). HspB1 also binds to and stabilizes the transcription factor Snail, and consequently induces EMT features (Wettstein et al. 2013). The phenomenon probably occurs via a Snail-induced transcriptional blockage of E-cadherin gene expression (Batlle et al. 2000). E-cadherin downregulation is necessary to trigger epithelial-to-mesenchymal transition and acquisition of metastatic potential at late stages of epithelial tumour progression. Concerning HspB5, a recent study mentions that its expression is associated with distant metastases formation in head and neck squamous cell carcinoma, a link that might relate to the chaperone function of HspB5 in mediating folding and secretion of VEGF and stimulating cell migration (van de Schootbrugge et al. 2013a). Thus, among the different small Hsps, at least HspB1 and HspB5 are considered as potent stimulators of tumor progression. However, we should be cautious before coming to a general conclusion on this topic, since, as indicated below (Sect. 2.2.3.2), in some tumors these Hsps have been recently shown to have an anti-tumor activity that counteracts tumor development.

Angiogenesis

Do small Hsps participate in the process triggering the excessive formation of blood vessels that irrigate cancer cells? Until recently, no answer could be given to this question since no data supported such a pro-angiogenic hypothesis. However, recent game-changing reports have clearly demonstrated that small Hsps indeed play a role in this process. First, it was shown that, in addition to their intracellular distribution, small Hsps can also be localized in plasma membrane and can be exported in the extracellular milieu (Chowdary et al. 2006; Rayner et al. 2008; Tsvetkova et al. 2002), a phenomenon that correlates with tumor growth and metastasis formation (Bausero et al. 2004). In addition to a possible immunological role for small Hsps, a first observation was that recombinant HspB1 added to the growth medium has a pro-angiogenic effect mediated by Toll-like receptor 3 (TLR3) at the surface of human microvascular endothelial cells (HMECs). The interaction stimulates NF-κB dependent vascular endothelial growth factor (VEGF) gene transcription and promotes secretion of VEGF-activating VEGF receptor type 2 and angiogenesis (Thuringer et al. 2013). Indeed, the production by endothelial cells of intracellular autocrine (intracrine) VEGF is critical for vasculature homeostasis. A more recent study showed that HspB1 is directly released from endothelial cells (ECs) and confirmed that it modulates angiogenesis via direct interaction with VEGF. However, these authors also showed that HspB1 can be cleaved by MMP9 (Matrix MetalloProteinase 9) and recovered as anti-angiogenic fragments which interfere with VEGF-induced ECs activation and tumor progression (Choi et al. 2014). Thus, it appears that the effect mediated by extracellular HspB1 in cancer pathologies may depend on the efficiency of its cleavage by MMP9. However, the first study used recombinant HspB1 added to culture medium, so that the cleavage activity of endogenous MMP9 could have been overwhelmed by an excess of HspB1 and thus a pro-angiogenic effect was observed. Thus, in vivo, HspB1 released from cells appears as an anti-angiogenic polypeptide. This is also supported by the fact that MMP inhibitors have failed in clinical trials, probably through their efficient knock out of HspB1 fragmentation.

Another small Hsp involved in angiogenesis is HspB5 since it is crucial for endothelial cell survival and is up regulated during vessel morphogenesis. For example, tumor vessels in HspB5 (–/–) mice showed signs of caspase-3 activation and apoptosis and tumors grown in such mice were significantly less vascularized than wild-type tumors and displayed increased areas of apoptosis/necrosis (Dimberg et al. 2008). Recently, it was shown that HspB5 is a VEGF chaperone that protects this growth factor against proteolytic degradation (Kerr and Byzova 2010; Ruan et al. 2011). HspB5 appears therefore strongly involved in the pathway maintaining intracrine VEGF signaling that sustains aberrant tumor angiogenesis (Dimberg et al. 2008; Ruan et al. 2011).

Gene Expression

The control by HspB1 of several crucial transcription factors (among them Snail, STAT3, NF-κB and HSF1) can have dramatic consequences particularly towards apoptosis inhibition and EMT promotion. HSF1 (heat shock factor 1), the transcription factor responsible for Hsps expression, has also been shown to play a crucial role in tumorogenesis (Mendillo et al. 2012). HSF1 is SUMO-2/3 modified by HspB1-Ubc9 complex (Brunet Simioni et al. 2009). This modification does not affect HSF1 DNA-binding ability but blocks its transactivation function suggesting that it could act, together with NuRD factors, as a transcriptional inhibitor that represses genes that oppose metastasis. Other hypotheses suggest that it could modulate energy metabolism or permit the development of polyploidy in cancer cells (Calderwood 2012; Mendillo et al. 2012).

HspB1, HspB7 and HspB8 can also favor the expression of pro-tumorigenic proteins though the control of mRNAs. Indeed, some clients of these Hsps regulate mRNA splicing, such as SAM68, Ddx20, EFTUD2 and SC35 (Badri et al. 2006; Hegele et al. 2012; Sun et al. 2010; Vos et al. 2009), while others play a role in translational initiation (eIF4G) (Andrieu et al. 2010) or mRNA stability (AUF1) (Sinsimer et al. 2008).

2.2.3.2 Anti-tumorigenic Effects

In contrast to the classical view described above favoring a pro-tumorigenic activity for HspB1 and HspB5, recent observations indicate that, in some cancer types, HspB1, HspB5 and HspB4 polypeptides display intriguing tumor suppressive activities. Moreover, recent studies dealing with HspB8 and HspB6 clearly show that these polypeptides promote tumor growth resistance and decrease cell survival.

Tumor Suppressive Role of HspB1

As mentioned above, HspB1 released from endothelial cells (ECs) regulates angiogenesis by interacting with VEGF (vascular endothelial growth factor). However, new observations have revealed that MMP9 (matrix metalloproteinase 9) can cleave HspB1 and release anti-angiogenic fragments that inhibit lung and liver tumor progression of B16F10 melanoma cells and lung tumor progression of CT26 colon carcinoma cells. The failure of MMP inhibitors in clinical trials could then be explained by their ability to decrease HspB1 fragmentation leading to pro-tumorigenic effects (Choi et al. 2014).

Tumor Suppressive Role of HspB5

In the case of nasopharyngeal carcinoma (NPC), an intriguing observation was that HspB5 downregulation is significantly associated with the progression of NPC while its overexpression interferes with NPC progression-associated phenotypes such as loss of cell adhesion, invasion, interaction with the tumor microenvironment, invasive protrusion formation and expression of epithelial-mesenchymal transition-associated markers. Molecular analysis revealed that HspB5 suppresses NPC progression by interacting with the cadherin/catenin adherens junction. This indirectly decreases the levels of expression of critical downstream targets such as cyclin-D1 and c-myc (Huang et al. 2012)

HspB4

The role of HspB4 in tumorigenesis appears rather equivocal (Deng et al. 2010). Indeed, depending of the tumor type the level of this protein is either up- or downregulated. In normal conditions, HspB4 is mainly expressed in the lens and is also detectable in the pancreas. Consequently, many of the lens tumor cells display high levels of HspB4 expression, such as those from retinoblastoma and eyelids with sebaceous carcinoma (Kase et al. 2009; Mahon et al. 1987; Rigas et al. 2009). In these cells, HspB4, like HspB5, can promote tumorigenesis since it bears an anti-apoptotic activity (Andley et al. 2000; Ciocca and Calderwood 2005) whose major property is to negatively regulate the pro-apoptotic members of the Bcl-2 family and caspase-3 (Hu et al. 2012). Contrasting with these observations, the moderate level of expression of HspB4 observed in normal human pancreas samples appears significantly reduced in many cases of pancreatic carcinoma of different types. Unfortunately, to date, the mechanism controlling HspB4 down-regulation in pancreatic carcinoma cells is not known. Another interesting point, as demonstrated by genetically forced expression of this protein, concerns the fact that, in the pancreas, HspB4 can act as a negative regulator that blocks cell transformation and retards cell migration (Deng et al. 2010). However, the mechanism by which HspB4 performs this pancreatic task is not yet solved. It may occur through a modulation of ERK MAP kinase activity regulating AP-1 expression and activity to halt cell transformation and retard cell migration (Chen et al. 2012; Deng et al. 2010). Thus, in spite of some common properties towards apoptosis, cell proliferation and tumor metastasis more work is needed to unravel the particular role of HspB4 in pancreatic carcinogenesis.

HspB8

It has been recently shown that in a large fraction of melanoma tumors, which are aggressive and drug-resistant cancers, HspB8 gene is silenced through aberrant DNA methylation. This phenomenon modulates Aza-C (5-Aza-2″-deoxycytidine) treatment efficiency (Smith et al. 2011). The anti-tumor property of HspB8 was then identified by experiments aimed at restoring its expression. Indeed, putting HspB8 back in cells inhibited tumor growth and induced the death of genetically diverse melanoma lines as a result of the activation of TAK1 (TGF-β activated kinase 1)-dependent death pathways (Li et al. 2007; Smith et al. 2012). Among the TAK1 putative down-stream pathways that could be involved is the inflammasome independent activation of caspase-1 resulting from the upregulation of ASC (apoptosis-associated speck-like protein containing a CARD). Apoptosis could then be caused by caspase-1-mediated cleavage of Beclin-1, a polypeptide upregulated in melanoma tumors as a result of mTOR (mammalian target of rapamycin) phosphorylation.

HspB6

Recent findings have shown that, in human hepatocellular carcinoma (HCC), HspB6 expression levels are inversely correlated with the progression of HCC. The negative effect mediated by HspB6 appears to result from its interaction with PI3K (phosphoinositide 3-kinase, an upstream kinase of Akt). This interaction suppresses PI3K activity, inhibits the AKT survival pathway and subsequently decreases HCC survival and growth (Matsushima-Nishiwaki et al. 2013).

Therapeutic Thoughts About Tumor Suppressive Small Hsps in Cancer

The examples presented above clearly indicate that, in some cancer cells, small Hsps can be associated with anti-tumorigenic activity. Hence, it is intriguing to note that cancer cells can devise strategies to improve their growth and dissemination by down-regulating the expression of these polypeptides. This may open up new therapeutic options aimed at restoring or up regulating the expression or activity of these proteins. However, restoring the specific expression of transcriptionally silenced genes is quite difficult. Moreover, as in the case of HspB8, the approach can be limited by the genetic diversity of the tumors. A better way to improve therapeutic strategies would be to mimic chemically the activation performed by small Hsps, as for example towards the TAK1 pathway in the case of melanoma. Similarly, restoring HspB4 or HspB5 level of expression, up-regulating HspB6 activity towards PI3K or stimulating HspB1 cleavage by MMP9 could be a challenge. In the meantime a better understanding of the role of HspB4 towards ERK MAP kinase activity and AP-1 expression as well as of HspB6 inhibitory interaction with PI3K may help in the discovery of new drugs effective against pancreatic and hepatic cells carcinogenesis.

2.2.4 *Extracellular Roles of Small Hsps*

Recently, a major discovery was that HspB1, HspB5 and HspB8 can localize in plasma membrane and be secreted in spite of their major intracellular localization (Chowdary et al. 2006; Rayner et al. 2008; Sreekumar et al. 2010; Tsvetkova et al. 2002). Thus, what could be the functions of these proteins at the cell surface or in the extracellular milieu? Do these circulating proteins share some of the properties of circulating Hsp70 (De Maio 2011)? For example, are they associated with immunogenic peptides which trigger an immune response (Delneste et al. 2002), or are they pro-immunosuppressive polypeptides (Chalmin et al. 2010). Are they involved in anti-inflammation, alarmone or other pathways by interacting with specific cellular receptors? Recent observations suggest that circulating HspB1 is not associated with immunogenic peptides but could have immunoregulatory activity. For example, circulating HspB5 stimulates macrophages through its ability to recognize CD14, TLR1 and TLR2 (Toll-like receptor 1 and 2) at their surface (van Noort et al. 2013). Similarly, HspB8 and HspB4 recognize TLR4 and induce dendritic cells activation (Roelofs et al. 2006). HspB1 was also found to activate NF-κB in macrophages (Salari et al. 2012). In addition, this protein recognizes several cell surface polypeptides such as CD10 (Dall'Era et al. 2007), Plasminogen, Angiostatin (Dudani et al. 2007) and TLR3 (Thuringer et al. 2013). In 4T1 breast adenocarcinoma cells, HspB1 cell surface expression appears correlated with tumor growth and metastasis formation (Bausero et al. 2004, 2006). Moreover, the angiogenic property of HspB1 is regulated by the cleavage efficiency of MMP9 (Choi et al. 2014; Thuringer et al. 2013) (see also Sect. Angiogenesis).

A key aspect of circulating small Hsps is that they can be either beneficial or harmful to patients suffering from different pathologies. In that regard they behave like intracellular small Hsps. For example, a major positive effect of circulating HspB1 is its impressive atheroprotective effect (Rayner et al. 2008; Salari et al. 2012). On the other hand, secreted HspB1 correlates with vascular complications in type 1 diabetic patients (Gruden et al. 2008) and is not a positive signal in cancers. Consequently, major care will have to be taken in case of therapeutic approaches targeting circulating Hsps. More studies are urgently needed to evaluate the multiple roles played by these extracellular proteins in normal and pathological physiological conditions.

2.2.5 Conclusions

As described here, small Hsps have immense cellular implications as a result of their interaction with many specific client polypeptides whose number is growing exponentially. Their ability to bind polypeptides and modulate their folding is a property that was originally discovered in heat shock treated cells where HspB1 was shown to interact with aberrantly folded polypeptides to prevent their aggregation. It is now well known that small Hsps can modulate folding or induce modifications in interacting clients. They also have the crucial ability to positively or negatively modulate their half-lifes. Taken together, these observations show that small Hsps can have a drastic influence on the level of expression as well as on the activity of interacting clients. Consequently, these Hsps indirectly appear to have a huge number of functions that allow cells to rest, grow or better adapt to changes in their physiology or pathological status. Moreover, by targeting specific clients, small Hsps can be protective and beneficial against cell degeneration. They can also have a disastrous effect by causing some cancer cells to proliferate and create metastasis.

The proteomic analysis presented here confirms our feeling that small Hsps, as Hsp90 (McClellan et al. 2007; Moulick et al. 2011; Taipale et al. 2010), are global regulators of cell systems that exert marked effects on normal biology and diseases through their chaperone/client interactome systems. Hence, we are now facing problems that are even more complex than those encountered by researchers working with Hsp90. The first of these illustrates the complexity associated with small Hsps and deals with the chimeric structures that can form between two small Hsps. These structures appear to have lost the properties associated with parental homo-oligomers, but do they have specific interactomes or are they inert? The second problem is common to small Hsps and Hsp90: what is the structural dynamic that acts on a diverse client repertoire in defined cellular conditions? In the case of HspB1, phosphorylation and oligomerization appear as key factors that dynamically react and provide a recognition platform for specific clients (Arrigo and Gibert 2013; Paul et al. 2010), however nothing is known about the molecular signaling mechanisms involved in this process. Thus, more in-depth structural work, signaling studies as well as analysis of the organization of small Hsps in living cells are necessary to unravel the problem of how these chaperones recognize client polypeptides. The third problem deals with therapeutic strategies aimed at modulating the level or activity of these chaperones. In the case of Hsp90, drugs interfering with its chaperone activity and broad interaction with clients have been clinically tested. Their modest effects and unsuspected side effects resulted in lack of FDA recognition (Whitesell et al. 2012). More specific drugs targeting only a subset of Hsp90-clients may prove more useful (Moulick et al. 2011). Similarly, the use of genetic techniques to invalidate the expression of small Hsps appears efficient (Gibert et al. 2012b; Wettstein et al. 2013) but in the long term they could be disappointing because of the complete disruption of small Hsps protein interactomes. Drugs or genetic techniques altering the structure of small Hsps can lead to interesting results (Gibert et al. 2011; Heinrich et al. 2011) but will require in-depth analysis of their effects on small Hsps interactomes. More work is needed to build comprehensive dynamic interactomes of small Hsps in specific pathologies. This will be necessary in characterizing both the good and pathological clients recognized by these Hsps. The discovery of new drugs or genetic techniques that preserve their interaction with the good clients and destroy those with the ugly ones will probably have a bright future.

References

Ackerley S, James PA, Kalli A, French S, Davies KE, Talbot K (2005) A mutation in the small heat shock protein HSPB1 leading to distal hereditary motor neuronopathy disrupts neurofilament assembly and the axonal transport of specific cellular cargoes. Hum Mol Genet 15(2):347–354

Adhikari AS, Singh BN, Rao KS, Rao Ch M (2011) alphaB-crystallin, a small heat shock protein, modulates NF-kappaB activity in a phosphorylation-dependent manner and protects muscle myoblasts from TNF-alpha induced cytotoxicity. Biochim Biophys Acta 1813(8):1532–1542

Agrawal P, Yu K, Salomon AR, Sedivy JM (2010) Proteomic profiling of Myc-associated proteins. Cell Cycle 9(24):4908–4921

Ahner A, Gong X, Schmidt BZ, Peters KW, Rabeh WM, Thibodeau PH, Lukacs GL, Frizzell RA (2012) Small heat shock proteins target mutant CFTR for degradation via a SUMO-dependent pathway. Mol Biol Cell 24(2):74–84

Alford KA, Glennie S, Turrell BR, Rawlinson L, Saklatvala J, Dean JL (2007) HSP27 functions in inflammatory gene expression and TAK1-mediated signalling. J Biol Chem 282:6232–6241
Al-Madhoun AS, Chen YX, Haidari L, Rayner K, Gerthoffer W, McBride H, O'Brien ER (2007) The interaction and cellular localization of HSP27 and ERbeta are modulated by 17beta-estradiol and HSP27 phosphorylation. Mol Cell Endocrinol 270(1–2):33–42
Andley UP (2007) Crystallins in the eye: function and pathology. Prog Retin Eye Res 26(1):78–98
Andley UP, Song Z, Wawrousek EF, Fleming TP, Bassnett S (2000) Differential protective activity of {alpha}A- and {alpha}B-crystallin in lens epithelial cells. J Biol Chem 275:36823–36831
Andrieu C, Taieb D et al (2010) Heat shock protein 27 confers resistance to androgen ablation and chemotherapy in prostate cancer cells through eIF4E. Oncogene 29(13):1883–1896
Aquilina JA, Shrestha S, Morris AM, Ecroyd H (2013) Structural and functional aspects of hetero-oligomers formed by the small heat-shock proteins alphaB crystallin and HSP27. J Biol Chem 288(19):13602–13609
Arany I, Clark JS, Reed DK, Ember I, Juncos LA (2012) Cisplatin enhances interaction between p66Shc and HSP27: its role in reorganization of the actin cytoskeleton in renal proximal tubule cells. Anticancer Res 32(11):4759–4763
Arrigo AP (1998) Small stress proteins: chaperones that act as regulators of intracellular redox state and programmed cell death. Biol Chem 379(1):19–26
Arrigo AP (2000) sHsp as novel regulators of programmed cell death and tumorigenicity. Pathol Biol (Paris) 48(3):280–288
Arrigo AP (2001) Hsp27: novel regulator of intracellular redox state. IUBMB Life 52(6):303–307
Arrigo AP (2005) Heat shock proteins as molecular chaperones. Med Sci (Paris) 21(6–7):619–625
Arrigo A-P (2007a) Anti-apoptotic, tumorigenic and metastatic potential of Hsp27 (HspB1) and alphaB-crystallin (HspB5): emerging targets for the development of new anti-cancer therapeutic strategies. In: Calderwood SK, Sherman M, Ciocca D (eds) Heat shock proteins in cancer. Springer, New-York, pp 73–92
Arrigo AP (2007b) The cellular "networking" of mammalian Hsp27 and its functions in the control of protein folding, redox state and apoptosis. Adv Exp Med Biol 594:14–26
Arrigo AP (2011) Structure-functions of HspB1 (Hsp27). Methods Mol Biol 787:105–119
Arrigo AP (2012a) Editorial: heat shock proteins in cancer. Curr Mol Med 12(9):1099–1101
Arrigo AP (2012b) Pathology-dependent effects linked to small heat shock proteins expression. Scientifica 2012:19 (Article ID 185641)
Arrigo AP (2013) Human small heat shock proteins: protein interactomes of homo- and hetero-oligomeric complexes: an update. FEBS Lett 587(13):1959–1969
Arrigo AP, Gibert B (2012) HspB1 dynamic phospho-oligomeric structure dependent interactome as cancer therapeutic target. Curr Mol Med 12:1151–1163
Arrigo AP, Gibert B (2013) Protein interactomes of three stress inducible small heat shock proteins: HspB1, HspB5 and HspB8. Int J Hyperthermia 29:409–422
Arrigo AP, Gibert B (2014) HspB1, HspB5 and HspB4 in human cancers: potent oncogenic role of some of their client proteins. Cancers (Basel) 6(1):333–365
Arrigo A-P, Simon S (2010) Dual, beneficial and deleterious, roles of small stress proteins in human diseases: implications for therapeutic strategies. In: Simon S, Arrigo A-P (eds) Book serie: protein science engineering. Nova Sciences, New York, pp 457–476
Arrigo A-P, Suhan JP, Welch WJ (1988) Dynamic changes in the structure and intracellular locale of the mammalian low-molecular-weight heat shock protein. Mol Cell Biol 8:5059–5071
Arrigo AP, Virot S, Chaufour S, Firdaus W, Kretz-Remy C, Diaz-Latoud C (2005) Hsp27 consolidates intracellular redox homeostasis by upholding glutathione in its reduced form and by decreasing iron intracellular levels. Antioxid Redox Signal 7(3–4):414–422
Arrigo AP, Simon S et al (2007) Hsp27 (HspB1) and alphaB-crystallin (HspB5) as therapeutic targets. FEBS Lett 581(19):3665–3674
Asthana A, Raman B, Ramakrishna T, Rao Ch M (2012) Structural aspects and chaperone activity of human HspB3: role of the "C-terminal extension". Cell Biochem Biophys 64(1):61–72. doi:10.1007/s12013-012-9366-x

Badri KR, Modem S, Gerard HC, Khan I, Bagchi M, Hudson AP, Reddy TR (2006) Regulation of Sam68 activity by small heat shock protein 22. J Cell Biochem 99(5):1353–1362

Barton KA, Hsu CD, Petrash JM (2009) Interactions between small heat shock protein alpha-crystallin and galectin-related interfiber protein (GRIFIN) in the ocular lens. Biochemistry 48(18):3956–3966

Basha E, O'Neill H, Vierling E (2011) Small heat shock proteins and alpha-crystallins: dynamic proteins with flexible functions. Trends Biochem Sci 37(3):106–117

Batlle E, Sancho E, Franci C, Dominguez D, Monfar M, Baulida J, Garcia De Herreros A (2000) The transcription factor snail is a repressor of E-cadherin gene expression in epithelial tumour cells. Nat Cell Biol 2(2):84–89

Bausero MA, Page DT, Osinaga E, Asea A (2004) Surface expression of Hsp25 and Hsp72 differentially regulates tumor growth and metastasis. Tumour Biol 25(5–6):243–251

Bausero MA, Bharti A et al (2006) Silencing the hsp25 gene eliminates migration capability of the highly metastatic murine 4T1 breast adenocarcinoma cell. Tumour Biol 27(1):17–26

Bellaye PS, Wettstein G et al (2014) The small heat-shock protein alphaB-crystallin is essential for the nuclear localization of Smad4: impact on pulmonary fibrosis. J Pathol 232(4):458–472

Bellomo G, Mirabelli F (1992) Oxidative stress and cytoskeletal alterations. Ann N Y Acad Sci 663:97–109

Bellyei S, Szigeti A, Pozsgai E, Boronkai A, Gomori E, Hocsak E, Farkas R, Sumegi B, Gallyas F Jr (2007) Preventing apoptotic cell death by a novel small heat shock protein. Eur J Cell Biol 86(3):161–171

Bennardini F, Wrzosek A, Chiesi M (1992) Alpha B-crystallin in cardiac tissue. Association with actin and desmin filaments. Circ Res 71(2):288–294

Benndorf R, Martin JL, Kosakovsky Pond SL, Wertheim JO (2014) Neuropathy- and myopathy-associated mutations in human small heat shock proteins: characteristics and evolutionary history of the mutation sites. Mutat Res. doi:10.1016/j.mrrev.2014.02.004

Beresford PJ, Jaju M, Friedman RS, Yoon MJ, Lieberman J (1998) A role for heat shock protein 27 in CTL-mediated cell death. J Immunol 161(1):161–167

Bharath S, Hsu M, Kaur D, Rajagopalan S, Andersen JK (2002) Glutathione, iron and Parkinson's disease. Biochem Pharmacol 64(5–6):1037–1048

Bhat SP, Nagineni CN (1989) αB subunit of lens-specific protein α-cristallin is present in other ocular and non-ocular tissues. Biochem Biophys Res Commun 158(1):319–325

Bjorkdahl C, Sjogren MJ, Zhou X, Concha H, Avila J, Winblad B, Pei JJ (2008) Small heat shock proteins Hsp27 or alphaB-crystallin and the protein components of neurofibrillary tangles: tau and neurofilaments. J Neurosci Res 86(6):1343–1352

Boelens WC, Croes Y, de Jong WW (2001) Interaction between alphaB-crystallin and the human 20S proteasomal subunit C8/alpha7. Biochim Biophys Acta 1544(1–2):311–319

Bova MP, Yaron O, Huang Q, Ding L, Haley DA, Stewart PL, Horwitz J (1999) Mutation R120G in alphaB-crystallin, which is linked to a desmin- related myopathy, results in an irregular structure and defective chaperone-like function. Proc Natl Acad Sci U S A 96(11):6137–6142

Browne SE, Ferrante RJ, Beal MF (1999) Oxidative stress in Huntington's disease. Brain Pathol 9(1):147–163

Bruey JM, Ducasse C et al (2000a) Hsp27 negatively regulates cell death by interacting with cytochrome c. Nat Cell Biol 2(9):645–652

Bruey JM, Paul C, Fromentin A, Hilpert S, Arrigo AP, Solary E, Garrido C (2000b) Differential regulation of HSP27 oligomerization in tumor cells grown in vitro and in vivo. Oncogene 19(42):4855–4863

Bruinsma IB, Bruggink KA et al (2011) Inhibition of alpha-synuclein aggregation by small heat shock proteins. Proteins 79(10):2956–2967

Brundel BJ, Ke L, Dijkhuis AJ, Qi X, Shiroshita-Takeshita A, Nattel S, Henning RH, Kampinga HH (2008) Heat shock proteins as molecular targets for intervention in atrial fibrillation. Cardiovasc Res 78(3):422–428

Brunet Simioni M, De Thonel A et al (2009) Heat shock protein 27 is involved in SUMO-2/3 modification of heat shock factor 1 and thereby modulates the transcription factor activity. Oncogene 28:3332–3344

Buchner J (1999) Hsp90 & Co. – a holding for folding. Trends Biochem Sci 24(4):136–141

Bukach OV, Glukhova AE, Seit-Nebi AS, Gusev NB (2009) Heterooligomeric complexes formed by human small heat shock proteins HspB1 (Hsp27) and HspB6 (Hsp20). Biochim Biophys Acta 1794(3):486–495

Bukau B, Horwich AL (1998) The Hsp70 and Hsp60 chaperone machines. Cell 92(3):351–366

Bullard B, Ferguson C et al (2004) Association of the chaperone alphaB-crystallin with titin in heart muscle. J Biol Chem 279(9):7917–7924

Calderwood SK (2012) HSF1, a versatile factor in tumorogenesis. Curr Mol Med 12(9):1102–1107

Calderwood SK, Khaleque MA, Sawyer DB, Ciocca DR (2006) Heat shock proteins in cancer: chaperones of tumorigenesis. Trends Biochem Sci 31(3):164–172

Carra S (2009) The stress-inducible HspB8-Bag3 complex induces the eIF2alpha kinase pathway: implications for protein quality control and viral factory degradation? Autophagy 5(3):428–429

Carra S, Sivilotti M, Chavez Zobel AT, Lambert H, Landry J (2005) HspB8, a small heat shock protein mutated in human neuromuscular disorders, has in vivo chaperone activity in cultured cells. Hum Mol Genet 14(12):1659–1669

Carra S, Seguin SJ, Lambert H, Landry J (2008a) HspB8 chaperone activity toward poly(Q)-containing proteins depends on its association with Bag3, a stimulator of macroautophagy. J Biol Chem 283(3):1437–1444

Carra S, Seguin SJ, Landry J (2008b) HspB8 and Bag3: a new chaperone complex targeting misfolded proteins to macroautophagy. Autophagy 4(2):237–239

Carra S, Rusmini P et al (2013) Different anti-aggregation and pro-degradative functions of the members of the mammalian sHSP family in neurological disorders. Philos Trans R Soc Lond B Biol Sci 368(1617):20110409

Cayado-Gutierrez N, Moncalero VL, Rosales EM, Beron W, Salvatierra EE, Alvarez-Olmedo D, Radrizzani M, Ciocca DR (2012) Downregulation of Hsp27 (HSPB1) in MCF-7 human breast cancer cells induces upregulation of PTEN. Cell Stress Chaperones 18(2):243–249

Chalmin F, Ladoire S et al (2010) Membrane-associated Hsp72 from tumor-derived exosomes mediates STAT3-dependent immunosuppressive function of mouse and human myeloid-derived suppressor cells. J Clin Invest 120(2):457–471

Charette SJ, Landry J (2000) The interaction of HSP27 with Daxx identifies a potential regulatory role of HSP27 in Fas-induced apoptosis. Ann N Y Acad Sci 926:126–131

Charette SJ, Lavoie JN, Lambert H, Landry J (2000) Inhibition of daxx-mediated apoptosis by heat shock protein 27. Mol Cell Biol 20(20):7602–7612

Chauhan D, Li G et al (2003) Hsp27 inhibits release of mitochondrial protein Smac in multiple myeloma cells and confers dexamethasone resistance. Blood 102(9):3379–3386

Chebotareva NA, Makeeva VF, Bazhina SG, Eronina TB, Gusev NB, Kurganov BI (2010) Interaction of Hsp27 with native phosphorylase kinase under crowding conditions. Macromol Biosci 10(7):783–789

Chen H, Zheng C, Zhang Y, Chang YZ, Qian ZM, Shen X (2006) Heat shock protein 27 downregulates the transferrin receptor 1-mediated iron uptake. Int J Biochem Cell Biol 38(8):1402–1416

Chen P, Ji W et al (2012) Alpha-crystallins and tumorigenesis. Curr Mol Med 12(9):1164–1173

Chen A, Karolczak-Bayatti M, Sweeney M, Treumann A, Morrissey K, Ulrich SM, Europe-Finner GN, Taggart MJ (2013) Lysine deacetylase inhibition promotes relaxation of arterial tone and C-terminal acetylation of HSPB6 (Hsp20) in vascular smooth muscle cells. Physiol Rep 1(6):e00127

Chernik IS, Seit-Nebi AS, Marston SB, Gusev NB (2007) Small heat shock protein Hsp20 (HspB6) as a partner of 14-3-3gamma. Mol Cell Biochem 295(1–2):9–17

Choi YW, Tan YJ, Lim SG, Hong W, Goh PY (2004) Proteomic approach identifies HSP27 as an interacting partner of the hepatitis C virus NS5A protein. Biochem Biophys Res Commun 318(2):514–519

Choi J, Rees HD, Weintraub ST, Levey AI, Chin LS, Li L (2005) Oxidative modifications and aggregation of Cu, Zn-superoxide dismutase associated with Alzheimer and Parkinson diseases. J Biol Chem 280(12):11648–11655

Choi SH, Lee HJ et al (2014) MMP9 processing of HSPB1 regulates tumor progression. PLoS One 9(1):e85509

Chowdary TK, Bakthisaran R, Tangirala R, Rao MC (2006) Interaction of mammalian Hsp22 with lipid membranes. Biochem J 401:437–445

Ciocca DR, Calderwood SK (2005) Heat shock proteins in cancer: diagnostic, prognostic, predictive, and treatment implications. Cell Stress Chaperones 10(2):86–103

Ciocca DR, Arrigo AP, Calderwood SK (2013) Heat shock proteins and heat shock factor 1 in carcinogenesis and tumor development: an update. Arch Toxicol 87(1):19–48

Cosentino C, Grieco D, Costanzo V (2011) ATM activates the pentose phosphate pathway promoting anti-oxidant defence and DNA repair. EMBO J 30(3):546–555

Crippa V, Sau D et al (2010) The small heat shock protein B8 (HspB8) promotes autophagic removal of misfolded proteins involved in amyotrophic lateral sclerosis (ALS). Hum Mol Genet 19(17):3440–3456

Cuesta R, Laroia G, Schneider RJ (2000) Chaperone Hsp27 inhibits translation during heat shock by binding eIF4G and facilitating dissociation of cap-initiation complexes. Genes Dev 14(12):1460–1470

Dalle-Donne I, Rossi R, Milzani A, Di Simplicio P, Colombo R (2001) The actin cytoskeleton response to oxidants: from small heat shock protein phosphorylation to changes in the redox state of actin itself. Free Radic Biol Med 31(12):1624–1632

Dall'Era MA, Oudes A, Martin DB, Liu AY (2007) HSP27 and HSP70 interact with CD10 in C4-2 prostate cancer cells. Prostate 67(7):714–721

Datskevich PN, Nefedova VV, Sudnitsyna MV, Gusev NB (2012) Mutations of small heat shock proteins and human congenital diseases. Biochemistry (Mosc) 77(13):1500–1514

De Maio A (2011) Extracellular heat shock proteins, cellular export vesicles, and the Stress Observation System: a form of communication during injury, infection, and cell damage. It is never known how far a controversial finding will go! Dedicated to Ferruccio Ritossa. Cell Stress Chaperones 16(3):235–249

de Thonel A, Vandekerckhove J et al (2010) HSP27 controls GATA-1 protein level during erythroid cell differentiation. Blood 116(1):85–96

de Wit NJ, Verschuure P, Kappe G, King SM, de Jong WW, van Muijen GN, Boelens WC (2004) Testis-specific human small heat shock protein HSPB9 is a cancer/testis antigen, and potentially interacts with the dynein subunit TCTEL1. Eur J Cell Biol 83(7):337–345

Del Vecchio PJ, MacElroy KS, Rosser MP, Church RL (1984) Association of alpha-crystallin with actin in cultured lens cells. Curr Eye Res 3(10):1213–1219

Delneste Y, Magistrelli G et al (2002) Involvement of LOX-1 in dendritic cell-mediated antigen cross-presentation. Immunity 17(3):353–362

den Engelsman J, Gerrits D, de Jong WW, Robbins J, Kato K, Boelens WC (2005) Nuclear import of {alpha}B-crystallin is phosphorylation-dependent and hampered by hyperphosphorylation of the myopathy-related mutant R120G. J Biol Chem 280(44):37139–37148

den Engelsman J, Boros S et al (2009) The small heat-shock proteins HSPB2 and HSPB3 form well-defined heterooligomers in a unique 3 to 1 subunit ratio. J Mol Biol 393(5):1022–1032

Deng M, Chen PC et al (2010) The small heat shock protein alphaA-crystallin is expressed in pancreas and acts as a negative regulator of carcinogenesis. Biochim Biophys Acta 1802(7–8):621–631

Devlin GL, Carver JA, Bottomley SP (2003) The selective inhibition of serpin aggregation by the molecular chaperone, alpha-crystallin, indicates a nucleation-dependent specificity. J Biol Chem 278(49):48644–48650

Diaz-Latoud C, Buache E, Javouhey E, Arrigo AP (2005) Substitution of the unique cysteine residue of murine hsp25 interferes with the protective activity of this stress protein through inhibition of dimer formation. Antioxid Redox Signal 7(3–4):436–445
Dierick I, Irobi J et al (2007) Genetic variant in the HSPB1 promoter region impairs the HSP27 stress response. Hum Mutat 28(8):830
Dieterich LC, Huang H, Massena S, Golenhofen N, Phillipson M, Dimberg A (2013) alphaB-crystallin/HspB5 regulates endothelial-leukocyte interactions by enhancing NF-kappaB-induced up-regulation of adhesion molecules ICAM-1, VCAM-1 and E-selectin. Angiogenesis 16(4):975–983 [In eng]
Dimberg A, Rylova S et al (2008) alphaB-crystallin promotes tumor angiogenesis by increasing vascular survival during tube morphogenesis. Blood 111(4):2015–2023
Djabali K, de Nechaud B, Landon F, Portier MM (1997) AlphaB-crystallin interacts with intermediate filaments in response to stress. J Cell Sci 110(Pt 21):2759–2769
Djabali K, Piron G, de Nechaud B, Portier MM (1999) alphaB-crystallin interacts with cytoplasmic intermediate filament bundles during mitosis. Exp Cell Res 253(2):649–662
Dodd SL, Hain B, Senf SM, Judge AR (2009) Hsp27 inhibits IKK{beta}-induced NF-{kappa}B activity and skeletal muscle atrophy. FASEB J 23:3415–3423
Doppler H, Storz P, Li J, Comb MJ, Toker A (2005) A phosphorylation state-specific antibody recognizes Hsp27, a novel substrate of protein kinase D. J Biol Chem 280(15):15013–15019
Dudani AK, Mehic J, Martyres A (2007) Plasminogen and angiostatin interact with heat shock proteins. Mol Cell Biochem 300(1–2):197–205
Duverger O, Paslaru L, Morange M (2004) HSP25 is involved in two steps of the differentiation of PAM212 keratinocytes. J Biol Chem 279(11):10252–10260
Eaton P, Awad WI, Miller JI, Hearse DJ, Shattock MJ (2000) Ischemic preconditioning: a potential role for constitutive low molecular weight stress protein translocation and phosphorylation? J Mol Cell Cardiol 32(6):961–971
Efthymiou CA, Mocanu MM, de Belleroche J, Wells DJ, Latchmann DS, Yellon DM (2004) Heat shock protein 27 protects the heart against myocardial infarction. Basic Res Cardiol 99(6):392–394. Epub 2004 Jul 13
Ehrnsperger M, Graber S, Gaestel M, Buchner J (1997) Binding of non-native protein to Hsp25 during heat shock creates a reservoir of folding intermediates for reactivation. EMBO J 16(2):221–229
Ehrnsperger M, Gaestel M, Buchner J (2000) Analysis of chaperone properties of small Hsp's. Methods Mol Biol 99:421–429
Elicker KS, Hutson LD (2007) Genome-wide analysis and expression profiling of the small heat shock proteins in zebrafish. Gene 403(1–2):60–69
Evgrafov OV, Mersiyanova I et al (2004) Mutant small heat-shock protein 27 causes axonal Charcot-Marie-Tooth disease and distal hereditary motor neuropathy. Nat Genet 36(6):602–606
Fanelli MA, Montt-Guevara M, Diblasi AM, Gago FE, Tello O, Cuello-Carrion FD, Callegari E, Bausero MA, Ciocca DR (2008) P-cadherin and beta-catenin are useful prognostic markers in breast cancer patients; beta-catenin interacts with heat shock protein Hsp27. Cell Stress Chaperones 13(2):207–220
Firdaus WJ, Wyttenbach A, Diaz-Latoud C, Currie RW, Arrigo AP (2006a) Analysis of oxidative events induced by expanded polyglutamine huntingtin exon 1 that are differentially restored by expression of heat shock proteins or treatment with an antioxidant. FEBS J 273(13):3076–3093
Firdaus WJ, Wyttenbach A, Giuliano P, Kretz-Remy C, Currie RW, Arrigo AP (2006b) Huntingtin inclusion bodies are iron-dependent centers of oxidative events. FEBS J 273(23):5428–5441
Fontaine JM, Sun X, Benndorf R, Welsh MJ (2005) Interactions of HSP22 (HSPB8) with HSP20, alphaB-crystallin, and HSPB3. Biochem Biophys Res Commun 337(3):1006–1011
Fontaine JM, Sun X, Hoppe AD, Simon S, Vicart P, Welsh MJ, Benndorf R (2006) Abnormal small heat shock protein interactions involving neuropathy-associated HSP22 (HSPB8) mutants. FASEB J 20:2168–2170

Forsman A, Ruetschi U, Ekholm J, Rymo L (2008) Identification of intracellular proteins associated with the EBV-encoded nuclear antigen 5 using an efficient TAP procedure and FT-ICR mass spectrometry. J Proteome Res 7(6):2309–2319
Fox JH, Kama JA et al (2007) Mechanisms of copper ion mediated Huntington's disease progression. PLoS One 2(3):e334
Freeman BC, Morimoto RI (1996) The human cytosolic molecular chaperones hsp90, hsp70 (hsc70) and hdj-1 have distinct roles in recognition of a non-native protein and protein refolding. EMBO J 15(12):2969–2979
Fu L, Liang JJ (2002) Detection of protein-protein interactions among lens crystallins in a mammalian two-hybrid system assay. J Biol Chem 277(6):4255–4260
Fu L, Liang JJ (2003) Enhanced stability of alpha B-crystallin in the presence of small heat shock protein Hsp27. Biochem Biophys Res Commun 302(4):710–714
Fuchs M, Poirier DJ, Seguin SJ, Lambert H, Carra S, Charette SJ, Landry J (2010) Identification of the key structural motifs involved in HspB8/HspB6-Bag3 interaction. Biochem J 425(1):245–255
Ganea E (2001) Chaperone-like activity of alpha-crystallin and other small heat shock proteins. Curr Protein Pept Sci 2(3):205–225
Gangalum RK, Bhat SP (2009) AlphaB-crystallin: a Golgi-associated membrane protein in the developing ocular lens. Invest Ophthalmol Vis Sci 50(7):3283–5290
Garrido C (2002) Size matters: of the small HSP27 and its large oligomers. Cell Death Differ 9(5):483–485
Garrido C, Fromentin A, Bonnotte B, Favre N, Moutet M, Arrigo AP, Mehlen P, Solary E (1998) Heat shock protein 27 enhances the tumorigenicity of immunogenic rat colon carcinoma cell clones. Cancer Res 58(23):5495–5499
Garrido C, Bruey JM, Fromentin A, Hammann A, Arrigo AP, Solary E (1999) HSP27 inhibits cytochrome c-dependent activation of procaspase-9. FASEB J 13(14):2061–2070
Garrido C, Brunet M, Didelot C, Zermati Y, Schmitt E, Kroemer G (2006) Heat shock proteins 27 and 70: anti-apoptotic proteins with tumorigenic properties. Cell Cycle 5:22
Gastmann O, Burfeind P, Gunther E, Hameister H, Szpirer C, Hoyer-Fender S (1993) Sequence, expression, and chromosomal assignment of a human sperm outer dense fiber gene. Mol Reprod Dev 36(4):407–418
Georgakis GV, Younes A (2005) Heat-shock protein 90 inhibitors in cancer therapy: 17AAG and beyond. Future Oncol 1(2):273–281
Gernold M, Knauf U, Gaestel M, Stahl J, Kloetzel P-M (1993) Development and tissue-specific distribution of mouse small heat shock protein hsp 25. Dev Genet 14:103–111
Ghosh JG, Houck SA, Clark JI (2007a) Interactive domains in the molecular chaperone human alphaB crystallin modulate microtubule assembly and disassembly. PLoS One 2(6):e498
Ghosh JG, Houck SA, Clark JI (2007b) Interactive sequences in the stress protein and molecular chaperone human alphaB crystallin recognize and modulate the assembly of filaments. Int J Biochem Cell Biol 39(10):1804–1815
Ghosh JG, Shenoy AK Jr, Clark JI (2007c) Interactions between important regulatory proteins and human alphaB crystallin. Biochemistry 46(21):6308–6317
Ghosh JG, Houck SA, Clark JI (2008) Interactive sequences in the molecular chaperone, human alphaB crystallin modulate the fibrillation of amyloidogenic proteins. Int J Biochem Cell Biol 40(5):954–967
Ghosh A, Lai C, McDonald S, Suraweera N, Sengupta N, Propper D, Dorudi S, Silver A (2013) HSP27 expression in primary colorectal cancers is dependent on mutation of KRAS and PI3K/AKT activation status and is independent of TP53. Exp Mol Pathol 94(1):103–108
Gibert B, Hadchity E, Czekalla A, Aloy MT, Colas P, Rodriguez-Lafrasse C, Arrigo AP, Diaz-Latoud C (2011) Inhibition of heat shock protein 27 (HspB1) tumorigenic functions by peptide aptamers. Oncogene 34:3672–3681
Gibert B, Eckel B et al (2012a) Knock down of heat shock protein 27 (HspB1) induces degradation of several putative client proteins. PLoS One 7(1):e29719

Gibert B, Eckel B et al (2012b) Targeting heat shock protein 27 (HspB1) interferes with bone metastasis and tumour formation in vivo. Br J Cancer 107(1):63–70
Gibert B, Simon S, Dimitrova V, Diaz-Latoud C, Arrigo A-P (2013) Peptide aptamers – tools to negatively or positively modulate HspB1 (27) function. Philos Trans R Soc B Biol Sci 368(1617):20120075. doi: 10.1098/rstb.2012.0075
Goldfarb LG, Vicart P, Goebel HH, Dalakas MC (2004) Desmin myopathy. Brain 127(Pt 4): 723–734
Golembieski WA, Thomas SL et al (2008) HSP27 mediates SPARC-induced changes in glioma morphology, migration, and invasion. Glia 56(10):1061–1075
Groenen P, Merck K, de Jong W, Bloemendal H (1994) Structure and modifications of the junior chaperone alpha-crystallin. From lens transparency to molecular pathology. Eur J Biochem 225:1–19
Gruden G, Bruno G et al (2008) Serum heat shock protein 27 and diabetes complications in the EURODIAB prospective complications study: a novel circulating marker for diabetic neuropathy. Diabetes 57(7):1966–1970
Gruvberger-Saal SK, Parsons R (2006) Is the small heat shock protein alphaB-crystallin an oncogene? J Clin Invest 116(1):30–32
Halliwell B (2001) Role of free radicals in the neurodegenerative diseases: therapeutic implications for antioxidant treatment. Drugs Aging 18(9):685–716
Halliwell B, Gutteridge J (1984) Role of iron in oxygen radical reactions. Methods Enzymol 105:47–56
Haslbeck M, Franzmann T, Weinfurtner D, Buchner J (2005) Some like it hot: the structure and function of small heat-shock proteins. Nat Struct Mol Biol 12(10):842–846
Hatters DM, Lindner RA, Carver JA, Howlett GJ (2001) The molecular chaperone, alpha-crystallin, inhibits amyloid formation by apolipoprotein C-II. J Biol Chem 276(36):33755–33761
Havugimana PC, Hart GT et al (2012) A census of human soluble protein complexes. Cell 150(5):1068–1081
Hayashi N, Peacock JW, Beraldi E, Zoubeidi A, Gleave ME, Ong CJ (2012) Hsp27 silencing coordinately inhibits proliferation and promotes Fas-induced apoptosis by regulating the PEA-15 molecular switch. Cell Death Differ 19(6):990–1002
Hegele A, Kamburov A et al (2012) Dynamic protein-protein interaction wiring of the human spliceosome. Mol Cell 45(4):567–5880
Heinrich JC, Tuukkanen A, Schroeder M, Fahrig T, Fahrig R (2011) RP101 (brivudine) binds to heat shock protein HSP27 (HSPB1) and enhances survival in animals and pancreatic cancer patients. J Cancer Res Clin Oncol 137(9):1349–1361
Hessling M, Richter K, Buchner J (2009) Dissection of the ATP-induced conformational cycle of the molecular chaperone Hsp90. Nat Struct Mol Biol 16(3):287–293
Hilditch-Maguire P, Trettel F, Passani LA, Auerbach A, Persichetti F, MacDonald ME (2000) Huntingtin: an iron-regulated protein essential for normal nuclear and perinuclear organelles. Hum Mol Genet 9(19):2789–2797
Hino M, Kurogi K, Okubo MA, Murata-Hori M, Hosoya H (2000) Small heat shock protein 27 (HSP27) associates with tubulin/microtubules in HeLa cells. Biochem Biophys Res Commun 271(1):164–169
Hishiya A, Salman MN, Carra S, Kampinga HH, Takayama S (2011) BAG3 directly interacts with mutated alphaB-crystallin to suppress its aggregation and toxicity. PLoS One 6(3):e16828
Hook D, Harding J (1996) Alpha-crystallin acting as a molecular chaperone protects catalase against steroid-induced inactivation. FEBS Lett 382:281–284
Horwitz J, Huang Q-L, Ding L-L (1992) Alpha-crystallin can function as a molecular chaperone. Proc Natl Acad Sci U S A 89:10449–10453
Hu WF, Gong L et al (2012) alphaA- and alphaB-crystallins interact with caspase-3 and Bax to guard mouse lens development. Curr Mol Med 12(2):177–187
Huang X, Moir RD, Tanzi RE, Bush AI, Rogers JT (2004) Redox-active metals, oxidative stress, and Alzheimer's disease pathology. Ann N Y Acad Sci 1012:153–163

Huang L, Min JN, Masters S, Mivechi NF, Moskophidis D (2007) Insights into function and regulation of small heat shock protein 25 (HSPB1) in a mouse model with targeted gene disruption. Genesis 45(8):487–501
Huang Z, Cheng Y et al (2012) Tumor suppressor Alpha B-crystallin (CRYAB) associates with the cadherin/catenin adherens junction and impairs NPC progression-associated properties. Oncogene 31(32):3709–3720
Jakob U, Gaestel M, Engels K, Buchner J (1993) Small heat shock proteins are molecular chaperones. J Biol Chem 268:1517–1520
Janue A, Olive M, Ferrer I (2007) Oxidative stress in desminopathies and myotilinopathies: a link between oxidative damage and abnormal protein aggregation. Brain Pathol 17(4):377–388
Jenner P, Olanow CW (1996) Oxidative stress and the pathogenesis of Parkinson's disease. Neurology 47(6):S161–S170
Jia Y, Ransom RF, Shibanuma M, Liu C, Welsh MJ, Smoyer WE (2001) Identification and characterization of hic-5/ARA55 as an hsp27 binding protein. J Biol Chem 276(43):39911–39918
Jiang T, Altman S (2001) Protein-protein interactions with subunits of human nuclear RNase P. Proc Natl Acad Sci U S A 98(3):920–925
Kammanadiminti SJ, Chadee K (2006) Suppression of NF-kappaB activation by Entamoeba histolytica in intestinal epithelial cells is mediated by heat shock protein 27. J Biol Chem 281(36):26112–26120
Kamradt MC, Lu M et al (2005) The small heat shock protein alpha B-crystallin is a novel inhibitor of TRAIL-induced apoptosis that suppresses the activation of caspase-3. J Biol Chem 280(12):11059–11066
Kang SH, Kang KW et al (2008) Upregulated HSP27 in human breast cancer cells reduces Herceptin susceptibility by increasing Her2 protein stability. BMC Cancer 8(1):286
Kase S, Parikh JG, Rao NA (2009) Expression of heat shock protein 27 and alpha-crystallins in human retinoblastoma after chemoreduction. Br J Ophthalmol 93(4):541–544
Ke L, Meijering RA, Hoogstra-Berends F, Mackovicova K, Vos MJ, Van Gelder IC, Henning RH, Kampinga HH, Brundel BJ (2011) HSPB1, HSPB6, HSPB7 and HSPB8 protect against RhoA GTPase-induced remodeling in tachypaced atrial myocytes. PLoS One 6(6):e20395
Keller JN, Dimayuga E, Chen Q, Thorpe J, Gee J, Ding Q (2004) Autophagy, proteasomes, lipofuscin, and oxidative stress in the aging brain. Int J Biochem Cell Biol 36(12):2376–2391
Kerr BA, Byzova TV (2010) alphaB-crystallin: a novel VEGF chaperone. Blood 115(16):3181–3183
Kiffin R, Bandyopadhyay U, Cuervo AM (2006) Oxidative stress and autophagy. Antioxid Redox Signal 8(1–2):152–162
Kijima K, Numakura C, Goto T, Takahashi T, Otagiri T, Umetsu K, Hayasaka K (2005) Small heat shock protein 27 mutation in a Japanese patient with distal hereditary motor neuropathy. J Hum Genet 50(9):473–476
Kim EH, Lee HJ et al (2007) Inhibition of heat shock protein 27-mediated resistance to DNA damaging agents by a novel PKC delta-V5 heptapeptide. Cancer Res 67(13):6333–6341
Klemenz R, Andres AC, Fröhli E, Schäfer R, Aoyama A (1993) Expression of the murine small heat shock proteins hsp25 and aB crystallin in the absence of stress. J Cell Biol 120(3):639–645
Knapinska AM, Gratacos FM, Krause CD, Hernandez K, Jensen AG, Bradley JJ, Wu X, Pestka S, Brewer G (2011) Chaperone Hsp27 modulates AUF1 proteolysis and AU-rich element-mediated mRNA degradation. Mol Cell Biol 31(7):1419–1431
Koch HB, Zhang R, Verdoodt B, Bailey A, Zhang CD, Yates JR 3rd, Menssen A, Hermeking H (2007) Large-scale identification of c-MYC-associated proteins using a combined TAP/MudPIT approach. Cell Cycle 6(2):205–217
Kurnellas MP, Brownell SE et al (2012) Chaperone activity of small heat shock proteins underlies therapeutic efficacy in experimental autoimmune encephalomyelitis. J Biol Chem 287(43):36423–36434

Kwok AS, Phadwal K, Turner BJ, Oliver PL, Raw A, Simon AK, Talbot K, Agashe VR (2011) HspB8 mutation causing hereditary distal motor neuropathy impairs lysosomal delivery of autophagosomes. J Neurochem 119(6):1155–1161

Latchman DS (2005) HSP27 and cell survival in neurones. Int J Hyperthermia 21(5):393–402

Lavoie JN, Hickey E, Weber LA, Landry J (1993) Modulation of actin microfilament dynamics and fluid phase pinocytosis by phosphorylation of heat shock protein 27. J Biol Chem 268(32):24210–24214

Lee HJ, Lee YS (2010) Repeated-dose toxicity of HSP27-binding heptapeptide in mice. Drug Chem Toxicol 33(3):284–290

Lee GJ, Vierling E (2000) A small heat shock protein cooperates with heat shock protein 70 systems to reactivate a heat-denatured protein. Plant Physiol 122(1):189–198

Lee GJ, Roseman AM, Saibil HR, Vierling E (1997) A small heat shock protein stably binds heat-denatured model substrates and can maintain a substrate in a folding-competent state. EMBO J 16:659–671

Lee S, Carson K, Rice-Ficht A, Good T (2006) Small heat shock proteins differentially affect Abeta aggregation and toxicity. Biochem Biophys Res Commun 347(2):527–533

Lee YS, Lim KH et al (2008) The cytoplasmic deacetylase HDAC6 is required for efficient oncogenic tumorigenesis. Cancer Res 68(18):7561–7569

Lee JS, Kim HY et al (2012) Expression of alphaB-crystallin overrides the anti-apoptotic activity of XIAP. Neuro Oncol 14(11):1332–1345

Lelj-Garolla B, Mauk AG (2005) Self-association of a small heat shock protein. J Mol Biol 345(3):631–642

Lelj-Garolla B, Mauk AG (2006) Self-association and chaperone activity of Hsp27 are thermally activated. J Biol Chem 281(12):8169–8174

Lemieux P, Oesterreich S, Lawrence JA, Steeg PS, Hilsenbeck SG, Harvey JM, Fuqua SA (1997) The small heat shock protein hsp27 increases invasiveness but decreases motility of breast cancer cells. Invasion Metastasis 17(3):113–123

Lewis SE, Mannion RJ, White FA, Coggeshall RE, Beggs S, Costigan M, Martin JL, Dillmann WH, Woolf CJ (1999) A role for HSP27 in sensory neuron survival. J Neurosci 19(20):8945–8953

Li DW, Liu JP et al (2005) Calcium-activated RAF/MEK/ERK signaling pathway mediates p53-dependent apoptosis and is abrogated by alpha B-crystallin through inhibition of RAS activation. Mol Biol Cell 16(9):4437–4453

Li B, Smith CC, Laing JM, Gober MD, Liu L, Aurelian L (2007) Overload of the heat-shock protein H11/HspB8 triggers melanoma cell apoptosis through activation of transforming growth factor-beta-activated kinase 1. Oncogene 26(24):3521–3531

Lin DI, Barbash O, Kumar KG, Weber JD, Harper JW, Klein-Szanto AJ, Rustgi A, Fuchs SY, Diehl JA (2006) Phosphorylation-dependent ubiquitination of cyclin D1 by the SCF(FBX4-alphaB crystallin) complex. Mol Cell 24(3):355–366

Liu BF, Liang JJ (2008) Confocal fluorescence microscopy study of interaction between lens MIP26/AQP0 and crystallins in living cells. J Cell Biochem 104(1):51–58

Liu C, Gilmont RR, Benndorf R, Welsh MJ (2000) Identification and characterization of a novel protein from Sertoli cells, PASS1, that associates with mammalian small stress protein hsp27. J Biol Chem 275(25):18724–18731

Liu JP, Schlosser R, Ma WY, Dong Z, Feng H, Lui L, Huang XQ, Liu Y, Li DW (2004) Human alphaA- and alphaB-crystallins prevent UVA-induced apoptosis through regulation of PKCalpha, RAF/MEK/ERK and AKT signaling pathways. Exp Eye Res 79(6):393–403

Liu J, Chen Q, Huang W, Horak KM, Zheng H, Mestril R, Wang X (2006) Impairment of the ubiquitin-proteasome system in desminopathy mouse hearts. FASEB J 20(2):362–364

Liu S, Li J, Tao Y, Xiao X (2007) Small heat shock protein alphaB-crystallin binds to p53 to sequester its translocation to mitochondria during hydrogen peroxide-induced apoptosis. Biochem Biophys Res Commun 354(1):109–114

Mahon KA, Chepelinsky AB, Khillan JS, Overbeek PA, Piatigorsky J, Westphal H (1987) Oncogenesis of the lens in transgenic mice. Science 235(4796):1622–1628
Mao YW, Liu JP, Xiang H, Li DW (2004) Human alphaA- and alphaB-crystallins bind to Bax and Bcl-X(S) to sequester their translocation during staurosporine-induced apoptosis. Cell Death Differ 11(5):512–526
Marin-Vinader L, Shin C, Onnekink C, Manley JL, Lubsen NH (2006) Hsp27 enhances recovery of splicing as well as rephosphorylation of SRp38 after heat shock. Mol Biol Cell 17(2):886–894
Markossian KA, Yudin IK, Kurganov BI (2009) Mechanism of suppression of protein aggregation by alpha-crystallin. Int J Mol Sci 10(3):1314–1345
Matsushima-Nishiwaki R, Kumada T et al (2013) Direct association of heat shock protein 20 (HSPB6) with phosphoinositide 3-kinase (PI3K) in human hepatocellular carcinoma: regulation of the PI3K activity. PLoS One 8(11):e78440
McClellan AJ, Xia Y, Deutschbauer AM, Davis RW, Gerstein M, Frydman J (2007) Diverse cellular functions of the Hsp90 molecular chaperone uncovered using systems approaches. Cell 131(1):121–135
McClung HM, Golembieski WA, Schultz CR, Jankowski M, Schultz LR, Rempel SA (2012) Deletion of the SPARC acidic domain or EGF-like module reduces SPARC-induced migration and signaling through p38 MAPK/HSP27 in glioma. Carcinogenesis 33(2):275–284
McDonough H, Patterson C (2003) CHIP: a link between the chaperone and proteasome systems. Cell Stress Chaperones 8(4):303–308
Mehlen P, Arrigo A-P (1994) The serum-induced phosphorylation of mammalian hsp27 correlates with changes in its intracellular localization and levels of oligomerization. Eur J Biochem 221:327–334
Mehlen P, Préville X, Chareyron P, Briolay J, Klemenz R, Arrigo A-P (1995) Constitutive expression of human hsp27, Drosophila hsp27, or human alpha B-crystallin confers resistance to TNF- and oxidative stress-induced cytotoxicity in stably transfected murine L929 fibroblasts. J Immunol 154(1):363–374
Mehlen P, Préville X, Kretz-Remy C, Arrigo A-P (1996a) Human hsp27, Drosophila hsp27 and human αB-crystallin expression-mediated increase in glutathione is essential for the protective activity of these protein against TNFα–induced cell death. EMBO J 15:2695–2706
Mehlen P, Schulze-Osthoff K, Arrigo AP (1996b) Small stress proteins as novel regulators of apoptosis. Heat shock protein 27 blocks Fas/APO-1- and staurosporine-induced cell death. J Biol Chem 271(28):16510–16514
Mehlen P, Hickey E, Weber L, Arrigo A-P (1997a) Large unphosphorylated aggregates as the active form of hsp27 which controls intracellular reactive oxygen species and glutathione levels and generates a protection against TNFα in NIH-3T3-ras cells. Biochem Biophys Res Commun 241:187–192
Mehlen P, Mehlen A, Godet J, Arrigo A-P (1997b) hsp27 as a switch between differentiation and apoptosis in murine embryonic stem cells. J Biol Chem 272:31657–31665
Mendez F, Sandigursky M, Franklin WA, Kenny MK, Kureekattil R, Bases R (2000) Heat-shock proteins associated with base excision repair enzymes in HeLa cells. Radiat Res 153(2):186–195
Mendillo ML, Santagata S et al (2012) HSF1 drives a transcriptional program distinct from heat shock to support highly malignant human cancers. Cell 150(3):549–562
Merendino AM, Paul C, Vignola AM, Costa MA, Melis M, Chiappara G, Izzo V, Bousquet J, Arrigo AP (2002) Heat shock protein-27 protects human bronchial epithelial cells against oxidative stress-mediated apoptosis: possible implication in asthma. Cell Stress Chaperones 7(3):269–280
Mickler M, Hessling M, Ratzke C, Buchner J, Hugel T (2009) The large conformational changes of Hsp90 are only weakly coupled to ATP hydrolysis. Nat Struct Mol Biol 16(3):281–286

Moulick K, Ahn JH et al (2011) Affinity-based proteomics reveal cancer-specific networks coordinated by Hsp90. Nat Chem Biol 7(11):818–826
Mounier N, Arrigo AP (2002) Actin cytoskeleton and small heat shock proteins: how do they interact? Cell Stress Chaperones 7(2):167–176
Moyano JV, Evans JR et al (2006) AlphaB-crystallin is a novel oncoprotein that predicts poor clinical outcome in breast cancer. J Clin Invest 116(1):261–270
Muchowski PJ (2002) Protein misfolding, amyloid formation, and neurodegeneration: a critical role for molecular chaperones? Neuron 35(1):9–12
Muchowski PJ, Wacker JL (2005) Modulation of neurodegeneration by molecular chaperones. Nat Rev Neurosci 6(1):11–22
Muchowski PJ, Valdez MM, Clark JI (1999) AlphaB-crystallin selectively targets intermediate filament proteins during thermal stress. Invest Ophthalmol Vis Sci 40(5):951–958
Mymrikov EV, Seit-Nebi AS, Gusev NB (2011) Large potentials of small heat shock proteins. Physiol Rev 91(4):1123–1159
Mymrikov EV, Seit-Nebi AS, Gusev NB (2012) Heterooligomeric complexes of human small heat shock proteins. Cell Stress Chaperones 17(2):157–169
Nagaraja GM, Kaur P, Neumann W, Asea EE, Bausero MA, Multhoff G, Asea A (2012a) Silencing Hsp25/Hsp27 gene expression augments proteasome activity and increases CD8+ T-cell-mediated tumor killing and memory responses. Cancer Prev Res (Phila) 5(1):122–137
Nagaraja GN, Kaur P, Asea A (2012b) Role of human and mouse HspB1 in metastasis. Curr Mol Med 12(9):1142–1150
Neckers L, Mimnaugh E, Schulte TW (1999) Hsp90 as an anti-cancer target. Drug Resist Updat 2(3):165–172
Nemes Z, Devreese B, Steinert PM, Van Beeumen J, Fesus L (2004) Cross-linking of ubiquitin, HSP27, parkin, and alpha-synuclein by gamma-glutamyl-epsilon-lysine bonds in Alzheimer's neurofibrillary tangles. FASEB J 18(10):1135–1137
Nivon M, Richet E, Codogno P, Arrigo AP, Kretz-Remy C (2009) Autophagy activation by NFkappaB is essential for cell survival after heat shock. Autophagy 5:766–783
Noh SJ, Jeong WJ et al (2008) Sensitization of RPE cells by alphaB-crystallin siRNA to SAHA-induced stage 1 apoptosis through abolishing the association of alphaB-crystallin with HDAC1 in SC35 speckles. Invest Ophthalmol Vis Sci 49(11):4753–4759
O'Callaghan-Sunol C, Gabai VL, Sherman MY (2007) Hsp27 modulates p53 signaling and suppresses cellular senescence. Cancer Res 67(24):11779–11788
Ohto-Fujita E, Fujita Y, Atomi Y (2007) Analysis of the alphaB-crystallin domain responsible for inhibiting tubulin aggregation. Cell Stress Chaperones 12(2):163–171
Ousman SS, Tomooka BH, van Noort JM, Wawrousek EF, O'Connor KC, Hafler DA, Sobel RA, Robinson WH, Steinman L (2007) Protective and therapeutic role for alphaB-crystallin in autoimmune demyelination. Nature 448(7152):474–479
Outeiro TF, Klucken J, Strathearn KE, Liu F, Nguyen P, Rochet JC, Hyman BT, McLean PJ (2006) Small heat shock proteins protect against alpha-synuclein-induced toxicity and aggregation. Biochem Biophys Res Commun 351(3):631–638
Pandey P, Farber R, Nakazawa A, Kumar S, Bharti A, Nalin C, Weichselbaum R, Kufe D, Kharbanda S (2000) Hsp27 functions as a negative regulator of cytochrome c-dependent activation of procaspase-3. Oncogene 19(16):1975–1981
Parcellier A, Schmitt E et al (2003) HSP27 is a ubiquitin-binding protein involved in I-kappaBalpha proteasomal degradation. Mol Cell Biol 23(16):5790–5802
Parcellier A, Brunet M et al (2006) HSP27 favors ubiquitination and proteasomal degradation of p27Kip1 and helps S-phase re-entry in stressed cells. FASEB J 20(8):1179–1181
Patil SB, Pawar MD, Bitar KN (2004) Direct association and translocation of PKC-alpha with calponin. Am J Physiol Gastrointest Liver Physiol 286(6):G954–G963
Paul C, Arrigo AP (2000) Comparison of the protective activities generated by two survival proteins: Bcl-2 and Hsp27 in L929 murine fibroblasts exposed to menadione or staurosporine. Exp Gerontol 35(6–7):757–766

Paul C, Manero F, Gonin S, Kretz-Remy C, Virot S, Arrigo AP (2002) Hsp27 as a negative regulator of cytochrome C release. Mol Cell Biol 22(3):816–834
Paul C, Simon S, Gibert B, Virot S, Manero F, Arrigo AP (2010) Dynamic processes that reflect anti-apoptotic strategies set up by HspB1 (Hsp27). Exp Cell Res 316(9):1535–1552
Perng MD, Cairns L, van den IP, Prescott A, Hutcheson AM, Quinlan RA (1999) Intermediate filament interactions can be altered by HSP27 and alphaB-crystallin. J Cell Sci 112(Pt 13):2099–2112
Perrin V, Regulier E, Abbas-Terki T, Hassig R, Brouillet E, Aebischer P, Luthi-Carter R, Deglon N (2007) Neuroprotection by Hsp104 and Hsp27 in lentiviral-based rat models of Huntington's disease. Mol Ther 15(5):903–911
Prabhu S, Raman B, Ramakrishna T, Rao Ch M (2012) HspB2/myotonic dystrophy protein kinase binding protein (MKBP) as a novel molecular chaperone: structural and functional aspects. PLoS One 7(1):e29810
Preville X, Mehlen P, Fabre-Jonca N, Chaufour S, Kretz-Remy C, Michel MR, Arrigo A-P (1996) Biochemical and immunofluorescence analysis of the constitutively expressed hsp27 stress protein in monkey CV-1 cells. J Biosci 21(2):1–14
Preville X, Gaestel M, Arrigo AP (1998a) Phosphorylation is not essential for protection of L929 cells by Hsp25 against H2O2-mediated disruption actin cytoskeleton, a protection which appears related to the redox change mediated by Hsp25. Cell Stress Chaperones 3(3):177–187
Preville X, Schultz H, Knauf U, Gaestel M, Arrigo AP (1998b) Analysis of the role of Hsp25 phosphorylation reveals the importance of the oligomerization state of this small heat shock protein in its protective function against TNFalpha- and hydrogen peroxide-induced cell death. J Cell Biochem 69(4):436–452
Preville X, Salvemini F, Giraud S, Chaufour S, Paul C, Stepien G, Ursini MV, Arrigo AP (1999) Mammalian small stress proteins protect against oxidative stress through their ability to increase glucose-6-phosphate dehydrogenase activity and by maintaining optimal cellular detoxifying machinery. Exp Cell Res 247(1):61–78
Qi S, Xin Y, Qi Z, Xu Y, Diao Y, Lan L, Luo L, Yin Z (2014) HSP27 phosphorylation modulates TRAIL-induced activation of Src-Akt/ERK signaling through interaction with beta-arrestin2. Cell Signal 26(3):594–602
Quraishe S, Asuni A, Boelens WC, O'Connor V, Wyttenbach A (2008) Expression of the small heat shock protein family in the mouse CNS: differential anatomical and biochemical compartmentalization. Neuroscience 153(2):483–491
Rane MJ, Coxon PY, Powell DW, Webster R, Klein JB, Pierce W, Ping P, McLeish KR (2001) p38 Kinase-dependent MAPKAPK-2 activation functions as 3- phosphoinositide-dependent kinase-2 for Akt in human neutrophils. J Biol Chem 276(5):3517–3523
Rane MJ, Pan Y, Singh S, Powell DW, Wu R, Cummins T, Chen Q, McLeish KR, Klein JB (2003) Heat shock protein 27 controls apoptosis by regulating Akt activation. J Biol Chem 278(30):27828–27835
Rayner K, Chen YX, McNulty M, Simard T, Zhao X, Wells DJ, de Belleroche J, O'Brien ER (2008) Extracellular release of the atheroprotective heat shock protein 27 is mediated by estrogen and competitively inhibits acLDL binding to scavenger receptor-A. Circ Res 103(2):133–141
Renkawek K, Bosman GJ, de Jong WW (1994) Expression of small heat-shock protein hsp 27 in reactive gliosis in Alzheimer disease and other types of dementia. Acta Neuropathol (Berl) 87(5):511–519
Rigas PK, Kase S, Rao NA (2009) Expression of alpha-crystallins in human sebaceous carcinoma of the eyelid. Eur J Ophthalmol 19(5):702–707
Robertson AL, Headey SJ, Saunders HM, Ecroyd H, Scanlon MJ, Carver JA, Bottomley SP (2010) Small heat-shock proteins interact with a flanking domain to suppress polyglutamine aggregation. Proc Natl Acad Sci U S A 107(23):10424–10429
Rocchi P, Beraldi E, Ettinger S, Fazli L, Vessella RL, Nelson C, Gleave M (2005) Increased Hsp27 after androgen ablation facilitates androgen-independent progression in prostate cancer via signal transducers and activators of transcription 3-mediated suppression of apoptosis. Cancer Res 65(23):11083–11093

Roelofs MF, Boelens WC, Joosten LA, Abdollahi-Roodsaz S, Geurts J, Wunderink LU, Schreurs BW, van den Berg WB, Radstake TR (2006) Identification of small heat shock protein B8 (HSP22) as a novel TLR4 ligand and potential involvement in the pathogenesis of rheumatoid arthritis. J Immunol 176(11):7021–7027
Rogalla T, Ehrnsperger M et al (1999) Regulation of Hsp27 oligomerization, chaperone function, and protective activity against oxidative stress/tumor necrosis factor alpha by phosphorylation. J Biol Chem 274(27):18947–18956
Rosenbaum EE, Brehm KS, Vasiljevic E, Liu CH, Hardie RC, Colley NJ (2011) XPORT-dependent transport of TRP and rhodopsin. Neuron 72(4):602–615
Rothbard JB, Kurnellas MP et al (2012) Therapeutic effects of systemic administration of chaperone alphaB-crystallin associated with binding proinflammatory plasma proteins. J Biol Chem 287(13):9708–9721
Rouse J, Cohen P, Trigon S, Morange M, Alonso-Llamazares A, Zamanillo D, Hunt T, Nebreda AR (1994) A novel kinase cascade triggered by stress and heat shock that stimulates MAPKAP kinase-2 and phosphorylation of the small heat shock proteins. Cell 78(6):1027–1037
Ruan Q, Han S, Jiang WG, Boulton ME, Chen ZJ, Law BK, Cai J (2011) alphaB-crystallin, an effector of unfolded protein response, confers anti-VEGF resistance to breast cancer via maintenance of intracrine VEGF in endothelial cells. Mol Cancer Res 9(12):1632–1643
Saha S, Das KP (2004) Relationship between chaperone activity and oligomeric size of recombinant human alphaA- and alphaB-crystallin: a tryptic digestion study. Proteins 57(3):610–617
Salari S, Seibert T, Chen YX, Hu T, Shi C, Zhao X, Cuerrier CM, Raizman JE, O'Brien ER (2012) Extracellular HSP27 acts as a signaling molecule to activate NF-kappaB in macrophages. Cell Stress Chaperones 18(1):53–63
Samali A, Cotter TG (1996) Heat shock proteins increase resistance to apoptosis. Exp Cell Res 223:163–170
Sayre LM, Perry G, Atwood CS, Smith MA (2000) The role of metals in neurodegenerative diseases. Cell Mol Biol 46:731–741
Schultz CR, Golembieski WA, King DA, Brown SL, Brodie C, Rempel SA (2012) Inhibition of HSP27 alone or in combination with pAKT inhibition as therapeutic approaches to target SPARC-induced glioma cell survival. Mol Cancer 11:20
Seit-Nebi AS, Gusev NB (2010) Versatility of the small heat shock protein HSPB6 (Hsp20). Cell Stress Chaperones 15(3):233–236
Shammas SL, Waudby CA, Wang S, Buell AK, Knowle TP, Ecroyd H, Welland ME, Carver JA, Dobson CM, Meehan S (2011) Binding of the molecular chaperone alphaB-crystallin to Abeta amyloid fibrils inhibits fibril elongation. Biophys J 101:1681–1689
Shimura H, Miura-Shimura Y, Kosik KS (2004) Binding of tau to heat shock protein 27 leads to decreased concentration of hyperphosphorylated tau and enhanced cell survival. J Biol Chem 279(17):17957–17962
Shinder GA, Lacourse MC, Minotti S, Durham HD (2001) Mutant Cu/Zn-superoxide dismutase proteins have altered solubility and interact with heat shock/stress proteins in models of amyotrophic lateral sclerosis. J Biol Chem 276(16):12791–12796
Shiota M, Bishop JL et al (2013) Hsp27 regulates epithelial mesenchymal transition, metastasis, and circulating tumor cells in prostate cancer. Cancer Res 73(10):3109–3119
Shoham S, Youdim MB (2000) Iron involvement in neural damage and microgliosis in models of neurodegenerative diseases. Cell Mol Biol (Noisy-le-Grand) 46(4):743–760
Simon S, Fontaine JM, Martin JL, Sun X, Hoppe AD, Welsh MJ, Benndorf R, Vicart P (2007) Myopathy-associated alpha B-crystallin mutants: abnormal phosphorylation, intracellular location, and interactions with other small heat shock proteins. J Biol Chem 82:34276–34287
Simon S, Dimitrova V et al (2013) Analysis of the dominant effects mediated by wild type or R120G mutant of alphaB-crystallin (HspB5) towards Hsp27 (HspB1). PLoS One 8(8):e70545
Singh BN, Rao KS, Ramakrishna T, Rangaraj N, Rao Ch M (2007) Association of alphaB-crystallin, a small heat shock protein, with actin: role in modulating actin filament dynamics in vivo. J Mol Biol 366(3):756–767

Sinsimer KS, Gratacos FM et al (2008) Chaperone Hsp27, a novel subunit of AUF1 protein complexes, functions in AU-rich element-mediated mRNA decay. Mol Cell Biol 28(17):5223–5237

Skouri-Panet F, Michiel M, Ferard C, Duprat E, Finet S (2012) Structural and functional specificity of small heat shock protein HspB1 and HspB4, two cellular partners of HspB5: role of the in vitro hetero-complex formation in chaperone activity. Biochimie 94(4):975–984

Sluchanko NN, Sudnitsyna MV, Chernik IS, Seit-Nebi AS, Gusev NB (2011) Phosphomimicking mutations of human 14-3-3zeta affect its interaction with tau protein and small heat shock protein HspB6. Arch Biochem Biophys 506(1):24–34

Smith CC, Li B, Liu J, Lee KS, Aurelian L (2011) The Levels of H11/HspB8 DNA methylation in human melanoma tissues and xenografts are a critical molecular marker for 5-Aza-2′-deoxycytidine therapy. Cancer Invest 29(6):383–395

Smith CC, Lee KS, Li B, Laing JM, Hersl J, Shvartsbeyn M, Aurelian L (2012) Restored expression of the atypical heat shock protein H11/HspB8 inhibits the growth of genetically diverse melanoma tumors through activation of novel TAK1-dependent death pathways. Cell Death Dis 3:e371

Solari E, Garrido C (2002) The forgotten chaperones. Nat Cell Biol 4:E125

Sreekumar PG, Kannan R, Kitamura M, Spee C, Barron E, Ryan SJ, Hinton DR (2010) alphaB crystallin is apically secreted within exosomes by polarized human retinal pigment epithelium and provides neuroprotection to adjacent cells. PLoS One 5(10):e12578

Sreelakshmi Y, Sharma KK (2006) The interaction between alphaA- and alphaB-crystallin is sequence-specific. Mol Vis 12:581–587

Srinivas PN, Reddy PY, Reddy GB (2008) Significance of alpha-crystallin heteropolymer with a 3:1 alphaA/alphaB ratio: chaperone-like activity, structure and hydrophobicity. Biochem J 414(3):453–460

Srinivasan A, Nagineni C, Bhat S (1992) alpha A-crystallin is expressed in non-ocular tissues. J Biol Chem 267:23337–23341

Stengel F, Baldwin AJ, Painter AJ, Jaya N, Basha E, Kay LE, Vierling E, Robinson CV, Benesch JL (2010) Quaternary dynamics and plasticity underlie small heat shock protein chaperone function. Proc Natl Acad Sci U S A 107(5):2007–2012

Stokoe D, Engel K, Campbell D, Cohen P, Gaestel M (1992) Identification of MAPKAP kinase 2 as a major enzyme responsible for the phosphorylation of the small mammalian heat shock proteins. FEBS Lett 313:307–313

Sugiyama Y, Suzuki A, Kishikawa M, Akutsu R, Hirose T, Waye MM, Tsui SK, Yoshida S, Ohno S (2000) Muscle develops a specific form of small heat shock protein complex composed of MKBP/HSPB2 and HSPB3 during myogenic differentiation. J Biol Chem 275(2):1095–1104

Sun X, Fontaine JM, Rest JS, Shelden EA, Welsh MJ, Benndorf R (2004) Interaction of human HSP22 (HSPB8) with other small heat shock proteins. J Biol Chem 279(4):2394–2402

Sun G, Guo M et al (2005) Bovine PrPC directly interacts with alphaB-crystalline. FEBS Lett 579(24):5419–5424

Sun Y, Yi H et al (2007) Identification of differential proteins in nasopharyngeal carcinoma cells with p53 silence by proteome analysis. FEBS Lett 581(1):131–139

Sun X, Fontaine JM et al (2010) Abnormal interaction of motor neuropathy-associated mutant HspB8 (Hsp22) forms with the RNA helicase Ddx20 (gemin3). Cell Stress Chaperones 15(5):567–582

Sun Y, Zhou M, Fu D, Xu B, Fang T, Ma Y, Chen J, Zhang J (2011) Ubiquitination of heat shock protein 27 is mediated by its interaction with Smad ubiquitination regulatory factor 2 in A549 cells. Exp Lung Res 37:568–573

Tabner BJ, Turnbull S, El-Agnaf O, Allsop D (2001) Production of reactive oxygen species from aggregating proteins implicated in Alzheimer's disease, Parkinson's disease and other neurodegenerative diseases. Curr Top Med Chem 1(6):507–517

Taipale M, Jarosz DF, Lindquist S (2010) HSP90 at the hub of protein homeostasis: emerging mechanistic insights. Nat Rev Mol Cell Biol 11(7):515–528

Tang G, Perng MD, Wilk S, Quinlan R, Goldman JE (2010) Oligomers of mutant glial fibrillary acidic protein (GFAP) Inhibit the proteasome system in alexander disease astrocytes, and the small heat shock protein alphaB-crystallin reverses the inhibition. J Biol Chem 285(14):10527–10537
Tanguay R, Wu Y, Khandjian E (1993) Tissue-specific expression of heat shock proteins of the mouse in the absence of stress. Dev Genet 14:112–118
Thedieck C, Kalbacher H, Kratzer U, Lammers R, Stevanovic S, Klein G (2008) alpha B-crystallin is a cytoplasmic interaction partner of the kidney-specific cadherin-16. J Mol Biol 378(1):145–153
Thuringer D, Jego G et al (2013) Extracellular HSP27 mediates angiogenesis through Toll-like receptor 3. FASEB J 27(10):4169–4183
Tsvetkova NM, Horvath I et al (2002) Small heat-shock proteins regulate membrane lipid polymorphism. Proc Natl Acad Sci U S A 99(21):13504–13509
Turnbull S, Tabner BJ, Brown DR, Allsop D (2003) Copper-dependent generation of hydrogen peroxide from the toxic prion protein fragment PrP106-126. Neurosci Lett 336(3):159–162
van de Schootbrugge C, Bussink J, Span PN, Sweep FC, Grenman R, Stegeman H, Pruijn GJ, Kaanders JH, Boelens WC (2013a) alphaB-crystallin stimulates VEGF secretion and tumor cell migration and correlates with enhanced distant metastasis in head and neck squamous cell carcinoma. BMC Cancer 13:128. doi:10.1186/1471-2407-13-128
van de Schootbrugge C, van Asten F et al (2013b) alphaB-crystallin expression is correlated with phospho-ERK1/2 expression in human breast cancer. Int J Biol Markers 28(4):e365–e370
van Noort JM, Bsibsi M et al (2013) Activation of an immune-regulatory macrophage response and inhibition of lung inflammation in a mouse model of COPD using heat-shock protein alpha B-crystallin-loaded PLGA microparticles. Biomaterials 34(3):831–840
Verschuure P, Croes Y, van den IPR, Quinlan RA, de Jong WW, Boelens WC (2002) Translocation of small heat shock proteins to the actin cytoskeleton upon proteasomal inhibition. J Mol Cell Cardiol 34(2):117–128
Vicart P, Caron A et al (1998) A missense mutation in the alphaB-crystallin chaperone gene causes a desmin-related myopathy. Nat Genet 20(1):92–95
Vos MJ, Kanon B, Kampinga HH (2009) HSPB7 is a SC35 speckle resident small heat shock protein. Biochim Biophys Acta 1793(8):1343–1353
Vos MJ, Zijlstra MP, Kanon B, van Waarde-Verhagen MA, Brunt ER, Oosterveld-Hut HM, Carra S, Sibon OC, Kampinga HH (2010) HSPB7 is the most potent polyQ aggregation suppressor within the HSPB family of molecular chaperones. Hum Mol Genet 19(23):4677–4693
Wang K, Spector A (1996) alpha-crystallin stabilizes actin filaments and prevents cytochalasin-induced depolymerization in a phosphorylation-dependent manner. Eur J Biochem 242(1):56–66
Wang J, Huo K et al (2011) Toward an understanding of the protein interaction network of the human liver. Mol Syst Biol 7:536
Warde-Farley D, Donaldson SL et al (2010) The GeneMANIA prediction server: biological network integration for gene prioritization and predicting gene function. Nucleic Acids Res 38(Web Server issue):W214–W220. doi:10.1093/nar/gkq537
Watanabe G, Kato S, Nakata H, Ishida T, Ohuchi N, Ishioka C (2009) alphaB-crystallin: a novel p53-target gene required for p53-dependent apoptosis. Cancer Sci 100(12):2368–2375
Wei L, Liu TT, Wang HH, Hong HM, Yu AL, Feng HP, Chang WW (2011) Hsp27 participates in the maintenance of breast cancer stem cells through regulation of epithelial-mesenchymal transition and nuclear factor-kappaB. Breast Cancer Res 13(5):R101
Welch WJ, Feramisco JR (1985) Disruption of the three cytoskeletal networks in mammalian cells does not affect transcription, translation, or protein translocation changes induced by heat shock. Mol Cell Biol 5(7):1571–1581
Wettstein G, Bellaye PS, Micheau O, Bonniaud P (2012) Small heat shock proteins and the cytoskeleton: an essential interplay for cell integrity? Int J Biochem Cell Biol 44(10):1680–1686
Wettstein G, Bellaye PS et al (2013) Inhibition of HSP27 blocks fibrosis development and EMT features by promoting Snail degradation. FASEB J 27(4):1549–1560
Whitesell L, Santagata S, Lin NU (2012) Inhibiting HSP90 to treat cancer: a strategy in evolution. Curr Mol Med 12(9):1108–1124

Wilhelmus MM, Boelens WC, Otte-Holler I, Kamps B, Kusters B, Maat-Schieman ML, de Waal RM, Verbeek MM (2006a) Small heat shock protein HspB8: its distribution in Alzheimer's disease brains and its inhibition of amyloid-beta protein aggregation and cerebrovascular amyloid-beta toxicity. Acta Neuropathol 111(2):139–149

Wilhelmus MM, Boelens WC, Otte-Holler I, Kamps B, de Waal RM, Verbeek MM (2006b) Small heat shock proteins inhibit amyloid-beta protein aggregation and cerebrovascular amyloid-beta protein toxicity. Brain Res 1089:67–78

Wu R, Kausar H, Johnson P, Montoya-Durango DE, Merchant M, Rane MJ (2007) Hsp27 regulates Akt activation and polymorphonuclear leukocyte apoptosis by scaffolding MK2 to Akt signal complex. J Biol Chem 282(30):21598–21608

Wu Y, Liu J, Zhang Z, Huang H, Shen J, Zhang S, Jiang Y, Luo L, Yin Z (2009) HSP27 regulates IL-1 stimulated IKK activation through interacting with TRAF6 and affecting its ubiquitination. Cell Signal 21(1):143–150

Wyttenbach A (2004) Role of heat shock proteins during polyglutamine neurodegeneration: mechanisms and hypothesis. J Mol Neurosci 23(1–2):69–96

Wyttenbach A, Sauvageot O, Carmichael J, Diaz-Latoud C, Arrigo AP, Rubinsztein DC (2002) Heat shock protein 27 prevents cellular polyglutamine toxicity and suppresses the increase of reactive oxygen species caused by huntingtin. Hum Mol Genet 11(9):1137–1151

Xi JH, Bai F, McGaha R, Andley UP (2006) Alpha-crystallin expression affects microtubule assembly and prevents their aggregation. FASEB J 20(7):846–857

Xu L, Chen S, Bergan RC (2006) MAPKAPK2 and HSP27 are downstream effectors of p38 MAP kinase-mediated matrix metalloproteinase type 2 activation and cell invasion in human prostate cancer. Oncogene 25:2987–2998

Yan LJ, Christians ES, Liu L, Xiao X, Sohal RS, Benjamin IJ (2002) Mouse heat shock transcription factor 1 deficiency alters cardiac redox homeostasis and increases mitochondrial oxidative damage. EMBO J 21(19):5164–5172

Yang Y, Ludwig RL et al (2005) Small molecule inhibitors of HDM2 ubiquitin ligase activity stabilize and activate p53 in cells. Cancer Cell 7(6):547–559

Yang K, Meinhardt A, Zhang B, Grzmil P, Adham IM, Hoyer-Fender S (2012) The small heat shock protein ODF1/HSPB10 is essential for tight linkage of sperm head to tail and male fertility in mice. Mol Cell Biol 32(1):216–225

Yerbury JJ, Gower D, Vanags L, Roberts K, Lee JA, Ecroyd H (2012) The small heat shock proteins alphaB-crystallin and Hsp27 suppress SOD1 aggregation in vitro. Cell Stress Chaperones 18(2):251–257

Zantema A, Vries MV-D, Maasdam D, Bol S, Avd E (1992) Heat shock protein 27 and αB-cristallin can form a complex, which dissociates by heat shock. J Biol Chem 267(18):12936–12941

Zha J, Harada H, Osipov K, Jockel J, Waksman G, Korsmeyer SJ (1997) BH3 domain of BAD is required for heterodimerization with BCL-XL and pro-apoptotic activity. J Biol Chem 272(39):24101–24104

Zhu Y, Tassi L, Lane W, Mendelsohn ME (1994) Specific binding of the transglutaminase, platelet factor XIII, to HSP27. J Biol Chem 269(35):22379–22384

Zhuang H, Jiang W et al (2009) Down-regulation of HSP27 sensitizes TRAIL-resistant tumor cell to TRAIL-induced apoptosis. Lung Cancer 68(1):27–38

Zoubeidi A, Zardan A, Beraldi E, Fazli L, Sowery R, Rennie P, Nelson C, Gleave M (2007) Cooperative interactions between androgen receptor (AR) and heat-shock protein 27 facilitate AR transcriptional activity. Cancer Res 67(21):10455–10465

Zoubeidi A, Zardan A, Wiedmann RM, Locke J, Beraldi E, Fazli L, Gleave ME (2010) Hsp27 promotes insulin-like growth factor-I survival signaling in prostate cancer via p90Rsk-dependent phosphorylation and inactivation of BAD. Cancer Res 70(6):2307–2317

Part II
Structure-Function

Chapter 3
Dynamics-Function Relationships of the Small Heat-Shock Proteins

Georg K.A. Hochberg and Justin L.P. Benesch

Abstract The Small Heat-Shock Proteins (sHSPs) are a widespread family of molecular chaperones that tend to populate ensembles of inter-converting conformational and oligomeric states at equilibrium. How this dynamic structure is linked to the sHSPs' ability to rapidly bind and sequester target proteins, intercepting them *en route* to aggregation and deposition during disease and cellular stress, is a controversial topic. Partly this is because the dynamics of the sHSPs pose challenges to all biophysical and structural biology techniques, rendering them difficult to study. Here we give a personal view on recent insights that have been obtained on the dynamic motions these proteins undergo, their regulation in the cell, and hypothesise on how they may directly underpin sHSP activity.

Keywords Small Heat-Shock Protein • Molecular chaperone • Protein dynamics • Subunit exchange • Polydispersity

3.1 Introduction

The paradigm underlying the field of structural molecular biology is that the function of proteins and other biomolecules can be directly related to their three-dimensional structure (Campbell 2002). Concomitantly, it has been known for some time that proteins are inherently highly dynamic molecules, undergoing fluctuations in their structure that can span a broad range of time- and length-scales (Karplus and McCammon 1983). However, it is only in the last decade or so that the consensus has emerged that these dynamic motions frequently underlie protein activity in the cell. As we will describe in this chapter, the Small Heat-Shock Proteins (sHSPs) undergo intrinsic structural motions on all levels of their protein organisation, and there is much evidence to suggest that these motions are integral to their activities

G.K.A. Hochberg • J.L.P. Benesch (✉)
Physical and Theoretical Chemistry Laboratory, Department of Chemistry, University of Oxford, South Parks Road, Oxford OX1 3QZ, UK
e-mail: justin.benesch@chem.ox.ac.uk

R.M. Tanguay, L.E. Hightower (eds.), *The Big Book on Small Heat Shock Proteins*, Heat Shock Proteins 8, DOI 10.1007/978-3-319-16077-1_3

in vivo. While many other protein families are known to undergo fluctuations in side-chain conformations and tertiary organization, sHSPs stand out in that they also display rapid and complex dynamics at the level of their oligomeric composition. This important group of molecular chaperones is therefore an interesting illustration of the dynamical modification to the structure-function paradigm.

The canonical function of sHSPs is to act as "paramedics" of the cell (Hilton et al. 2013b), intercepting target proteins that are in a non-native state and on the pathway towards forming non-functional and potentially harmful aggregates, brought about by conditions of cellular stress (Basha et al. 2011). This function requires neither binding nor hydrolysis of ATP, meaning therefore that the energy required for the sHSPs to perform any conformational contortions necessary for their function comes from the ambient environment (McHaourab et al. 2009). This is very different from the mechanism of energy transfer used by most other molecular chaperones, and reveals that the sHSPs populate a relatively shallow free-energy surface (Papoian 2008). From a biophysical perspective, therefore, a picture emerges of a remarkable family of proteins whose thermodynamic and kinetic properties are delicately balanced to allow functional regulation by subtle changes in cellular conditions. Here we consider both the structure and dynamics of the sHSPs, the mechanism of their regulation, and how this may underpin their function in the cell.

3.2 Measuring sHSP Structural Dynamics

Despite the importance of the sHSPs in the proteostasis 'network' of the cell (Powers et al. 2009) and as part of the heat-shock response (Richter et al. 2010), the sHSPs remain relatively poorly characterised at the atomic level. This is largely due to their tendency to adopt a variety of alternative and rapidly interconverting states at equilibrium, both in terms of the conformation of individual monomers and the organisation of monomers into oligomers (Basha et al. 2011; Hilton et al. 2013b; McHaourab et al. 2009). The majority, but certainly not all, of sHSPs studied assemble into oligomeric structures, typically comprising six or more dimeric building blocks.

A major step forward in the last 5 years has been the crystallisation of truncated constructs of mammalian sHSPs, providing high-resolution information on the sHSP core (Bagnéris et al. 2009). This has provided a framework for employing ingenious selective labelling schemes, enabling electron paramagnetic resonance spectroscopy (Koteiche et al. 1998) and nuclear magnetic resonance spectroscopy (NMR) experiments (Baldwin et al. 2011a; Jehle et al. 2010) to reveal interatomic distances and associated fluctuations. To complement these high-resolution data, which are ensemble-averaged onto the level of monomers, with information on quaternary structure, various research groups have implemented advanced transmission electron microscopy (EM) (Braun et al. 2011), small-angle X-ray scattering (SAXS) (Jehle et al. 2010) or ion mobility mass spectrometry (IM-MS) approaches (Baldwin et al. 2011b). As a result, many of the recent gains in understanding have come from a synthesis of structural and dynamical information obtained on multiple levels of sHSP organisation.

A particularly notorious aspect of sHSP self-assembly is the propensity of the many members of the family to exist as a polydisperse ensemble of oligomers, rather than as a single stoichiometry. Such structural plasticity, while likely important to sHSP cellular function, renders the application of biophysical techniques challenging as it requires techniques that act to separate, or interrogate individually, the oligomeric states that contribute to the ensemble. This challenge is typically met either through the use of mass transfer approaches that are inherently dispersive (e.g. mass spectrometry, centrifugation, chromatography), or by subsequent data partitioning *in silico*. At present, the low resolution of separation, afforded by these techniques is not yet high enough to remove all ambiguity from the data (Delbecq and Klevit 2013), and is compounded by the complexity introduced by the timescale of analysis being slow relative to the equilibria within the analyte (Hochberg and Benesch 2014). As a result, while in recent years the study of the sHSPs on the atomic level has been invigorated by the application of novel and integrated structural biology approaches, on-going technical developments are needed to resolve the remaining inconsistencies in our understanding of these chaperones.

3.3 Assembly, Disassembly, and Subunit Exchange

One of the most intriguing aspects of the sHSPs' self-assembly is their capacity for hetero-oligomerisation, both *in vitro* and *in vivo*. In fact, with multiple members frequently expressed in the same cellular compartment, in many cases it is therefore more appropriate to consider the proteins as a 'multi-dimensional' sHSP ensemble. In the event of no kinetic or thermodynamic preference for different hetero-oligomers, their abundance will be given by a multinomial distribution, according to the expression ratios of the constituent sHSPs. This has been observed for some sHSPs (Sobott et al. 2002), but in other cases it appears that certain oligomeric compositions are defined (den Engelsman et al. 2009), or thermodynamically somewhat favoured (Mymrikov et al. 2012).

The process whereby the equilibrium distribution of these mixed oligomers is reached, and maintained, is known as "subunit exchange". This occurs via the continual dissociation and association from the assembled oligomers of individual subunits, typically monomers, dimers, or a combination of both. For instance, a pair of plant sHSPs from *Arabidopsis* were found to exchange predominantly via dimers (Painter et al. 2008), whereas two closely related sHSPs from wheat and pea displayed significant exchange via monomers (Sobott et al. 2002). The rate of exchange is a direct measure of the lability of the sHSP assembly involved, while the proportion of the process associated with the movement of dimers and monomers provides information as to the relative ease of breaking the inter- and intra-dimer interfaces. Such quaternary dynamics, i.e. a fluctuation in the subunits comprising the assembled oligomeric forms, while not limited to the sHSPs, have been reported for relatively few other proteins. It appears therefore that this is a behaviour characteristic of these molecular chaperones that acts to establish a hetero-oligomeric sHSP ensemble (Stengel et al. 2010).

The sheer variety of known sHSP assemblies raises questions about how individual stoichiometries and architectures evolve in this protein family. It is, for example, not known whether polydispersity, i.e. the ability to populate more than one stoichiometry at equilibrium, represents the ancestral state of all oligomeric sHSPs, or whether it is a derived characteristic, evolved only in some organisms to serve a particular functional need. This issue is further complicated by the fact that some sHSPs switch their oligomeric organization in response to heat stress or post-translational modification (see below). Answering such questions will require a much more detailed understanding of how quaternary dynamics are underpinned by the primary, secondary and tertiary structure and associated fluctuations of the protomeric building blocks of dynamic oligomers.

3.4 Dynamic Structure, Labile Interfaces, and Allosteric Communication

sHSPs are identified through the presence of an 'α-crystallin' domain, a region of sequence rich in β-structure that adopts an immunoglobulin-type fold (Poulain et al. 2010). In general, the α-crystallin domain is found to dimerise, forming the basic building block of higher order sHSP structure, in either one of two ways (Basha et al. 2011; Hilton et al. 2013b). In the structures determined for mammalian sHSPs, extended "β6+7" strands on each monomer pair-up in an anti-parallel orientation (Fig. 3.1). Contrastingly, it appears that in plants and lower organisms the sHSP dimers are formed through reciprocal donation of the β6 strand, located in a loop, into the β-sandwich of the partner (Fig. 3.2). While the latter dimerisation mode has only been observed in one form, in studies on mammalian sHSPs three distinct states have been observed, brought about by shifts in the register between the paired β6+7 strands (Clark et al. 2011). In the case of αB-crystallin, all of these three registration states have been observed under different crystallisation conditions (Fig. 3.1b, Hochberg et al. 2014).

These polymorphs, termed AP_{I}, AP_{II} and AP_{III} in order of decreasing overlap of anti-parallel β6+7 strands (Laganowsky et al. 2010), have to date only been observed directly in X-ray structures obtained from truncated constructs. Measurements in solution, while not ruling out the presence of multiple co-populated registers, certainly suggest that they are not equally populated (Fig. 3.1c, Hochberg et al. 2014). While it is therefore not yet clear whether mammalian sHSPs can ratchet along the dimer interface, there is good evidence that the dimer interface is relatively weak and dynamic, with a dissociation constant on the order of a few micromolar (Hilton et al. 2013a; Jehle et al. 2009). Indeed, this lability of the dimer

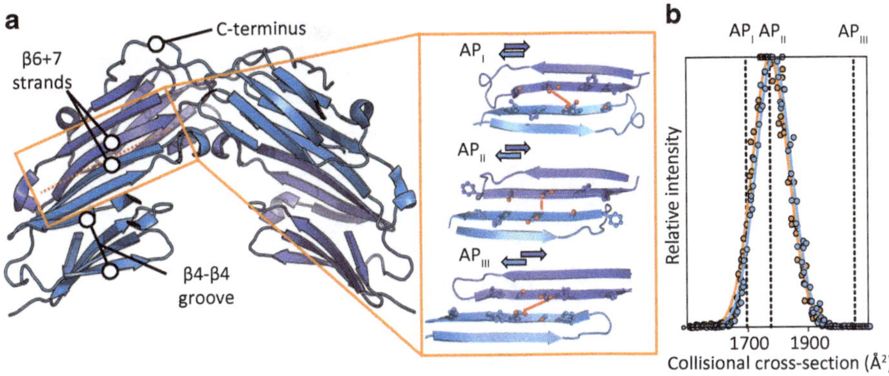

Fig. 3.1 Polymorphic dimer interfaces in αB crystallin. (**a**) *Left*: crystal structure (PDB ID: 3L1G) of a truncated αB-crystallin construct, illustrating how two monomers assemble into a dimer along their respective β6+7 strands (dimer interface indicated by *red line*). Dimers assemble into higher order structures via inter-dimer C-terminal interactions. *Right*: different registrations of the dimer interface observed crystallographically, termed from upper to lower, AP_I, AP_{II}, and AP_{III}, respectively. In the AP_{II} form a central glutamic acid (*red line*) forms the centre of a twofold axis, whereas the two monomers are shifted outward (AP_{III}) and inward (AP_I) relative to each other in the other two registrations. (**b**) Ion mobility mass spectrometry allows the measurement of physical size in the gas phase, in terms of a collisional cross-section (Benesch and Ruotolo 2011). Analysis of the truncated αB-crystallin construct is consistent with the majority of the dimers populating the AP_{II} register (*blue*). Theoretical collisional cross-sections for the different AP states are indicated with *dashed lines*. Engineering a cysteine on the central twofold axis to create a dimer that is conformationally restricted to AP_{II} results in a similar distribution of sizes (*orange*). This indicates that for αB-crystallin the AP_{II} state appears to be predominantly populated in solution (Figure adapted from Hilton et al. 2013a; Hochberg et al. 2014)

interface is such that, for the polydisperse mammalian sHSPs, oligomers containing an odd number of subunits (and hence, necessarily containing at least one un-dimerised monomer) are clearly observed (Aquilina et al. 2004, 2013).

In general, the α-crystallin domain is flanked by N-terminal and C-terminal regions, the lengths of which vary considerably across the sHSP family, and can contain further subdivisions (Fig. 3.2a). The C-terminal region is typically relatively short, and acts to bridge between dimers, such that a conserved I/VXI/V motif may bind into a groove between strands β4 and β8 of a neighbouring monomer (Figs. 3.1a and 3.2b). This "cross-linking" is enabled by a notable flexibility of the C-terminal region, appearing to give it complete freedom in the trajectory it takes towards its binding partner (Hilton et al. 2013b). As a result, even within the same oligomer, monomers can adopt strikingly different C-terminal orientations (Fig. 3.2c, Hanazono et al. 2013; van Montfort et al. 2001). Interestingly, for mammalian sHSPs it appears that the I/VXI/V motif does not bind inter-monomerically all of the time, but rather can also bind to the β4-β8 groove intra-monomerically, and under many conditions can be partly or even entirely detached (Hochberg and Benesch 2014). The relative populations of these different states can be facilely regulated by variations in solution condition, or through modification of the protein

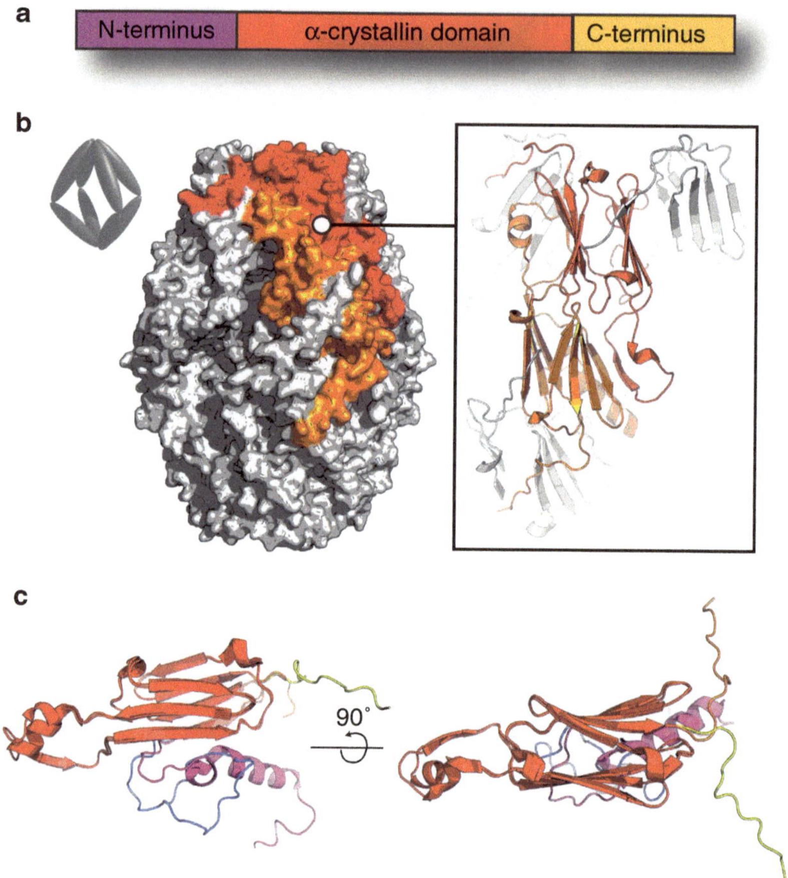

Fig. 3.2 Heterogeneity of monomer conformations within a sHSP oligomer. (**a**) Geometry of the sHSP assembly of HSP16.0 from *Schizosaccharomyces pombe* (PDB ID: 3W1Z) (Hanazono et al. 2013). Each ellipsoid represents a dimer, collectively assembling into a 16mer resembling an octahedron lacking the equatorial edges. The oligomer is comprised of eight dimers tightly packed such that each dimer donates and receives two C-termini to and from neighbouring dimers. Two monomers comprising a single dimer are highlighted in *orange* and *red*. Inset: cartoon representation of the dimer, illustrating a mode of dimerisation very different to that seen in the mammalian sHSPs (Fig. 3.1). (**b**) Non-equivalent conformations of the N- and C-termini (*yellow* and *orange*) are observed within the same dimer. Two monomers are overlaid, illustrating how in one the N-terminus is largely in a loop conformation (*blue*), whereas in the other it adopts a partially helical fold (*purple*). The C-termini (*yellow* and *orange*) protrude from the aligned monomers in orthogonal directions, illustrating the conformational flexibility of this region

sequence. An attractive hypothesis is that these C-terminal fluctuations act to regulate access to the β4-β8 groove, which may serve as a binding site for target proteins. In other words the C-terminus could be crucial to chaperone function, acting through a form of auto-inhibitory regulation.

The N-terminal region is the major source of variation in sHSP sequences across nature (Kriehuber et al. 2010). Unfortunately, structural and dynamical information on this part of the proteins is however currently very sparse. The limited crystallographic data that does exist comes from sHSPs with a comparatively short N-terminal region, and is consistent with it being located within the central portion of the oligomer (Hanazono et al. 2013; van Montfort et al. 2001). Notably, the data also point towards structural heterogeneity, as not all monomers display the same orientations within the oligomers (Fig. 3.2c). Similarly, molecular dynamics investigations suggest that the N-termini can explore multiple different conformations (Patel et al. 2014; Weeks et al. 2014). Nonetheless, the majority of the structural data that exists for the N-terminus points towards at least part of this region being involved in making inter-monomer contacts, and therefore contributing to the assembly of the oligomers. Complementing this data, hydrogen/deuterium exchange experiments reveal that N-terminal regions tend to be more solvent accessible than the majority of the sequence, suggesting a significant presence of dynamics at equilibrium, and revealing the relative lability of the interfaces they make (Cheng et al. 2008).

The sHSPs therefore contain a number of inter-subunit contacts, including one that serves to construct the protomeric dimer. One of the most exciting discoveries of the last few years is that for αB-crystallin there appears to be an allosteric coupling between these interfaces, such that changes at the dimer interface have a knock-on effect on how those dimers assemble (Baldwin et al. 2011c). In other words, the intra- and inter-dimer interfaces are linked such that energy can flow between them. Evidence from mass spectrometry and NMR experiments have narrowed down this effect to, at least in part, a coupling between C-terminal binding and dimer dissociation (Delbecq et al. 2012; Hilton et al. 2013a), however more details are required to characterise the atomic details of this allosteric behaviour, and to ascertain its generality across the sHSP family. Whatever the precise relationship between these various interactions, the combined evidence reveals the sHSPs to be constructed from a robust α-crystallin core, flanked by dynamic regions of structure via a series of labile interfaces.

3.5 Regulation of sHSP Dynamics

All the dynamical fluctuations in sHSPs described in the previous two sections of this chapter are affected by variations in solution and cellular conditions, as well as through modifications of protein sequence. Perhaps obviously, due to the temperature factor in the entropy contribution to free energy ($\Delta G = \Delta H - T\Delta S$), variations in temperature affect sHSP dynamics. For instance, subunit exchange proceeds faster at higher temperature, the equilibrium position between oligomeric and dissociated forms shifts towards the latter, and flexible regions of the protein tumble more freely. These phenomena are not unrelated: for αB-crystallin the temperature dependence of the transitions made by the C-termini could be quantitatively linked to the dissociation of monomers from the oligomer, which in turn were free to engage in subunit exchange (Baldwin et al. 2011a).

While it is tempting to evoke this simple thermodynamic influence on the oligomeric ensemble as being functionally relevant, it is not intuitively obvious that this must represent a means of regulation. Evidence that it is the case comes from the fact that, for sHSPs that have been studied in detail in this regard, the temperature dependence of rate constants can be particularly acute (Baldwin et al. 2011a), or involve multiple pathways (Benesch et al. 2010; Franzmann et al. 2008). Moreover, other changes in solution conditions, such as a drop in pH, also appear to cause a similar cascade in the dynamics (Baldwin et al. 2011c; Jehle et al. 2010). Regulation through pH is likely functional, as differences in chaperone activity have been noted *in vitro* as a function of pH, and in both cardiac and eye lens tissue there are important pH changes that act to recruit αB-crystallin. That different modes of activation appear to function via a similar pathway to changes caused by heating, provides strong evidence for the thermally induced changes observed being functional. This conclusion is probably unsurprising, considering that the sHSPs are, in the main, called into action under conditions of cellular thermal stress.

Dynamical changes caused by variations in primary structure have been reported. For example, the R120G mutation of αB-crystallin is associated with chaperone hyper-activity (Bova et al. 1999), and exchanges subunits more rapidly than the wild type (Michiel et al. 2009). Many post-translational modifications (PTMs) of mammalian sHSPs have been observed, though not so in plants and lower organisms. While some of those reported for the α-crystallins are likely just an inevitable consequence of their longevity in the eye lens, other PTMs appear to be specific regulators of activity. Chief amongst these are phosphorylation, which is prevalent in several mammalian sHSPs, and in particular HSP27 and αB-crystallin. On each of these two proteins there are three major phosphosites that fall within the N-terminal portion of the protein. HSP27 undergoes dissociation into sub-olgomeric species upon phosphorylation (Hayes et al. 2009; McDonald et al. 2012; Rogalla et al. 1999), and, while more subtle, a destabilisation of the inter-subunit interfaces is observed for αB-crystallin (Aquilina et al. 2004; Peschek et al. 2013).

Notwithstanding any effect of PTMs on downstream signalling pathways, for example, it is possible that these effects on the interfaces are important for chaperone function, possibly by exposing target-protein binding regions either on the sub-oligomeric species, or along the sHSP disassembly pathway. It appears therefore that the dynamics of sHSPs can be regulated in several different ways, depending on the organism and particular sHSP. Post-translational modification, and phosphorylation in particular, is likely to play a key role in controlling sHSP activity in homeotherms. In other organisms, such as bacteria and plants, environmental temperature changes may be sufficient to regulate activity, taking advantage of a finely tuned free-energy surface populated by the sHSPs.

3.6 Dynamic Formation of sHSP: Target Complexes

Not only are the sHSPs dynamic entities, but so too are the complexes they form with target proteins. In the first instance, self-assembly of these sHSP:target complexes appears to follow biphasic kinetics (Fig. 3.3). This can be explained by the target first being bound by the sHSP, following which additional sHSP subunits are recruited by the complex (Stengel et al. 2010). Considering the continual dissociation and reassociation of sHSP subunits mediating subunit exchange of the complexes

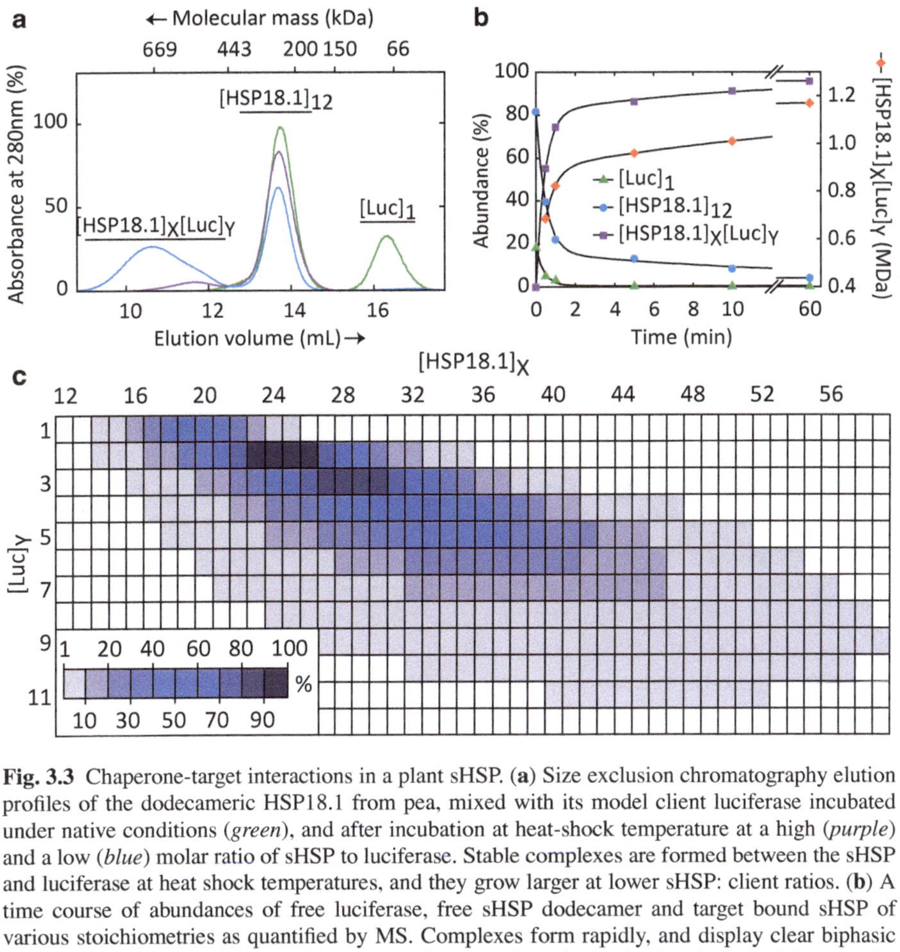

Fig. 3.3 Chaperone-target interactions in a plant sHSP. (**a**) Size exclusion chromatography elution profiles of the dodecameric HSP18.1 from pea, mixed with its model client luciferase incubated under native conditions (*green*), and after incubation at heat-shock temperature at a high (*purple*) and a low (*blue*) molar ratio of sHSP to luciferase. Stable complexes are formed between the sHSP and luciferase at heat shock temperatures, and they grow larger at lower sHSP: client ratios. (**b**) A time course of abundances of free luciferase, free sHSP dodecamer and target bound sHSP of various stoichiometries as quantified by MS. Complexes form rapidly, and display clear biphasic kinetics, indicative of two separate processes. These are hypothesised to be binding of the target by the sHSP, followed by recruitment of additional sHSP subunits to the complex. (**c**) Two-dimensional histogram of the relative abundance of different HSP18.1:luciferase stoichiometries formed during protection, obtained by a combination of tandem mass spectrometry and size exclusion chromatography analysis. The resulting mixture is extremely polydisperse, with over 300 different stoichiometries populated to an appreciable amount (Figure adapted from Stengel et al. 2010, 2012)

(Friedrich et al. 2004), the second phase can be thought of biophysically as being brought about by a decrease in dissociation rate of the sHSP subunit from the complex relative to from the oligomer. This would be expected if the sHSP subunits make somewhat stronger interactions with the target protein than with themselves, in line with the idea of sHSP oligomerisation in the absence of destabilized targets serving to shield their target binding surfaces from making non-specific interactions. As progressively more sHSP get incorporated into a complex, the dissociation rate will increase until it reaches that of the sHSP oligomer on its own, a situation that will be achieved when the additional sHSP subunit no longer makes any interaction with the bound target. This amounts to an elegant mechanism, as it will naturally coincide with the moment at which the bound target no longer has any hydrophobic surfaces exposed in solution.

Intrinsic sHSP dynamics are therefore potentially a useful means for regulating the binding of targets, and the formation of the resultant complexes. At the same time, the plasticity of the complex architectures formed by sHSPs, as evidenced by their polydispersity, also provides the potential for great flexibility in self-assembly. For instance, the hypothesis that sHSP self-assembly is based on polygonal shapes (Baldwin et al. 2011b) would in theory lead to the possibility of all manner of architectures, not just polyhedra and rings, but also sheets, or tubes. It is indeed possible that the sHSPs could employ this ability to encapsulate aggregated protein, for instance such as amyloid fibres (Knowles et al. 2007).

However, there is also data that demonstrates that *in vitro* it is not necessary for the sHSPs to actually be assembly competent, or even dynamic. Our recent work demonstrated that the structured core domain dimer of αB-crystallin is sufficient to prevent both amyloid and amorphous protein aggregation (Hochberg et al. 2014). This remained true even when the dimer was covalently linked to prevent disassembly into monomers or a change of registration state (Fig. 3.2). In other words, even when making the protein as un-dynamic as possible, it remained an effective chaperone. Taken in context of the other data, what this finding highlights is that sHSPs can act at different points of the protein unfolding pathway (Binger et al. 2013). It has been proposed for some time that the sHSPs can act in two modes: a low affinity mode with high capacity, and a high affinity mode with low capacity (Shashidharamurthy et al. 2005). It is quite possible that the former requires only transient interactions with the core domain and does not result in complex formation, whereas the former results in stable complexes necessitating the involvement of the flanking sequences. Nonetheless, elucidating the contributions of different sHSP sub-structures, and their associated dynamics, still requires considerable investigation.

3.7 Beyond the Structure-Function Paradigm

Many questions remain about the dynamics-function relationships of sHSPs. The major bottleneck is real paucity of structural insight into the nature of the sHSP:target interaction. Whilst structures for other classes of chaperones bound to destabilised proteins have become available in recent years, such attempts have thus far failed for sHSPs. Detailed knowledge of how exactly sHSPs interact with targets on the atomic level, and over what time-scale this interaction is stable, are crucial for rationalising the increasingly well-understood dynamical behaviour of sHSPs from a functional standpoint.

It is attractive to hypothesise that the polydispersity of many sHSPs, and their ability to assemble as hetero-oligomers, reflects somehow the diversity of targets that sHSPs can protect from aggregation. One can envisage that a hyper-variable binding surface on sHSP oligomers allows the fine-tuned binding of a large variety of clients (Stengel et al. 2010). However, until we better understand the features of destabilised proteins that sHSPs recognize, and with what parts of their sequence they do so, this remains just one of several explanations for sHSP structural heterogeneity. sHSPs that constitute major components of vertebrate eye-lenses have a more obvious use for heterogeneous and dynamic assemblies. These proteins are present in the eye-lens at exceptionally high concentrations and must resist crystallization for nearly the whole of the organism's life-span (Bloemendal et al. 2004). Although this goes some way to explaining polydispersity, particularly in mammalian sHSPs, several non-lenticular sHSPs also assemble into similarly heterogeneous ensembles. As such, it remains to be seen whether the plasticity afforded by the variable sHSP quaternary structure is of general utility, in terms of binding targets, remodelling into stable sHSP:target complexes, both, or neither.

An additional problem in the study of sHSP dynamics is that it is not known how common rapid quaternary dynamics and polydispersity are outside the family. Strikingly however, the subunit exchange rates observed for the sHSPs is fast on the timescale of typical cellular stress conditions. Nonetheless, since such dynamics pose serious problems to high-resolution structural biology techniques, it is possible that selection for well-behaved targets has created a bias against heterogeneous and dynamic protein assemblies in the protein databank and other popular databases. Such dynamics might be a more general feature of large protein assemblies, and assigning any one particular function to them might fall victim to the fallacy illustrated with Voltaire's parody of Leibnitzian optimism: "all is for the best in the best of all possible worlds" (Gould and Lewontin 1979). In this context, sHSPs remain a fascinating example to study the interplay between physical, functional, and evolutionary constraints in the shaping of protein assemblies.

Acknowledgments JLPB would like to thank all members of his group, past and present, for their enthusiasm, hard work, and intellectual insights that have driven our research programme in the study of this fascinating family of proteins. Similarly, we are extremely grateful to our collaborators, many of which have also contributed chapters to this book, who continue to galvanise our work, and make this field such an exciting and stimulating one to work in. GKAH is funded by the Doctoral Training Centre in System Biology that is funded by the Engineering and Physical Sciences Research Council, and JLPB by the Royal Society.

References

Aquilina JA, Benesch JLP, Ding LL, Yaron O, Horwitz J, Robinson CV (2004) Phosphorylation of alphaB-crystallin alters chaperone function through loss of dimeric substructure. J Biol Chem 279:28675–28680

Aquilina JA, Shrestha S, Morris AM, Ecroyd H (2013) Structural and functional aspects of hetero-oligomers formed by the small heat shock proteins alphaB-crystallin and HSP27. J Biol Chem 288:13602–13609

Bagnéris C, Bateman OA, Naylor CE, Cronin N, Boelens WC, Keep NH, Slingsby C (2009) Crystal structures of α-crystallin domain dimers of αB-crystallin and Hsp20. J Mol Biol 392:1242–1252

Baldwin AJ, Hilton GR, Lioe H, Bagneris C, Benesch JLP, Kay LE (2011a) Quaternary dynamics of aB-crystallin as a direct consequence of localised tertiary fluctuations in the C-terminus. J Mol Biol 413:310–320

Baldwin AJ, Lioe H, Hilton GR, Kay LE, Benesch JLP (2011b) The polydispersity of αB-crystallin is rationalised by an inter-converting polyhedral architecture. Structure 19:1855–1863

Baldwin AJ, Lioe H, Robinson CV, Kay LE, Benesch JLP (2011c) αB-crystallin polydispersity is a consequence of unbiased quaternary dynamics. J Mol Biol 413:297–309

Basha E, O'Neill H, Vierling E (2011) Small heat shock proteins and alpha-crystallins: dynamic proteins with flexible functions. Trends Biochem Sci 37(3):106–117

Benesch JLP, Ruotolo BT (2011) Mass spectrometry: an approach come-of-age for structural and dynamical biology. Curr Opin Struct Biol 21:641–649

Benesch JLP, Aquilina JA, Baldwin AJ, Rekas A, Stengel F, Lindner RA, Basha E, Devlin GL, Horwitz J, Vierling E et al (2010) The quaternary organization and dynamics of the molecular chaperone HSP26 are thermally regulated. Chem Biol 17:1008–1017

Binger KJ, Ecroyd H, Yang S, Carver JA, Howlett GJ, Griffin MD (2013) Avoiding the oligomeric state: alphaB-crystallin inhibits fragmentation and induces dissociation of apolipoprotein C-II amyloid fibrils. FASEB J 27:1214–1222

Bloemendal H, de Jong W, Jaenicke R, Lubsen NH, Slingsby C, Tardieu A (2004) Ageing and vision: structure, stability and function of lens crystallins. Prog Biophys Mol Biol 86:407–485

Bova MP, Yaron O, Huang QL, Ding LL, Haley DA, Stewart PL, Horwitz J (1999) Mutation R120G in alpha B-crystallin, which is linked to a desmin-related myopathy, results in an irregular structure and defective chaperone-like function. Proc Natl Acad Sci U S A 96:6137–6142

Braun N, Zacharias M, Peschek J, Kastenmuller A, Zou J, Hanzlik M, Haslbeck M, Rappsilber J, Buchner J, Weinkauf S (2011) Multiple molecular architectures of the eye lens chaperone alphaB-crystallin elucidated by a triple hybrid approach. Proc Natl Acad Sci U S A 108:20491–20496

Campbell ID (2002) Timeline: the march of structural biology. Nat Rev Mol Cell Biol 3:377–381

Cheng GL, Basha E, Wysocki VH, Vierling E (2008) Insights into small heat shock protein and substrate structure during chaperone action derived from hydrogen/deuterium exchange and mass spectrometry. J Biol Chem 283:26634–26642

Clark AR, Naylor CE, Bagneris C, Keep NH, Slingsby C (2011) Crystal structure of R120G disease mutant of human alphaB-crystallin domain dimer shows closure of a groove. J Mol Biol 408:118–134

Delbecq SP, Klevit RE (2013) One size does not fit all: the oligomeric states of alphaB crystallin. FEBS Lett 587:1073–1080

Delbecq S, Jehle S, Klevit R (2012) Binding determinants of the small heat shock protein, [alpha] B-crystallin: recognition of the 'IxI' motif. EMBO J 31:4587–4594

den Engelsman J, Boros S, Dankers PY, Kamps B, Vree Egberts WT, Bode CS, Lane LA, Aquilina JA, Benesch JLP, Robinson CV et al (2009) The small heat-shock proteins HSPB2 and HSPB3 form well-defined heterooligomers in a unique 3 to 1 subunit ratio. J Mol Biol 393:1022–1032

Franzmann TM, Menhorn P, Walter S, Buchner J (2008) Activation of the chaperone Hsp26 is controlled by the rearrangement of its thermosensor domain. Mol Cell 29:207–216

Friedrich KL, Giese KC, Buan NR, Vierling E (2004) Interactions between small heat shock protein subunits and substrate in small heat shock protein-substrate complexes. J Biol Chem 279:1080–1089

Gould SJ, Lewontin RC (1979) The spandrels of San Marco and the Panglossian paradigm: a critique of the adaptationist programme. Proc R Soc Lond B Biol Sci 205:581–598

Hanazono Y, Takeda K, Oka T, Abe T, Tomonari T, Akiyama N, Aikawa Y, Yohda M, Miki K (2013) Nonequivalence observed for the 16-meric structure of a small heat shock protein, SpHsp16.0, from Schizosaccharomyces pombe. Structure 21:220–228

Hayes D, Napoli V, Mazurkie A, Stafford WF, Graceffa P (2009) Phosphorylation dependence of hsp27 multimeric size and molecular chaperone function. J Biol Chem 284:18801–18807

Hilton GR, Hochberg GKA, Laganowsky A, McGinnigle SI, Baldwin AJ, Benesch JLP (2013a) C-terminal interactions mediate the quaternary dynamics of αB-crystallin. Philos Trans R Soc Lond B Biol Sci 368:20110405

Hilton GR, Lioe H, Stengel F, Baldwin AJ, Benesch JLP (2013b) Small heat-shock proteins: paramedics of the cell. Top Curr Chem 328:69–98

Hochberg GKA, Benesch JLP (2014) Dynamical structure of alphaB-crystallin. Prog Biophys Mol Biol 115:11–20

Hochberg GKA, Ecroyd H, Liu C, Cox D, Cascio D, Sawaya MR, Collier MP, Stroud J, Carver JA, Baldwin AJ et al (2014) The structured core domain of alphaB-crystallin can prevent amyloid fibrillation and associated toxicity. Proc Natl Acad Sci U S A 111:E1562–E1570

Jehle S, van Rossum B, Stout JR, Noguchi SM, Falber K, Rehbein K, Oschkinat H, Klevit RE, Rajagopal P (2009) alphaB-crystallin: a hybrid solid-state/solution-state NMR investigation reveals structural aspects of the heterogeneous oligomer. J Mol Biol 385:1481–1497

Jehle S, Rajagopal P, Bardiaux B, Markovic S, Kuhne R, Stout JR, Higman VA, Klevit RE, van Rossum BJ, Oschkinat H (2010) Solid-state NMR and SAXS studies provide a structural basis for the activation of alphaB-crystallin oligomers. Nat Struct Mol Biol 17:1037–1042

Karplus M, McCammon JA (1983) Dynamics of proteins: elements and function. Annu Rev Biochem 52:263–300

Knowles TP, Shu W, Devlin GL, Meehan S, Auer S, Dobson CM, Welland ME (2007) Kinetics and thermodynamics of amyloid formation from direct measurements of fluctuations in fibril mass. Proc Natl Acad Sci U S A 104:10016–10021

Koteiche HA, Berengian AR, McHaourab HS (1998) Identification of protein folding patterns using site-directed spin labeling. Structural characterization of a beta-sheet and putative substrate binding regions in the conserved domain of alpha A-crystallin. Biochemistry 37:12681–12688

Kriehuber T, Rattei T, Weinmaier T, Bepperling A, Haslbeck M, Buchner J (2010) Independent evolution of the core domain and its flanking sequences in small heat shock proteins. FASEB J 24:3633–3642

Laganowsky A, Benesch JLP, Landau M, Ding L, Sawaya M, Cascio D, Huang Q, Robinson CV, Horwitz J, Eisenberg D (2010) Crystal structures of truncated alphaA and alphaB crystallins reveal structural mechanisms of polydispersity important for eye lens function. Protein Sci 19:1031–1043

McDonald ET, Bortolus M, Koteiche HA, McHaourab HS (2012) Sequence, structure, and dynamic determinants of Hsp27 (HspB1) equilibrium dissociation are encoded by the N-terminal domain. Biochemistry 51:1257–1268

McHaourab HS, Godar JA, Stewart PL (2009) Structure and mechanism of protein stability sensors: chaperone activity of small heat shock proteins. Biochemistry 48:3828–3837

Michiel M, Skouri-Panet F, Duprat E, Simon S, Ferard C, Tardieu A, Finet S (2009) Abnormal assemblies and subunit exchange of alpha B-crystallin R120 mutants could be associated with destabilization of the dimeric substructure. Biochemistry 48:442–453

Mymrikov EV, Seit-Nebi AS, Gusev NB (2012) Heterooligomeric complexes of human small heat shock proteins. Cell Stress Chaperones 17:157–169

Painter AJ, Jaya N, Basha E, Vierling E, Robinson CV, Benesch JLP (2008) Real-time monitoring of protein complexes reveals their quaternary organization and dynamics. Chem Biol 15:246–253

Papoian GA (2008) Proteins with weakly funneled energy landscapes challenge the classical structure-function paradigm. Proc Natl Acad Sci U S A 105:14237–14238

Patel S, Vierling E, Tama F (2014) Replica exchange molecular dynamics simulations provide insight into substrate recognition by small heat shock proteins. Biophys J 106:2644–2655

Peschek J, Braun N, Rohrberg J, Back K, Kriehuber T, Kastenmüller A, Weinkauf S, Buchner J (2013) Regulated structural transitions unleash the chaperone activity of αB-crystallin. Proc Natl Acad Sci U S A 110(40):E3780–E3789, 201308898

Poulain P, Gelly JC, Flatters D (2010) Detection and architecture of small heat shock protein monomers. PLoS ONE 5:e9990

Powers ET, Morimoto RI, Dillin A, Kelly JW, Balch WE (2009) Biological and chemical approaches to diseases of proteostasis deficiency. Annu Rev Biochem 78:959–991

Richter K, Haslbeck M, Buchner J (2010) The heat shock response: life on the verge of death. Mol Cell 40:253–266

Rogalla T, Ehrnsperger M, Preville X, Kotlyarov A, Lutsch G, Ducasse C, Paul C, Wieske M, Arrigo AP, Buchner J et al (1999) Regulation of Hsp27 oligomerization, chaperone function, and protective activity against oxidative stress tumor necrosis factor alpha by phosphorylation. J Biol Chem 274:18947–18956

Shashidharamurthy R, Koteiche HA, Dong J, McHaourab HS (2005) Mechanism of chaperone function in small heat shock proteins: dissociation of the HSP27 oligomer is required for recognition and binding of destabilized T4 lysozyme. J Biol Chem 280:5281–5289

Sobott F, Benesch JLP, Vierling E, Robinson CV (2002) Subunit exchange of multimeric protein complexes. Real-time monitoring of subunit exchange between small heat shock proteins by using electrospray mass spectrometry. J Biol Chem 277:38921–38929

Stengel F, Baldwin AJ, Painter AJ, Jaya N, Basha E, Kay LE, Vierling E, Robinson CV, Benesch JLP (2010) Quaternary dynamics and plasticity underlie small heat shock protein chaperone function. Proc Natl Acad Sci U S A 107:2007–2012

Stengel F, Baldwin AJ, Bush MF, Hilton GR, Lioe H, Basha E, Jaya N, Vierling E, Benesch JLP (2012) Dissecting heterogeneous molecular chaperone complexes using a mass spectrum deconvolution approach. Chem Biol 19:599–607

van Montfort RL, Basha E, Friedrich KL, Slingsby C, Vierling E (2001) Crystal structure and assembly of a eukaryotic small heat shock protein. Nat Struct Biol 8:1025–1030

Weeks SD, Baranova EV, Heirbaut M, Beelen S, Shkumatov AV, Gusev NB, Strelkov SV (2014) Molecular structure and dynamics of the dimeric human small heat shock protein HSPB6. J Struct Biol 185:342–354

Chapter 4
Insights into How Small Heat Shock Proteins Bind a Great Diversity of Substrate Proteins: A Super-Transformer Model

Xinmiao Fu

Abstract Small heat shock proteins (sHSPs) are ubiquitous, ATP-independent molecular chaperones that play crucial roles in the cellular protein quality control and whose dysfunction is also linked to various diseases. They are able to suppress the aggregation of non-native substrate proteins and assist the refolding of these substrates in cooperation with ATP-dependent chaperones. Substrate recognition and binding by sHSPs are essential for their chaperone functions. In this review, I focus on how a sHSP recognizes and binds various substrate proteins that are greatly diversified in both structures and functions. Such a broad substrate spectrum is bestowed not only by the well-known dynamics of sHSP oligomeric structures, but also by their characteristic disordered substrate-binding regions that may interact with disordered substrate proteins via a way of disorder-adapting-disorder. Further, utilization of numerous substrate-binding residues is also crucial, as these residues include both hydrophobic and polar amino acids equally and are also spatially chaotic throughout the sHSP oligomers. In keeping with these details, a novel super-transformer model is presented to account for the structural characteristics and broad substrate spectrum of sHSPs.

Keywords Protein quality control • Molecular chaperone • Small heat shock protein • Protein aggregation • Protein dynamics • Intrinsically disordered structures • Oligomer • Transformer

Abbreviations

Bpa	*p*-benzoyl-L-phenylalanine
sHSPs	Small heat shoc k proteins

X. Fu (✉)
State Key Laboratory of Protein and Plant Gene Research, School of Life Sciences, Peking University, Beijing 100871, China
e-mail: fuxinmiao@pku.edu.cn

R.M. Tanguay, L.E. Hightower (eds.), *The Big Book on Small Heat Shock Proteins*, Heat Shock Proteins 8, DOI 10.1007/978-3-319-16077-1_4

4.1 Introduction

Proteins have to fold into defined three-dimensional structures (i.e., native conformations) to gain functional activity. But in the cellular environment, many newly synthesized proteins are unable to spontaneously fold into such native conformations. Rather, they need the assistance of molecular chaperone proteins, which are defined as proteins assisting other proteins to acquire functionally active structures but not being the components of these final structures (Ellis 1993; Hartl et al. 2011). Molecular chaperones are usually classified (according to their molecular size) as five families: Hsp100s, Hsp90s, Hsp70s, Hsp60s and small heat shock proteins (sHSPs), and they play essential roles in the folding, refolding, oligomeric assembly, translocation, and/or degradation of cellular proteins (Hartl et al. 2011). Among these five molecular chaperone families, only sHSPs are ATP-independent ones (Horwitz 1992; Jakob et al. 1993), but have been found to play crucial roles in a wide range of physiological processes, including cell differentiation (Favet et al. 2001), apoptosis (Salinthone et al. 2007), and animal longevity (Morrow et al. 2004). And their dysfunction was found to be linked to a variety of diseases (Laskowska et al. 2010), including cancer development (Richards et al. 1996), cardiovascular disease (Hollander et al. 2004), cataract (Litt et al. 1998), myopathy (Vicart et al. 1998), and neuron disorder (Irobi et al. 2004; Evgrafov et al. 2004).

Under in vitro conditions, sHSPs are able to effectively interact with unfolded model substrate proteins and keep them in a folding-competent state for subsequent refolding in cooperation with such ATP-dependent chaperones as Hsp70s and Hsp100s (Veinger et al. 1998; Mogk et al. 2003; Lee et al. 1997; Cashikar et al. 2005; Ehrnsperger et al. 1997). As such, heterologous over-expression of sHSPs was repeatedly found to increase the tolerance of host cells against various stresses (Balogi et al. 2008; Sun et al. 2012; Montero-Barrientos et al. 2007; Kim et al. 2012; Seo et al. 2006; Yeh et al. 1997; Kitagawa et al. 2000; Landry et al. 1989). In terms of primary structure, sHSPs are characterized by the presence of a conserved α-crystallin domain of approximately 100 amino acids (de Jong et al. 1993) that is flanked by a highly variable N-terminal arm (van Montfort et al. 2001; Kim et al. 1998; Fu et al. 2005; Cheng et al. 2008) and a short flexible C-terminal region (Blakytny et al. 1997; Carver and Lindner 1998; Kim et al. 1998). Interestingly, most sHSPs exist as large oligomers of 12–40 subunits (Kim et al. 1998; van Montfort et al. 2001; Sobott et al. 2002; Hanazono et al. 2012, 2013) while a few of them present as small oligomers of 2–4 subunits (Kokke et al. 1998; Basha et al. 2013; Leroux et al. 1997a). Each subunit of sHSP oligomers folds as immunoglobin-like β-sandwiches in its α-crystallin domain, with a few α-helices being formed in the N-terminal arm and C-terminal region (Kim et al. 1998).

My colleagues and I have focused on sHSPs for years and investigated their chaperone function and mechanism under both in vitro (Chang et al. 1996; Yang et al. 1999; Gu et al. 2002; Fu et al. 2005; Jiao et al. 2005b, 2008; Shi et al. 2011) and in vivo (Jiao et al. 2005a; Ezemaduka et al. 2014; Fu et al. 2013b) conditions. One intriguing feature of sHSPs, in our opinion, is that they are able to recognize and bind a great diversity of substrate proteins (Bepperling et al. 2012; Basha et al.

2004; Fu 2014). For instances, *Deinococcus radiodurans* Hsp20.2 and *Escherichia coli* IbpB were found to respectively interact with at least 89 and 110 substrate proteins in cells (Bepperling et al. 2012; Fu et al. 2013a). These substrate proteins appear to participate in a variety of cellular processes, including metabolism, DNA replication and recombination, protein translation, and so forth (Fu 2014). Here I propose a super-transformer model to illustrate the underlying mechanism for their broad substrate spectrum. I mainly focus on the structural characteristics of sHSPs (from quaternary, tertiary, secondary to primary structures) bestowing them this capability.

4.2 Dynamic Oligomeric Structures of sHSPs

4.2.1 General Significance of Oligomers for Substrate-Binding

The most important structural feature of sHSPs is that they exclusively exist as multi-subunit oligomers, which had been observed 40 years ago on mammalian sHSPs (α-crystallin and Hsp27) (Spector et al. 1971; Arrigo and Welch 1987; Longoni et al. 1990) and later on many sHSPs from different organisms (Chang et al. 1996; Haley et al. 2000; Kim et al. 1998; van Montfort et al. 2001; de Miguel et al. 2009; den Engelsman et al. 2009; Haslbeck et al. 1999, 2008; Kennaway et al. 2005; Shi et al. 2011; Leroux et al., 1997b; Giese and Vierling 2004; Sun et al. 2004; Hilario et al. 2011; Hockertz et al. 1991; Merck et al. 1993b). Although subunits within a sHSP oligomer are theoretically identical in their substrate-binding residues, the architecture of such multi-subunits would enable the sHSP oligomer to incorporate substrate proteins at multiple independent sites. Further, it is conceivable that the basic function units within the oligomer could be multiplied such that monomers, dimers and/or tetramers are all suitable to bind the substrate polypeptide chain. In this token, there are numerous types of substrate-binding patches in a single sHSP oligomer. It follows that each molecule of sHSP oligomers could simultaneously accommodate more than one types of substrate proteins as we recently suggested (Fu et al. 2013a). In support of this, it was observed that pre-existing sHSP-substrate complexes could incorporate additional substrate proteins (Friedrich et al. 2004; Stromer et al. 2003). Nevertheless, it remains controversial whether the formation of large oligomers per se is required for the functionality of sHSPs (Lentze et al. 2003; Basha et al. 2013; Franzmann et al. 2005; Takeda et al. 2011; Bukach et al. 2004; Kappe et al. 2004).

4.2.2 Oligomeric Polydispersity

Existence as an ensemble of multiple oligomeric states for one type of sHSP (Haley et al. 2000; Braun et al. 2011; Shearstone and Baneyx 1999; Haley et al. 1998; Aquilina et al. 2003. 2005; Stengel et al. 2012; Lelj-Garolla and Mauk 2005), termed as oligomeric polydispersity, is an important feature for many sHSPs, particularly of those from metazoan and certain bacteria as we reported earlier (Fu et al. 2006). The polydispersity of sHSP oligomers, accordingly, will enable the chaperone to cope with much more types of substrate proteins, assuming that different oligomers differ in the substrate-binding efficiency, selectivity and capacity. On the other hand, it reflects that sHSP oligomers indeed undergo rapid transformations between different oligomeric states, which are supposed to be nearly isoenergetic and thus inter-changeable without overcoming significant energetic barrier. In this respect, the oligomeric polydispersity of sHSPs reflects protein dynamics in general (Frauenfelder et al. 1991; Henzler-Wildman and Kern 2007) in the level of protein quaternary structures.

4.2.3 Oligomeric Plasticity

The oligomeric plasticity of sHSPs is reflected by the different distributions of variable oligomers in equilibrium under different conditions. In reality, sHSP oligomers undergo rapid re-organization in response to the disturbance of environmental factors and/or chemicals. For instance, temperature, the most biologically relevant environmental factor, substantially impacts the oligomeric states of sHSPs (Ehrnsperger et al. 1999; Gu et al. 2002; Haslbeck et al. 1999; van Montfort et al. 2001; Raman and Rao 1997). Since temperature also substantially affects the folding status and abundance of substrate proteins in cells, temperature-induced structural changes of sHSPs would enable them to act as robust chaperones to cope with temperature-mediated alterations on substrate proteins, as we recently observed on IbpB in living *E. coli* cells (Fu et al. 2013b), where hundreds of natural substrate proteins of IbpB are susceptible to heat shock-induced unfolding/misfolding.

4.2.4 Subunit Exchange

Lastly, subunit exchange is commonly observed to take place between sHSP oligomers (Bova et al. 1997, 2002; Gu et al. 2002; Aquilina et al. 2005; Friedrich et al. 2004; Fu and Chang 2004; Sobott et al. 2002; Basha et al. 2010), presumably via reversible oligomeric dissociation and association (Gu et al. 2002; Sobott et al. 2002). It even occurs between different types of but structurally similar and/or functionally related sHSPs (Merck et al. 1993a; Sobott et al. 2002; den Engelsman et al. 2009; Leroux et al. 1997b; Fu et al. 2006), as well as between free sHSP oligomers and sHSP-substrate complexes (Friedrich et al. 2004). Conceivably, such subunit exchange will dynamically change the "morphology" of sHSP oligomers and thus creates much more conformational resembles to cope with substrate diversity, in a similar way with the polydispersity of sHSP oligomers as discussed above. Further, it may also serve as an activation mechanism, by which the sHSP oligomers release sub-oligomers that are able to efficiently capture substrate proteins as we suggested before (Fu et al. 2003b). In support of this, the subunit exchange rate for many sHSPs was reported to positively correlate with chaperone activities (Bova et al. 1997, 2002; Fu and Chang 2004; Raman and Rao 1997; Santhoshkumar et al. 2009).

The sHSP oligomeric dynamics is believed to be largely resulted from the conformational flexibility or the even structural disorder of the N-terminal arm and C-terminal extension (Basha et al. 2012; Sudnitsyna et al. 2012; Baldwin et al. 2011). In sum, the dynamic oligomers of sHSPs can be viewed as a kind of plastic nanostructures that are able to accommodate structurally diversified substrate proteins. This concept was nicely demonstrated by the observation that Hsp18.1 was able to form complexes of over 300 different stoichiometries with a single substrate protein (luciferase) (Stengel et al. 2010). By contrast, it is imaginable that monomeric molecular chaperones such as Hsp70 would be unable to do so.

4.3 Disorder-Adapting-Disorder for sHSP-Substrate Recognition and Interaction

Protein-protein interactions are generally considered to occur through ordered, specific interfaces of two folded proteins. Nevertheless, in many cases protein-protein interactions were reported to be mediated by an disordered interface of one interacting protein (as reviewed (Wright and Dyson 1999)), which in turn becomes folded upon interaction with its targets (Sugase et al. 2007). Such disordered interfaces apparently confer functional advantages on a protein, including the ability to bind to several different targets (Wright and Dyson 1999), i.e., the broad client spectrum. Hinted by this, we propose that sHSP-substrate interactions undergo possibly through a way of disorder-adapting-disorder. In other words, the disordered substrate-binding region of a sHSP interacts with a globally or locally disordered substrate protein.

It is hard to obtain the high-resolution structures of the substrate proteins of sHSPs due to their structural heterogeneity. It appears that they are neither folded nor fully unfolded. Specifically, these substrate proteins, when complexed with sHSPs, were found to be characterized by native-like secondary structures but compromised tertiary structures and somehow exist in molten globule states, as revealed by fluorescence spectroscopy, NMR, CD spectroscopy and spin labeling (Das et al. 1996, 1999; Lindner et al. 1997; Rawat and Rao 1998; Sathish et al. 2003). Nevertheless, the exact folding status of substrate proteins, before they are recognized and bound by sHSPs, remains unclear. Given that they are aggregation-prone intermediates during unfolding and/or refolding (Carver et al. 1995, 2002; Carver and Lindner 1998; Treweek et al. 2000; Farahbakhsh et al. 1995; Lindner et al. 1997, 2001; Das et al. 1996, 1999), it is most likely that sHSP substrate proteins are largely disordered in structures before being bound by sHSPs. One excellent example in support of this assumption comes from melittin, which is a random coil peptide under physiological condition and was reported to be bound by α-crystallin (Farahbakhsh et al. 1995). In another critical study, Carver et al. reported that a disordered intermediate of α-lactalbumin was more efficiently bound by α-crystallin than a less disordered intermediate (Carver et al. 2002). Possibly, upon binding to sHSPs, such largely disordered structures in the substrate proteins are stabilized by sHSPs and thus transformed into more ordered structures. In support of this, unfolded insulin B chain towards aggregation, after complexed with α-crystallin, was reported to be strongly immobilized along much of its length and also adopt to its native fold (Farahbakhsh et al. 1995).

On the other hand, the N-terminal arm and C-terminal extension of sHSPs were reported to be extensively flexible in conformations or even structurally disordered (Kim et al. 1998; van Montfort et al. 2001; Fu et al. 2005; Jiao et al. 2005b; Cheng et al. 2008; Blakytny et al. 1997; Carver and Lindner 1998; Hartl et al. 2011; Baldwin et al. 2011). Whereas the exact role of the flexible C-terminal extension in binding substrate proteins is still in debate (Carver and Lindner 1998; Smulders et al. 1996; Jehle et al. 2010; Saji et al. 2008; Treweek et al. 2007; Jaya et al. 2009; Jiao et al. 2005b; Fu et al. 2013b), it is well established that the structurally disordered N-terminal arm serves as a principal substrate-binding region (Fu et al. 2005, 2013b; Sharma et al. 1997; Jaya et al. 2009; Lambert et al. 2013). In this context, sHSPs appear to dominantly utilize their disordered N-terminal arms for substrate-binding, and thus allow them to present diverse geometries for substrate-binding. Further, we previously reported that *M. tuberculosis* Hsp16.3, when losing its native structures became an unstructured (disordered) protein either due to a point mutation or at high temperature, even exhibited much higher chaperone activity than the wild type protein (Chen et al. 2005; Fu and Chang 2006a; Fu et al. 2003a), indicating the importance of disordered structures of sHSPs for substrate-binding. Taken together, the sHSP-substrate interaction could be considered to undergo in a way of "disorder-adapting-disorder" as proposed by us recently (Fu et al. 2013b). This unique interaction way, unlike the general way of order-fitting-order specifically occurring between structurally ordered proteins, would allow both sides to fit with each other in a nearly infinite conformational space and thus dramatically bestow the broad substrate spectrum of sHSPs.

The principle governing such "disorder-adapting-disorder" is currently unknown. It was found that molecular chaperone proteins are overall more disordered in their structures than most of other functional proteins (Tompa and Csermely 2004). Such disordered structures are considered to play a fundamental mechanistic role via a way of "entropy transfer", by which the transition of disorder to order in the chaperone enables the misfolded substrate to be locally unfolded for subsequent refolding in a limited conformational space (Tompa and Csermely 2004). With respect to sHSPs, the flexible C-terminal extension may act in this way, as in αB-crystallin this region was reported to lose its flexibility upon binding to substrate proteins (Carver et al. 1995; Carver and Lindner 1998). Nevertheless, regarding the disorder-adapting-disorder interaction between the disordered substrate-binding regions of sHSPs (e.g., the N- terminal arm) and the disordered element of substrate proteins, the free energy would be presumably stemmed from the multiple non-covalent interactions between them. It follows that during disorder-adapting-disorder the total entropy is reduced and both disordered structures are transiently stabilized. In line with this, synthesized peptides respectively corresponding to the N-terminal arm of Hsp16.3 and a substrate-binding sequence of αA-crystallin were both found to be able to directly interact with unfolded substrate proteins under in vitro conditions (Sharma et al. 2000; Fu et al. 2005). The disorder-adapting-disorder scenario may be applicable to other molecular chaperones (e.g., Hsp33 and HdeA) that utilize disordered structures to bind unfolded substrate proteins (Reichmann et al. 2012; Bardwell and Jakob 2012; Hong et al. 2005).

4.4 Structural Diversity and Spatial Chaos of Substrate-Binding Residues of sHSPs

Last but not least, the broad substrate spectrum of sHSPs is bestowed from diversified amino acids that are utilized by sHSPs for substrate-binding. Studies by truncation or by chemical crosslinking showed that the three characteristic domains of sHSPs are all crucial for the chaperone functional integrity (Merck et al. 1993b; Fu and Chang 2006a; Fu et al. 2005; Basha et al. 2006; Giese and Vierling 2004; Leroux et al. 1997b; Jiao et al. 2005b; Lentze et al. 2003; Ghosh et al. 2006; Saji et al. 2008; Strozecka et al. 2012; Studer et al. 2002; Treweek et al. 2007; Fu and Chang 2006b; Lambert et al. 2013; Shemetov and Gusev 2011). However, these studies cannot reveal the exact role of each amino acid residue for substrate-binding. An elegant approach using unnatural amino acid-mediated photo-crosslinking has been recently utilized to characterize the substrate-binding residues of sHSPs (Fu et al. 2013b; Jaya et al. 2009). Interestingly, both hydrophobic and charged residues in pea Hsp18.1 were found to participate in binding model substrate proteins (luciferase and malate dehydrogenase) under in vitro conditions (Jaya et al. 2009), in line with conventional point mutation studies on αA crystalline and αB-crystallin (Plater et al. 1996; Smulders et al. 1995). Importantly, we found by in vivo photo-crosslinking that the 48 substrate-binding residues of IpbB are composed of nearly equal amount of polar amino acids (including 13 charged residues) and hydrophobic amino acids. Together, these results thus demonstrate that not only hydrophobic interactions but also electrostatic forces and hydrogen bonds are involved in the interaction of sHSPs with substrate proteins. Accordingly, such multi-types of non-covalent interactions could enable sHSPs to recognize structurally diversified substrate proteins carrying both non-native hydrophobic and hydrophilic amino acids.

Another noteworthy point is that the substrate-binding residues of both Hsp18.1 and IbpB appear spatially chaotic throughout the sHSP oligomers. First, similarly with those of Hsp18.1 that did not form discrete binding surfaces (Jaya et al. 2009), the substrate-binding residues of IbpB appear to be largely dispersed (Fig. 4.1a). Such global dispersion of multi-types of numerous substrate-binding residues in the oligomeric structures of Hsp18.1 and IbpB strongly support the notion that each sHSP oligomer binds the substrate proteins at multiple independent sites (Cheng

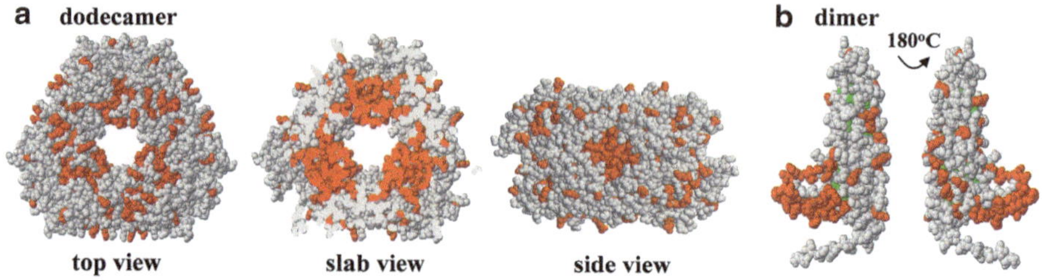

Fig. 4.1 IbpB's substrate-binding residues are spatially chaotic within the modeled dodecameric and dimeric structures. 48 substrate-binding residues of IbpB as identified at 50 °C are mapped into the modeled IbpB dodecameric (panel **a**) or dimeric (panel **b**) structures. The dodecameric structure of IbpB was modeled by referring to the determined 3-D structure of Hsp16.9 as described in our recent study (Fu et al. 2013b). The amino-terminal 31 residues in half of the 12 subunits of IbpB are assumed to be structurally disordered by referring to Hsp16.9 (van Montfort et al. 2001) and are thus disregarded during modeling. In panel **b**, fully or partially exposed residues are colored in *red* (38 residues) while largely buried ones are in *green* (10 residues)

et al. 2008). On the other hand, this topology would allow the chaperones to create unlimited number of patches for substrate-binding. Nevertheless, it is noted that many of the substrate-binding residues in IbpB also form a few discrete patches (Fig. 4.1a). It is conceivable that each of the globally dispersed residues has low binding affinity while those spatially enriched ones exhibit high binding affinity to substrates. In support of this scenario, in vitro studies revealed that Hsp16.5 binds lysozyme with both low and high affinities (Sathish et al. 2003; Shi et al. 2013). It follows that both the globally dispersed and locally enriched substrate-binding residues are crucial for the recognition and interaction of sHSPs towards substrate proteins. In addition, mapping analysis reveals that, although a large portion of substrate-binding residues are present on the surface, many are deeply buried in the core structure of both Hsp18.1 (Jaya et al. 2009) and IbpB dimers (Fig. 4.1b). These observations indicate that sHSP dimers have to undergo significant structural readjustment before utilizing these deeply buried sites for substrate-binding.

In sum, the substrate-binding residues of sHSPs are composed of nearly equal amount of polar (charged or uncharged) and hydrophobic amino acids, and are spatially chaotic within the context of oligomers. These characteristics will also bestow the broad substrate spectrum of sHSPs.

4.5 sHSPs Act as Super-Transformers in Interacting with Diversified Substrate Proteins

So far no single model is considered to sufficiently describe the structure, function or mechanism of sHSPs (Basha et al. 2012) due to the diversity in their primary sequences, oligomeric structures, substrate-binding sites and substrate proteins. It appears that sHSPs are more divergent than the ATP-dependent molecular chaperone families (e.g., Hsp70, Hsp60 and Hsp90) that are usually relatively conserved in structure and function (de Jong et al. 1993, 1998). In our opinion, the dynamic structures of sHSPs should be considered as the nature of this chaperone family and is also the key to underlie their broad substrate spectrum. In keeping with all the details related to sHSPs as discussed above, a novel "super-transformer" model is proposed, as illustrated in Fig. 4.2, to account for the structural characteristics and broad substrate spectrum of sHSPs.

In this model, each sHSP monomer behaves like a transformer, which is able to adapt to different conformations (the upper part in the left frame of Fig. 4.2). Multiple monomers, presumably via forming the build block dimer per se, easily assemble into various transformable and dynamic oligomers, termed as super-transformer (the lower part in the left frame of Fig. 4.2). Each super-transformer is able to capture structurally distinct substrate proteins using its numerous multi-type substrate-binding sites that are spatially chaotic and somehow disordered. Disorder-adapting-disorder may play important roles for the initial recognition and interaction of sHSPs with substrate proteins. In the end, large sHSP-substrate complexes are formed (the right frame in Fig. 4.2) and stabilized via numerous multi-type weak interactions between sHSPs and substrates. However, such complexes are also dynamic, given that subunit exchange occurs between sHSP subunits in the complexes and free sHSP subunits in the solution (Friedrich et al. 2004), that free substrate proteins could be incorporated into the pre-existing sHSP-substrate complexes (Stromer et al. 2003; Friedrich et al. 2004), and that the local conformational flexibility of sHSPs in the complexes is not decreased

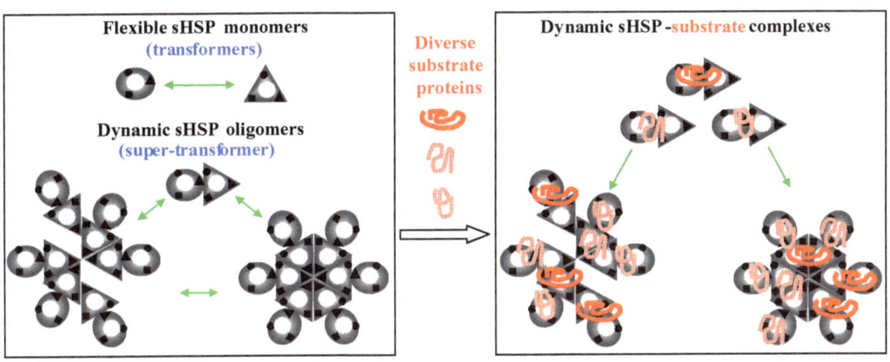

Fig. 4.2 A super-transformer model illustrating both the dynamic structures of sHSPs and the variable sHSP-substrate interactions. Each sHSP monomer, like a transformer, is able to adapt to different conformations, and easily assembles into various transformable and dynamic oligomers (super-transformer). The super-transformer is able to utilize its numerous multi-type substrate-binding sites (indicated by *black circles*, *squares* and *triangles*) that are spatially chaotic and potentially disordered (*white circles*) to recognize and capture multi-type substrate proteins, possibly via a way of disorder-adapting-disorder. In the end, apparently stable and large sHSP-substrate complexes are formed, with the sHSP subunits in the complexes still being dynamic

(Cheng et al. 2008). It follows that the dynamics of sHSP-substrate complexes would facilitate the subsequent substrate release in cooperation with other protein quality control components.

4.6 Perspectives

Here we have emphasized the common features of sHSPs and introduced some new terms such as disorder-adapting-disorder, super-transformer and spatial chaos to define their structural characteristics or interactions with substrate proteins. The super-transformer model provides a simple, somehow universal explanation for understanding the broad substrate spectrum of sHSPs. In this context, it is of great importance to probe how such sHSP super-transformers cooperate ATP-dependent molecular chaperones and/or proteases for protein quality control. In addition, it is of general interest to explore the significance of sHSP structural dynamics with respect to the general protein structural dynamics and to examine how the disorder-adapting-disorder for sHSP-substrate interaction differs from the general protein-protein interaction. Determination of high-resolution structures of sHSP-substrate complexes by such methods as cryo-electron microscopy may advance to these issues.

Acknowledgements This work was supported by research grants from the National Basic Research Program of China (973 Program) (No. 2012CB917300 to X.F.) and the National Natural Science Foundation of China (No. 31100559 and No. 31270804 to X.F.). The author is grateful to Professor Zengyi Chang (Peking University School of Life Sciences) for his full support and useful discussions.

References

Aquilina JA, Benesch JL, Bateman OA, Slingsby C, Robinson CV (2003) Polydispersity of a mammalian chaperone: mass spectrometry reveals the population of oligomers in alphaB-crystallin. Proc Natl Acad Sci U S A 100:10611–10616

Aquilina JA, Benesch JL, Ding LL, Yaron O, Horwitz J, Robinson CV (2005) Subunit exchange of polydisperse proteins: mass spectrometry reveals consequences of alphaA-crystallin truncation. J Biol Chem 280:14485–14491

Arrigo AP, Welch WJ (1987) Characterization and purification of the small 28,000-dalton mammalian heat shock protein. J Biol Chem 262:15359–15369

Baldwin AJ, Hilton GR, Lioe H, Bagneris C, Benesch JL, Kay LE (2011) Quaternary dynamics of alphaB-crystallin as a direct consequence of localised tertiary fluctuations in the C-terminus. J Mol Biol 413:310–320

Balogi Z, Cheregi O, Giese KC, Juhasz K, Vierling E, Vass I, Vigh L, Horvath I (2008) A mutant small heat shock protein with increased thylakoid association provides an elevated resistance against UV-B damage in synechocystis 6803. J Biol Chem 283:22983–22991

Bardwell JC, Jakob U (2012) Conditional disorder in chaperone action. Trends Biochem Sci 37:517–525

Basha E, Lee GJ, Breci LA, Hausrath AC, Buan NR, Giese KC, Vierling E (2004) The identity of proteins associated with a small heat shock protein during heat stress in vivo indicates that these chaperones protect a wide range of cellular functions. J Biol Chem 279:7566–7575

Basha E, Friedrich KL, Vierling E (2006) The N-terminal arm of small heat shock proteins is important for both chaperone activity and substrate specificity. J Biol Chem 281:39943–39952
Basha E, Jones C, Wysocki V, Vierling E (2010) Mechanistic differences between two conserved classes of small heat shock proteins found in the plant cytosol. J Biol Chem 285:11489–11497
Basha E, O'neill H, Vierling E (2012) Small heat shock proteins and alpha-crystallins: dynamic proteins with flexible functions. Trends Biochem Sci 37:106–117
Basha E, Jones C, Blackwell AE, Cheng G, Waters ER, Samsel KA, Siddique M, Pett V, Wysocki V, Vierling E (2013) An unusual dimeric small heat shock protein provides insight into the mechanism of this class of chaperones. J Mol Biol 425:1683–1696
Bepperling A, Alte F, Kriehuber T, Braun N, Weinkauf S, Groll M, Haslbeck M, Buchner J (2012) Alternative bacterial two-component small heat shock protein systems. Proc Natl Acad Sci U S A 109:20407–20412
Blakytny R, Carver JA, Harding JJ, Kilby GW, Sheil MM (1997) A spectroscopic study of glycated bovine alpha-crystallin: investigation of flexibility of the C-terminal extension, chaperone activity and evidence for diglycation. Biochim Biophys Acta 1343:299–315
Bova MP, Ding LL, Horwitz J, Fung BK (1997) Subunit exchange of alphaA-crystallin. J Biol Chem 272:29511–29517
Bova MP, Huang Q, Ding L, Horwitz J (2002) Subunit exchange, conformational stability, and chaperone-like function of the small heat shock protein 16.5 from Methanococcus jannaschii. J Biol Chem 277:38468–38475
Braun N, Zacharias M, Peschek J, Kastenmuller A, Zou J, Hanzlik M, Haslbeck M, Rappsilber J, Buchner J, Weinkauf S (2011) Multiple molecular architectures of the eye lens chaperone alphaB-crystallin elucidated by a triple hybrid approach. Proc Natl Acad Sci U S A 108:20491–20496
Bukach OV, Seit-Nebi AS, Marston SB, Gusev NB (2004) Some properties of human small heat shock protein Hsp20 (HspB6). Eur J Biochem 271:291–302
Carver JA, Lindner RA (1998) NMR spectroscopy of alpha-crystallin. Insights into the structure, interactions and chaperone action of small heat-shock proteins. Int J Biol Macromol 22:197–209
Carver JA, Guerreiro N, Nicholls KA, Truscott RJ (1995) On the interaction of alpha-crystallin with unfolded proteins. Biochim Biophys Acta 1252:251–260
Carver JA, Lindner RA, Lyon C, Canet D, Hernandez H, Dobson CM, Redfield C (2002) The interaction of the molecular chaperone alpha-crystallin with unfolding alpha-lactalbumin: a structural and kinetic spectroscopic study. J Mol Biol 318:815–827
Cashikar AG, Duennwald M, Lindquist SL (2005) A chaperone pathway in protein disaggregation. Hsp26 alters the nature of protein aggregates to facilitate reactivation by Hsp104. J Biol Chem 280:23869–23875
Chang Z, Primm TP, Jakana J, Lee IH, Serysheva I, Chiu W, Gilbert HF, Quiocho FA (1996) Mycobacterium tuberculosis 16-kDa antigen (Hsp16.3) functions as an oligomeric structure in vitro to suppress thermal aggregation. J Biol Chem 271:7218–7223
Chen X, Fu X, Ma Y, Chang Z (2005) Chaperone-like activity of Mycobacterium tuberculosis Hsp16.3 does not require its intact (native) structures. Biochemistry (Mosc) 70:913–919
Cheng G, Basha E, Wysocki VH, Vierling E (2008) Insights into small heat shock protein and substrate structure during chaperone action derived from hydrogen/deuterium exchange and mass spectrometry. J Biol Chem 283:26634–26642
Das KP, Petrash JM, Surewicz WK (1996) Conformational properties of substrate proteins bound to a molecular chaperone alpha-crystallin. J Biol Chem 271:10449–10452
Das KP, Choo-Smith LP, Petrash JM, Surewicz WK (1999) Insight into the secondary structure of non-native proteins bound to a molecular chaperone alpha-crystallin. An isotope-edited infrared spectroscopic study. J Biol Chem 274:33209–33212
De Jong WW, Leunissen JA, Voorter CE (1993) Evolution of the alpha-crystallin/small heat-shock protein family. Mol Biol Evol 10:103–126
De Jong WW, Caspers GJ, Leunissen JA (1998) Genealogy of the alpha-crystallin–small heat-shock protein superfamily. Int J Biol Macromol 22:151–162

De Miguel N, Braun N, Bepperling A, Kriehuber T, Kastenmuller A, Buchner J, Angel SO, Haslbeck M (2009) Structural and functional diversity in the family of small heat shock proteins from the parasite Toxoplasma gondii. Biochim Biophys Acta 1793:1738–1748

Den Engelsman J, Boros S, Dankers PY, Kamps B, Vree Egberts WT, Bode CS, Lane LA, Aquilina JA, Benesch JL, Robinson CV, De Jong WW, Boelens WC (2009) The small heat-shock proteins HSPB2 and HSPB3 form well-defined heterooligomers in a unique 3 to 1 subunit ratio. J Mol Biol 393:1022–1032

Ehrnsperger M, Graber S, Gaestel M, Buchner J (1997) Binding of non-native protein to Hsp25 during heat shock creates a reservoir of folding intermediates for reactivation. EMBO J 16:221–229

Ehrnsperger M, Lilie H, Gaestel M, Buchner J (1999) The dynamics of Hsp25 quaternary structure. Structure and function of different oligomeric species. J Biol Chem 274:14867–14874

Ellis RJ (1993) The general concept of molecular chaperones. Philos Trans R Soc Lond B Biol Sci 339:257–261

Evgrafov OV, Mersiyanova I, Irobi J, van Den Bosch L, Dierick I, Leung CL, Schagina O, Verpoorten N, van Impe K, Fedotov V, Dadali E, Auer-Grumbach M, Windpassinger C, Wagner K, Mitrovic Z, Hilton-Jones D, Talbot K, Martin JJ, Vasserman N, Tverskaya S, Polyakov A, Liem RK, Gettemans J, Robberecht W, de Jonghe P, Timmerman V (2004) Mutant small heat-shock protein 27 causes axonal Charcot-Marie-Tooth disease and distal hereditary motor neuropathy. Nat Genet 36:602–606

Ezemaduka AN, Yu J, Shi X, Zhang K, Yin CC, Fu X, Chang Z (2014) A small heat shock protein enables Escherichia coli to grow at a lethal temperature of 50{degrees}C conceivably by maintaining cell envelope integrity. J Bacteriol 196:2004–2011

Farahbakhsh ZT, Huang QL, Ding LL, Altenbach C, Steinhoff HJ, Horwitz J, Hubbell WL (1995) Interaction of alpha-crystallin with spin-labeled peptides. Biochemistry 34:509–516

Favet N, Duverger O, Loones MT, Poliard A, Kellermann O, Morange M (2001) Overexpression of murine small heat shock protein HSP25 interferes with chondrocyte differentiation and decreases cell adhesion. Cell Death Differ 8:603–613

Franzmann TM, Wuhr M, Richter K, Walter S, Buchner J (2005) The activation mechanism of Hsp26 does not require dissociation of the oligomer. J Mol Biol 350:1083–1093

Frauenfelder H, Sligar SG, Wolynes PG (1991) The energy landscapes and motions of proteins. Science 254:1598–1603

Friedrich KL, Giese KC, Buan NR, Vierling E (2004) Interactions between small heat shock protein subunits and substrate in small heat shock protein-substrate complexes. J Biol Chem 279:1080–1089

Fu X (2014) Chaperone function and mechanism of small heat-shock proteins. Acta Biochim Biophys Sin (Shanghai) 46:347–356

Fu X, Chang Z (2004) Temperature-dependent subunit exchange and chaperone-like activities of Hsp16.3, a small heat shock protein from Mycobacterium tuberculosis. Biochem Biophys Res Commun 316:291–299

Fu X, Chang Z (2006a) Identification of a highly conserved pro-gly doublet in non-animal small heat shock proteins and characterization of its structural and functional roles in Mycobacterium tuberculosis Hsp16.3. Biochemistry (Mosc) 71(Suppl 1):S83–S90

Fu X, Chang Z (2006b) Identification of bis-ANS binding sites in Mycobacterium tuberculosis small heat shock protein Hsp16.3: evidences for a two-step substrate-binding mechanism. Biochem Biophys Res Commun 349:167–171

Fu X, Li W, Mao Q, Chang Z (2003a) Disulfide bonds convert small heat shock protein Hsp16.3 from a chaperone to a non-chaperone: implications for the evolution of cysteine in molecular chaperones. Biochem Biophys Res Commun 308:627–635

Fu X, Liu C, Liu Y, Feng X, Gu L, Chen X, Chang Z (2003b) Small heat shock protein Hsp16.3 modulates its chaperone activity by adjusting the rate of oligomeric dissociation. Biochem Biophys Res Commun 310:412–420

Fu X, Zhang H, Zhang X, Cao Y, Jiao W, Liu C, Song Y, Abulimiti A, Chang Z (2005) A dual role for the N-terminal region of Mycobacterium tuberculosis Hsp16.3 in self-oligomerization and binding denaturing substrate proteins. J Biol Chem 280:6337–6348

Fu X, Jiao W, Chang Z (2006) Phylogenetic and biochemical studies reveal a potential evolutionary origin of small heat shock proteins of animals from bacterial class A. J Mol Evol 62:257–266

Fu X, Shi X, Yan L, Zhang H, Chang Z (2013a) In vivo substrate diversity and preference of small heat shock protein IbpB as revealed by using a genetically incorporated photo-cross-linker. J Biol Chem 288:31646–31654

Fu X, Shi X, Yin L, Liu J, Joo K, Lee J, Chang Z (2013b) Small heat shock protein IbpB acts as a robust chaperone in living cells by hierarchically activating its multi-type substrate-binding residues. J Biol Chem 288:11897–11906

Ghosh JG, Shenoy AK, Clark JI (2006) N- and C-terminal motifs in human alpha B crystallin play an important role in the recognition, selection, and solubilization of substrates. Biochemistry 45:13847–13854

Giese KC, Vierling E (2004) Mutants in a small heat shock protein that affect the oligomeric state. Analysis and allele-specific suppression. J Biol Chem 279:32674–32683

Gu L, Abulimiti A, Li W, Chang Z (2002) Monodisperse Hsp16.3 nonamer exhibits dynamic dissociation and reassociation, with the nonamer dissociation prerequisite for chaperone-like activity. J Mol Biol 319:517–526

Haley DA, Horwitz J, Stewart PL (1998) The small heat-shock protein, alphaB-crystallin, has a variable quaternary structure. J Mol Biol 277:27–35

Haley DA, Bova MP, Huang QL, Mchaourab HS, Stewart PL (2000) Small heat-shock protein structures reveal a continuum from symmetric to variable assemblies. J Mol Biol 298:261–272

Hanazono Y, Takeda K, Yohda M, Miki K (2012) Structural studies on the oligomeric transition of a small heat shock protein, StHsp14.0. J Mol Biol 422:100–108

Hanazono Y, Takeda K, Oka T, Abe T, Tomonari T, Akiyama N, Aikawa Y, Yohda M, Miki K (2013) Nonequivalence observed for the 16-meric structure of a small heat shock protein, SpHsp16.0, from Schizosaccharomyces pombe. Structure 21:220–228

Hartl FU, Bracher A, Hayer-Hartl M (2011) Molecular chaperones in protein folding and proteostasis. Nature 475:324–332

Haslbeck M, Walke S, Stromer T, Ehrnsperger M, White HE, Chen S, Saibil HR, Buchner J (1999) Hsp26: a temperature-regulated chaperone. EMBO J 18:6744–6751

Haslbeck M, Kastenmuller A, Buchner J, Weinkauf S, Braun N (2008) Structural dynamics of archaeal small heat shock proteins. J Mol Biol 378:362–374

Henzler-Wildman K, Kern D (2007) Dynamic personalities of proteins. Nature 450:964–972

Hilario E, Martin FJ, Bertolini MC, Fan L (2011) Crystal structures of Xanthomonas small heat shock protein provide a structural basis for an active molecular chaperone oligomer. J Mol Biol 408:74–86

Hockertz MK, Clark-Lewis I, Candido EP (1991) Studies of the small heat shock proteins of Caenorhabditis elegans using anti-peptide antibodies. FEBS Lett 280:375–378

Hollander JM, Martin JL, Belke DD, Scott BT, Swanson E, Krishnamoorthy V, Dillmann WH (2004) Overexpression of wild-type heat shock protein 27 and a nonphosphorylatable heat shock protein 27 mutant protects against ischemia/reperfusion injury in a transgenic mouse model. Circulation 110:3544–3552

Hong W, Jiao W, Hu J, Zhang J, Liu C, Fu X, Shen D, Xia B, Chang Z (2005) Periplasmic protein HdeA exhibits chaperone-like activity exclusively within stomach pH range by transforming into disordered conformation. J Biol Chem 280:27029–27034

Horwitz J (1992) Alpha-crystallin can function as a molecular chaperone. Proc Natl Acad Sci U S A 89:10449–10453

Irobi J, van Impe K, Seeman P, Jordanova A, Dierick I, Verpoorten N, Michalik A, de Vriendt E, Jacobs A, van Gerwen V, Vennekens K, Mazanec R, Tournev I, Hilton-Jones D, Talbot K, Kremensky I, van den Bosch L, Robberecht W, van Vandekerckhove J, van Broeckhoven C,

Gettemans J, de Jonghe P, Timmerman V (2004) Hot-spot residue in small heat-shock protein 22 causes distal motor neuropathy. Nat Genet 36:597–601

Jakob U, Gaestel M, Engel K, Buchner J (1993) Small heat shock proteins are molecular chaperones. J Biol Chem 268:1517–1520

Jaya N, Garcia V, Vierling E (2009) Substrate binding site flexibility of the small heat shock protein molecular chaperones. Proc Natl Acad Sci U S A 106:15604–15609

Jehle S, Rajagopal P, Bardiaux B, Markovic S, Kuhne R, Stout JR, Higman VA, Klevit RE, van Rossum BJ, Oschkinat H (2010) Solid-state NMR and SAXS studies provide a structural basis for the activation of alphaB-crystallin oligomers. Nat Struct Mol Biol 17:1037–1042

Jiao W, Li P, Zhang J, Zhang H, Chang Z (2005a) Small heat-shock proteins function in the insoluble protein complex. Biochem Biophys Res Commun 335:227–231

Jiao W, Qian M, Li P, Zhao L, Chang Z (2005b) The essential role of the flexible termini in the temperature-responsiveness of the oligomeric state and chaperone-like activity for the polydisperse small heat shock protein IbpB from Escherichia coli. J Mol Biol 347:871–884

Jiao W, Hong W, Li P, Sun S, Ma J, Qian M, Hu M, Chang Z (2008) The dramatically increased chaperone activity of small heat-shock protein IbpB is retained for an extended period of time after the stress condition is removed. Biochem J 410:63–70

Kappe G, Aquilina JA, Wunderink L, Kamps B, Robinson CV, Garate T, Boelens WC, de Jong WW (2004) Tsp36, a tapeworm small heat-shock protein with a duplicated alpha-crystallin domain, forms dimers and tetramers with good chaperone-like activity. Proteins 57:109–117

Kennaway CK, Benesch JL, Gohlke U, Wang L, Robinson CV, Orlova EV, Saibil HR, Keep NH (2005) Dodecameric structure of the small heat shock protein Acr1 from Mycobacterium tuberculosis. J Biol Chem 280:33419–33425

Kim KK, Kim R, Kim SH (1998) Crystal structure of a small heat-shock protein. Nature 394:595–599

Kim KH, Alam I, Kim YG, Sharmin SA, Lee KW, Lee SH, Lee BH (2012) Overexpression of a chloroplast-localized small heat shock protein OsHSP26 confers enhanced tolerance against oxidative and heat stresses in tall fescue. Biotechnol Lett 34:371–377

Kitagawa M, Matsumura Y, Tsuchido T (2000) Small heat shock proteins, IbpA and IbpB, are involved in resistances to heat and superoxide stresses in Escherichia coli. FEMS Microbiol Lett 184:165–171

Kokke BP, Leroux MR, Candido EP, Boelens WC, de Jong WW (1998) Caenorhabditis elegans small heat-shock proteins Hsp12.2 and Hsp12.3 form tetramers and have no chaperone-like activity. FEBS Lett 433:228–232

Lambert W, Rutsdottir G, Hussein R, Bernfur K, Kjellstrom S, Emanuelsson C (2013) Probing the transient interaction between the small heat-shock protein Hsp21 and a model substrate protein using crosslinking mass spectrometry. Cell Stress Chaperones 18:75–85

Landry J, Chretien P, Lambert H, Hickey E, Weber LA (1989) Heat shock resistance conferred by expression of the human HSP27 gene in rodent cells. J Cell Biol 109:7–15

Laskowska E, Matuszewska E, Kuczynska-Wisnik D (2010) Small heat shock proteins and protein-misfolding diseases. Curr Pharm Biotechnol 11:146–157

Lee GJ, Roseman AM, Saibil HR, Vierling E (1997) A small heat shock protein stably binds heat-denatured model substrates and can maintain a substrate in a folding-competent state. EMBO J 16:659–671

Lelj-Garolla B, Mauk AG (2005) Self-association of a small heat shock protein. J Mol Biol 345:631–642

Lentze N, Studer S, Narberhaus F (2003) Structural and functional defects caused by point mutations in the alpha-crystallin domain of a bacterial alpha-heat shock protein. J Mol Biol 328:927–937

Leroux MR, Ma BJ, Batelier G, Melki R, Candido EP (1997a) Unique structural features of a novel class of small heat shock proteins. J Biol Chem 272:12847–12853

Leroux MR, Melki R, Gordon B, Batelier G, Candido EP (1997b) Structure-function studies on small heat shock protein oligomeric assembly and interaction with unfolded polypeptides. J Biol Chem 272:24646–24656

Lindner RA, Kapur A, Carver JA (1997) The interaction of the molecular chaperone, alpha-crystallin, with molten globule states of bovine alpha-lactalbumin. J Biol Chem 272:27722–27729
Lindner RA, Treweek TM, Carver JA (2001) The molecular chaperone alpha-crystallin is in kinetic competition with aggregation to stabilize a monomeric molten-globule form of alpha-lactalbumin. Biochem J 354:79–87
Litt M, Kramer P, Lamorticella DM, Murphey W, Lovrien EW, Weleber RG (1998) Autosomal dominant congenital cataract associated with a missense mutation in the human alpha crystallin gene CRYAA. Hum Mol Genet 7:471–474
Longoni S, James P, Chiesi M (1990) Cardiac alpha-crystallin. I. Isolation and identification. Mol Cell Biochem 99:113–120
Merck KB, Groenen PJ, Voorter CE, de Haard-Hoekman WA, Horwitz J, Bloemendal H, de JONG WW (1993a) Structural and functional similarities of bovine alpha-crystallin and mouse small heat-shock protein. A family of chaperones. J Biol Chem 268:1046–1052
Merck KB, Horwitz J, Kersten M, Overkamp P, Gaestel M, Bloemendal H, de Jong WW (1993b) Comparison of the homologous carboxy-terminal domain and tail of alpha-crystallin and small heat shock protein. Mol Biol Rep 18:209–215
Mogk A, Deuerling E, Vorderwulbecke S, Vierling E, Bukau B (2003) Small heat shock proteins, ClpB and the DnaK system form a functional triade in reversing protein aggregation. Mol Microbiol 50:585–595
Montero-Barrientos M, Cardoza RE, Gutierrez S, Monte E, Hermosa R (2007) The heterologous overexpression of hsp23, a small heat-shock protein gene from Trichoderma virens, confers thermotolerance to T. harzianum. Curr Genet 52:45–53
Morrow G, Battistini S, Zhang P, Tanguay RM (2004) Decreased lifespan in the absence of expression of the mitochondrial small heat shock protein Hsp22 in Drosophila. J Biol Chem 279:43382–43385
Plater ML, Goode D, Crabbe MJ (1996) Effects of site-directed mutations on the chaperone-like activity of alphaB-crystallin. J Biol Chem 271:28558–28566
Raman B, Rao CM (1997) Chaperone-like activity and temperature-induced structural changes of alpha-crystallin. J Biol Chem 272:23559–23564
Rawat U, Rao M (1998) Interactions of chaperone alpha-crystallin with the molten globule state of xylose reductase. Implications for reconstitution of the active enzyme. J Biol Chem 273:9415–9423
Reichmann D, Xu Y, Cremers CM, Ilbert M, Mittelman R, Fitzgerald MC, Jakob U (2012) Order out of disorder: working cycle of an intrinsically unfolded chaperone. Cell 148:947–957
Richards EH, Hickey E, Weber L, Master JR (1996) Effect of overexpression of the small heat shock protein HSP27 on the heat and drug sensitivities of human testis tumor cells. Cancer Res 56:2446–2451
Saji H, Iizuka R, Yoshida T, Abe T, Kidokoro S, Ishii N, Yohda M (2008) Role of the IXI/V motif in oligomer assembly and function of StHsp14.0, a small heat shock protein from the acidothermophilic archaeon, Sulfolobus tokodaii strain 7. Proteins 71:771–782
Salinthone S, Ba M, Hanson L, Martin JL, Halayko AJ, Gerthoffer WT (2007) Overexpression of human Hsp27 inhibits serum-induced proliferation in airway smooth muscle myocytes and confers resistance to hydrogen peroxide cytotoxicity. Am J Physiol Lung Cell Mol Physiol 293:L1194–L1207
Santhoshkumar P, Murugesan R, Sharma KK (2009) Deletion of (54)FLRAPSWF(61) residues decreases the oligomeric size and enhances the chaperone function of alphaB-crystallin. Biochemistry 48:5066–5073
Sathish HA, Stein RA, Yang G, Mchaourab HS (2003) Mechanism of chaperone function in small heat-shock proteins. Fluorescence studies of the conformations of T4 lysozyme bound to alphaB-crystallin. J Biol Chem 278:44214–44221
Seo JS, Lee YM, Park HG, Lee JS (2006) The intertidal copepod Tigriopus japonicus small heat shock protein 20 gene (Hsp20) enhances thermotolerance of transformed Escherichia coli. Biochem Biophys Res Commun 340:901–908

Sharma KK, Kaur H, Kester K (1997) Functional elements in molecular chaperone alpha-crystallin: identification of binding sites in alpha B-crystallin. Biochem Biophys Res Commun 239:217–222
Sharma KK, Kumar RS, Kumar GS, Quinn PT (2000) Synthesis and characterization of a peptide identified as a functional element in alphaA-crystallin. J Biol Chem 275:3767–3771
Shearstone JR, Baneyx F (1999) Biochemical characterization of the small heat shock protein IbpB from Escherichia coli. J Biol Chem 274:9937–9945
Shemetov AA, Gusev NB (2011) Biochemical characterization of small heat shock protein HspB8 (Hsp22)-Bag3 interaction. Arch Biochem Biophys 513:1–9
Shi X, Wang Z, Yan L, Ezemaduka AN, Fan G, Wang R, Fu X, Yin C, Chang Z (2011) Small heat shock protein AgsA forms dynamic fibrils. FEBS Lett 585:3396–3402
Shi J, Koteiche HA, Mcdonald ET, Fox TL, Stewart PL, Mchaourab HS (2013) Cryoelectron microscopy analysis of small heat shock protein 16.5 (hsp16.5) complexes with t4 lysozyme reveals the structural basis of multimode binding. J Biol Chem 288:4819–4830
Smulders RH, Merck KB, Aendekerk J, Horwitz J, Takemoto L, Slingsby C, Bloemendal H, de Jong WW (1995) The mutation Asp69-->Ser affects the chaperone-like activity of alpha A-crystallin. Eur J Biochem 232:834–838
Smulders R, Carver JA, Lindner RA, van Boekel MA, Bloemendal H, van Jong WW (1996) Immobilization of the C-terminal extension of bovine alphaA-crystallin reduces chaperone-like activity. J Biol Chem 271:29060–29066
Sobott F, Benesch JL, Vierling E, Robinson CV (2002) Subunit exchange of multimeric protein complexes. Real-time monitoring of subunit exchange between small heat shock proteins by using electrospray mass spectrometry. J Biol Chem 277:38921–38929
Spector A, Li LK, Augusteyn RC, Schneider A, Freund T (1971) -Crystallin. The isolation and characterization of distinct macromolecular fractions. Biochem J 124:337–343
Stengel F, Baldwin AJ, Painter AJ, Jaya N, Basha E, Kay LE, Vierling E, Robinson CV, Benesch JL (2010) Quaternary dynamics and plasticity underlie small heat shock protein chaperone function. Proc Natl Acad Sci U S A 107:2007–2012
Stengel F, Baldwin AJ, Bush MF, Hilton GR, Lioe H, Basha E, Jaya N, Vierling E, Benesch JL (2012) Dissecting heterogeneous molecular chaperone complexes using a mass spectrum deconvolution approach. Chem Biol 19:599–607
Stromer T, Ehrnsperger M, Gaestel M, Buchner J (2003) Analysis of the interaction of small heat shock proteins with unfolding proteins. J Biol Chem 278:18015–18021
Strozecka J, Chrusciel E, Gorna E, Szymanska A, Zietkiewicz S, Liberek K (2012) Importance of N- and C-terminal regions of IbpA, Escherichia coli small heat shock protein, for chaperone function and oligomerization. J Biol Chem 287:2843–2853
Studer S, Obrist M, Lentze N, Narberhaus F (2002) A critical motif for oligomerization and chaperone activity of bacterial alpha-heat shock proteins. Eur J Biochem 269:3578–3586
Sudnitsyna MV, Mymrikov EV, Seit-Nebi AS, Gusev NB (2012) The role of intrinsically disordered regions in the structure and functioning of small heat shock proteins. Curr Protein Pept Sci 13:76–85
Sugase K, Dyson HJ, Wright PE (2007) Mechanism of coupled folding and binding of an intrinsically disordered protein. Nature 447:1021–1025
Sun Y, Mansour M, Crack JA, Gass GL, Macrae TH (2004) Oligomerization, chaperone activity, and nuclear localization of p26, a small heat shock protein from Artemia franciscana. J Biol Chem 279:39999–40006
Sun L, Liu Y, Kong X, Zhang D, Pan J, Zhou Y, Wang L, Li D, Yang X (2012) ZmHSP16.9, a cytosolic class I small heat shock protein in maize (Zea mays), confers heat tolerance in transgenic tobacco. Plant Cell Rep 31:1473–1484
Takeda K, Hayashi T, Abe T, Hirano Y, Hanazono Y, Yohda M, Miki K (2011) Dimer structure and conformational variability in the N-terminal region of an archaeal small heat shock protein, StHsp14.0. J Struct Biol 174:92–99
Tompa P, Csermely P (2004) The role of structural disorder in the function of RNA and protein chaperones. FASEB J 18:1169–1175

Treweek TM, Lindner RA, Mariani M, Carver JA (2000) The small heat-shock chaperone protein, alpha-crystallin, does not recognize stable molten globule states of cytosolic proteins. Biochim Biophys Acta 1481:175–188

Treweek TM, Ecroyd H, Williams DM, Meehan S, Carver JA, Walker MJ (2007) Site-directed mutations in the C-terminal extension of human alphaB-crystallin affect chaperone function and block amyloid fibril formation. PLoS One 2:e1046

van Montfort RL, Basha E, Friedrich KL, Slingsby C, Vierling E (2001) Crystal structure and assembly of a eukaryotic small heat shock protein. Nat Struct Biol 8:1025–1030

Veinger L, Diamant S, Buchner J, Goloubinoff P (1998) The small heat-shock protein IbpB from Escherichia coli stabilizes stress-denatured proteins for subsequent refolding by a multichaperone network. J Biol Chem 273:11032–11037

Vicart P, Caron A, Guicheney P, Li Z, Prevost MC, Faure A, Chateau D, Chapon F, Tome F, Dupret JM, Paulin D, Fardeau M (1998) A missense mutation in the alphaB-crystallin chaperone gene causes a desmin-related myopathy. Nat Genet 20:92–95

Wright PE, Dyson HJ (1999) Intrinsically unstructured proteins: re-assessing the protein structure-function paradigm. J Mol Biol 293:321–331

Yang H, Huang S, Dai H, Gong Y, Zheng C, Chang Z (1999) The Mycobacterium tuberculosis small heat shock protein Hsp16.3 exposes hydrophobic surfaces at mild conditions: conformational flexibility and molecular chaperone activity. Protein Sci 8:174–179

Yeh CH, Chang PF, Yeh KW, Lin WC, Chen YM, Lin CY (1997) Expression of a gene encoding a 16.9-kDa heat-shock protein, Oshsp16.9, in Escherichia coli enhances thermotolerance. Proc Natl Acad Sci U S A 94:10967–10972

Chapter 5
Model Chaperones: Small Heat Shock Proteins from Plants

Indu Santhanagopalan, Eman Basha, Keith N. Ballard, Nathen E. Bopp, and Elizabeth Vierling

Abstract Small heat shock proteins (sHSPs) are ubiquitous stress proteins proposed to act as ATP-independent molecular chaperones to prevent irreversible aggregation of stress-labile proteins. sHSPs range in size from ~12 to 42 kDa, but typically assemble into 12 to >32 subunit oligomers. The monomers are defined by a conserved α-crystallin domain flanked by divergent and flexible N-terminal and C-terminal arms. In higher plants sHSPs have evolved independently of metazoan and bacterial homologs and comprise multiple families of cytosolic proteins, along with proteins targeted to the nucleus, chloroplasts, mitochondria, endoplasmic reticulum and peroxisomes. This diversity of sHSPs is unique to land plants and likely arose as a result of their frequent exposure to stress due to their sessile nature. The availability of the high resolution structure of a dodecameric cytosolic class I sHSP from wheat, Ta16.9 (PDB ID: 1GME; 2.7 Å resolution), has facilitated detailed in vitro studies of sHSP chaperone action. A working model proposes that sHSP oligomers dissociate into dimers during heat stress, revealing hydrophobic patches that interact with exposed hydrophobic regions on denaturing substrates, maintaining them in a soluble, folding-competent state. sHSP-substrate complexes are then acted on by ATP-dependent chaperones to restore substrates to their native state. However, much remains to be done to connect this model with the function of the many different sHSPs found in plants. Further genetic and biochemical studies are needed to identify sHSP substrates and to define the mechanism by which sHSPs function, not only during stress, but also during specific developmental stages in plants.

I. Santhanagopalan • K.N. Ballard • N.E. Bopp • E. Vierling (✉)
Life Science Laboratories, Department of Biochemistry and Molecular Biology, University of Massachusetts, 240 Thatcher Road, Amherst, MA 01003, USA
e-mail: indu@biochem.umass.edu; kballard@mcb.umass.edu; Nbopp@umass.edu; vierling@biochem.umass.edu

E. Basha
Botany Department, College of Science, Tanta University, Tanta, Egypt

Biology Department, Faculty of Science, Taif University, Taif, Saudi Arabia
e-mail: basha_eman@hotmail.com

R.M. Tanguay, L.E. Hightower (eds.), *The Big Book on Small Heat Shock Proteins*, Heat Shock Proteins 8, DOI 10.1007/978-3-319-16077-1_5

Keywords α-crystallin domain • Hydrogen-deuterium exchange • Nanoelectrospray mass spectrometry • IxI motif • Chaperone efficiency • Replica exchange molecular dynamics • *Arabidopsis* thaliana • *Synechocystis sp.* PCC 6803 • N-terminal arm • Stress response • Chaperone network

Abbreviations

ACD	Alpha-crystallin domain
AKRA2	Ankyrin repeat containing protein 2A (UniProt accession: Q9SAR5)
At14.7	*Arabidopsis thaliana* heat shock protein 14.7 (UnitProt accession: Q6NLV0)
At15.7	*Arabidopsis thaliana* heat shock protein 15.7 (UniProt accession: Q9FHQ3)
At17.4	*Arabidopsis thaliana* heat shock protein 17.4 (SwissProt accession: Q9SYG1)
At17.7	*Arabidopsis thaliana* heat shock protein 17.7 (UniProt accession: O81822)
At17.8	*Arabidopsis thaliana* heat shock protein 17.8 (UniProt accession: Q9LNW0)
At18.5	*Arabidopsis thaliana* heat shock protein 18.5 (UniProt accession: O64564)
At21.7	*Arabidopsis thaliana* heat shock protein 21.7 (UniProt accession: Q9FIT9)
At22.0	*Arabidopsis thaliana* heat shock protein 22.0 (UniProt accession: Q38806)
At23.5	*Arabidopsis thaliana* heat shock protein 23.5 (UniProt accession: Q9FGM9)
At23.6	*Arabidopsis thaliana* heat shock protein 23.6 (UniProt accession: Q96331)
At26.5	*Arabidopsis thaliana* heat shock protein 26.5 (UnitProt accession: Q9SSQ8)
C. elegans	*Caenorhabditis elegans*
Dr17.7	*Deinococcus radiodurans* heat shock protein 17.7 (UniProt accession: Q9RTR5)
Dr20.2	*Deinococcus radiodurans* heat shock protein 20.2 (UniProt accession: Q9RVB5)
EPR	Electron paramagnetic resonance
ER	Endoplasmic reticulum
FRET	Förster (or fluorescence) resonance energy transfer
GST	Glutathione-S-transferase
HSP	Heat shock protein
MALDI-TOF	Matrix assisted laser desorption ionization – time of flight
MI	Mitochondrial class I small heat shock protein

MII	Mitochondrial class II small heat shock protein
Mj16.5	*Methanocaldococcus jannaschii* heat shock protein 16.5 (UniProt accession: Q57733)
NtHsp24.6	*Nicotiana tabacum* heat shock protein 24.6
OEP7	Outer envelope protein 7 (chloroplast) (UnitProt accession: Q9SVC4)
Ps17.7	*Pisum sativum* heat shock protein 17.7 (UniProt accession: P19242)
Ps18.1	*Pisum sativum* heat shock protein 18.1 (UniProt accession: P19243)
Sc26	*Saccharomyces cerevisiae* heat shock protein 26 (UniProt accession: P15992)
sHSP	Small heat shock protein
Sp16.0	*Schizosaccharomyces pombe* heat shock protein 16.0 (UniProt accession: O14368)
Syn16.6	*Synechocystis* heat shock protein 16.6 (Unit Prot accession: M1LDX9)
Ta16.9	*Triticum asetivum* heat shock protein 16.9 (UniProt accession: P12810)
Ta17.8	*Triticum asetivum* heat shock protein 17.8 (UniProt accession: Q94KM0)

5.1 Defining the Small Heat Shock Proteins in Plants

Over three decades ago it was recognized that all organisms respond to heat stress by elevating the transcription and translation of a set of genes that were designated as encoding "heat shock proteins" (HSPs). The first observations of HSP synthesis in higher plants were reported for soybean and tobacco (Barnett et al. 1980; Key et al. 1981). It was already obvious from these initial studies that, compared to *Drosophila*, *Saccharomyces cerevisiae*, bacteria and humans, plants synthesize a very large number of HSPs with molecular weights between 15 and 25 kDa ("small" HSPs or sHSPs). In plants, sHSPs were found to be not only numerous, but the corresponding mRNAs were also produced at such high levels during heat stress (estimated >20,000 copies per cell), that cDNAs encoding these sHSPs were among the first protein coding sequences cloned from plants (Schöffl and Key 1982). The production of major classes of HSP mRNAs following heat stress in plants is illustrated in Fig. 5.1. The accumulation of mRNAs encoding members of the HSP100/ClpB, Hsp90 and Hsp70 families of HSPs along with numerous sHSP mRNAs is apparent. The abundance and diversity of plant sHSPs has made their function and evolution of considerable interest.

The first sequences of sHSPs from plants were obtained from soybean in 1985, and the data revealed that the proteins were homologous to sHSPs that had already been characterized from *Drosophila*, *C. elegans* and *Xenopus* (Nagao et al. 1985). The common characteristic of all sHSPs is a 90–100 amino acid, central domain (the α-crystallin domain (ACD)) (Figs. 5.2 and 5.3) recognized as homologous to the eye lens α-crystallins by Ingolia and Craig (1982). sHSPs are now known to be found in

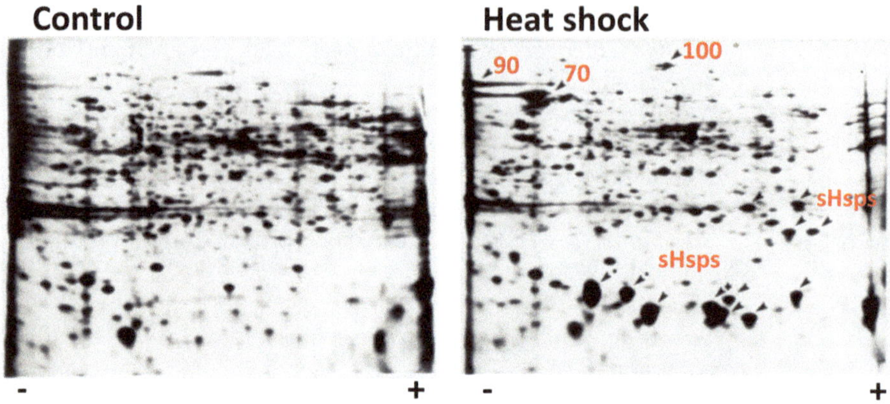

Fig. 5.1 Heat stress induces the accumulation of HSP mRNAs in plants. Autoradiogram of in vitro translation products of mRNA isolated from control (22 °C) or heat stressed (38 °C, 90 min) pea (*Pisum sativum*) seedlings. PolyA RNA was extracted after treatment and used for in vitro translation in wheat germ extract in the presence of ^{35}S-methionine. The translation products were separated by two-dimensional polyacrylamide gel electrophoresis and imaged by autoradiography. *Arrowheads* point to the indicated HSPs. sHSPs at the *middle right* represent the precursors of mitochondrial and chloroplast targeted sHSPs (Data from Vierling, unpublished)

all domains of life. An understanding of the origins and diversity of the sHSPs in plants has been gained in subsequent decades of research. While sHSPs in metazoans are found primarily in the cytosol and nucleus, plant sHSPs are present in these compartments, and in addition, nuclear-encoded plant sHSPs have been characterized that are targeted to chloroplasts (Van Aken et al. 2009; Vierling et al. 1988; Waters and Vierling 1999a), the endoplasmic reticulum (ER) (Helm et al. 1995), mitochondria (Lenne et al. 1995; Van Aken et al. 2009), the nucleus, and peroxisomes (Ma et al. 2006; Scharf et al. 2001). Outside of plants, *Drosophila* is the only organism known to have an organelle specific sHSP, which is reported to localize to mitochondria (Wadhwa et al. 2010). Furthermore, angiosperm cytosolic sHSPs group into five or more families that originated hundreds of millions of years ago and show evidence of continued diversification (Lopes-Caitar et al. 2013; Siddique et al. 2008; Waters et al. 2008; Waters and Vierling 1999b). The sequence relationship of plant sHSPs is illustrated for all sHSPs from *Arabidopsis thaliana* in Fig. 5.2. A general conclusion is that the evolution of the diverse land plant sHSPs has been driven by the sessile nature of plants, which subjects them to inescapable environmental stresses potentially counteracted by sHSP function.

Although all sHSPs are grouped together as ATP-independent molecular chaperones that are proposed to prevent irreversible aggregation of stress-sensitive proteins, exactly how they function and what they protect remains far from determined (Basha et al. 2012). Here we will first review basic data that have helped formulate a model for sHSP function from studies of plant sHSPs, with reference to data from other organisms. We then go on to discuss further studies that both support, but also call into question the generality of the current model. Finally, we will pose some of the major outstanding questions that remain to be tackled.

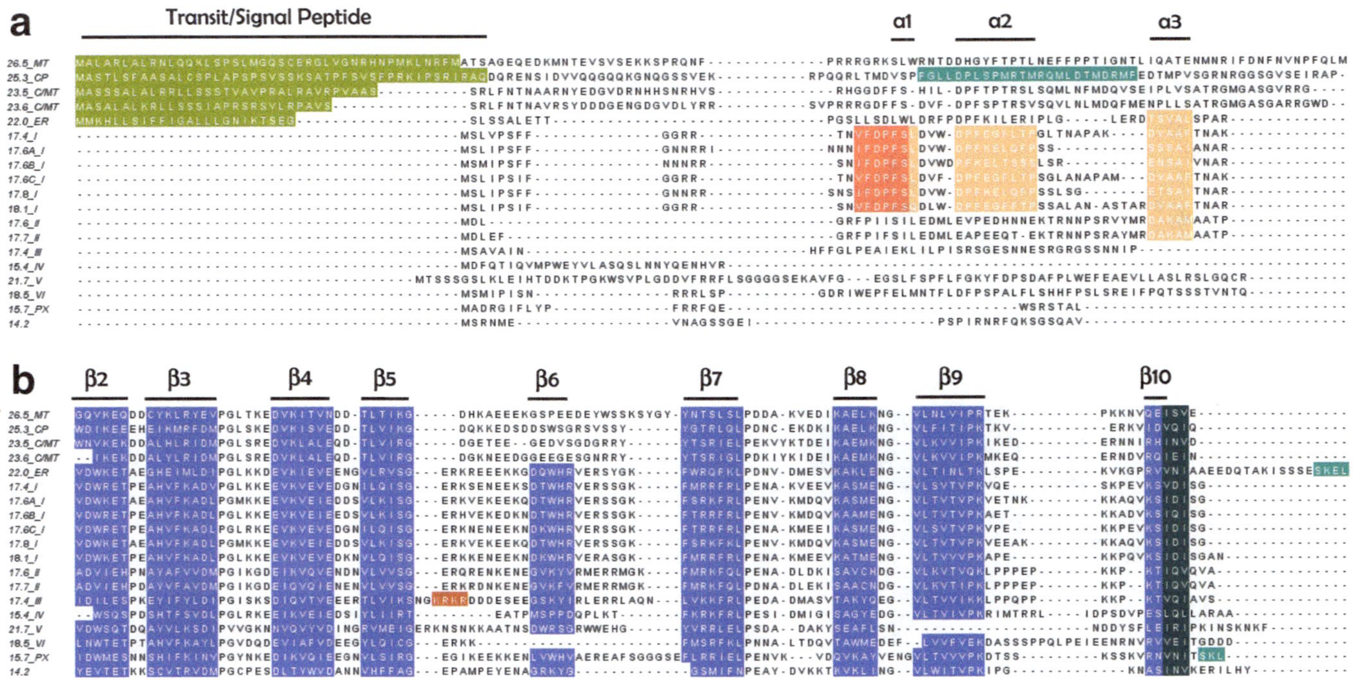

Fig. 5.2 Amino acid sequence alignment of the 19 sHSPs from *Arabidopsis thaliana*. (**a**) The N-terminal arms, and (**b**) the ACDs and C-terminal extensions. Alignment was performed with Jalview (Waterhouse et al. 2009) and adjusted based on secondary structures predicted using Jpred (Cole et al. 2008) and alignment with the Ta16.9 sequence. Secondary structure is numbered according to the Ta16.9 structure (1GME). In (**a**) the transit (or signal) peptides required for localization to intracellular organelles are highlighted at the N-terminus in *green*. Predicted alpha helices are highlighted in *pink*, with the conserved V/IFDPFS motif of the class I proteins partially overlapping α-helix 1 in *orange*. The Methione-Bristle motif (Chen and Vierling 1991) in chloroplast-targeted At25.3 is highlighted in *turquoise*. In **b** the characteristic β-strands of the ACD are shown in *blue* where predictions were possible. The C-terminal IxI motif (*black*) overlaps by one residue with β10. Additional features indicated are: the nuclear-localization signal in class III At17.4 (*red*), the C-terminal ER-retention signal in At22.0 (*turquoise*) and the C-terminal peroxisomal type 1 targeting signal in At15.7 (*turquoise*). Arabidopsis genome identification numbers for the sHSPs are: 26.5-MT, At1g52560; 25.3-CP, At4g27670; 23.5-C/MT, At5g51440; At23.6-C/MT, At4g25200;At22.0-ER, At4g10250; At17.4-I, At3g46230; At17.6A-I, At1g59860; At17.6B-I, At2g29500; At17.6C-I,At1g53540; At17.8-I, At1g07400; At18.1-I, At5g59720; At17.6-II,At5g12020; At17.7-II, At5g12030; At17.4-III, At1g54050; At15.4-IV, At4g21870; At21.7-V,At5g54660; At18.5-VI, At2g19310; At15.7-PX, At5g37670; At14.2, At5g47600

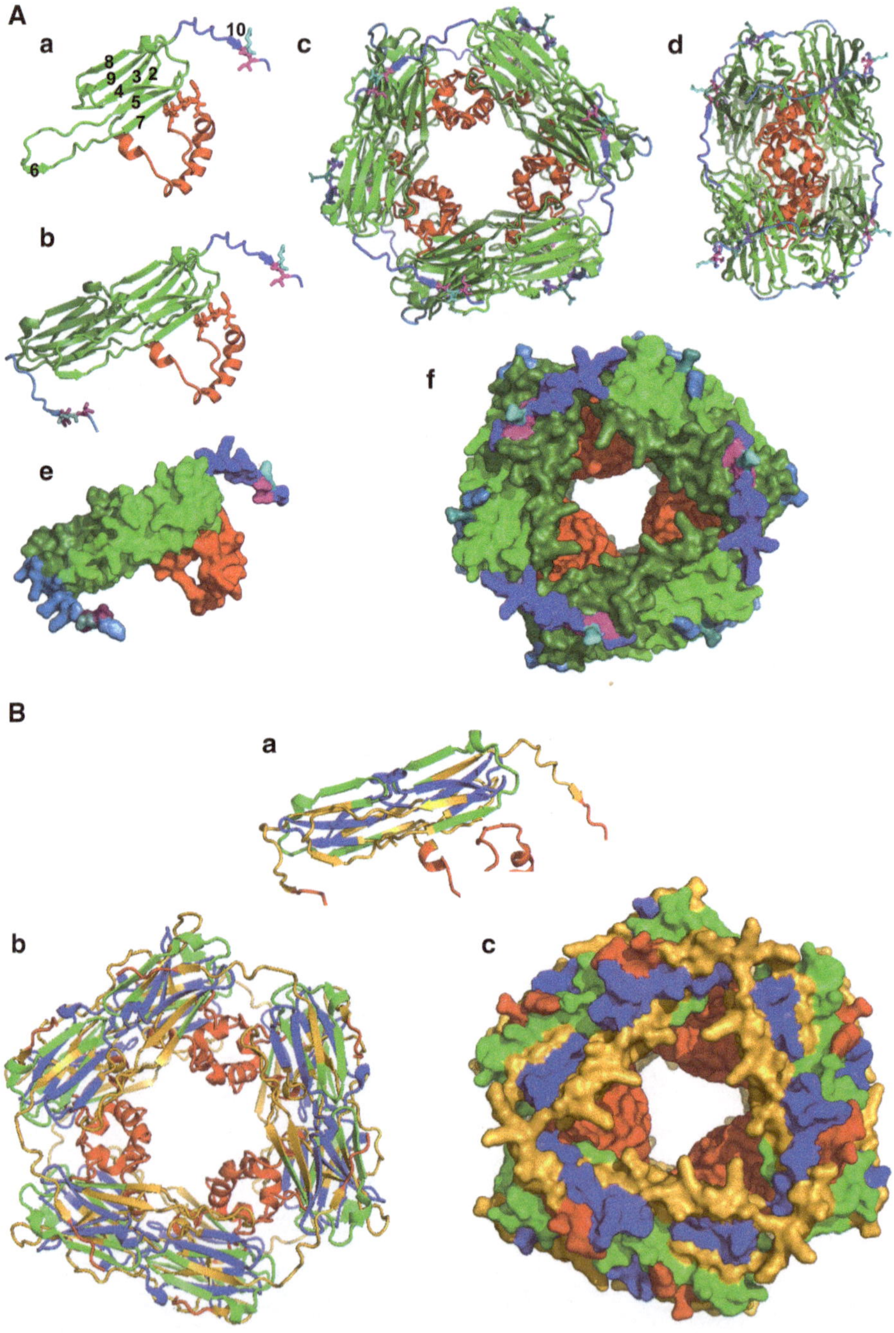

Fig. 5.3 Structural features and dynamics of wheat (*Triticum aestivum*) Ta16.9 (1GME). (**a**) Color-coding used is *red*=N-terminal arm, *green*=ACD, *blue*=C-terminal extension and the IxI motif is colored *magenta-cyan-magenta*. One monomer of each dimer is shown in *darker shades*. **Fig. 5.3** (continued) Monomer, with numbered ACD β-strands (*a*), dimer (*b*) and dodecamer (*c*

5.2 Insights into sHSP Structure Gained from Plant sHSPs

Like any other protein, one key to understanding sHSPs lies in gaining insights about their structures. While the monomeric masses of sHSPs from diverse organisms range from 12 to 42 kDa, the majority assemble into multimers of 12 to over 32 monomers (Basha et al. 2012; Van Montfort et al. 2001a). The oligomeric size and polydispersity of many sHSPs have made obtaining high resolution structures difficult (Delbecq and Klevit 2013). A summary of structural data available for sHSPs is provided in Table 5.1. Studies of a single evolutionary class of plant cytosolic sHSPs, referred to as cytosolic class I (Fig. 5.2), have made major contributions to the current model for sHSP chaperone function, in part due to the availability of the 2.7 Å structure of the native class I dodecamer from wheat (1GME) (van Montfort et al. 2001b) (Fig. 5.3). Dodecameric wheat Hsp16.9 (Ta16.9) and the recent X-ray structure of the 16-mer Hsp16.0 from *Schizosaccharomyces pombe* (Hanazono et al. 2013) remain the only eukaryotic, and the most complete high resolution sHSP structures available, but limited activity data are available for the latter protein. In this section, we focus on structural information gained from Ta16.9, with reference to other sHSPs.

The Ta16.9 and *Methanocaldococcus janachii* Mj16.5 structures defined the conserved ACD as comprising a seven or eight strand β-sandwich, which is a structural hallmark of sHSPs (Fig. 5.3). In addition, both these sHSPs are built from a dimeric substructure that involves swapping of β-strand 6 between monomers. The dimers of most plant, bacterial and yeast sHSPs comprise this type of strand-swapped dimer (Hanazono et al. 2013; Kim et al. 1998; van Montfort et al. 2001b) (Table 5.1). Note that the secondary and tertiary structure of the ACD is much more conserved within the sHSP family than the amino acid sequence of this domain. High resolution structures of the ACD from several metazoans (Table 5.1) have defined a second type of dimer that does not involve strand swapping. These metazoan proteins, including αA- and αB-crystallin, do not have a β6 strand, but rather possess an elongated β7 strand that forms the dimeric interface by aligning in an antiparallel fashion (Bagneris et al. 2009; Baranova et al. 2011; Clark et al. 2011; Laganowsky et al. 2010; Laganowsky and Eisenberg 2010). The two types of dimers were highlighted in recent reviews (Basha et al. 2012; Delbecq and Klevit 2013). The extent to which these differences in sHSP dimer structure may determine differences in mechanism is unknown.

The characteristic ACD is preceded by a variable N-terminal arm. The variability in N-terminal arms of different sHSPs is well-illustrated by the alignment of

Fig. 5.3 (continued) and *d*) are shown in cartoon form, with the IxI and N-terminal V/IFDPFS motifs in stick representation. Dodecamer in (*d*) is rotated 90° in the plane of the page from (**c**). Surface representations of the dimer (*e*) and dodecamer (*f*). (**b**) Ta16.9 structure colored according to percent of amide hydrogen-deuterium exchange within 5 s at 25 °C (Cheng et al. 2008) with *red* = 80–100 %, *orange* = 60–80 %, *green* = 40–60 % and *blue* = 20–40 %. Note that the N-terminal arm and C-terminus are the most dynamic, followed by the rest of the C-terminal extension and the β6-containing loop. The ACD is the least dynamic. Cartoon representation of the dimer (*a*) and dodecamer (*b*) and surface representation of the dodecamer (*c*) are shown. Pictures were created with PyMol

Table 5.1 Structural data available for sHSPs

High resolution structures							
Protein	**Organism**	**Length**	**PDB**	**Residues in structure**	**Resolution**	**Subunits in oligomer**[a]	**Reference**
Hsp 16.5	*Methanocaldococcus jannaschii*	147	1SHS	33–147	2.90	24(24)	Kim et al. (1998)
Hsp 16.9	Wheat (*Triticum aestivum*)	151	1GME	2/43–151[b]	2.70	12(12)	van Montfort et al. (2001b)
Tsp36	Tapeworm (*Taenia saginata*)	314	2BOL	2–314	2.50	2(2)	Stamler et al. (2005)
HspA	*Xanthomonas axonopodis pv. citri*	158	3GLA	40–139	1.65	2(12)	Hilario et al. (2006)
			3GT6	39/40–138/139	2.15	2(12)	Hilario et al. (2011)
			3GUF	37–139	2.28	2(12)	Hilario et al. (2011)
Hsp 14.0	*Sulfolobus tokodaii*	123	3VQK	17–123	4.50	24(24)	Hanazono et al. (2012)
Hsp 14.0(I120F/I122F)	*Sulfolobus tokodaii*	123	3AAB	14–123/9–114	1.85	2(24)	Takeda et al. (2011)
Hsp 16.0	Yeast (*Schizoscaccharomyces pombe*)	143	3W1Z	1/10/14/34–143[b]	2.40	2(16)	Hanazono et al. (2013)
Hsp 17.7	*Deinococcus radiodurans*	166	4FEI	46–147	2.40	2(2)	Bepperling et al. (2012)
HspB1(Hsp 27)	Human (*Homo sapiens*)	205	3Q9P/3Q9Q	90–171	2.00/2.20	6(<24–32)[c]	Baranova et al. (2011)
HspB1(Hsp 27) with C-terminal peptide	Human (*Homo sapiens*)	205	4MJH	84–176	2.60	2(<24–32)	Hochberg et al. (2014)
HspB5(αB-crystallin)	Human (*Homo sapiens*)	175	2WJ7	67–157	2.63	2(<24–32)	Bagneris et al. (2009)
			3L1G	68–162	3.32	2(<24–32)	Laganowsky et al. (2010)
			2KLR	69–150	1.53	2(<24–32)[d]	Jehle et al. (2010)
HspB5(αB-crystallin) R120G/L137M	Human (*Homo sapiens*)	175	2Y1Z	67–157	2.50	2(<24–32)	Clark et al. (2011)
HspB5(αB-crystallin) core with C-terminal peptide	Human (*Homo sapiens*)	175	4M5S	68–153	1.37	2(<24–32)	Hochberg et al. (2014)

HspB5(αB-crystallin) core E117C with C-terminal peptide	Human (*Homo sapiens*)	175	4M5T	68–153	2.00	2(<24–32)	Hochberg et al. (2014)
HspB4 (αA-crystallin) with zinc	Cow (*Bos taurus*)	173	3L1E	59–163	1.15	2(24–32)	Laganowsky et al. (2010)
HspB4(αA-crystallin)	Cow (*Bos taurus*)	173	3L1F	62–163	1.53	2(24–32)	Laganowsky et al. (2010)
HspB6(Hsp20)	Rat (*Rattus norvegicus*)	162	2WJ5	65–162	1.12	2(2)	Bagneris et al. (2009)
HspB4(αA-crystallin)	Zebrafish (*Danio rerio*)	173	3N3E	61–166	1.75	2(24–32)	Laganowsky and Eisenberg (2010)
Low resolution structures							
Protein	**Organism**	**Length**	**PDB**[e]	**EM file**	**Resolution**	**Subunits in oligomer**	**Reference**
ACR1(Hsp 16.3)	*Mycobacterium tuberculosis*	144	2BYU	1,149	16.5	2(12)	Kennaway et al. 2005)
Hsp 26	Yeast (*Saccharomyces cerevisiae*)	214	2H50	1,221	10.8	24(24–32)	White et al. (2006)
			2H53	1,126	11.5	24(24–32)	White et al. (2006)
HspB5(αB-crystallin)	Human (*Homo sapiens*)	175	n.a	1,121	20.0	24(24–32)	Peschek et al. (2009)

[a]The first number indicates the number of subunits in the determined structure. Number in parentheses indicates the number of monomers in the native protein

[b]The residues in the N-terminal arms for which coordinates are available, are not the same for all chains in the structures-1GME and 3W1Z. IGME has coordinate information beginning from residue 2 for six of the 12 chains, while the other six have information beginning from residue 43. For the structure 3W1Z, coordinates are available from residues 1, 10, 14 and 34 for four sets of chains comprising four chains each

[c]The hexamer observed for human HspB1 is non-native

[d]The structure was determined by solid state NMR

[e]Coordinates for these files are for the wheat Hsp16.9 ACD domain fitted into the determined EM density

n.a. not available

Arabidopsis sHSPs in Fig. 5.2. The overall diversity of sHSP N-terminal arms in all kingdoms of life has been nicely summarized in recent studies using extensive genomic data (Kriehuber et al. 2010; Poulain et al. 2010). These and other characteristics can be accessed in public databases that have compiled and analyzed sHSP sequences (http://forge.info.univ-angers.fr/~gh/Shspdb/index.php; http://pdslab.biochem.iisc.ernet.in/hspir/). The N-terminal arm is poorly conserved compared to the ACD and variable in length, but contains recognizable motifs in related sHSP families. For example, 237 class I proteins from 84 different plant species have N-terminal arms ranging from 24 to 62 amino acids (average 49) and contain a conserved motif V/IFDPFS, in which the proline residue is >97 % conserved (Fig. 5.2). This N-terminal motif can be considered a signature of class I proteins. In the Ta16.9 structure this motif appears to function to patch one end of the ACD β-sandwich (Fig. 5.3), likely explaining its conservation.

Among the available high resolution structures of sHSPs, there are limited data on the N-terminal arms. Many of the structures were determined using a truncated ACD dimer, and in the three oligomeric sHSP structures the arms are either completely missing (e.g. Mj16.5), or only a subset are fully resolved (Ta16.9, 6 out of 12; Sp16.0, 4 complete and 8 partial out of 16) (Table 5.1). The disorder of many N-terminal arms in the crystal structures, as well as dynamic features of this domain as observed by NMR, have led to the suggestion that they are intrinsically disordered (Uversky and Dunker 2010). However, most arms that are resolved have defined α- and 3_{10} helices, although these structures may well be very dynamic in solution (Fig. 5.3). The latter conclusion is supported by hydrogen-deuterium exchange experiments with Ta16.9 and Ps18.1 that show amide hydrogens in the N-terminal arms are fully exchanged within the first 5 s of exposure to deuterium at room temperature (Cheng et al. 2008; Wintrode et al. 2003) (Fig. 5.3b). In contrast, ACD amide hydrogens exchange much more slowly, consistent with a more stable β-sandwich structure for this signature domain (Fig. 5.3b). We and others have proposed that the N-terminal arms are a major substrate binding domain, as discussed in a later section of this review.

Following the ACD sHSPs have a short C-terminal extension. In Ta16.9 and other plant class I proteins the entire extension is only 14–21 (average 15) amino acids, in comparison to the longer extension found in αA- and αB-crystallins (28 and 26 amino acids, respectively, in humans). Notably, the C-terminal extension of a majority of all sHSPs contains a conserved I/V/L-x-I/V/L motif (IxI motif) first recognized in 1998 (de Jong et al. 1998) (Figs. 5.2 and 5.3). In plant class I proteins, IxI occurs ~87 % of the time, and the remainder are either VxI or IxV. In Ta16.9 the motif is IQI and in Ps18.1 it is IEI, which are the two most frequent variants of the IxI sequence in class I proteins (25 % and 54 %, respectively from 207 proteins), while IDI occurs in 20 % of these sHSPs. The IxI motif is significant because it makes an important contact linking sHSP dimers into higher order oligomers. The motif of one monomer binds to a hydrophobic patch on a monomer of another dimer, which is created by β-strands 4 and 8 on one face of the ACD β-sandwich. Although there are no specific reports on the effect of mutation or truncation of this motif in higher plant sHSPs, truncation of the IxI in Syn16.6 from the cyanobacterium *Synechocystis sp.* PCC 6803 converts the native oligomer to dimers (Giese et al.

2005), and destabilization of sHSPs oligomers is observed in mutants of archael sHSPs (Quinlan et al. 2013; Saji et al. 2008) and mammalian sHSPs (Delbecq and Klevit 2013). The C-terminal extension can also adopt different angles in relation to the ACD, which facilitates the assembly of oligomers with different geometries. This flexibility is well-illustrated by the Ta16.9 structure (van Montfort et al. 2001b) and by the fission yeast Sp16.0 structure (Hanazono et al. 2013).

5.3 Class I Cytosolic sHSPs from Plants Are "Gold-Standard" Chaperones

In our current model for sHSP function, denatured substrates bound to sHSPs can be refolded by the ATP-dependent Hsp70/DnaK chaperones, potentially with help from the Hsp100/ClpB AAA+ protein disaggregases in those cells and organelles in which these proteins are found (Basha et al. 2012) (Fig. 5.4). This model for sHSP

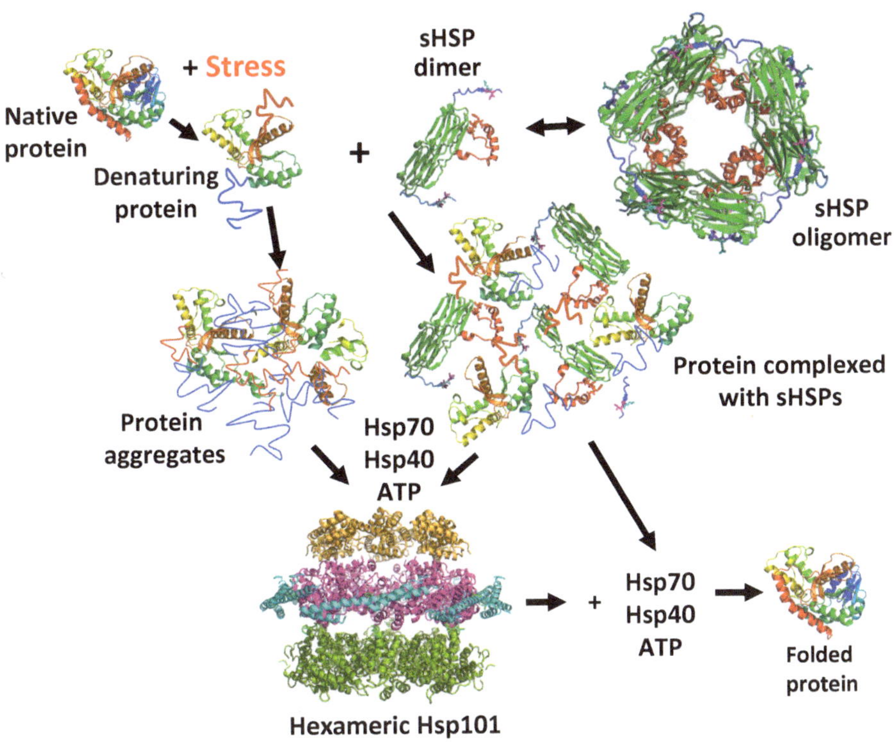

Fig. 5.4 Model for the action of class I sHSPs in the plant cytosol. Native protein (e.g. MDH, 1MLD) is partially denatured by stress and either captured by sHSP dimers to form soluble heterogeneous sHSP-substrate complexes or forms aggregates. sHSP-substrate complexes can be acted on by Hsp70 and co-chaperones using ATP to effect release and substrate refolding. Refolding can be enhanced by the action of the cytosolic protein disaggregase Hsp101, a hexameric ATPase (structure shown is the homologous *E. coli* ClpB, 4D2U), which can act on protein aggregates, but more efficiently works on sHSP-substrate complexes (Mogk et al. 2003b). sHSP structures (1GME) are colored as in Fig. 5.3a

chaperone function has been developed through extensive in vitro studies reconstituting these reactions using different sHSPs from plants, cyanobacteria, *E. coli* and humans (Giese et al. 2005; Lee et al. 1997; Lee and Vierling 2000; Van Montfort et al. 2001a) and has support from a variety of in vivo studies (Haslbeck et al. 2005; Lee et al. 2005; Mogk et al. 2003a; Tripp et al. 2009).

Much of the biochemical work with plant sHSPs has utilized Ta16.9 and a homologous class I protein from *Pisum sativum* (pea) Hsp18.1 (Ps18.1). Both Ta16.9 and Ps18.1 are dodecamers in solution at room temperature, consistent with the Ta16.9 crystal structure, and all available data indicate that sHSPs in the plant class I family most likely all assemble into dodecamers (Basha et al. 2010; Kirschner et al. 2000; Tiroli and Ramos 2007; Yoon et al. 2005). It has been suggested that sHSP oligomers act as reservoirs of the active dimeric units of the chaperone (Stengel et al. 2010; van Montfort et al. 2001b). As illustrated in Fig. 5.4, the dimeric units of the sHSP are proposed to become available to 'stressed' cellular proteins upon activation of the sHSP. Stress can activate the sHSPs by shifting the equilibrium to the dimeric form, which can then bind unfolded/misfolded proteins. In this model, the oligomers are essentially the storage form of the chaperone in which binding sites for denaturing proteins are sequestered from the cellular contents until needed. Indeed, sHSPs are an excellent illustration of the idea that protein dynamics are able to regulate and control protein function.

The existence of an equilibrium between sHSP oligomers and dimers has been shown by several experiments. Analysis of recombinant proteins on non-denaturing gels showed that Ta16.9 and Ps18.1 could exchange subunits to form hetero-oligomers when simply incubated together at room temperature (van Montfort et al. 2001b). The same results have been obtained with a class I sHSP from Arabidopsis (Basha et al. 2010). Friedrich et al. (2004) used exchange between a C-terminal affinity tagged Ps18.1 and wild type Ps18.1 incubated in a molar ratio of 1:1 (12 μM each) to demonstrate that hetero-oligomers were formed in less than 15 min at room temperature and no exchange occurred when proteins were incubated together on ice. More accurate kinetic data along with information about the units being exchanged between oligomers was determined using nanospray mass spectrometry. In these experiments the mass of the non-covalent dodecamers could be measured, and the stoichiometry of the hetero-oligomers was directly monitored. Results demonstrated that the subunit exchange between Ps18.1 and Ta16.9 was rapid and that exchange preferentially involved dimers (Sobott et al. 2002). Further exchange studies with other class I sHSPs confirmed that movement of species with an even number of subunits between dodecamers was more prevalent than movement of monomers (Painter et al. 2008). A rate of 0.40 min^{-1} at 30 °C was measured for exchange of dimers between Ps18.1 and an Arabidopsis class I sHSPs (Painter et al. 2008). Dynamic behavior of sHSP oligomeric assemblies appears to be a common property of these proteins. Exchange of dimers between α-crystallin molecules has been observed by fluorescence resonance energy transfer (FRET) and mass spectrometric studies (Baldwin et al. 2011; Bova et al. 2000). Mixtures of labeled and unlabeled sHSP from *Saccharomyces cerevisiae* (Sc26) also showed exchange of dimeric species at elevated temperatures (Benesch et al. 2010). Evidence that the equilibrium between oligomers and dimers shifts towards dimers for class I plant sHSPs at elevated temperatures, comes from several experiments. Analytical size exclusion chromatography and native gel electrophoresis studies with purified Ta16.9, Ps18.1 and a class I sHSP from *Arabidopsis*, At17.6, show that these dodecameric proteins reversibly dissociate to smaller oligomers, predominantly dimers at higher temperatures (Basha et al. 2010; van Montfort et al. 2001b). Similar observations were made with class I proteins from sugarcane (Tiroli-Cepeda and Ramos 2010). Since the oligomers rapidly reassemble and are dominant even at very low micromolar concentrations at room temperature, it is only possible to observe the dimeric forms when biochemical analysis is performed at elevated temperature. In total these data are consistent with the model that at elevated temperatures the dimer is the active form (Fig. 5.4). However, because dimers are constantly exchanging, they are essentially always available, and the equilibrium towards dimers would be enhanced in the presence of 'dimer-binding' unfolded substrates, even at ambient temperatures. Thus, due to the dynamic release of dimers the oligomeric 'storage' form can still act as a chaperone even in the absence of high temperature stress. This has been nicely illustrated with mammalian sHSPs and destabilized forms of T4 lysozyme by the McHaourab group (McHaourab et al. 2009).

To date the majority of work examining sHSP protection of substrates has involved model, heat-sensitive proteins, as opposed to stress labile proteins from the same organism as the sHSP. This is due to our general lack of knowledge of specific in vivo substrates protected by sHSPs. The proteins most commonly used as model substrates in sHSP chaperone assays for protection from heat denaturation are firefly luciferase, porcine heart mitochondrial malate dehydrogenase, and porcine liver citrate synthase. These proteins are all heat sensitive at moderate temperatures (40 °C), can be readily assayed for enzymatic activity and are available commercially. The complexes formed when sHSPs interact with these denaturing substrates at elevated temperatures are large (from a few hundred to a few thousand kDa) and highly polydisperse. As characterized in detail with plant class I sHSPs, the sizes of the complexes vary depending on the ratio of sHSP to the substrate, the identity of the substrate, and the temperature at which substrate denaturation is performed (Basha et al. 2004b, 2006, 2013; Lee et al. 1997; Mogk et al. 2003b). Similar observations have been made with ScHsp26 from yeast (Stromer et al. 2003). When sHSPs are limiting, sHSP-substrate complexes are larger, presumably because there is insufficient sHSP to block self-interaction of the denaturing substrate. Conversely, sHSP-substrate complexes are smaller at higher ratios of sHSP to substrate because there is less self-aggregation of substrate due to relatively more sHSP being available to form contacts with substrate (Friedrich et al. 2004). The complexes formed in vitro between certain sHSPs and substrates are quite stable and essentially do not release free substrate. Addition of more Ps18.1 to pre-formed complexes of luciferase did not alter their size since the sHSP cannot bring about disaggregation of proteins to smaller aggregates (Friedrich et al. 2004). However, adding luciferase to pre-formed complexes increased their size, and exchange between tagged and untagged Ps18.1 in sHSP-substrate complexes was observed, indicating that the complexes are dynamic (Friedrich et al. 2004).

Defining the stoichiometry and mode of sHSP interaction with substrates remains difficult. In mass spectrometry studies of Ps18.1 with the substrate luciferase, more than 300 stoichiometries of sHSP to substrate were observed (Stengel et al. 2010). This polydispersity may be due to sHSPs capturing substrates in different degrees of unfolding, as well as there being no single specific substrate binding site on the sHSPs (Jaya et al. 2009). In complexes of Ps18.1 with different substrates, it was also observed that there was a predominance of one or two molecules of monomeric or dimeric substrates, respectively (Stengel et al. 2010). The finding that two molecules of dimeric substrates were captured in complexes is consistent with the conclusion that partially unfolded substrates, retaining some degree of native conformational characteristics, are recognized by sHSPs. Related work found that sHSP-substrate complexes that have an even number of sHSP monomers are predominant, supporting the sHSP dimer as the major substrate binding species (Stengel et al. 2012). However, the dimeric interface is also labile, and the dimers dissociate under stress conditions, and sHSP-substrate complexes carrying an odd number of sHSP monomers have also been observed, although to a lesser extent than those with even numbered species. Dissociation of the Ta16.9 and Ps18.1 dodecamers appears to be the rate limiting step (Sobott et al. 2002) and coincides with substrate binding. This is then followed by a slower step of augmentation of the complex with addition of more sHSP molecules (Stengel et al. 2010). The dynamic interaction of sHSPs with substrate was further observed by hydrogen-deuterium exchange experiments followed by mass spectrometry, which typically can identify protein-protein interfaces (Cheng et al. 2008). Amide hydrogen exchange from the substrate malate dehydrogenase revealed that N- and C-terminal regions of the protein, which have stable secondary structure in the native state, were rapidly exchanged in Ps18.1 or Ta16.9-substrate complexes, but that the core of the protein was much slower to exchange. These observations also support the conclusion that sHSPs bind substrates at an early stage of unfolding. The surprising result was that no change in the pattern of amide hydrogen exchange was observed for the sHSPs, whether or not they were in complex with the substrate. Thus, no new stable sHSP secondary structure forms upon interaction of these sHSPs with heat-denatured malate dehydrogenase, although they very efficiently protect it, and the complexes are stable. We suggest that these data can be explained by the existence of multiple low affinity contacts between sHSP and substrate that when combined create a stable interaction.

Dissociation into dimers presumably makes surfaces that are normally buried in the sHSP oligomers available for binding substrates. Interaction between substrates and sHSPs are proposed to occur through exposed hydrophobic surfaces (Basha et al. 2012; Lee et al. 1997; Van Montfort et al. 2001a). As indicated by a number of experiments, these binding surfaces appear to reside primarily on the flexible N-terminal arm of sHSPs. The high flexibility and exposed nature of the N-terminal arms, particularly at elevated temperatures when the equilibrium is shifted to the dimeric form, makes this protein region highly suitable for substrate binding (Cheng et al. 2008). The role of the N-terminal arm in the efficiency of substrate protection (mole sHSP required per mole substrate) of Ps18.1 and Ta16.9 were studied by creating chimeric proteins. Although Ps18.1 and Ta16.9 are very similar (68 %

sequence similarity, 79 % sequence identity), Ps18.1 displays up to a sixfold higher efficiency in protecting some substrates in vitro. The chaperone activity of chimeric proteins, in which different portions of the N-termini were swapped, was tested with citrate synthase, firefly luciferase and malate dehydrogenase. The composition of the N-terminus clearly determined the efficiency of protection for citrate synthase and luciferase, although both the N-terminal arm and the ACD impacted interaction with malate dehydrogenase (Basha et al. 2006). In vitro cross-linking studies with Ps18.1 mutants carrying the photo-crosslinker p-benzoylphenylalanine at different positions, further indicated that interaction with substrates mostly involved the N-terminal arm (Jaya et al. 2009). Deletion and mutation studies with non-plant sHSPs also support this conclusion (Basha et al. 2012). The importance of the N-terminal arm for function in vivo was demonstrated for the sHSP from cyanobacterial Syn16.6. Cells expressing Syn16.6 with certain point mutations in the N-terminal arm were unable to survive high temperature treatment, although in vitro these same mutants displayed normal oligomer formation and chaperone activity with model substrates (Giese et al. 2005). It is possible that these N-terminal mutations disrupted the ability of Syn16.6 to protect one or more specific, essential in vivo substrates. Altogether, support for a critical role of the N-terminal arm in substrate protection is strong.

The study of isolated dimeric subunits from the sHSP oligomers has not been experimentally feasible owing to the dynamic, polydisperse nature of these proteins at elevated temperatures and the absence of other conditions under which the dimers are stable. Potential hydrophobic substrate binding surfaces on the dimers have been explored by a replica exchange molecular dynamics simulation study of Ps18.1 and Ta16.9 (Patel et al. 2014). The difference in chaperone efficiency of these two proteins, as mentioned above, helped address the dynamics of the dimers in relation to activity. Open and closed conformational ensembles for the dimers were defined based on larger and smaller distances separating the two N-terminal arms. The simulations showed that for the more efficient Ps18.1, the population of open dimers is larger and the increase in open conformations at higher temperatures (27 °C vs. 46 °C) is enhanced. It has previously been suggested that the N-terminal arms are intrinsically disordered (Sudnitsyna et al. 2012; Uversky and Dunker 2010). However, the simulations indicate that these regions are not completely disordered, but rather retain flexible helical segments with poorly packed side chains, indicating a more molten globule-like state. The composition of N-terminal arms from a majority of sHSPs shows enrichment of Pro, Ser and Phe residues (Kriehuber et al. 2010), which is suggested to provide increased flexibility and possibility for increased hydrophobic interactions with denatured substrates. Examination of the accessible surface area of the N-terminal arms in open and closed conformations also showed the presence of hydrophobic patches. These patches are more exposed in open forms, and the surface areas exposed (1,100–1,700 $Å^2$) compare well with the area buried in inter-protein interactions (Chakrabarti and Janin 2002). These large surface areas on N-terminal arms are in contrast to smaller patches observed in the β4–β8 groove of the ACD (100–300 $Å^2$), which binds the C-terminal IXI motif and has been suggested to be a major substrate binding region. The much larger size of hydrophobic patches on the N-terminal arms argues that they are the major substrate binding region.

The refolding of substrates bound to sHSPs requires the activity of ATP-dependent chaperones; the majority of substrates do not spontaneously release and refold from sHSP-substrate complexes formed with heat-denaturing substrates (Friedrich et al. 2004; Mogk et al. 2003b). Refolding of firefly luciferase in complex with Ps18.1 was accomplished in vitro by addition of eukaryotic Hsc70 and DnaJ (*S. cerevisiae* Hdj1 plus Ydj1) or prokaryotic DnaK/DnaJ/GrpE, plus ATP (Lee and Vierling 2000). The ability of both, prokaryotic and mamallian Hsp70 systems to bring about refolding of luciferase in complex with a plant sHSP hints towards the absence of specific interactions between sHSPs and the Hsp70 machinery. In another study with Syn16.6 and Ps18.1 using three different substrates, the role of different ATP-dependent chaperones was examined in substrate refolding (Mogk et al. 2003b). When sHSP-substrate complexes were relatively small (600 kDa or less) due to a higher ratio of sHSP to substrate, active substrate was readily recovered by the DnaK machinery alone. However, dissociation and refolding of the substrates was even more efficient with DnaK plus the protein disaggregase ClpB, particularly when complexes were larger, having been formed with less sHSP. The GroEL/GroES machinery could not bring about refolding of sHSP-bound substrates, although it could provide some enhancement of recovery of active substrate. The sHSPs, therefore, appear to enhance the availability of substrates for the DnaK machinery, holding them in a form that can readily be refolded.

5.4 The Diversity of Plant sHSPs

Plant class I cytosolic sHSPs have been excellent models for mechanistic studies of how denaturing proteins can be protected from irreversible aggregation, but as mentioned previously, class I sHSPs are only 1 of at least 11 different families of higher plant sHSPs that have a domain organization comprising an N-terminal arm, ACD, and C-terminal extension (Basha et al. 2012) (Fig. 5.2). In addition to the class I cytosolic sHSPs, there are four other classes of cytosolic sHSPs, designated CII through CV, that are common to monocots and dicots (Siddique et al. 2008; Waters et al. 2008), and additional groups that appear to be specific to either dicots or monocots, or to even more restricted taxonomic groups (Lopes-Caitar et al. 2013; Waters et al. 2008). Genes encoding organelle-targeted proteins (chloroplast, peroxisomes, mitochondria and ER) are also found in all higher plants. Extensive evolutionary analysis indicates that the different classes of plant sHSPs arose after divergence of plants from other organisms (Waters 2013; Waters and Rioflorido 2007; Waters and Vierling 1999b). A case in point is that the mitochondrial and chloroplast-targeted sHSPs are more closely related to their class I counterparts than to α-proteobacterial or cyanobacterial sHSPs (Waters 2013; Waters and Vierling 1999a). The ER targeted sHSP also appears to be derived from a class I protein. This is a significantly different evolutionary history than the Hsp70 chaperones, in which the chloroplast and plant mitochondrial Hsp70s are bacterial in character (DnaK-like), and the ER forms arose early in eukaryotic evolution (Boorstein et al. 1994;

Lin et al. 2001). Although plants have these multiple sHSP gene classes, it is of note that class I proteins accumulate to the highest levels in heat stressed cells (estimated close to 1 % of total cell protein (Derocher et al. 1991)), and genes encoding class I sHSPs are the most numerous. In poplar (*Populus tremuloides*), rice (*Oryza sativa*), *Arabidopsis thaliana*, and soybean (*Glycine max*), for which comprehensive genome analysis of sHSP genes has been performed, there are 18, 8, 6, and 18 genes encoding class I proteins, respectively (Lopes-Caitar et al. 2013; Waters et al. 2008). In contrast, there are only one or a few genes in each of the other sHSP classes, such that the corresponding total number of sHSP genes for poplar, rice, Arabidopsis and soybean is 39, 22, 19 and 51, respectively. As more plant genomes are sequenced, a similar pattern of sHSP gene abundance is expected, as well as the recognition of more sHSPs with limited taxonomic distribution, indicative of recent evolution.

The diversity of plant sHSP sequences is well-illustrated by an alignment of all of the sHSPs from *Arabidopsis thaliana* (Fig. 5.2). The conservation of the ACD (β2–β9), compared to the N-terminal arm is evident. The first five aligned sequences have N-terminal extensions that direct them to organelles, and related sHSPs have been found in all higher plants (Fragkostefanakis et al. 2015; Lopes-Caitar et al. 2013; Waters et al. 2008). At26.5 has been designated as one of two classes of mitochondrion-targeted proteins (MII class). The plastid-localized At25.3, was the first organelle-targeted sHSP recognized (Vierling et al. 1986, 1988), and has been considered to be the only chloroplast targeted protein. However, At23.5 and At23.6, which were originally defined as mitochondrial proteins (MI class), have recently been reported to be targeted to chloroplasts as well (Van Aken et al. 2009). The ER-targeted At22.0 has a C-terminal SKEL sequence, presumably acting to retain it in the ER; the C-termini are predicted to be exposed, based on the structure of Ta16.9 (Fig. 5.3). Two other proteins, At17.4 and At15.7, contain specific subcellular targeting signals. At17.4 localizes to the nucleus with a nuclear targeting signal (KRKR) in the loop region involved in strand exchange in the formation of sHSP dimers (Siddique et al. 2003). Modeling the position of this sequence on the Ta16.9 structure indicates it could be readily accessible even if the sHSP is a large oligomer. At15.7 has a C-terminal peroxisomal type 1 targeting motif (SKL), which is also expected to be exposed in the oligomeric form of the sHSP.

For the N-terminal arms, the sequence alignment appears meaningful only for proteins within the same class, which in Arabidopsis are the six class I proteins, two class II proteins and the two mitochondrial/chloroplast-targeted proteins. The diversity of the N-terminal arms between classes within a species contrasts with the conserved motifs that can be identified within classes across evolutionarily distant higher plants. In addition to the conserved motif in the N-terminal arm of the class I proteins noted above, the chloroplast sHSP (exemplified by At25.3) has a highly conserved sequence predicted to form an amphipathic alpha helix with multiple methionines on the hydrophobic face (a "methionine bristle") (Fig. 5.2) (Chen and Vierling 1991; Waters and Vierling 1999a). Other class-specific N-terminal arm motifs were noted by Scharf et al. (2001), but these were based on a relatively small number of sequences. Defining the structural and functional significance of conserved N-terminal arm features of plant sHSPs requires further research.

The alignment of Arabidopsis sHSPs shows the conservation of predicted β strands in the ACD (Fig. 5.2). Analysis of plant sHSP sequences prior to the availability of the high resolution structures recognized the regions comprising β2–β5 and β7–β9 as conserved and designated these as "region I" and "region II" (Vierling 1991). These β-strands are now known to form the ACD β-sandwich (Fig. 5.3a). The sequence between β5 and β7 is more variable, although most of the Arabidopsis proteins contain predicted β structure (β6) and the sequence is of sufficient length in all but At18.5 to allow for dimer formation through strand-swapping with β6 as shown for Ta16.9 in Fig. 5.3a.

There are limited data on the oligomeric stoichiometry of most plant sHSPs. Specific Class II and chloroplast sHSPs have been characterized as well-defined dodecamers (Basha et al. 2010; Lambert et al. 2011). Recombinant peroxisomal and mitochondrial proteins form large oligomers as assessed by crosslinking, but the exact stoichiometry is unknown (Siddique et al. 2008). One plant sHSP, At18.5, is dimeric, and will be discussed in a later section (Basha et al. 2013).

Although sHSPs were first discovered as induced by heat stress, the patterns of sHSP expression are complex and vary between organisms and developmental stages. Sixteen of the 19 sHSP genes in Arabidopsis (including all six class I genes) show dramatic increases in transcription at elevated temperatures, while three (At14.7, At15.4 and At21.7), show no response at all to heat (Siddique et al. 2008; Waters et al. 2008). Many studies, along with the gene expression data document increases in specific sHSP transcripts under other abiotic stress conditions, in response to biotic stress, and in normal development and in organ specific patterns (Giorno et al. 2010; Lopes-Caitar et al. 2013; Siddique et al. 2008; Waters et al. 2008). One of the best characterized patterns of sHSP expression in the absence of stress, is the transcription and accumulation of class I proteins during seed development (Almoguera and Jordano 1992; Coca et al. 1994; Reddy et al. 2014; Wehmeyer et al. 1996; Wehmeyer and Vierling 2000). Unfortunately to date it has been difficult to link sHSPs directly to any form of stress tolerance in seeds (Personat et al. 2014; Tejedor-Cano et al. 2010).

5.5 Mechanistic Differences Between sHSPs

Class II sHSPs are the second major class of sHSP genes cloned and are defined not only in higher plants, but also in early land plants (Ruibal et al. 2013; Vierling 1991; Waters 2013; Waters et al. 2008). Like class I sHSPs, class II proteins accumulate to significant levels during heat stress and can make up 0.25 % of the total cellular protein (Basha et al. 2010). Considerably fewer studies have examined the biochemistry and function of these plant sHSPs. As mentioned above, nano-electrospray ionization mass spectrometry of recombinant class II sHSPs from wheat (Ta17.8), pea (Ps17.7) and Arabidopsis (At17.7), along with other data indicate that they form dodecamers like their class I counterparts (Basha et al. 2004b, 2010; Lee et al. 1995). Despite forming homo-oligomers with the same number of subunits and co-localizing to heat shock granules during heat stress in vivo (Kirschner et al. 2000; Weber et al. 2008), class I and II sHSPs do not directly interact to form hetero-oligomers (Basha et al. 2010; Helm et al. 1997). However, class I and class II sHSPs do form hetero-oligomers with other proteins within their own class (Basha et al. 2010; van Montfort et al 2001b). What structural features prevent the formation of class I/II hetero-oligomers remain to be determined, but this property, along with their evolutionary divergence and conservation argue that the two classes serve distinct functions.

The behavior of class II proteins in vitro also distinguishes them from class I sHSPs. The class II proteins do not dissociate into stable dimers upon heat stress, as appears to occur with the class I proteins and is proposed to be important for substrate protection (Basha et al. 2004b, 2010) (Fig. 5.4). However, non-denaturing polyacrylamide gel electrophoresis and mass spectrometry show that class II sHSPs do exchange subunits at elevated temperatures, which could expose hydrophobic substrate binding sites. Consistent with this idea, binding of the hydrophobic probe bis (8-anilinonaphthalene-1-sulfonate) also increases at high temperatures (Basha et al. 2010). What oligomeric form of class II sHSPs is the active species remains to be determined. Class II sHSPs do act as efficient chaperones in vitro, being able to protect various model substrates from aggregation by forming large, soluble heterogeneous complexes, analogous to what has been observed with class I sHSPs, although there are notable differences. Heating luciferase and class II Ps17.7 at a high sHSP:Luc molar ratio (12:1 and 24:1) resulted in a decrease in solubility for both proteins, indicating co-aggregation rather than protection. Class II Ta17.8 (from wheat) fully protects luciferase at low molar ratios (sHSP:luc), while class I Ta16.9 is unable to prevent aggregation of luciferase, suggesting that for this substrate the CII protein is a more efficient chaperone than the CI protein (Basha et al. 2006, 2010). Perhaps the differences between class I and class II proteins reflect specific mechanisms by which sHSPs recognize distinctively different substrates. Given the complexity of the plant proteome it makes sense that plants would evolve multiple chaperones to protect labile substrates from aggregation during stress. However, the existing in vitro assays do not fully capture these differences.

Studies with sHSPs from other organisms also suggest mechanistic differences from the model presented in Fig. 5.4. Oligomeric Sc26 from *S. cerevisiae* was originally reported to dissociate before interacting with substrate (Haslbeck et al. 1999; Stromer et al. 2004), but subsequent studies indicated that dissociation was not essential for chaperone activity (Franzmann et al. 2005). A disulfide bond between the N-terminal arms, which was introduced by mutation (S4C), was reported to prevent normal dissociation of the protein into dimers at higher temperatures as shown by size exclusion chromatography. However, the oxidized protein, which could not dissociate, was reported to protect citrate synthase as efficiently as the reduced protein, which could dissociate. In further studies the authors suggest that heating activates the Sc26 oligomer by altering the conformation and accessibility of part of the N-terminal arm such that dissociation is not required (Franzmann et al. 2008).

Experiments with the archael Mj16.5 have also been interpreted to support a "static mechanism" for sHSP chaperone function, in which the oligomers do not dissociate. Cryoelectron microscopy and EPR studies of MjHsp16.5 in complex with a destabilized mutant of T4 lysozyme showed that the substrate was within the cavity of this 24 subunit sHSP making contacts with both the N-terminal arms and ACD (Shi et al. 2013). The sHSP-substrate complexes examined were prepared to obtain homogenous particles, which are not typically observed with heat-denatured substrates and other sHSPs. To accommodate substrate, a conformational change was observed that expanded the sHSP cavity. Whether this mechanism can be generalized to other sHSPs remains to be shown. Under the moderate temperatures examined in this study, Mj16.5 does not appear to undergo subunit exchange. However, *M. jannaschii* is a hyperthermophile and, thus, at temperatures optimum for its growth, Mj16.5 might dissociate and function as a suboligomer.

In addition to heat-induced conformational changes, phosphorylation appears to be a mode of sHSP activation, particularly in the case of the α-crystallins. In HspB5 and HspB4, phosphorylation sites reside in the N-terminal arm (Peschek et al. 2013). An HspB5 phosphomimetic mutant, in which glutamate residues were substituted at three phosphorylation sites in the N-terminal arm, showed a predominance of hexameric rings, dimers and dodecamers as opposed to the native higher order oligomers (Peschek et al. 2013). Although the N-terminal arm interactions were disrupted, leading to formation of smaller oligomers, the introduction of glutamate residues did not disrupt the ACD and C-terminal arm packing in the oligomers. This mutant showed higher efficiency than wild type in protecting the substrate MDH, providing further evidence that suboligomeric sHSP species are active chaperones.

5.6 Dimeric sHSPs

As already described, the majority of sHSPs form large homo-oligomeric complexes, and oligomeric structure has often been considered essential for activity (Basha et al. 2012; McHaourab et al. 2009). However, not all sHSPs form higher-order oligomeric assemblies. From plants, recombinant cytosolic At18.5 from Arabidopsis is dimeric (Basha et al. 2013; Siddique et al. 2008). At18.5 was designated as class CIV by Siddique et al. (2008) and CVI by (Waters et al. 2008), and is not found in monocots. Unlike other plant sHSPs, the amino acid sequence comprising the dimerization loop containing the β6 strand is absent in At18.5 (Fig. 5.2), and the basic dimer structure is likely different from that of Ta16.9. One possibility is that the dimer interface involves anti-parallel interactions of β7 strands, similar to vertebrate α-crystallins, which also lack a β6 strand (Basha et al. 2012; Delbecq and Klevit 2013). This unusual dimeric sHSP displays a robust chaperone activity in protecting luciferase and malate dehydrogenase, and it forms large, soluble heterogeneous complexes when bound to these substrates, as seen with class I and II sHSPs. Like the class I proteins, the N-terminal arm of At18.5 is important for its chaperone activity. Removal of 24 N-terminal residues of At18.5 led to complete loss or 50 % decrease in chaperone activity with model substrates. The fact that this dimeric sHSP functions as efficiently with some model substrate as the oligomeric class I and II proteins is consistent with the model that sHSP dimers are the functional form that captures substrates.

Another plant sHSP, recombinant mitochondrial NtHsp24.6 from tobacco has been reported to be a dimer (Kim et al. 2011b). However, the protein was purified from *E. coli* using a hexahistidine tag fused to the N-terminus. As the N-termini pack together in the oligomeric state in all sHSPs for which data are available, it is possible that this recombinant protein does not reflect the native state. None the less, the purified NtHsp24.6 displayed protection of model substrates (luciferase and citrate synthase) against heat-induced aggregation, again indicating that sHSP dimers can be effective chaperones.

Dimeric sHSPs have also been described from other organisms. Dr17.7 from the prokaryote *Deinococcus radiodurans* is dimeric and can suppress aggregation of luciferase, citrate synthase, lysozyme and insulin, although no stable sHSP-substrate complexes could be detected (Bepperling et al. 2012). Crystal structure of this protein was solved at 2.4 Å resolution (Table 5.1), revealing the same type of swapped-strand dimer interface as in Ta16.9; electron density for the flexible N- and C-terminal regions was lacking. *D. radiodurans* has a second sHSP, Dr20.2, which is a large oligomer (~36 subunits) that also has chaperone activity, but forms stable complexes with substrates. A stoichiometry of 3:1 for a 550 kDa complex of Dr20.2:lysozyme was estimated by analytical ultracentrifugation. Dr17.7 was not able to exchange subunits with Dr20.2. The authors hypothesized that Dr17.7 forms transient complexes with unfolded substrates, perhaps making them available to Dr20.2.

Recent structural studies of a human sHSP, HspB6, revealed that it forms stable, compact dimers in solution (Weeks et al. 2014). The multiple different quaternary states adopted by sHSPs, from dimers to oligomers with 36 or more subunits, emphasizes how the ACD can be assembled in multiple geometries depending on the flanking sequences.

5.7 Identification of Endogenous sHSP-Interacting Proteins

Despite the amount of biochemical information describing the interaction of sHSPs with model substrates, the identities of endogenous sHSP substrates or potential interacting partners remains largely unknown, particularly in plants. Studies of ANKYRIN REPEAT-CONTAINING PROTEIN 2A (AKR2A) in Arabidopsis suggest it may interact with one or more class I sHSPs to facilitate targeting of chloroplast outer membrane proteins (Kim et al. 2011a). AKR2A (At4g35450) is a 350 amino acid protein with four C-terminal ankyrin repeats and an N-terminal PEST domain. AKR2A has been suggested to function as a molecular chaperone that binds the membrane-targeted peroxisomal protein ascorbate peroxidase 3 and the chloroplast outer envelope protein 7 (OEP7), preventing their aggregation to allow them to reach their target membranes (Shen et al. 2010). Incubation of glutathione

S-transferase (GST)-tagged AKR2A with total soluble leaf protein extracts followed by affinity isolation recovered At17.8, a class I sHSP (Fig. 5.2). Interactions of At17.8 and AKR2A were further probed using pull-down experiments with GST fusion proteins, and through over-expression of tagged proteins in isolated protoplasts. Data suggest N-terminally His-tagged At17.8 associated with chloroplasts in a dimeric state and enhanced targeting of OEP7. GST pull-downs with three other Arabidopsis class I proteins also showed interaction with AKR2A. Experiments testing the interaction of purified, native AKR2A with native class I proteins have not been reported. It is interesting to speculate that class I proteins could aid normal membrane protein targeting, even though they are present at very low levels, and that the major increase in their expression during heat stress copes specifically with increased difficulty in targeting membrane proteins. Studies of sHSP targets during heat stress are warranted.

Another specific sHSP interaction has recently been reported. Arabidopsis chloroplast sHSP At25.3 (also called Hsp21, based on the mature size of the protein after import into chloroplasts) was found associated with the plastid nucleoid protein pTAC5 under heat stress conditions (Zhong et al. 2013). A knockout mutant of At25.3 grows normally in the absence of stress, but produces yellow cotyledons when seedlings are treated at 30 °C in the light after germinating in the dark at optimal temperatures. Affinity isolation of proteins interacting with At25.3, using a His-tagged variant expressed in the knockout background, identified pTAC5 (plastid transcriptionally active chromatin protein 5; At4g13670) as the major associated protein. pTAC5 is a 387 amino acid protein with a peptidoglycan binding-like domain and a DnaJ domain with zinc-dependent disulfide isomerase activity. It is one of 18 pTAC proteins that purify with an insoluble preparation of chloroplast RNA polymerase. Zhong et al. (2013) present colocalization, biomolecular fluorescence complementation, and in vitro At25.3-GST pull-down assays with pTAC-MBP, all supporting interaction of these two proteins. Molecular phenotypes, including plastid RNA and protein levels, led the authors to suggest that At25.3 works with pTAC5 to protect plastid-encoded RNA polymerase activity during heat stress. No heat stress phenotypes were reported for plants beyond the cotyledon stage, and direct interaction of native pTAC5 and At25.3 has not been tested. Whether At25.3 protects pTAC by the chaperone mechanism proposed here, or what other proteins the chloroplast sHSP may protect remain to be tested.

Early work on the heat stress response in tomato cells reported that sHSPs and Hsp70 were present in structures termed "heat stress granules" (HSGs), which were characterized as large cytoplasmic complexes composed of 40 nm particles (Nover et al. 1983, 1989). By expressing class I and class II sHSPs in tobacco protoplasts, Kirchner et al. (2000) demonstrated that class II proteins were required for formation of HSG-like structures and that class I sHSPs were recruited to these complexes. The chloroplast sHSP and sHSPs from non-plant species were not incorporated into HSGs, indicating there was specificity of the interactions observed. While initially HSGs were reported to store mRNAs of housekeeping genes during heat stress, this result is now thought to be an artifact of co-isolation of "stress granules", which are mRNA-containing particles that are distinct, but often associated with HSGs (Weber et al. 2008). Thus, the function and composition of plant HSGs remains unresolved. One possibility is that they represent class I and II sHSPs bound to heat labile cellular substrates, consistent with the chaperone model of their function.

Attempts to identify a comprehensive set of putative sHSP substrates have not yet been reported in higher plants, but experiments of this type have been performed in the cyanobacterium *Synechocystis sp*. PCC 6803 with its single sHSP, Syn16.6 (Basha et al. 2004a). Syn16.6 deletion mutants grow normally in the absence of heat stress, but are several orders of magnitude more sensitive than wild type cells to 44 °C (Giese et al. 2005; Giese and Vierling 2002, 2004). A C-terminally tagged Syn16.6, which complemented the heat- sensitive phenotype of the deletion mutant, was used to recover proteins specifically associated with the sHSP during heat stress, and 13 polypeptides were identified by MALDI-TOF mass spectrometry. Most of these putative Syn16.6 substrates were also present in the insoluble cell fraction, as well as associated with the sHSP after heat stress, suggesting that they are thermally labile. Notably, the proteins were released from the Syn16.6 affinity isolates when incubated with ATP, DnaK, DnaJ and GrpE. Among the proteins identified were components of the translational machinery, specifically elongation factor Ts (a GDP/GTP exchange factor), ribosomal protein S1, a putative tRNA/rRNA methyltransferase and GTPase TypA/BipA. The other potential sHSP targets had diverse functions and included metabolic enzymes (heme oxygenase, shikimate kinase, serine esterase), ferredoxin $NADP^+$ reductase, a peptidyl-prolyl isomerase, and a urease accessory protein. The serine esterase was purified and found to be heat sensitive and to interact with the Syn16.6 in vitro, consistent with it being protected by the sHSP. However, it has not yet been shown that the stability or activity of any of these proteins depend on the presence of Syn16.6 in vivo. In summary, considerable work remains to demonstrate that proteins found associated with sHSPs are in fact protected by sHSPs through the chaperone mechanism as proposed in Fig. 5.4.

5.8 Studies of In Vivo Function of Plant sHSPs

Experiments have been performed to test if plant sHSP expression can provide stress tolerance in vivo, both in *E. coli* and in plants. As listed in Table 5.2, cytosolic class I proteins from a variety of species are reported to enhance survival of *E. coli* cells treated at 50 °C for varying periods of time. Other studies have extended these observations to include some degree of tolerance to salinity, osmotic and cold stress. Stress protection in some studies correlated with the protection of total *E. coli* proteins from insolubility during the stress. In addition to experiments using class I proteins, there is a recent report that expression of a mitochondrial sHSP confers tolerance to salt and arsenic in *E. coli* (Table 5.2).

Several different plant species have been transformed to constitutively express specific sHSPs using the 35S promoter to drive expression of the transgene (Table 5.3). Some reports indicate that the resulting increased levels of sHSPs in the

Table 5.2 Phenotypes of *Escherichia coli* cells expressing plant sHSPs

Plant species	sHSP	Class	Phenotype	Reference
Oryza sativa (rice)	Os16.9	I	Heat tolerance	Yeh et al. (1997)
Castanea sativa (chestnut)	Cs17.5	I	Tolerance at 50 °C and 4 °C	Soto et al. (1999)
Nicotiana tabacum (tobacco)	Nt18.3	I	Tolerance to 50 °C	Joe et al. (2000)
Nicotiana tabacum (tobacco)	Nt18.0	I	Tolerance to 50 °C	Smykal et al. (2000)
Oryza sativa (rice)	Os16.9	I	Heat tolerance	Yeh et al. (2002)
Rosa chinensis (rose)	Rc17.8	I	Heat, salt, oxidative stress tolerance	Jiang et al. (2009)
Daucus carota (carrot)	Dc17.7	I	Enhanced viability at 2 °C	Song (2010, #1670)
Daucus carota (carrot)	Dc17.7	I	Salt tolerance	Song and Ahn (2011)
Medicago sativa (alfalfa)	Ms23.0	Mito	Salt and arsenic tolerance	Lee et al. (2012)

absence of stress have no major effect on growth of the plants, although specific data are not always provided. The largest number of experiments have been performed with class I cytosolic proteins, with reports of enhanced tolerance to heat, drought, NaCl, mannitol and H_2O_2. Over-expression of the chloroplast sHSP has been linked to cold, heat and oxidative stress tolerance, as well as to enhanced differentiation of chromoplasts from chloroplasts. A few studies indicate that class II cytosolic sHSPs could contribute to drought, salt, heat and oxidative stress tolerance (Table 5.3). In almost all the over-expression studies, the tests of stress tolerance have been restricted to young seedlings or to germination, and to a very limited range of stress conditions. None of these plants has been assessed outside of the laboratory, and little consideration has been given to mimic heat stress conditions experienced by plants in the field or to assess effects on yield.

Studies of sHSP mutants or plants in which sHSP expression has been inhibited by antisense RNA or other methods have been limited. As noted in Table 5.3, antisense inhibition of class I sHSP expression in carrot cells led to increased electrolyte leakage after 50 °C heat treatment (Malik et al. 1999). Antisense inhibition of the chloroplast sHSP in Arabidopsis was also reported to decrease viability after 2 h at 39 °C (Chauhan et al. 2012), although no similar phenotype was reported in the Arabidopsis At25.3 mutant (Zhong et al. 2013). Reduced survival of tobacco seedlings after a 2 h stress at 46 °C was associated with antisense inhibition of the mitochondrial sHSP (Sanmiya et al. 2004). Each of these studies involving inhibition of sHSPs with antisense RNA reports results from only a single stress condition. Experiments with Arabidopsis mutants include the At25.3 knockout mutant discussed above, and one report analyzing class I knockout mutants (Dafny-Yelin et al. 2008). In the latter study, T-DNA insertion lines for At17.4, At.17.6A and At18.1 class I proteins were isolated. The mutants were reported to grow identically to wild type at 22 °C and under 34 °C day/28 °C night conditions. However, a heat stress

Table 5.3 Studies of sHSP overexpression in transgenic plants

Species transformed	sHSP	Class	Phenotypes	Reference
Daucus carota (carrot)	Carrot Dc17.7	I	Reduced electrolyte leakage of leaves incubated at 50 °C (AS[a] – increased electrolyte leakage)	Malik et al. (1999)
Solanum tuberosum (potato)	Carrot HisDc17.7[b]	I	Reduced electrolyte leakage of leaf pieces at 47 °C, 4 h; more tuber production in vitro at 29 °C	Ahn and Zimmerman (2006)
Oryza sativa (rice)	Rice Os17.7	I	10 day seedlings could regrow after 6 days no H_2O (soil: −15 MPa), or 3 days in 30 % PEG	Sato and Yokoya (2008)
Arabidopsis thaliana	Rose Rc17.8	I	10 day seedlings, enhanced survival after 45 °C, 2 h; 4 week old seedlings less electrolyte leakage (45 °C,2 h)	Jiang et al. (2009)
Arabidopsis thaliana	Rose Rc17.8	I	9 day old seedlings had longer roots on mannitol; 4 weeks old plants 15 days no H_2O had more siliques, higher fresh wt on recovery	Jiang et al. (2009)
Arabidopsis thaliana	*Opuntia streptacantha* Os18.0	I	Enhanced germination rate in NaCl (125–175 mM), glucose or mannitol (6 or 7 %), ABA(3–9 uM); enhanced plant survival of NaCl (150 mM, 14 days), mannitol (5 %, 21 day)	Salas-Munoz et al. (2012)
Nicotiana tabacum (tobacco)	*Zea mays* Zm16.9	I	Enhanced germination after 40 °C, 10 days or H_2O_2 (5 mM, 20 days); increased root length in 7 day old seedlings 40 °C, 9 h or H_2O_2; minor decrease in oxidative stress markers	Sun et al. (2012)
Oryza sativa (rice)	Rice Os17	I	Enhanced germination, longer roots on NaCl (100–150 mM) or mannitol (50–150 mM); enhanced seedling survival to NaCl (200 mM, 24 h) and air exposure (9.5 h), less oxidative stress on PEG (3 days, 20 %) NaCl (200 mM)	Zou et al. (2012)
Arabidopsis thaliana	Arabidopsis At17.8-HA[c]	I	Sensitivity of germination and reduced root growth on ABA (3 uM; 12 uM); reduced water loss of detached leaves	Kim et al. (2013)
Lactuca sativa (lettuce)	Arabidopsis At17.8-HA[c]	I	Sensitivity of germination and reduced root growth on ABA (3 uM, 12 uM) increased dehydration (3 weeks plants, 4 weeks no H_2O) and NaCl tolerance (5 weeks plants, 500 mM NaCl)	Kim et al. (2013)
Arabidopsis thaliana	Arabidopsis At17.7A	II	Increased plant fresh weight after 17 days w/o H_2O or with NaCl (75 mM)	Sun et al. (2001)

(continued)

Table 5.3 (continued)

Species transformed	sHSP	Class	Phenotypes	Reference
Arabidopsis thaliana	*Nelumbo nucifere* Nn17.5	II	5 day old seedlings enhanced survival 44 °C, 60–75 min; better germination, growth and higher SOD after accelerated ageing (41 °C, 72 h, 100 % humidity)	Zhou et al. (2012)
Arabidopsis thaliana	*Lillium davidii* Ld16.45	II	Increased germination after 45 °C (1–2 h) or 1–7 days on NaCl (100–150 mM). Constitutive increase in SOD and CAT in absence of stress. Longer roots after 14 days on H_2O_2 (1–2 mM)	Mu et al. (2013)
Lycopersicon esculentum (tomato)	Tomato Le21.0	Cplast	Enhanced PSII activity after 4 °C then high light; no PSII protection from heat; enhanced chromoplast differentiation	Neta-Sharir et al. (2005)
Nicotiana tabacum (tobacco)	*Capsicum annuum* Cp26.0	CPlast	20 % higher O_2 evolution after 42 °C, 2 h or 4 °C, 6 h. ~15 % higher P700 after 4 °C 2 h	Guo et al. (2007)
Festuca arundinacea (fescue)	Rice Os26	CPlast	6 week old plants, methyl viologen or 42 °C for 12–48 h, 42 °C, reduced electrolyte leakage and lipid oxidation, enhanced PS activity	Kim et al. (2012)
Arabidopsis thaliana	Wheat Ta26.0	CPlast	35 °C 2 weeks constant, enhanced fresh wt, germination, seed wt, PS. (AS – 2 h 39 °C lethal)	Chauhan et al. (2012)
Nicotiana tabacum (tobacco)	*Capsicum annuum* Cp26.0	CPlast	4 week old greenhouse plants, 4 °C for 6 h; enhanced PS activities	Li et al. (2012)
Nicotiana tabacum (tobacco)	Tomato Le25.0	Mito[d]	Increased seedling survival on plates at 48 °C, 2 h. (AS – reduced survival 46 °C, 2 h)	Sanmiya et al. (2004)
Lycopersicon esculentum (tomato)	Tomato Le21.5	ER	14 day old liquid-grown seedlings recovered from 24 h 10 ug/ml tunicamycin	Zhao et al. (2007)
Arabidopsis thaliana	*S. cerevisiae* Hsp26	n/a	3 week old seedlings on plates −10 °C, 2.5 h: higher survival, >proline, >soluble sugars, >freezing response gene expression	Xue et al. (2009)

[a]AS – results from anti-sense inhibition of sHSP expression reported in the same publication
[b]sHSP was introduced with an N-terminal histidine affinity tag
[c]sHSP introduced with a C-terminal HA affinity tag
[d]Correspondence of this protein to Arabidopsis organelle-localized proteins in Fig. 5.2 was not reported

phenotype was detected for dark grown seedlings adapted at 38 °C prior to heat stress at 45 °C; the mutant seedlings elongated 50 % or less than wild type. This result was interpreted to indicate that the individual class I sHSPs had non-redundant functions in acquired heat tolerance. In order to determine if these sHSPs had overlapping functions in other processes, attempts were made to generate double mutants, but none could be obtained. Phenotypic analysis led the authors to propose that class I sHSP function is required in seed development either during or after fertilization. The developmental roles of sHSPs will no doubt prove to be as interesting as their functions in stress tolerance.

A recent study in the moss *Physcomitrella patens*, which can be manipulated by homologous recombination to knock out or replace specific genes, examined the phenotype of plants deleted for two class II sHSPs (both Pp16.4) (Ruibal et al. 2013). These sHSPs were discovered as being regulated by ABA and also by heat and other stresses. Growth of the sHSP knockout moss was assessed by measuring recovery from different treatments, including 37 °C for 7 days or growth for 10 days on 900 mM mannitol or 500 mM NaCl. The lack of these class II sHSPs resulted in a 50 % or greater reduction in growth after these treatments.

In total, these in vivo studies, along with the evolutionary conservation of the sHSPs and their diverse patterns of expression indicate that these proteins have diverse roles in plants. Unfortunately, these studies do not provide any further insight into potential mechanisms that could be responsible for the protective effects observed in plants or cells that over-express sHSPs, or the loss-of-function phenotypes in sHSP mutants.

5.9 Future Challenges

The major challenge in plant sHSP research is to develop a mechanistic understanding of sHSP function in vivo. Despite elegant in vitro studies of sHSP properties and chaperone activity in protecting a range of model substrates from irreversible denaturation and aggregation, the connection to in vivo function is still missing. Fragmented data suggest different classes of higher plant sHSPs are required for stress tolerance in vivo (e.g. Tables 5.2 and 5.3), but the essential targets they might protect are unknown. Extensive work with class I sHSPs has been instrumental in developing the chaperone model for sHSP function presented in Fig. 5.4, and different in vivo studies can be fit to this model, but whether all sHSPs work the same way is far from demonstrated (Basha et al. 2012). One issue is the relative abundance of the different sHSPs. For example, the chloroplast-localized protein does not accumulate to very high levels, arguing that it likely protects one or a few critical substrates, rather than acting on many denaturing proteins. The in vitro chaperone model also does not explain the conservation of the class I and II sHSPs, which first appeared in the mosses ~400 million years ago (Waters 2013; Waters and Vierling 1999b). These two classes of proteins can protect the same model substrates in a test tube (Basha et al. 2010), but must serve different functions within the cell. Similarly, N-terminal arm mutants of Syn16.6 can protect model substrates in vitro, but do not protect cells from heat stress, although they accumulate to the same levels as wild type in the soluble cell fraction during stress (Giese et al. 2005). This incongruity between in vitro and in vivo data indicates that we have a very limited understanding of the specific roles of sHSPs in cells, particularly in plants. It will require continued biochemical studies, coupled with plant genetics to unravel the plant sHSP puzzle. Identifying specific interactions of sHSP with other cellular components, either functional partners or substrates, is the next key step.

Acknowledgements E. V. would like to acknowledge long term support for work on sHSP biochemistry and function from the National Institutes of Health GM RO142762, as well as support from the National Science Foundation and United States Department of Agriculture competitive grants program for studies of sHSPs in plants and the Department of Energy Basic Bioenergy Sciences program for work on the interaction of plant sHSPs and Hsp101. We also thank many lab members and colleagues for critical and productive collaborations over many years.

References

Ahn YJ, Zimmerman JL (2006) Introduction of the carrot HSP17.7 into potato (Solanum tuberosum L.) enhances cellular membrane stability and tuberization in vitro. Plant Cell Environ 29:95–104

Almoguera C, Jordano J (1992) Developmental and environmental concurrent expression of sunflower dry-seed-stored low-molecular-weight heat-shock protein and Lea mRNAs. Plant Mol Biol 19:781–792

Bagneris C, Bateman OA, Naylor CE, Cronin N, Boelens WC, Keep NH, Slingsby C (2009) Crystal structures of alpha-crystallin domain dimers of alphaB-crystallin and Hsp20. J Mol Biol 392:1242–1252

Baldwin AJ, Lioe H, Robinson CV, Kay LE, Benesch JL (2011) alphaB-crystallin polydispersity is a consequence of unbiased quaternary dynamics. J Mol Biol 413:297–309

Baranova EV, Weeks SD, Beelen S, Bukach OV, Gusev NB, Strelkov SV (2011) Three-dimensional structure of alpha-crystallin domain dimers of human small heat shock proteins HSPB1 and HSPB6. J Mol Biol 411:110–122

Barnett T, Altschuler M, McDaniel CN, Mascarenhas JP (1980) Heat shock induced proteins in plant cells. Dev Genet 1:331–340

Basha E, Lee GJ, Breci LA, Hausrath AC, Buan NR, Giese KC, Vierling E (2004a) The identity of proteins associated with a small heat shock protein during heat stress in vivo indicates that these chaperones protect a wide range of cellular functions. J Biol Chem 279:7566–7575

Basha E, Lee GJ, Demeler B, Vierling E (2004b) Chaperone activity of cytosolic small heat shock proteins from wheat. Eur J Biochem 271:1426–1436

Basha E, Friedrich KL, Vierling E (2006) The N-terminal arm of small heat shock proteins is important for both chaperone activity and substrate specificity. J Biol Chem 281:39943–39952

Basha E, Jones C, Wysocki V, Vierling E (2010) Mechanistic differences between two conserved classes of small heat shock proteins found in the plant cytosol. J Biol Chem 285:11489–11497

Basha E, O'Neill H, Vierling E (2012) Small heat shock proteins and alpha-crystallins: dynamic proteins with flexible functions. Trends Biochem Sci 37:106–117

Basha E, Jones C, Blackwell AE, Cheng G, Waters ER, Samsel KA, Siddique M, Pett V, Wysocki V, Vierling E (2013) An unusual dimeric small heat shock protein provides insight into the mechanism of this class of chaperones. J Mol Biol 425:1683–1696

Benesch JL, Aquilina JA, Baldwin AJ, Rekas A, Stengel F, Lindner RA, Basha E, Devlin GL, Horwitz J, Vierling E, Carver JA, Robinson CV (2010) The quaternary organization and dynamics of the molecular chaperone HSP26 are thermally regulated. Chem Biol 17:1008–1017

Bepperling A, Alte F, Kriehuber T, Braun N, Weinkauf S, Groll M, Haslbeck M, Buchner J (2012) Alternative bacterial two-component small heat shock protein systems. Proc Natl Acad Sci U S A 109:20407–20412

Boorstein WR, Ziegelhoffer T, Craig EA (1994) Molecular evolution of the HSP70 multigene family. J Mol Evol 38:1–17

Bova MP, McHaourab HS, Han Y, Fung BK (2000) Subunit exchange of small heat shock proteins. Analysis of oligomer formation of alphaA-crystallin and Hsp27 by fluorescence resonance energy transfer and site-directed truncations. J Biol Chem 275:1035–1042

Chakrabarti P, Janin J (2002) Dissecting protein-protein recognition sites. Proteins 47:334–343

Chauhan H, Khurana N, Nijhavan A, Khurana JP, Khurana P (2012) The wheat chloroplastic small heat shock protein (sHSP26) is involved in seed maturation and germination and imparts tolerance to heat stress. Plant Cell Environ 35:1912–1931

Chen Q, Vierling E (1991) Analysis of conserved domains identifies a unique structural feature of a chloroplast heat shock protein. Mol Gen Genet 228:328

Cheng G, Basha E, Wysocki VH, Vierling E (2008) Insights into small heat shock protein and substrate structure during chaperone action derived from hydrogen/deuterium exchange and mass spectrometry. J Biol Chem 283:26634–26642

Clark AR, Naylor CE, Bagneris C, Keep NH, Slingsby C (2011) Crystal structure of R120G disease mutant of human alphaB-crystallin domain dimer shows closure of a groove. J Mol Biol 408:118–134

Coca MA, Almoguera C, Jordano J (1994) Expression of sunflower low-molecular-weight heat-shock proteins during embryogenesis and persistence after germination: localization and possible functional implications. Plant Mol Biol 25:479–492

Cole C, Barber JD, Barton GJ (2008) The Jpred 3 secondary structure prediction server. Nucleic Acids Res 36:W197–W201

Dafny-Yelin M, Tzfira T, Vainstein A, Adam Z (2008) Non-redundant functions of sHSP-CIs in acquired thermotolerance and their role in early seed development in Arabidopsis. Plant Mol Biol 67:363–373

de Jong WW, Caspers GJ, Leunissen JA (1998) Genealogy of the alpha-crystallin – small heat-shock protein superfamily. Int J Biol Macromol 22:151–162

Delbecq SP, Klevit RE (2013) One size does not fit all: the oligomeric states of alphaB crystallin. FEBS Lett 587:1073–1080

Derocher AE, Helm KW, Lauzon LM, Vierling E (1991) Expression of a conserved family of cytoplasmic low molecular weight heat shock proteins during heat stress and recovery. Plant Physiol 96:1038–1047

Fragkostefanakis S, Simm S, Paul P, Bublak D, Scharf KD, Schleiff E (2015) Chaperone network composition in Solanum lycopersicum explored by transcriptome profiling and microarray meta-analysis. Plant Cell Environ 38:693–709

Franzmann TM, Wuhr M, Richter K, Walter S, Buchner J (2005) The activation mechanism of Hsp26 does not require dissociation of the oligomer. J Mol Biol 350:1083–1093

Franzmann TM, Menhorn P, Walter S, Buchner J (2008) Activation of the chaperone Hsp26 is controlled by the rearrangement of its thermosensor domain. Mol Cell 29:207–216

Friedrich KL, Giese KC, Buan NR, Vierling E (2004) Interactions between small heat shock protein subunits and substrate in small heat shock protein-substrate complexes. J Biol Chem 279:1080–1089

Giese KC, Vierling E (2002) Changes in oligomerization are essential for the chaperone activity of a small heat shock protein in vivo and in vitro. J Biol Chem 277:46310–46318

Giese KC, Vierling E (2004) Mutants in a small heat shock protein that affect the oligomeric state. Analysis and allele-specific suppression. J Biol Chem 279:32674–32683

Giese KC, Basha E, Catague BY, Vierling E (2005) Evidence for an essential function of the N terminus of a small heat shock protein in vivo, independent of in vitro chaperone activity. Proc Natl Acad Sci U S A 102:18896–18901

Giorno F, Wolters-Arts M, Grillo S, Scharf KD, Vriezen WH, Mariani C (2010) Developmental and heat stress-regulated expression of HsfA2 and small heat shock proteins in tomato anthers. J Exp Bot 61:453–462

Guo SJ, Zhou HY, Zhang XS, Li XG, Meng QW (2007) Overexpression of CaHSP26 in transgenic tobacco alleviates photoinhibition of PSII and PSI during chilling stress under low irradiance. J Plant Physiol 164:126–136

Hanazono Y, Takeda K, Yohda M, Miki K (2012) Structural studies on the oligomeric transition of a small heat shock protein, StHsp14.0. J Mol Biol 422:100–108

Hanazono Y, Takeda K, Oka T, Abe T, Tomonari T, Akiyama N, Aikawa Y, Yohda M, Miki K (2013) Nonequivalence observed for the 16-meric structure of a small heat shock protein, SpHsp16.0, from Schizosaccharomyces pombe. Structure 21:220–228

Haslbeck M, Walke S, Stromer T, Ehrnsperger M, White HE, Chen S, Saibil HR, Buchner J (1999) Hsp26: a temperature-regulated chaperone. EMBO J 18:6744–6751

Haslbeck M, Miess A, Stromer T, Walter S, Buchner J (2005) Disassembling protein aggregates in the yeast cytosol. The cooperation of Hsp26 with Ssa1 and Hsp104. J Biol Chem 280:23861–23868

Helm KW, Schmeits J, Vierling E (1995) An endomembrane-localized small heat-shock protein from Arabidopsis thaliana. Plant Physiol 107:287–288

Helm KW, Lee GJ, Vierling E (1997) Expression and native structure of cytosolic class II small heat-shock proteins. Plant Physiol 114:1477–1485

Hilario E, Teixeira EC, Pedroso GA, Bertolini MC, Medrano FJ (2006) Crystallization and preliminary X-ray diffraction analysis of XAC1151, a small heat-shock protein from Xanthomonas axonopodis pv. citri belonging to the alpha-crystallin family. Acta Crystallogr Sect F Struct Biol Cryst Commun 62:446–448

Hilario E, Martin FJ, Bertolini MC, Fan L (2011) Crystal structures of Xanthomonas small heat shock protein provide a structural basis for an active molecular chaperone oligomer. J Mol Biol 408:74–86

Hochberg GK, Ecroyd H, Liu C, Cox D, Cascio D, Sawaya MR, Collier MP, Stroud J, Carver JA, Baldwin AJ, Robinson CV, Eisenberg DS, Benesch JL, Laganowsky A (2014) The structured core domain of alphaB-crystallin can prevent amyloid fibrillation and associated toxicity. Proc Natl Acad Sci U S A 111:E1562–E1570

Ingolia TD, Craig EA (1982) Four small Drosophila heat shock proteins are related to each other and to mammalian alpha-crystallin. Proc Natl Acad Sci U S A 79:2360–2364

Jaya N, Garcia V, Vierling E (2009) Substrate binding site flexibility of the small heat shock protein molecular chaperones. Proc Natl Acad Sci U S A 106:15604–15609

Jehle S, Rajagopal P, Bardiaux B, Markovic S, Kuhne R, Stout JR, Higman VA, Klevit RE, van Rossum BJ, Oschkinat H (2010) Solid-state NMR and SAXS studies provide a structural basis for the activation of alphaB-crystallin oligomers. Nat Struct Mol Biol 17:1037–1042

Jiang C, Xu J, Zhang H, Zhang X, Shi J, Li M, Ming F (2009) A cytosolic class I small heat shock protein, RcHSP17.8, of Rosa chinensis confers resistance to a variety of stresses to Escherichia coli, yeast and Arabidopsis thaliana. Plant Cell Environ 32:1046–1059

Joe MK, Park SM, Lee YS, Hwang DS, Hong CB (2000) High temperature stress resistance of Escherichia coli induced by a tobacco class I low molecular weight heat-shock protein. Mol Cells 10:519–524

Kennaway CK, Benesch JL, Gohlke U, Wang L, Robinson CV, Orlova EV, Saibil HR, Keep NH (2005) Dodecameric structure of the small heat shock protein Acr1 from Mycobacterium tuberculosis. J Biol Chem 280:33419–33425

Key JL, Lin CY, Chen YM (1981) Heat shock proteins of higher plants. Proc Natl Acad Sci U S A 78:3526–3530

Kim KK, Kim R, Kim SH (1998) Crystal structure of a small heat-shock protein. Nature 394:595–599

Kim DH, Xu ZY, Na YJ, Yoo YJ, Lee J, Sohn EJ, Hwang I (2011a) Small heat shock protein Hsp17.8 functions as an AKR2A cofactor in the targeting of chloroplast outer membrane proteins in Arabidopsis. Plant Physiol 157:132–146

Kim KP, Yu JH, Park SM, Koo HJ, Hong CB (2011b) Tobacco mitochondrial small heat shock protein NtHSP24.6 adopts a dimeric configuration and has a broad range of substrates. BMB Rep 44:816–820

Kim KH, Alam I, Kim YG, Sharmin SA, Lee KW, Lee SH, Lee BH (2012) Overexpression of a chloroplast-localized small heat shock protein OsHSP26 confers enhanced tolerance against oxidative and heat stresses in tall fescue. Biotechnol Lett 34:371–377

Kim DH, Xu ZY, Hwang I (2013) AtHSP17.8 overexpression in transgenic lettuce gives rise to dehydration and salt stress resistance phenotypes through modulation of ABA-mediated signaling. Plant Cell Rep 32:1953–1963

Kirschner M, Winkelhaus S, Thierfelder JM, Nover L (2000) Transient expression and heat-stress-induced co-aggregation of endogenous and heterologous small heat-stress proteins in tobacco protoplasts. Plant J 24:397–411

Kriehuber T, Rattei T, Weinmaier T, Bepperling A, Haslbeck M, Buchner J (2010) Independent evolution of the core domain and its flanking sequences in small heat shock proteins. FASEB J 24:3633–3642

Laganowsky A, Eisenberg D (2010) Non-3D domain swapped crystal structure of truncated zebrafish alphaA crystallin. Protein Sci 19:1978–1984

Laganowsky A, Benesch JL, Landau M, Ding L, Sawaya MR, Cascio D, Huang Q, Robinson CV, Horwitz J, Eisenberg D (2010) Crystal structures of truncated alphaA and alphaB crystallins reveal structural mechanisms of polydispersity important for eye lens function. Protein Sci 19:1031–1043

Lambert W, Koeck PJ, Ahrman E, Purhonen P, Cheng K, Elmlund D, Hebert H, Emanuelsson C (2011) Subunit arrangement in the dodecameric chloroplast small heat shock protein Hsp21. Protein Sci 20:291–301

Lee GJ, Vierling E (2000) A small heat shock protein cooperates with heat shock protein 70 systems to reactivate a heat-denatured protein. Plant Physiol 122:189–198

Lee GJ, Pokala N, Vierling E (1995) Structure and in vitro molecular chaperone activity of cytosolic small heat shock proteins from pea. J Biol Chem 270:10432–10438

Lee GJ, Roseman AM, Saibil HR, Vierling E (1997) A small heat shock protein stably binds heat-denatured model substrates and can maintain a substrate in a folding-competent state. EMBO J 16:659–671

Lee U, Wie C, Escobar M, Williams B, Hong SW, Vierling E (2005) Genetic analysis reveals domain interactions of Arabidopsis Hsp100/ClpB and cooperation with the small heat shock protein chaperone system. Plant Cell 17:559–571

Lee KW, Cha JY, Kim KH, Kim YG, Lee BH, Lee SH (2012) Overexpression of alfalfa mitochondrial HSP23 in prokaryotic and eukaryotic model systems confers enhanced tolerance to salinity and arsenic stress. Biotechnol Lett 34:167–174

Lenne C, Block MA, Garin J, Douce R (1995) Sequence and expression of the mRNA encoding HSP22, the mitochondrial small heat-shock protein in pea leaves. Biochem J 311(Pt 3):805–813

Li M, Ji L, Yang X, Meng Q, Guo S (2012) The protective mechanisms of CaHSP26 in transgenic tobacco to alleviate photoinhibition of PSII during chilling stress. Plant Cell Rep 31:1969–1979

Lin BL, Wang JS, Liu HC, Chen RW, Meyer Y, Barakat A, Delseny M (2001) Genomic analysis of the Hsp70 superfamily in Arabidopsis thaliana. Cell Stress Chaperones 6:201–208

Lopes-Caitar VS, de Carvalho MC, Darben LM, Kuwahara MK, Nepomuceno AL, Dias WP, Abdelnoor RV, Marcelino-Guimarães FC (2013) Genome-wide analysis of the Hsp20 gene family in soybean: comprehensive sequence, genomic organization and expression profile analysis under abiotic and biotic stresses. BMC Genomics 14:577. doi:10.1186/1471-2164-14-577

Ma C, Haslbeck M, Babujee L, Jahn O, Reumann S (2006) Identification and characterization of a stress-inducible and a constitutive small heat-shock protein targeted to the matrix of plant peroxisomes. Plant Physiol 141:47–60

Malik MK, Slovin JP, Hwang CH, Zimmerman JL (1999) Modified expression of a carrot small heat shock protein gene, hsp17. 7, results in increased or decreased thermotolerancedouble dagger. Plant J 20:89–99

McHaourab HS, Godar JA, Stewart PL (2009) Structure and mechanism of protein stability sensors: chaperone activity of small heat shock proteins. Biochemistry 48:3828–3837

Mogk A, Deuerling E, Vorderwulbecke S, Vierling E, Bukau B (2003a) Small heat shock proteins, ClpB and the DnaK system form a functional triade in reversing protein aggregation. Mol Microbiol 50:585–595

Mogk A, Schlieker C, Friedrich KL, Schonfeld HJ, Vierling E, Bukau B (2003b) Refolding of substrates bound to small Hsps relies on a disaggregation reaction mediated most efficiently by ClpB/DnaK. J Biol Chem 278:31033–31042

Mu C, Zhang S, Yu G, Chen N, Li X, Liu H (2013) Overexpression of small heat shock protein LimHSP16.45 in Arabidopsis enhances tolerance to abiotic stresses. PLoS One 8:e82264

Nagao RT, Czarnecka E, Gurley WB, Schoffl F, Key JL (1985) Genes for low-molecular-weight heat shock proteins of soybeans: sequence analysis of a multigene family. Mol Cell Biol 5:3417–3428

Neta-Sharir I, Isaacson T, Lurie S, Weiss D (2005) Dual role for tomato heat shock protein 21: protecting photosystem II from oxidative stress and promoting color changes during fruit maturation. Plant Cell 17:1829–1838

Nover L, Scharf KD, Neumann D (1983) Formation of cytoplasmic heat shock granules in tomato cell cultures and leaves. Mol Cell Biol 3:1648–1655

Nover L, Scharf KD, Neumann D (1989) Cytoplasmic heat shock granules are formed from precursor particles and are associated with a specific set of mRNAs. Mol Cell Biol 9:1298–1308

Painter AJ, Jaya N, Basha E, Vierling E, Robinson CV, Benesch JL (2008) Real-time monitoring of protein complexes reveals their quaternary organization and dynamics. Chem Biol 15:246–253

Patel S, Vierling E, Tama F (2014) Replica exchange molecular dynamics simulations provide insight into substrate recognition by small heat shock proteins. Biophys J 106:2644–2655

Personat JM, Tejedor-Cano J, Prieto-Dapena P, Almoguera C, Jordano J (2014) Co-overexpression of two heat shock factors results in enhanced seed longevity and in synergistic effects on seedling tolerance to severe dehydration and oxidative stress. BMC Plant Biol 14:56

Peschek J, Braun N, Franzmann TM, Georgalis Y, Haslbeck M, Weinkauf S, Buchner J (2009) The eye lens chaperone alpha-crystallin forms defined globular assemblies. Proc Natl Acad Sci U S A 106:13272–13277

Peschek J, Braun N, Rohrberg J, Back KC, Kriehuber T, Kastenmuller A, Weinkauf S, Buchner J (2013) Regulated structural transitions unleash the chaperone activity of alphaB-crystallin. Proc Natl Acad Sci U S A 110:E3780–E3789

Poulain P, Gelly JC, Flatters D (2010) Detection and architecture of small heat shock protein monomers. PLoS One 5:e9990

Quinlan RA, Zhang Y, Lansbury A, Williamson I, Pohl E, Sun F (2013) Changes in the quaternary structure and function of MjHSP16.5 attributable to deletion of the IXI motif and introduction of the substitution, R107G, in the alpha-crystallin domain. Philos Trans R Soc Lond B Biol Sci 368:20120327

Reddy PS, Kavi Kishor PB, Seiler C, Kuhlmann M, Eschen-Lippold L, Lee J, Reddy MK, Sreenivasulu N (2014) Unraveling regulation of the small heat shock proteins by the heat shock factor HvHsfB2c in barley: its implications in drought stress response and seed development. PLoS One 9:e89125

Ruibal C, Castro A, Carballo V, Szabados L, Vidal S (2013) Recovery from heat, salt and osmotic stress in Physcomitrella patens requires a functional small heat shock protein PpHsp16.4. BMC Plant Biol 13:174

Saji H, Iizuka R, Yoshida T, Abe T, Kidokoro S, Ishii N, Yohda M (2008) Role of the IXI/V motif in oligomer assembly and function of StHsp14.0, a small heat shock protein from the acido-thermophilic archaeon, Sulfolobus tokodaii strain 7. Proteins 71:771–782
Salas-Munoz S, Gomez-Anduro G, Delgado-Sanchez P, Rodriguez-Kessler M, Jimenez-Bremont JF (2012) The opuntia streptacantha OpsHSP18 gene confers salt and osmotic stress tolerance in Arabidopsis thaliana. Int J Mol Sci 13:10154–10175
Sanmiya K, Suzuki K, Egawa Y, Shono M (2004) Mitochondrial small heat-shock protein enhances thermotolerance in tobacco plants. FEBS Lett 557:265–268
Sato Y, Yokoya S (2008) Enhanced tolerance to drought stress in transgenic rice plants overexpressing a small heat-shock protein, sHSP17.7. Plant Cell Rep 27:329–334
Scharf KD, Siddique M, Vierling E (2001) The expanding family of Arabidopsis thaliana small heat stress proteins and a new family of proteins containing alpha-crystallin domains (Acd proteins). Cell Stress Chaperones 6:225–237
Schoffl F, Key JL (1982) An analysis of mRNAs for a group of heat shock proteins of soybean using cloned cDNAs. J Mol Appl Genet 1:301–314
Shen G, Kuppu S, Venkataramani S, Wang J, Yan J, Qiu X, Zhang H (2010) ANKYRIN REPEAT-CONTAINING PROTEIN 2A is an essential molecular chaperone for peroxisomal membrane-bound ASCORBATE PEROXIDASE3 in Arabidopsis. Plant Cell 22:811–831
Shi J, Koteiche HA, McDonald ET, Fox TL, Stewart PL, McHaourab HS (2013) Cryoelectron microscopy analysis of small heat shock protein 16.5 (Hsp16.5) complexes with T4 lysozyme reveals the structural basis of multimode binding. J Biol Chem 288:4819–4830
Siddique M, Port M, Tripp J, Weber C, Zielinski D, Calligaris R, Winkelhaus S, Scharf KD (2003) Tomato heat stress protein Hsp16.1-CIII represents a member of a new class of nucleocytoplasmic small heat stress proteins in plants. Cell Stress Chaperones 8:381–394
Siddique M, Gernhard S, von Koskull-Doring P, Vierling E, Scharf KD (2008) The plant sHSP superfamily: five new members in Arabidopsis thaliana with unexpected properties. Cell Stress Chaperones 13:183–197
Smykal P, Masin J, Hrdy I, Konopasek I, Zarsky V (2000) Chaperone activity of tobacco HSP18, a small heat-shock protein, is inhibited by ATP. Plant J 23:703–713
Sobott F, Benesch JL, Vierling E, Robinson CV (2002) Subunit exchange of multimeric protein complexes. Real-time monitoring of subunit exchange between small heat shock proteins by using electrospray mass spectrometry. J Biol Chem 277:38921–38929
Song NH, Ahn YJ (2010) DcHsp17.7, a small heat shock protein from carrot, is upregulated under cold stress and enhances cold tolerance by functioning as a molecular chaperone. HortSci 45:469–474
Song NH, Ahn YJ (2011) DcHsp17.7, a small heat shock protein in carrot, is tissue-specifically expressed under salt stress and confers tolerance to salinity. N Biotechnol 28:698–704
Soto A, Allona I, Collada C, Guevara MA, Casado R, Rodriguez-Cerezo E, Aragoncillo C, Gomez L (1999) Heterologous expression of a plant small heat-shock protein enhances Escherichia coli viability under heat and cold stress. Plant Physiol 120:521–528
Stamler R, Kappe G, Boelens W, Slingsby C (2005) Wrapping the alpha-crystallin domain fold in a chaperone assembly. J Mol Biol 353:68–79
Stengel F, Baldwin AJ, Painter AJ, Jaya N, Basha E, Kay LE, Vierling E, Robinson CV, Benesch JL (2010) Quaternary dynamics and plasticity underlie small heat shock protein chaperone function. Proc Natl Acad Sci U S A 107:2007–2012
Stengel F, Baldwin AJ, Bush MF, Hilton GR, Lioe H, Basha E, Jaya N, Vierling E, Benesch JL (2012) Dissecting heterogeneous molecular chaperone complexes using a mass spectrum deconvolution approach. Chem Biol 19:599–607
Stromer T, Ehrnsperger M, Gaestel M, Buchner J (2003) Analysis of the interaction of small heat shock proteins with unfolding proteins. J Biol Chem 278:18015–18021
Stromer T, Fischer E, Richter K, Haslbeck M, Buchner J (2004) Analysis of the regulation of the molecular chaperone Hsp26 by temperature-induced dissociation: the N-terminal domail is important for oligomer assembly and the binding of unfolding proteins. J Biol Chem 279:11222–11228

Sudnitsyna MV, Mymrikov EV, Seit-Nebi AS, Gusev NB (2012) The role of intrinsically disordered regions in the structure and functioning of small heat shock proteins. Curr Protein Pept Sci 13:76–85
Sun W, Bernard C, van de Cotte B, Van Montagu M, Verbruggen N (2001) At-HSP17.6A, encoding a small heat-shock protein in Arabidopsis, can enhance osmotolerance upon overexpression. Plant J 27:407–415
Sun L, Liu Y, Kong X, Zhang D, Pan J, Zhou Y, Wang L, Li D, Yang X (2012) ZmHSP16.9, a cytosolic class I small heat shock protein in maize (Zea mays), confers heat tolerance in transgenic tobacco. Plant Cell Rep 31:1473–1484
Takeda K, Hayashi T, Abe T, Hirano Y, Hanazono Y, Yohda M, Miki K (2011) Dimer structure and conformational variability in the N-terminal region of an archaeal small heat shock protein, StHsp14.0. J Struct Biol 174:92–99
Tejedor-Cano J, Prieto-Dapena P, Almoguera C, Carranco R, Hiratsu K, Ohme-Takagi M, Jordano J (2010) Loss of function of the HSFA9 seed longevity program. Plant Cell Environ 33:1408–1417
Tiroli AO, Ramos CH (2007) Biochemical and biophysical characterization of small heat shock proteins from sugarcane. Involvement of a specific region located at the N-terminus with substrate specificity. Int J Biochem Cell Biol 39:818–831
Tiroli-Cepeda AO, Ramos CH (2010) Heat causes oligomeric disassembly and increases the chaperone activity of small heat shock proteins from sugarcane. Plant Physiol Biochem 48:108–116
Tripp J, Mishra SK, Scharf KD (2009) Functional dissection of the cytosolic chaperone network in tomato mesophyll protoplasts. Plant Cell Environ 32:123–133
Uversky VN, Dunker AK (2010) Understanding protein non-folding. Biochim Biophys Acta 1804:1231–1264
Van Aken O, Zhang B, Carrie C, Uggalla V, Paynter E, Giraud E, Whelan J (2009) Defining the mitochondrial stress response in Arabidopsis thaliana. Mol Plant 2:1310–1324
Van Montfort R, Slingsby C, Vierling E (2001a) Structure and function of the small heat shock protein/alpha-crystallin family of molecular chaperones. Adv Protein Chem 59:105–156
van Montfort RL, Basha E, Friedrich KL, Slingsby C, Vierling E (2001b) Crystal structure and assembly of a eukaryotic small heat shock protein. Nat Struct Biol 8:1025–1030
Vierling E (1991) The roles of heat-shock proteins in plants. Annu Rev Plant Physiol 42:579–620
Vierling E, Mishkind ML, Schmidt GW, Key JL (1986) Specific heat shock proteins are transported into chloroplasts. Proc Natl Acad Sci U S A 83:361–365
Vierling E, Nagao RT, DeRocher AE, Harris LM (1988) A heat shock protein localized to chloroplasts is a member of a eukaryotic superfamily of heat shock proteins. EMBO J 7:575–581
Wadhwa R, Ryu J, Gao R, Choi IK, Morrow G, Kaur K, Kim I, Kaul SC, Yun CO, Tanguay RM (2010) Proproliferative functions of Drosophila small mitochondrial heat shock protein 22 in human cells. J Biol Chem 285:3833–3839
Waterhouse AM, Procter JB, Martin DM, Clamp M, Barton GJ (2009) Jalview version 2 – a multiple sequence alignment editor and analysis workbench. Bioinformatics 25:1189–1191
Waters ER (2013) The evolution, function, structure, and expression of the plant sHSPs. J Exp Bot 64:391–403
Waters ER, Rioflorido I (2007) Evolutionary analysis of the small heat shock proteins in five complete algal genomes. J Mol Evol 65:162–174
Waters ER, Vierling E (1999a) Chloroplast small heat shock proteins: evidence for atypical evolution of an organelle-localized protein. Proc Natl Acad Sci U S A 96:14394–14399
Waters ER, Vierling E (1999b) The diversification of plant cytosolic small heat shock proteins preceded the divergence of mosses. Mol Biol Evol 16:127–139
Waters ER, Aevermann BD, Sanders-Reed Z (2008) Comparative analysis of the small heat shock proteins in three angiosperm genomes identifies new subfamilies and reveals diverse evolutionary patterns. Cell Stress Chaperones 13:127–142
Weber C, Nover L, Fauth M (2008) Plant stress granules and mRNA processing bodies are distinct from heat stress granules. Plant J 56:517–530

Weeks SD, Baranova EV, Heirbaut M, Beelen S, Shkumatov AV, Gusev NB, Strelkov SV (2014) Molecular structure and dynamics of the dimeric human small heat shock protein HSPB6. J Struct Biol 185:342–354

Wehmeyer N, Vierling E (2000) The expression of small heat shock proteins in seeds responds to discrete developmental signals and suggests a general protective role in desiccation tolerance. Plant Physiol 122:1099–1108

Wehmeyer N, Hernandez LD, Finkelstein RR, Vierling E (1996) Synthesis of small heat-shock proteins is part of the developmental program of late seed maturation. Plant Physiol 112:747–757

White HE, Orlova EV, Chen S, Wang L, Ignatiou A, Gowen B, Stromer T, Franzmann TM, Haslbeck M, Buchner J, Saibil HR (2006) Multiple distinct assemblies reveal conformational flexibility in the small heat shock protein Hsp26. Structure 14:1197–1204

Wintrode PL, Friedrich KL, Vierling E, Smith JB, Smith DL (2003) Solution structure and dynamics of a heat shock protein assembly probed by hydrogen exchange and mass spectrometry. Biochemistry 42:10667–10673

Xue Y, Peng R, Xiong A, Li X, Zha D, Yao Q (2009) Yeast heat-shock protein gene HSP26 enhances freezing tolerance in Arabidopsis. J Plant Physiol 166:844–850

Yeh CH, Chang PF, Yeh KW, Lin WC, Chen YM, Lin CY (1997) Expression of a gene encoding a 16.9-kDa heat-shock protein, Oshsp16.9, in Escherichia coli enhances thermotolerance. Proc Natl Acad Sci U S A 94:10967–10972

Yeh CH, Chen YM, Lin CY (2002) Functional regions of rice heat shock protein, Oshsp16.9, required for conferring thermotolerance in Escherichia coli. Plant Physiol 128:661–668

Yoon HJ, Kim KP, Park SM, Hong CB (2005) Functional mode of NtHSP17.6. a cytosolic small heat-shock protein from Nicotiana tabacum. J Plant Biol 48:120–127

Zhao C, Shono M, Sun A, Yi S, Li M, Liu J (2007) Constitutive expression of an endoplasmic reticulum small heat shock protein alleviates endoplasmic reticulum stress in transgenic tomato. J Plant Physiol 164:835–841

Zhong L, Zhou W, Wang H, Ding S, Lu Q, Wen X, Peng L, Zhang L, Lu C (2013) Chloroplast small heat shock protein HSP21 interacts with plastid nucleoid protein pTAC5 and is essential for chloroplast development in Arabidopsis under heat stress. Plant Cell 25:2925–2943

Zhou Y, Chen H, Chu P, Li Y, Tan B, Ding Y, Tsang EW, Jiang L, Wu K, Huang S (2012) NnHSP17.5, a cytosolic class II small heat shock protein gene from Nelumbo nucifera, contributes to seed germination vigor and seedling thermotolerance in transgenic Arabidopsis. Plant Cell Rep 31:379–389

Zou J, Liu C, Liu A, Zou D, Chen X (2012) Overexpression of OsHsp17.0 and OsHsp23.7 enhances drought and salt tolerance in rice. J Plant Physiol 169:628–635

Chapter 6
Regulation of the Chaperone Function of Small Hsps

Martin Haslbeck, Sevil Weinkauf, and Johannes Buchner

Abstract Small heat shock proteins (sHsps) are ubiquitous molecular chaperones that represent a first line of defense against proteotoxic stress and prevent the aggregation of unfolding proteins. The most striking feature of sHsps is their ability to form higher-order oligomers. Within the last decade, especially the dynamic ensembles with a broad distribution of different oligomers, the diversity of assembly types and the regulation of their activity were in the focus of research. Interestingly, the activity of sHsps directly correlates to these structural features as it is regulated by changes in the composition of the ensembles. In this chapter, we describe the mechanisms known so far which are responsible for the activation of chaperone function and their linkage to changes in the structure of sHsps.

Keywords Molecular chaperone • Stress response • Protein aggregation • Alpha crystallin • Protein dynamics • Flexible quaternary structure • Chaperone networks • Regulation of chaperone activity

6.1 Small Heat Shock Proteins Form Dynamic Ensembles of Interconverting Oligomers

sHsps are ATP-independent molecular chaperones which act within the proteostasis network of the cell as a buffer system to bind unfolding proteins upon stress, protect them from irreversible aggregation and keep them in a folding-competent state until their refolding by downstream ATP-dependent chaperone systems (Haslbeck et al. 2005a; Basha et al. 2012; Delbecq and Klevit 2013; Garrido et al. 2012). The most striking feature of most sHsps, also the key to their function, is their ability to assemble into large oligomeric structures of a unique hierarchal arrangement mediated by different protein domains at different levels.

M. Haslbeck (✉) • S. Weinkauf • J. Buchner
Department Chemie, Center for Integrated Protein Science, Technische Universität München, 85747 Garching, Germany
e-mail: martin.haslbeck@tum.de; sevil.weinkauf@tum.de; johannes.buchner@tum.de

R.M. Tanguay, L.E. Hightower (eds.), *The Big Book on Small Heat Shock Proteins*, Heat Shock Proteins 8, DOI 10.1007/978-3-319-16077-1_6

sHsps share a conserved domain organization (Delbecq and Klevit 2013; Haslbeck et al. 2005a; Garrido et al. 2012; De Jong et al. 1998; Kriehuber et al. 2010; Maaroufi and Tanguay 2013). The primary structure can be dissected in three major parts of functional importance: an N-terminal region followed by the conserved α-crystallin domain and a C-terminal region (Fig. 6.1a) (Caspers et al. 1995; De Jong et al. 1998; Kriehuber et al. 2010).

The α-crystallin domain represents the signature motif of sHsps, albeit the overall amino acid sequence conservation of this domain is rather low. Multiple sequence alignments including sHsps from different organisms revealed only very few consensus residues with the A-x-x-x-n-G-v-L motif at the end of the α-crystallin domain being the most significant indicator (Fig. 6.1a) (Narberhaus 2002; Caspers et al. 1995; De Jong et al. 1998). Although not a single amino acid within the domain is invariant and shared by all family members, the length and structure of the α-crystallin domain are well conserved (Kriehuber et al. 2010). This domain is on average 94 amino acids long and forms a compact β-sheet sandwich similar to the immunoglobulin fold (Fig. 6.1b). The β-sheet sandwich is composed of two antiparallel layers of three and four β-strands, respectively, connected by a short interdomain loop (Fig. 6.1b) (van Montfort et al. 2001, 2002; Delbecq and Klevit 2013).

A growing number of structures of isolated, recombinantly produced α-crystallin domains demonstrates that these usually form stable dimers (Fig. 6.1c) and that the α-crystallin domain on its own is not sufficient for oligomer formation (Bagneris et al. 2009; Baranova et al. 2011; Clark et al. 2011; Laganowsky et al. 2010). Within

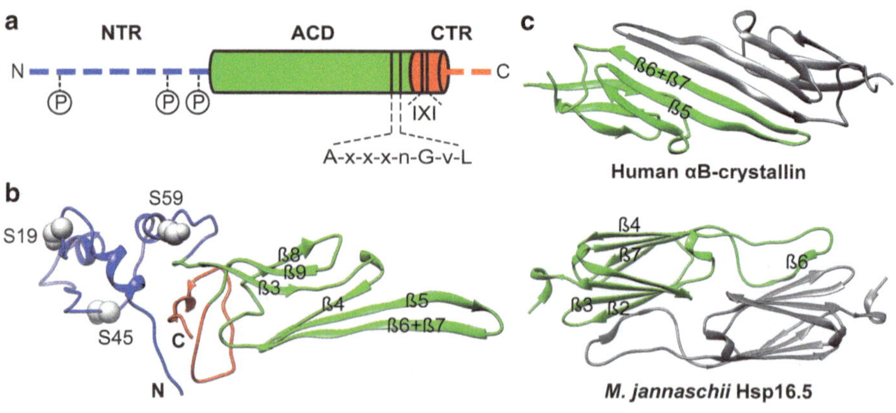

Fig. 6.1 (**a**) Domain organization of sHsps. N-terminal domain (NTR) (*blue*), α-crystallin domain (ACD) (*green*), C-terminal extension (CTR) (*red*). The motifs common to all sHsps and the three major phosphorylation sites of αB-crystallin are indicated. (**b**) Monomer structure of full-length human αB-crystallin according to Braun et al. (2011) (PDB: 2YGD). The domains are color coded according to the domain organization shown in (**a**). The three major in vivo phosphorylation sites Ser19, Ser45, and Ser59 are depicted as grey spheres. (**c**) Structures of the ACD-dimers of human αB-crystallin (solid state NMR, PDB: 2KLR) (*top*) and of *M. jannaschii* Hsp16.5 (X-ray crystallography, PDB: 1SHS) (*bottom*). The ACDs of individual protomers are colored *green* and *gray*. Note the differences in metazoan-type (human αB-crystallin) and non-metazoan-type (*M. jannaschii* Hsp16.5) dimerization modes

the oligomeric assemblies of the sHsps, the dimer seems to be the basic building block and the first level of internal order (Fig. 6.2a) (Delbecq and Klevit 2013). Dimerization in plants, yeast, archaeal and bacterial sHsps occurs via reciprocal strand swapping of the ß6 strands into the ß-sandwich of the neighboring monomer (Fig. 6.1c) (Kim et al. 1998; van Montfort et al. 2001; Hanazono et al. 2013; Bepperling et al. 2012). This "non-metazoan" dimer structure is different from the "metazoan" type of dimer assembly found for example in human αB-crystallin (Jehle et al. 2011; Delbecq and Klevit 2013). Here, the β6- and β7-strands are fused into an elongated strand which forms the dimer interface with its counterpart from the neighboring monomer in an anti-parallel orientation (Fig. 6.1c).

The α-crystallin domain is flanked by N- and C-terminal regions. The C-terminal region following the α-crystallin domain varies substantially. It is on average 10 amino acids long and typically does not exceed 20 amino acids (Kriehuber et al. 2010). Most importantly, the C-terminal region contains the highly conserved I-x-I motif which is essential for the association of the α-crystallin domain to higher oligomers (Fig. 6.2). Some sHsp family members like human Hsp20 completely

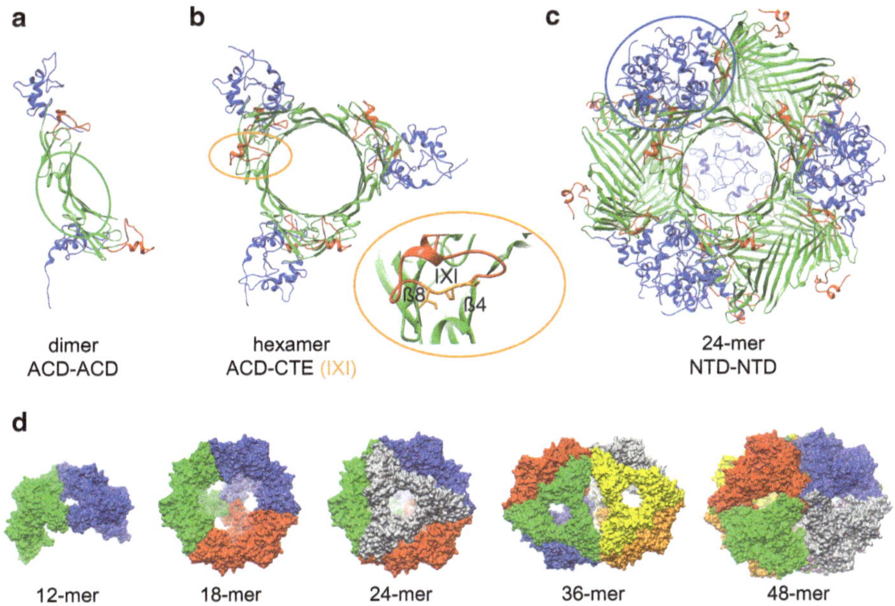

Fig. 6.2 The hierarchy of the human αB-crystallin 24-meric assembly (Braun et al. 2011) (PDB: 2YGD). (**a**) The dimeric building block assembled through the interactions between the ß6+ß7-strands located within the ACDs of the adjacent protomers. (**b**) The hexameric sub-assembly formed by inter-dimer interactions, i.e., through the binding of C-terminal IXI motifs (I159-P160-I161) to the hydrophobic groove formed by the ß4- and ß8-strands of the adjacent protomer (*inset*). (**c**) 24-mer assembled from hexameric blocks through contacts within N-terminal regions. For the color coding of the domains across columns a–c see Fig. 6.1a. (**d**) Pseudo-atomic models of αB-crystallin oligomers composed of multiples of hexameric blocks depicted in different colors (Braun et al. 2011). Note the modular architecture of the assemblies

lack a C-terminal region and thus also the conserved I-x-I motif necessary for inter-dimer contacts (Chen et al. 2010; Basha et al. 2012; Delbecq and Klevit 2013).

The N-terminal region is very diverse in sequence and length, e.g., 24 residues in Hsp12.2 from *Caenorhabtidis elegans* (Candido 2002) and 247 residues in Hsp42 from *Saccharomyces cerevisiae* (Haslbeck et al. 2004; Kriehuber et al. 2010; Wotton et al. 1996). Especially in fungi, a general increase of the length of the N-termini is observed (Kriehuber et al. 2010). Interestingly, the N-terminal region seems to be highly flexible, as it is not or only partially resolved in all yet available sHsp crystal structures.

Due to the limited number of oligomeric sHsp structures, it is still not unambiguously understood which inter-subunit interactions are responsible for the formation of the defined, higher-order oligomeric species. However, the emerging picture from recent structural studies is that residues in all three regions of sHsps are required for oligomerization (Delbecq and Klevit 2013). While the α-crystallin domain is necessary for dimer formation and thus assembles the basic building block, both flanking regions promote the formation of higher-order structures. Thus, the three domains contribute to the assembly process in a hierarchical way (Fig. 6.2a–c). For example, in the case of the human αB-crystallin 24-mer, three dimers, each formed through interactions between the β6+β7-strands of two adjacent monomers (Fig. 6.2a), assemble into a hexamer via binding of the C-terminal I-x-I motif into the hydrophobic groove formed by the β4- and β8-strands of the neighboring monomer (Fig. 6.2b). Four such hexamers further associate into the 24-mer through contacts within N-terminal regions (Fig. 6.2c) (Jehle et al. 2011; Braun et al. 2011).

This type of hierarchy seems to be conserved among sHsps and indicates that the total number of subunits in the oligomers can be modulated by variations especially in the N-terminal region. This is, for example, highlighted by studies on a variant of Hsp16.5 from *M. jannaschii* (McHaourab et al. 2012). Here, the insertion of a 14 amino acid sequence of the N-terminal region of human Hsp27 at the junction between the N-terminal region and the α-crystallin domain of Hsp16.5 resulted in the increase of the number of subunits in the oligomer from 24 to 48 subunits. Furthermore, the expanded oligomer showed an increased chaperone activity.

Another key feature of sHsps is their tendency to populate a range of oligomeric states at equilibrium. The oligomers constantly exchange subunits and thus form polydisperse and dynamic ensembles (Fig. 6.2d). The degree of structural plasticity and heterogeneity appears, however, variable for different members of the family. In this context, it should be noted that also sHsps which were earlier described as "monodisperse" do not exist in one oligomeric state, but are rather ensembles of different oligomers with one highly abundant species within the size distribution. For example, Hsp26 from *S. cerevisiae* shows a single peak when analyzed at physiological temperatures (25 °C) by size exclusion chromatography, indicating a monodisperse population of 24-mers (Haslbeck et al. 1999). Mass spectrometry, however, reveals the heterogeneity of the Hsp26 ensemble that actually consists of a range of oligomeric states from monomers to higher-order species such as 40-mers (Benesch et al. 2010; White et al. 2006). Hsp16.5 from the hyper-thermophilic organism *M. jannaschii* which crystallizes as a 24-mer (Kim et al. 1998) and displays a single oligomeric species in mass spectrometry experiments at low temperatures (Hilton and Benesch 2012), forms a dynamic ensemble when incubated at physiological (Bova et al. 2002; Haslbeck et al. 2008). Similarly, wheat Hsp16.9 which crystallizes as a dodecamer and represents only dodecamers in mass spectrometry experiments at different conditions (van Montfort et al. 2001; Hilton and Benesch 2012), frequently exchanges subunits at physiological temperatures and thus forms a dynamic ensemble of different species (Sobott et al. 2002; Wintrode et al. 2003). sHsps which were described as "polydisperse", on the other hand, show generally a broader distribution of different species within the ensemble. In this context, human αB-crystallin populates a wide variety of inter-converting oligomers ranging from monomers to 48-mers (Fig. 6.2d) (Baldwin et al. 2011; Braun et al. 2011). Even more complicated, *Escherichia coli* IbpB and IbpA present complex mixtures from dimers to roughly 50-mers or fibrils and it is still enigmatic if any major species exists in the ensemble (Ratajczak et al. 2010; Matuszewska et al. 2005; Shearstone and Baneyx 1999).

The ability of sHsps to generate oligomers of different sizes from a single sequence is linked to their modular architecture. In human αB-crystallin, heterogeneity is achieved by addition or subtraction of subunits from oligomeric complexes while the overall structure of the constituent monomers and the interface of the dimeric building block remain unaltered (Braun et al. 2011). Thus, the distribution of oligomers seems to be governed mostly by fluctuations of the inter-subunit contacts mediated by the C- and N-terminal regions. Conditions that destabilize interactions at subunit interfaces lead to an enhanced rate of dissociation of subunits, concomitant with an enhanced subunit exchange kinetics, increasing the populations of smaller oligomers (and vice versa). This seemingly simple mechanism of establishing a delicate balance between different oligomer populations by adjusting the dissociation/association rates of building blocks is unique to sHsps and is tightly correlated with the regulation of their chaperone activity according to the needs of the cell (see Part III).

Although the data in the literature on subunit exchange rates of different sHsps under physiological and stress conditions are rather limited, so far known values for the half times of subunit exchange for different sHsps at the respective physiological temperatures are usually between ~5 and 500 s. Intriguingly, "polydisperse" and "monodisperse" representatives seem to differ in their exchange rates by at least one order of magnitude with highly similar values for "monodisperse" members at the upper end and for "polydisperse" members at the lower end of the above scale (Bova et al. 2000, 2002; Franzmann et al. 2008; deMiguel et al. 2009). In general, the subunit exchange rate of sHsps increases with increasing temperature and chaperone activity, while substrate binding seems to reduce subunit exchange rates.

6.2 Small Heat Shock Proteins Are Promiscuous Molecular Chaperones

The cellular function of sHsps remained elusive for a very long time after their discovery. In the early 1990s, in vitro studies demonstrated that sHsps bind denatured proteins and prevent them from irreversible aggregation (Jakob et al. 1993; Horwitz 1992; De Jong et al. 1993; Merck et al. 1993). Meanwhile, many sHsps from different species have been shown to bind a variety of non-native proteins in vitro (Fig. 6.3) (Stege et al. 1999; Jakob et al. 1993; Studer and Narberhaus 2000; Lee et al. 1995, 1997; Basha et al. 2004b; Mchaourab et al. 2002). sHsps are generally ATP-independent molecular chaperones which bind their substrate proteins tightly and form stable complexes (Basha et al. 2012; Haslbeck et al. 2005a; Garrido et al. 2012; Horwitz 1992; Bova et al. 2000; McHaourab et al. 2009) which also seem to exist as a polydisperse ensemble of different species (Fig. 6.3) (Stengel et al. 2010). Intriguingly, analyses by electron microscopy (Haslbeck et al. 1999; Stromer et al. 2003) and mass spectrometry (Stengel et al. 2010) have revealed that the substrate complexes of many sHsps also contain highly-ordered species that are larger than the substrate-free sHsp oligomers. Although their structures are not yet resolved, these species convey the impression that they have re-assembled from small sHsps oligomers, presumably dimers, which re-associate to a new form of complex upon substrate binding. For some sHsps, this re-assembling process upon substrate binding seems to be even cooperative (Stromer et al. 2003; Haslbeck et al. 1999). Thus, within the proteostasis network of the cell, sHsps function as a buffer system to bind and complex unfolding proteins upon stress and to protect them from irreversible aggregation (Fig. 6.3). Notably, ratios of up to one non-native substrate protein per sHsp dimer have been reported (Lee et al. 1997; Haslbeck et al. 1999) indicating the high binding capacity of sHsps. In vitro experiments revealed that the non-native protein trapped in sHsp/substrate complexes remain folding-competent (Ehrnsperger et al. 1997; Lee et al. 1997) suggesting that additional ATP-dependent chaperones

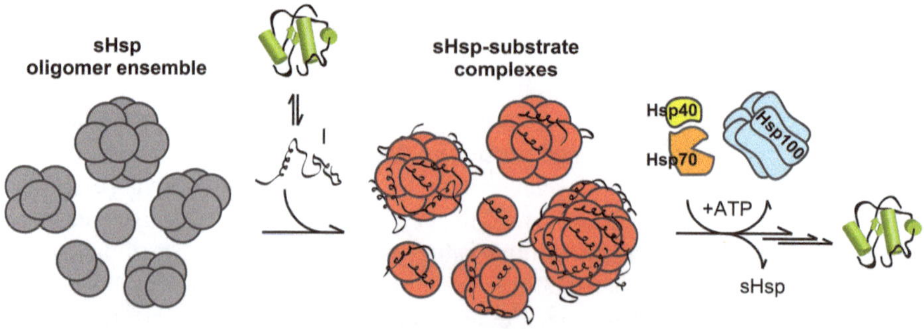

Fig. 6.3 Mechanism of the chaperone function of sHsps. Under stress conditions sHsps bind partially unfolded proteins in an energy-independent manner and keep them in a folding-competent state in sHps/substrate complexes. The substrates are subsequently refolded by downstream ATP-dependent chaperone systems

might trigger the release of the substrate (Fig. 6.3). Meanwhile, it has been shown that in mammalian cells, the Hsp70/Hsp40 system is required for the refolding of substrate proteins bound to sHsps (Ehrnsperger et al. 1997; Stege et al. 1994; Lee and Vierling 2000). Similarly, for bacteria such as *E. coli*, it has been demonstrated that bound non-native proteins were specifically transferred to the DnaK-DnaJ-GrpE chaperone system and subsequently re-activated (Mogk et al. 2003; Veinger et al. 1998). However, this re-activation mechanism appears to be largely dependent on the ratio of sHsp to substrate and the Hsp70/Hsp40 system alone is effective in refolding the substrate proteins only if sHsps are present at stoichiometric or excess concentrations, where soluble, defined sHsp/substrate complexes form. At excess levels of the substrate protein, sHsps become incorporated into aggregates of the substrate protein. For the refolding of substrates from these aggregate-like sHsp/substrate complexes, a mechanism involving Hsp70/Hsp40 and members of the Hsp100 family is necessary. This mechanism seems to be conserved from bacteria to lower eukaryotes, involving the Hsp100 family members ClpB in *E. coli* and Hsp104 in *S. cerevisiae* (Mogk et al. 2003; Cashikar et al. 2005; Haslbeck et al. 2005b; Bepperling et al. 2012).

Taken together, these results suggest that sHsps provide organisms with the capability to separate the prevention of aggregation from ATP-dependent refolding steps. Furthermore, the incorporation of sHsps in non-native substrate aggregates is necessary for efficient refolding of the substrates (Fig. 6.3).

The substrate specificity of sHsps is still a matter of debate. Several proteomic approaches in different organisms have shown that a significant number of cytosolic proteins is maintained in a soluble state by sHsps under heat shock conditions (Basha et al. 2004a; Haslbeck et al. 2004; Bepperling et al. 2012; Fu et al. 2013). Interestingly, a preference for translation-related proteins (e.g., ribosomal proteins and amino-acyl tRNA synthases) and for metabolic enzymes has been observed in prokaryotes (Fu et al. 2013; Bepperling et al. 2012; Fu 2014). Nevertheless, it still remains to be explored whether these preferences also exist for eukaryotic sHsps and if they are due to preferred recognition motifs or if these proteins are simply the ones representing the largest fraction of aggregation-prone proteins under heat stress conditions.

Intriguingly, in some organisms containing multiple cytosolic sHsps, the substrate spectra of individual members show substantial overlaps (Haslbeck et al. 2004) and it remains enigmatic whether specific substrates exist for a given sHsps when multiple sHsps are found in one cellular compartment. However, when sHsps are expressed in different tissues or developmental stages, the respective substrate spectra seem to be optimized and adjusted to the proteome of the respective cells. The most prominent example in this respect is αA-crystallin which is specific to the eye lens and which seems to be needed to specifically recognize, as a hetero-oligomer with αB-crystallin, β- and γ-crystallins (Fu and Liang 2002). αB-crystallin, on the other hand, recognizes many more substrates in other tissues (Arrigo 2013; Arrigo and Gibert 2013). As a further example, the testis-specific human HspB9 seems to be specifically needed to bind one substrate or, in this case rather a partner protein, the dynein subunit TCTEL1 (de Wit et al. 2004).

In summary, sHsps recognize a wide spectrum of in vivo substrates as well as various artificial model substrates demonstrating that they act as highly promiscuous chaperones. However, it is still unclear which regions of the substrate proteins are bound by sHsps and whether common recognition motifs exist. sHsps are typically less effective in suppressing the aggregation of larger proteins, in terms of the ratios of sHsps:substrate needed, indicating that the interaction depends on the mass ratio rather than on the molar ratio and hints to a charge- and/or hydrophobicity-driven capturing of the substrates (Basha et al. 2012).

The residues and the respective surface patches on the sHsps which are involved in substrate interaction are still largely unknown. The emerging picture is that there are several binding sites throughout the molecule which act together, presumably in a different manner for different substrates. Studies utilizing the incorporation of hydrophobic dyes suggested that substrates might bind to short segments in the N-terminal region (Sharma et al. 1998; Lee et al. 1997). Recent evidence arising from cross-linking experiments and analyses by mass spectrometry or peptide libraries support this tendency (Ghosh et al. 2007; Ahrman et al. 2007b; Jaya et al. 2009; Cheng et al. 2008). Nevertheless, an exchange of the highly conserved G in the AxxxGVL motif of HspH from *Bradyrhizobium japonicum* specifically impaired chaperone activity without interfering with other properties of the protein indicating also the involvement of the α-crystallin domain (Lentze and Narberhaus 2004; Lentze et al. 2004; Hochberg et al. 2014). Additionally, mutations of amino acid residues in the C-terminal region of αB-crystallin affect chaperone activity and indicate that the flexibility of the C-terminus is necessary for substrate recognition (Treweek et al. 2007, 2010). Together with the observation that the I-x-I motif is bound into the groove between the β4- and β8-strands within the α-crystallin domain of the adjacent monomer, this might hint to a potential binding pocket for unfolded proteins at the hydrophobic patches which becomes exposed when the C-terminal contacts fall apart during oligomer disassembly (Treweek et al. 2010; Laganowsky et al. 2010; van Montfort et al. 2001; Delbecq et al. 2012). However, no direct interactions of the α-crystallin domain with the substrates have been detected in cross-linking experiments so far (Jaya et al. 2009; Ahrman et al. 2007b). Altogether, substrate binding seems to be accomplished most likely by the non-conserved variable sequences outside the α-crystallin domain which in turn might explain variations in the substrate specificity of different sHsps and why the smallest known representatives of the sHsp family, Hsp12.1 Hsp12.2, 12.3 and 12.6 of *C. elegans*, which solely consist of an α-crystallin domain with only very short N- and C-terminal sequences, have failed to show chaperone-like properties in vitro (Kokke et al. 1998; Leroux et al. 1997). Furthermore, phylogenetic comparison of these sequences showed that they evolved independently and in parallel (Kriehuber et al. 2010), making variations in the substrate recognition profiles of sHsps from evolutionally far apart species most likely. But at present, it is still not possible to state which exact sequence parts or motifs of sHsps are necessary for their chaperone function and which are involved in oligomer interactions – or if there is a substantial crosstalk between those two functions.

Studies on the interaction site on diverse model substrates by cross-linking demonstrated that in general a limited number of sequence segments are bound to the sHsps. Only a single α-crystallin-binding site was identified in yeast alcohol dehydrogenase (ADH) by a cross-linking approach (Santhoshkumar and Sharma 2002). For citrate synthase (CS), a specific region in the N-terminal part was shown to bind to *Arabidopsis thaliana* Hsp21 (Ahrman et al. 2007a) and in malate dehydrogenase (MDH) only two core regions were identified to bind to pea Hsp18.1 or wheat Hsp16.9 (Cheng et al. 2008). Commonly, no cross-links to regions or sequences stretches located in the interior of the folded substrate proteins have been detected. Together, these findings indicate that upon stress conditions the substrate proteins in the cell may be bound as early unfolding intermediates rather than as completely unfolded peptide chains as only regions which are presumably exposed early during unfolding of the substrate proteins bind to the sHsps (Ahrman et al. 2007a, b; Cheng et al. 2008). This is further evident in experiments using a library of fluorescently labeled T4-lysozyme variants with different thermodynamic stabilities (McHaourab et al. 2009). Here, the experimental setup allowed equilibrium measurements of sHsp-substrate interactions and demonstrated that already very weakly destabilized T4-lysozyme variants bind to human Hsp27 and that the binding affinity increases with stronger destabilization of the T4 lysozyme.

In summary, the described chaperone function of sHsps and their incorporation in the protein homeostasis network is consistent with a general protective function in the dynamic environment of a living cell to combat proteotoxic stress. In addition, sHsps might assume tissue-specific functions related to protein aggregation. For example, in the static environment of the eye-lens, irreversible binding of lens proteins by α-crystallin, which occurs in the aging lens, prevents aggregation and light scattering (Horwitz 1992, 2000). However, as the most likely natural substrate proteins in the eye-lens, the β– and γ-crystallins are more stable than α-crystallin, the key function of α-crystallin in the eye lens is probably not aggregation suppression by forming stable complexes with the substrates but rather the adjustment and regulation of the refractive index gradient (Clark et al. 2012; Slingsby et al. 2013; Slingsby and Wistow 2014; Bloemendal et al. 2004). Moreover, their polydispersity allows them to resist phase separation, aggregation and crystallization at the high protein concentrations present in the lens and the adjustment of their oligomeric size in accordance to the actual, local needs.

6.3 Regulation of Activity of Small Heat Shock Proteins

For molecular chaperones in general, the existence of low- and high-affinity states for non-native proteins represents a conserved trait (Fig. 6.4a) (Walter and Buchner 2002). In all families of molecular chaperones except for sHsps, the transition between the two functional states is governed by ATP binding and hydrolysis (Buchner and Walter 2005). As sHsps act in an ATP-independent manner, their activity is regulated by intrinsic mechanisms that are strictly related to their

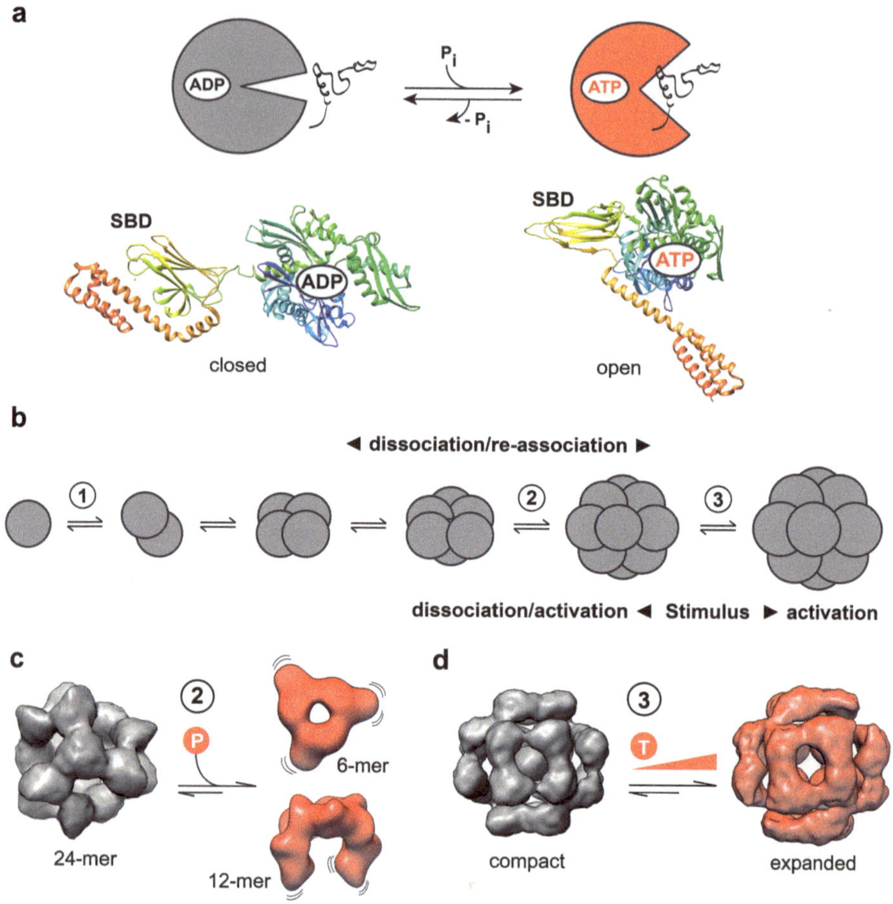

Fig. 6.4 (**a**) General activation mechanism for ATP-dependent molecular chaperones. The transition between the low-affinity and high-affinity states of the chaperone for non-native proteins is accomplished by conformational changes mediated by ATP-binding and – hydrolysis. Example: ADP-bound, "closed" conformation of Hsp70-DnaK with hidden substrate-binding domain (SBD) (*left*, PDB: 2KHO) (Bertelsen et al. 2009), ATP-bound, "open" conformation with accessible (SBD) (*right*, PDB: 4JN4) (Qi et al. 2013). (**b**) Activation mechanism of sHsps. sHsps populate at equilibrium a wide variety of inter-converting oligomers with different substrate affinities. The transition into a state of increased chaperone activity may occur through (i) remodeling of the ensemble composition in favor of species with enhanced substrate binding affinity (1 and 2), (ii) slight conformational changes within the monomer without perturbation of the oligomeric state (3), or (iii) a combination of both pathways. The association/dissociation equilibria are modulated by triggers such as elevated temperatures or post-translational modifications. (**c**) Phosphorylation of human αB-crystallin in NTD triggers the dissociation of larger oligomers, e.g., 24-mers, into smaller species, e.g., 12-mers and hexamers, in which the "free" N-termini gain flexibility and show higher substrate binding affinity. (**d**) Heat stress leads to structural rearrangements of the middle domain and expansion of yeast Hsp26, resulting in activation of the chaperone activity

structural features and ability to populate different oligomeric forms. Several lines of evidence indicate that these mechanisms, which are initiated by various triggers, involve either conformational changes within the subunit or remodeling of the ensemble composition, most likely, however, a combination of both (Fig. 6.4b–d).

For some sHsps, e.g., for Hsp26 from *S. cerevisiae*, the transition to a state of increased substrate binding affinity and chaperone activity occurs through slight conformational changes within the monomer without perturbation of the oligomeric state (Franzmann et al. 2005, 2008). According to analyses by cryo-electron microscopy, these subtle structural rearrangements lead to a transfer of electron density from the interior of the oligomer to the surface accompanied by a slight increase in particle diameter (Fig. 6.4d) (White et al. 2006; Haslbeck et al. 2008). The switch between the "compact" and "expanded" oligomer most likely results in the presentation of substrate recognizing surfaces which are otherwise buried in the interior of the compact oligomer. This oligomer "breathing" concomitant with an increase in substrate binding affinity was also observed with the archaeal sHsp Hsp20.2 from *Archaeoglobus fulgidus* (Haslbeck et al. 2008). A similar model was proposed for the exposure of substrate binding sites on IbpB oligomers, where heat-stress temperatures are proposed to induce the step-wise and hierarchical exposure of binding sites (Fu 2014). It is conceivable that both the breathing and the exposure of binding sites present only one, and presumably the first step in the activation of these sHsps, as they dissociate upon stress (see below) (Stromer et al. 2004; Haslbeck et al. 1999, 2008; White et al. 2006).

A majority of sHsps requires the dissociation of the oligomer and in consequence, the remodeling of the ensemble composition for the efficient recognition and binding of client proteins. Dissociation leads to an ensemble with a higher content of smaller species, correlating to the "active state", while the "inactive state" shows a higher fraction of larger oligomers which seem to correspond to "dormant" storage species (Fig. 6.4b) (Basha et al. 2012; Haslbeck et al. 2005a; Garrido et al. 2012; Delbecq and Klevit 2013; McHaourab et al. 2009). For example, class I sHsps in plants show a shift of the oligomer ensemble towards dimers upon heat stress, accompanied by an increase in chaperone activity. Plant class II sHsps, on the other hand, remain oligomeric but undergo structural rearrangements and start to exchange subunits (Basha et al. 2010). The increase of subunit exchange by a factor of 10–100 is a commonly observed phenomenon that correlates with the activation of sHsps (deMiguel et al. 2009; Franzmann et al. 2008; Bova et al. 2000, 2002; Cheng et al. 2008; Wintrode et al. 2003; Basha et al. 2010). In most cases, the exchanging unit seems to be a dimer or even a monomer (Sobott et al. 2002; Aquilina et al. 2005) but in some cases the exchanging units might be also tetramers or hexamers (Peschek et al. 2013; Basha et al. 2010). Currently, it is unclear whether these larger subunits are always involved in this process but disassemble too fast to be recognized or whether they are bypassed in the disassembling process of sHsps where only dimers and monomers are observed (Fig. 6.4b).

A further variation of the activation pathway becomes evident from the existence of solely dimeric sHsps (Basha et al. 2013; Bepperling et al. 2012). An example is Hsp17.7 from *Deinocaldococcus radiodurans* which interacts with its substrates only transiently without forming stable sHsp/substrate complexes (Bepperling et al. 2012). On the other hand, the dimeric Hsp18.5 from *Arabidopsis thaliana* is in equilibrium with a monomeric species and can be activated at heat stress temperatures where the subunit exchange of the monomers is significantly increased (Basha et al. 2013). Additionally, Hsp18.5 is still able to form stable, larger sHsp/substrate complexes, presumably including several dimer/substrate-complexes.

In summary, these findings underline that the transition of sHsps between active and inactive states is not a simple, two-state mechanisms, but a complex process which might involve several transitions. Some sHsps apparently make use of the full scale of possible transitions while others focus on only parts of it (Fig. 6.4b). Intriguingly, the activation by a hierarchical, most likely step-wise dissociation mechanism might explain why ATP consumption is not needed as already subtle changes in the energy profiles of the subunit interfaces would modulate the association/dissociation equilibria. This might also explain why sHsps are constitutively active as the ensemble always contains a certain amount of active species due to the flat equilibria of the association/dissociation processes.

Which regulatory stimuli affect the association/dissociation equilibria of sHsps and lead to activation? According to our current knowledge, four different stimuli exist: (i) the presence of unfolded or partially folded substrates; (ii) changes in the environmental temperature; (iii) phosphorylation or more general post- translational modifications and, (iv) the formation of hetero-oligomers.

Especially the first regulatory principle is eye catching and establishes sHsps as the first line of defense of the cell to guarantee the stability of the proteome at physiological as well as upon stress conditions. The dynamic assembly/disassembly behavior allows the substrate binding sites, which seem to be buried in the oligomeric complexes, to become exposed by dissociation (Giese and Vierling 2002; Shashidharamurthy et al. 2005; Yang et al. 1999; Lindner et al. 1998; McHaourab et al. 2009). Thus, in the ensemble there will always be some molecules with exposed binding sites which can bind unfolding substrate proteins. This would titrate the active species out of the equilibrium leading to a gradual shift of the ensemble to more active species. Thus, the spontaneous dissociation/re-association process may be a sensing process monitoring the presence of non-native proteins in the cellular environment (McHaourab et al. 2009; Shi et al. 2013; Benesch et al. 2008). Especially, the above-mentioned studies on human Hsp27, in which the equilibrium dissociation is encoded in the N-terminal region that interacts with substrates dependent on their stability, support this sensing hypothesis (Shashidharamurthy et al. 2005; McDonald et al. 2012).

Besides this substrate-based regulation mechanism, temperature is a further general trigger that acts on the association/dissociation equilibria of sHsps as all sHsps described so far can be specifically activated upon (heat) stress temperatures. Intriguingly, the temperature range of activation is adapted to the respective physiological temperature of the respective organism. While the mesophilic yeast Hsp26 is activated in a temperature range from 20 to 43 °C, the hyper-thermophilic *M. jannaschii* Hsp16.5 is activated in the range from 60 to 95 °C and sHsps from fish living below 15 °C are activated at ~10–25 °C (Haslbeck et al. 1999, 2008; Posner et al. 2012). First studies on yeast Hsp26 recognizing temperature as a regulatory mechanism of their chaperone activity also established the close correlation with the disassembly of the oligomer (Haslbeck et al. 1999). Later on, it was recognized that the dissociation was not essential because a cross-linked oligomer unable to dissociate into smaller species showed chaperone activity which was further enhanced by a shift to higher temperatures (Franzmann et al. 2005). This indicates that the observed "breathing" mechanism of the Hsp26 oligomer (Fig. 6.4d), representing the first step in the activation cascade, is already sufficient for substrate recognition. Nevertheless, also yeast Hsp26 undergoes further activation steps in form of dissociation under physiological conditions. Interestingly, the "breathing" mechanism observed for Hsp26 involves the structural rearrangement of a so-called middle domain, a temperature sensing part of the N-terminal sequence localized next to the α-crystallin domain (White et al. 2006; Franzmann et al. 2008). Currently, it is unclear whether such a thermo-sensing segment also exists in the *A. fulgidus* or *M. jannaschii* sHsps with a similar "breathing" behavior (Haslbeck et al. 2008). While the "breathing" mechanism as a first step in the activation cascade seems to be specific for only a subset of sHsps, the response to temperature appears to be common and temperature-induced enhanced dissociation seems to be the general scheme. First studies attempting to investigate the thermodynamics of the assembly process in more detail showed that the overall energy needed for dissociation rather low in comparison to the unfolding of the α-crystallin domain (Stromer et al. 2004). Thus, the slight shift from physiological to heat stress temperatures, which usually comprises only a few degrees, would provide sufficient activation energy to shift the equilibria towards enhanced dissociation. However, a detailed biophysical characterization of individual steps in the hierarchical assembly/disassembly process is still lacking and such studies will require separating different oligomeric species.

At first sight, the above considerations are contradicted by the observation that human Hsp27 shows no enhanced dissociation but an increase in its overall oligomer size under heat stress conditions (40–43 °C) (Lelj-Garolla and Mauk 2005; Mymrikov et al. 2012). Nevertheless, Hsp27 still is activated in this temperature range and additionally, according to substrate interaction studies, the Hsp27 dimer seems to be the substrate-interacting species (McDonald et al. 2012). The current models of dissociation-dependent substrate recognition and activation cannot fully explain the observations on Hsp27. Further experimental evidence is needed to clarify whether the behavior of Hsp27 corresponds to an additional activation mechanism or whether a "breathing" mechanism also exists for Hsp27. Another observation that remains enigmatic from a mechanistic point of view is that in the case of sHsps from the parasite *T. gondii* temperature-induced activation is achieved not by "heat shock" but by "cold shock" temperatures (deMiguel et al. 2009).

The importance of temperature stress itself as an activation mechanism is additionally highlighted on the level of translation control. Studies on mRNAs encoding sHsps from mesophilic and thermophilic cyanobacteria indicate "RNA thermometers" (RNAT) where internal hairpins in the 5′-untranslated region are inhibiting effective translation of the sHsp RNAs at physiological temperatures (Cimdins et al. 2014). At heat-stress temperatures, these hairpins (RNATs) melt and effective translation of the sHsp mRNA is achieved. Thus, organisms seem to control the activity and amount of sHsps on the level of transcription by heat-shock factor dependent promotors, translation by RNATs and on the protein level by temperature-induced dissociation/activation.

Similar to the presence of substrate, the temperature stimulus represents a very effective and even faster trigger for the activation of sHsps when the amount of unfolded protein increases within a cell. It is tempting to speculate that, in comparison to phosphorylation and hetero-oligomer formation, this activation stimulus of sHsps might be the evolutionarily oldest regulatory mechanism. It allows the organisms to react rapidly to changes in the surrounding temperature in stabilizing its proteome without the need for protein synthesis, giving the cell time to reorganize its gene expression profiles to withstand prolonged heat-stress situations.

Besides temperature, other stress situations, e.g., changes in pH might represent further triggers for activation. Preliminary observations indicate that a change in pH seems to induce structural changes in Hsp27 and αB-crystallin which, in turn, affects the assembly of the oligomers and/or dimers, e.g., Hsp27 oligomer equilibrium shifts towards larger species upon slight acidification (Chernik et al. 2004). In the case of αB-crystallin, the dimer interface is destabilized upon acidification leading to a higher content of monomers (Baldwin et al. 2011; Jehle et al. 2011; Delbecq and Klevit 2013). However, the correlation with a coincident change in the activity is still lacking.

The third mechanism, i.e., regulation of the chaperone activity by phosphorylation or more general post-translational modifications, seems to be specific to eukaryotes. It allows efficient fine-tuning of sHsp activity. The oligomeric state of mammalian sHsps can be altered in particular by serine-specific phosphorylation (Kantorow and Piatigorsky 1998; Gaestel 2002; Derham and Harding 1999). All human sHsp investigated so far are phosphorylated and the state of phosphorylation is regulated in response to stress, cytokines or growth factors (Ito et al. 1997; van den IJssel et al. 1998; Voorter et al. 1989) and phosphatase-mediated dephosphorylation counter-regulates the phosphorylation state (Gaestel et al. 1992; Cairns et al. 1994; Moroni and Garland 2001). Human Hsp27, for example, possesses three phosphorylation sites, S15, S78 and S82, whose modification via a MAP-kinase cascade leads to eight possible isoforms (Kemp and Pearson 1990). Thr143 comprises an alternative phosphorylation site in Hsp27 that is phosphorylated by a cGMP-dependent protein kinase (Butt et al. 2001).

Phosphorylation of sHsps, similar to the other already described stimuli, leads to a shift in the ensemble towards smaller species by dissociation of the larger oligomers (Fig. 6.4c). Interestingly, for most of the sHsps which have been described to react to phosphorylation, the smaller species are enriched in tetramers and hexamers indicating that some of the equilibria within the assembly process are affected stronger than others. For example, phosphorylation by MAPKAP2/3 kinase in case of Hsp27 leads first to an enrichment of tetramers which further dissociate into dimers (Hayes et al. 2009; Ehrnsperger et al. 1997; Rogalla et al. 1999; Kato et al. 1994). Similarly, studies using phosphorylation mimicking variants of αB-crystallin reveal an oligomer ensemble mainly consisting of 12-mers, hexamers and dimers (Fig. 6.4c)

(Peschek et al. 2013; Ito et al. 2001). The predominance of these species indicates that mainly the N-terminal contacts in the oligomer are influenced by phosphorylation which is also in accordance with the localization of the phosphorylation sites (Fig. 6.1a). Mechanistically, the negative charges incorporated into the oligomer upon phosphorylation seems to destabilize subunit interfaces. Interestingly, in the pseudo-atomic model of αB-crystallin, the three phosphorylation sites co-localize in the same region indicating a contact patch that seems to be destroyed by increasing phosphorylation in a titrable manner (Braun et al. 2011; Peschek et al. 2013).

While phosphorylation of αB-crystallin results for all substrate proteins studied so far in an increase in chaperone activity, the situation turns out to be more complicated for Hsp27. Depending on the experimental conditions and the model substrate used, Hsp27 was shown to be either active or inactive upon phosphorylation (Lelj-Garolla and Mauk 2005; Shashidharamurthy et al. 2005; Hayes et al. 2009; Rogalla et al. 1999; Kato et al. 1994; Ito et al. 2001). Generally, the interaction of Hsp27 with a specific substrate protein is dependent on its phosphorylation state (Arrigo 2013). Glucose-6-phosphate dehydrogenase is stabilized by phosphorylated Hsp27 in a much more efficient manner (Arrigo 2013). In addition, phosphorylation allows the cell to regulate and adjust the appropriate substrate spectra of the sHsps. In mammalian cells, for example, the phosphorylation-dependent regulation of sHsp-substrate interactions seems to be utilized to modulate the organization of the cytoskeleton (Arrigo 2013).

It should be mentioned that also other, non-mammalian sHsps can be phosphorylated. For example, in Hsp22 from maize mitochondria, the covalent modification occurs at a serine residue by an unspecified kinase activity (Lund et al. 2001). Similarly, for yeast Hsp26 phosphorylated species were described already in the early 1990s which were meanwhile confirmed by phosphoproteomic studies utilizing mass spectrometry (Bentley et al. 1992; Bodenmiller et al. 2010; Ficarro et al. 2002). However, it remains elusive whether phosphorylation of non-mammalian sHsps has a similar impact on their activity and substrate spectra as it is the case for mammalian sHsps.

Besides phosphorylation, other post-translational modifications have been described to have regulatory influences on the chaperone activity of sHsps. Deamidation, as described for α-crystallin, introduces a negative charge at the respective site and affects oligomerization (Gupta and Srivastava 2004). Similarly, modifications by oxidation (Chen et al. 2001; Chalova et al. 2014), glycation (Satish Kumar et al. 2004) and the attachment of methylglyoxal (MG), a dicarbonyl side product of glycolysis, have been suggested to influence the oligomerization of sHsps (Oya-Ito et al. 2006). MG was described earlier to provoke the formation of Hsp27 oligomers (Oya-Ito et al. 2006). Additionally, redox-induced modifications were suggested mainly for Hsp27 and αB-crystallin (Arrigo 2013; Chalova et al. 2014; Garrido et al. 2012; Zavialov et al. 1998a, b). However, one should bear in mind that sHsps are usually in a reducing environment and it is unclear whether oxidation of cysteines resulting in disulfide-bridges inter-connecting sHsp subunits really represents a regulatory mechanism, or if this is a side reaction only taking place when the sHsps are released from dying cells. Nevertheless, human Hsp27 and αB-crystallin stabilize the cell against oxidative stress conditions, shows anti-apoptotic activity (Arrigo 2007; Pasupuleti et al. 2010b) and seems to be involved in the regulation of the glutathione pool of the cell (Christians et al. 2012). On the structural level, oxidation of Hsp27 influences the size distribution of the oligomers "freezing" the ensemble in a more homogenous state and results in a decrease of the overall chaperone activity (Chalova et al. 2014). On the other hand, in the closely homologous murine Hsp25, only the dimer seems to be cross-linked by a disulfide bond upon oxidation and oligomer assembly and chaperone activity are not affected (Zavialov et al. 1998a). Mutation of the cysteine of human Hsp27, however, decreases its anti-apoptotic activity under oxidative stress conditions by inhibiting the interaction with cytochrome-c (Bruey et al. 2000; Pasupuleti et al. 2010a).

The flexible nature of sHsp structure and the constant exchange of subunits open up the possibility to form hetero-oligomeric assemblies with other sHsps present in the same compartment. This hetero-oligomer formation represents the fourth stimulus which allows the activation of sHsps. The most prominent sHsp hetero-oligomer is formed by αA- and αB-crystallin in the human eye lens, although hetero-oligomer formation was also shown for several other human sHsp (for a comprehensive survey see: Arrigo 2013). Indeed, cytosolic, human sHsps can be roughly divided into two classes based on the ability to interact with each other, with Hsp27 (HspB1), Hsp20 (HspB6), and αB-crystallin (HspB5) in one class and HspB2 and HspB3 in the second class (Sugiyama et al. 2000). While the members of these two classes only form hetero-oligomers with sHsps from the same class, Hsp22 (HspB8) is able to interact with members of both classes (Fontaine et al. 2005; Sun et al. 2004). In contrast to the stimuli described before, the formation of hetero-oligomers does not per se lead to a shift of the ensemble to smaller species. For example, Hsp27/αB-crystallin hetero-oligomers possess the same size as the isolated proteins (Mymrikov et al. 2012; Skouri-Panet et al. 2012). Nevertheless, the Hsp27/αB-crystallin hetero-oligomers show higher chaperone activities than the parental homo-oligomers and the presence of Hsp27 seems to stimulate the chaperone activity of αB-crystallin (HspB5). In case of hetero-oligomer formation, the regulation of activity probably occurs through the variation of the oligomer dynamics by incorporation of slower or faster exchanging hetero-subunits and/or through the modulation of the surface properties of the sHsps (Skouri-Panet et al. 2012). One intriguing observation with sHsp hetero-oligomers is that the substrate spectra of the homo-oligomeric, parental sHsps are modified when they are incorporated in hetero-oligomers (Arrigo 2013). Furthermore, as the phosphorylation of the parental sHsps additionally influences their hetero-oligomer formation properties as well as, the two stimuli (phosphorylation and hetero-oligomer formation) in combination allow a highly complex fine-tuning and adjusting of the chaperone properties to specific substrates, e.g., Hsp27 (HspB1) found in complex with αB-crystallin (HspB5) is differently and less phosphorylated than Hsp27 found in homo-oligomers and the interactomes of hetero- and homo-oligomers vary significantly (Arrigo and Gibert 2013). In context of human, cytosolic sHsps this fine-tuning of the substrate spectra is further apparent in the rather specific, seemingly regulatory interactions they show with some substrates (Arrigo 2013). Prominent examples are the regulatory interactions of sHsps with cytoskeleton components.

In the human cytosol, Hsp20 which forms only dimers, does not change its oligomeric state upon phosphorylation (Bukach et al. 2004). However, when it forms hetero-oligomers with Hsp27 or αB-crystallin, the size of the resulting hetero-oligomers is significantly decreased compared to the parental Hsp27 or αB-crystallin homo-oligomers (Bukach et al. 2009; Mymrikov et al. 2012). Thus, Hsp20 seems to be a real modulator changing the properties of Hsp27 and αB-crystallin oligomers upon association and hetero-oligomer formation. Similarly, *E. coli* IbpA itself shows only low chaperone activity, but it enhances the chaperone activity of IbpB when both form a hetero-oligomer, also indicating that one sHsps is used as the modulator of the activity of a "partner" sHsp (Ratajczak et al. 2009).

It is of special interest that not all sHsps simultaneously present in the same compartment form hetero-oligomers or exchange subunits. In *T. gondii*, for example, only two of four cytosolic sHsp interact with each other (deMiguel et al. 2009). Similarly, in plants the class I and II sHsps do not cross-interact (Basha et al. 2012) and most intriguingly, the two yeast sHsps Hsp26 and Hsp42 also do not form hetero-oligomers (Haslbeck et al. 2004). The situation in the human cytosol where Hsp27 and αB-crystallin form hetero-oligomers appears even more complicated. While only a fraction of the parental Hsp27 is incorporated into the hetero-oligomers that co-exist with Hsp27 homo-oligomers, all αB-crystallin oligomers seem to associate with Hsp27 (Arrigo 2013). Usually, the fraction of Hsp27 homo-oligomers is ~10 % independent of the ratio of Hsp27 to αB-crystallin but seems to be controlled by the phosphorylation status of Hsp27.

Overall, these observations suggest that the formation of hetero-oligomers is tightly regulated and the co-existence of homo-oligomeric and hetero-oligomeric ensembles is possible. Due to the close homology of most sHsps within one compartment of an organism, it will be highly interesting, nevertheless challenging, to find out how such tremendous influences on the assembly mechanisms can be achieved on the structural level by only slight variations in the amino acid composition.

In summary, the mechanistic correlation of the chaperone activity of sHsps and their structural dynamics has become evident in the last years. Especially, the increased number of available sHsp 3D-structures obtained by hybrid approaches helped for a better understanding of the mechanistic features. Nevertheless, we are still far from understanding the mechanistic details and differences of the various members of the sHsp family and key aspects of their function are still to be resolved. It remains elusive how different assembly types can be achieved by variations in the N-terminal sequence. Furthermore, it is enigmatic whether a maximum number of subunits for oligomers exists, what this number might be and what the structurally limiting parameters are. Finally, the mechanisms of substrate recognition, the corresponding binding sites and how the re-assembly to organized, structured and stable sHsps-substrate complexes is achieved, is still a mystery. All these open questions leave exciting perspectives for the years to come.

Acknowledgments We thank the Deutsche Forschungsgemeinschaft (SFB 1035) and CIPS^{M} for financial support.

References

Ahrman E, Gustavsson N, Hultschig C, Boelens WC, Emanuelsson CS (2007a) Small heat shock proteins prevent aggregation of citrate synthase and bind to the N-terminal region which is absent in thermostable forms of citrate synthase. Extremophiles 11:659–666

Ahrman E, Lambert W, Aquilina JA, Robinson CV, Emanuelsson CS (2007b) Chemical cross-linking of the chloroplast localized small heat-shock protein, Hsp21, and the model substrate citrate synthase. Protein Sci 16:1464–1478

Aquilina JA, Benesch JL, Ding LL, Yaron O, Horwitz J, Robinson CV (2005) Subunit exchange of polydisperse proteins: mass spectrometry reveals consequences of alphaA-crystallin truncation. J Biol Chem 280:14485–14491

Arrigo AP (2007) The cellular "networking" of mammalian Hsp27 and its functions in the control of protein folding, redox state and apoptosis. Adv Exp Med Biol 594:14–26

Arrigo AP (2013) Human small heat shock proteins: protein interactomes of homo- and hetero-oligomeric complexes: an update. FEBS Lett 587:1959–1969

Arrigo AP, Gibert B (2013) Protein interactomes of three stress inducible small heat shock proteins: HspB1, HspB5 and HspB8. Int J Hyperth 29:409–422

Bagneris C, Bateman OA, Naylor CE, Cronin N, Boelens WC, Keep NH, Slingsby C (2009) Crystal structures of alpha-crystallin domain dimers of alphaB-crystallin and Hsp20. J Mol Biol 392:1242–1252

Baldwin AJ, Lioe H, Robinson CV, Kay LE, Benesch JL (2011) alphaB-crystallin polydispersity is a consequence of unbiased quaternary dynamics. J Mol Biol 413:297–309

Baranova EV, Weeks SD, Beelen S, Bukach OV, Gusev NB, Strelkov SV (2011) Three-dimensional structure of alpha-crystallin domain dimers of human small heat shock proteins HSPB1 and HSPB6. J Mol Biol 411:110–122

Basha E, Lee GJ, Breci LA, Hausrath AC, Buan NR, Giese KC, Vierling E (2004a) The identity of proteins associated with a small heat shock protein during heat stress in vivo indicates that these chaperones protect a wide range of cellular functions. J Biol Chem 279:7566–7575

Basha E, Lee GJ, Demeler B, Vierling E (2004b) Chaperone activity of cytosolic small heat shock proteins from wheat. Eur J Biochem 271:1426–1436

Basha E, Jones C, Wysocki V, Vierling E (2010) Mechanistic differences between two conserved classes of small heat shock proteins found in the plant cytosol. J Biol Chem 285:11489–11497

Basha E, O'Neill H, Vierling E (2012) Small heat shock proteins and alpha-crystallins: dynamic proteins with flexible functions. Trends Biochem Sci 37:106–117

Basha E, Jones C, Blackwell AE, Cheng G, Waters ER, Samsel KA, Siddique M, Pett V, Wysocki V, Vierling E (2013) An unusual dimeric small heat shock protein provides insight into the mechanism of this class of chaperones. J Mol Biol 425:1683–1696

Benesch JL, Ayoub M, Robinson CV, Aquilina JA (2008) Small heat shock protein activity is regulated by variable oligomeric substructure. J Biol Chem 283:28513–28517

Benesch JL, Aquilina JA, Baldwin AJ, Rekas A, Stengel F, Lindner RA, Basha E, Devlin GL, Horwitz J, Vierling E, Carver JA, Robinson CV (2010) The quaternary organization and dynamics of the molecular chaperone HSP26 are thermally regulated. Chem Biol 17:1008–1017

Bentley NJ, Fitch IT, Tuite MF (1992) The small heat-shock protein Hsp26 of *Saccharomyces cerevisiae* assembles into a high molecular weight aggregate. Yeast 8:95–106

Bepperling A, Alte F, Kriehuber T, Braun N, Weinkauf S, Groll M, Haslbeck M, Buchner J (2012) Alternative bacterial two-component small heat shock protein systems. Proc Natl Acad Sci U S A 109:20407–20412

Bertelsen EB, Chang L, Gestwicki JE, Zuiderweg ER (2009) Solution conformation of wild-type *E. coli* Hsp70 (DnaK) chaperone complexed with ADP and substrate. Proc Natl Acad Sci U S A 106:8471–8476

Bloemendal H, de Jong W, Jaenicke R, Lubsen NH, Slingsby C, Tardieu A (2004) Ageing and vision: structure, stability and function of lens crystallins. Prog Biophys Mol Biol 86:407–485

Bodenmiller B, Wanka S, Kraft C, Urban J, Campbell D, Pedrioli PG, Gerrits B, Picotti P, Lam H, Vitek O, Brusniak MY, Roschitzki B, Zhang C, Shokat KM, Schlapbach R, Colman-Lerner A, Nolan GP, Nesvizhskii AI, Peter M, Loewith R, von Mering C, Aebersold R (2010) Phosphoproteomic analysis reveals interconnected system-wide responses to perturbations of kinases and phosphatases in yeast. Sci Signal 3:rs4

Bova MP, McHaourab HS, Han Y, Fung BKK (2000) Subunit exchange of small heat shock proteins – analysis of oligomer formation of αA-crystallin and Hsp27 by fluorescence resonance energy transfer and site-directed truncations. J Biol Chem 275:1035–1042

Bova MP, Huang Q, Ding L, Horwitz J (2002) Subunit exchange, conformational stability, and chaperone-like function of the small heat shock protein 16.5 from *Methanococcus jannaschii*. J Biol Chem 277:38468–38475

Braun N, Zacharias M, Peschek J, Kastenmuller A, Zou J, Hanzlik M, Haslbeck M, Rappsilber J, Buchner J, Weinkauf S (2011) Multiple molecular architectures of the eye lens chaperone alphaB-crystallin elucidated by a triple hybrid approach. Proc Natl Acad Sci U S A 108:20491–20496

Bruey JM, Ducasse C, Bonniaud P, Ravagnan L, Susin SA, Diaz-Latoud C, Gurbuxani S, Arrigo AP, Kroemer G, Solary E, Garrido C (2000) Hsp27 negatively regulates cell death by interacting with cytochrome c. Nat Cell Biol 2:645–652

Buchner J, Walter S (2005) Analysis of chaperone function in vitro. In: Buchner J, Kiefhaber T (eds) Protein folding handbook. Wiley-VCH, Weinheim, pp 162–196

Bukach OV, Seit-Nebi AS, Marston SB, Gusev NB (2004) Some properties of human small heat shock protein Hsp20 (HspB6). Eur J Biochem 271:291–302

Bukach OV, Glukhova AE, Seit-Nebi AS, Gusev NB (2009) Heterooligomeric complexes formed by human small heat shock proteins HspB1 (Hsp27) and HspB6 (Hsp20). Biochim Biophys Acta 1794:486–495

Butt E, Immler D, Meyer HE, Kotlyarov A, Laass K, Gaestel M (2001) Heat shock protein 27 is a substrate of cGMP-dependent protein kinase in intact human platelets: phosphorylation-induced actin polymerization caused by HSP27 mutants. J Biol Chem 276:7108–7113

Cairns J, Qin S, Philp R, Tan YH, Guy GR (1994) Dephosphorylation of the small heat shock protein Hsp27 in vivo by protein phosphatase 2A. J Biol Chem 269:9176–9183

Candido EP (2002) The small heat shock proteins of the nematode *Caenorhabditis elegans*: structure, regulation and biology. Prog Mol Subcell Biol 28:61–78

Cashikar AG, Duennwald ML, Lindquist SL (2005) A chaperone pathway in protein disaggregation: Hsp26 alters the nature of protein aggregates to facilitate reactivation by Hsp104. J Biol Chem 280:23869–23875

Caspers GJ, Leunissen JA, De Jong WW (1995) The expanding small heat-shock protein family, and structure predictions of the conserved "alpha-crystallin domain". J Mol Evol 40:238–248

Chalova AS, Sudnitsyna MV, Semenyuk PI, Orlov VN, Gusev NB (2014) Effect of disulfide cross-linking on thermal transitions and chaperone-like activity of human small heat shock protein HspB1. Cell Stress Chaperones 19:963–972

Chen SJ, Sun TX, Akhtar NJ, Liang JJ (2001) Oxidation of human lens recombinant alphaA-crystallin and cysteine-deficient mutants. J Mol Biol 305:969–976

Chen J, Feige MJ, Franzmann TM, Bepperling A, Buchner J (2010) Regions outside the alpha-crystallin domain of the small heat shock protein Hsp26 are required for its dimerization. J Mol Biol 398:122–131

Cheng G, Basha E, Wysocki VH, Vierling E (2008) Insights into small heat shock protein and substrate structure during chaperone action derived from hydrogen/deuterium exchange and mass spectrometry. J Biol Chem 283:26634–26642

Chernik IS, Panasenko OO, Li Y, Marston SB, Gusev NB (2004) pH-induced changes of the structure of small heat shock proteins with molecular mass 24/27 kDa (HspB1). Biochem Biophys Res Commun 324(4):1199–1203

Christians ES, Ishiwata T, Benjamin IJ (2012) Small heat shock proteins in redox metabolism: implications for cardiovascular diseases. Int J Biochem Cell Biol 44:1632–1645

Cimdins A, Klinkert B, Aschke-Sonnenborn U, Kaiser FM, Kortmann J, Narberhaus F (2014) Translational control of small heat shock genes in mesophilic and thermophilic cyanobacteria by RNA thermometers. RNA Biol 11:594–608

Clark AR, Naylor CE, Bagneris C, Keep NH, Slingsby C (2011) Crystal structure of R120G disease mutant of human alphaB-crystallin domain dimer shows closure of a groove. J Mol Biol 408:118–134

Clark AR, Lubsen NH, Slingsby C (2012) sHSP in the eye lens: crystallin mutations, cataract and proteostasis. Int J Biochem Cell Biol 44:1687–1697

De Jong WW, Leunissen JA, Voorter CE (1993) Evolution of the alpha-crystallin/small heat-shock protein family. Mol Biol Evol 10:103–126

De Jong WW, Caspers GJ, Leunissen JA (1998) Genealogy of the alpha-crystallin–small heat-shock protein superfamily. Int J Biol Macromol 22:151–162

de Wit NJ, Verschuure P, Kappe G, King SM, de Jong WW, van Muijen GN, Boelens WC (2004) Testis-specific human small heat shock protein HSPB9 is a cancer/testis antigen, and potentially interacts with the dynein subunit TCTEL1. Eur J Cell Biol 83:337–345

Delbecq SP, Klevit RE (2013) One size does not fit all: the oligomeric states of alphaB crystallin. FEBS Lett 587:1073–1080

Delbecq SP, Jehle S, Klevit R (2012) Binding determinants of the small heat shock protein, alphaB-crystallin: recognition of the 'IxI' motif. EMBO J 31:4587–4594

deMiguel N, Braun N, Bepperling A, Kriehuber T, Kastenmuller A, Buchner J, Angel SO, Haslbeck M (2009) Structural and functional diversity in the family of small heat shock proteins from the parasite *Toxoplasma gondii*. Biochim Biophys Acta 1793:1738–1748

Derham BK, Harding JJ (1999) α-crystallin as a molecular chaperone. Prog Retin Eye Res 18:463–509

Ehrnsperger M, Graber S, Gaestel M, Buchner J (1997) Binding of non-native protein to Hsp25 during heat shock creates a reservoir of folding intermediates for reactivation. EMBO J 16:221–229

Ficarro SB, McCleland ML, Stukenberg PT, Burke DJ, Ross MM, Shabanowitz J, Hunt DF, White FM (2002) Phosphoproteome analysis by mass spectrometry and its application to *Saccharomyces cerevisiae*. Nat Biotechnol 20:301–305

Fontaine JM, Sun X, Benndorf R, Welsh MJ (2005) Interactions of HSP22 (HSPB8) with HSP20, alphaB-crystallin, and HSPB3. Biochem Biophys Res Commun 337:1006–1011

Franzmann TM, Wuhr M, Richter K, Walter S, Buchner J (2005) The activation mechanism of Hsp26 does not require dissociation of the oligomer. J Mol Biol 350:1083–1093

Franzmann TM, Menhorn P, Walter S, Buchner J (2008) Activation of the chaperone Hsp26 is controlled by the rearrangement of its thermosensor domain. Mol Cell 29:207–216

Fu X (2014) Chaperone function and mechanism of small heat-shock proteins. Biochim Biophys Acta 46:347–356

Fu L, Liang JJ (2002) Detection of protein-protein interactions among lens crystallins in a mammalian two-hybrid system assay. J Biol Chem 277:4255–4260

Fu X, Shi X, Yan L, Zhang H, Chang Z (2013) In vivo substrate diversity and preference of small heat shock protein IbpB as revealed by using a genetically incorporated photo-cross-linker. J Biol Chem 288:31646–31654

Gaestel M (2002) sHsp-phosphorylation: enzymes, signaling pathways and functional implications. Prog Mol Subcell Biol 28:151–169

Gaestel M, Benndorf R, Hayess K, Priemer E, Engel K (1992) Dephosphorylation of the small heat shock protein hsp25 by calcium/calmodulin-dependent (type 2B) protein phosphatase. J Biol Chem 267:21607–21611

Garrido C, Paul C, Seigneuric R, Kampinga HH (2012) The small heat shock proteins family: the long forgotten chaperones. Int J Biochem Cell Biol 44:1588–1592

Ghosh JG, Shenoy AK Jr, Clark JI (2007) Interactions between important regulatory proteins and human alphaB crystallin. Biochemistry 46:6308–6317

Giese KC, Vierling E (2002) Changes in oligomerization are essential for the chaperone activity of a small heat shock protein in vivo and in vitro. J Biol Chem 277:46310–46318

Gupta R, Srivastava OP (2004) Deamidation affects structural and functional properties of human alphaA-crystallin and its oligomerization with alphaB-crystallin. J Biol Chem 279:44258–44269

Hanazono Y, Takeda K, Oka T, Abe T, Tomonari T, Akiyama N, Aikawa Y, Yohda M, Miki K (2013) Nonequivalence observed for the 16-meric structure of a small heat shock protein, SpHsp16.0, from *Schizosaccharomyces pombe*. Structure 21:220–228

Haslbeck M, Walke S, Stromer T, Ehrnsperger M, White HE, Chen S, Saibil HR, Buchner J (1999) Hsp26: a temperature-regulated chaperone. EMBO J 18:6744–6751

Haslbeck M, Braun N, Stromer T, Richter B, Model N, Weinkauf S, Buchner J (2004) Hsp42 is the general small heat shock protein in the cytosol of *Saccharomyces cerevisiae*. EMBO J 23:638–649

Haslbeck M, Franzmann T, Weinfurtner D, Buchner J (2005a) Some like it hot: the structure and function of small heat-shock proteins. Nat Struct Mol Biol 12:842–846

Haslbeck M, Miess A, Stromer T, Walter S, Buchner J (2005b) Disassembling protein aggregates in the yeast cytosol. The cooperation of Hsp26 with Ssa1 and Hsp104. J Biol Chem 280:23861–23868

Haslbeck M, Kastenmuller A, Buchner J, Weinkauf S, Braun N (2008) Structural dynamics of archaeal small heat shock proteins. J Mol Biol 378:362–374

Hayes D, Napoli V, Mazurkie A, Stafford WF, Graceffa P (2009) Phosphorylation dependence of hsp27 multimeric size and molecular chaperone function. J Biol Chem 284:18801–18807

Hilton GR, Benesch JL (2012) Two decades of studying non-covalent biomolecular assemblies by means of electrospray ionization mass spectrometry. J R Soc Interface 9:801–816

Hochberg GK, Ecroyd H, Liu C, Cox D, Cascio D, Sawaya MR, Collier MP, Stroud J, Carver JA, Baldwin AJ, Robinson CV, Eisenberg DS, Benesch JL, Laganowsky A (2014) The structured core domain of alphaB-crystallin can prevent amyloid fibrillation and associated toxicity. Proc Natl Acad Sci U S A 111:1562–1570

Horwitz J (1992) Alpha-crystallin can function as a molecular chaperone. Proc Natl Acad Sci U S A 89:10449–10453

Horwitz J (2000) The function of alpha-crystallin in vision. Semin Cell Dev Biol 11(1):53–60

Ito H, Okamoto K, Nakayama H, Isobe T, Kato K (1997) Phosphorylation of αB-crystallin in response to various types of stress. J Biol Chem 272:29934–29941

Ito H, Kamei K, Iwamoto I, Inaguma Y, Nohara D, Kato K (2001) Phosphorylation-induced change of the oligomerization state of alpha B-crystallin. J Biol Chem 276:5346–5352

Jakob U, Gaestel M, Engel K, Buchner J (1993) Small heat shock proteins are molecular chaperones. J Biol Chem 268:1517–1520

Jaya N, Garcia V, Vierling E (2009) Substrate binding site flexibility of the small heat shock protein molecular chaperones. Proc Natl Acad Sci U S A 106:15604–15609

Jehle S, Vollmar BS, Bardiaux B, Dove KK, Rajagopal P, Gonen T, Oschkinat H, Klevit RE (2011) N-terminal domain of alphaB-crystallin provides a conformational switch for multimerization and structural heterogeneity. Proc Natl Acad Sci U S A 108:6409–6414

Kantorow M, Piatigorsky J (1998) Phosphorylations of alpha A- and alpha B-crystallin. Int J Biol Macromol 22:307–314

Kato K, Hasegawa K, Goto S, Inaguma Y (1994) Dissociation as a result of phosphorylation of an aggregated form of the small stress protein, hsp27. J Biol Chem 269:11274–11278

Kemp BE, Pearson RB (1990) Protein kinase recognition sequence motifs. Trends Biochem Sci 15:342–346

Kim KK, Kim R, Kim SH (1998) Crystal structure of a small heat-shock protein. Nature 394:595–599

Kokke BP, Leroux MR, Candido EP, Boelens WC, De Jong WW (1998) *Caenorhabditis elegans* small heat-shock proteins Hsp12.2 and Hsp12.3 form tetramers and have no chaperone-like activity. FEBS Lett 433:228–232

Kriehuber T, Rattei T, Weinmaier T, Bepperling A, Haslbeck M, Buchner J (2010) Independent evolution of the core domain and its flanking sequences in small heat shock proteins. FASEB J 24:3633–3642

Laganowsky A, Benesch JL, Landau M, Ding L, Sawaya MR, Cascio D, Huang Q, Robinson CV, Horwitz J, Eisenberg D (2010) Crystal structures of truncated alphaA and alphaB crystallins reveal structural mechanisms of polydispersity important for eye lens function. Protein Sci 19:1031–1043

Lee GJ, Vierling E (2000) A small heat shock protein cooperates with heat shock protein 70 systems to reactivate a heat-denatured protein. Plant Physiol 122:189–198

Lee GJ, Pokala N, Vierling E (1995) Structure and in vitro molecular chaperone activity of cytosolic small heat shock proteins from pea. J Biol Chem 270:10432–10438

Lee GJ, Roseman AM, Saibil HR, Vierling E (1997) A small heat shock protein stably binds heat-denatured model substrates and can maintain a substrate in a folding-competent state. EMBO J 16:659–671

Lelj-Garolla B, Mauk AG (2005) Self-association of a small heat shock protein. J Mol Biol 345:631–642

Lentze N, Narberhaus F (2004) Detection of oligomerisation and substrate recognition sites of small heat shock proteins by peptide arrays. Biochem Biophys Res Commun 325:401–407

Lentze N, Aquilina JA, Lindbauer M, Robinson CV, Narberhaus F (2004) Temperature and concentration-controlled dynamics of rhizobial small heat shock proteins. Eur J Biochem 271:2494–2503

Leroux MR, Ma BJ, Batelier G, Melki R, Candido EP (1997) Unique structural features of a novel class of small heat shock proteins. J Biol Chem 272:12847–12853

Lindner RA, Kapur A, Mariani M, Titmuss SJ, Carver JA (1998) Structural alterations of alpha-crystallin during its chaperone action. Eur J Biochem 25:170–183

Lund AA, Rhoads DM, Lund AL, Cerny RL, Elthon TE (2001) In vivo modifications of the maize mitochondrial small heat stress protein, HSP22. J Biol Chem 276:29924–29929

Maaroufi H, Tanguay RM (2013) Analysis and phylogeny of small heat shock proteins from marine viruses and their cyanobacteria host. PLoS ONE 8:e81207

Matuszewska M, Kuczynska-Wisnik D, Laskowska E, Liberek K (2005) The small heat shock protein IbpA of *Escherichia coli* cooperates with IbpB in stabilization of thermally aggregated proteins in a disaggregation competent state. J Biol Chem 280:12292–12298

McDonald ET, Bortolus M, Koteiche HA, McHaourab HS (2012) Sequence, structure, and dynamic determinants of Hsp27 (HspB1) equilibrium dissociation are encoded by the N-terminal domain. Biochemistry 51:1257–1268

Mchaourab HS, Dodson EK, Koteiche HA (2002) Mechanism of chaperone function in small heat shock proteins. Two-mode binding of the excited states of T4 lysozyme mutants by alphaA-crystallin. J Biol Chem 277:40557–40566

McHaourab HS, Godar JA, Stewart PL (2009) Structure and mechanism of protein stability sensors: chaperone activity of small heat shock proteins. Biochemistry 48:3828–3837

McHaourab HS, Lin YL, Spiller BW (2012) Crystal structure of an activated variant of small heat shock protein Hsp16.5. Biochemistry 51:5105–5112

Merck KB, Horwitz J, Kersten M, Overkamp P, Gaestel M, Bloemendal H, De Jong WW (1993) Comparison of the homologous carboxy-terminal domain and tail of alpha-crystallin and small heat shock protein. Mol Biol Rep 18:209–215

Mogk A, Deuerling E, Vorderwulbecke S, Vierling E, Bukau B (2003) Small heat shock proteins, ClpB and the DnaK system form a functional triade in reversing protein aggregation. Mol Microbiol 50:585–595

Moroni M, Garland D (2001) In vitro dephosphorylation of α-crystallin is dependent on the state of oligomerization. Biochim Biophys Acta 1546:282–290

Mymrikov EV, Seit-Nebi AS, Gusev NB (2012) Heterooligomeric complexes of human small heat shock proteins. Cell Stress Chaperones 17:157–169

Narberhaus F (2002) α-Crystallin-type heat shock proteins: socializing minichaperones in the context of a multichaperone network. Microbiol Mol Biol Rev 66:64–93

Oya-Ito T, Liu BF, Nagaraj RH (2006) Effect of methylglyoxal modification and phosphorylation on the chaperone and anti-apoptotic properties of heat shock protein 27. J Cell Biochem 99:279–291

Pasupuleti N, Gangadhariah M, Padmanabha S, Santhoshkumar P, Nagaraj RH (2010a) The role of the cysteine residue in the chaperone and anti-apoptotic functions of human Hsp27. J Cell Biochem 110:408–419

Pasupuleti N, Matsuyama S, Voss O, Doseff AI, Song K, Danielpour D, Nagaraj RH (2010b) The anti-apoptotic function of human alphaA-crystallin is directly related to its chaperone activity. Cell Death Dis 1:e31

Peschek J, Braun N, Rohrberg J, Back KC, Kriehuber T, Kastenmuller A, Weinkauf S, Buchner J (2013) Regulated structural transitions unleash the chaperone activity of alphaB-crystallin. Proc Natl Acad Sci U S A 110:3780–3789

Posner M, Kiss AJ, Skiba J, Drossman A, Dolinska MB, Hejtmancik JF, Sergeev YV (2012) Functional validation of hydrophobic adaptation to physiological temperature in the small heat shock protein alphaA-crystallin. PLoS ONE 7:e34438

Qi R, Sarbeng EB, Liu Q, Le KQ, Xu X, Xu H, Yang J, Wong JL, Vorvis C, Hendrickson WA, Zhou L, Liu Q (2013) Allosteric opening of the polypeptide-binding site when an Hsp70 binds ATP. Nat Struct Mol Biol 20:900–907

Ratajczak E, Zietkiewicz S, Liberek K (2009) Distinct activities of *Escherichia coli* small heat shock proteins IbpA and IbpB promote efficient protein disaggregation. J Mol Biol 386:178–189

Ratajczak E, Strozecka J, Matuszewska M, Zietkiewicz S, Kuczynska-Wisnik D, Laskowska E, Liberek K (2010) IbpA the small heat shock protein from *Escherichia coli* forms fibrils in the absence of its cochaperone IbpB. FEBS Lett 584:2253–2257

Rogalla T, Ehrnsperger M, Preville X, Kotlyarov A, Lutsch G, Ducasse C, Paul C, Wieske M, Arrigo AP, Buchner J, Gaestel M (1999) Regulation of Hsp27 oligomerization, chaperone function, and protective activity against oxidative stress/tumor necrosis factor α by phosphorylation. J Biol Chem 274:18947–18956

Santhoshkumar P, Sharma KK (2002) Identification of a region in alcohol dehydrogenase that binds to *a*-crystallin during chaperone action. Biochim Biophys Acta 1598:115–121

Satish Kumar M, Mrudula T, Mitra N, Bhanuprakash Reddy G (2004) Enhanced degradation and decreased stability of eye lens alpha-crystallin upon methylglyoxal modification. Exp Eye Res 79:577–583

Sharma KK, Kumar GS, Murphy AS, Kester K (1998) Identification of 1,1′-bi(4-anilino)naphthalene-5,5′-disulfonic acid binding sequences in alpha-crystallin. J Biol Chem 273:15474–15478

Shashidharamurthy R, Koteiche HA, Dong J, Mchaourab HS (2005) Mechanism of chaperone function in small heat shock proteins: dissociation of the HSP27 oligomer is required for recognition and binding of destabilized T4 lysozyme. J Biol Chem 280:5281–5289

Shearstone JR, Baneyx F (1999) Biochemical characterization of the small heat shock protein IbpB from *Escherichia coli*. J Biol Chem 274:9937–9945

Shi J, Koteiche HA, McDonald ET, Fox TL, Stewart PL, McHaourab HS (2013) Cryoelectron microscopy analysis of small heat shock protein 16.5 (Hsp16.5) complexes with T4 lysozyme reveals the structural basis of multimode binding. J Biol Chem 288:4819–4830

Skouri-Panet F, Michiel M, Ferard C, Duprat E, Finet S (2012) Structural and functional specificity of small heat shock protein HspB1 and HspB4, two cellular partners of HspB5: role of the in vitro hetero-complex formation in chaperone activity. Biochimie 94:975–984

Slingsby C, Wistow GJ (2014) Functions of crystallins in and out of lens: roles in elongated and post-mitotic cells. Prog Biophys Mol Biol 115:52–67

Slingsby C, Wistow GJ, Clark AR (2013) Evolution of crystallins for a role in the vertebrate eye lens. Protein Sci 22:367–380

Sobott F, Benesch JL, Vierling E, Robinson CV (2002) Subunit exchange of multimeric protein complexes. Real-time monitoring of subunit exchange between small heat shock proteins by using electrospray mass spectrometry. J Biol Chem 277:38921–38929

Stege GJ, Li GC, Li L, Kampinga HH, Konings AW (1994) On the role of hsp72 in heat-induced intranuclear protein aggregation. Int J Hyperth 10:659–674

Stege GJ, Renkawek K, Overkamp PS, Verschuure P, van Rijk AF, Reijnen-Aalbers A, Boelens WC, Bosman GJ, De Jong WW (1999) The molecular chaperone alphaB-crystallin enhances amyloid beta neurotoxicity. Biochem Biophys Res Commun 262:152–156

Stengel F, Baldwin AJ, Painter AJ, Jaya N, Basha E, Kay LE, Vierling E, Robinson CV, Benesch JL (2010) Quaternary dynamics and plasticity underlie small heat shock protein chaperone function. Proc Natl Acad Sci U S A 107:2007–2012

Stromer T, Ehrnsperger M, Gaestel M, Buchner J (2003) Analysis of the interaction of small heat shock proteins with unfolding proteins. J Biol Chem 278:18015–18021

Stromer T, Fischer E, Richter K, Haslbeck M, Buchner J (2004) Analysis of the regulation of the molecular chaperone Hsp26 by temperature-induced dissociation – the N-terminal domain is important for oligomer assembly and the binding of unfolding proteins. J Biol Chem 279:11222–11228

Studer S, Narberhaus F (2000) Chaperone activity and homo- and hetero-oligomer formation of bacterial small heat shock proteins. J Biol Chem 275:37212–37218

Sugiyama Y, Suzuki A, Kishikawa M, Akutsu R, Hirose T, Waye MM, Tsui SK, Yoshida S, Ohno S (2000) Muscle develops a specific form of small heat shock protein complex composed of MKBP/HSPB2 and HSPB3 during myogenic differentiation. J Biol Chem 275:1095–1104

Sun X, Fontaine JM, Rest JS, Shelden EA, Welsh MJ, Benndorf R (2004) Interaction of human HSP22 (HSPB8) with other small heat shock proteins. J Biol Chem 279:2394–2402

Treweek TM, Ecroyd H, Williams DM, Meehan S, Carver JA, Walker MJ (2007) Site-directed mutations in the C-terminal extension of human alphaB-crystallin affect chaperone function and block amyloid fibril formation. PLoS ONE 2:e1046

Treweek TM, Rekas A, Walker MJ, Carver JA (2010) A quantitative NMR spectroscopic examination of the flexibility of the C-terminal extensions of the molecular chaperones, alphaA- and alphaB-crystallin. Exp Eye Res 91:691–699

van den IJssel PR, Overkamp P, Bloemendal H, de Jong WW (1998) Phosphorylation of αB-crystallin and HSP27 is induced by similar stressors in HeLa cells. Biochem Biophys Res Commun 247:518–523

van Montfort RL, Basha E, Friedrich KL, Slingsby C, Vierling E (2001) Crystal structure and assembly of a eukaryotic small heat shock protein. Nat Struct Biol 8:1025–1030

van Montfort R, Slingsby C, Vierling E (2002) Structure and function of the small heat shock protein/alpha-crystallin family of molecular chaperones. Adv Protein Chem 59:105–156

Veinger L, Diamant S, Buchner J, Goloubinoff P (1998) The small heat-shock protein IbpB from *Escherichia coli* stabilizes stress-denatured proteins for subsequent refolding by a multichaperone network. J Biol Chem 273:11032–11037

Voorter CE, de Haard-Hoekman WA, Roersma ES, Meyer HE, Bloemendal H, de Jong WW (1989) The in vivo phosphorylation sites of bovine αB-crystallin. FEBS Lett 259:50–52

Walter S, Buchner J (2002) Molecular chaperones-cellular machines for protein folding. Angew Chem Int Ed 41:1098–1113

White HE, Orlova EV, Chen S, Wang L, Ignatiou A, Gowen B, Stromer T, Franzmann TM, Haslbeck M, Buchner J, Saibil HR (2006) Multiple distinct assemblies reveal conformational flexibility in the small heat shock protein Hsp26. Structure 14:1197–1204

Wintrode PL, Friedrich KL, Vierling E, Smith JB, Smith DL (2003) Solution structure and dynamics of a heat shock protein assembly probed by hydrogen exchange and mass spectrometry. Biochemistry 42:10667–10673

Wotton D, Freeman K, Shore D (1996) Multimerization of Hsp42p, a novel heat shock protein of *Saccharomyces cerevisiae*, is dependent on a conserved carboxyl-terminal sequence. J Biol Chem 271:2717–2723

Yang H, Huang S, Dai H, Gong Y, Zheng C, Chang Z (1999) The *Mycobacterium tuberculosis* small heat shock protein Hsp16.3 exposes hydrophobic surfaces at mild conditions: conformational flexibility and molecular chaperone activity. Protein Sci 8:174–179

Zavialov A, Benndorf R, Ehrnsperger M, Zav'yalov V, Dudich I, Buchner J, Gaestel M (1998a) The effect of the intersubunit disulfide bond on the structural and functional properties of the small heat shock protein Hsp25. Int J Biol Macromol 22:163–173

Zavialov AV, Gaestel M, Korpela T, Zav'yalov VP (1998b) Thiol/disulfide exchange between small heat shock protein 25 and glutathione. Biochim Biophys Acta 1388:123–132

Chapter 7
Redefining the Chaperone Mechanism of sHsps: Not Just Holdase Chaperones

Heath Ecroyd

Abstract The small heat-shock proteins (sHsps) are molecular chaperones that play a fundamental role in maintaining cellular protein homeostasis (proteostasis) by preventing the aggregation of destabilised proteins. They are generally described as 'holdase' type chaperones since they have the ability to bind partially folded intermediate states of target proteins, in an ATP-independent manner, and, in doing so, they can form high molecular weight complexes with some of them. However, recent work has shown that the ability of sHsps to interact with target proteins is multi-faceted. This review highlights the mechanisms by which sHsps can interact with aggregation-prone target proteins and proposes that they should be considered as protein 'stabilisers' rather than 'holdase' chaperones.

Keywords Protein aggregation • Heat shock proteins • Amyloid fibrils • Proteostasis • Crystallin

Abbreviations

RCM	Reduced and carboxymethylated
RCMα-LA	Reduced and carboxymethylated α-lactalbumin
sHsp(s)	small Heat shock protein(s).
αA-C	αA-crystallin
αB-C	αB-crystallin
α-LA	α-lactalbumin

H. Ecroyd (✉)
Illawarra Health and Medical Research Institute and School of Biological Sciences, University of Wollongong, Wollongong 2522, NSW, Australia
e-mail: heath_ecroyd@uow.edu.au

R.M. Tanguay, L.E. Hightower (eds.), *The Big Book on Small Heat Shock Proteins*, Heat Shock Proteins 8, DOI 10.1007/978-3-319-16077-1_7

7.1 Introduction

Proteostasis refers to the functional maintenance of the proteome and encompasses all the processes that balance protein synthesis, protein folding, protein trafficking and the degradation of proteins in an organism (Balch et al. 2008; Kikis et al. 2010; Vendruscolo et al. 2011; Wyatt et al. 2013). This balance can be impacted by a number of factors, including protein mutations (either inherited or the result of errors during protein synthesis), protein misfolding, protein unfolding, or modifications to proteins induced by stress. Protein misfolding can lead to subsequent aggregation, an often irreversible process which may result in proteotoxicity due to the accumulation of cytotoxic protein species (Hightower 1991; Guijarro et al. 1998; Bucciantini et al. 2002; Kayed et al. 2003). Protein misfolding and aggregation has been implicated in various diseases ranging from neurodegenerative disorders to cancer (Bucciantini et al. 2002; Kayed et al. 2003; Morimoto and Cuervo 2009). All cells (and organisms) have various control pathways in place that act to maintain proteostasis, and therefore reduce the accumulation of damaged, dysfunctional and/or toxic proteins (Cohen et al. 2006; Vendruscolo et al. 2011; Wyatt et al. 2013).

Of the quality control mechanisms responsible for combating protein aggregation and disease, molecular chaperones are arguably the most important, as they form a cell's first line of defence against protein misfolding (Hohfeld et al. 2001; Macario and Conway de Macario 2005; Muchowski and Wacker 2005; Voisine et al. 2010; Hartl et al. 2011). Molecular chaperones represent a structurally diverse class of proteins that support correct protein folding by preventing unproductive interactions (Barral et al. 2004; Hartl et al. 2011). More specifically, chaperones recognise and interact with the exposed hydrophobic surfaces of aggregation-prone protein intermediates so as to bind and sequester them from the crowded cell milieu. Chaperones are often classified as either 'holdases' or 'foldases' based on their mechanism of action. This chapter focusses on the small heat-shock proteins (sHsps), one such class of molecular chaperones that have been shown to be capable of suppressing protein aggregation. The sHsps are generally described as 'holdase' type chaperones (Ehrnsperger et al. 1997; Beissinger and Buchner 1998), however, after considering the multi-faceted nature by which sHsp chaperones can interact with aggregating proteins, it is proposed that protein 'stabilisers' is a more fitting term to describe their chaperone action.

7.2 The Protein Folding and Unfolding Pathways

Before considering the mechanism of chaperone action of sHsps, it is pertinent to discuss the salient features of protein misfolding leading to aggregation and precipitation. Immediately following their synthesis on the ribosome, proteins are unfolded (U) and many must first fold into their native (functional) state (N) before gaining their biological function (Fig. 7.1) (Carver et al. 2003). An exception to this are the class of intrinsically unfolded/disordered proteins that do not fold, however, some

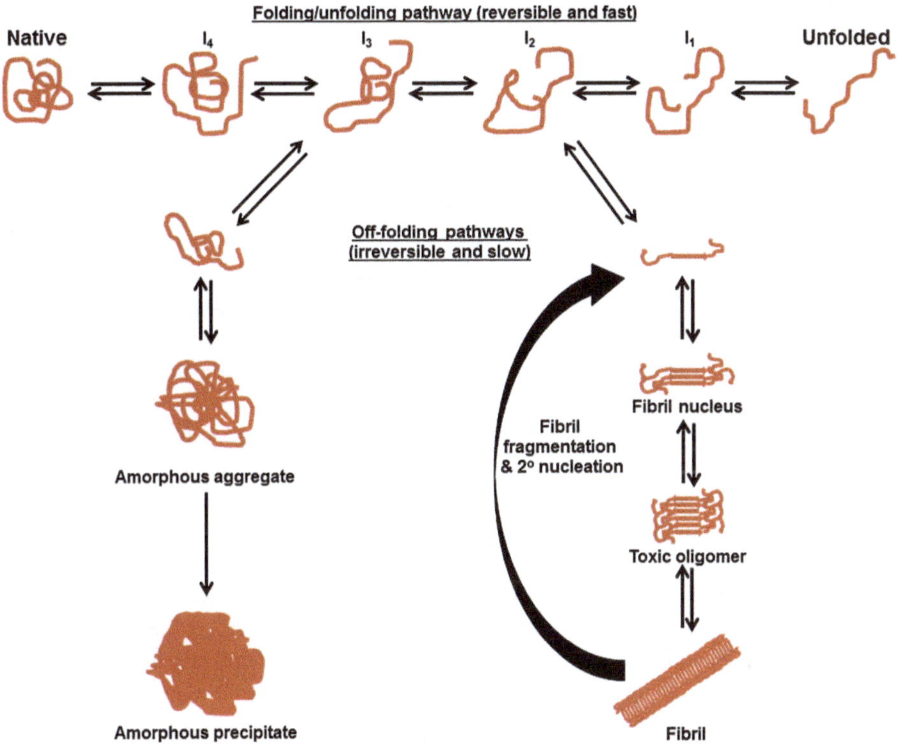

Fig. 7.1 The protein folding/unfolding and off-folding pathways. An unfolded protein (U) folds via a variety of intermediate states (I_1, I_2, I_3, …) into its native state (N) via the fast and reversible folding/unfolding pathway. Under conditions of stress or when required (e.g. to cross a membrane), folded proteins can unfold along the same pathway. However, if the partially folded intermediate states linger for too long, they can self-associate and enter the off-folding pathways, comprising either the disordered (amorphous) aggregation pathway, which produces irreversibly precipitated, amorphously aggregated species, or the amyloid fibril-forming pathway which, via the formation of small, soluble nuclei, leads to insoluble, highly ordered cross β-sheet fibrils (Adapted from (Carver et al. 2003))

of these may adopt structure once they bind to targets (Dunker et al. 2002; Uversky 2002). Proteins that fold do so quickly (in the range of a few μsec to msec for most proteins) via the folding/unfolding pathway. Folding typically involves the formation of a series of partially folded intermediate states (I_1, I_2, I_3,…) that have varying degrees of structure. These partially folded intermediate states represent proteins in which the process of hydrophobic collapse (that dictates their folding) has commenced (Dobson 2001, 2003; Hartl et al. 2011). However, since folding is not complete, these partially folded intermediate states expose hydrophobic regions to solution and are therefore inherently unstable and prone to self-association (via hydrophobic-mediated interactions), the first step towards aggregation. The prolonged presence of partially folded intermediates can be exacerbated if conditions are not conducive to folding (e.g. in the presence of stressors such as increases in temperature, changes in pH or the presence of oxidants or reductants). Moreover, such stressors can also break the bonds that maintain the native structure of proteins, causing them to unfold (via the folding/unfolding pathway) and populate these partially folded states.

Protein aggregation occurs when partially folded intermediate states mutually associate (via hydrophobic interactions) causing the protein to leave the folding/unfolding pathway and enter an off-folding pathway (Fig. 7.1) (Carver et al. 2003; Ecroyd and Carver 2008). There are two off-folding pathways that a protein may enter: the disordered (amorphous) or ordered (amyloid fibril) aggregation pathways (Chiti and Dobson 2006; Ecroyd and Carver 2008). Disordered aggregation occurs when intermediates randomly associate, through a three-dimensional process, producing unstructured (amorphous) aggregates which become insoluble when a critical size is reached (Stranks et al. 2009). Alternatively, intermediates that aggregate via an ordered mechanism (mediated though β-strand interactions) form soluble protofibrils, which then laterally associate to form highly structured, β-sheet rich amyloid fibrils (Dobson 2001, 2002; Jiménez et al. 2002). In both cases aggregation occurs through a nucleation-dependent mechanism in which the formation of the nucleus is the rate-determining step in most cases (Harper and Lansbury 1997; Knowles et al. 2009; Stranks et al. 2009), but not all (Colon and Kelly 1992; Quintas et al. 1999; Ecroyd et al. 2008). The aggregates that are formed eventually precipitate from solution. Specific factors that govern whether a protein aggregates to form amorphous or fibrillar aggregates include (1) the rate at which aggregation occurs (with faster aggregation typically favouring the formation of amorphous aggregates), and (2) the nature of the precursor partially folded intermediate (the more destabilised and less structured intermediates typically being precursors to amorphous type aggregates).

Both forms of protein aggregates are associated with disease; however, amyloid fibril formation has garnered the most attention of late due to its links with neurological disorders, such as Alzheimer's disease, Parkinson's disease, Huntington's disease, Creutzfeldt-Jakob disease and Amytrophic Lateral Sclerosis, and cancers (Dobson 1999; Stefani and Dobson 2003; Barral et al. 2004; Lee et al. 2011). It is generally accepted that small, soluble, prefibrillar oligomers formed as a result of fibrillar aggregation are cytotoxic and therefore responsible for the neuron loss seen in the neurodegenerative disorders (Walsh et al. 2002; Caughey and Lansbury 2003; Kayed et al. 2003; Stefani 2012). Of particular interest, recent work has indicated that the majority of these cytotoxic species arise from the mature fibrils themselves, due to secondary nucleation events and fibril fragmentation, rather than from the association of monomers during the initial aggregation process (Knowles et al. 2009; Cremades et al. 2012). Thus, whilst mature fibrils themselves are considered less cytotoxic, they can be the greatest source of toxic oligomers.

7.3 The Mechanism of Action of sHsp Chaperone Proteins

7.3.1 Interactions with Partially Folded Intermediate States

The original description of the chaperone action of α-crystallin by Horwitz (1992), was an experimental verification of Ingolia and Craig's earlier classification of α–crystallin as a member of the sHsp molecular chaperone family (Ingolia and Craig 1982). Since then at least five of the ten individual human sHsps, i.e. Hsp27 (HSB1), αA-c (HSP4), αB-c (HSB5), Hsp20 (HSPB6), and Hsp22 (HSPB8), have been reported to have the ability to act in a chaperone-like manner to suppress the aggregation of target proteins in vitro (e.g. (Jakob et al. 1993; Carver et al. 1994; Smulders et al. 1996; van de Klundert et al. 1998; Bukach et al. 2004; Kim et al. 2004). However, this chaperone activity has not been definitively demonstrated for all human sHsps (e.g. HSPB10) and some sHsps may play roles in other proteostasis pathways (e.g. apoptosis, protein degradation) independent of acting as chaperones (e.g. Carra et al. 2008, 2012; Crippa et al. 2010; Ahner et al. 2013).

With regards to those sHsps that have been shown to act as chaperones, whilst most studies have used amorphous aggregation models to assess their chaperone activity, work using targets that undergo fibrillar aggregation are becoming more prevalent. Many studies have indicated that, in preventing aggregation, sHsp chaperones form stable complexes with target proteins (e.g. (Rao et al. 1993; Carver et al. 2002; Basha et al. 2004; Regini et al. 2010). Based on such findings the most widely accepted mechanism of sHsp chaperone action is that they bind to exposed hydrophobic residues on partially folded protein intermediates to sequester them into soluble high molecular mass complexes, in a 'holdase' type activity (Carver et al. 1995; Beissinger and Buchner 1998). Since sHsps have no inherent capacity to refold bound substrates (i.e. their activity is ATP-independent) this complex has been described as a reservoir-of-intermediates (Ehrnsperger et al. 1997). The sHsps are thought to maintain bound intermediates in a refolding competent state, until cellular conditions are permissive for ATP-dependent chaperones (e.g. Hsp70) to re-fold them back to their native conformation (Ehrnsperger et al. 1997). There is also evidence that some sHsps, such as Hsp27 (Ahner et al. 2013) and Hsp22 (Crippa et al. 2010), may shuttle intermediates to the proteosome/autophagosome system to facilitate their degradation.

The concept of sHsps as "holdase" chaperones now pervades most general reviews and textbooks that discuss their chaperone mechanism; however, this description does not fully describe the multi-faceted nature by which these chaperones can interact with target proteins to prevent their aggregation. It is well recognised that sHsp chaperones can also interact through highly dynamic, albeit weak and transient, interactions with target proteins to prevent their aggregation. For example, in preventing fibril formation by apolipoprotein C-II (apoC-II), αB-c acts (at very low sub-stoichiometric ratios) transiently and dynamically with the partially folded intermediates of the aggregation-prone monomer rather than by forming stable complexes with it (Hatters et al. 2001). The interaction of αB-c with aggregation-prone monomeric apoC-II stabilizes it, presumably so that it may refold back to its native state via the reversible on-folding pathway. Thus, taking into account that sHsps do not always form complexes with target proteins in preventing their aggregation, a more apt description of sHsp chaperones is as protein 'stabilisers' rather than 'holdases'.

Rather than the mode of interaction between sHsp chaperones and target proteins being dependent on the target protein per se, we have recently shown that it is dictated by the nature of the partially folded intermediate that is the precursor to aggregation (Kulig and Ecroyd 2012). Using bovine α-lactalbumin (α-LA) as a model target protein, we developed conditions whereby it could be induced to aggregate down either the amorphous or fibrillar off-folding pathway under conditions of physiological pH and temperature (i.e. pH 7.0 and 37 °C) and in comparable buffers. Thus, when α-LA was reduced it aggregated amorphously, whereas a reduced and carboxymethylated form (RCMα-LA) aggregated to form amyloid fibrils (Kulig and Ecroyd 2012). In doing so, we demonstrated that in preventing aggregation, αB-c forms high molecular mass complexes with highly disordered, precipitation bound intermediates that expose a high degree of hydrophobicity to solution (Fig. 7.2). In contrast, αB-c acts via weak, dynamic and transient interactions to suppress the aggregation of more stable, structured intermediates which expose less hydrophobicity to solution (Kulig and Ecroyd 2012).

As an alternative approach to evaluating sHsp binding to target proteins, in a series of elegant papers, McHaourab and colleagues assessed the ability of the sHsps αB-c, αA-c and Hsp27 to bind destabilised targets using a series of T4 lysozyme mutants (McHaourab et al. 2002; Koteiche and McHaourab 2003; Sathish et al. 2003; Shashidharamurthy et al. 2005). These studies led to a proposed mechanism of sHsp chaperone action in which the equilibrium between the folding and unfolding of the target protein (in this case T4 lysozyme) is coupled to the equilibrium between dissociated (dimeric) and oligomeric states of the sHsp. Thus, they concluded that these sHsps bind destabilised T4 lysozyme in two modes, either with high or low affinity. These may be analogous to the strong (holdase) and weak (stabiliser) binding mechanisms seen using other target proteins.

In the context of the cell, these dynamic interactions are clearly of benefit as it enables the levels of chaperone-competent sHsps to be maintained (rather than depleted) under basal conditions. Thus, it is considered that the weak transient interactions with target proteins is the mechanism that likely predominates in cells not subjected to stress, since these conditions are not conducive to large-scale protein destabilisation. Upon exposure to a stress, activation of the heat-shock response leads to dramatic up-regulation of the expression of some sHsps (de Thonel et al. 2012), and would also lead to complex formation (and stabilization) with more destabilised target proteins. Accordingly, it is envisaged that sHsps, acting as stability sensors in the cell, only mediate stable high-molecular-mass complex formation with protein intermediates when binding is more energetically favourable than refolding.

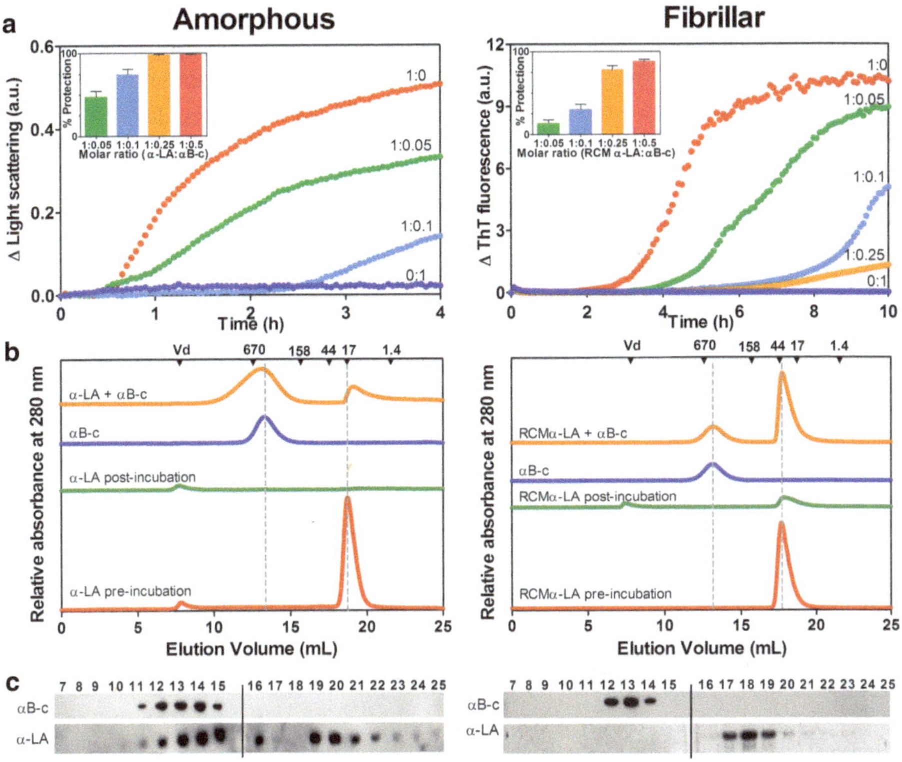

Fig. 7.2 αB-c forms a stable complex with a-lactalbumin to prevent its amorphous aggregation, but uses weak transient interactions to prevent its fibril formation. (**a**) To induce amorphous aggregation α-LA (100 μM) was incubated at 37 °C in 50 mM phosphate buffer (pH 7.0) containing 100 mM KCl, 5 mM EDTA and 20 mM DTT. To induce fibril formation a reduced and carboxymethylated form of a-lactalbumin (RCMα-LA) incubated at 37 °C in 50 mM phosphate buffer containing 100 mM KCl and 10 mM $MgCl_2$. The formation of aggregates was monitored by an increase in relative light scatter at 360 nm (amorphous aggregation) or ThT fluorescence at 490 nm (fibrillar aggregation) in the absence or presence of αB-c. The change in light scattering or ThT fluorescence for each molar ratio tested (reported as α-LA: αB-c) was monitored and is shown in arbitrary units (a.u.). The results shown are representative of three independent experiments. Insets show the percentage protection afforded by αB-c against α-LA aggregation at various molar ratios (α-LA:αB-c). Mean ± SEM ($n=3$). (**b**) Size exclusion profiles of α-LA/RCMα-LA (100 μM), αB-c (50 μM) and soluble post aggregation samples of α-LA/RCMα-LA (100 μM) in the absence and presence of αB-c (50 μM). The elution volume for standards (kDa) used during calibration are shown at the top of the graph. (**c**) Immunoblot of the eluate fractions collected from the size exclusion column. A sample from every second fraction (fraction size was 0.5 mL) collected between 7 and 25 mL was loaded onto a SDS-PAGE gel, transferred to nitrocellulose and blotted with an anti-αB-c or anti-α-LA antibody. Both αB-c and α-LA elute in a high molecular mass complex when α-LA is incubated under conditions that induce its amorphous aggregation, but no complex is evident when a-LA is incubated under conditions which induce it to form amyloid fibrils (Adapted from Kulig and Ecroyd 2012)

With respect to the overall efficacy of sHsp chaperones to inhibit protein aggregation, it is well established that the kinetics of the aggregation process are a significant factor, i.e. these sHsps are most effective against slowly-aggregating target proteins (Lindner et al. 2001; Carver et al. 2002). It is concluded this is because sHsps interact with partially folded states of target proteins that have left the folding pathway (Treweek et al. 2000). Thus, the longer these states persist, the more likely that sHsps are able to interact and stabilise them. In support of this, the limited work that has been done in this area suggests that the longer the lag phase of fibril formation (representing the time taken to form stable nuclei), the more effective the chaperone ability of αB-c. For example, αB-c is more effective at inhibiting fibril formation by wild-type α-synuclein compared to A53Tα-synuclein (the latter forms fibrils more rapidly) (Rekas et al. 2004). As well, when the rate of α-synuclein fibril formation is increased through the addition of dextran (which acts as a molecular crowding agent), the effectiveness of αB-c as a chaperone decreases (Ecroyd, unpublished data). Thus, on the basis of aggregation kinetics, it is concluded that sHsps are better suited to prevent fibril formation as compared to amorphous aggregation, since the formation of fibrils is typically a much slower process than amorphous aggregation.

7.3.2 Interactions with Amyloid Fibrils

The vast majority of studies that have tested the in vitro chaperone action of sHsps (and indeed other chaperone proteins) have involved addition of the protein prior to aggregation commencing. However, the design of such studies does not address the effect(s) sHsps have on the latter stages of aggregation. This aspect is important to consider since, in vivo, levels of sHsps in the cell increase after aggregation has commenced as a result of the activation of the stress response (Shinohara et al. 1993; Renkawek et al. 1994). Initial studies investigating the ability of αB-c to inhibit α-synuclein fibril formation (which is associated with Parkinson's disease) demonstrated that, when the chaperone was added after aggregation had commenced (i.e. during the elongation phase of aggregation), fibril formation was halted (Rekas et al. 2004). The authors' interpretation of these findings were that this was due to αB-c interacting with monomeric species to prevent their association with the growing fibril (Rekas et al. 2004); however, subsequent studies have demonstrated that αB-c can bind directly to pre-formed α-synuclein fibrils and, in doing so, inhibit their further growth (Waudby et al. 2010). It is now well established that αB-c binds along the lateral surface of mature amyloid fibrils formed by α-synuclein (Rekas et al. 2007; Waudby et al. 2010), Aβ peptides (Shammas et al. 2011), and apoC-II (Binger et al. 2013) and therefore appears to have a generic fibril-binding activity (Fig. 7.3).

So, why do sHsps bind to mature fibrils? We have recently conducted work to address this question using αB-c and fibrils formed from apoC-II (Binger et al. 2013). We found that by binding to apoC-II fibrils, αB-c stabilises them, preventing

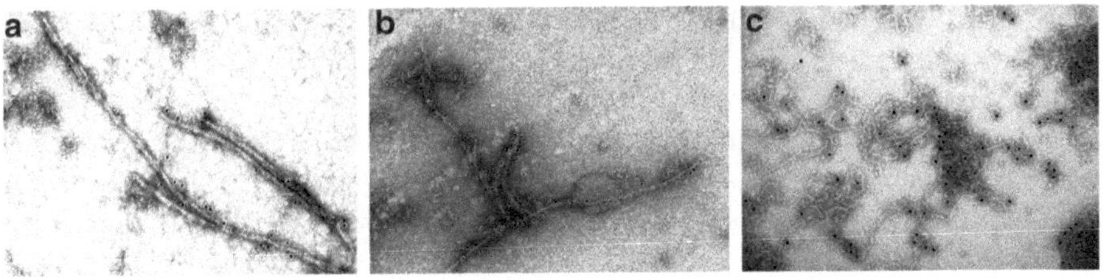

Fig. 7.3 The sHsp αB-c binds to mature amyloid fibrils. Transmission electron micrographs of fibrils formed from (**a**) α-syn, (**b**) Aβ1-42, and (**c**) apoC-II following incubation with αB-c and subsequent immunogold labelling of the bound αB-c

their (dilution-induced) fragmentation, and causes them to associate (tangle) into larger species, reminiscent of protein inclusions (Binger et al. 2013). Thus, by coating these fibrils αB-c may inhibit fibril fragmentation and prevent secondary nucleation events that can occur on the fibril surface. This is significant because both of these processes (i.e. fragmentation and secondary nucleation) can be the main source of cytotoxic oligomers that are associated with amyloid fibril formation (Knowles et al. 2009). αB-c's fibril binding-activity therefore appears to represent another mechanism by which it (and other chaperone-active sHsps) minimises the toxic effects of protein aggregation in cells. Moreover, it helps to rationalise why some sHsps are found in the protein deposits associated with diseases such as Alzheimer's disease and Parkinson's disease (Lowe et al. 1992; Shinohara et al. 1993; Renkawek et al. 1994; McLean et al. 2002; Pountney et al. 2005; Wilhelmus et al. 2006); i.e. their co-localization with amyloid deposits likely results from a specific interaction with the fibrillar aggregate. Furthermore, this interaction may actually help mediate the coalescing of the aggregates into larger protein deposits. There is evidence that protein inclusions are protective in the cell; studies have reported a lack of correlation between amyloid deposition in the brain and disease progression (Schmitz et al. 2004; Rabinovici and Jagust 2009), and in cell culture models of protein aggregation multiple inclusions can be present in otherwise 'healthy' cells (Arrasate et al. 2004; Gong et al. 2008; Ormsby et al. 2013). Thus, by promoting the formation of inclusions sHsps may assist in shielding the cell from toxic oligomers. Lastly, it was recently shown that αB-c promotes the dissociation of potentially toxic β2-microglobulin oligomers into monomers, highlighting yet another role these chaperones may have in cells to directly protect them from the adverse effects of protein aggregation (Esposito et al. 2013).

Of course, apart from their ability to inhibit aggregation, sHsps may protect cells from the toxicity associated with fibrillar protein aggregation in other ways. For example, both Hsp27 and αB-c increase the resistance of cells to oxidative stress (e.g. Mehlen et al. 1995; Preville et al. 1999; Shin et al. 2009; Xu et al. 2013) and can inhibit apoptosis via a range of interactions with partner proteins involved in cell death pathways (Acunzo et al. 2012).

In summary, recent work on the interaction of sHsps with amyloid fibrils has provided new insights into the potential roles of sHsps in preventing amyloid toxicity, i.e. stabilizing fibrils to prevent their fragmentation into toxic oligomeric species, mediating amyloid plaque formation, and driving the dissociation of fibrils to non-toxic monomers. It is early days in this area of research, however, the prospect that there are likely to be multiple mechanisms through which sHsps can interact with target protein undergoing amyloid fibril type aggregation is exciting. Much remains to be established, but further work in this area promises to uncover some fundamental activities of these chaperone proteins that, up until recently, had not been fully appreciated. Two of the most pressing questions in this area are (i) can all sHsps capable of binding to fibrils? and (ii) are larger oligomers or dissociated species, or a combination of these, responsible for this fibril-binding activity?

7.3.3 A Revised Model of sHsp Chaperone Action

In light of the mechanisms discussed above, a revised model of sHsp activity is proposed here which extends previous models that have focused solely on the interaction of sHsps with partially folded intermediate states of target proteins (Fig. 7.4). The revised model highlights the multi-faceted role that sHsps play in preventing the toxic effects of protein aggregation by incorporating their interaction with soluble, partially folded states and their ability to bind to fibrillar species. The interaction with partially folded intermediates can be via high affinity binding, resulting in the formation of a stable sHsp-target protein complex, or through low affinity, transient and dynamic interactions that facilitate the intermediate re-entering the folding/unfolding pathway and thus reattaining its native state.

Whilst this revised model of sHsps chaperone action (Fig. 7.3) recognises our growing appreciation of the mechanisms by which sHsps interact with aggregating proteins, critical aspects still remain to be established. It is amazing that some of these key gaps in our knowledge were first raised by Horwitz in his seminal paper describing the chaperone action of α-crystallin (Horwitz 1992). For example, we still do not know whether the chaperone activity of sHsps depends on its oligomeric size or sub-unit exchange. Whilst chaperone active species are normally depicted as dissociated (dimeric) subunits (and large oligomers as 'reservoirs' of these species), there is no definitive evidence to-date that these alone are exclusively chaperone active. Moreover, some studies have indicated that there is no correlation between the concentration of sub-oligomeric species (or sub-unit exchange rate) and apparent chaperone activity (Aquilina et al. 2005). Also, glutaraldehyde cross-linked α-crystallin, which is incapable of subunit exchange, retains significant chaperone activity in vitro (Augusteyn 2004). Thus, the relationship between oligomeric size and chaperone activity remains enigmatic.

Little is known about the stoichiometry of mammalian sHsp-target protein complexes. To-date, efforts to study, in precise detail, the manner by which sHsps bind to target proteins to form complexes have been hampered by the large, polydisperse and dynamic nature of sHsp oligomers and the low abundance of individual species in these heterogeneous samples. This problem is further confounded by the use of

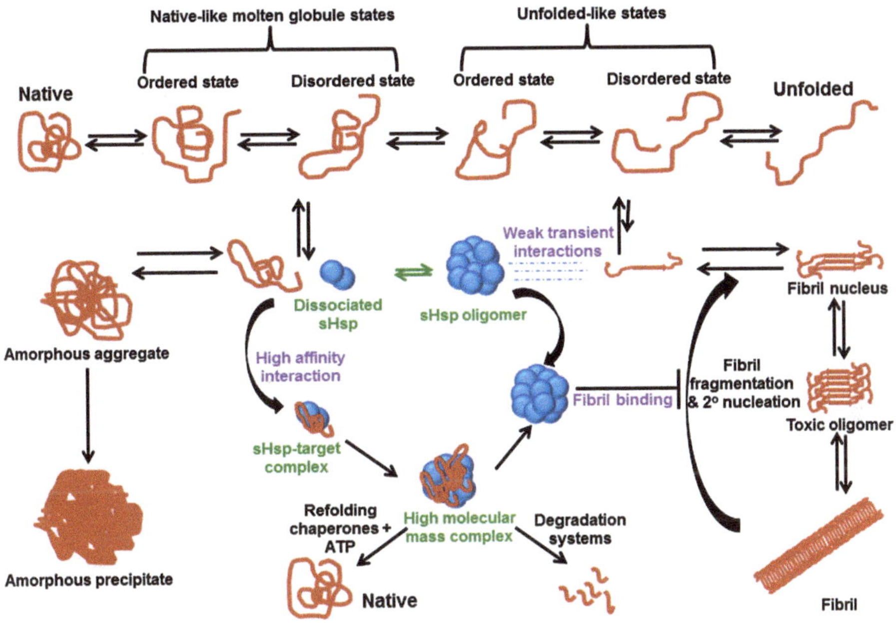

Fig. 7.4 The mechanism of action of sHsp chaperones. Multiple partially folded protein intermediate states populate the folding/unfolding pathway of a protein. The mechanism by which sHsps, such as αB-c, prevent target protein aggregation (either amorphous or fibrillar) is dictated by the conformational stability and exposed hydrophobicity of the precursor protein intermediates. High affinity interactions occur with highly destabilised intermediates (which exceed the threshold of binding) and these are sequestered into stable high molecular mass complexes. Target proteins in these complexes can be re-folded through the action of other ATP-dependent chaperones or shuttled for degradation. Alternatively, weak, transient interactions occur with more stable protein intermediates, which re-directs them back to the folding pathway so as to facilitate their re-folding. sHsps can also interact with pre-fibrillar and fibrillar aggregates formed by target proteins. By binding to these species sHsps stabilise them, preventing their further elongation, fibril fragmentation and secondary nucleation events along their surface

'bulk' averaging techniques that can mask the presence of rare species formed in such dynamic systems. Work using native mass spectrometry and a monodisperse (dodecameric) sHsp from pea (Hsp18.1) demonstrated the remarkable heterogeneity of the complexes it forms with target proteins (over 300 different stoichiometries of interaction!) (Stengel et al. 2010). Thus, the possible number of stoichiometries of sHsp-target protein complexes formed by polydisperse mammalian sHsps is expected to be enormous. Recent advances in methods that enable the study of dynamic heterogeneous protein ensembles, such as nanospray electrospray ionisation mass spectrometry (Benesch et al. 2006; Benesch and Ruotolo 2011; Stengel et al. 2012) and single molecule fluorescence techniques (Laurence et al. 2007; van Oijen 2011; Loveland et al. 2012), augur fundamental advances in our understanding of the mechanisms by which sHsps interact with their target proteins.

Finally, the models of sHsp chaperone action are predominately based on in vitro studies; little is known about how they relate to the chaperone activity of sHsps inside the crowded environment of the cell. For example, nothing is known about the precise concentrations of sHsps inside a cell, let alone cellular compartments such as the nucleus versus the cytoplasm. Since the oligomeric state of sHsps can be concentration dependent, e.g. Hsp27 dissociates into smaller dimers at low concentrations in vitro (Kato et al. 1994; Rogalla et al. 1999; Shashidharamurthy et al. 2005), it therefore remains to be elucidated which oligomeric forms are relevant in a cellular context. Moreover, the relative importance of each of these chaperone mechanisms in the cellular function of sHsp chaperones remains to be established.

7.4 Conclusions

Whilst the sHsps are often described as holdase types chaperones, this does not fully describe the multi-faceted nature in which they interact with aggregating target proteins. For this reason it is proposed that they should be considered protein 'stabilisers' rather than 'holdase' chaperones. Whilst we have made substantial advances in our understanding of how these (often forgotten) chaperones act to protect cells in times of cellular stress, much remains to be done. Considering the instrumental role the sHsp molecular chaperones play in preserving proteostasis and suppressing protein aggregation, ongoing research to further elucidate the mechanisms by which they interact with aggregating target proteins is essential. This is particularly true if the full potential of these chaperone proteins is to be exploited as an avenue for developing novel therapeutics against protein aggregation-based diseases.

References

Acunzo J, Katsogiannou M, Rocchi P (2012) Small heat shock proteins HSP27 (HspB1), alphaB-crystallin (HspB5) and HSP22 (HspB8) as regulators of cell death. Int J Biochem Cell Biol 44(10):1622–1631

Ahner A, Gong X, Schmidt BZ, Peters KW, Rabeh WM, Thibodeau PH, Lukacs GL, Frizzell RA (2013) Small heat shock proteins target mutant cystic fibrosis transmembrane conductance regulator for degradation via a small ubiquitin-like modifier-dependent pathway. Mol Biol Cell 24(2):74–84

Aquilina JA, Benesch JL, Ding LL, Yaron O, Horwitz J, Robinson CV (2005) Subunit exchange of polydisperse proteins: mass spectrometry reveals consequences of alphaA-crystallin truncation. J Biol Chem 280(15):14485–14491

Arrasate M, Mitra S, Schweitzer ES, Segal MR, Finkbeiner S (2004) Inclusion body formation reduces levels of mutant huntingtin and the risk of neuronal death. Nature 431(7010):805–810

Augusteyn RC (2004) Dissociation is not required for alpha-crystallin's chaperone function. Exp Eye Res 79(6):781–784

Balch WE, Morimoto RI, Dillin A, Kelly JW (2008) Adapting proteostasis for disease intervention. Science 319(5865):916–919

Barral JM, Broadley SA, Schaffar G, Hartl FU (2004) Roles of molecular chaperones in protein misfolding diseases. Semin Cell Dev Biol 15(1):17–29

Basha E, Lee GJ, Breci LA, Hausrath AC, Buan NR, Giese KC, Vierling E (2004) The identity of proteins associated with a small heat shock protein during heat stress in vivo indicates that these chaperones protect a wide range of cellular functions. J Biol Chem 279(9):7566–7575

Beissinger M, Buchner J (1998) How chaperones fold proteins. Biol Chem 379(3):245–259

Benesch JL, Ruotolo BT (2011) Mass spectrometry: come of age for structural and dynamical biology. Curr Opin Struct Biol 21(5):641–649

Benesch JLP, Aquilina JA, Ruotolo BT, Sobott F, Robinson CV (2006) Tandem mass spectrometry reveals the quaternary organization of macromolecular assemblies. Chem Biol 13(6):597–605

Binger KJ, Ecroyd H, Yang S, Carver JA, Howlett GJ, Griffin MD (2013) Avoiding the oligomeric state: AlphaB-crystallin inhibits fragmentation and induces dissociation of apolipoprotein C-II amyloid fibrils. FASEB J 27(3):1214–1222

Bucciantini M, Giannoni E, Chiti F, Baroni F, Formigli L, Zurdo J, Taddei N, Ramponi G, Dobson CM, Stefani M (2002) Inherent toxicity of aggregates implies a common mechanism for protein misfolding diseases. Nature 416(6880):507–511

Bukach OV, Seit-Nebi AS, Marston SB, Gusev NB (2004) Some properties of human small heat shock protein Hsp20 (HspB6). Eur J Biochem 271(2):291–302

Carra S, Seguin SJ, Lambert H, Landry J (2008) HspB8 chaperone activity toward poly(Q)-containing proteins depends on its association with Bag3, a stimulator of macroautophagy. J Biol Chem 283(3):1437–1444

Carra S, Crippa V, Rusmini P, Boncoraglio A, Minoia M, Giorgetti E, Kampinga HH, Poletti A (2012) Alteration of protein folding and degradation in motor neuron diseases: Implications and protective functions of small heat shock proteins. Prog Neurobiol 97(2):83–100

Carver JA, Aquilina JA, Cooper PG, Williams GA, Truscott RJ (1994) Alpha-crystallin: molecular chaperone and protein surfactant. Biochim Biophys Acta 1204(2):195–206

Carver JA, Guerreiro N, Nicholls KA, Truscott RJ (1995) On the interaction of alpha-crystallin with unfolded proteins. Biochim Biophys Acta 1252(2):251–260

Carver JA, Lindner RA, Lyon C, Canet D, Hernandez H, Dobson CM, Redfield C (2002) The interaction of the molecular chaperone alpha-crystallin with unfolding alpha-lactalbumin: a structural and kinetic spectroscopic study. J Mol Biol 318(3):815–827

Carver JA, Rekas A, Thorn DC, Wilson MR (2003) Small heat-shock proteins and clusterin: intra- and extracellular molecular chaperones with a common mechanism of action and function? IUBMB Life 55(12):661–668

Caughey B, Lansbury PT (2003) Protofibrils, pores, fibrils, and neurodegeneration: separating the responsible protein aggregates from the innocent bystanders. Annu Rev Neurosci 26:267–298

Chiti F, Dobson CM (2006) Protein misfolding, functional amyloid, and human disease. Annu Rev Biochem 75:333–366

Cohen E, Bieschke J, Perciavalle RM, Kelly JW, Dillin A (2006) Opposing activities protect against age-onset proteotoxicity. Science 313(5793):1604–1610

Colon W, Kelly JW (1992) Partial denaturation of transthyretin is sufficient for amyloid fibril formation in vitro. Biochemistry 31(36):8654–8660

Cremades N, Cohen SI, Deas E, Abramov AY, Chen AY, Orte A, Sandal M, Clarke RW, Dunne P, Aprile FA, Bertoncini CW, Wood NW, Knowles TP, Dobson CM, Klenerman D (2012) Direct observation of the interconversion of normal and toxic forms of alpha-synuclein. Cell 149(5):1048–1059

Crippa V, Sau D, Rusmini P, Boncoraglio A, Onesto E, Bolzoni E, Galbiati M, Fontana E, Marino M, Carra S, Bendotti C, De Biasi S, Poletti A (2010) The small heat shock protein B8 (HspB8) promotes autophagic removal of misfolded proteins involved in amyotrophic lateral sclerosis (ALS). Hum Mol Genet 19(17):3440–3456

de Thonel A, Le Mouel A, Mezger V (2012) Transcriptional regulation of small HSP-HSF1 and beyond. Int J Biochem Cell Biol 44(10):1593–1612

Dobson CM (1999) Protein misfolding, evolution and disease. Trends Biochem Sci 24(9):329–332

Dobson CM (2001) The structural basis of protein folding and its links with human disease. Philos Trans R Soc Lond B Biol Sci 356(1406):133–145
Dobson CM (2002) Getting out of shape. Nature 418(6899):729–730
Dobson CM (2003) Protein folding and misfolding. Nature 426(6968):884–890
Dunker AK, Brown CJ, Lawson JD, Iakoucheva LM, Obradovic Z (2002) Intrinsic disorder and protein function. Biochemistry 41(21):6573–6582
Ecroyd H, Carver JA (2008) Unraveling the mysteries of protein folding and misfolding. IUBMB Life 60(12):769–774
Ecroyd H, Koudelka T, Thorn DC, Williams DM, Devlin G, Hoffmann P, Carver JA (2008) Dissociation from the oligomeric state is the rate-limiting step in amyloid fibril formation by kappa-casein. J Biol Chem 283(14):9012–9022
Ehrnsperger M, Graber S, Gaestel M, Buchner J (1997) Binding of non-native protein to Hsp25 during heat shock creates a reservoir of folding intermediates for reactivation. EMBO J 16(2):221–229
Esposito G, Garvey M, Alverdi V, Pettirossi F, Corazza A, Fogolari F, Polano M, Mangione PP, Giorgetti S, Stoppini M, Rekas A, Bellotti V, Heck AJ, Carver JA (2013) Monitoring the interaction between beta2-microglobulin and the molecular chaperone alphaB-crystallin by NMR and mass spectrometry: AlphaB-crystallin dissociates beta2-microglobulin oligomers. J Biol Chem 288(24):17844–17858
Gong B, Lim MC, Wanderer J, Wyttenbach A, Morton AJ (2008) Time-lapse analysis of aggregate formation in an inducible PC12 cell model of Huntington's disease reveals time-dependent aggregate formation that transiently delays cell death. Brain Res Bull 75(1):146–157
Guijarro JI, Sunde M, Jones JA, Campbell ID, Dobson CM (1998) Amyloid fibril formation by an SH3 domain. Proc Natl Acad Sci U S A 95(8):4224–4228
Harper JD, Lansbury PT Jr (1997) Models of amyloid seeding in Alzheimer's disease and scrapie: mechanistic truths and physiological consequences of the time-dependent solubility of amyloid proteins. Annu Rev Biochem 66:385–407
Hartl FU, Bracher A, Hayer-Hartl M (2011) Molecular chaperones in protein folding and proteostasis. Nature 475(7356):324–332
Hatters DM, Lindner RA, Carver JA, Howlett GJ (2001) The molecular chaperone, alpha-crystallin, inhibits amyloid formation by apolipoprotein C-II. J Biol Chem 276(36):33755–33761
Hightower LE (1991) Heat shock, stress proteins, chaperones, and proteotoxicity. Cell 66(2):191–197
Hohfeld J, Cyr DM, Patterson C (2001) From the cradle to the grave: molecular chaperones that may choose between folding and degradation. EMBO Rep 2(10):885–890
Horwitz J (1992) Alpha-crystallin can function as a molecular chaperone. Proc Natl Acad Sci U S A 89(21):10449–10453
Ingolia TD, Craig EA (1982) Four small Drosophila heat shock proteins are related to each other and to mammalian alpha-crystallin. Proc Natl Acad Sci U S A 79(7):2360–2364
Jakob U, Gaestel M, Engel K, Buchner J (1993) Small heat shock proteins are molecular chaperones. J Biol Chem 268(3):1517–1520
Jiménez JL, Nettleton EJ, Bouchard M, Robinson CV, Dobson CM, Saibil HR (2002) The protofilament structure of insulin amyloid fibrils. Proc Natl Acad Sci U S A 99(14):9196–9201
Kato K, Hasegawa K, Goto S, Inaguma Y (1994) Dissociation as a result of phosphorylation of an aggregated form of the small stress protein, hsp27. J Biol Chem 269(15):11274–11278
Kayed R, Head E, Thompson JL, McIntire TM, Milton SC, Cotman CW, Glabe CG (2003) Common structure of soluble amyloid oligomers implies common mechanism of pathogenesis. Science 300(5618):486–489
Kikis EA, Gidalevitz T, Morimoto RI (2010) Protein homeostasis in models of aging and age-related conformational disease. Adv Exp Med Biol 694:138–159
Kim MV, Seit-Nebi AS, Marston SB, Gusev NB (2004) Some properties of human small heat shock protein Hsp22 (H11 or HspB8). Biochem Biophys Res Commun 315(4):796–801

Knowles TP, Waudby CA, Devlin GL, Cohen SI, Aguzzi A, Vendruscolo M, Terentjev EM, Welland ME, Dobson CM (2009) An analytical solution to the kinetics of breakable filament assembly. Science 326(5959):1533–1537

Koteiche HA, McHaourab HS (2003) Mechanism of chaperone function in small heat-shock proteins. Phosphorylation-induced activation of two-mode binding in alphaB-crystallin. J Biol Chem 278(12):10361–10367

Kulig M, Ecroyd H (2012) The small heat-shock protein alphaB-crystallin uses different mechanisms of chaperone action to prevent the amorphous versus fibrillar aggregation of alpha-lactalbumin. Biochem J 448(3):343–352

Laurence TA, Kwon Y, Yin E, Hollars CW, Camarero JA, Barsky D (2007) Correlation spectroscopy of minor fluorescent species: signal purification and distribution analysis. Biophys J 92(6):2184–2198

Lee SJ, Lim HS, Masliah E, Lee HJ (2011) Protein aggregate spreading in neurodegenerative diseases: problems and perspectives. Neurosci Res 70(4):339–348

Lindner RA, Treweek TM, Carver JA (2001) The molecular chaperone alpha-crystallin is in kinetic competition with aggregation to stabilize a monomeric molten-globule form of alpha-lactalbumin. Biochem J 354(1):79–87

Loveland AB, Habuchi S, Walter JC, van Oijen AM (2012) A general approach to break the concentration barrier in single-molecule imaging. Nat Methods 9(10):987–992

Lowe J, McDermott H, Pike I, Spendlove I, Landon M, Mayer RJ (1992) AlphaB-crystallin expression in non-lenticular tissues and selective presence in ubiquitinated inclusion bodies in human disease. J Pathol 166(1):61–68

Macario AJ, Conway de Macario E (2005) Sick chaperones, cellular stress, and disease. N Engl J Med 353(14):1489–1501

McHaourab HS, Dodson EK, Koteiche HA (2002) Mechanism of chaperone function in small heat shock proteins. Two-mode binding of the excited states of T4 lysozyme mutants by alphaA-crystallin. J Biol Chem 277(43):40557–40566

McLean PJ, Kawamata H, Shariff S, Hewett J, Sharma N, Ueda K, Breakefield XO, Hyman BT (2002) TorsinA and heat shock proteins act as molecular chaperones: suppression of alpha-synuclein aggregation. J Neurochem 83(4):846–854

Mehlen P, Preville X, Chareyron P, Briolay J, Klemenz R, Arrigo AP (1995) Constitutive expression of human hsp27, Drosophila hsp27, or human alphaB-crystallin confers resistance to TNF- and oxidative stress-induced cytotoxicity in stably transfected murine L929 fibroblasts. J Immunol 154(1):363–374

Morimoto RI, Cuervo AM (2009) Protein homeostasis and aging: taking care of proteins from the cradle to the grave. J Gerontol A Biol Sci Med Sci 64(2):167–170

Muchowski PJ, Wacker JL (2005) Modulation of neurodegeneration by molecular chaperones. Nat Rev Neurosci 6(1):11–22

Ormsby AR, Ramdzan YM, Mok YF, Jovanoski KD, Hatters DM (2013) A platform to view huntingtin exon 1 aggregation flux in the cell reveals divergent influences from chaperones hsp40 and hsp70. J Biol Chem 288(52):37192–37203

Pountney DL, Treweek TM, Chataway T, Huang Y, Chegini F, Blumbergs PC, Raftery MJ, Gai WP (2005) AlphaB-crystallin is a major component of glial cytoplasmic inclusions in multiple system atrophy. Neurotox Res 7(1–2):77–85

Preville X, Salvemini F, Giraud S, Chaufour S, Paul C, Stepien G, Ursini MV, Arrigo AP (1999) Mammalian small stress proteins protect against oxidative stress through their ability to increase glucose-6-phosphate dehydrogenase activity and by maintaining optimal cellular detoxifying machinery. Exp Cell Res 247(1):61–78

Quintas A, Saraiva MJ, Brito RM (1999) The tetrameric protein transthyretin dissociates to a non-native monomer in solution. A novel model for amyloidogenesis. J Biol Chem 274(46):32943–32949

Rabinovici GD, Jagust WJ (2009) Amyloid imaging in aging and dementia: testing the amyloid hypothesis in vivo. Behav Neurol 21(1):117–128

Rao PV, Horwitz J, Zigler JS Jr (1993) Alpha-crystallin, a molecular chaperone, forms a stable complex with carbonic anhydrase upon heat denaturation. Biochem Biophys Res Commun 190(3):786–793

Regini JW, Ecroyd H, Meehan S, Bremmell K, Clarke MJ, Lammie D, Wess T, Carver JA (2010) The interaction of unfolding alpha-lactalbumin and malate dehydrogenase with the molecular chaperone alphaB-crystallin: a light and X-ray scattering investigation. Mol Vis 16:2446–2456

Rekas A, Adda CG, Aquilina JA, Barnham KJ, Sunde M, Galatis D, Williamson NA, Masters CL, Anders RF, Robinson CV, Cappai R, Carver JA (2004) Interaction of the molecular chaperone alphaB-crystallin with alpha-synuclein: effects on amyloid fibril formation and chaperone activity. J Mol Biol 340(5):1167–1183

Rekas A, Jankova L, Thorn DC, Cappai R, Carver JA (2007) Monitoring the prevention of amyloid fibril formation by alpha-crystallin. Temperature dependence and the nature of the aggregating species. FEBS J 274(24):6290–6304

Renkawek K, Bosman GJ, de Jong WW (1994) Expression of small heat-shock protein hsp 27 in reactive gliosis in Alzheimer disease and other types of dementia. Acta Neuropathol 87(5):511–519

Rogalla T, Ehrnsperger M, Preville X, Kotlyarov A, Lutsch G, Ducasse C, Paul C, Wieske M, Arrigo AP, Buchner J, Gaestel M (1999) Regulation of Hsp27 oligomerization, chaperone function, and protective activity against oxidative stress/tumor necrosis factor alpha by phosphorylation. J Biol Chem 274(27):18947–18956

Sathish HA, Stein RA, Yang G, McHaourab HS (2003) Mechanism of chaperone function in small heat-shock proteins. Fluorescence studies of the conformations of T4 lysozyme bound to alphaB-crystallin. J Biol Chem 278(45):44214–44221

Schmitz C, Rutten BP, Pielen A, Schafer S, Wirths O, Tremp G, Czech C, Blanchard V, Multhaup G, Rezaie P, Korr H, Steinbusch HW, Pradier L, Bayer TA (2004) Hippocampal neuron loss exceeds amyloid plaque load in a transgenic mouse model of Alzheimer's disease. Am J Pathol 164(4):1495–1502

Shammas SL, Waudby CA, Wang S, Buell AK, Knowles TP, Ecroyd H, Welland ME, Carver JA, Dobson CM, Meehan S (2011) Binding of the molecular chaperone alphaB-crystallin to Abeta amyloid fibrils inhibits fibril elongation. Biophys J 101(7):1681–1689

Shashidharamurthy R, Koteiche HA, Dong J, McHaourab HS (2005) Mechanism of chaperone function in small heat shock proteins: dissociation of the HSP27 oligomer is required for recognition and binding of destabilized T4 lysozyme. J Biol Chem 280(7):5281–5289

Shin JH, Kim SW, Lim CM, Jeong JY, Piao CS, Lee JK (2009) AlphaB-crystallin suppresses oxidative stress-induced astrocyte apoptosis by inhibiting caspase-3 activation. Neurosci Res 64(4):355–361

Shinohara H, Inaguma Y, Goto S, Inagaki T, Kato K (1993) AlphaB-crystallin and HSP28 are enhanced in the cerebral cortex of patients with Alzheimer's disease. J Neurol Sci 119(2):203–208

Smulders R, Carver JA, Lindner RA, van Boekel MA, Bloemendal H, de Jong WW (1996) Immobilization of the C-terminal extension of bovine alphaA-crystallin reduces chaperone-like activity. J Biol Chem 271(46):29060–29066

Stefani M (2012) Structural features and cytotoxicity of amyloid oligomers: implications in Alzheimer's disease and other diseases with amyloid deposits. Prog Neurobiol 99(3):226–245

Stefani M, Dobson CM (2003) Protein aggregation and aggregate toxicity: new insights into protein folding, misfolding diseases and biological evolution. J Mol Med 81(11):678–699

Stengel F, Baldwin AJ, Painter AJ, Jaya N, Basha E, Kay LE, Vierling E, Robinson CV, Benesch JL (2010) Quaternary dynamics and plasticity underlie small heat shock protein chaperone function. Proc Natl Acad Sci U S A 107(5):2007–2012

Stengel F, Baldwin AJ, Bush MF, Hilton GR, Lioe H, Basha E, Jaya N, Vierling E, Benesch JL (2012) Dissecting heterogeneous molecular chaperone complexes using a mass spectrum deconvolution approach. Chem Biol 19(5):599–607

Stranks SD, Ecroyd H, Van Sluyter S, Waters EJ, Carver JA, von Smekal L (2009) Model for amorphous aggregation processes. Phys Rev E 80(5):13

Treweek TM, Lindner RA, Mariani M, Carver JA (2000) The small heat-shock chaperone protein, alpha-crystallin, does not recognize stable molten globule states of cytosolic proteins. Biochim Biophys Acta 1481(1):175–188

Uversky VN (2002) What does it mean to be natively unfolded? Eur J Biochem 269(1):2–12

van de Klundert FA, Smulders RH, Gijsen ML, Lindner RA, Jaenicke R, Carver JA, de Jong WW (1998) The mammalian small heat-shock protein Hsp20 forms dimers and is a poor chaperone. Eur J Biochem 258(3):1014–1021

van Oijen AM (2011) Single-molecule approaches to characterizing kinetics of biomolecular interactions. Curr Opin Biotechnol 22(1):75–80

Vendruscolo M, Knowles TP, Dobson CM (2011) Protein solubility and protein homeostasis: a generic view of protein misfolding disorders. Cold Spring Harb Perspect Biol 3(12):a010454

Voisine C, Pedersen JS, Morimoto RI (2010) Chaperone networks: tipping the balance in protein folding diseases. Neurobiol Dis 40(1):12–20

Walsh DM, Klyubin I, Fadeeva JV, Cullen WK, Anwyl R, Wolfe MS, Rowan MJ, Selkoe DJ (2002) Naturally secreted oligomers of amyloid beta protein potently inhibit hippocampal long-term potentiation in vivo. Nature 416(6880):535–539

Waudby CA, Knowles TP, Devlin GL, Skepper JN, Ecroyd H, Carver JA, Welland ME, Christodoulou J, Dobson CM, Meehan S (2010) The interaction of alphaB-crystallin with mature alpha-synuclein amyloid fibrils inhibits their elongation. Biophys J 98(5):843–851

Wilhelmus MM, Otte-Holler I, Wesseling P, de Waal RM, Boelens WC, Verbeek MM (2006) Specific association of small heat shock proteins with the pathological hallmarks of Alzheimer's disease brains. Neuropathol Appl Neurobiol 32(2):119–130

Wyatt AR, Yerbury JJ, Ecroyd H, Wilson MR (2013) Extracellular chaperones and proteostasis. Annu Rev Biochem 82:295–322

Xu F, Yu H, Liu J, Cheng L (2013) AlphaB-crystallin regulates oxidative stress-induced apoptosis in cardiac H9c2 cells via the PI3K/AKT pathway. Mol Biol Rep 40(3):2517–2526

Chapter 8
Everything but the ACD, Functional Conservation of the Non-conserved Terminal Regions in sHSPs

Michelle Heirbaut, Sergei V. Strelkov, and Stephen D. Weeks

Abstract At the primary level small heat shock proteins are commonly described as a conserved α-crystallin domain flanked by regions that have disparate sequence content. While this holds true when analysing simple pairwise alignments, it belittles the importance of these N-terminal and C-terminal extensions. Careful examination of their sequences, combined with an improved understanding of the structure and activity of these proteins, yields an alternative view where the N- and C-terminal arms play an important role in function. In this chapter we shall describe the current understanding of these two regions and highlight that they both demonstrate structural and functional properties that are highly conserved across all kingdoms of life.

Keywords Sequence analysis • Structure • Oligomerisation • Chaperone-like activity • Truncation

8.1 Introduction

Viewed with the crudest of lenses small heat shock proteins (sHSPs) can be considered as having a tripartite architecture composed of an α-crystallin domain (ACD) flanked by, what are typically considered to be, highly variable N- and C-terminal regions herein termed the NTR and CTR, respectively (Fig. 8.1). Most of our structural understanding of this chaperone family lies with the ACD as this particular domain has proven most amenable to X-ray crystallographic studies (Bagnéris et al. 2009; Baranova et al. 2011; Laganowsky et al. 2010). Characterization of the N- and C-terminal arms has long frustrated researchers primarily due to their apparent lack of sequence conservation and structure. Conversely the fact that across all kingdoms of life these regions are maintained in this state speaks to their importance and hints

M. Heirbaut • S.V. Strelkov • S.D. Weeks (✉)
Laboratory for Biocrystallography, Department of Pharmaceutical and Pharmacological Sciences, KU Leuven, Leuven, Belgium
e-mail: stephen.weeks@pharm.kuleuven.be

R.M. Tanguay, L.E. Hightower (eds.), *The Big Book on Small Heat Shock Proteins*, Heat Shock Proteins 8, DOI 10.1007/978-3-319-16077-1_8

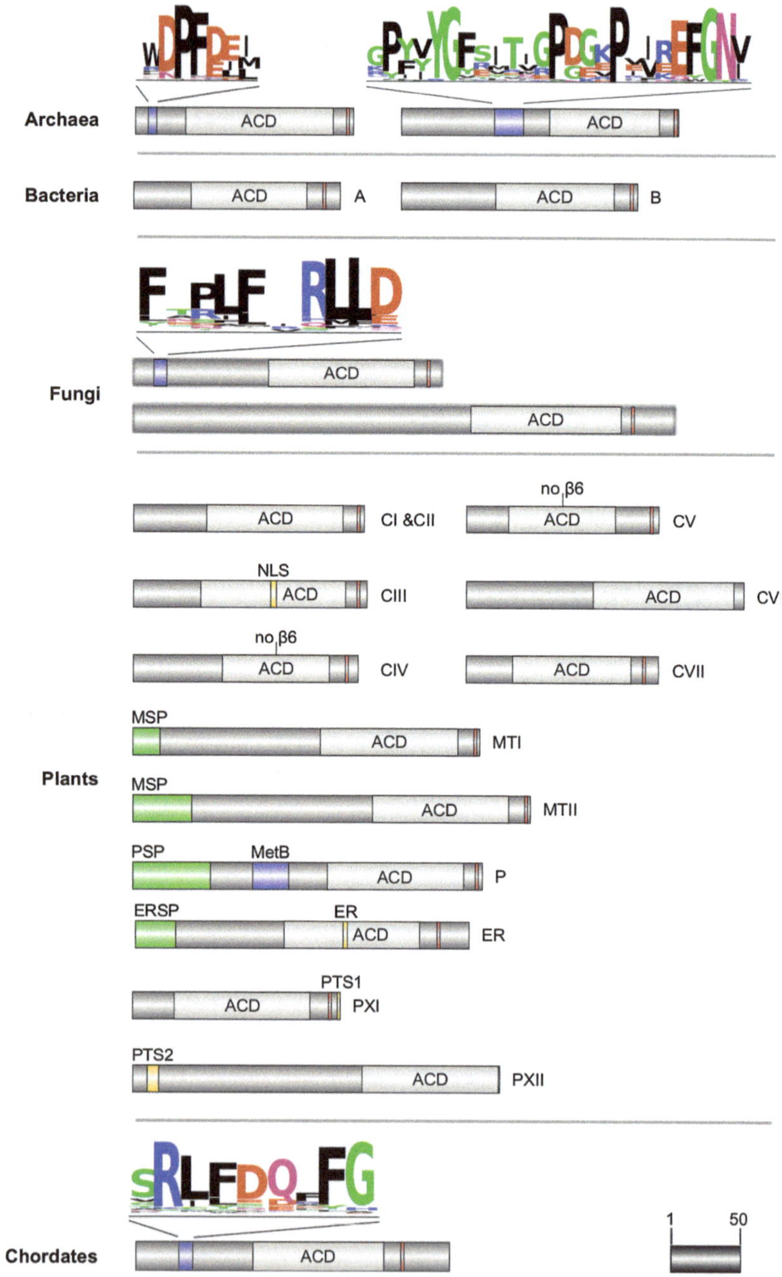

Fig. 8.1 Cartoon representation of sHSP domain architecture from different phyla. The ACD (*light gray*), the conserved IxI-motif (*red*), conserved N-terminal sequences (*blue*) with their corresponding sequence logo, cleavable signalling peptide sequences (*green*) and targeting sequences (*yellow*) are shown. For archaea, two representatives from each class are depicted; for bacteria the two conserved classes: A and B are shown. The shorter HSP26 and longer HSP42 from *S. cerevisiae*

at conserved functional roles. This is probably best demonstrated by the identification of congenital mutations within these two regions in a number of human homologues that are linked to the same disease state as those localized in the structured ACD (Benndorf et al. 2014).

Despite the challenge in understanding the function of the NTR and the CTR, decades of research employing deletion and mutational studies, combined with the application of modern biophysical techniques have begun to shed some light into their roles. In this chapter we shall explore this rich pool of studies in an attempt to show how these regions appear to have overlapping structural and functional properties. In particular examining their role in the two defining characteristics of this protein family: self-oligomerisation and chaperoning. Additionally we shall also highlight some studies that show specialisation within these regions that have allowed for differentiation of sHSP activity both within and between species.

Although there are a number of ways of defining the end point of the NTR and the start of the CTR we shall base our boundaries on the tertiary structure. This model assumes that the ACD across species is delineated by a conserved seven stranded β-sandwich, where the strands are numbered from β2 to β9. An additional strand is observed in non-metazoan sHSPs, termed β6 that is formed by domain swapping in the ACD dimer (Kim et al. 1998). Although this strand is absent in higher eukaryotes, resulting in a different dimer interface, it does not alter the terminal definitions of this domain. Notably, the β2-β9 demarcation differs from the sequence based PFAM classification (PF00011, Finn et al. 2014a), which omits the β2-strand of the ACD but includes a large portion of what we consider to be the CTR. We stress this point because in the literature both boundary descriptions have been employed for analysing the full sequence space available in public databases, giving rise to differences in the reported mean domain lengths (Kriehuber et al. 2010; Poulain et al. 2010). Any differences in values mentioned below to those published are thus a result of the definition of the specific domain borders.

Fig. 8.1 (continued) were chosen as representatives for the fungal kingdom. In plants, the average lengths of the domains per sHSP sub-class are shown based on values from Bondino et al. (2011) and Siddique et al. (2008). For chordates, the cartoon is based on the human HSPB1. MetB: methionine bristle found in plant chloroplast sHSPs. ER: signalling sequence for endoplasmatic reticulum localization; PTS1 and PTS2: peroxisomal targeting sequences 1 and 2. Targeting peptides; MSP: mitochondrial signalling peptide, PSP: chloroplast signalling peptide, ERSP: endoplasmatic reticulum signalling peptide. A scale bar corresponding to an amino acid length of 50 residues is shown on the bottom right (Cartoon representations were created using DOG (Ren et al. 2009). Sequence logos were generated online using WebLogo (Crooks et al. 2004))

8.2 Sequence Properties

Even with a limited number of sHSP sequences it was recognised very early on that the N- and C-terminal arms showed considerably less conservation than the central ACD (de Jong et al. 1998). The explosion in data resulting from various genome sequencing projects supports this initial observation although detailed analysis suggests a more nuanced view should be taken. While comparison between the different taxonomic kingdoms shows significant differences, at the level of phyla and subphyla a more complex picture of conservation can be presented. An example of this is shown for vertebrates in the conservation matrix in Fig. 8.2. While

		HUMAN				MOUSE				CHICKEN				ZEBRA FISH			
		B1	B2	B5	B8	B1	B2	B5	B8	B1	B2	B5	B8	B1	B2	B5	B8
HUMAN	B1		41	41	14	87	30	43	14	57	27	38	16	41	32	27	31
	B2	40		34	26	19	94	31	26	40	31	26	23	20	43	23	39
	B5	38	47		7	32	29	96	11	48	38	78	41	41	37	52	36
	B8	46	34	33		22	29	7	96	22	28	7	63	33	37	19	28
MOUSE	B1	85	37	38	45		19	32	16	51	30	38	19	38	35	27	28
	B2	38	98	47	34	35		29	29	37	31	37	20	29	46	23	26
	B5	39	47	97	34	40	47		11	48	38	82	37	37	41	52	33
	B8	45	35	33	95	43	37	30		22	28	11	67	30	37	19	28
CHICKEN	B1	73	40	39	48	72	40	40	49		45	44	22	56	37	44	39
	B2	37	74	45	30	35	75	48	31	41		35	24	38	59	41	28
	B5	39	38	76	30	34	39	76	28	35	48		15	41	44	56	44
	B8	46	32	32	68	44	31	33	66	39	31	29		37	33	30	36
ZEBRA FISH	B1	64	37	38	42	70	37	40	43	66	39	35	40		52	33	31
	B2	37	64	39	31	33	64	42	33	38	81	43	33	38		36	28
	B5	29	36	68	24	31	38	68	22	35	46	62	27	29	44		26
	B8	40	26	28	56	44	29	24	54	32	28	31	59	36	26	24	

Fig. 8.2 Matrix of percentage sequence similarity values calculated from the pairwise alignment of the terminal extensions of select vertebrate sHSPs. Shown are the values for the N-terminal (*below diagonal*) and C-terminal (*above diagonal*) regions. Initial sequences were identified using public databases. Regions were defined by multiple sequence alignment of the full sequences and removal of the sequence section corresponding to strands β2 to β9 of the α-crystallin domain. The sequence similarity was then determined for each sequence pair with the BLOSUM62 matrix using the MatGAT program (Campanella et al. 2003). For the N-terminal region background colors are based on sequence percentage similarity threshold values where <50 % (*white*), 50–75 % (*light gray*), ≥75 % (*dark gray*). For the C-terminal region background colors are based on percentage similarity threshold values where <50 % (*dark gray*), 50–75 % (*light gray*), ≥75 % (*white*)

comparison of any one sHSP to orthologues both within and between species shows weak conservation for the two ACD flanking arms, homologues from different species demonstrate reasonably high values: for example HSPB5 (αB-crystallin) shows 68 and 52 % sequence similarity between *Homo sapiens* and *Danio rerio* in the NTR and CTR, respectively. In this case it suggests early branching and specialisation of this family of chaperones in a common ancestor of the vertebrate lineage. In addition to this homologue based conservation, careful inspection at the sequence level across all species also reveals a clearly conserved residue bias in these two domains (Kriehuber et al. 2010). Furthermore a number of highly conserved sequence epitopes which in the NTR appear to be phyla specific are also observed whilst the CTR contains a motif conserved across all kingdoms (de Jong et al. 1998; Poulain et al. 2010).

8.2.1 The N-Terminal Region

Analysis of a non-redundant set of sequences belonging to the sHSP family (PF00011) shows a wide variety in the NTR, not only with regard to residue conservation but also the sequence length between different kingdom representatives (Table 8.1). On average the NTR contains 55.4 (±43.6) residues, which is close to the mean number of residues of this domain in bacteria (42.7 residues), a result due to the over-representation of sequences from this group. Viruses (Maaroufi and Tanguay 2013) and archaea have a relatively short N-terminal stretch, with average lengths of 46.8 and 52.4 respectively. All eukaryotes encode sHSPs that have longer NTRs than prokaryotes, but they also show a much larger variability in overall sequence length. The greatest differences are seen in fungi and plants, having average sequence lengths of 103.0 (±85.1) and 82.8 (±85.3), respectively. This increased

Table 8.1 Average sequence lengths of sHSP regions for the different taxonomic kingdoms

	Full-length	NTR	ACD	CTR
Viruses (26)	150.1±9.2	46.8±6.4	85.4±1.0	17.9±6.3
Archaea (365)	152.0±27.3	52.4±27.7	84.2±8.3	15.4±9.7
Bacteria (5,721)	148.2±15.0	42.7±12.7	86.5±3.9	19.0±8.1
Funghi (367)	248.1±110.3	103.0±85.1	121.2±39.7	23.9±39.7
Plants (1,121)	214.5±105.9	82.8±85.3	92.7±20.2	39.0±60.3
Metazoans[a] (721)	205.1±84.6	75.2±49.9	84.8±10.5	45.1±51.2
Chordates (690)	183.1±30.4	70.5±21.1	84.2±7.8	28.3±12.8
All (9,011)	167.9±60.2	55.4±43.6	88.2±13.9	24.3±29.5

Non-redundant sequences for each kingdom were extracted from PFAM-family PF0011. The number of representative sequences is indicated between brackets. For each group the sequences were divided into three regions based on multiple sequence alignment. The average sequence length ± standard deviation is shown

NTR N-terminal region, *ACD* α-crystallin domain, *CTR* C-terminal region

[a]Metazoans excluding chordate sequences

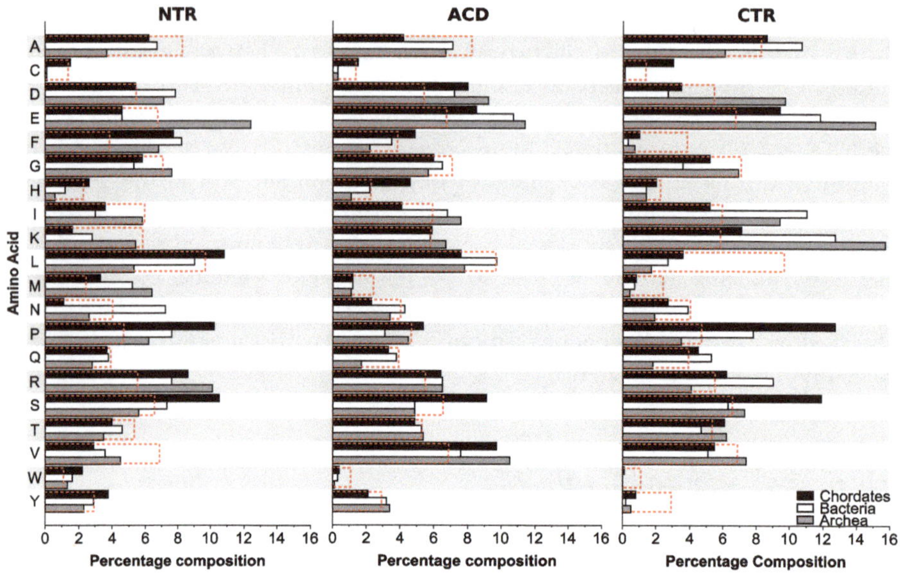

Fig. 8.3 Cross kingdom comparison of the sequence composition of the three sHSP domains. sHSP sequences for each phyla were downloaded from the PFAM-database (Finn et al. 2014b) and manually defined using Jalview (Waterhouse et al. 2009) using the boundaries described in Sect. 8.1. The sequence statistics were calculated for chordates (*black bars*), bacteria (*white bars*) and archaea (*gray bars*) in SeaView (Gouy et al. 2010) and compared to the mean composition for whole genomic datasets (*dashed lines*)

variability arises from the existence of different distinct sHSP classes, especially within the plant kingdom where the NTR can contain additional domains dictating localization and possibly specialized functions (Fig. 8.1) (Waters and Vierling 1999a, b). Metazoan sHSPs have slightly shorter NTRs when compared to both fungi and plants, with the average number of amino acids being 70.5 for chordates and 75.2 (±49.9) for examples that fall outside this phyla. Both also have a decreased variability when compared to the other eukaryotic phyla (Table 8.1).

Examining the sequences in closer detail reveals a specific residue bias within the NTR that exists across all species (Kriehuber et al. 2010). In particular there is an over-representation in the number of arginines, phenylalanines and prolines, and an apparent lower number of cysteines and lysines, when compared to the average number of these residues in whole genomic datasets (Fig. 8.3). Comparison of the sequence content between the different sHSP regions highlights the specific bias of the arginine, phenylalanine and lysine residues, but also shows that the NTR on average contains more tryptophans than the following domains (Fig. 8.3). Phenylalanines have been found to be important for assembly and chaperone activity in a number of sHSPs, suggesting a conserved role for these aromatic hydrophobic residues in interacting with either substrate or self molecules (Hanazono et al. 2013; Horwitz et al. 1998; Kelley and Abraham 2003; Plater et al. 1996). Likewise, similar structural and functional properties can be speculated for the higher abundance of tryptophan residues in this region. For the arginines it is possible to postulate that these positively charged residues may also be involved in self-assembly, mediated by cation-π interactions with the phenylalanine residues. At the same time, the chaotropic properties of the side chain guanidinium group have been shown to stabilise proteins in solution (Arakawa et al. 2007; Shah et al. 2012) thus a possible role in chaperoning can also be hypothesised. For the lower abundance of lysines and cysteines, it can be argued that these are kept to a minimum in the extended and likely solvent-exposed NTRs to avoid post-translational modifications (PTMs). Lysines in particular have been shown to be highly reactive and can undergo a wide variety of PTMs, including non-enzymatic glycation and carbamylation leading to non-disulfide crosslinking (Soskić et al. 2008). Cysteine residues on the other hand are the prime candidates for oxidation. Many of these PTMs have been shown to decrease the chaperone-like activity (van Boekel et al. 1996) and the increase of modified α-crystallins in the eye lens has been linked to cataract (Sharma and Santhoshkumar 2009).

Careful analysis of multiple sequence alignments of the NTR alone also suggests the presence of a number of conserved sequence motifs. These are typically restricted to phyla and subphyla but there are a number of examples that cross these taxonomic classifications. In archaea, two motifs can be identified in the two different orthologues commonly present in most species. The first, found in sHSPs with longer NTRs, encompasses approximately 20 residues (Fig. 8.2). The second sequence is found close to the N-terminus of the smaller orthologues and has the pattern (D/N)PF(D/E)(D/E). Interestingly a similar sequence to the latter motif was identified in vertebrate HSPB1(Franck et al. 2004), the deletion of which impaired chaperone-like activity and oligomerization in *Cricetulus griseus* HSPB1 (Lambert et al. 1999; Theriault 2004), although equivalent studies with the human orthologue encompassing this sequence failed to protect insulin from aggregating but did not affect the assembly (Lelj-Garolla and Mauk 2005, 2012) and deletion of this sequence in mouse HSPB1 had no effect on either chaperoning or assembly (Guo and Cooper 2000). Cytosolic class I sHSPs from plants also have a sequence resembling this short motif residing within a larger conserved region, whether it has a functional role though is not yet known (Waters et al. 2008; Waters and Vierling 1999a).

HSP26 and HSP42, the main cytosolic sHSPs found in the budding yeasts, are probably the best-studied members from this taxonomic group. Comparison of their architecture explains the large variance seen in the NTR lengths observed in fungi (Fig. 8.1 and Table 8.1). In particular the *Saccharomycetales* specific HSP42, although poorly conserved even within this order, consistently contains long stretches of repetitive sequence rich in the acidic residues aspartate and glutamate or their amide containing equivalents asparagine and glutamine. Looking at the sHSPs with shorter NTRs a conserved region can be found (Fig. 8.1) although its function remains unclear.

In the special case of plants, where sHSPs demonstrate organelle-specific localization, homologous proteins across species clearly contain conserved sequences necessary for translocation to the target organelle (Fig. 8.1). Examination of the

NTR outside these targeting peptides shows little conservation between subfamilies in a single species, although these sequences are conserved when comparing equivalent homologues within this kingdom. However to what extent these regions are involved in specific functions, is less understood (Scharf et al. 2001). One particular region found in chloroplast sHSPs, the so-called methionine bristle, has been studied in more detail and has been found to be involved in chaperone-like activity and sensing redox state in chloroplasts (Gustavsson et al. 2001; Härndahl et al. 2001; Waters 2013).

Many vertebrate sHSPs contain a 'SRLFDQxFG' motif, originally described as the phenylalanine-rich region in the α-crystallins (Crabbe and Goode 1994; Van Der Ouderaa et al. 1974). Mutations of this motif, which is typically found in the centre of the NTR (Fig. 8.1), appear to effect assembly and subunit turnover (see Sect. 8.4) (Derham et al. 2001; Horwitz et al. 1998; Pasta et al. 2003; Plater et al. 1996). A highly similar sequence has also been identified in the sHSPs of some arthropods (Liu et al. 2014; Heirbaut et al. 2014). These particular family representatives, that show highest overall sequence similarity to the vertebrate α-crystallins, suggest that a common ancestor containing this motif appeared more than 500 million years ago. The conservation of this motif, particular between the numerous homologues found encoded in vertebrate genomes underlines an important functional role not yet fully understood.

A second shorter and less conserved motif (P/G)YY(I/V)R has also been described. Found in the end of the NTR of some vertebrate sHSPs, deletion of this sequence impairs chaperone activity (Ghosh et al. 2006; Heirbaut et al. 2014). In addition, and somewhat reminiscent of the methionine-rich regions found in plant chloroplast sHSPs, the vertebrate member called HSPB7 (also known as cardiovascular HSP; cvHSP) contains a poly-serine stretch in the first half of the NTR (Franck et al. 2004). This region, ranging from 6 to 18 serine residues depending on the species examined, has been found to be involved in localization of this sHSP to nuclear splicing speckles, suggesting it has acquired a specialized function that may not necessarily be related to chaperone activity (Vos et al. 2009).

8.2.2 The C-Terminal Region

For all sequences analysed the CTR is smaller than the extension N-terminal to the ACD (Table 8.1). On average this region is approximately one third of the length of the corresponding NTR in each kingdom. Like the NTR the sequence length of the CTR differs considerably between the domains, with bacteria and archaea demonstrating some of the shortest lengths with 19.0 and 15.4 residues, respectively (Table 8.1).

Relative to the NTR, the CTR shows greater sequence divergence between homologues when comparing within a phylum (Fig. 8.2). Even so this region shows a conserved sequence bias across all species that is even more apparent than the NTR (Fig. 8.3). In particular, relative to the mean amino acid composition of

UniProtKB/Swiss-Prot data bank (April 2013 release), there is a greater presence of the charged residues glutamate and lysine. At the same time the CTR has a consistently lower number of the aromatic amino acids; tryptophan, phenylalanine and tyrosine as well as the hydrophobic residue leucine. These latter residues reveal the most pointed differences between this domain and both the NTR and ACD (Fig. 8.3) and suggests that this region would preferentially be solvent exposed. Similar to the NTR there is a lower than average number of cysteines while prolines have an above average representation in bacteria and chordates.

The most characteristic sequence feature present in the CTR was originally identified as the RxIxI motif (de Jong et al. 1998), later truncated to just the last three residues IxI or IxI/V. This tripeptide, also named the C-terminal anchoring motif (CAM; Poulain et al. 2010), appears to play an important role in assembly (see Sect. 8.3.2). An extensive genome-wide survey showed some species based bias for the residues in this motif, in particular the central amino acid is typically a proline in animals (Poulain et al. 2010). In the same survey the CAM sequence was found to be on average 14 (±3) residues away from the final β9-strand of the ACD. This means it is generally found at the C-terminus in bacteria and archaea while in eukaryotes it is commonly found in the first half of the CTR (Fig. 8.1). Closer examination of the residues surrounding the CAM in vertebrates also suggests that it sits in a conserved palindromic region (Hilton et al. 2013; Laganowsky et al. 2010), for example in humans: HSPB1, EITIPITFE; αA-cystallin, ERAIPVSRE; and αB-crystallin, ERTIPITRE.

8.3 Structure

Independent of the species of origin, prediction algorithms typically suggest that the NTR contains little to no secondary structure although, other than the immediate N-terminus, it appears not to contain sequence stretches typical for intrinsically disordered regions (IDR; Weeks et al. 2014). The CTR is similarly barren in secondary structure but, with its penchant to contain charged residues and lack hydrophobic ones, this region fulfils the standard criteria for being an IDR (Vucetic et al. 2003). It is these specific properties that have made structural characterisation of either region a challenge. Nevertheless, the concerted efforts from many individual laboratories have begun to reveal some interesting details about them.

8.3.1 The N-Terminal Region

Initial structure predictions of the NTR were first confirmed by the X-ray crystal structure of HSP16.5 from *Methanocaldococcus jannaschii* (MjHSP16.5, Kim et al. 1998). The solved structure contains eight subunits in the asymmetric unit (ASU) yielding the biologically relevant 24-mer after applying the crystallographic threefold symmetry operations. The resultant oligomer resembles a hollow sphere with

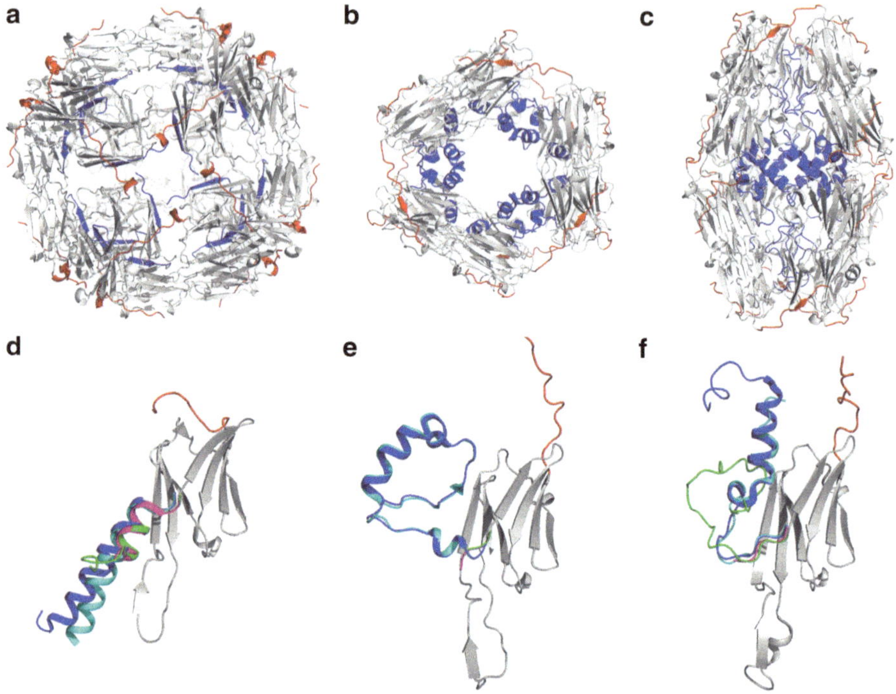

Fig. 8.4 Crystal structures of small heat shock proteins. (**a–c**) Cartoon representations of the structures of the oligomeric assemblies of (**a**) *M. jannaschii* HSP16.5 (PDB 1SHS), (**b**) *T. aestivum* HSP16.9 (PDB 1GME) and **c** *S. pombe* HSP16.0 (PDB 3W1Z). The NTR (*blue*), CTR (*red*) and ACD (*gray*) (**d–f**) Comparison of the structure of the N-terminal extension from various sHSPs. Individual chains from each crystal structure (*blue, cyan, magenta and green*) were superposed via the ACD region. For clarity only one ACD and C-terminal domain are shown per panel (*gray*). (**d**) *S. tokodaii* HSP14 (PDBs 3AAB and 3AAC), (**e**) *T. aestivum* HSP16.9 and (**f**) *S. pombe* HSP16.0

an internal apparent solvent filled volume of approximately 1.8×10^5 Å^3 (Fig. 8.4a). An additional β-strand (β1) is observed for all monomers in the ASU, corresponding to residues 33–40 of the NTR, that pairs with the β7-strand of the ACD. This strand appears to be unique to MjHSP16.5, even amongst other prokaryotic sHSPs (Hanazono et al. 2012; Hilario et al. 2011; Takeda et al. 2011). Despite this fact, the numbering of β-strands in other sHSPs is based on this structure. Importantly though, there is no observable electron density for the first 32 residues of each subunit, or a total of 768 residues in the 24-mer. The first N-terminal residue observed for each monomer, and indeed the β1-strand, is found on a fourfold non-crystallographic symmetry (NCS) axis in the interior of the 24-mer (Fig. 8.4a). The absent residues would equate to a volume of approximately 1.1×10^5 Å^3 (assuming a partial specific volume of 0.73 ml g^{-1}), equivalent to just under two-thirds of the internal volume of the large assembly. Presumably the remainder of the NTRs are localised in this voluminous cavity but, displaying multiple conformations in the crystal, the electron density is averaged out.

Similar to MjHSP16.5 the structure of the full-length *Sulfolobus tokodaii* HSP14.0 (StHSP14.0) also yielded a hollow 24-mer where only a part of the NTR, found again on the inside of the oligomeric assembly, was observed (Hanazono et al. 2012). By mutating or deleting a portion of the CTR the same group also managed to obtain structures of the dimeric form of this sHSP at higher resolutions than that of the oligomer (Hanazono et al. 2012; Takeda et al. 2011). In these structures the full NTR was only partially resolved and, differing to the MjHSP16.5 structure, an α-helix was consistently present in the region closest to the ACD (Fig. 8.4d). Interestingly the fragments of the NTR that were modelled in the different crystal forms show considerable variability. Superimposition of the ACD of the available structures highlights differences in both the length and the relative position of the α-helix (Fig. 8.4d), confirming an inherent propensity for movement and a lack of structure within the whole domain. The results also suggest that parts of the NTR can exist in multiple structural conformations even within the highly ordered arrangement found in a crystal.

The schizophrenic properties of the NTR are probably best illustrated by the structure of wheat HSP16.9 (*Triticum aestivum*; TaHSP16.9). This eukaryotic sHSP is an oblate dodecamer, composed of a stack of two rings, where each ring is trimer of three ACD-mediated dimers (Fig. 8.4b, van Montfort et al. 2001). For each dimer in the assembly the electron density of the full NTR, corresponding to the first 42 residues of this sHSP, is only seen for one of the component protomers while in the other it is completely absent. Alone, the modelled NTR shows an extended conformation containing some α-helical segments (Fig. 8.4e). This structure intertwines with a symmetry-related NTR originating from a monomer on the opposing trimer of dimers and, stabilised by a combination of hydrophobic and polar interactions, holds the assembly together (van Montfort et al. 2001). Crucially the observed NTRs are only found at the vertices formed between the dimers in the oligomer (Fig. 8.4b). Although the position of the first N-terminal residue of the chains without an observable NTR are in symmetry related positions to those where the full sequence is present, the limited space at the vertices of this assembly excludes the possibility of the missing residues from interacting in an equivalent manner. Therefore although association between the NTRs of different monomers is important for assembly of TaHSP16.9 it appears that this same region selectively destabilises the structure of equivalent domains from other monomers within the same oligomer.

This non-equivalence of the NTR was also seen in the structure of HSP16.0 from the fission yeast *Schizosaccharomyces pombe* (SpHSP16.0; Hanazono et al. 2013). In this prolate 16-mer the full NTR is observed in the interior of the oligomeric assembly at the equatorial interface of the protein subunits (Fig. 8.4c). At this position the partially helical NTR of eight monomers make a number of inter-subunit interactions that stabilise the interface. In particular a number of phenylalanines were shown to be crucial in this interaction (Hanazono et al. 2013). However, within the same assembly, the NTRs at the poles are either partially observed but unstructured or completely absent (Fig. 8.4f). The structure of this eukaryotic sHSP therefore illustrates that a subtle difference in the local environment can have a potent influence on the structure of the NTR.

Although some moderate success has been achieved in obtaining atomic resolution structures of full-length prokaryotic, yeast and plant sHSPs, to date structures of homologues from vertebrates are limited to the ACD and portions of the CTR (Bagnéris et al. 2009; Baranova et al. 2011; Hochberg et al. 2014; Laganowsky et al. 2010). These latter results have been achieved by removing the majority of the NTR. Hence very little is known about the structural capacity of this region in this important taxonomic group. Only a short portion of the NTR immediately preceding the strand β2 was seen in the crystals of the N-terminally truncated human HSPB6 (ΔN56). The retained residues of the NTR showed no secondary structure. Interestingly, the residues corresponding to strand β2 were involved in interactions with the ACD of another dimer, resulting in tetramer formation, which was however an artefact of truncation (Weeks et al. 2014). Solid-state NMR studies of full-length human αB-crystallin suggest the presence of a mix of α and β secondary structure in the NTR (Jehle et al. 2010). The limited number of restraints and heterogeneity of the chemical shifts led the authors to state that these structures may not simultaneously exist in one chain or, for that matter, be present in all chains of the oligomer.

8.3.2 The C-Terminal Region

Unlike the NTR, structures of the shorter CTR at an atomic resolution are more abundant. In the first structure of a sHSP, MjHSP16.5, this region was clearly observed making inter-protomer contacts (Fig. 8.4a, Kim et al. 1998). In particular the interaction was mediated by the highly conserved IxI motif patching a groove formed between the β4 and β8-strands on one side of the ACD β-sandwich structure. The equivalent interaction has been observed in structures of full-length prokaryote, plant and yeast sHSPs (Fig. 8.4b, c, Bepperling et al. 2012), as well as in truncated constructs of vertebrate homologues (Hochberg et al. 2014; Laganowsky et al. 2010; Laganowsky and Eisenberg 2010). In the latter cases parts of the CTD downstream of the CAM are not observed (Fig. 8.5a) as this region was intentionally removed from the initial construct to enable crystallization (Laganowsky et al. 2010; Laganowsky and Eisenberg 2010).

Independent of the kingdom under study the patching interface is the same. The side-chains of the two outer conserved residues act as knobs filling two small hydrophobic depressions on the β4/β8-face of the ACD (Fig. 8.5a). In all species the interaction is commonly stabilised by van der Waals forces and a limited number of backbone hydrogen bonds (Fig. 8.5b). As stated in Sect. 8.2.2 the CAM sequence is palindromic, a property that extends to the surrounding residues of some vertebrate sHSPs. This has led to the hypothesis that the orientation of the patching residues could occur in either direction relative to the ACD β4/β8-face, partially explaining the polydispersity typically common in this family of proteins (Delbecq et al. 2012;

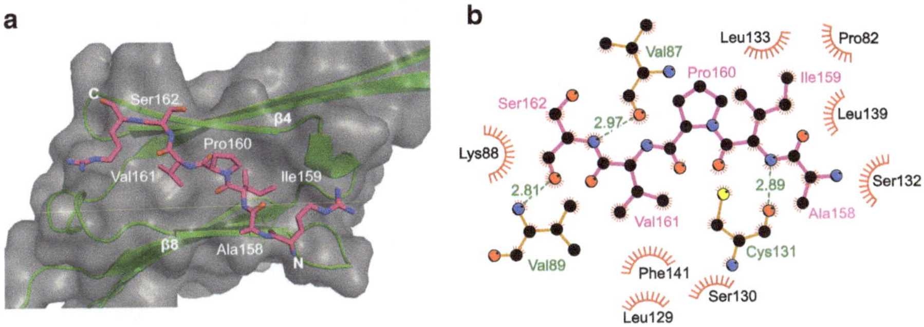

Fig. 8.5 Patching of the β4/β8 face of the ACD by the conserved IxI motif. (**a**) Structure of the patching sequence in αA-crystallin (PDB 3L1E). The ACD is shown as a cartoon and surface representation simultaneously. The patching sequence is shown using stick representation of the connected atoms, (**b**) Interaction diagram of the patching sequence with the ACD. Hydrogen bonds are shown as *dashed lines*. Hydrophobic interactions mediated by the ACD are shown with *spoked arcs*. Interacting atoms of the peptide are also sunbeamed (Image generated with LigPlot+ (Laskowski and Swindells 2011))

Laganowsky et al. 2010). Despite this possibility, examination of available structures of full-length sHSPs always shows one preferred orientation. In addition mutational studies covering the extended palindromic region of human αB-crystallin demonstrated an asymmetric effect on oligomer stability, suggesting that a unique direction also exists for vertebrate sHSPs (Hilton et al. 2013).

Although the patching of the ACD by the IxI-motif is unambiguously shown in a number of crystal structures, there is some debate over the physiological relevance of this interaction. While solid-state NMR experiments of human αB-crystallin performed at 0 °C demonstrated patching equivalent to that seen in crystal structures (Jehle et al. 2010), solution NMR studies at higher temperature showed that the nuclear spin of the knob residues relaxed much more quickly than immobilized residues in the large oligomeric assembly, suggesting that they are highly dynamic (Baldwin et al. 2011a). Calculations predict that at physiological temperatures only 2 % of the IxI-motifs are bound at any given time for this sHSP, although the CTR is believed to be in close proximity to the β4/β8-face of ACD within the oligomeric assembly (Baldwin et al. 2011a). Whether this phenomenon is also true for sHSPs from other kingdoms has yet to be tested, although it is interesting to note that in the crystal structure of HSPA from the bacterium *Xanthomonas axonopodis* patching was not seen despite the presence of the full CTR (Hilario et al. 2011).

8.4 Functional Roles of the N- and C-Terminal Regions

When describing the properties of small heat shock proteins we think mostly in terms of their ability to assemble into large oligomers, and their capacity to chaperone partially unfolded proteins. While there are examples of sHSPs that have additional functions that fall outside this classical model (Dreiza et al. 2010; Vos et al.

2009), here we will focus on the conserved roles that the ACD flanking arms play in defining these prototypical attributes.

The formation of large oligomers is a corner stone of the majority of studied sHSPs. While a small number form monodisperse preparations that have made them suitable for structural studies (Hanazono et al. 2012, 2013; Kim et al. 1998; van Montfort et al. 2001) generally the isolated proteins, independent of the kingdom of origin, yield polydisperse material with oligomers containing a differing numbers of subunits. In the best characterised system, the vertebrate αB-crystallin, the purified oligomers can contain between 10 and 40 protomers (Baldwin et al. 2011b), a value that can vary with temperature (Clauwaert et al. 1989). In addition to changes in the subunit number, the individual αB-crystallin protomers can also assemble into a variety of different oligomeric shapes, yielding a further dimension in sample polydispersity (Baldwin et al. 2011b). An added complexity to the study of these assemblies is also the highly dynamic turnover of the constituent monomers. In the case of αB-crystallin a complete stochastic exchange of the component subunits occurs within 40 min at physiological pH and temperature (Hilton et al. 2013). This dynamic behaviour is believed to be necessary for chaperone activity. In this model a released protomer is the chaperoning competent state capable of recognizing an aberrantly folded substrate (Haslbeck et al. 2005). Crucially this model of activity suggests that these two properties of sHSPs are likely intertwined at the sequence level – a supposition supported by the various studies that will be outlined below – but for simplicity we shall describe the role of the NTR and CTR in defining them independently.

8.4.1 Assembly in Non-metazoans

The crystal structures of the full-length sHSPs hint at a possible role of the N- and C-terminal regions in assembly (Fig. 8.4a–c). One of the most compelling arguments supporting this hypothesis can be found in solution studies of the isolated α-crystallin domain from the sHSPs of a variety of species that clearly show that this conserved region is only capable of forming a dimer – the basic building block of the large oligomeric assemblies. Deletion and point mutation studies have attempted to delineate the importance of the flanking domains in oligomerization, the results of which show a picture of varying influence when comparing results across the kingdoms.

In archaea the role of the CTR, and in particular the IxI-motif, in inter-subunit patching of the component dimers of the oligomer was clearly seen in the crystal structure of the HSP16.5 from *M. jannaschii* (Fig. 8.4a). A C-terminal truncation of this sHSP, that eliminated the conserved motif, resulted in collapse of the oligomer (Kim et al. 2003; Quinlan et al. 2013). Similar results were also obtained for StHSP14. Deletion of the CAM yielded dimers, a result that could be replicated by mutating both isoleucines in the IKI sequence to either alanine, serine or phenylalanine (Hanazono et al. 2012; Saji et al. 2008). Interestingly, a double mutation where both isoleucines were replaced by tryptophans led to oligomers with the same overall size as the wild type but that showed an increase in dissociation to smaller entities at lower temperatures (Saji et al. 2008).

Although the NTR was not observed in the MjHSP16.5 crystal structure deletion of the first 12 residues also halted assembly at the level of a tetramer. Complete removal of this domain resulted in dimers (Kim et al. 2003). Conversely truncation of the NTR in StHSP14 did not disrupt oligomeric structure but led to a decreased subunit exchange. In addition negative stain electron microscopy revealed that the complete removal of the NTR resulted in oligomeric assemblies containing hollow centres (Usui et al. 2004).

In bacteria several studies have implicated both N- and C-terminal sequences in regulating oligomerization. Removal of the first 11 residues from the NTR or the last 11 of the CTR, the latter containing the canonical CAM sequence, prevented *E. coli* IbpB from forming assemblies bigger than the ACD-mediated dimers (Jiao et al. 2005). The same effect was seen for a C-terminal mutation in the cyanobacterial HSP16.6 from *Synechocystis* where mutation or removal of the VKV-motif limited this sHSP to a dimeric form (Giese et al. 2005; Giese and Vierling 2004). Interestingly, NTR localised suppressor mutations (S2Y and N7Y) were identified that could rescue the VKV mutation, suggesting that both domains can play a role in oligomerization (Giese and Vierling 2004).

Truncations and site-specific mutations in both the class A and B sHSPs from the plant symbiote *Bradyrhizobium japonicum* resulted in an inability to form oligomers (Studer et al. 2002). In particular alanine substitutions of the patching isoleucine residues (single or double) of the class B HSPH yielded only dimers in solution (Studer et al. 2002). Truncation of the C-terminus, leaving the IxI-motif intact in class A HSPF led to larger oligomers. Similarly, truncation of the very N-terminus did not disrupt oligomerization, however removal of 9 or more residues led to a loss of multimerization for the class B sHSP, the same effect was seen in the class A sHSP upon deletion of 30 or more residues from the NTR (Studer et al. 2002).

Deletion of the full CTR in AsgA from *Salmonella typhimurium* led to a non-functional sHSP in vivo hinting at a role for the CTR in conferring thermotolerance, whereas removal of 11 residues – leaving the IxI-motif intact – resulted in the formation of smaller oligomers (Tomoyasu et al. 2010). The same phenomenon was also observed by truncating the NTR, where removal of 11 or 17 residues gave smaller multimers that dissociated at lower temperatures (Tomoyasu et al. 2010). In *Mycobacterium tuberculosis* HSP16.3 (MtHSP16.3), deletion of the CAM led to a construct that could only form dimers. This truncated protein was however still able to assemble into oligomers when mixed with the wild-type protein, clearly showing a certain redundancy in the involvement of the CTR in regulating and modulating assembly, as presence of only a few protomers containing the CTR appears to be sufficient. Deletion of the first 35 residues from the NTR, led to the formation of dimers and trimers that undergo reversible self-association dependent on concentration, seen as an increase in the relative amount of trimers (Fu et al. 2005).

In comparison to prokaryotes only a limited number of studies report characterization of the sequence determinants necessary for oligomerization in fungal sHSPs. For the *Saccharomyces cerevisiae* HSP26, the NTR has been implicated in assembly as deletion of this whole region yielded dimers. Shorter truncations, removing only the first 30 residues of this domain, did not disrupt oligomer formation but an increase in temperature-induced dissociation was seen (Haslbeck et al. 2004; Stromer et al. 2004). Residues 30–90, corresponding to a globular middle domain in this sHSP, have been shown to be involved in assembly by cryo-EM analysis (White et al. 2006). In HSP16.0 from the fission yeast *Schizosaccharomyces pombe*, an essential role of phenylalanine residues in the NTR in oligomerization was inferred from the crystal structure of the hexadecameric assembly (Hanazono et al. 2013). The mutation of either Phe6 or Phe7 of SpHSP16.0 to alanine resulted in a similar phenotype as the ScHSP26 truncations, showing an increased readiness to dissociate into smaller oligomers at lower temperatures (Hanazono et al. 2013). The role of the CTR in fungi sHSPs in assembly is less understood, while the structure of the SpHSP16.0 shows inter-subunit patching mediated by the conserved IxI-motif no attempts, to the best of our knowledge, have been made to assess the true role in assembly.

Information on the roles of the two ACD-flanking domains on assembly is equally sparse in plants. Some studies do suggest a possible role of the IxI-motif in assembly as removal of the CAM in *Pisum sativum* HSP17 (PsHSP17) resulted in a smaller species with a reduced capacity to interact with other sHSPs when expressed in tobacco protoplasts (Kirschner et al. 2000). A study where chimeras of TaHSP16.9 and PsHSP18.1, both class I sHSPs, were created showed a concentration-dependent oligomerisation for the chimeric construct containing the HSP16.9 NTR. This suggests that the assemblies formed were less stable, whereas the construct carrying the HSP18.1 NTR behaved like the WT HSP18.1 (Basha et al. 2006). Both observations allude to a role for the NTR in determining the propensity to oligomerise.

8.4.2 Assembly in Metazoans

The role of the CTR in assembly of metazoan sHSPs seems to differ from that in bacteria and archaea. While in the latter two kingdoms the IxI-motif is necessary for oligomerization, in animals the story is more complicated. A study with HSP26 of the arthropod *Artemia franciscana* showed that removal of the full CTR impaired oligomerization. At the same time truncation of the CTR by only ten residues – leaving the IxI intact – had little effect on the size of the protein (Crack et al. 2002; Sun et al. 2004). Studies with HSP16.2 from *Caenorhabditis elegans* (CeHSP16.2) revealed that mutating the CAM gave rise to larger oligomers with reduced stability, as determined by its aggregative behaviour upon freeze-thawing (Leroux et al. 1997). When the IxI-motif of human αA-crystallin was mutated to GxG, large and highly polydisperse oligomers were formed that were approximately three times the size of the wild-type protein. The same mutation in human αB-crystallin however only led to a marginal increase in size and polydispersity (Pasta et al. 2004). Increase in oligomer size is also seen for the Pro182Leu mutant of human HSPB1. This disease causing autosomal dominant mutation in the centre of the CAM sequence leads to the formation of large aggregates of the sHSP in vivo (Ackerley et al. 2006;

Almeida-Souza et al. 2010), a result we have also seen with the recombinantly produced and isolated mutant (unpublished data).

A study on porcine αB-crystallin that investigated the role of the C-terminal lysines on the structure and function found that truncating the CTR by ten or more residues reduced the solubility and thermostability of the protein, whereas mutation or deletion of the terminal lysines increased the thermostability (Liao et al. 2009). Equivalent work with human αB-crystallin also led to similar results. Mutation of both Lys174 and 175 to alanine or the IxI-motif to AxA led to the formation of slightly larger assemblies. Mutation of both glutamates at positions 164 and 165 dramatically reduced the thermostability of the protein, suggesting that the charged residues found in higher abundance in the CTR may be important in defining solubility of sHSPs and their assemblies (Treweek et al. 2007). These results were confirmed by a similar study on mouse HSPB1, where deletion of the C-terminal region downstream of the IxI-motif showed a decrease in thermostability without affecting the size of the oligomers (Lindner et al. 2000). Conversely, the same truncation in human HSPB1, increased the thermostability of this protein and eliminated the temperature-dependent increase in subunit exchange without affecting its overall size, suggesting a role for the flexible C-terminal tail in oligomer dissociation and thermal stability (Lelj-Garolla and Mauk 2012).

Overall mutations in the polar CTR of vertebrate sHSPs increase oligomer size, albeit only marginally in most cases. These results are completely opposite to the studies with archaea and eubacteria sHSPs, where mutation or deletion blocks higher order assembly. In the former species it thus appears that the CTR regulates the solubility and stability of the protein as well as subunit exchange rather than assembly (Andley et al. 1996; Aquilina et al. 2005; Kallur et al. 2008; Plater et al. 1996; Smulders et al. 1996). This is supported by recent biophysical characterisation using NMR and native MS of human αB-crystallin that suggest that the CAM, although conserved, is not patching the ACD at physiological temperatures (Baldwin et al. 2011a; Hilton et al. 2013).

While the CTR appears to have a limited role in assembly of animal sHSPs, removal of the full NTR consistently leads to disruption of oligomerization (Baranova et al. 2011; Feil et al. 2001; Kundu et al. 2007; Laganowsky et al. 2010). Even though most of our understanding of this taxonomic group comes from studies on vertebrate sHSPs, the crucial role of the NTR in assembly has also been confirmed in other metazoan species. Removal of the full N-terminal arm in HSP26 of *A. fransiscana* led to monomers and dimers, where the length of truncation correlated positively with the extent of dissociation (Crack et al. 2002; Sun et al. 2004). Deletion of the NTR in HSP16.2 of *C. elegans* also impaired oligomerization with the largest assembly being tetramers (Leroux et al. 1997).

Although complete removal of the NTR prevents oligomerization, truncations involving just the N-terminal end of this region usually have little effect on oligomer assembly. This was clearly demonstrated by trypsin digestion studies and truncation analysis on the human α-crystallins, where removal of the first 20 residues did not influence assembly (Kundu et al. 2007; Saha and Das 2004). Studies on HSPB1 from different species show mixed results. Truncation of the first 14 and 24 residues of human HSPB1 led to a minor decrease in size of the oligomers formed by this sHSP. Both deletions also abrogated the temperature-dependent increase in oligomer size seen for the wild-type protein, but had minimal effect on subunit exchange (Lelj-Garolla and Mauk 2012). Deletion of residues 5–23 from hamster HSPB1 limited self-association at the level of dimers (Lambert et al. 1999), while a study on mouse HSPB1 where the first 33 amino acids were deleted, had no effect on oligomerization (Guo and Cooper 2000). Even though the exact cause of these discrepancies remains unclear, one possible explanation could lie within either the difference in purification protocols or the methods used to assess the size of the assemblies.

Regions within the NTR have also been identified as involved in assembly albeit no single deletion has been reported to result in the same effects as complete removal of the NTR. In human αB-crystallin deletion of residues 54–61 led to a decrease in overall size (Santhoshkumar et al. 2009). The same effects were seen when deleting the conserved SRLFDQxFG-motif in α-crystallin of the same species – residues 20–28 in αA and 21–29 in αB – with a concomitant increase in subunit exchange (Pasta et al. 2003). Residues 37–54 were found to be involved in subunit-subunit interaction through peptide pin array studies on αB-crystallin from humans (Ghosh and Clark 2005). Finally deletion of residues 58–70 in human HSPB1 shifted the equilibrium, for both a wild-type and a phospho-mimicking mutant of this sHSP, towards large assemblies at protein concentrations that typically result in full dissociation of these complexes to the dimer (Shashidharamurthy et al. 2005).

While large sequence deletions have provided some insight into the role of the NTR in assembly, results from single point mutations have traditionally proven to be unsatisfactory in pinpointing key regions (Derham et al. 2001; Horwitz et al. 1998; Plater et al. 1996). The residue Phe28 of bovine αB-crystallin when mutated to a serine resulted in increased dissociation of the assembly at elevated temperatures (Kelley and Abraham 2003) Mutation of Phe27 of the same sHSP from mouse on the other hand had no observable effect on the size and assembly in another study (Plater et al. 1996), although it had a slight increase in size in separate work that employed the human orthologue (Horwitz et al. 1998). Pro39 of human αB-crystallin was also subjected to investigation (Numoto et al. 2012) which, according to the authors, is conserved in the archaeal MjHSP16.5 NTR. However when this residue was mutated to arginine, it led to only a minor increase in oligomer size and polydispersity. Similarly, mutation of another conserved proline in the same protein, Pro20 to serine – a mutation associated with cataract – had no observable effect on assembly, nor the size of the hetero-oligomeric complex with αA-crystallin (Li et al. 2008).

Information gathered on post-translational modifications, mostly phosphorylation on serines found in the NTR, has also implied a regulatory role for the NTR in the assembly and subunit dynamics of vertebrate sHSPs. A phospho-mimicking mutant of HSPB1 has been found to shift the equilibrium between oligomers and dimers in favor of the smaller assemblies (Shashidharamurthy et al. 2005). A second study by the same group confirmed these results and carried out a detailed analysis of the NTR in terms of assembly and substrate binding. Cysteine mutations in the NTR, introduced to conduct EPR analysis also affected the size and assembly of the oligomers, again accentuating the role of this domain in both properties

(McDonald et al. 2012). Hetero-oligomer formation of this phospho-mimicking mutant of HSPB1 with HSPB6 was increased when compared to wild type (Bukach et al. 2004). Furthermore, PTMs occurring with age have been investigated for the α-crystallins, where racemization of Asp58, Ser59 and Ser62 is seen in ageing cataractous lenses (Fujii et al. 2003; Hooi et al. 2013). Glycation and proteolysis have also been linked to ageing and development of cataract, but if and how these specific PTMs modify the assembly and structure of the lens crystallins remains to be elucidated (Groenen et al. 1994; Horwitz 2003; van Boekel et al. 1996).

8.4.3 Chaperone Activity in Non-metazoans

With few exceptions the NTR and CTR appear to have a dual role in oligomerisation and chaperone activity, therefore the results from the deletion and mutation studies presented in the previous section also overlap with the discussion here. Although the published data can at times appear to be contradictory, possibly reflecting the use of different substrates and experimental conditions between laboratories when testing the activity of these proteins, the general consensus is that the NTR plays a major role in chaperoning.

Truncation of the CTR, removing the IxI-motif of the archaeal MjHSP16.5 did not significantly alter its chaperoning capabilities (Kim et al. 2003; Quinlan et al. 2013), whereas mutation of the CAM in StHSP14.0 resulted in a non-functional chaperone (Saji et al. 2008). However, truncation of the NTR in both archaeal sHSPs impaired chaperone activity, with the length of the deletion positively correlating to the severity of the impairment (Kim et al. 2003; Usui et al. 2004).

Similar results were observed with bacterial sHSPs, where mutation of the IxI-motif impaired chaperone-like activity along with assembly (Giese et al. 2005; Giese and Vierling 2004; Jiao et al. 2005; Studer et al. 2002; Tomoyasu et al. 2010). A single exception to these results was seen with MtHSP16.3, where deletion of the CAM led to a more active chaperone (Fu et al. 2005). Deletion of the NTR again negatively affected activity in all studies. Removal of nine or more residues in HSPF or 30 or more residues from HSPH from *Bradyrhizobium japonicum* led to a measurable decrease in the ability to prevent aggregation of citrate synthase (Studer et al. 2002). Likewise, the first 35 residues from MtHSP16.3 and the first 27 residues of AsgA from *S. typhimurium* seem to be essential for chaperoning (Fu et al. 2005; Tomoyasu et al. 2010). Random mutagenesis of the *Synechocystis* HSP16.6 also highlighted a functional role for both the CTR and NTR. Single mutations in either domain impaired activity in vivo, as assessed by cell survival after heat treatment, where mutations in the NTR showed the least viability. Surprisingly, in vitro these mutants were still capable of keeping firefly luciferase in a folding-competent state, clearly underlining the limitations of the typical biochemical assays used to asses sHSP activity in vitro (Giese et al. 2005).

Truncation of the full NTR in yeast HSP26 completely abolished chaperone-like activity, whereas removal of the first 29 residues only slightly impaired it (Haslbeck et al. 2004). Mutation of a number of phenylalanines in the shorter NTR of SpHSP16.0 resulted in a reduction in the ability to prevent citrate synthase aggregation (Hanazono et al. 2013). Studies on the possible role of the CTR in activity are currently missing in this group of sequences.

The implication of the NTR in chaperoning was most clearly demonstrated in a study that incorporated a photo-activatable crosslinking analog of phenylalanine, *p*-benzoyl-L-phenylalanine (Bpa), into the pea PsHSP18.1 which showed preferential crosslinking of the NTR to two different substrates (Jaya et al. 2009). Interestingly, three of the mutations – that incorporated the unnatural amino acid – in the NTR fully protected malate dehydrogenase but not firefly luciferase from heat dependent aggregation. Combined with the fact that the crosslinking patterns seen for the two different substrates were not identical, these results hint at a possible discrimination of N-terminal sequences between substrates (Jaya et al. 2009) Substrate specificity was also seen in a study where chimeras of two different plant sHSPs were created. Substitution of the first four residues of TaHSP16.9 with the first ten of PsHSP18.1 proved sufficient to increase the chaperone-like activity of this chimera towards luciferase and citrate synthase when compared to wild-type TaHSP16.9 (Basha et al. 2006). Results from H/D exchange studies on both wild-type plant proteins indicated that the NTR is solvent exposed as measured by a high exchange. Surprisingly no change in the H/D exchange rate was seen between bound and unbound forms, indicating that the NTR does not adopt a new inaccessible conformation upon association with a substrate (Cheng et al. 2008). The NTR of HSP21 from *Arabidopsis thaliana* was also implicated in chaperoning via chemical crosslinking of lysine residues with heat-treated citrate synthase. Multiple crosslinks were observed between the substrate and the NTR, and a few with the CTR, supporting the idea that multiple binding sites exist (Ahrman et al. 2007). Substitution of the conserved methionines to leucines in the Met-bristle of the same protein did not result in a significant change in chaperone-like activity (Gustavsson et al. 2001). However sulfoxidation of the methionines by H_2O_2 did impair the ability to prevent both citrate synthase and insulin from aggregating, leading the authors to hypothesise a possible role for this conserved region as a redox sensor in the chloroplast (Härndahl et al. 2001).

8.4.4 Chaperone Activity in Metazoans

Involvement of both the NTR and the CTR in chaperoning has also been demonstrated for metazoan sHSPs, although again the role of the CTR in activity is not conclusive. In HSP26 from *A. franciscana*, removal of the full CTR severely hampered activity both in vivo and in vitro (Crack et al. 2002; Sun et al. 2004). Deletion of residues 169–186 from this sHSP, identified as a sequence with high serine/threonine content, also impaired chaperoning capabilities, but to a lesser extent (Sun and MacRae 2005). Removal of the full NTR of the same protein completely abrogated in vitro chaperone activity (Sun et al. 2004), although a second study showed no impairment to confer thermotolerance in vivo (Crack et al. 2002). Deletion of residues 36–45 in the NTR, designated as the arginine-rich region, resulted in a decrease in chaperone activity, albeit not fully abolished (Sun and MacRae 2005), supporting the hypothesis that these kingdom wide conserved residues may be important within this role.

Mutation of the IxI-motif in the nematode CeHSP16.2 had no observable effect on activity, but reduced the stability of the protein. As expected, truncation of the NTR reduced the chaperoning capabilities (Leroux et al. 1997). Blocking the NTR of *Drosophila melanogaster* HSP22 with a histidine-tag also reduced the ability to prevent aggregation of citrate synthase (Morrow et al. 2006). Removal of the first 17 amino acids in HSP30C from the amphibian *Xenopus laevis* did not result in a change in activity, whereas removal of 25 amino acids from the CTR impaired the ability to protect citrate synthase and keep luciferase in a folding-competent state, a result likely due to a decreased stability of the protein and its substrate-complexes (Fernando and Heikkila 2000). In a follow-up study, the negatively charged aspartate residues in the CTR were identified as main players in this phenomenon (Fernando et al. 2002).

The ample pool of studies on vertebrate sHSPs all support a conserved role of the NTR in chaperone activity, whereas results examining the CTR differ. Mutation of the IxI-motif in both human α-crystallins did not abolish their chaperone activity, on the contrary both mutants were active at lower temperatures than the wild-type proteins (Pasta et al. 2004). Truncation of the CTR of porcine αB-crystallin by 11 or more residues, leaving the CAM intact, greatly reduced the activity towards β_L-crystallin and insulin whereas smaller truncations increased the activity, a result also seen by mutating the terminal lysines to alanines. The largest improvement of activity was seen when both lysine 174 and 175 were mutated (Liao et al. 2009). Oppositely, mutating both lysine 174 and 175 in murine αB-crystallin to glycines or alanines suppressed the ability to protect insulin and γ-crystallin from aggregating (Plater et al. 1996). In a separate study, mutation of glutamate 164 and 165 to alanines reduced the chaperone-like activity of human αB-crystallin towards both β_L-crystallin and insulin, whereas mutation of the CAM diminished activity towards β_L-crystallin but enhanced the activity towards insulin. Such contradicting results were also seen when lysine 174 and 175 were mutated to alanines, although the changes in activity were not as dramatic (Treweek et al. 2007). Removal of the last 23 residues from the CTR in human HSPB1, that left the CAM intact, resulted in diminished activity at 20 °C but no difference at 40 °C (Lelj-Garolla and Mauk 2012).

Truncation of the NTR in human αA-crystallin with an increasing number of residues led to reduction of the chaperone-like activity, with the length of the truncation correlating with the decline in activity (Kundu et al. 2007). Again a different chaperoning profile was observed depending on the substrate, for β_L-crystallin and insulin as much as 35 residues could be deleted before a reduction in activity was seen, whereas it took only the first 10 residues to diminish the activity against carbonic anhydrase (Kundu et al. 2007). More or less the same results were seen when truncating the NTR of the dimeric human sHSP HSPB6, deletion of the first

56 residues resulted in a construct with no ability to chaperone insulin, whereas removal of the first 23 residues only slightly diminished the activity (Weeks et al. 2014). A short truncation has a stronger effect on human HSPB1, where removal of the first 14 residues was enough to limit the ability to chaperone reduced insulin (Lelj-Garolla and Mauk 2012). Interaction of the NTR with substrate proteins has also been studied using peptide pin-arrays, in which overlapping peptides corresponding to part of the NTR of human αB-crystallin were assessed for binding different substrates. Two main interacting domains were found spanning residues 9–20 and 43–58. Nonetheless, when looking at the identified peptides that protected β_H-crystallin, γD-crystallin, alcohol dehydrogenase or citrate synthase, the whole NTR lit up as being necessary for chaperoning, albeit with different regions necessary to protect the different substrates (Ghosh et al. 2005, 2006). In a similar study where peptides of mouse HSPB1 were assessed for their ability to inhibit actin polymerization, an isolated peptide corresponding to NTR residues 43–57 was also found to be enough, suggesting the possibility of the involvement of these residues in binding and regulating cytoskeleton proteins (Wieske et al. 2001).

A number of groups have also observed an increase in chaperone function upon deletion of sections of the NTR. Deletion of the phenylalanine-rich region in both αA- and αB-crystallin from humans showed higher activity than their wild-type counterparts (Pasta et al. 2003). Deletion of residues 31–35, encompassing the second half of this conserved region, showed the same effect on human HSPB6 activity (Heirbaut et al. 2014). Another study, deleting residues 54–61 in human αB-crystallin also enhanced the activity of this sHSP (Santhoshkumar et al. 2009). Even though counter-intuitive, these observations should not be considered surprising. These sequences may act as negative regulators of chaperoning, preventing unnecessary and unwanted binding. If the NTR would so much as recognize every single exposed hydrophobic site without some form of discrimination, one could imagine detrimental effects within the cell from simply binding any protein with such an exposed surface. The conservation of SRLFDQxFG motif within the animal kingdom suggests a similar fashion of self-regulation of sHSP activity within orthologues of different species.

As described in Sect. 8.4.1, single point mutation studies of vertebrate sHSPs have not only proven inadequate for pinpointing residues required for assembly, but also for chaperoning. All of the aforementioned studies also observed changes in activity although contradictory results have been published (Derham et al. 2001; Kelley and Abraham 2003; Numoto et al. 2012; Plater et al. 1996). Even though these results may prove disappointing in gaining molecular insights into the key residues necessary for chaperoning, they clearly underline the pivotal role of the NTR in regulating activity. The inconsistencies seen when using single point mutations may also reflect a certain level of redundancy in the N-terminal sequences or can be explained by differences in assay set-up. In addition, one could postulate that most likely multiple regions of the NTR are required for activity and that these regions may even have preferential binding partners, hence the observed disparities.

Studies on the effect of PTMs on chaperone-like activity also emphasize the role of the NTR. The phospho-mimicking mutant of human HSPB1 had increased affinity towards a destabilizing mutant of T4 lysozyme (Shashidharamurthy et al. 2005), likely an effect of the exposure of the NTR, as seen by a loss of restricted spin label motion in electron paramagnetic resonance studies (McDonald et al. 2012). Phosphorylation of mouse HSPB1 was found to abolish its actin-polymerization inhibiting activity (Benndorf et al. 1994). The opposite was seen for HSPB6, where the phosphomimicking mutant showed a decrease in chaperone-like activity (Bukach et al. 2004), but an increase in the ability to disrupt the actin cytoskeleton (Dreiza et al. 2005). In addition, in vivo phosphorylated α-crystallin purified from calf lenses, was found to have better capacity to prevent the aggregation of β_L-crystallin, with the observed effect being more pronounced for αA-crystallin than αB-crystallin (van Boekel et al. 1996).

8.5 Conclusions

Since their discovery sHSPs have been the subject of many investigations, both structurally and functionally. Even though the importance of the arms flanking the ACD was recognised early on, in the literature these regions are commonly described as being poorly conserved. We argue that this statement, to a general reader, undermines the understanding of their influence on the biology of these proteins. While this description holds true when comparing the sequence conservation of this group of chaperones with that of other HSP families, the functions of the NTR and the CTR clearly have been conserved across all kingdoms, a finding underlined by the many studies outlined here.

Analysis of the relative length of both domains shows an increase that is correlated with the complexity of the organism (Table 8.1). It is possible to suggest that this may simply reflect the fact that the proteome of eukaryotes contains larger multi-domained proteins. Despite the obvious difference in lengths across species there appears to be a conserved bias in the residue content of both the NTR and CTR that likely is important structurally and functionally (Fig. 8.3).

As outlined in Sect. 8.3, three-dimensional structural information on the regions flanking the ACD is limited, likely a result of their inherent propensity to be unstructured and flexible. In structures where part of the NTR was resolved, some secondary structure could be seen, but it remains likely that not all sequences within the oligomeric complex adopt the same structure at the same time, a finding supported by NMR and H/D exchange experiments (Cheng et al. 2008; Jehle et al. 2011). Even though structural knowledge has been obtained from crystal structures of a few monodisperse sHSPs, it is important to recognise that the majority of these proteins form polydisperse oligomers not amendable for high resolution structure determination. The use of techniques such as cryo-EM and native MS have recently begun to provide information about the structure of such entities, although the role of both the NTR and CTR in these assemblies is not fully resolved (Baldwin et al. 2011b; Braun et al. 2011; Jehle et al. 2011).

An important question still remaining is what the structure of a sHSP bound to a substrate looks like. One interesting strategy beginning to answer this challenging problem involves the detailed biophysical characterisation of the interaction of a sHSPs with a library of destabilised mutants of T4 lysozyme (McDonald et al. 2012; Shi et al. 2013). The use of this model substrate, supported by parallel studies of other proteins with known thermosensitive mutants, will be essential to achieve an understanding of the underlying mechanism and structural changes necessary for recognition and binding of a partially unfolded species.

While we have addressed the roles of both the N- and C-terminal regions in assembly and chaperoning separately, these characteristics are intertwined at the sequence level. This is especially clear in the prokaryotic sHSPs, where deletion or even a single point mutation in either domain has a concerted and profound effect on assembly and chaperoning. In higher eukaryotes, a more complex picture arises: even though the CTR contains the highly conserved IxI-motif the latter appears to have lost its indispensable function in stabilizing large inter-subunit assemblies, and seems to have evolved further as a stabilizer and/or solubiliser of sHSP-substrate complexes (Hilton et al. 2013; Kallur et al. 2008; Liao et al. 2009; Pasta et al. 2004). Indeed, counter-intuitively, recent studies on human αB-crystallin suggest that the CAM sequence is rarely bound at physiological temperatures, but actually acts as a switch region regulating dissociation of the individual protomers from the oligomeric assemblies (Baldwin et al. 2011a; Hilton et al. 2013). The majority of studies of the NTR have found that mutation or truncation is accompanied by a change in both activity and assembly. Large truncations all favour the formation of smaller assemblies usually with no capacity to chaperone, clearly advocating the crucial role of this region across all kingdoms of life.

Attempts to understand the exact role of the NTR in chaperoning are principally limited to deletion studies. The results typically underline the importance of sequence stretches rather than single amino acids in defining activity. In addition it seems that there is not a single region solely responsible for chaperoning and that multiple sites may act mutually on the same substrate. In vitro isolated sHSPs are capable of chaperoning a multitude of different test substrates, albeit with somewhat different effectiveness and specificity (Basha et al. 2006; Braun et al. 2011; Jaya et al. 2009), a finding also underlined by in vivo studies identifying natural substrates (Andley et al. 2014; Basha et al. 2004; Fu et al. 2013). This promiscuity of the NTR towards the very diverse array of proteins that are commonly employed can be explained by the specific residue bias observed in this region that would presumably permit a similar mode of recognition and binding of an unstable species. The reported substrate specificity could in part be defined by the remaining differences in sequence that fine-tune the activity and specificity.

It is crucial though to point out that the substrates commonly used to assess chaperone activity have some specific limitations. These problems include method-specific drawbacks such as the all-or-nothing induced aggregation, via either incubation at high temperatures or the reduction of disulfide bonds, and substrate-related issues such as the use of model proteins, for example insulin, that are highly unlikely to ever come into contact with a sHSP in a host. These issues can lead to false read-outs that possibly gave rise to the comparative discrepancies described in Sects. 8.4.3 and 8.4.4. A clear example of these failings can be found in the study by Giese et al. (2005) where mutations in the NTR that negatively affected the viability of bacteria, following a short incubation at heat-shock temperature, had no observable effect in vitro on the ability to chaperone the commonly employed substrate luciferase. Future efforts to delineate the roles of the ACD flanking regions should thus take this into consideration. In absence of an ideal substrate, chaperone assays looking at the prevention of amyloid formation may be more suitable (Bruinsma et al. 2011; Raman et al. 2005; Waudby et al. 2010). It is also increasingly recognised that some members of the sHSP family appear to interact with specific proteins for example, HSPB1 with pro-caspase 3 (Voss et al. 2007) and HSPB6 with protein phosphatase 1 (Qian et al. 2011). Although unique for these particular sHSPs, studies of the sequence determinants necessary for activity may yield a better picture of the role of the NTR and CTR in function.

Despite the exponential growth in our understanding of the biology of sHSPs, their story has not been fully told. Crucially, further details of the role of the NTR and CTR in defining function are essential. The application of new biophysical techniques and improved assays, together with established methods, are set to begin to explain in molecular detail the kingdom wide importance of these 'non-conserved' regions.

Acknowledgements M.H. is funded by the Department of Pharmaceutical and Pharmacological Sciences, KU Leuven. S.D.W. acknowledges the support of Marie Curie Reintegration Grant. This research was further supported by the grant OT13/097 from the KU Leuven, and the grant G.0936.15 from the Research Foundation Flanders/FWO, both awarded to S.V.S. The authors would like to thank Drs. Alexander Shkumatov and Nikolai Sluchanko for their careful review of the chapter and useful suggestions.

References

Ackerley S, James PA, Kalli A, French S, Davies KE, Talbot K (2006) A mutation in the small heat-shock protein HSPB1 leading to distal hereditary motor neuronopathy disrupts neurofilament assembly and the axonal transport of specific cellular cargoes. Hum Mol Genet 15:347–354

Ahrman E, Lambert W, Aquilina JA, Robinson CV, Emanuelsson CS (2007) Chemical cross-linking of the chloroplast localized small heat-shock protein, Hsp21, and the model substrate citrate synthase. Protein Sci 16:1464–1478

Almeida-Souza L, Goethals S, de Winter V, Dierick I, Gallardo R, Van Durme J, Irobi J, Gettemans J, Rousseau F, Schymkowitz J, Timmerman V, Janssens S (2010) Increased monomerization of mutant HSPB1 leads to protein hyperactivity in Charcot-Marie-Tooth neuropathy. J Biol Chem 285:12778–12786

Andley UP, Mathur S, Griest TA, Petrash JM (1996) Cloning, expression, and chaperone-like activity of human alphaA-crystallin. J Biol Chem 271:31973–31980

Andley UP, Malone JP, Townsend RR (2014) In vivo substrates of the lens molecular chaperones αA-crystallin and αB-crystallin. PLoS One 9:e95507

Aquilina JA, Benesch JLP, Ding LL, Yaron O, Horwitz J, Robinson CV (2005) Subunit exchange of polydisperse proteins: mass spectrometry reveals consequences of alphaA-crystallin truncation. J Biol Chem 280:14485–14491

Arakawa T, Ejima D, Tsumoto K, Obeyama N, Tanaka Y, Kita Y, Timasheff SN (2007) Suppression of protein interactions by arginine: a proposed mechanism of the arginine effects. Biophys Chem 127:1–8

Bagnéris C, Bateman OA, Naylor CE, Cronin N, Boelens WC, Keep NH, Slingsby C (2009) Crystal structures of α-crystallin domain dimers of αB-crystallin and Hsp20. J Mol Biol 392:1242–1252

Baldwin AJ, Hilton GR, Lioe H, Bagnéris C, Benesch JLP, Kay LE (2011a) Quaternary dynamics of αB-crystallin as a direct consequence of localised tertiary fluctuations in the C-terminus. J Mol Biol 413:310–320

Baldwin AJ, Lioe H, Hilton GR, Baker LA, Rubinstein JL, Kay LE, Benesch JLP (2011b) The polydispersity of αB-crystallin is rationalized by an interconverting polyhedral architecture. Structure 19:1855–1863

Baranova EV, Weeks SD, Beelen S, Bukach OV, Gusev NB, Strelkov SV (2011) Three-dimensional structure of α-crystallin domain dimers of two human small heat shock proteins, HSPB1 and HSPB6. J Mol Biol 411:110–122

Basha E, Lee GJ, Breci LA, Hausrath AC, Buan NR, Giese KC, Vierling E (2004) The identity of proteins associated with a small heat shock protein during heat stress in vivo indicates that these chaperones protect a wide range of cellular functions. J Biol Chem 279:7566–7575

Basha E, Friedrich KL, Vierling E (2006) The N-terminal arm of small heat shock proteins is important for both chaperone activity and substrate specificity. J Biol Chem 281:39943–39952

Benndorf R, Hayess K, Ryazantsev S, Wieske M, Behlke J, Lutsch G (1994) Phosphorylation and supramolecular organization of murine small heat shock protein HSP25 abolish its actin polymerization-inhibiting activity. J Biol Chem 269:20780–20784

Benndorf R, Martin JL, Kosakovsky Pond SL, Wertheim JO (2014) Neuropathy- and myopathy-associated mutations in human small heat shock proteins: characteristics and evolutionary history of the mutation sites. Mutat Res Rev Mutat Res. doi:10.1016/j.mrrev.2014.02.004

Bepperling A, Alte F, Kriehuber T, Braun N, Weinkauf S, Groll M, Haslbeck M, Buchner J (2012) Alternative bacterial two-component small heat shock protein systems. Proc Natl Acad Sci U S A 109:20407–20412

Bondino HG, Valle EM, Ten Have A (2011) Evolution and functional diversification of the small heat shock protein/α-crystallin family in higher plants. Planta 235:1299–1313

Braun N, Zacharias M, Peschek J, Kastenmüller A, Zou J, Hanzlik M, Haslbeck M, Rappsilber J, Buchner J, Weinkauf S (2011) Multiple molecular architectures of the eye lens chaperone αB-crystallin elucidated by a triple hybrid approach. Proc Natl Acad Sci U S A 108:20491–20496

Bruinsma IB, Bruggink KA, Kinast K, Versleijen AAM, Segers-Nolten IMJ, Subramaniam V, Kuiperij HB, Boelens W, de Waal RMW, Verbeek MM (2011) Inhibition of α-synuclein aggregation by small heat shock proteins. Proteins 79:2956–2967

Bukach OV, Seit-Nebi AS, Marston SB, Gusev NB (2004) Some properties of human small heat shock protein Hsp20 (HspB6). Eur J Biochem 271:291–302

Campanella JJ, Bitincka L, Smalley J (2003) MatGAT: an application that generates similarity/identity matrices using protein or DNA sequences. BMC Bioinformatics 4:29

Cheng G, Basha E, Wysocki VH, Vierling E (2008) Insights into small heat shock protein and substrate structure during chaperone action derived from hydrogen/deuterium exchange and mass spectrometry. J Biol Chem 283:26634–26642

Clauwaert J, Ellerton HD, Koretz JF, Thomson K, Augusteyn RC (1989) The effect of temperature on the renaturation of alpha-crystallin. Curr Eye Res 8:397–403

Crabbe MJ, Goode D (1994) Alpha-crystallin: chaperoning and aggregation. Biochem J 297(Pt 3):653–654

Crack JA, Mansour M, Sun Y, MacRae TH (2002) Functional analysis of a small heat shock/alpha-crystallin protein from Artemia franciscana. Oligomerization and thermotolerance. Eur J Biochem 269:933–942

Crooks GE, Hon G, Chandonia J-M, Brenner SE (2004) WebLogo: a sequence logo generator. Genome Res 14:1188–1190

De Jong WW, Caspers GJ, Leunissen JA (1998) Genealogy of the alpha-crystallin–small heat-shock protein superfamily. Int J Biol Macromol 22:151–162

Delbecq SP, Jehle S, Klevit R (2012) Binding determinants of the small heat shock protein, αB-crystallin: recognition of the "IxI" motif. EMBO J 31:4587–4594

Derham BK, van Boekel MA, Muchowski PJ, Clark JI, Horwitz J, Hepburne-Scott HW, de Jong WW, Crabbe MJ, Harding JJ (2001) Chaperone function of mutant versions of alpha A- and alpha B-crystallin prepared to pinpoint chaperone binding sites. Eur J Biochem 268:713–721

Dreiza CM, Brophy CM, Komalavilas P, Furnish EJ, Joshi L, Pallero MA, Murphy-Ullrich JE, von Rechenberg M, Ho YJ, Richardson B, Xu N, Zhen Y, Peltier JM, Panitch A (2005) Transducible heat shock protein 20 (HSP20) phosphopeptide alters cytoskeletal dynamics. FASEB J 19:261–263

Dreiza C, Komalavilas P, Furnish E, Flynn C, Sheller M, Smoke C, Lopes L, Brophy C (2010) The small heat shock protein, HSPB6, in muscle function and disease. Cell Stress Chaperones 15:1–11

Feil IK, Malfois M, Hendle J, van Der Zandt H, Svergun DI (2001) A novel quaternary structure of the dimeric alpha-crystallin domain with chaperone-like activity. J Biol Chem 276:12024–12029

Fernando P, Heikkila JJ (2000) Functional characterization of Xenopus small heat shock protein, Hsp30C: the carboxyl end is required for stability and chaperone activity. Cell Stress Chaperones 5:148–159

Fernando P, Abdulle R, Mohindra A, Guillemette JG, Heikkila JJ (2002) Mutation or deletion of the C-terminal tail affects the function and structure of Xenopus laevis small heat shock protein, hsp30. Comp Biochem Physiol B Biochem Mol Biol 133:95–103

Finn RD, Bateman A, Clements J, Coggill P, Eberhardt RY, Eddy SR, Heger A, Hetherington K, Holm L, Mistry J, Sonnhammer ELL, Tate J, Punta M (2014) Pfam: the protein families database. Nucleic Acids Res 42:D222–D230

Franck E, Madsen O, Rheede T, Ricard G, Huynen MA, Jong WW (2004) Evolutionary diversity of vertebrate small heat shock proteins. J Mol Evol 59:792–805

Fu X, Zhang H, Zhang X, Cao Y, Jiao W, Liu C, Song Y, Abulimiti A, Chang Z (2005) A dual role for the N-terminal region of Mycobacterium tuberculosis Hsp16.3 in self-oligomerization and binding denaturing substrate proteins. J Biol Chem 280:6337–6348

Fu X, Shi X, Yin L, Liu J, Joo K, Lee J, Chang Z (2013) Small heat shock protein IbpB acts as a robust chaperone in living cells by hierarchically activating its multi-type substrate-binding residues. J Biol Chem 288:11897–11906

Fujii N, Awakura M, Takemoto L, Inomata M, Takata T, Fujii N, Saito T (2003) Characterization of alphaA-crystallin from high molecular weight aggregates in the normal human lens. Mol Vis 9:315–322

Ghosh JG, Clark JI (2005) Insights into the domains required for dimerization and assembly of human alphaB crystallin. Protein Sci 14:684–695

Ghosh JG, Estrada MR, Clark JI (2005) Interactive domains for chaperone activity in the small heat shock protein, human alphaB crystallin. Biochemistry 44:14854–14869

Ghosh JG, Shenoy AK, Clark JI (2006) N- and C-Terminal motifs in human alphaB crystallin play an important role in the recognition, selection, and solubilization of substrates. Biochemistry 45:13847–13854

Giese KC, Vierling E (2004) Mutants in a small heat shock protein that affect the oligomeric state. Analysis and allele-specific suppression. J Biol Chem 279:32674–32683

Giese KC, Basha E, Catague BY, Vierling E (2005) Evidence for an essential function of the N terminus of a small heat shock protein in vivo, independent of in vitro chaperone activity. Proc Natl Acad Sci U S A 102:18896–18901

Gouy M, Guindon S, Gascuel O (2010) SeaView version 4: a multiplatform graphical user interface for sequence alignment and phylogenetic tree building. Mol Biol Evol 27:221–224

Groenen PJ, Merck KB, de Jong WW, Bloemendal H (1994) Structure and modifications of the junior chaperone alpha-crystallin. From lens transparency to molecular pathology. Eur J Biochem 225:1–19

Guo Z, Cooper LF (2000) An N-terminal 33-amino-acid-deletion variant of hsp25 retains oligomerization and functional properties. Biochem Biophys Res Commun 270:183–189

Gustavsson N, Kokke BP, Anzelius B, Boelens WC, Sundby C (2001) Substitution of conserved methionines by leucines in chloroplast small heat shock protein results in loss of redox-response but retained chaperone-like activity. Protein Sci 10:1785–1793

Hanazono Y, Takeda K, Yohda M, Miki K (2012) Structural studies on the oligomeric transition of a small heat shock protein, StHsp14.0. J Mol Biol 422:100–108

Hanazono Y, Takeda K, Oka T, Abe T, Tomonari T, Akiyama N, Aikawa Y, Yohda M, Miki K (2013) Nonequivalence observed for the 16-meric structure of a small heat shock protein, SpHsp16.0, from Schizosaccharomyces pombe. Structure 21:220–228

Härndahl U, Kokke BP, Gustavsson N, Linse S, Berggren K, Tjerneld F, Boelens WC, Sundby C (2001) The chaperone-like activity of a small heat shock protein is lost after sulfoxidation of conserved methionines in a surface-exposed amphipathic alpha-helix. Biochim Biophys Acta 1545:227–237

Haslbeck M, Ignatiou A, Saibil H, Helmich S, Frenzl E, Stromer T, Buchner J (2004) A domain in the N-terminal part of Hsp26 is essential for chaperone function and oligomerization. J Mol Biol 343:445–455

Haslbeck M, Franzmann T, Weinfurtner D, Buchner J (2005) Some like it hot: the structure and function of small heat-shock proteins. Nat Struct Mol Biol 12:842–846

Heirbaut M, Beelen S, Strelkov SV, Weeks SD (2014) Dissection the functional role of the N-terminal domain of the human small heat shock protein HSPB6. PLoS One 9:e105892

Hilario E, Martin FJM, Bertolini MC, Fan L (2011) Crystal structures of xanthomonas small heat shock protein provide a structural basis for an active molecular chaperone oligomer. J Mol Biol 408:74–86

Hilton GR, Hochberg GKA, Laganowsky A, McGinnigle SI, Baldwin AJ, Benesch JLP (2013) C-terminal interactions mediate the quaternary dynamics of αB-crystallin. Philos Trans R Soc Lond B Biol Sci 368:20110405

Hochberg GKA, Ecroyd H, Liu C, Cox D, Cascio D, Sawaya MR, Collier MP, Stroud J, Carver JA, Baldwin AJ, Robinson CV, Eisenberg DS, Benesch JLP, Laganowsky A (2014) The structured core domain of αB-crystallin can prevent amyloid fibrillation and associated toxicity. Proc Natl Acad Sci U S A 111:E1562–E1570

Hooi MYS, Raftery MJ, Truscott RJW (2013) Age-dependent racemization of serine residues in a human chaperone protein. Protein Sci 22:93–100

Horwitz J (2003) Alpha-crystallin. Exp Eye Res 76:145–153

Horwitz J, Bova M, Huang QL, Ding L, Yaron O, Lowman S (1998) Mutation of alpha B-crystallin: effects on chaperone-like activity. Int J Biol Macromol 22:263–269

Jaya N, Garcia V, Vierling E (2009) Substrate binding site flexibility of the small heat shock protein molecular chaperones. Proc Natl Acad Sci U S A 106:15604–15609

Jehle S, Rajagopal P, Bardiaux B, Markovic S, Kühne R, Stout JR, Higman VA, Klevit RE, van Rossum B-J, Oschkinat H (2010) Solid-state NMR and SAXS studies provide a structural basis for the activation of alphaB-crystallin oligomers. Nat Struct Mol Biol 17:1037–1042

Jehle S, Vollmar BS, Bardiaux B, Dove KK, Rajagopal P, Gonen T, Oschkinat H, Klevit RE (2011) N-terminal domain of alphaB-crystallin provides a conformational switch for multimerization and structural heterogeneity. Proc Natl Acad Sci U S A 108:6409–6414

Jiao W, Qian M, Li P, Zhao L, Chang Z (2005) The essential role of the flexible termini in the temperature-responsiveness of the oligomeric state and chaperone-like activity for the polydisperse small heat shock protein IbpB from Escherichia coli. J Mol Biol 347:871–884

Kallur LS, Aziz A, Abraham EC (2008) C-Terminal truncation affects subunit exchange of human alphaA-crystallin with alphaB-crystallin. Mol Cell Biochem 308:85–91

Kelley PB, Abraham EC (2003) Thermally induced disintegration of the oligomeric structure of alphaB-crystallin mutant F28S is associated with diminished chaperone activity. Mol Cell Biochem 252:273–278

Kim KK, Kim R, Kim SH (1998) Crystal structure of a small heat-shock protein. Nature 394:595–599

Kim R, Lai L, Lee H-H, Cheong G-W, Kim KK, Wu Z, Yokota H, Marqusee S, Kim S-H (2003) On the mechanism of chaperone activity of the small heat-shock protein of Methanococcus jannaschii. Proc Natl Acad Sci U S A 100:8151–8155

Kirschner M, Winkelhaus S, Thierfelder JM, Nover L (2000) Transient expression and heat-stress-induced co-aggregation of endogenous and heterologous small heat-stress proteins in tobacco protoplasts. Plant J 24:397–411

Kriehuber T, Rattei T, Weinmaier T, Bepperling A, Haslbeck M, Buchner J (2010) Independent evolution of the core domain and its flanking sequences in small heat shock proteins. FASEB J 24:3633–3642

Kundu M, Sen PC, Das KP (2007) Structure, stability, and chaperone function of alphaA-crystallin: role of N-terminal region. Biopolymers 86:177–192

Laganowsky A, Eisenberg D (2010) Non-3D domain swapped crystal structure of truncated zebrafish alphaA crystallin. Protein Sci 19:1978–1984

Laganowsky A, Benesch JLP, Landau M, Ding L, Sawaya MR, Cascio D, Huang Q, Robinson CV, Horwitz J, Eisenberg D (2010) Crystal structures of truncated alphaA and alphaB crystallins reveal structural mechanisms of polydispersity important for eye lens function. Protein Sci 19:1031–1043

Lambert H, Charette SJ, Bernier AF, Guimond A, Landry J (1999) HSP27 multimerization mediated by phosphorylation-sensitive intermolecular interactions at the amino terminus. J Biol Chem 274:9378–9385

Laskowski RA, Swindells MB (2011) LigPlot+: multiple ligand-protein interaction diagrams for drug discovery. J Chem Inf Model 51:2778–2786

Lelj-Garolla B, Mauk AG (2005) Self-association of a small heat shock protein. J Mol Biol 345:631–642

Lelj-Garolla B, Mauk AG (2012) Roles of the N- and C-terminal sequences in Hsp27 self-association and chaperone activity. Protein Sci 21:122–133

Leroux MR, Melki R, Gordon B, Batelier G, Candido EP (1997) Structure-function studies on small heat shock protein oligomeric assembly and interaction with unfolded polypeptides. J Biol Chem 272:24646–24656

Li H, Li C, Lu Q, Su T, Ke T, Li DW-C, Yuan M, Liu J, Ren X, Zhang Z, Zeng S, Wang QK, Liu M (2008) Cataract mutation P20S of alphaB-crystallin impairs chaperone activity of alphaA-crystallin and induces apoptosis of human lens epithelial cells. Biochim Biophys Acta 1782:303–309

Liao J-H, Lee J-S, Wu S-H, Chiou S-H (2009) COOH-terminal truncations and site-directed mutations enhance thermostability and chaperone-like activity of porcine alphaB-crystallin. Mol Vis 15:1429–1444

Lindner RA, Carver JA, Ehrnsperger M, Buchner J, Esposito G, Behlke J, Lutsch G, Kotlyarov A, Gaestel M (2000) Mouse Hsp25, a small shock protein. The role of its C-terminal extension in oligomerization and chaperone action. Eur J Biochem 267:1923–1932

Liu Z, Yao P, Guo X, Xu B (2014) Two small heat shock protein genes in Apis cerana cerana: characterization, regulation, and developmental expression. Gene 545:205–214

Maaroufi H, Tanguay RM (2013) Analysis and phylogeny of small heat shock proteins from marine viruses and their cyanobacteria host. PLoS One 8:e81207

McDonald ET, Bortolus M, Koteiche HA, McHaourab HS (2012) Sequence, structure, and dynamic determinants of Hsp27 (HspB1) equilibrium dissociation are encoded by the N-terminal domain. Biochemistry 51:1257–1268

Morrow G, Heikkila JJ, Tanguay RM (2006) Differences in the chaperone-like activities of the four main small heat shock proteins of Drosophila melanogaster. Cell Stress Chaperones 11:51–60

Numoto N, Kita A, Fujii N, Miki K (2012) A P39R mutation at the N-terminal domain of human αB-crystallin regulates its oligomeric state and chaperone-like activity. Biochem Biophys Res Commun 425:601–606

Pasta SY, Raman B, Ramakrishna T, Rao CM (2003) Role of the conserved SRLFDQFFG region of alpha-crystallin, a small heat shock protein. Effect on oligomeric size, subunit exchange, and chaperone-like activity. J Biol Chem 278:51159–51166

Pasta SY, Raman B, Ramakrishna T, Rao CM (2004) The IXI/V motif in the C-terminal extension of alpha-crystallins: alternative interactions and oligomeric assemblies. Mol Vis 10:655–662
Plater ML, Goode D, Crabbe MJ (1996) Effects of site-directed mutations on the chaperone-like activity of alphaB-crystallin. J Biol Chem 271:28558–28566
Poulain P, Gelly J-C, Flatters D (2010) Detection and architecture of small heat shock protein monomers. PLoS One 5:e9990
Qian J, Vafiadaki E, Florea SM, Singh VP, Song W, Kung Lam C, Wang Y, Yuan Q, Pritchard TJ, Cai W, Haghighi K, Rodriguez P, Wang H-S, Sanoudou D, Fan G-C, Kranias EG (2011) Small heat shock protein 20 interacts with protein phosphatase-1 and enhances sarcoplasmic reticulum calcium cycling. Circ Res 108:1429–1438
Quinlan RA, Zhang Y, Lansbury A, Williamson I, Pohl E, Sun F (2013) Changes in the quaternary structure and function of MjHSP16.5 attributable to deletion of the IXI motif and introduction of the substitution, R107G, in the α-crystallin domain. Philos Trans R Soc Lond B Biol Sci 368:20120327
Raman B, Ban T, Sakai M, Pasta SY, Ramakrishna T, Naiki H, Goto Y, Rao CM (2005) AlphaB-crystallin, a small heat-shock protein, prevents the amyloid fibril growth of an amyloid beta-peptide and beta2-microglobulin. Biochem J 392:573–581
Ren J, Wen L, Gao X, Jin C, Xue Y, Yao X (2009) DOG 1.0: illustrator of protein domain structures. Cell Res 19:271–273
Saha S, Das KP (2004) Relationship between chaperone activity and oligomeric size of recombinant human alphaA- and alphaB-crystallin: a tryptic digestion study. Proteins 57:610–617
Saji H, Iizuka R, Yoshida T, Abe T, Kidokoro S-I, Ishii N, Yohda M (2008) Role of the IXI/V motif in oligomer assembly and function of StHsp14.0, a small heat shock protein from the acidothermophilic archaeon, Sulfolobus tokodaii strain 7. Proteins 71:771–782
Santhoshkumar P, Murugesan R, Sharma KK (2009) Deletion of (54)FLRAPSWF(61) residues decreases the oligomeric size and enhances the chaperone function of alphaB-crystallin. Biochemistry 48:5066–5073
Scharf KD, Siddique M, Vierling E (2001) The expanding family of Arabidopsis thaliana small heat stress proteins and a new family of proteins containing alpha-crystallin domains (Acd proteins). Cell Stress Chaperones 6:225–237
Shah D, Li J, Shaikh AR, Rajagopalan R (2012) Arginine-aromatic interactions and their effects on arginine-induced solubilization of aromatic solutes and suppression of protein aggregation. Biotechnol Prog 28:223–231
Sharma KK, Santhoshkumar P (2009) Lens aging: effects of crystallins. Biochim Biophys Acta 1790:1095–1108
Shashidharamurthy R, Koteiche HA, Dong J, McHaourab HS (2005) Mechanism of chaperone function in small heat shock proteins: dissociation of the HSP27 oligomer is required for recognition and binding of destabilized T4 lysozyme. J Biol Chem 280:5281–5289
Shi J, Koteiche HA, McDonald ET, Fox TL, Stewart PL, McHaourab HS (2013) Cryoelectron microscopy analysis of small heat shock protein 16.5 (Hsp16.5) complexes with T4 lysozyme reveals the structural basis of multimode binding. J Biol Chem 288:4819–4830
Siddique M, Gernhard S, von Koskull-Döring P, Vierling E, Scharf K-D (2008) The plant sHSP superfamily: five new members in Arabidopsis thaliana with unexpected properties. Cell Stress Chaperones 13:183–197
Smulders RH, Carver JA, Lindner RA, van Boekel MA, Bloemendal H, de Jong WW (1996) Immobilization of the C-terminal extension of bovine alphaA-crystallin reduces chaperone-like activity. J Biol Chem 271:29060–29066
Soskić V, Groebe K, Schrattenholz A (2008) Nonenzymatic posttranslational protein modifications in ageing. Exp Gerontol 43:247–257
Stromer T, Fischer E, Richter K, Haslbeck M, Buchner J (2004) Analysis of the regulation of the molecular chaperone Hsp26 by temperature-induced dissociation: the N-terminal domail is important for oligomer assembly and the binding of unfolding proteins. J Biol Chem 279:11222–11228
Studer S, Obrist M, Lentze N, Narberhaus F (2002) A critical motif for oligomerization and chaperone activity of bacterial alpha-heat shock proteins. Eur J Biochem 269:3578–3586

Sun Y, MacRae TH (2005) Characterization of novel sequence motifs within N- and C-terminal extensions of p26, a small heat shock protein from Artemia franciscana. FEBS J 272:5230–5243

Sun Y, Mansour M, Crack JA, Gass GL, MacRae TH (2004) Oligomerization, chaperone activity, and nuclear localization of p26, a small heat shock protein from Artemia franciscana. J Biol Chem 279:39999–40006

Takeda K, Hayashi T, Abe T, Hirano Y, Hanazono Y, Yohda M, Miki K (2011) Dimer structure and conformational variability in the N-terminal region of an archaeal small heat shock protein, StHsp14.0. J Struct Biol 174:92–99

Theriault JR (2004) Essential role of the NH2-terminal WD/EPF motif in the phosphorylation-activated protective function of mammalian Hsp27. J Biol Chem 279:23463–23471

Tomoyasu T, Tabata A, Nagamune H (2010) Investigation of the chaperone function of the small heat shock protein-AgsA. BMC Biochem 11:27

Treweek TM, Ecroyd H, Williams DM, Meehan S, Carver JA, Walker MJ (2007) Site-directed mutations in the C-terminal extension of human alphaB-crystallin affect chaperone function and block amyloid fibril formation. PLoS One 2:e1046

Usui K, Hatipoglu OF, Ishii N, Yohda M (2004) Role of the N-terminal region of the crenarchaeal sHsp, StHsp14.0, in thermal-induced disassembly of the complex and molecular chaperone activity. Biochem Biophys Res Commun 315:113–118

Van Boekel MA, Hoogakker SE, Harding JJ, de Jong WW (1996) The influence of some post-translational modifications on the chaperone-like activity of alpha-crystallin. Ophthalmic Res 28(Suppl 1):32–38

Van Der Ouderaa FJ, De Jong WW, Hilderink A, Bloemendal H (1974) The amino-acids sequence of the alphaB2 chain of bovine alpha-crystallin. Eur J Biochem 49:157–168

Van Montfort RL, Basha E, Friedrich KL, Slingsby C, Vierling E (2001) Crystal structure and assembly of a eukaryotic small heat shock protein. Nat Struct Biol 8:1025–1030

Vos MJ, Kanon B, Kampinga HH (2009) HSPB7 is a SC35 speckle resident small heat shock protein. Biochim Biophys Acta 1793:1343–1353

Voss OH, Batra S, Kolattukudy SJ, Gonzalez-Mejia ME, Smith JB, Doseff AI (2007) Binding of caspase-3 prodomain to heat shock protein 27 regulates monocyte apoptosis by inhibiting caspase-3 proteolytic activation. J Biol Chem 282:25088–25099

Vucetic S, Brown CJ, Dunker AK, Obradovic Z (2003) Flavors of protein disorder. Proteins 52:573–584

Waterhouse AM, Procter JB, Martin DMA, Clamp M, Barton GJ (2009) Jalview version 2–a multiple sequence alignment editor and analysis workbench. Bioinformatics 25:1189–1191

Waters ER (2013) The evolution, function, structure, and expression of the plant sHSPs. J Exp Bot 64:391–403

Waters ER, Vierling E (1999a) Chloroplast small heat shock proteins: evidence for atypical evolution of an organelle-localized protein. Proc Natl Acad Sci U S A 96:14394–14399

Waters ER, Vierling E (1999b) The diversification of plant cytosolic small heat shock proteins preceded the divergence of mosses. Mol Biol Evol 16:127–139

Waters ER, Aevermann BD, Sanders-Reed Z (2008) Comparative analysis of the small heat shock proteins in three angiosperm genomes identifies new subfamilies and reveals diverse evolutionary patterns. Cell Stress Chaperones 13:127–142

Waudby CA, Knowles TPJ, Devlin GL, Skepper JN, Ecroyd H, Carver JA, Welland ME, Christodoulou J, Dobson CM, Meehan S (2010) The interaction of alphaB-crystallin with mature alpha-synuclein amyloid fibrils inhibits their elongation. Biophys J 98:843–851

Weeks SD, Baranova EV, Heirbaut M, Beelen S, Shkumatov AV, Gusev NB, Strelkov SV (2014) Molecular structure and dynamics of the dimeric human small heat shock protein HSPB6. J Struct Biol 185:342–354

White HE, Orlova EV, Chen S, Wang L, Ignatiou A, Gowen B, Stromer T, Franzmann TM, Haslbeck M, Buchner J, Saibil HR (2006) Multiple distinct assemblies reveal conformational flexibility in the small heat shock protein Hsp26. Structure 14:1197–1204

Wieske M, Benndorf R, Behlke J, Dölling R, Grelle G, Bielka H, Lutsch G (2001) Defined sequence segments of the small heat shock proteins HSP25 and alphaB-crystallin inhibit actin polymerization. Eur J Biochem 268:2083–2090

Chapter 9
HSPB6 (Hsp20) as a Versatile Molecular Regulator

Maria V. Sudnitsyna, Nikolai N. Sluchanko, and Nikolai B. Gusev

Abstract HspB6 (Hsp20) is a member of the large family of human small heat shock proteins. In contrast to other human small heat shock proteins (HspB1, HspB5) forming large oligomers, HspB6 predominantly forms stable dimers which tend to self-association. HspB6 is ubiquitously expressed in practically all human tissues, undergoes posttranslational modifications (such as phosphorylation and acetylation) and forms heterooligomeric complexes with two other human small heat shock proteins, HspB1 and HspB5. Possessing chaperone-like activity, HspB6 prevents aggregation of amyloid-β and α-synuclein, decreases cytotoxicity induced by accumulation of amyloids and, interacting with Bag3, modulates autophagosomal degradation of misfolded proteins. HspB6 protects cardiomyocytes from ischemia/reperfusion injuries, prevents cardiac hypertrophy and, possessing anti-apoptotic activity, protects cardiomyocytes from different unfavorable conditions. Phosphorylation of HspB6 catalyzed by cyclic nucleotide-dependent protein kinases induces relaxation of different smooth muscle. Exact molecular mechanism underlying relaxation effect of phosphorylated HspB6 in smooth muscle remains enigmatic, but seems to be dependent on the remodeling of actin cytoskeleton. HspB6 is not a genuine actin-binding protein, but interacting with universal adapter protein 14-3-3 seems to be able indirectly affect activity of certain regulatory actin-binding proteins thus inducing cytoskeleton remodeling. Penetrating phosphorylated peptides of HspB6 were successfully used for relaxation of airway smooth muscle and prevention of vasospasm in human blood vessels. Full-size HspB6 and its short peptides modulate platelet aggregation. Versatility of HspB6 is awaiting further investigation, however it can be at least partially explained by the ability of phosphorylated HspB6 to interact with the universal adapter protein 14-3-3 and to displace different target proteins from their complexes with 14-3-3. This displacement may result in modulation of target protein activity and consequently can induce multiple and diverse effects.

M.V. Sudnitsyna • N.B. Gusev (✉)
Department of Biochemistry, School of Biology, Moscow State University,
Moscow 119991, Russian Federation
e-mail: NBGusev@mail.ru

N.N. Sluchanko
A. N. Bach Institute of Biochemistry, Russian Academy of Sciences,
Moscow 119071, Russian Federation

R.M. Tanguay, L.E. Hightower (eds.), *The Big Book on Small Heat Shock Proteins*, Heat Shock Proteins 8, DOI 10.1007/978-3-319-16077-1_9

Keywords Oligomeric structure • Protein-protein interaction • Phosphorylation • Homeostasis • Protective activity

9.1 Introduction

The small heat shock protein with apparent molecular weight of about 20 kDa, and therefore initially named Hsp20 (or p20), was firstly described by Kato et al. (1994). This protein was co-purified with αB-crystallin and another small heat shock protein, Hsp27 (HspB1), and was detected in practically all rat tissues. It has been shown that this protein belongs to the large family of small heat shock proteins (sHsp) and, according to the new classification, it was renamed to HspB6 (Kappe et al. 2001, 2003; Fontaine et al. 2003).

9.2 Expression and Intracellular Localization

HspB6, together with three other members of sHsp family, namely HspB1, HspB5 (αB-crystallin) and HspB8, belongs to the group of the small heat shock proteins demonstrating ubiquitous expression (Taylor and Benjamin 2005; Kirbach and Golenhofen 2011). The level of HspB6 expression is especially high in smooth and skeletal muscle and in the heart (Kirbach and Golenhofen 2011; Kim et al. 2004; Cross et al. 2007; Dohke et al. 2006). Expression of HspB6 seems not to be heat-inducible (Taylor and Benjamin 2005; Kirbach and Golenhofen 2011; Arrigo 2013), however sodium arsenite, oxidative and hyperosmotic stresses were accompanied by increase of HspB6 expression in cultured hyppocampal neurons (Bartelt-Kirbach and Golenhofen 2013). The level of HspB6 in skeletal and cardiac muscle was also increased in the course of ontogenesis (Verschuure et al. 2003; Inaguma et al. 1996). Expression of HspB6 can vary under different physiological conditions. For instance, exercise training (Boluyt et al. 2006) and congestive heart failure (Dohke et al. 2006) were accompanied by increased level of HspB6 in the heart, whereas spinal cord isolation and denervation resulted in decreased level of HspB6 in different skeletal muscles (Inaguma et al. 1996; Huey et al. 2004).

HspB6 can undergo translocation and therefore its exact intracellular localization still remains controversial. Under normal unstressed conditions HspB6 was predominantly detected in cytosol of cardiomyocytes, however part of HspB6 was colocalized with actin (Pipkin et al. 2003; Fan et al. 2004). Under ischemic stress HspB6 was reported to translocate to the cardiac myofilaments (Golenhofen et al. 2004). Similar translocation was reported after isoproterenol stimulation of rat cardiomyocytes (Fan et al. 2004). At the same time, HspB6 was predominantly detected in cytoplasm of carotid artery cells (Rembold and Zhang 2001) and stimulation of vascular smooth muscle did not lead to dramatic changes in HspB6 distribution

(Brophy et al. 1999a). However, according to other reports, hormone-induced phosphorylation induces binding of HspB6 to actin (Rembold et al. 2000). Endothelial injury leading to the damage of the outer cellular membrane is accompanied by releasing of HspB6 in the blood stream (Kozawa et al. 2002) where it can function as potential regulator of platelet aggregation. Moreover, the recently published data indicate that the small heat shock proteins can be secreted by a number of different cells and can be involved in cell-cell communication and immune response (van Noort et al. 2012; Arrigo 2012). At present this phenomenon was described for HspB1, HspB4, HspB5 and HspB8, and very recently confirmed for HspB6 (Zhang et al. 2012).

9.3 Molecular Structure and Physico-chemical Properties of HspB6

The primary structure of HspB6 is homologous to the structure of all other small heat shock proteins and contains conservative α-crystallin domain flanked by N- and C-terminal extensions (Kappe et al. 2003; Fontaine et al. 2003). Human HspB6 (Uniprot ID O14558) contains 160 amino acid residues and is one of the smallest members of the family of human small heat shock proteins (Fig. 9.1a).

Up to now all attempts to crystallize full-size HspB6 were unsuccessful probably because of its flexible and highly dynamic termini which could be easily cleaved off and fragmented by proteases (Sluchanko et al. 2012; Weeks et al. 2013). However, differently truncated fragments of HspB6 containing α-crystallin domain have been crystallized and their three-dimensional structure has been solved (Bagneris et al. 2009; Weeks et al. 2013). The structure of truncated fragments of HspB6 seems to be very similar to the structure of the corresponding fragments of other vertebrate small heat shock proteins (HspB4 and HspB5) (Clark et al. 2011; Laganowsky et al. 2010; Laganowsky and Eisenberg 2010). In all these structures the α-crystallin domain is composed of six (or seven) β-strands packed in two sheets forming β-sandwich. Two monomers interact via antiparallel packed β7 strands which can be paired in at least three different registers (Laganowsky et al. 2010; Bagneris et al. 2009; Weeks et al. 2013). In the case of HspB6 the β7 strands are in the so-called APII register (Fig. 9.1b) (Weeks et al. 2013) and β4 and β8 strands form hydrophobic groove that can be occupied by hydrophobic peptides belonging to different small heat shock proteins (Weeks et al. 2013; Delbecq et al. 2012; Delbecq and Klevit 2013), protein partners interacting with HspB6 (for instance, Bag3 (Fuchs et al. 2010)) or protein substrates tending to aggregate and recognized by HspB6. Although at present high resolution data on the structure of full-size HspB6 are unavailable, recent study utilizing combination of small angle X-ray scattering (SAXS), X-ray diffraction on crystallized proteolytic fragments of HspB6 and size-exclusion chromatography were used for building a hybrid model of HspB6 dimer structure (Fig. 9.1b) (Weeks et al. 2013).

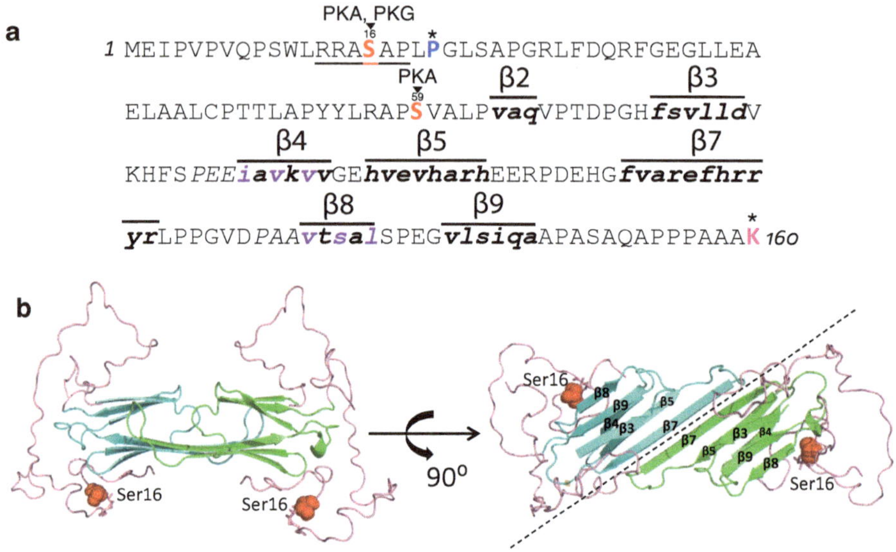

Fig. 9.1 (**a**) The primary structure of human HspB6. Residues involved in formation of β-strands are indicated by *italic lowercase letters* and putative location of β-strands is marked by the *lines* above the sequence. Residues of β4 and β8 participating in formation of hydrophobic cleft are marked *violet*. Ser16 and Ser59 phosphorylated by cyclic nucleotide-dependent protein kinases (protein kinase A (PKA) and protein kinase G (PKG)) are marked *red*. Consensus sequence containing Ser16 and recognized by 14-3-3 is *underlined*. The C-terminal Lys undergoing acetylation (*magenta*) and Pro20 mutated in dilated cardiomyopathy (*blue*) are marked by an *asterisk*. (**b**) Hybrid model of the HspB6 dimer structure in solution showing the β-sheet α-crystallin domain dimer flanked by flexible termini (only one position of the termini from the model 7AAA in Protein Ensemble Database (http://pedb.vib.be/query.php) is presented). β-Strands (with the corresponding numbers) of two monomers are marked *cyan* and *green*, position of Ser16 is marked *red*. HspB6 dimer interface plane is shown as *a dashed line*

The data on the quaternary structure of HspB6 still remain controversial. Early investigations performed on rat HspB6 indicated that this protein forms interconvertible oligomers with molecular weights of 43–67 and 300–470 kDa (Kato et al. 1994; van de Klundert et al. 1998). Analysis of crude muscle extract by size-exclusion chromatography or sucrose density gradient detected HspB6 in fractions with apparent molecular weight in the range of 68–540 kDa (Kato et al. 1994; Brophy et al. 1999a). Increase of intracellular concentration of cyclic AMP was accompanied by shifting HspB6 distribution towards fractions with smaller molecular weight (Brophy et al. 1999a). Therefore it was postulated that phosphorylation or a heat shock induce dissociation of HspB6 oligomers and can affect its interaction with target proteins (Salinthone et al. 2008; Dreiza et al. 2010; Fan and Kranias 2011). It is postulated that HspB6 interacts with many protein partners (see Table 9.1 and discussions below). Therefore the data obtained on crude extracts might indicate that phosphorylation or a heat shock affected the structure or stability of hetero-complexes formed by HspB6, but does not unequivocally mean that these treatments

Table 9.1 Postulated potential protein partners of HspB6

Partner	Ascribed function	Methods used for establishing of protein-protein interaction	Reference
Adapter and heat shock proteins			
HspB1	Cell homeostasis	Immunochemical and biochemical methods	Bukach et al. (2009)
HspB5	Cell homeostasis		Mymrikov et al. (2012)
HspB8	Uncertain		Mymrikov et al. (2011)
HspB2	Unknown		Boros et al. (2004)
14-3-3	Regulation of smooth muscle and cytoskeleton	Immunochemical and biochemical methods	Chernik et al. (2007)
			Dreiza et al. (2005)
			Sluchanko et al. (2011a, 2012)
Proteins involved in neurodegenerative diseases			
α-Synuclein	Prevention of targets aggregation	Biochemical and cytochemical methods	Bruinsma et al. (2011)
Amyloid-β			Wilhelmus et al. (2006b)
Proteins involved in regulation of apoptosis and autophagy			
Bag3	Autophagy	Immunochemistry	Fuchs et al. (2010)
		Size-exclusion chromatography Chemical crosslinking	Shemetov and Gusev (2011)
Beclin1	Autophagy	Coimmunoprecipitation	Qian et al. (2009)
Bax	Apoptosis	Coimmunoprecipitation	Fan et al. (2005b)
Protein kinases			
ASK1 (apoptosis signal-regulating kinase 1)	Cardioprotection (inhibition of ASK1-signaling cascade)	Immunostaining	Fan et al. (2006)
PI3K (phosphoinositide 3-kinase)	Inhibition of hepatocellular carcinoma progression	Coimmunoprecipitation	Matsushima-Nishiwaki et al. (2013)
AKT	Cardioprotection	Coimmunoprecipitation	Fan et al. (2008)
Other enzymes			
Protein phosphatase 1 (PP-1)	Cardioprotection	Affinity chromatography immunochemistry	Damer et al. (1998)
			Qian et al. (2011)
cAMP phosphodiesterase-4 (PDE4)	Modulation of cAMP level	Peptide array immunocytochemistry, coimmunoprecipitation	Sin et al. (2011)
Transglutaminase 2	Cell survival upon excitotoxic stress	Coimmunoprecipitation	Caccamo et al. (2013)
Cytoskeletal proteins			
Actin	Regulation of smooth muscle contraction and cytoskeleton	Coimmunoprecipitation	Tessier et al. (2003)
		Cosedimentation	Rembold et al. (2000)
			Flynn et al. (2003)
			Bukach et al. (2005)
α-Actinin	Vasorelaxation	Coimmunoprecipitation	Tessier et al. (2003)
Receptors			
VEGF receptor 2	HspB6-mediated angiogenesis	Immunochemistry	Zhang et al. (2012)

affected the quaternary structure of HspB6 itself. Indeed, in vitro phosphorylation or phosphomimicking mutation did not affect the apparent molecular weight of isolated HspB6 which predominantly formed dimers (Bukach et al. 2004; Chernik et al. 2007; Sluchanko et al. 2011a). Inability to form large oligomers characteristic for many other small heat shock proteins can be at least partially due to the absence of conservative (I/V/L)-X-(I/V/L) sequence in the short C-terminal sequence of HspB6. This sequence is involved in inter-dimer interaction and formation of large oligomers of αA- and αB-crystallin (Bagneris et al. 2009; Delbecq et al. 2012; Delbecq and Klevit 2013; Sudnitsyna et al. 2012a). The absence of this sequence in the C-terminus of HspB6 can be one of the reasons making formation of large oligomers of this protein less probable. However, at a very high concentration human HspB6 demonstrates attractive self-association (Weeks et al. 2013). Taking into account high HspB6 content in certain tissues and crowding conditions inside the cell increasing its effective concentration, we cannot exclude that in vivo HspB6 can form higher order homooligomeric complexes formed by its self-associated dimers. Further investigations are required in order to analyze effect of different conditions and posttranslational modifications on formation of these complexes.

9.4 Mutations and Posttranslational Modifications of HspB6

To our knowledge only one point mutation (P20L) of human HspB6 was described in the literature, despite the fact that at least several other validated missense mutations could be found in NCBI Gene Database (e.g., S10F, V92A, G94S). Mutation P20L was detected in one patient suffering from dilated cardiomyopathy (Nicolaou et al. 2008). This mutation decreases phosphorylation of Ser16 and abrogates anti-apoptotic and cardioprotective effect of HspB6 (Nicolaou et al. 2008).

HspB6 can undergo multisite phosphorylation. PI3-kinase dependent pathway seems to be involved in phosphorylation of Ser157 of rat HspB6 and this event is supposed to be dependent on insulin stimulation (Wang et al. 1999). The same residue was phosphorylated in swine carotid HspB6 and this phosphorylation was independent on forskolin treatment (Meeks et al. 2005). The primary structure at this site is not absolutely conservative and bovine and human HspB6 are not phosphorylated at the corresponding residues. At least three other sites of HspB6 phosphorylation were reported in the literature. The first site is located somewhere in peptide restricted by residues 123–140 of bovine HspB6 and becomes phosphorylated upon stimulation of carotid artery by phorbol ester or application of forskolin and phosphodiesterase inhibitors (Beall et al. 1999). The second residue, Ser59, which is highly conservative in human, macaque, bovine, rat and mouse HspB6, is phosphorylated in vitro by cAMP-dependent protein kinase (Beall et al. 1999) and was recently reported to be phosphorylated in human skeletal muscle (Hojlund et al. 2009) (Fig. 9.1a). The third site, containing Ser16, with the primary structure strictly conserved in human, bovine, rat and mouse HspB6, is phosphorylated by cyclic nucleotide-dependent protein kinases both in vitro and in vivo (Beall et al. 1997,

1999; Rembold et al. 2000) (Fig. 9.1a). Phosphorylation of this site seems to be very important and affects physiologically important properties of HspB6 and its interaction with many target proteins (see below).

Recently published data indicate that HspB6 can undergo acetylation (Karolczak-Bayatti et al. 2011; Chen et al. 2013). It was shown that histone deacetylase 8 (HDAC8) is non-nuclear lysine deacetylase being able to interact with HspB6 which could be then recognized by anti-acetylated ε-lysine antibodies. Selective inhibition of deacetylase is accompanied by decrease of spontaneous or oxytocin-augmented contraction of myometrial strips (Karolczak-Bayatti et al. 2011). The very C-terminal Lys of HspB6 (Fig. 9.1a) can undergo acetylation and inhibition of deacetylase promotes HspB6 phosphorylation (Chen et al. 2013). This means that two different posttranslational modifications are somehow connected with each other. However, we assume that acetylation of HspB6 under physiological conditions in vivo requires further detailed investigation.

9.5 Interaction of HspB6 with Other Members of sHsp Family

As already mentioned, the structure of HspB6 is similar to that of other members of sHsp family (Basha et al. 2012) and therefore one can tentatively suggest that HspB6 will be able to interact with other small heat shock proteins. HspB6 was co-purified and formed tight complexes with HspB5 (αB-crystallin) (Kato et al. 1994; Sugiyama et al. 2000). These complexes can be crosslinked by transglutaminase (Boros et al. 2004). The so-called Cys mutants of HspB6 and HspB5 containing single Cys residue engineered in the middle of the β7 strand can form disulfide crosslinked heterodimers (Mymrikov et al. 2012) (Table 9.1). Finally, fluorescent chimeras consisting of the corresponding small heat shock proteins and different fluorescent proteins interact with each other both in vitro (Datskevich et al. 2012a) and ex vivo (Simon et al. 2007).

Interaction of HspB6 with HspB1 is also well-established (Table 9.1). According to the data of size-exclusion chromatography, these proteins form two types of complexes with apparent molecular weight of ~100 and 300 kDa (Bukach et al. 2004). Formation of heterooligomeric complexes of HspB1 and HspB6 was also demonstrated by means of disulfide crosslinking (Mymrikov et al. 2012) and Förster resonance energy transfer (Datskevich et al. 2012a). Formation of these complexes is modulated by HspB1 and HspB6 phosphorylation (Bukach et al. 2009) and mutations of HspB1 correlating with peripheral distal neuropathies affect its interaction with HspB6 (Nefedova et al. 2013a, b).

A number of different methods indicate that under certain conditions HspB6 can interact with another ubiquitously expressed small heat shock protein, HspB8 (Fontaine et al. 2005, 2006). However, other experimental approaches do not confirm tight interaction between HspB6 and HspB8 (Mymrikov et al. 2012; Datskevich et al. 2012a). Therefore further investigations are required to validate direct interaction of these two proteins.

Transglutaminase induces crosslinking of HspB6 and HspB2 (Boros et al. 2004). HspB2 does not belong to the group of ubiquitously expressed sHsp and even if expressed, the concentration of this protein is rather low (Kirbach and Golenhofen 2011). Therefore, an importance of HspB2-HspB6 interaction remains uncertain.

Many tissues simultaneously express large quantities of different small heat shock proteins and this increases the probability of formation of heterooligomeric complexes. These complexes can differ from the corresponding homooligomeric complexes in their intracellular location, interaction with target proteins and other physiologically important properties (Arrigo 2013; Aquilina et al. 2013). Although there are many methodological problems in analyzing the structure and properties of heterooligomeric complexes of sHsp in the living cell, this line of investigation seems to be very important and perspective.

9.6 Chaperone-Like Activity and Probable Participation of HspB6 in Selective Elimination of Unfolded Proteins

Chaperone-like activity, i.e. ability to prevent aggregation of partially denatured or unfolded proteins, is common for all small heat shock proteins (McHaourab et al. 2009; Vos et al. 2008; Basha et al. 2012). However, originally in vitro experiments performed with rat HspB6 indicated that the chaperone-like activity of this protein was lower than that of α-crystallin (van de Klundert et al. 1998). Later studies performed on human HspB6 have shown that depending on the nature of model protein substrate the chaperone-like activity of human HspB6 is comparable to that of αB-crystallin (Bukach et al. 2004; Mymrikov et al. 2010; Datskevich et al. 2012b). On the cell level, thermoprotective effect produced by overexpression of HspB6 was comparable with that of αB-crystallin (van de Klundert et al. 1999). HspB6 associates with non-fibrillar amyloid-β (Aβ) peptides and inhibits or completely prevents aggregation of Aβ (1–42) significantly decreasing cerebrovascular toxicity of these peptides (Wilhelmus et al. 2006a, c). Furthermore, small heat shock proteins including HspB6 were reported to transiently interact with the wild-type and mutant α-synuclein (αSYN) and inhibit its fibril formation (Bruinsma et al. 2011). At the same time, HspB6 was not detected in neurofibrillar tangles predominantly formed by incorrectly folded tau protein (Wilhelmus et al. 2006c). It is important to mention that in Alzheimer disease brains HspB6 was detected both inside and outside of cells (Wilhelmus et al. 2006c). This can be due either to necrosis of neurons or to specific secretion of HspB6 from the living cells as a protective reaction against extracellular accumulation of amyloid peptides.

Small heat shock proteins can prevent accumulation of aggregates of improperly folded proteins by at least three means: (1) by preventing aggregation (the so-called holdase or chaperone-like activity), (2) by transferring improperly folded proteins to ATP-dependent chaperone Hsp70, or (3) by promoting degradation of improperly folded proteins in autophagosomes or proteasomes (Basha et al. 2012; Vos et al. 2011). The proteins of Bag family are co-chaperones involved in regulation of Hsp70 ATPase activity (Takayama and Reed 2001; Rosati et al. 2011). One of the members of Bag family, namely Bag3, contains several different domains (McCollum et al. 2010). The N-terminal part of Bag3 contains the so-called WW region participating in the interaction with PXXP regions of different proteins (or Bag3 itself) and two IPV-motifs. The C-terminal part of Bag3 contains PXXP domain responsible for interaction with SH3 domain-containing proteins (and among them with γ-subunit of phospholipase C) and Bag-domain providing for interaction of Bag3 with Hsp70 and Bcl2 (McCollum et al. 2010). The primary structure of IPV motifs of Bag3 is similar to the corresponding sequence located in the C-terminal extension of many sHsp (such as HspB1, HspB4 and HspB5) and lacking in the structure of other small heat shock proteins (HspB6 and HspB8). In the case of sHsp tending to form large oligomers (HspB1, HspB4, HspB5) the (I/V/L)-P-(I/V/L) sequence of the C-terminal extension of one dimer interacts with the hydrophobic groove formed by β4 and β8 strands of another dimer and thus participates in stabilization of oligomeric structure (Delbecq et al. 2012; Delbecq and Klevit 2013). HspB8 and HspB6, lacking this motif in their own structure, interact with IPV motifs of Bag3 (Fuchs et al. 2010; Carra et al. 2008). The stoichiometry of complexes formed by Bag3 and HspB8 (or HspB6) is variable and binding of small heat shock proteins stabilizes the structure of predominantly disordered Bag3 (Shemetov and Gusev 2011). Bag3 seems to be able to simultaneously interact with Hsp70 and the small heat shock proteins, playing a role of adapter providing contacts between two different groups of heat shock proteins. Moreover, HspB8 can participate in formation of large protein complex consisting of Bag3, Hsp70 and chaperone-associated ubiquitin ligase (CHIP) thus promoting ubiquitination of target proteins leading to its autosomal proteolytic degradation (Arndt et al. 2010; Crippa et al. 2010). HspB6 weaker interacts with Bag3, however it is assumed that if the ratio HspB6/HspB8 in the given tissue is high, HspB6 can also provide selective Bag3-dependent autophagosomal degradation of misfolded proteins (Fuchs et al. 2010). It is interesting to mention, that P209L mutation located in one of HspB6/HspB8-binding sites of Bag3 correlates with severe dominant childhood muscular dystrophy (Selcen et al. 2009). This indirectly indicates that interaction with the small heat shock proteins somehow affects autosomal degradation of improperly folded proteins or modulates antiapoptotic activity of Bag3.

9.7 HspB6 Induced Cardioprotection

Ischemia followed by reperfusion induces severe heart injury. Cardiac-specific overexpression of HspB6 in transgenic mouse lines produces improved functional recovery and reduction in infarct size after ischemia/reperfusion (Fan et al. 2005b). The level of HspB6 expression in the heart can be regulated by micro-RNA and micro-RNA mir-320 negatively regulates HspB6 translation (Ren et al. 2009).

Overexpression of mir-320 was accompanied by severe damages induced by ischemia/reperfusion, whereas its knockdown was cytoprotective (Ren et al. 2009). Protective effect of HspB6 is dependent on its phosphorylation and the animals expressing nonphosphorylatable mutant (S16A) of HspB6 demonstrated lower recovery after ischemia/reperfusion and increased necrotic damage than nontransgenic animals (Qian et al. 2009). It is worthwhile to mention, that S16A mutation of HspB6 correlates with decreased autophagosomal activity and drug-induced increase of autophagosomal activity improves functional recovery (Qian et al. 2009).

Cardioprotective effect is at least partially due to the antiapoptotic activity of HspB6. Prolonged treatment with isoproterenol induced apoptosis of cardiomyocytes and overexpression of HspB6 protected cultured cardiac myocytes from apoptosis and antiapoptotic effect of HspB6 was again dependent on Ser16 phosphorylation (Fan et al. 2004). Overexpression of HspB6 is accompanied by increase Bcl-2/Bax ratio and reduced caspase-3 activity in cardiomyocytes and neuroblastoma cells (Fan et al. 2005b; Zeng et al. 2013). It is believed that HspB6 induces its antiapoptotic effect by direct binding of pro-apoptotic protein Bax and preventing its interaction with mitochondria membranes (Fan et al. 2005b) (Table 9.1). In addition, HspB6 seems to be able to regulate activity of several protein kinases. HspB6 interacts and activates phosphorylated Akt, i.e. protein kinase possessing strong antiapoptotic activity (Fan et al. 2008) and at the same time inhibits the apoptosis signal-regulating/c-jun-NH_2-terminal kinase/p38 pathway thus inhibiting initiation of apoptosis (Fan et al. 2006). In addition, it is postulated that HspB6 can inhibit caspase-3 activity inducing additional antiapoptotic effect (Wang et al. 2009). Chaperone-like activity can also be important for cardioprotection and antiapoptotic effect of HspB6. Deletion of the C-terminal extension, which is believed to be important for chaperone-like activity, decreases cardioprotective effect of HspB6 (Islamovic et al. 2007).

Physical exercises and pathological conditions like pressure/volume overload are accompanied by cardiac hypertrophy. Cardiac hypertrophy is accompanied by increased expression of HspB6 (Willis and Patterson 2010) and vice versa, overexpression of HspB6 decreases cardiac hypertrophy (Fan et al. 2006). Molecular mechanisms of cardiac remodeling by phosphorylated HspB6 remain enigmatic. However, interesting experimental results appeared recently. Members of cAMP-specific phosphodiesterase family PDE4 interact with HspB6 (Sin et al. 2011) (Table 9.1). The site of interaction is close to the catalytic domain of PDE and therefore the local concentration of cyclic nucleotides in vicinity of HspB6 is very low. This prevents phosphorylation of HspB6 by cAMP-dependent protein kinase and leaves HspB6 in unphosphorylated dormant state (Sin et al. 2011). Disrupting peptides with the sequence of the HspB6-docking site of PDE4 can dissociate PDE4-HspB6 complex, and liberated HspB6 is phosphorylated and activated by cAMP-dependent protein kinase. It is suggested that the multi-purpose scaffold protein AKAP-Lbc interacts with HspB6 and three protein kinases (protein kinase A, protein kinase C and protein kinase D). This can increase the rate of HspB6 phosphorylation (Edwards et al. 2012a) and induce protein kinase D phosphorylation. Phosphorylated protein kinase D migrates to nucleus where it can depress transcription factors mediating hypertrophic phenotype (Edwards et al. 2012b).

9.8 HspB6 and Regulation of Smooth Muscle Contraction

Activation of cAMP- or cGMP-dependent protein kinases and/or inhibition of protein phosphatases is accompanied by effective phosphorylation of Ser16 of HspB6 which correlates with relaxation of trachealis and carotid artery smooth muscle (Rembold et al. 2000; Beall et al. 1997, 1999). Relaxation induced by HspB6 phosphorylation seems to be independent of the level of myosin regulatory light chains phosphorylation (Rembold et al. 2000; Woodrum et al. 1999). Therefore it was suggested that phosphorylated HspB6 somehow affects actin filaments or modulates their interaction with phosphorylated myosin.

Two hypotheses trying to explain molecular mechanism of relaxation induced by HspB6 phosphorylation were put forward. According to the first hypothesis, there is a troponin I homologous peptide in the structure of HspB6 (Rembold et al. 2000). Phosphorylation induces translocation of HspB6 to actin filament and the corresponding peptide inhibits interaction of phosphorylated myosin with actin like troponin I. Thus, HspB6 phosphorylation induced by activation of cyclic-nucleotide dependent protein kinases (Rembold et al. 2000), hypoxia (Frobert et al. 2005), or heat pretreatment (O'Connor and Rembold 2002) results in inactivation of actin filament regions containing phosphorylated HspB6.

According to the second hypothesis, unphosphorylated HspB6 interacts with F-actin and α-actinin and by this means stabilizes actin cytoskeleton providing effective contraction. Phosphorylated HspB6 dissociates from actin cytoskeleton causing disruption of actin filaments, their dissociation from α-actinin, destabilization and reorganization of cytoskeleton thus leading to smooth muscle relaxation (Brophy et al. 1999b; Tessier et al. 2003; Woodrum et al. 2003). Detailed description and comparison of both hypotheses were presented earlier (Mymrikov et al. 2011).

Both hypotheses had their supporters and after long discussion it was postulated that phosphorylated HspB6 could reorganize actin cytoskeleton disrupting force transmission as well as disrupting actin-myosin binding (Ba et al. 2009). For direct realization of both these activities HspB6 should be a genuine actin-binding protein and should be (at least under certain conditions) bound to actin. At the same time the total concentration of actin in smooth muscle is more than 1,000 μM (Rembold et al. 2000), whereas the total concentration of HspB6 monomer is in the range of 100–200 μM (Kato et al. 1994). As already mentioned, HspB6 forms stable dimers or is involved in formation of large heterooligomers with other small heat shock proteins having apparent molecular weight in the range of 65–500 kDa. This means that the real concentration of HspB6 homo- or heterooligomers would be less than 50 μM, thus being more than 20 times smaller than that of actin. In reality the HspB6/actin ratio should be even smaller since HspB6 is predominantly located in cytosol with only small portion being bound to the cytoskeleton (Rembold and Zhang 2001; Tyson et al. 2008). We analyzed interaction of the wild type HspB6 and its S16D mutant mimicking phosphorylation with actin filaments and myofibrils (Bukach et al. 2005). Even at high HspB6 concentration, the HspB6/actin molar ratio in the pellet formed after co-sedimentation was less than 0.04. Our experimental results together with theoretical calculations indicate that HspB6 is not a genuine actin-binding protein and being presented at a very low stoichiometry on actin filaments, HspB6 cannot directly regulate interaction of actin with myosin. Therefore one should theorize that HspB6 indirectly regulates actin filaments.

Penetrating HspB6 peptide containing phosphorylated Ser16 within the motif RRA**S16**APLP (resembling typical binding site for universal adapter protein 14-3-3 (RXXpSXP) (Yaffe et al. 1997)) induced the loss of actin stress fibers and focal adhesion complexes in Swiss 3T3 cells (Dreiza et al. 2005). It was proposed that HspB6 phosphopeptide interacts with 14-3-3 and displaces phosphorylated cofilin from its complex with 14-3-3. Displaced phosphorylated cofilin undergoes rapid dephosphorylation catalyzed by slingshot protein phosphatase, regains its F-actin severing activity and induces depolymerization or detachment of actin filament from focal adhesion complexes (Dreiza et al. 2005). We analyzed interaction of intact full-size HspB6 with 14-3-3 and found that phosphorylated (but not unphosphorylated) HspB6 indeed forms tight complexes with 14-3-3 (Chernik et al. 2007). Interaction of these proteins seems to be finely regulated. For instance, the point mutations mimicking phosphorylation of 14-3-3 modulate its interaction with phosphorylated HspB6. In the case of ζ-isoform of 14-3-3, mutation S58E decreases, whereas mutation S184E increases interaction of 14-3-3 with HspB6 (Sluchanko et al. 2011a) (Fig. 9.2). Under certain conditions usually dimeric 14-3-3 can dissociate forming monomers having distinct properties (Sluchanko and Gusev 2012). We found that the dimer-deficient mutants of 14-3-3 also tightly interact with HspB6 (Sluchanko et al. 2011b, 2012). This interaction was again strictly dependent on HspB6 phosphorylation and binding of phosphorylated HspB6 induced stabilization of dimer-deficient mutant of 14-3-3 (Sluchanko et al. 2012). Interaction of HspB6 and 14-3-3 is modulated by physiological concentrations of phosphate (Sluchanko et al. 2013) and therefore oscillation of phosphate concentration appearing in the course of muscle contraction can affect interaction of HspB6 and 14-3-3 (Fig. 9.2).

Thus, at present the interaction of phosphorylated full-size HspB6 or its phosphorylated peptide with γ- and ζ-isoforms of 14-3-3 is well-established. Moreover, fluorescently labeled phosphorylated peptide of HspB6 was successfully used for the large scale screening of low molecular weight pharmacological compounds which can be potentially used for relaxation of airway smooth muscle (An et al. 2011). The principle of the screening was based on the ability of pharmacological compounds to affect fluorescence polarization of the complex formed by fluorescently labeled phosphorylated HspB6 peptide and 14-3-3. Effective method was developed for obtaining preparative quantities of cell-penetrating phosphorylated form of HspB6 (Flynn et al. 2007) and it was shown that the phosphorylated peptide or phosphorylated full-size HspB6 effectively prevented vasospasm of human saphenous vein segments (McLemore et al. 2004). Penetrating phosphorylated peptide of HspB6 prevents vasospasm without significant effect on cell proliferation and therefore seems to be very promising for peripheral vascular reconstruction (Tessier et al. 2004).

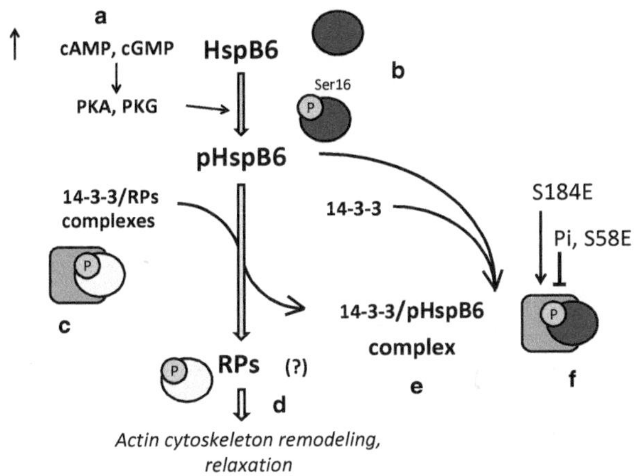

Fig. 9.2 Proposed model of HspB6 participation in regulation of smooth muscle contraction. Upon increasing concentration of cyclic nucleotides protein kinase A and protein kinase G are activated (**a**) and phosphorylate HspB6 at Ser16 (**b**). Phosphorylated HspB6 binds to 14-3-3 and displaces different regulatory proteins (*RPs*) from their complexes with 14-3-3 (**c**). Liberated regulatory proteins of incompletely characterized nature induce cytoskeleton remodeling and smooth muscle relaxation (**d**). Stability of the complex between phosphorylated HspB6 and 14-3-3 (**e**) might be regulated by 14-3-3 phosphorylation (at Ser58, Ser184 and, probably, other sites) and/or concentration of inorganic phosphate (**f**)

Although the HspB6/14-3-3 interaction is well-established, the detailed mechanism of smooth muscle relaxation induced by phosphorylated HspB6 remains enigmatic. As already mentioned, at the beginning it was suggested that phosphorylated HspB6 competes with phosphorylated cofilin for interaction with 14-3-3 (Dreiza et al. 2005, 2010). After phosphorylation HspB6 displaces phosphorylated cofilin from its complex with 14-3-3, the liberated cofilin undergoes rapid dephosphorylation and regains its actin-severing activity. All these events lead to disruption of actin filaments and induce smooth muscle relaxation. This hypothesis, based on experimental data of Gohla and Bokoch (2002), postulated direct tight interaction of phosphorylated cofilin and 14-3-3, remains popular till now (Edwards et al. 2011; Karolczak-Bayatti et al. 2011; Hocking et al. 2013). However, the recently published data indicate that cofilin only very weakly interacts with 14-3-3 in a phosphorylation-independent manner (Sudnitsyna et al. 2012b; Grebenova et al. 2012) and therefore it cannot directly compete with phosphorylated HspB6 for 14-3-3 binding (Sudnitsyna et al. 2012b). This means that there should be other mechanisms underlying the relaxation effect induced by binding of phosphorylated HspB6 to 14-3-3.

The recently published data indicate that the universal adapter protein 14-3-3 can in theory interact with more than 700 protein clients (Uhart and Bustos 2013) and among them there are many proteins directly or indirectly involved in regulation of cytoskeleton (Sluchanko and Gusev 2010). For instance, phosphorylated slingshot protein phosphatase interacts with 14-3-3 and this leads to dissociation of slingshot from actin filaments and inhibition of its activity (Freeman and Morrison 2011; Mizuno 2013). Phosphorylated HspB6 can displace phosphorylated slingshot from its complex with 14-3-3 thus leading to the activation of protein phosphatase, cofilin dephosphorylation and fragmentation of actin filaments resulting in smooth muscle relaxation (Sudnitsyna et al. 2012b; Mizuno 2013; Sluchanko and Gusev 2010). In addition, 14-3-3 interacts with phosphorylated cortactin (the protein participating in regulation of actin polymerization) and phosphorylated IRSp53 (the protein involved in the interaction of actin filaments with plasma membrane) (Freeman and Morrison 2011) and if phosphorylated HspB6 displaces these (or other) regulatory proteins (RPs) from their complexes with 14-3-3, it will indirectly regulate the cytoskeleton and smooth muscle relaxation (Fig. 9.2). Among other proteins which are potential targets of 14-3-3, we can mention α- and β-chains of integrin and β-catenin (reviewed in (Mymrikov et al. 2011; Sluchanko and Gusev 2010)) which also participate in formation and reorganization of cytoskeleton. Unfortunately, direct competition of phosphorylated HspB6 with the above mentioned proteins (except of phosphorylated cofilin) for their binding to 14-3-3 has not been analyzed yet. However, taking into account high concentration of HspB6 in smooth muscle and high affinity of phosphorylated HspB6 to 14-3-3, we assume that the relaxation effect of HspB6 can be explained by its ability to displace certain proteins involved in regulation of cytoskeleton from their complexes with 14-3-3.

9.9 HspB6 and Platelet Aggregation

HspB6 in a concentration-dependent manner inhibits thrombin-induced platelet aggregation (Matsuno et al. 1998). HspB6 binds platelet with high affinity and prevents thrombin-induced calcium entrance from outside the cell (Niwa et al. 2000). Detailed mechanism of antiaggregation effect of HspB6 remains obscure, however, it is believed that HspB6 somehow inhibits thrombin-induced activation of phospholipase C and by this means prevents activation of protein kinase C which is involved in regulation of calcium entrance (Kozawa et al. 2002). In this case HspB6 exerts its effect outside of the cell, however penetrating phosphorylated peptide of HspB6 is also able to inhibit platelet aggregation (McLemore et al. 2004). It is interesting to mention that both the full-size αB-crystallin and HspB6 and even short peptides of αB-crystallin (^{9}WIRRPFFPF^{17}) and of HspB6 (^{11}WLRRASAPL^{19}) prevent thrombin-induced platelet aggregation (Kanno and Matsuno 2006; Matsuno et al. 2003). To our knowledge it is unknown whether these peptides produce their effects outside or inside the cell. Carotid artery injury is accompanied by decrease of HspB6 level in endothelium and increased level of HspB6 in the blood thus suggesting that it can be secreted from vascular wall or simply released from damaged cells and participates in regulation of platelet function in circulation (Kozawa et al. 2002). Therefore HspB6 and its peptides are considered as important potential targets for therapy of thrombosis and cardiovascular diseases (Fan and Kranias 2011; Edwards et al. 2011).

9.10 HspB6 as a Therapeutic Target

The recently published data indicate that the promoter region of *HspB6* gene is hypermethylated in melanoma cells and the level of hypermethylation is increased in advanced stages of melanoma (Koga et al. 2009). Increased methylation of *HspB6* gene promoter is accompanied by a decreased level of HspB6 expression implying interrelation between progression of melanoma and HspB6 expression (Koga et al. 2009). This is in line with the finding that HspB6 expression was significantly reduced in hepatocellular carcinoma cell lines, where the level of HspB6 correlated inversely with a tumor stage and size (Noda et al. 2007). Protective effect of HspB6 in hepatocellular carcinoma can be at least partially explained by its ability to directly interact with both 85 and 110 kDa subunits of phosphoinositide 3-kinase (PI3K) and to inhibit activity of this protein kinase which is upstream of Akt kinase involved in carcinogenesis (Matsushima-Nishiwaki et al. 2013). Significant decrease of *HspB6* gene transcript was detected in lung cell adenocarcinomas (Yap et al. 2005) and HspB6 is one of the proteins that were significantly and reproducibly downregulated in gliomas (Deighton et al. 2010). It seems very probable that in future HspB6 can be used as diagnostic or prognostic biomarker in oncology.

As already mentioned, short penetrating peptides of HspB6 can be successfully used under certain disease-related situation. For instance, these peptides were used against excessive keloid scarring (Lopes et al. 2009), for relaxation of airway smooth muscle (Flynn et al. 2007; Komalavilas et al. 2008), prevention of vasospasm of human umbilical artery (Flynn et al. 2007) and subarachnoid haemorrhage induced by delayed vasospasm (Furnish et al. 2010) as well as prevention of platelet aggregation (McLemore et al. 2004). Special peptides were developed in order to modulate and disrupt protein complexes formed by HspB6 and phosphodiesterase of cAMP (Sin et al. 2011). Disruption of these complexes increases concentration of cAMP in vicinity of HspB6 and by this means increases the level of HspB6 phosphorylation which is important for many physiological effects of this protein.

Low molecular weight compounds, being able to modulate interaction of phosphorylated HspB6 and 14-3-3, were developed and were proven to be effective in relaxation of human airway smooth muscle (An et al. 2011). Understanding structural determinants of interaction between 14-3-3 and phosphorylated full-size HspB6 will help in future attempts to develop compounds modulating 14-3-3/HspB6 complexes which could be used as a complementary therapy along with existing phosphopeptides, like AZX100 (Lopes et al. 2009).

As already mentioned, HspB6 effectively prevents aggregation of amyloid-β peptides (Wilhelmus et al. 2006a) and α-synuclein (Bruinsma et al. 2011) and therefore can probably affect progression of Alzheimer's and Parkinson's disease providing a platform for therapy of these important disorders. In addition, HspB6 possesses multifunctional cardioprotective activity (Fan and Kranias 2011). Genetic manipulations with viral vectors and/or permeabilization were able to increase the intracellular level of HspB6 which was accompanied by cardioprotective effects (Pipkin et al. 2003; Fan et al. 2005a). However, up to now these approaches have significant limitations in clinical application.

9.11 Conclusions and Perspectives

HspB6 belongs to the family of the small heat shock proteins and is ubiquitously expressed in human tissues. This protein is unusually versatile and prevents aggregation of improperly folded proteins, participates in selective autophagosomal deletion of misfolded proteins, participates in regulation of apoptosis and proliferation, possesses multifunctional cardioprotective activity, participates in regulation of smooth muscle contraction and regulation of platelet aggregation (Fig. 9.3). This versatility can be at least partially explained by the ability of HspB6 to interact and to modify activity and intracellular location of a large number of different proteins (Table 9.1). Among them are other members of the sHsp family (αB-crystallin (HspB5) and HspB1), protein phosphatase I, cAMP phosphodiesterase, different protein kinases (Akt, Ask1, PI3K), pro- and anti- apoptotic proteins Bax, Beclin-1 and Bag3 and universal adapter protein 14-3-3 (Fan et al. 2005a; Fan and Kranias 2011; Edwards et al. 2011; Mymrikov et al. 2011). We believe that for understanding of HspB6 versatility the interaction of this protein with 14-3-3 is of special importance. More than 700 potential partners of 14-3-3 are already described in the literature (Uhart and Bustos 2013) and the interaction of 14-3-3 with these partners depends on their phosphorylation, intracellular location and affinity for 14-3-3. Among potential 14-3-3 targets are many protein kinases, protein phosphatases, transcription factors, membrane and cytoskeletal proteins (Pozuelo Rubio et al. 2004). Activation of cyclic nucleotide-dependent protein kinases results in rapid phosphorylation of HspB6 which is highly expressed in certain tissues. Phosphorylated HspB6 has a high affinity to 14-3-3 and accumulation of phosphorylated HspB6 will inevitably affect interaction between 14-3-3 and other potential protein targets leading to different intracellular effects. If this suggestion is correct, the versatility of HspB6 could at least partially be explained by the well-known versatility of 14-3-3. Further investigations are required in order to analyze probable effect of phosphorylated HspB6 on the interaction of 14-3-3 with different target proteins.

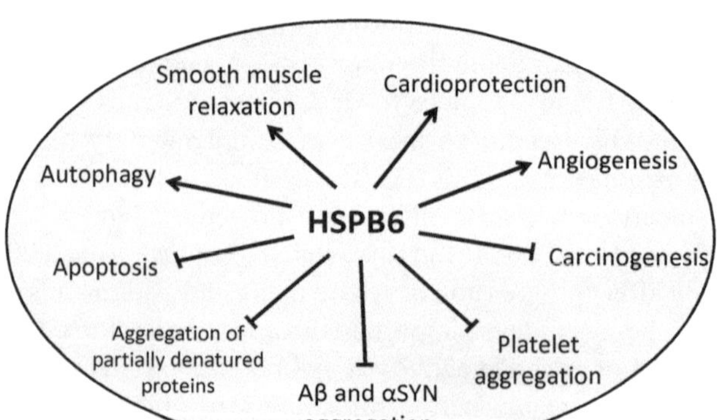

Fig. 9.3 Versatility of HspB6 action (See text for details)

The data presented indicate that HspB6 is a very promising therapeutic target. Therefore, further development of special methods of targeted transportation of HspB6 or its peptides to certain organs or selective regulation of HspB6 expression seems to be very important from a medical point of view. Construction of new pharmacological agents able to modulate interaction of HspB6 with its protein targets seems very promising for the development of novel therapies of different human diseases. Acquiring detailed structural information on the complexes of HspB6 with its partner proteins will provide bases for the new drug development.

Since the number of publications on HspB6 is very large, we were unable to discuss all problems concerning this protein. The interested reader is addressed to a number of reviews dealing with this small heat shock protein and other members of sHsp family (Basha et al. 2012; McHaourab et al. 2009; Vos et al. 2008; Fan et al. 2005a; Fan and Kranias 2011; Mymrikov et al. 2011; Edwards et al. 2011; Dreiza et al. 2010; Seit-Nebi and Gusev 2010; Dubinska-Magiera et al. 2014; Delbecq and Klevit 2013).

Acknowledgement This investigation was supported by Russian Foundation for Basic Research (14-04-00146 to NNS and 13-04-00015 to NBG).

Note The recently published data indicate that under crowding conditions HspB6 tends to self-association and that phosphorylation abrogates this tendency (Sluchanko NN, Chebotareva NA, Gusev NB (2015) Quaternary structure of human small heat shock protein HSPB6 (Hsp20) in crowded media modeled by trimethylamine N-oxide (TMAO): Effect of protein phosphorylation. Biochimie 108:68–75. doi:10.1016/j.biochi.2014.11.001).

References

An SS, Askovich PS, Zarembinski TI, Ahn K, Peltier JM, von Rechenberg M, Sahasrabudhe S, Fredberg JJ (2011) A novel small molecule target in human airway smooth muscle for potential treatment of obstructive lung diseases: a staged high-throughput biophysical screening. Respir Res 12:8. doi:10.1186/1465-9921-12-8

Aquilina JA, Shrestha S, Morris AM, Ecroyd H (2013) Structural and functional aspects of hetero-oligomers formed by the small heat shock proteins alphaB-crystallin and HSP27. J Biol Chem 288(19):13602–13609. doi:10.1074/jbc.M112.443812

Arndt V, Dick N, Tawo R, Dreiseidler M, Wenzel D, Hesse M, Furst DO, Saftig P, Saint R, Fleischmann BK, Hoch M, Hohfeld J (2010) Chaperone-assisted selective autophagy is essential for muscle maintenance. Curr Biol 20(2):143–148, S0960-9822(09)02000-4

Arrigo AP (2012) Pathology-dependent effects linked to small heat shock proteins expression: an update. Scientifica 2012:185641. doi:10.6064/2012/185641

Arrigo AP (2013) Human small heat shock proteins: protein interactomes of homo- and hetero-oligomeric complexes: an update. FEBS Lett 587(13):1959–1969, S0014-5793(13)00353-0

Ba M, Singer CA, Tyagi M, Brophy C, Baker JE, Cremo C, Halayko A, Gerthoffer WT (2009) HSP20 phosphorylation and airway smooth muscle relaxation. Cell Health Cytoskeleton 2009(1):27–42

Bagneris C, Bateman OA, Naylor CE, Cronin N, Boelens WC, Keep NH, Slingsby C (2009) Crystal structures of alpha-crystallin domain dimers of alphaB-crystallin and Hsp20. J Mol Biol 392(5):1242–1252, S0022-2836(09)00936-X

Bartelt-Kirbach B, Golenhofen N (2013) Reaction of small heat-shock proteins to different kinds of cellular stress in cultured rat hippocampal neurons. Cell Stress Chaperones. doi:10.1007/s12192-013-0452-9

Basha E, O'Neill H, Vierling E (2012) Small heat shock proteins and alpha-crystallins: dynamic proteins with flexible functions. Trends Biochem Sci 37(3):106–117, S0968-0004(11)00190-3

Beall AC, Kato K, Goldenring JR, Rasmussen H, Brophy CM (1997) Cyclic nucleotide-dependent vasorelaxation is associated with the phosphorylation of a small heat shock-related protein. J Biol Chem 272(17):11283–11287

Beall A, Bagwell D, Woodrum D, Stoming TA, Kato K, Suzuki A, Rasmussen H, Brophy CM (1999) The small heat shock-related protein, HSP20, is phosphorylated on serine 16 during cyclic nucleotide-dependent relaxation. J Biol Chem 274(16):11344–11351

Boluyt MO, Brevick JL, Rogers DS, Randall MJ, Scalia AF, Li ZB (2006) Changes in the rat heart proteome induced by exercise training: increased abundance of heat shock protein hsp20. Proteomics 6(10):3154–3169. doi:10.1002/pmic.200401356

Boros S, Kamps B, Wunderink L, de Bruijn W, de Jong WW, Boelens WC (2004) Transglutaminase catalyzes differential crosslinking of small heat shock proteins and amyloid-beta. FEBS Lett 576(1–2):57–62, S0014579304010816

Brophy CM, Dickinson M, Woodrum D (1999a) Phosphorylation of the small heat shock-related protein, HSP20, in vascular smooth muscles is associated with changes in the macromolecular associations of HSP20. J Biol Chem 274(10):6324–6329

Brophy CM, Lamb S, Graham A (1999b) The small heat shock-related protein-20 is an actin-associated protein. J Vasc Surg 29(2):326–333, S0741521499000427

Bruinsma IB, Bruggink KA, Kinast K, Versleijen AA, Segers-Nolten IM, Subramaniam V, Kuiperij HB, Boelens W, de Waal RM, Verbeek MM (2011) Inhibition of alpha-synuclein aggregation by small heat shock proteins. Proteins 79(10):2956–2967. doi:10.1002/prot.23152

Bukach OV, Seit-Nebi AS, Marston SB, Gusev NB (2004) Some properties of human small heat shock protein Hsp20 (HspB6). Eur J Biochem 271(2):291–302, 3928

Bukach OV, Marston SB, Gusev NB (2005) Small heat shock protein with apparent molecular mass 20 kDa (Hsp20, HspB6) is not a genuine actin-binding protein. J Muscle Res Cell Motil 26(4–5):175–181. doi:10.1007/s10974-005-9008-7

Bukach OV, Glukhova AE, Seit-Nebi AS, Gusev NB (2009) Heterooligomeric complexes formed by human small heat shock proteins HspB1 (Hsp27) and HspB6 (Hsp20). Biochim Biophys Acta 1794(3):486–495, S1570-9639(08)00372-5

Caccamo D, Condello S, Ferlazzo N, Curro M, Griffin M, Ientile R (2013) Transglutaminase 2 interaction with small heat shock proteins mediate cell survival upon excitotoxic stress. Amino Acids 44(1):151–159. doi:10.1007/s00726-011-1083-z

Carra S, Seguin SJ, Landry J (2008) HspB8 and Bag3: a new chaperone complex targeting misfolded proteins to macroautophagy. Autophagy 4(2):237–239, 5407

Chen A, Karolczak-Bayatti M, Sweeney M, Treumann A, Morrissey K, Ulrich SM, Nicholas Europe-Finner G, Taggart MJ (2013) Lysine deacetylase inhibition promotes relaxation of arterial tone and C-terminal acetylation of HSPB6 (Hsp20) in vascular smooth muscle cells. Physiol Rep 1(6):e00127. doi:10.1002/phy2.127

Chernik IS, Seit-Nebi AS, Marston SB, Gusev NB (2007) Small heat shock protein Hsp20 (HspB6) as a partner of 14-3-3 gamma. Mol Cell Biochem 295(1–2):9–17. doi:10.1007/s11010-006-9266-8

Clark AR, Naylor CE, Bagneris C, Keep NH, Slingsby C (2011) Crystal structure of R120G disease mutant of human alphaB-crystallin domain dimer shows closure of a groove. J Mol Biol 408(1):118–134, S0022-2836(11)00164-1

Crippa V, Carra S, Rusmini P, Sau D, Bolzoni E, Bendotti C, De Biasi S, Poletti A (2010) A role of small heat shock protein B8 (HspB8) in the autophagic removal of misfolded proteins responsible for neurodegenerative diseases. Autophagy 6(7):958–960, 13042

Cross BE, O'Dea HM, MacPhee DJ (2007) Expression of small heat shock-related protein 20 (HSP20) in rat myometrium is markedly decreased during late pregnancy and labour. Reproduction 133(4):807–817, 133/4/807

Damer CK, Partridge J, Pearson WR, Haystead TA (1998) Rapid identification of protein phosphatase 1-binding proteins by mixed peptide sequencing and data base searching. Characterization of a novel holoenzymic form of protein phosphatase 1. J Biol Chem 273(38):24396–24405

Datskevich PN, Mymrikov EV, Gusev NB (2012a) Utilization of fluorescent chimeras for investigation of heterooligomeric complexes formed by human small heat shock proteins. Biochimie 94(8):1794–1784, S0300-9084(12)00146-0

Datskevich PN, Mymrikov EV, Sluchanko NN, Shemetov AA, Sudnitsyna MV, Gusev NB (2012b) Expression, purification and some properties of fluorescent chimeras of human small heat shock proteins. Protein Expr Purif 82(1):45–54, S1046-5928(11)00313-5

Deighton RF, McGregor R, Kemp J, McCulloch J, Whittle IR (2010) Glioma pathophysiology: insights emerging from proteomics. Brain Pathol 20(4):691–703. doi:10.1111/j.1750-3639.2010.00376.x

Delbecq SP, Klevit RE (2013) One size does not fit all: the oligomeric states of alphaB crystallin. FEBS Lett 587(8):1073–1080, S0014-5793(13)00052-5

Delbecq SP, Jehle S, Klevit R (2012) Binding determinants of the small heat shock protein, alphaB-crystallin: recognition of the 'IxI' motif. EMBO J 31(24):4587–4594, emboj2012318

Dohke T, Wada A, Isono T, Fujii M, Yamamoto T, Tsutamoto T, Horie M (2006) Proteomic analysis reveals significant alternations of cardiac small heat shock protein expression in congestive heart failure. J Card Fail 12(1):77–84, S1071-9164(05)00703-7

Dreiza CM, Brophy CM, Komalavilas P, Furnish EJ, Joshi L, Pallero MA, Murphy-Ullrich JE, von Rechenberg M, Ho YS, Richardson B, Xu N, Zhen Y, Peltier JM, Panitch A (2005) Transducible heat shock protein 20 (HSP20) phosphopeptide alters cytoskeletal dynamics. FASEB J 19(2):261–263, 04-2911fje

Dreiza CM, Komalavilas P, Furnish EJ, Flynn CR, Sheller MR, Smoke CC, Lopes LB, Brophy CM (2010) The small heat shock protein, HSPB6, in muscle function and disease. Cell Stress Chaperones 15(1):1–11. doi:10.1007/s12192-009-0127-8

Dubinska-Magiera M, Jablonska J, Saczko J, Kulbacka J, Jagla T, Daczewska M (2014) Contribution of small heat shock proteins to muscle development and function. FEBS Lett. doi:10.1016/j.febslet.2014.01.005

Edwards HV, Cameron RT, Baillie GS (2011) The emerging role of HSP20 as a multifunctional protective agent. Cell Signal 23(9):1447–1454, S0898-6568(11)00146-X

Edwards HV, Scott JD, Baillie GS (2012a) The A-kinase anchoring protein AKAP-Lbc facilitates cardioprotective PKA phosphorylation of Hsp20 on Serine 16. Biochem J 446(3):437–443, BJ20120570

Edwards HV, Scott JD, Baillie GS (2012b) PKA phosphorylation of the small heat-shock protein Hsp20 enhances its cardioprotective effects. Biochem Soc Trans 40(1):210–214, BST20110673

Fan GC, Kranias EG (2011) Small heat shock protein 20 (HspB6) in cardiac hypertrophy and failure. J Mol Cell Cardiol 51(4):574–577, S0022-2828(10)00344-5

Fan GC, Chu G, Mitton B, Song Q, Yuan Q, Kranias EG (2004) Small heat-shock protein Hsp20 phosphorylation inhibits beta-agonist-induced cardiac apoptosis. Circ Res 94(11):1474–1482. doi:10.1161/01.RES.0000129179.66631.00

Fan GC, Chu G, Kranias EG (2005a) Hsp20 and its cardioprotection. Trends Cardiovasc Med 15(4):138–141, S1050-1738(05)00058-7

Fan GC, Ren X, Qian J, Yuan Q, Nicolaou P, Wang Y, Jones WK, Chu G, Kranias EG (2005b) Novel cardioprotective role of a small heat-shock protein, Hsp20, against ischemia/reperfusion injury. Circulation 111(14):1792–1799, 01.CIR.0000160851.41872.C6

Fan GC, Yuan Q, Song G, Wang Y, Chen G, Qian J, Zhou X, Lee YJ, Ashraf M, Kranias EG (2006) Small heat-shock protein Hsp20 attenuates beta-agonist-mediated cardiac remodeling through apoptosis signal-regulating kinase 1. Circ Res 99(11):1233–1242, 01.RES.0000251074.19348.af

Fan GC, Zhou X, Wang X, Song G, Qian J, Nicolaou P, Chen G, Ren X, Kranias EG (2008) Heat shock protein 20 interacting with phosphorylated Akt reduces doxorubicin-triggered oxidative stress and cardiotoxicity. Circ Res 103(11):1270–1279, CIRCRESAHA.108.182832

Flynn CR, Komalavilas P, Tessier D, Thresher J, Niederkofler EE, Dreiza CM, Nelson RW, Panitch A, Joshi L, Brophy CM (2003) Transduction of biologically active motifs of the small heat shock-related protein HSP20 leads to relaxation of vascular smooth muscle. FASEB J 17(10):1358–1360. doi:10.1096/fj.02-1028fje

Flynn CR, Smoke CC, Furnish E, Komalavilas P, Thresher J, Yi Z, Mandarino LJ, Brophy CM (2007) Phosphorylation and activation of a transducible recombinant form of human HSP20 in Escherichia coli. Protein Expr Purif 52(1):50–58, S1046-5928(06)00263-4

Fontaine JM, Rest JS, Welsh MJ, Benndorf R (2003) The sperm outer dense fiber protein is the 10th member of the superfamily of mammalian small stress proteins. Cell Stress Chaperones 8(1):62–69

Fontaine JM, Sun X, Benndorf R, Welsh MJ (2005) Interactions of HSP22 (HSPB8) with HSP20, alphaB-crystallin, and HSPB3. Biochem Biophys Res Commun 337(3):1006–1011, S0006-291X(05)02180-7

Fontaine JM, Sun X, Hoppe AD, Simon S, Vicart P, Welsh MJ, Benndorf R (2006) Abnormal small heat shock protein interactions involving neuropathy-associated HSP22 (HSPB8) mutants. FASEB J 20(12):2168–2170, fj.06-5911fje

Freeman AK, Morrison DK (2011) 14-3-3 Proteins: diverse functions in cell proliferation and cancer progression. Semin Cell Dev Biol 22(7):681–687. doi:10.1016/j.semcdb.2011.08.009

Frobert O, Buus CL, Rembold CM (2005) HSP20 phosphorylation and interstitial metabolites in hypoxia-induced dilation of swine coronary arteries. Acta Physiol Scand 184(1):37–44, APS1426

Fuchs M, Poirier DJ, Seguin SJ, Lambert H, Carra S, Charette SJ, Landry J (2010) Identification of the key structural motifs involved in HspB8/HspB6-Bag3 interaction. Biochem J 425(1):245–255, BJ20090907

Furnish EJ, Brophy CM, Harris VA, Macomson S, Winger J, Head GA, Shaver EG (2010) Treatment with transducible phosphopeptide analogues of the small heat shock-related protein, HSP20, after experimental subarachnoid hemorrhage: prevention and reversal of delayed decreases in cerebral perfusion. J Neurosurg 112(3):631–639. doi:10.3171/2009.7.JNS09730

Gohla A, Bokoch GM (2002) 14-3-3 regulates actin dynamics by stabilizing phosphorylated cofilin. Curr Biol 12(19):1704–1710, S0960982202011843

Golenhofen N, Perng MD, Quinlan RA, Drenckhahn D (2004) Comparison of the small heat shock proteins alphaB-crystallin, MKBP, HSP25, HSP20, and cvHSP in heart and skeletal muscle. Histochem Cell Biol 122(5):415–425. doi:10.1007/s00418-004-0711-z

Grebenova D, Roselova P, Pluskalova M, Halada P, Rosel D, Suttnar J, Brodska B, Otevrelova P, Kuzelova K (2012) Proteins implicated in the increase of adhesivity induced by suberoylanilide hydroxamic acid in leukemic cells. J Proteomics 77:406–422. doi:10.1016/j.jprot.2012.09.014

Hocking KM, Baudenbacher FJ, Putumbaka G, Venkatraman S, Cheung-Flynn J, Brophy CM, Komalavilas P (2013) Role of cyclic nucleotide-dependent actin cytoskeletal dynamics: Ca(2+)](i) and force suppression in forskolin-pretreated porcine coronary arteries. PLoS One 8(4):e60986. doi:10.1371/journal.pone.0060986

Hojlund K, Bowen BP, Hwang H, Flynn CR, Madireddy L, Geetha T, Langlais P, Meyer C, Mandarino LJ, Yi Z (2009) In vivo phosphoproteome of human skeletal muscle revealed by phosphopeptide enrichment and HPLC-ESI-MS/MS. J Proteome Res 8(11):4954–4965. doi:10.1021/pr9007267

Huey KA, Thresher JS, Brophy CM, Roy RR (2004) Inactivity-induced modulation of Hsp20 and Hsp25 content in rat hindlimb muscles. Muscle Nerve 30(1):95–101. doi:10.1002/mus.20063

Inaguma Y, Hasegawa K, Kato K, Nishida Y (1996) cDNA cloning of a 20-kDa protein (p20) highly homologous to small heat shock proteins: developmental and physiological changes in rat hindlimb muscles. Gene 178(1–2):145–150

Islamovic E, Duncan A, Bers DM, Gerthoffer WT, Mestril R (2007) Importance of small heat shock protein 20 (hsp20) C-terminal extension in cardioprotection. J Mol Cell Cardiol 42(4):862–869, S0022-2828(07)00004-1

Kanno Y, Matsuno H (2006) The possibility of novel antiplatelet peptides: the physiological effects of low molecular weight HSPs on platelets. Curr Pharm Des 12(7):887–892

Kappe G, Verschuure P, Philipsen RL, Staalduinen AA, Van de Boogaart P, Boelens WC, De Jong WW (2001) Characterization of two novel human small heat shock proteins: protein kinase-related HspB8 and testis-specific HspB9. Biochim Biophys Acta 1520(1):1–6

Kappe G, Franck E, Verschuure P, Boelens WC, Leunissen JA, de Jong WW (2003) The human genome encodes 10 alpha-crystallin-related small heat shock proteins: HspB1-10. Cell Stress Chaperones 8(1):53–61

Karolczak-Bayatti M, Sweeney M, Cheng J, Edey L, Robson SC, Ulrich SM, Treumann A, Taggart MJ, Europe-Finner GN (2011) Acetylation of heat shock protein 20 (Hsp20) regulates human myometrial activity. J Biol Chem 286(39):34346–34355. doi:10.1074/jbc.M111.278549

Kato K, Goto S, Inaguma Y, Hasegawa K, Morishita R, Asano T (1994) Purification and characterization of a 20-kDa protein that is highly homologous to alpha B crystallin. J Biol Chem 269(21):15302–15309

Kim NK, Joh JH, Park HR, Kim OH, Park BY, Lee CS (2004) Differential expression profiling of the proteomes and their mRNAs in porcine white and red skeletal muscles. Proteomics 4(11):3422–3428. doi:10.1002/pmic.200400976

Kirbach BB, Golenhofen N (2011) Differential expression and induction of small heat shock proteins in rat brain and cultured hippocampal neurons. J Neurosci Res 89(2):162–175. doi:10.1002/jnr.22536

Koga Y, Pelizzola M, Cheng E, Krauthammer M, Sznol M, Ariyan S, Narayan D, Molinaro AM, Halaban R, Weissman SM (2009) Genome-wide screen of promoter methylation identifies novel markers in melanoma. Genome Res 19(8):1462–1470. doi:10.1101/gr.091447.109

Komalavilas P, Penn RB, Flynn CR, Thresher J, Lopes LB, Furnish EJ, Guo M, Pallero MA, Murphy-Ullrich JE, Brophy CM (2008) The small heat shock-related protein, HSP20, is a cAMP-dependent protein kinase substrate that is involved in airway smooth muscle relaxation. Am J Physiol Lung Cell Mol Physiol 294(1):L69–L78. doi:10.1152/ajplung.00235.2007

Kozawa O, Matsuno H, Niwa M, Hatakeyama D, Oiso Y, Kato K, Uematsu T (2002) HSP20, low-molecular-weight heat shock-related protein, acts extracellularly as a regulator of platelet functions: a novel defense mechanism. Life Sci 72(2):113–124, S0024320502021446

Laganowsky A, Eisenberg D (2010) Non-3D domain swapped crystal structure of truncated zebrafish alphaA crystallin. Protein Sci. doi:10.1002/pro.471

Laganowsky A, Benesch JL, Landau M, Ding L, Sawaya MR, Cascio D, Huang Q, Robinson CV, Horwitz J, Eisenberg D (2010) Crystal structures of truncated alphaA and alphaB crystallins reveal structural mechanisms of polydispersity important for eye lens function. Protein Sci 19(5):1031–1043. doi:10.1002/pro.380

Lopes LB, Furnish EJ, Komalavilas P, Flynn CR, Ashby P, Hansen A, Ly DP, Yang GP, Longaker MT, Panitch A, Brophy CM (2009) Cell permeant peptide analogues of the small heat shock protein, HSP20, reduce TGF-beta1-induced CTGF expression in keloid fibroblasts. J Invest Dermatol 129(3):590–598, jid2008264

Matsuno H, Kozawa O, Niwa M, Usui A, Ito H, Uematsu T, Kato K (1998) A heat shock-related protein, p20, plays an inhibitory role in platelet activation. FEBS Lett 429(3):327–329, S0014-5793(98)00626-7

Matsuno H, Ishisaki A, Nakajima K, Kato K, Kozawa O (2003) A peptide isolated from alpha B-crystallin is a novel and potent inhibitor of platelet aggregation via dual prevention of PAR-1 and GPIb/V/IX. J Thromb Haemost 1(12):2636–2642

Matsushima-Nishiwaki R, Kumada T, Nagasawa T, Suzuki M, Yasuda E, Okuda S, Maeda A, Kaneoka Y, Toyoda H, Kozawa O (2013) Direct association of heat shock protein 20 (HSPB6) with phosphoinositide 3-kinase (PI3K) in human hepatocellular carcinoma: regulation of the PI3K activity. PLoS One 8(11):e78440. doi:10.1371/journal.pone.0078440

McCollum AK, Casagrande G, Kohn EC (2010) Caught in the middle: the role of Bag3 in disease. Biochem J 425(1):e1–e3, BJ20091739

McHaourab HS, Godar JA, Stewart PL (2009) Structure and mechanism of protein stability sensors: chaperone activity of small heat shock proteins. Biochemistry 48(18):3828–3837. doi:10.1021/bi900212j

McLemore EC, Tessier DJ, Flynn CR, Furnish EJ, Komalavilas P, Thresher JS, Joshi L, Stone WM, Fowl RJ, Brophy CM (2004) Transducible recombinant small heat shock-related protein, HSP20, inhibits vasospasm and platelet aggregation. Surgery 136(3):573–578. doi:10.1016/j.surg.2004.04.024

Meeks MK, Ripley ML, Jin Z, Rembold CM (2005) Heat shock protein 20-mediated force suppression in forskolin-relaxed swine carotid artery. Am J Physiol Cell Physiol 288(3):C633–C639, 00269.2004

Mizuno K (2013) Signaling mechanisms and functional roles of cofilin phosphorylation and dephosphorylation. Cell Signal 25(2):457–469. doi:10.1016/j.cellsig.2012.11.001

Mymrikov EV, Bukach OV, Seit-Nebi AS, Gusev NB (2010) The pivotal role of the beta 7 strand in the intersubunit contacts of different human small heat shock proteins. Cell Stress Chaperones 15(4):365–377. doi:10.1007/s12192-009-0151-8

Mymrikov EV, Seit-Nebi AS, Gusev NB (2011) Large potentials of small heat shock proteins. Physiol Rev 91(4):1123–1159, 91/4/1123

Mymrikov EV, Seit-Nebi AS, Gusev NB (2012) Heterooligomeric complexes of human small heat shock proteins. Cell Stress Chaperones 17(2):157–169. doi:10.1007/s12192-011-0296-0

Nefedova VV, Datskevich PN, Sudnitsyna MV, Strelkov SV, Gusev NB (2013a) Physico-chemical properties of R140G and K141Q mutants of human small heat shock protein HspB1 associated with hereditary peripheral neuropathies. Biochimie 95(8):1582–1592. doi:10.1016/j.biochi.2013.04.014

Nefedova VV, Sudnitsyna MV, Strelkov SV, Gusev NB (2013b) Structure and properties of G84R and L99M mutants of human small heat shock protein HspB1 correlating with motor neuropathy. Arch Biochem Biophys 538(1):16–24. doi:10.1016/j.abb.2013.07.028

Nicolaou P, Knoll R, Haghighi K, Fan GC, Dorn GW 2nd, Hasenfub G, Kranias EG (2008) Human mutation in the anti-apoptotic heat shock protein 20 abrogates its cardioprotective effects. J Biol Chem 283(48):33465–33471, M802307200

Niwa M, Kozawa O, Matsuno H, Kato K, Uematsu T (2000) Small molecular weight heat shock-related protein, HSP20, exhibits an anti-platelet activity by inhibiting receptor-mediated calcium influx. Life Sci 66(1):PL7–PL12, S0024320599005664

Noda T, Kumada T, Takai S, Matsushima-Nishiwaki R, Yoshimi N, Yasuda E, Kato K, Toyoda H, Kaneoka Y, Yamaguchi A, Kozawa O (2007) Expression levels of heat shock protein 20 decrease in parallel with tumor progression in patients with hepatocellular carcinoma. Oncol Rep 17(6):1309–1314

O'Connor MJ, Rembold CM (2002) Heat-induced force suppression and HSP20 phosphorylation in swine carotid media. J Appl Physiol 93(2):484–488. doi:10.1152/japplphysiol.00009.2002

Pipkin W, Johnson JA, Creazzo TL, Burch J, Komalavilas P, Brophy C (2003) Localization, macromolecular associations, and function of the small heat shock-related protein HSP20 in rat heart. Circulation 107(3):469–476

Pozuelo Rubio M, Geraghty KM, Wong BH, Wood NT, Campbell DG, Morrice N, Mackintosh C (2004) 14-3-3-affinity purification of over 200 human phosphoproteins reveals new links to regulation of cellular metabolism, proliferation and trafficking. Biochem J 379(Pt 2):395–408. doi:10.1042/BJ20031797

Qian J, Ren X, Wang X, Zhang P, Jones WK, Molkentin JD, Fan GC, Kranias EG (2009) Blockade of Hsp20 phosphorylation exacerbates cardiac ischemia/reperfusion injury by suppressed autophagy and increased cell death. Circ Res 105(12):1223–1231, CIRCRESAHA.109.200378

Qian J, Vafiadaki E, Florea SM, Singh VP, Song W, Lam CK, Wang Y, Yuan Q, Pritchard TJ, Cai W, Haghighi K, Rodriguez P, Wang HS, Sanoudou D, Fan GC, Kranias EG (2011) Small heat shock protein 20 interacts with protein phosphatase-1 and enhances sarcoplasmic reticulum calcium cycling. Circ Res 108(12):1429–1438, CIRCRESAHA.110.237644

Rembold CM, Zhang E (2001) Localization of heat shock protein 20 in swine carotid artery. BMC Physiol 1:10

Rembold CM, Foster DB, Strauss JD, Wingard CJ, Eyk JE (2000) cGMP-mediated phosphorylation of heat shock protein 20 may cause smooth muscle relaxation without myosin light chain dephosphorylation in swine carotid artery. J Physiol 524(Pt 3):865–878, PHY_0041

Ren XP, Wu J, Wang X, Sartor MA, Qian J, Jones K, Nicolaou P, Pritchard TJ, Fan GC (2009) MicroRNA-320 is involved in the regulation of cardiac ischemia/reperfusion injury by targeting heat-shock protein 20. Circulation 119(17):2357–2366. doi:10.1161/CIRCULATIONAHA.108.814145

Rosati A, Graziano V, De Laurenzi V, Pascale M, Turco MC (2011) BAG3: a multifaceted protein that regulates major cell pathways. Cell Death Dis 2:e141, cddis201124

Salinthone S, Tyagi M, Gerthoffer WT (2008) Small heat shock proteins in smooth muscle. Pharmacol Ther 119(1):44–54, S0163-7258(08)00080-6

Seit-Nebi AS, Gusev NB (2010) Versatility of the small heat shock protein HSPB6 (Hsp20). Cell Stress Chaperones 15(3):233–236. doi:10.1007/s12192-009-0141-x

Selcen D, Muntoni F, Burton BK, Pegoraro E, Sewry C, Bite AV, Engel AG (2009) Mutation in BAG3 causes severe dominant childhood muscular dystrophy. Ann Neurol 65(1):83–89. doi:10.1002/ana.21553

Shemetov AA, Gusev NB (2011) Biochemical characterization of small heat shock protein HspB8 (Hsp22)-Bag3 interaction. Arch Biochem Biophys 513(1):1–9, S0003-9861(11)00249-9

Simon S, Fontaine JM, Martin JL, Sun X, Hoppe AD, Welsh MJ, Benndorf R, Vicart P (2007) Myopathy-associated alphaB-crystallin mutants: abnormal phosphorylation, intracellular location, and interactions with other small heat shock proteins. J Biol Chem 282(47):34276–34287, M703267200

Sin YY, Edwards HV, Li X, Day JP, Christian F, Dunlop AJ, Adams DR, Zaccolo M, Houslay MD, Baillie GS (2011) Disruption of the cyclic AMP phosphodiesterase-4 (PDE4)-HSP20 complex attenuates the beta-agonist induced hypertrophic response in cardiac myocytes. J Mol Cell Cardiol 50(5):872–883. doi:10.1016/j.yjmcc.2011.02.006

Sluchanko NN, Gusev NB (2010) 14-3-3 proteins and regulation of cytoskeleton. Biochemistry 75(13):1528–1546

Sluchanko NN, Gusev NB (2012) Oligomeric structure of 14-3-3 protein: what do we know about monomers? FEBS Lett 586(24):4249–4256. doi:10.1016/j.febslet.2012.10.048

Sluchanko NN, Sudnitsyna MV, Chernik IS, Seit-Nebi AS, Gusev NB (2011a) Phosphomimicking mutations of human 14-3-3zeta affect its interaction with tau protein and small heat shock protein HspB6. Arch Biochem Biophys 506(1):24–34, S0003-9861(10)00463-7

Sluchanko NN, Sudnitsyna MV, Seit-Nebi AS, Antson AA, Gusev NB (2011b) Properties of the monomeric form of human 14-3-3zeta protein and its interaction with tau and HspB6. Biochemistry 50(45):9797–9808. doi:10.1021/bi201374s

Sluchanko NN, Artemova NV, Sudnitsyna MV, Safenkova IV, Antson AA, Levitsky DI, Gusev NB (2012) Monomeric 14-3-3zeta has a chaperone-like activity and is stabilized by phosphorylated HspB6. Biochemistry 51(31):6127–6138. doi:10.1021/bi300674e

Sluchanko NN, Chebotareva NA, Gusev NB (2013) Modulation of 14-3-3/phosphotarget interaction by physiological concentrations of phosphate and glycerophosphates. PLoS One 8(8):e72597. doi:10.1371/journal.pone.0072597

Sudnitsyna MV, Mymrikov EV, Seit-Nebi AS, Gusev NB (2012a) The role of intrinsically disordered regions in the structure and functioning of small heat shock proteins. Curr Protein Pept Sci 13(1):76–85, BSP/CPPS/E-Pub/163

Sudnitsyna MV, Seit-Nebi AS, Gusev NB (2012b) Cofilin weakly interacts with 14-3-3 and therefore can only indirectly participate in regulation of cell motility by small heat shock protein HspB6 (Hsp20). Arch Biochem Biophys 521(1–2):62–70, S0003-9861(12)00101-4

Sugiyama Y, Suzuki A, Kishikawa M, Akutsu R, Hirose T, Waye MM, Tsui SK, Yoshida S, Ohno S (2000) Muscle develops a specific form of small heat shock protein complex composed of MKBP/HSPB2 and HSPB3 during myogenic differentiation. J Biol Chem 275(2):1095–1104

Takayama S, Reed JC (2001) Molecular chaperone targeting and regulation by BAG family proteins. Nat Cell Biol 3(10):E237–E241. doi:10.1038/ncb1001-e237

Taylor RP, Benjamin IJ (2005) Small heat shock proteins: a new classification scheme in mammals. J Mol Cell Cardiol 38(3):433–444, S0022-2828(05)00003-9

Tessier DJ, Komalavilas P, Panitch A, Joshi L, Brophy CM (2003) The small heat shock protein (HSP) 20 is dynamically associated with the actin cross-linking protein actinin. J Surg Res 111(1):152–157, S0022480403001136

Tessier DJ, Komalavilas P, Liu B, Kent CK, Thresher JS, Dreiza CM, Panitch A, Joshi L, Furnish E, Stone W, Fowl R, Brophy CM (2004) Transduction of peptide analogs of the small heat shock-related protein HSP20 inhibits intimal hyperplasia. J Vasc Surg 40(1):106–114. doi:10.1016/j.jvs.2004.03.028

Tyson EK, Macintyre DA, Smith R, Chan EC, Read M (2008) Evidence that a protein kinase A substrate, small heat-shock protein 20, modulates myometrial relaxation in human pregnancy. Endocrinology 149(12):6157–6165, en.2008-0593

Uhart M, Bustos DM (2013) Human 14-3-3 paralogs differences uncovered by cross-talk of phosphorylation and lysine acetylation. PLoS One 8(2):e55703. doi:10.1371/journal.pone.0055703

van de Klundert FA, Smulders RH, Gijsen ML, Lindner RA, Jaenicke R, Carver JA, de Jong WW (1998) The mammalian small heat-shock protein Hsp20 forms dimers and is a poor chaperone. Eur J Biochem 258(3):1014–1021

van de Klundert F, van den Ijssel P, Stege GJ, de Jong WW (1999) Rat Hsp20 confers thermoresistance in a clonal survival assay, but fails to protect coexpressed luciferase in Chinese hamster ovary cells. Biochem Biophys Res Commun 254(1):164–168, S0006291X9899917X

van Noort JM, Bsibsi M, Nacken P, Gerritsen WH, Amor S (2012) The link between small heat shock proteins and the immune system. Int J Biochem Cell Biol 44(10):1670–1679. doi:10.1016/j.biocel.2011.12.010

Verschuure P, Tatard C, Boelens WC, Grongnet JF, David JC (2003) Expression of small heat shock proteins HspB2, HspB8, Hsp20 and cvHsp in different tissues of the perinatal developing pig. Eur J Cell Biol 82(10):523–530

Vos MJ, Hageman J, Carra S, Kampinga HH (2008) Structural and functional diversities between members of the human HSPB, HSPH, HSPA, and DNAJ chaperone families. Biochemistry 47(27):7001–7011. doi:10.1021/bi800639z

Vos MJ, Zijlstra MP, Carra S, Sibon OC, Kampinga HH (2011) Small heat shock proteins, protein degradation and protein aggregation diseases. Autophagy 7(1):101–103, 13935

Wang Y, Xu A, Pearson RB, Cooper GJ (1999) Insulin and insulin antagonists evoke phosphorylation of P20 at serine 157 and serine 16 respectively in rat skeletal muscle. FEBS Lett 462(1–2):25–30, S0014-5793(99)01496-9

Wang X, Zingarelli B, O'Connor M, Zhang P, Adeyemo A, Kranias EG, Wang Y, Fan GC (2009) Overexpression of Hsp20 prevents endotoxin-induced myocardial dysfunction and apoptosis via inhibition of NF-kappaB activation. J Mol Cell Cardiol 47(3):382–390, S0022-2828(09)00227-2

Weeks SD, Baranova EV, Heirbaut M, Beelen S, Shkumatov AV, Gusev NB, Strelkov SV (2013) Molecular structure and dynamics of the dimeric human small heat shock protein HSPB6. J Struct Biol. doi:10.1016/j.jsb.2013.12.009

Wilhelmus MM, Boelens WC, Otte-Holler I, Kamps B, de Waal RM, Verbeek MM (2006a) Small heat shock proteins inhibit amyloid-beta protein aggregation and cerebrovascular amyloid-beta protein toxicity. Brain Res 1089(1):67–78, S0006-8993(06)00762-1

Wilhelmus MM, Boelens WC, Otte-Holler I, Kamps B, Kusters B, Maat-Schieman ML, de Waal RM, Verbeek MM (2006b) Small heat shock protein HspB8: its distribution in Alzheimer's disease brains and its inhibition of amyloid-beta protein aggregation and cerebrovascular amyloid-beta toxicity. Acta Neuropathol 111(2):139–149. doi:10.1007/s00401-005-0030-z

Wilhelmus MM, Otte-Holler I, Wesseling P, de Waal RM, Boelens WC, Verbeek MM (2006c) Specific association of small heat shock proteins with the pathological hallmarks of Alzheimer's disease brains. Neuropathol Appl Neurobiol 32(2):119–130, NAN689

Willis MS, Patterson C (2010) Hold me tight: role of the heat shock protein family of chaperones in cardiac disease. Circulation 122(17):1740–1751. doi:10.1161/CIRCULATIONAHA.110.942250

Woodrum DA, Brophy CM, Wingard CJ, Beall A, Rasmussen H (1999) Phosphorylation events associated with cyclic nucleotide-dependent inhibition of smooth muscle contraction. Am J Physiol 277(3 Pt 2):H931–H939

Woodrum D, Pipkin W, Tessier D, Komalavilas P, Brophy CM (2003) Phosphorylation of the heat shock-related protein, HSP20, mediates cyclic nucleotide-dependent relaxation. J Vasc Surg 37(4):874–881. doi:10.1067/mva.2003.153

Yaffe MB, Rittinger K, Volinia S, Caron PR, Aitken A, Leffers H, Gamblin SJ, Smerdon SJ, Cantley LC (1997) The structural basis for 14-3-3:phosphopeptide binding specificity. Cell 91(7):961–971

Yap YL, Lam DC, Luc G, Zhang XW, Hernandez D, Gras R, Wang E, Chiu SW, Chung LP, Lam WK, Smith DK, Minna JD, Danchin A, Wong MP (2005) Conserved transcription factor binding sites of cancer markers derived from primary lung adenocarcinoma microarrays. Nucleic Acids Res 33(1):409–421. doi:10.1093/nar/gki188

Zeng L, Tan J, Lu W, Hu Z (2013) Blockade of Ser16-Hsp20 phosphorylation attenuates neuroprotection dependent upon Bcl-2 and Bax. Curr Neurovasc Res 10(3):208–215

Zhang X, Wang X, Zhu H, Kranias EG, Tang Y, Peng T, Chang J, Fan GC (2012) Hsp20 functions as a novel cardiokine in promoting angiogenesis via activation of VEGFR2. PLoS One 7(3):e32765. doi:10.1371/journal.pone.0032765

Chapter 10
The Chloroplast-Localized Plant sHsp in *Arabidopsis Thaliana:* Role of Its Oligomeric Conformation and Its Translocation into Membranes

Katja Bernfur, Gudrun Rutsdottir, Cecilia Månsson, and Cecilia Emanuelsson

Abstract Recent advances in quantitative proteomics show that small heat shock proteins (sHsps) are among the most highly upregulated proteins in cellular stress response. In plants there are numerous paralogous sHsps expressed in various cellular compartments. The chloroplast-localized sHsp, named Hsp21 with reference to its monomeric size, has an N-terminal region that is some 40 amino acids longer compared to the cytosolic sHsps, and increases plant stress tolerance as shown by *Arabidopsis thaliana* plants which overexpress Hsp21. Recombinantly expressed and purified Hsp21-protein shows the features expected for a chaperone protein in rescuing temperature-sensitive model substrate proteins from aggregation. Hsp21 is a dodecameric protein, with C-terminal tails that keep the dodecamer together but are highly flexible in solution. The N-terminal domains, which resemble intrinsically disordered proteins and are located in the interior of the dodecamer, contain conserved methionine residues of crucial importance for function. Methionine sulfoxidation abolishes the chaperone activity in vitro, but in vivo such oxidized methioniones appear to be continuously re-reduced thanks to a chloroplast-localized form of peptide methionine sulfoxide reductase, an important enzyme expressed ubiquitously and belonging to the minimal gene set for life. There is not a clear picture of which the endogenous substrate proteins of Hsp21 are. For more than a decade it has been observed that cyanobacterial sHsps enter the membranes at increased temperatures. In a quantitative proteomics approach we have recently analyzed and compared Hsp21 with hundreds of other chloroplast proteins, and found that Hsp21 is fairly unique with respect to its translocation into membrane in heat-stressed plants. One reason for the oligomericity of Hsp21 is suggested to be the possibility to rapidly supply hydrophobic surfaces with chaperoning capacity in response to stress, while still preventing membrane lysis by such hydrophobic surfaces under non-stress conditions.

K. Bernfur • G. Rutsdottir • C. Månsson • C. Emanuelsson (✉)
Department of Biochemistry and Structural Biology, Center for Molecular Protein Science, Lund University, PO Box 124, SE-22100 Lund, Sweden
e-mail: Cecilia.Emanuelsson@biochemistry.lu.se

R.M. Tanguay, L.E. Hightower (eds.), *The Big Book on Small Heat Shock Proteins*, Heat Shock Proteins 8, DOI 10.1007/978-3-319-16077-1_10

Keywords Arabidopis • Chloroplast • Heat stress • N15-labelling • Quantitative proteomics

10.1 The sHsps as Actors in Preventive Healthcare in Cells

The small heat shock protein (sHsp) chaperones play an important role as "paramedics in the cell" (Hilton et al. 2013), rapidly sequestering destabilized target proteins and overcoming the kinetic competition with aggregation, until other members of the chaperone proteostasis network can refold or degrade the target proteins. As such they act as "Nature's sponges" to avoid protein aggregation (Eyles and Gierasch 2010). They act on early unfolding intermediates (McHaourab et al. 2009), as concluded from the finding that T4 lysozyme mutants bind sHsps with higher affinity the lower the folding equilibrium constants, showing that sHsps can capture unfolded states of proteins even present for only a small fraction of the time. As such sHsps may even be regarded as workers in preventive healthcare rather than as paramedics. A number of human diseases are associated with malfunctioning sHsps (Carra et al. 2013; Garrido et al. 2012; Kampinga and Garrido 2012). In heat-stressed cells, the sHsps can be among the most highly upregulated proteins, as detected by recently developed proteomic techniques where quantitative mass spectrometry is combined with cryo electron tomography to determine protein concentrations in the human pathogen *Leptospira interrogans*, where the sHsp concentration was found to increase from 0.06 μM in non-stimulated cells to 30 μM (~45 μg/ml) in heat-stressed cells (Beck et al. 2009; Malmstrom et al. 2009). In plants the sHsps confer an especially pronounced part of the stress response, with several paralogous sHsps in different cell compartments, often heat-induced or developmentally expressed (Van Montfort et al. 2001a; Waters et al. 1996).

10.2 The Chloroplast-Localized sHsp, Hsp21

A chloroplast-localized sHsp evolved when the land-plants developed, through gene duplication from the cytosolic sHsps and not from the sHsp in the endosymbiont that gave rise to the chloroplast, in contrast to the organellar Hsp70s that are of bacterial origin (Waters and Vierling 1999). The interpretation of this finding is that the endosymbiont sHsp may not have been needed in the early eukaryotes and instead a later selection pressure, unique to the plant lineage, led to the diversification of cytosolic sHsps into the various organellar sHsps, in the chloroplasts, mitochondria and ER. It was recognized by Vierling and coworkers who cloned the gene for the chloroplast-localized sHsp, Hsp21, named with reference to its monomer size, that its N-terminal domain contained a unique set of conserved methionines in an amphipathic α-helix-motif (Chen and Vierling 1991). Genomic sequence data from *Physcomitrella patens* indicate that mosses do have an amphipathic α-helix in the N-terminal domain of their Hsp21-orthologues, whereas the methionines appear to have evolved later in evolution, in response to a specific selection pressure on higher land plants (Waters 2013).

By genetically engineering plants of *Arabidopsis thaliana* that constitutively overexpressed Hsp21 to 20 times the level on heat-induced wildtype plants, we found that Hsp21 plays a role in tolerance against oxidative stress (Fig. 10.1), and that the Hsp21 oligomer undergoes a conformational change in response to oxidation (Harndahl et al. 1999). Using recombinantly expressed and purified Hsp21 protein we also found that the conformational change is due to methionine oxidation (Gustavsson et al. 1999), and that substitution of conserved methionines by leucines results in loss of such redox-response in Hsp21, yet the chaperone activity is retained (Gustavsson et al. 2001). In plants, reversible methionine oxidation of Hsp21 is presumably part of a scavenging cycle for reactive oxygen species (ROS) since methionine sulfoxidation in Hsp21 is reversible (Gustavsson et al. 2002), due to a chloroplast-localized peptide methionine sulfoxide-reductase (PMSR), a crucial enzyme universally present in all organisms and belonging to the minimal gene set for life (Kim et al. 2014; Mushegian and Koonin 1996). Five PMSR-genes were detected in *Arabidopsis thaliana* by Murphy and coworkers, who cloned the chloroplast-localized form, which is highly expressed in photosynthetic tissues (Sadanandom et al. 2000).

When the wildtype and the Hsp21-overexpressing plants of *Arabidopsis thaliana* were analyzed by non-denaturing PAGE and Western blotting it was clear that the Hsp21 dodecamers become less visible during stress, especially repeated heat stress, as the antibodies decorated the whole lane as a smear, presumably showing "Hsp21 at work" (Harndahl et al. 1999). The Hsp21 dodecamers again became visible as distinct bands several days after the heat-stress. As a recombinantly

Fig. 10.1 The chloroplast-localized sHsp, Hsp21, increases plant stress tolerance. Plants of *Arabidopsis thaliana* (ecotype RLD, 21 days old), wildtype plants (*left*) and transgenic plants (*right*) which constitutively overexpressed Hsp21 to 20× normal levels, were exposed to repeated daily heat stress treatment (4 h, 40 °C) at high light-intensity (1,600 μE/m^{2}s). After 1 week, the wildtype plants were half the height of the Hsp21-overexpressors as seen in the picture, and the dry weight per plant had dropped to one third (30 %) for wildtype plants but only to half (53 %) for Hsp21-overexpressors, compared to plants given no heat stress treatment (Härndahl et al. 1999)

expressed and purified protein, Hsp21 is dodecameric and stable at high temperatures provided that DTT is present to keep the methionines reduced, whereas the dodecamer dissociates in presence of unfolding model substrate protein, as concluded from experiments with chemical crosslinking and SDS-PAGE (Ahrman et al. 2007).

The structure of Hsp21 is quite similar to the dodecameric wheat Hsp16.9, whose structure has been resolved to atomic resolution (Van Montfort et al. 2001b). Hsp 16.9 however is smaller and lacks the methionine-rich motif in the N-terminal domain, which is 41 amino acids longer in Hsp21. With Hsp16.9 as a template we generated a structure model for Hsp21, by homology modeling, single-particle electron microscopy and lysine-specific chemical crosslinking (Lambert et al. 2011). The model showed that the Hsp21 subunits were arranged in two hexameric discs that were rotated by 25° in relation to each other. Possibly, the protein density maps has caught a glimpse of a more dynamic conformation compared to a more symmetric and stable conformation required for crystallization. The rapid subunit exchange is functionally important (Bova et al. 1997, 2000; Giese and Vierling 2002) whereby hydrophobic parts hidden in the oligomer become exposed for interactions with the unfolding client proteins, forming polydisperse ensembles of sHsp-client complexes (Basha et al. 2011; Stengel et al. 2012). We have repeatedly seen in proteolysis experiments that the N-terminal domain in Hsp21 is rapidly degraded, behaving as intrinsically disordered proteins (Bardwell and Jakob 2012; Tompa and Csermely 2004) and in various functional in vitro assays Hsp21 displays the typical properties of a molecular chaperone, suppressing the aggregation of heat-sensitive model substrate proteins and of fibrillation-prone peptides (unpublished data). Chemical crosslinking and mass spectrometry suggest an interaction between the disordered N-terminal region of Hsp21 and the C-terminal presumably unfolding part of the temperature-sensitive substrate protein malate dehydrogenase (Lambert et al. 2013).

Which the endogenous target proteins for Hsp21 are is not clear, although an interesting, single target was recently identified, a transcription regulator, the plastid nucleoid protein pTAC5, that is required to maintain the plastid-encoded RNA-polymerase under heat stress (Zhong et al. 2013). In case of the 11 different mammalian sHsps, of which many are constitutively expressed, there are several known cytoskeletal target proteins, and a connection between certain protein aggregates and sHsps (Carra et al. 2013; Garrido et al. 2012; Kampinga and Garrido 2012). A cyanobacterial homologue to Hsp21, *Synechocystis* Hsp16.6, was investigated for target proteins using co-immunoprecipitation (Basha et al. 2004), and a large number of various cellular proteins with various cellular functions were identified as target proteins. Using the same *Synechocystis* Hsp16.6 system, a screen was also performed for mutations affecting activity, and it was concluded that mutations in the N-terminal domain could be detected that hampered the thermotolerance in vivo, yet without reducing the in vitro assay activity (Giese et al. 2005). Importantly, the fact that the N-terminal mutations were impaired in terms of their function in vivo, yet active as chaperones in vitro, indicates that chaperone activity assays in vitro should be interpreted with some precaution. They may reflect rather a very general chaperoning property in suppressing aggregation in vitro, rather than the physiologically relevant and essential features of the sHsp mechanism of action.

10.3 Quantification of Translocation into Membrane in Heat-Stressed Plants

Another avenue to follow regarding the sHsp mechanism of action is that Vigh and coworkers have discovered and described that cyanobacterial sHsps translocate into membranes under heat stress (Balogi et al. 2008; Torok et al. 2001), suggesting dual roles for sHsps in stress-protection, as molecular chaperones and as membrane stabilizers. Membranes are both targets and sensors of stress in all organisms but especially so in photosynthetic organisms (Horvath et al. 2012; Vigh et al. 2007).

We therefore addressed the question of translocation into membranes; to what extent the chloroplast-localized Hsp21 can translocate into membranes during heat stress and to what extent other proteins in the chloroplast proteome also translocate (Bernfur et al. unpublished data). We applied a quantitative proteomics procedure previously developed for *Arabidopsis thaliana* plants (Bernfur et al. 2013) as outlined in Fig. 10.2.

Plants were grown hydroponically with stable isotope labeling as indicated in Fig. 10.2a. Growth on ^{15}N-medium does not affect the protein composition of the plants compared to growth on normal (^{14}N) medium. All plants were pre-treated to induce expression of Hsp21. The ^{14}N plants were kept as control plants and the ^{15}N

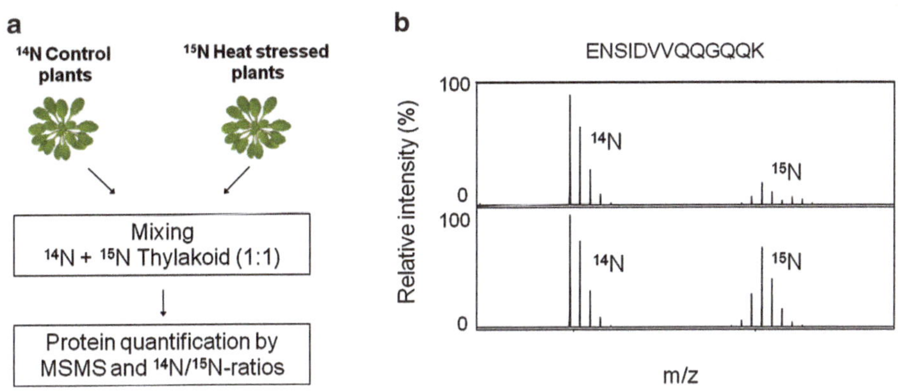

Fig. 10.2 Quantitative proteomics approach to assess the relative abundance of various proteins in the thylakoid membrane in heat-stressed compared to control plants. Plants of *Arabidopsis thaliana* were isotope-labelled and exposed to control and heat-stress conditions, after which they were processed to calculate the relative abundance of protein present in the thylakoid membranes of heat-stressed compared to control plants. (**a**) After heat stress, chloroplasts were immediately isolated, from the heat-stressed plants and from control plants, and subfractionated into soluble and thylakoid membrane fraction. After 1:1 mixing of thylakoid fraction from the heat-stressed (^{15}N) and control (^{15}N) plants, proteins in this mixture were analyzed by Orbitrap nano-LC MSMS and Mascot Distiller Quantitation Toolbox used to determine the signal intensities of the isotope peak pairs, and the relative abundance of the isotope forms for each protein was calculated. (**b**) Mass spectra show example of an isotope peak pair for one peptide (peptide ENSIDVVQQGQQK, amino acid 6–18 in the Hsp21-protein. The relative abundance was calculated as presented for in total 169 different proteins in Fig. 10.3

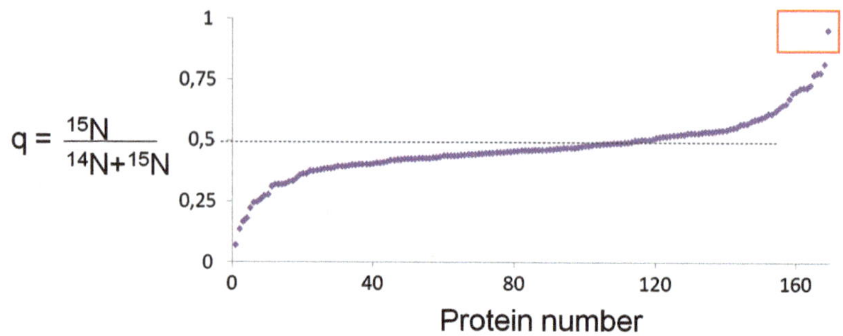

Fig. 10.3 Relative abundance of proteins detected in the thylakoid membrane in heat-stressed plants compared to control plants. The relative amount of protein present in the thylakoid membranes in heat-stressed plants compared to control plants is shown for 169 proteins. Protein quantification based on $^{14}N/^{15}N$-ratios was used to assess the relative abundance of proteins in the thylakoid membranes in heat-stressed plants compared to control plants, which is expressed as the q-value of the protein which is the average of the q-values of the peptides determined as $q = {}^{15}N / {}^{14}N + {}^{15}N$ according to Fig. 10.2. A value of $q > 0.5$ means larger amount of protein in thylakoid membrane in heat-stressed plants compared to control plants, a value of $q < 0.5$ means lower amount in thylakoid membrane in heat-stressed plants compared to control plants, and a value of $q = 0.5$ means that the protein is distributed similarly between control and heat-stressed plants

plants were heat-stressed for 1 h and immediately thereafter the thylakoid membranes were isolated, mixed 1:1 and processed for proteomic protein quantification with the Mascot Distiller Quantitation Toolbox software (www.matrixscience.com). The output gives, for several peptides in each protein, the L/H (light/heavy) isotope ratios from the signal intensities in the mass spectra, where peptides appear as peak pairs with one ^{14}N-peak (light) and one ^{15}N-peak (heavy), as exemplified in Fig. 10.2b. This gives the relative abundance, i.e. the amount of each protein present, in the thylakoid membranes in the heat-stressed plants compared to the control plants.

The relative abundance, defined as the q-value (from the sum of the ^{15}N isotope signal divided by the sum of the signals of the ^{14}N and ^{15}N), was determined for 169 proteins in total. If a protein has the same relative abundance in the membranes from heat-stressed and control plants the q-value is 0.5. We found that approximately 95 % of 169 proteins did not deviate strongly from $q = 0.5$ (Fig. 10.3). Contrastingly, Hsp21 had a q-value of 0.95, i.e. translocated into membrane in heat-stressed plants.

Among the 169 proteins, approximately one third were well-known high-abundance thylakoid membrane proteins belonging to the four multisubunit complexes in photosynthetic light reaction (photosystem I and II, the cytochrome b/f-complex, the ATP-synthase), with q-values between 0.36 and 0.55 (average $q = 0.44$), showing, reasonably enough, that their abundance in the thylakoid membrane did not change in heat-stressed plants compared to control plants. A few proteins showed very low values of $q < 0.2$, one of them was the chloroplast form of lipoxygenase ($q = 0.14$), involved in the biosynthesis of the stress signaling lipid compound jasmonic acid and the chloroplast RNA-binding protein ($q = 0.24$), involved in transcription. These two proteins appear to be degraded in heat-stressed plants compared to control plants, since the corresponding q-values in the soluble fraction also were low (q=0.13 and 0.32, respectively). Members of the Hsp70 subfamily and transketolase were detected in thylakoid membranes with q-values >0.77, and with the corresponding q-values in soluble stroma lower than 0.5 (q=0.41 and 0.5, respectively), suggesting translocation from the soluble stroma to the thylakoid membrane, albeit less pronounced than for Hsp21. Other proteins belonging to the family of heat shock proteins, two members of the DnaJ/Hsp40 subfamily and one of the Hsp60 subfamily, were detected in the thylakoid fractions with q-values at or slightly above 0.5 (0.46, 0.57 and 0.68, respectively). The data altogether suggest that Hsp21 translocates into the membrane in heat-stressed plants whereas, with few exceptions, other proteins largely do not.

10.4 Quantification of Translocation into Isolated Membranes of Hsp21-Protein

A temperature-dependent translocation of Hsp21 into membrane also occurred when the phenomenon was investigated in reconstituted systems in vitro, using isolated thylakoid membranes which were incubated with either isolated stroma that contained Hsp21 and all other proteins in the soluble chloroplast stroma, or with recombinantly expressed and purified Hsp21 protein. We used the same quantification procedure based on L/H-isotope ratios by using ^{15}N-labelled Hsp21, obtained by growing the bacterial host in minimal medium containing $^{15}N\text{-}NH_4Cl$ as nitrogen source as described in (Söderberg et al. 2012).

Moreover, we have generated a mutant of Hsp21 that is dimeric, as judged by gel filtration and non-denaturing PAGE (Bernfur et al. unpublished data). In this Hsp21 V181A-substitution mutant the hydrophobic interaction is reduced between the conserved the IXI/V-motif in the C-terminal tails (which in Hsp21 and its orthologues is extended to IX**V**XI, where the **V** in bold corresponds to V181) and the hydrophobic grooves in the adjacent monomers, i.e. the interactions which are so important in keeping sHsp oligomers together. When we have grown this Hsp21 V181A-substitution mutant protein, we have repeatedly observed that the *E. coli* cells expressing this mutant protein grow less well, are very easily lysed and that the Hsp21 V181A-substitution mutant protein has a tendency to leak out into the medium more than the wildtype protein. We hypothesized that this reflects the functional importance of why Hsp21 is oligomeric, and why sHsps are oligomeric, and used our quantitative proteomics approach with L/H-isotope ratios to precisely determine to what extent and under which conditions Hsp21 dodecamer and Hsp21 dimer translocate into membrane.

Thylakoid membranes were incubated with a mixture of wildtype dodecameric Hsp21-protein (^{15}N) and mutant dimeric Hsp21 (^{14}N). After incubation at certain time and temperature the thylakoid membranes were spun down and processed for

Table 10.1 Quantification of temperature-dependent translocation into membrane of purified Hsp21 proteins using ^{15}N-labelled Hsp21 dodecamer and Hsp21 V181A mutant dimer

	L/H	q	Comment:
Mix of ^{15}N dodecamer and ^{14}N dimer	0.24	0.81	Mixture with 80 % ^{15}N-labelled dodecamer
Membrane	–	–	No Hsp21 in membrane
Membrane + mix, 20 °C 15 min	0.55	0.64	Mostly ^{14}N dimer into the membrane
Membrane + mix, 45 °C 15 min	0.07	0.93	Mostly ^{15}N dodecamer into the membrane
Membrane + mix, 20 °C, 30 min	1.03	0.49	Mostly ^{14}N dimer into the membrane

Recombinantly expressed Hsp21 dodecamer protein was purified was labelled with ^{15}N by growing the bacterial host in minimal medium containing ^{15}N-NH_4Cl as nitrogen source as described in Söderberg et al. (2012). The recombinantly expressed Hsp21 dimer V181A substitution mutant protein was purified by ion exchange chromatography and gel filtration (unpublished results)
L/H = (light/heavy) ratio is the output from Mascot Distiller Quantitation Toolbox software
$q = {}^{15}N / {}^{14}N + {}^{15}N$ as described in Figs. 10.2 and 10.3

determination of relative abundance of proteins a summarized in Table 10.1, the data obtained indicated that wildtype Hsp21-protein (^{15}N), which was in surplus in the mixture, is preferentially inserted into the thylakoid membranes after incubation at increased temperature (45 °C). However, after incubation at lower temperature (20 °C) the mutant dimeric Hsp21-protein (^{14}N) is preferentially inserted compared to wildtype dodecameric protein.

At present it is not known in what form Hsp21 is inserted or associated with the thylakoid membrane. Presumably dissociated subunits of Hsp21 (monomers or dimers) become inserted into the membrane, but it remains to be investigated whether it is the N-terminal domain with the amphipathic α-helices that bring about the translocation into membrane, as in amphitropic enzymes (Huang et al. 2013). Also it is yet not known whether Hsp21 is just inserted into the lipid bilayer or whether it interacts with membrane-associated proteins or with integral membrane proteins, whose soluble parts also may unfold in response to increased temperature. These data also suggest a partly new explanation as to why sHsps need to be oligomeric: to hide the hydrophobic regions that are exposed in the dimers to avoid membrane lysis under non-stressed conditions.

The conserved amphipathic N-terminal domain of Hsp21 (Chen and Vierling 1991; Waters et al. 1996), which shows features of an intrinsically disordered protein (Bardwell and Jakob 2012; Tompa and Csermely 2004), could interact either with the membrane bilayer, proteins therein, or with soluble proteins, as outlined in Fig. 10.4. Due to the reversible methionine oxidation cycle (Gustavsson et al. 2002) involving Hsp21 and the chloroplast-localized methionine sulfoxide reductase, the translocation of Hsp21 into thylakoid membranes could also protect the membrane from lipid peroxidation by scavenging reactive oxygen species, as demonstrated for the Hsp21-homologue in *Mycobacterium tuberculosis* (Abulimiti et al. 2003).

10.5 Conclusions

Hsp21, the chloroplast-localized sHsp, protects plants and the photosynthesis machinery, which is the driver of plant growth, against oxidative stress and the conserved methionines in the N-terminal domain of Hsp21 act in scavenging reactive oxygen species (Figs. 10.1 and 10.4). Using stable isotope labelling of *Arabidopsis thaliana* plants and quantitative proteomics based on $^{14}N/^{15}N$ isotope ratios, we have shown that most chloroplast proteins have the same relative abundance in the thylakoid membranes in heat-stressed plants compared to control plants, whereas Hsp21 translocates into the thylakoid membrane in heat-stressed plants (Figs. 10.2 and 10.3). Also in reconstituted systems with isolated thylakoid membranes, Hsp21 enters the membrane at increased temperature, and the dimeric mutant Hsp21 does so already at lower temperature (Table 10.1). The picture is emerging for several

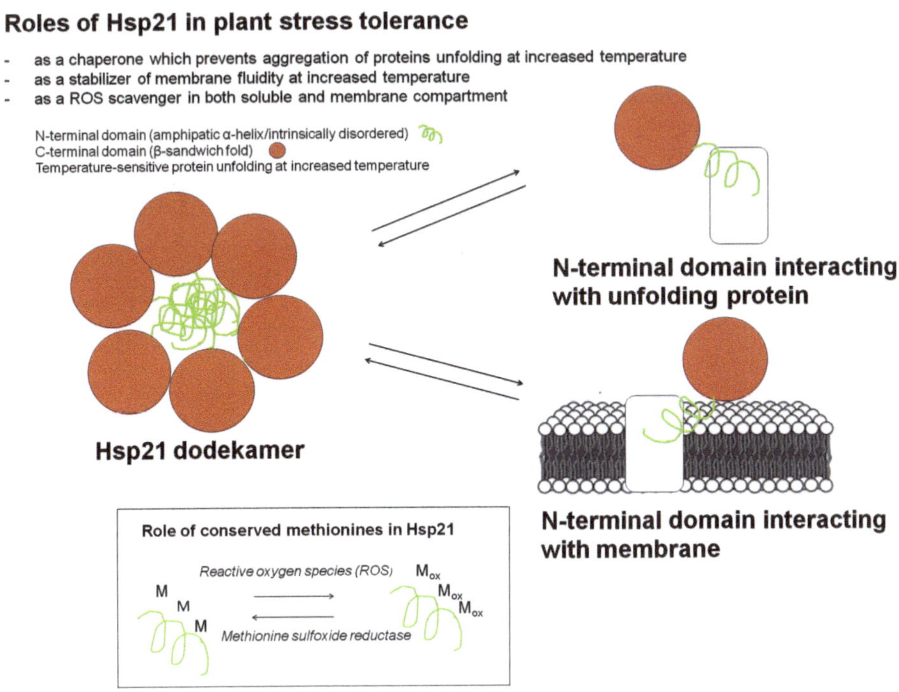

Fig. 10.4 Role of the chloroplast-localized Hsp21 in plant stress tolerance. Hsp21 can interact as a chaperone with partly unfolded substrate proteins to prevent aggregation, and/or interact with the thylakoid membrane. The interior of the Hsp21 dodecamer holds 12N-terminal domains, with a propensity for forming amphipathic α-helices while also showing properties resembling intrinsically disordered proteins, and which may interact with unfolding substrate proteins as well as with the membrane. Insert: The conserved methionines in the N-terminal domains of Hsp21 may add a ROS-scavenger function to the chaperone, with reversible oxidation-reduction of the methionines in a cycle including a chloroplast-localized form of a peptide methionine sulfoxide reductase (Gustavsson et al. 2002). Such a cycle might also counteract lipid peroxidation in the membranes

sHsps that the disordered regions are kept safely ready to be "unleashed" upon oligomer disassembly in order to perform the chaperoning activity (Basha et al. 2013; Peschek et al. 2014). Were it not for the oligomericity the high expression level of sHsps might have caused membrane lysis under non-stress conditions, as happened with the non-dodecameric Hsp21 mutant that entered the membrane also at lower temperature. Questions that remain to be answered are what parts of Hsp21 that becomes inserted into membrane, if Hsp21 also translocates into the chloroplast envelope membrane and whether the cytosolic sHsps enter into the plasma membrane and/or the chloroplast envelope membrane. The here described quantitative proteomics approach (Bernfur et al. unpublished data), especially in combination with chemical crosslinking, are ideal tools to address these questions.

Our data altogether suggest that temperature-dependent translocation into membranes is a general and unique property of sHsps and that attempts to understand the physiological role of Hsp21 in plant stress resistance must take into account this translocation into membranes.

Acknowledgement Thanks to Elizabeth Vierling who cloned the gene for Hsp21 and gave it to us, Wilbert C Boelens for collaboration on sHsp chaperone activity measurements, Denis Murphy for discovery and purification of the chloroplast-localized peptide methionine sulfoxide reductase and the former PhD students Ulrika Härndahl, Niklas Gustavsson, Emma Åhrman and Wietske Lambert who contributed to the knowledge on the chloroplast-localized sHsp in *Arabidopsis thaliana*.

References

Abulimiti A, Qiu X, Chen J, Liu Y, Chang Z (2003) Reversible methionine sulfoxidation of Mycobacterium tuberculosis small heat shock protein Hsp16.3 and its possible role in scavenging oxidants. Biochem Biophys Res Commun 305:87–93

Ahrman E, Lambert W, Aquilina JA, Robinson CV, Emanuelsson CS (2007) Chemical cross-linking of the chloroplast localized small heat-shock protein, Hsp21, and the model substrate citrate synthase. Protein Sci 16:1464–1478

Balogi Z, Cheregi O, Giese KC, Juhasz K, Vierling E, Vass I, Vigh L, Horvath I (2008) A mutant small heat shock protein with increased thylakoid association provides an elevated resistance against UV-B damage in synechocystis 6803. J Biol Chem 283:22983–22991

Bardwell JC, Jakob U (2012) Conditional disorder in chaperone action. Trends Biochem Sci 37:517–525

Basha E, Lee GJ, Breci LA, Hausrath AC, Buan NR, Giese KC, Vierling E (2004) The identity of proteins associated with a small heat shock protein during heat stress in vivo indicates that these chaperones protect a wide range of cellular functions. J Biol Chem 279:7566–7575

Basha E, O'Neill H, Vierling E (2011) Small heat shock proteins and alpha-crystallins: dynamic proteins with flexible functions. Trends Biochem Sci 37:106–117

Basha E, Jones C, Blackwell AE, Cheng G, Waters ER, Samsel KA, Siddique M, Pett V, Wysocki V, Vierling E (2013) An unusual dimeric small heat shock protein provides insight into the mechanism of this class of chaperones. J Mol Biol 425:1683–1696

Beck M, Malmstrom JA, Lange V, Schmidt A, Deutsch EW, Aebersold R (2009) Visual proteomics of the human pathogen Leptospira interrogans. Nat Methods 6:817–823

Bernfur K, Larsson O, Larsson C, Gustavsson N (2013) Relative abundance of integral plasma membrane proteins in Arabidopsis leaf and root tissue determined by metabolic labeling and mass spectrometry. PLoS One 8:e71206

Bova MP, Ding LL, Horwitz J, Fung BK (1997) Subunit exchange of alphaA-crystallin. J Biol Chem 272:29511–29517

Bova MP, McHaourab HS, Han Y, Fung BK (2000) Subunit exchange of small heat shock proteins. Analysis of oligomer formation of alphaA-crystallin and Hsp27 by fluorescence resonance energy transfer and site-directed truncations. J Biol Chem 275:1035–1042

Carra S, Rusmini P, Crippa V, Giorgetti E, Boncoraglio A, Cristofani R, Naujock M, Meister M, Minoia M, Kampinga HH, Poletti A (2013) Different anti-aggregation and pro-degradative functions of the members of the mammalian sHSP family in neurological disorders. Philos Trans R Soc Lond B Biol Sci 368:20110409

Chen Q, Vierling E (1991) Analysis of conserved domains identifies a unique structural feature of a chloroplast heat shock protein. Mol Gen Genet 226:425–431

Eyles SJ, Gierasch LM (2010) Nature's molecular sponges: small heat shock proteins grow into their chaperone roles. Proc Natl Acad Sci U S A 107:2727–2728

Garrido C, Paul C, Seigneuric R, Kampinga HH (2012) The small heat shock proteins family: the long forgotten chaperones. Int J Biochem Cell Biol 44:1588–1592

Giese KC, Vierling E (2002) Changes in oligomerization are essential for the chaperone activity of a small heat shock protein in vivo and in vitro. J Biol Chem 277:46310–46318

Giese KC, Basha E, Catague BY, Vierling E (2005) Evidence for an essential function of the N terminus of a small heat shock protein in vivo, independent of in vitro chaperone activity. Proc Natl Acad Sci U S A 102:18896–18901

Gustavsson N, Harndahl U, Emanuelsson A, Roepstorff P, Sundby C (1999) Methionine sulfoxidation of the chloroplast small heat shock protein and conformational changes in the oligomer. Protein Sci 8:2506–2512

Gustavsson N, Kokke BP, Anzelius B, Boelens WC, Sundby C (2001) Substitution of conserved methionines by leucines in chloroplast small heat shock protein results in loss of redox-response but retained chaperone-like activity. Protein Sci 10:1785–1793

Gustavsson N, Kokke BP, Harndahl U, Silow M, Bechtold U, Poghosyan Z, Murphy D, Boelens WC, Sundby C (2002) A peptide methionine sulfoxide reductase highly expressed in photosynthetic tissue in Arabidopsis thaliana can protect the chaperone-like activity of a chloroplast-localized small heat shock protein. Plant J 29:545–553

Harndahl U, Hall RB, Osteryoung KW, Vierling E, Bornman JF, Sundby C (1999) The chloroplast small heat shock protein undergoes oxidation-dependent conformational changes and may protect plants from oxidative stress. Cell Stress Chaperones 4:129–138

Hilton GR, Lioe H, Stengel F, Baldwin AJ, Benesch JL (2013) Small heat-shock proteins: paramedics of the cell. Top Curr Chem 328:69–98

Horvath I, Glatz A, Nakamoto H, Mishkind ML, Munnik T, Saidi Y, Goloubinoff P, Harwood JL, Vigh L (2012) Heat shock response in photosynthetic organisms: membrane and lipid connections. Prog Lipid Res 51:208–220

Huang HK, Taneva SG, Lee J, Silva LP, Schriemer DC, Cornell RB (2013) The membrane-binding domain of an amphitropic enzyme suppresses catalysis by contact with an amphipathic helix flanking its active site. J Mol Biol 425:1546–1564

Kampinga HH, Garrido C (2012) HSPBs: small proteins with big implications in human disease. Int J Biochem Cell Biol 44:1706–1710

Kim G, Weiss SJ, Levine RL (2014) Methionine oxidation and reduction in proteins. Biochim Biophys Acta 1840:901–905

Lambert W, Koeck PJ, Ahrman E, Purhonen P, Cheng K, Elmlund D, Hebert H, Emanuelsson C (2011) Subunit arrangement in the dodecameric chloroplast small heat shock protein Hsp21. Protein Sci 20:291–301

Lambert W, Rutsdottir G, Hussein R, Bernfur K, Kjellstrom S, Emanuelsson C (2013) Probing the transient interaction between the small heat-shock protein Hsp21 and a model substrate protein using crosslinking mass spectrometry. Cell Stress Chaperones 18:75–85

Malmstrom J, Beck M, Schmidt A, Lange V, Deutsch EW, Aebersold R (2009) Proteome-wide cellular protein concentrations of the human pathogen Leptospira interrogans. Nature 460:762–765

McHaourab HS, Godar JA, Stewart PL (2009) Structure and mechanism of protein stability sensors: chaperone activity of small heat shock proteins. Biochemistry 48:3828–3837

Mushegian AR, Koonin EV (1996) A minimal gene set for cellular life derived by comparison of complete bacterial genomes. Proc Natl Acad Sci U S A 93:10268–10273

Peschek J, Braun N, Rohrberg J, Back KC, Kriehuber T, Kastenmuller A, Weinkauf S, Buchner J (2014) Regulated structural transitions unleash the chaperone activity of alphaB-crystallin. Proc Natl Acad Sci U S A 110:E3780–E3789

Sadanandom A, Poghosyan Z, Fairbairn DJ, Murphy DJ (2000) Differential regulation of plastidial and cytosolic isoforms of peptide methionine sulfoxide reductase in Arabidopsis. Plant Physiol 123:255–264

Söderberg C, Lambert W, Kjellström S, Wiegandt A, Peterson Wulff R, Rutsdottir G, Emanuelsson C (2012) Detection of crosslinks within and between proteins by LC-MALDI-TOFTOF and the software FINDX to reduce the MSMS-data to acquire for validation. PLoS One 7(6):e38927

Stengel F, Baldwin AJ, Bush MF, Hilton GR, Lioe H, Basha E, Jaya N, Vierling E, Benesch JL (2012) Dissecting heterogeneous molecular chaperone complexes using a mass spectrum deconvolution approach. Chem Biol 19:599–607

Tompa P, Csermely P (2004) The role of structural disorder in the function of RNA and protein chaperones. FASEB J 18:1169–1175

Torok Z, Goloubinoff P, Horvath I, Tsvetkova NM, Glatz A, Balogh G, Varvasovszki V, Los DA, Vierling E, Crowe JH, Vigh L (2001) Synechocystis HSP17 is an amphitropic protein that stabilizes heat-stressed membranes and binds denatured proteins for subsequent chaperone-mediated refolding. Proc Natl Acad Sci U S A 98:3098–3103

Van Montfort R, Slingsby C, Vierling E (2001a) Structure and function of the small heat shock protein/alpha-crystallin family of molecular chaperones. Adv Protein Chem 59:105–156

Van Montfort RL, Basha E, Friedrich KL, Slingsby C, Vierling E (2001b) Crystal structure and assembly of a eukaryotic small heat shock protein. Nat Struct Biol 8:1025–1030

Vigh L, Nakamoto H, Landry J, Gomez-Munoz A, Harwood JL, Horvath I (2007) Membrane regulation of the stress response from prokaryotic models to mammalian cells. Ann N Y Acad Sci 1113:40–51

Waters ER (2013) The evolution, function, structure, and expression of the plant sHSPs. J Exp Bot 64:391–403

Waters ER, Vierling E (1999) Chloroplast small heat shock proteins: evidence for atypical evolution of an organelle-localized protein. Proc Natl Acad Sci U S A 96:14394–14399

Waters ER, Lee GJ, Vierling E (1996) Evolution, structure and function of the small heat shock proteins in plants. J Exp Bot 47:325–338

Zhong L, Zhou W, Wang H, Ding S, Lu Q, Wen X, Peng L, Zhang L, Lu C (2013) Chloroplast small heat shock protein HSP21 interacts with plastid nucleoid protein pTAC5 and is essential for chloroplast development in Arabidopsis under heat stress. Plant Cell 25:2925–2943

Part III
sHsps in the Clinic

Chapter 11
Multifunctional Roles of αB-Crystallin in Skeletal and Cardiac Muscle Homeostasis and Disease

Katie A. Mitzelfelt and Ivor J. Benjamin

Abstract αB-Crystallin, or HspB5, is a small molecular-weight heat shock protein expressed highly in cardiac and skeletal muscle with multifaceted cellular roles including, chaperone function towards essential myofibrillar components. Insights into protective roles played by αB-crystallin, as well as mutations in the gene encoding αB-crystallin, *CRYAB*, which resulted in human pathologies, have highlighted the critical functions of αB-crystallin in both skeletal and cardiac muscle, *inter alia*. Various human mutations in *CRYAB* appear to have tissue-specific effects, with loss of αB-crystallin only impacting skeletal muscle under basal conditions. This review aims to highlight the roles of αB-crystallin in skeletal and cardiac muscle homeostasis as well as under conditions of stress and disease, drawing insights from human pathologies resulting from *CRYAB* mutations, and to discuss the potential of using induced pluripotent stem cells to model αB-crystallin-opathies in vitro.

Keywords Cardiac muscle • Skeletal muscle • Myopathy • *CRYAB*, HspB5 • Induced pluripotent stem cells (iPSCs) • Protein aggregation

11.1 Introduction

αB-Crystallin, also known as HspB5, is a member of the small molecular-weight heat shock family (sHSPs) of molecular chaperones, a diverse family of proteins that are characterized by the presence of a conserved α-crystallin domain (Ignolia and Craig 1982; Kappé et al. 2002). The α-crystallin domain in αB-crystallin interacts with an adjacent monomer to form a dimeric building block, and further assembles into higher order oligomers (Bagneris et al. 2009; Jehle et al. 2010;

K.A. Mitzelfelt
Department of Biochemistry, University of Utah, Salt Lake City, UT, USA

I.J. Benjamin, M.D. (✉)
Cardiovascular Center, The Medical College of Wisconsin, Milwaukee, WI, USA
e-mail: ibenjamin@mcw.edu

R.M. Tanguay, L.E. Hightower (eds.), *The Big Book on Small Heat Shock Proteins*, Heat Shock Proteins 8, DOI 10.1007/978-3-319-16077-1_11

Laganowsky et al. 2010). The *CRYAB* gene encodes αB-crystallin expression and primarily functions as a molecular chaperone (Horwitz 1992), among other things. Though its expression was originally thought to be confined to the lens, it was later realized that αB-crystallin is relatively ubiquitously and abundantly expressed in cardiac and skeletal muscle (Dubin et al. 1989). Enhancer elements modulating the *CRYAB* promoter account for its tissue specificity (Gopal-Srivastava et al. 1995; Gopal-Srivastava and Piatigorsky 1993). As described in this review, αB-crystallin plays essential functions in tissue maintenance during homeostasis and in stressed or pathological states of cardiac and skeletal muscle. Additionally, mutations in human *CRYAB* result in cardiac and/or skeletal myopathies, with causal mechanisms of the apparent tissue-specific effects of the mutations remaining unresolved. This review discusses the implications and requirements for αB-crystallin in both cardiac and skeletal muscle during homeostasis as well as under stressed or pathological conditions. We also invoke studies involving mutant forms of *CRYAB* present in patients to make inferences into tissue-specific requirements.

11.2 Structure and Function of αB-Crystallin

11.2.1 Constitutive and Inducible Expression

αB-Crystallin is classified as a Class I sHSP due to its ubiquitous expression (Taylor and Benjamin 2005). First discovered in 1894, it was originally thought to be a lens-specific protein (Morner 1894), until murine *CRYAB* was cloned and found to be expressed at high levels in the heart, skeletal muscle, kidney, and lung and low levels in the brain and spleen (Dubin et al. 1989). Muscle-specific expression of *CRYAB* is conferred by upstream enhancer elements that regulate promoter activity. Murine *CRYAB* cis-acting enhancer elements, identified by reporter expression driven by the *CRYAB* promoter, show regions required for expression in skeletal muscle, termed αBE-1, αBE-2, αBE-3, and MRF (Gopal-Srivastava and Piatigorsky 1993), which are also required for cardiac muscle expression along with an additional, unique element, αBE-4 (Gopal-Srivastava et al. 1995). The MRF site contains an E-box that in skeletal muscle is bound and activated by the bHLH myogenic regulatory factors (MRFs), including MyoD and myogenin (Gopal-Srivastava and Piatigorsky 1993). In cardiac muscle, the E-box of the MRF site may be bound by upstream stimulatory factor (USF) or an antigenically similar factor (Gopal-Srivastava et al. 1995). The cardiac-specific element, αBE-4, contains a reverse CArG box, which may be bound by serum response factor (SRF) or an alternate, antigenically similar protein (Gopal-Srivastava et al. 1995).

In addition to its constitutive expression, *CRYAB* is also inducible in response to multiple forms of stress including heat, oxidative stress, and inflammation, which are sensed by heat shock factor 1 (HSF1), a transcriptional activator that homo-trimerizes to interact with heat shock elements in the promoters of stress-response genes, including *CRYAB* (reviewed in (Morimoto 1998; Christians et al. 2002)). HSF1 deficient mice exhibit decreased basal levels of αB-crystallin in the normal heart, indicating HSF1 may be involved in regulation of *CRYAB* expression under non-stress conditions as well (Yan et al. 2002).

Continuous motor nerve stimulation of rabbit tibialis anterior muscle increases αB-crystallin levels, and this upregulation may involve interaction of the MRFs with the E-box in the *CRYAB* enhancer region (Neufer and Benjamin 1996). Eccentric contraction (lengthening contraction) in skeletal muscle also increases the levels of *CRYAB* along with other HSPs (Kostek et al. 2007; Thompson et al. 2001) resulting from mechanical and/or oxidative stress (Koh 2002). Mechanical load and nerve innervation of skeletal muscle also regulate the level of *CRYAB* expression as shown in experiments where suspension of the rat hindlimb decreases *CRYAB* mRNA levels, and denervation has differential effects on *CRYAB* mRNA levels in different types of skeletal muscle (Atomi et al. 1991). Passive stretch increases *CRYAB* mRNA levels in skeletal muscle (Atomi et al. 1991).

αB-Crystallin is the most abundant sHSP in the heart, making up 3 % of cardiac homogenates and its expression in the heart is limited to cardiomyocytes (Lutsch et al. 1997). Stress including ischemia/reperfusion (I/R) injury upregulates αB-crystallin in the heart (Martin et al. 1997; Ray et al. 2001).

11.2.2 Structure and Oligomerization

sHSPs such as αB-crystallin contain an 80–100 residue, conserved α-crystallin domain (de Jong et al. 1998) comprised of an immunoglobulin-like β-sandwich, including strands β2–β9, which is critical for dimer formation through interaction of the extended β6+7 strand on adjacent monomers (Bagneris et al. 2009). Homo- or hetero-dimers assemble into large, oligomeric structures in the inactive state (Mymrikov et al. 2011) with an average of 40 subunits and a molecular mass of 800 kDa (Bloemendal 1981). Heat stimulates the rapid exchange of subunits, as determined by fluorescence resonance energy transfer experiments, and binding to large, denatured substrates greatly reduced the rate of subunit exchange (Bova et al. 1997). The variable c-terminal region of αB-crystallin contains a conserved I/V/L-X-I/V/L motif that binds the β4/8 groove of the adjacent monomer (Delbecq et al. 2012), contributing to dimer formation, and may also modulate chaperone activity (Bagneris et al. 2009; Laganowsky et al. 2010; Ghosh et al. 2006). The hydrophobic groove at the dimer interface is essential for substrate binding (Clark et al. 2011). The function of the variable N-terminus is less well-defined, though it is thought to impact oligomerization, containing three phosphorylation sites that can modulate the polydispersity of oligomers (Ecroyd et al. 2007) and reduce their size (Peschek et al. 2013).

11.2.3 Subcellular Localization

11.2.3.1 Cytoplasm

αB-Crystallin has disperse cytoplasmic localization and can bind to the intermediate filament protein desmin (Bennardini et al. 1992), as well as contractile proteins, including actin (Bennardini et al. 1992) and titin (Golenhofen et al. 2002; Bullard et al. 2004; Kotter et al. 2014). These interactions are stimulated in response to stress including I/R stress (Golenhofen et al. 1999, 2002) and stretch caused by eccentric contraction in skeletal muscle (Koh and Escobedo 2004; Kotter et al. 2014). Transgenic mice expressing the R120G mutant form of αB-crystallin in cardiomyocytes exhibit defects in mitochondrial organization, architecture, and respiration (Maloyan et al. 2005), which is attributable to breakdown of the desmin network (Wang et al. 2001). Specific mitochondrial alignment along adjacent sarcomeres is critical for maximal respiratory function in striated muscle, dependent upon cytoskeletal components, with cardiac and different types of skeletal muscle having differential arrangements and amounts of mitochondria (Milner et al. 2000; Reipert et al. 1999; Rambourg and Segretain 1980; Rappaport et al. 1998; Ogata and Yamaksak 1997). These data indicate that αB-crystallin, through interactions with cytoskeletal elements including desmin, among other things, may be crucial for maintaining organization and proper function of mitochondria.

11.2.3.2 Mitochondrial Interaction

Mitochondria are vital for both cardiac and skeletal muscle function, and as mentioned previously, the organization of mitochondria along myofibrils, which varies based on muscle type and is maintained through cytoskeletal interactions, is essential for maintaining maximal functional capacity (Milner et al. 2000; Reipert et al. 1999; Rambourg and Segretain 1980; Rappaport et al. 1998; Ogata and Yamaksak 1997). Maintaining mitochondrial integrity is crucial for inhibiting apoptosis and αB-crystallin has been shown to play a key role in this process during stress. αB-Crystallin associates with mitochondria, and this association increases with I/R stress, shown in rats (Mitra et al. 2013), and after exposure to hydrogen peroxide, shown in culture of mouse neonatal cardiomyocytes (Chis et al. 2012). Both phosphorylated and unphosphorylated αB-crystallin interact with the voltage-dependent anion channel (VDAC) and translocase of outer mitochondrial membrane 20 kDa (TOM 20) at the mitochondrial membrane as well as with caspase 3 and caspase 12 (Chis et al. 2012). It is thought that the interaction with VDAC is critical for preventing apoptosis through the mitochondrial pathway during I/R injury (Mitra et al. 2013). Temporal studies with I/R performed ex vivo in mouse hearts demonstrated that αB-crystallin translocates to the mitochondria during ischemia, becomes phosphorylated at Ser-59, and aids in maintaining mitochondrial membrane potential, thereby preventing apoptosis (Whittaker et al. 2009).

11.2.3.3 Nuclear Speckles

Nuclear speckles, located within the interchromatin space of the nucleoplasm, are proposed to be storage sites for splicing components and are associated with transcriptional activity (Spector and Lamond 2011). αB-Crystallin and other sHSPs, including Hsp27 (HspB1) localize to nuclear speckles in a variety of unstressed, transcriptionally active cell lines and primary cells (Vandenijssel et al. 2003). Overexpression of the R120G mutant form of αB-crystallin inhibited localization to nuclear speckles (Vandenijssel et al. 2003). These data suggest a nuclear role for αB-crystallin in modulating transcript processing, which may be important in R120G αB-crystallin-opathy (Vandenijssel et al. 2003).

A study utilizing mouse C2C12 myoblasts and differentiated myotubes showed differential nuclear speckle localization of αB-crystallin in response to heat shock (Adhikari et al. 2004). In the basal state, myoblasts showed expression of αB-crystallin in the cytoplasm and nucleus, with co-localization with lamin A/C and splicing factor SC-35, indicating nuclear speckle localization, and upon heat stress, myoblasts exhibited almost complete re-localization of αB-crystallin to the nucleus in nuclear speckles, returning to basal state following 3 h recovery. Interestingly, myotubes have αB-crystallin expression exclusively in the cytoplasm under basal and heat stressed conditions. This data may speak for a differential role of αB-crystallin in transcriptional modulation or cytoskeletal protection during differentiation and myogenesis.

11.2.3.4 Posttranslational Modifications

αB-Crystallin contains three phosphorylation sites (Ser-19, Ser-45, Ser-59), identified through experiments exposing cultured cells to various forms of stress (Kato et al. 1998; Ito et al. 1997). These sites are located within the n-terminal region, which can impact both structural and functional aspects of the sHSP, though the impact of phosphorylation on protection of the cell seems to be context dependent. Through studies employing phosphomimetic forms of αB-crystallin, it has been shown that phosphorylation at all three residues decreases oligomer size, with the dominant species being 12-mers and 6-mers, resulting in a more flexible n-terminal domain and increased chaperone activity toward malate dehydrogenase and p53 (Peschek et al. 2013). Additional studies have shown both increases and decreases in chaperone activity of αB-crystallin with phosphorylation depending on the substrate and conditions (Ahmad et al. 2008; Ecroyd et al. 2007), and the ratio of phosphorylated to non-phosphorylated forms of αB-crystallin present in oligomers may be key to regulating its wide range of chaperone activity (Ahmad et al. 2008).

Phosphorylation at Ser-45 may disrupt dimer formation, resulting in an increase in oligomers with an odd number of subunits (Aquilina et al. 2004). One study showed that phosphorylation at Ser-59 is required for nuclear import, and localization to nuclear speckles required phosphorylation at Ser-45 in HeLa cells (den Engelsman et al. 2005). Though a different study employing transient transfection of a phospho-defective form of αB-crystallin, demonstrated that phosphorylation was not required for nuclear localization (Vandenijssel et al. 2003). Phosphorylation at Ser-59 occurs in response to I/R stress in cardiomyocytes (Ito et al. 1997). A phosphomimetic at Ser-59 was shown to be necessary and sufficient for attenuating hyperosmotic or hypoxic stress-induced apoptosis in rat neonatal cardiomyocytes through inhibition of caspase 3 activation (Morrison et al. 2003). In differentiating C2C12 myoblasts, overexpression of wildtype αB-crystallin inhibits caspase 3 activation thereby blocking differentiation-induced apoptosis, and pseudophosphorylation of αB-crystallin at all three serine residues prevents αB-crystallin from performing this function (Kamradt et al. 2002). In light of these data showing varying effects of phosphorylation of αB-crystallin on its structure, localization, and function, it is likely that phosphorylation may be a mechanism for fine-tuning the activity of αB-crystallin, which is context dependent. In addition to phosphorylation, αB-crystallin undergoes O-GlcNAcylation at threonine 170, which has been linked to effects on αB-crystallin subcellular localization, protein-protein interactions, chaperone function, and degradation (Krishnamoorthy et al. 2013).

11.2.4 Roles

11.2.4.1 Chaperone

The major role of αB-crystallin is as a molecular chaperone. Chaperones bind hydrophobic regions of unfolded proteins, stabilizing the protein and preventing it from precipitating out of solution (Ellis and van der Vies 1991). αB-Crystallin, functions in an ATP-independent manner to bind severely compromised, aggregation-prone, late unfolding protein intermediates that are about to precipitate out of solution, and forms a stable, soluble complex that is resistant to aggregation (Carver et al. 1995; Rajaraman et al. 2001). Additionally, αB-crystallin transiently interacts with early unfolding intermediates and promotes their refolding (Rajaraman et al. 2001). Substrate recognition by αB-crystallin is not fully defined, but occurs through interaction of its hydrophobic regions, possibly various regions throughout the protein with exposed hydrophobic regions of unfolded proteins. This provides αB-crystallin with the ability to recognize a wide range of substrates (Basha et al. 2012).

The importance of αB-crystallin in maintaining cytoskeletal integrity is exemplified by its ability to bind and stabilize actin microfilaments, intermediate filaments, including desmin, vimentin, and GFAP, as well as microtubules reviewed in (Liang and MacRae 1997). In addition to binding cytoskeletal elements, αB-crystallin also has the ability to bind and prevent the aggregation of some growth factors, including FGF-2 and VEGF (Ghosh et al. 2007a, b). An in vitro study demonstrated the transient appearance of dimeric αB-crystallin interacting with the model destabilized substrate, α-lactalbumin, indicating that dimer release from the larger oligomers may be important for the chaperone function of αB-crystallin (Smirnova et al. 2013).

Desmin, a major intermediate filament protein in striated muscle, is responsible for connecting myofibrils together through Z-discs (Lockard and Bloom 1993). αB-Crystallin interacts with and functions as a chaperone for desmin (Perng et al. 1999a), and a mutant form of αB-crystallin, R120G, can increase its affinity for desmin resulting in aberrant desmin aggregation (Perng et al. 2004; Vicart et al. 1998). A study identifying that the interaction between αB-crystallin with desmin involved the β3 and β8 strands as well as c-terminal residues 155–165, also suggests that binding or cosedimentation of αB-crystallin with desmin does not necessarily increase αB-crystallin chaperone function towards desmin (Houck et al. 2011).

The muscle contractile protein titin, responsible for the elastic properties of muscle, contains spring-like immunoglobulin (Ig) domains that can unfold and allow contracted muscles to stretch (Minajeva et al. 2001), which can result in aggregation of the unfolded titin domains (Minajeva et al. 2001; Rief 1997). Titin and αB-crystallin were co-immunoprecipitated in pig hearts following ischemia and αB-crystallin was shown to bind to the I-band region of titin, with extraction of actin, in ischemic rat cardiomyocytes (Golenhofen et al. 2002). A following study more specifically identified that αB-crystallin binds to the N2B region of the cardiac titin isoform in physiologically stretched cardiomyocytes, and that higher stretching forces are required to unfold the titin domains in the presence of αB-crystallin (Bullard et al. 2004). Whereas binding of αB-crystallin appears to be specific to the N2B element of titin in cardiac muscle, in skeletal muscle, αB-crystallin shows scattered binding along the length of the I-band in rats, indicating it may bind to other Ig domains (Golenhofen et al. 2004). In vitro work shows that unfolded Ig domains, including cardiac N2B and skeletal N2A, of titin aggregate with increased aggregation under acidic conditions, and the presence of αB-crystallin prevents this aggregation (Kotter et al. 2014). αB-Crystallin was also localized to the Ig domain regions of titin in diseased, human cardiomyocytes and skeletal muscle, suggesting that this localization is necessary to prevent aggregation of unfolded Ig domains in overstretched muscle cells, and this interaction may also be relevant in skeletal muscle during exercise (Kotter et al. 2014). Additionally, in I/R experiments using isolated papillary muscles from *CRYAB/HSPB2* double knockout (DKO) mouse hearts, it was speculated that more pronounced contracture in mutant muscles may be due to loss of αB-crystallin chaperone function of titin, thereby causing increased stiffness in response to ischemia (Golenhofen et al. 2006).

αB-Crystallin interacts with an f-box protein, FBX4, an important component of the SCF complex, which carries out ubiquitination of proteins, marking them for degradation (den Engelsman et al. 2003). This interaction is increased with phosphomimetics at Ser-19 and Ser-45 and also by the R120G mutation. These data suggest a potential role for αB-crystallin in directing substrate ubiquitination in the ubiquitin proteasome pathway, which may assist in maintaining cellular homeostasis by degrading proteins that will not refold properly. Activation of autophagy or the inflammatory response may also be a role for αB-crystallin in protein quality control as has been suggested for other sHSPs (Carra et al. 2008, 2009; Bruinsma et al. 2011).

11.2.4.2 Anti-apoptotic

Exogenous αB-crystallin expression in human lens epithelial cells prevented UVA-induced apoptosis through inhibition of the RAF/MEK/ERK pathway (Liu et al. 2004) via inhibition of RAS activation (Li et al. 2005). αB-Crystallin binds and modulates caspase 3 and Bax in the lens, regulating differentiation (Hu et al. 2012). It was also shown in retinal pigment epithelial cells that αB-crystallin is able to inhibit the translocation of Bax and Bcl-2 to the mitochondria (Mao et al. 2004). In cancer cell lines, αB-crystallin was identified as a target gene directly trans-activated by p53. αB-Crystallin binds to the DNA binding domain of p53 and reduced p53-induced apoptosis (Watanabe et al. 2009). In response to hydrogen peroxide treatment of C2C12 myogenic cells, αB-crystallin interacts with p53 in the cytoplasm, possibly preventing its translocation to the mitochondria and abrogating oxidative stress-induced apoptosis (Liu et al. 2007).

When myoblasts are stimulated to differentiate by growth factor deprivation, they will undergo apoptosis if they do not develop resistance to apoptosis, allowing differentiation to ensue. In differentiating C2C12 myoblasts, αB-crystallin is induced in those myoblasts that develop resistance to apoptosis and overexpression of αB-crystallin prevents differentiation-induced apoptosis through inhibition of caspase 3 activation (Kamradt et al. 2001, 2002). An independent experiment confirmed that overexpression of αB-crystallin decreased the levels of cleaved, activated caspase 3 in differentiating C2C12 myoblasts (Singh et al. 2010). These results suggest a critical role for αB-crystallin in the prevention of apoptosis during normal differentiation of myoblasts to myotubes. Additionally, DKO mice exhibit increases in both apoptosis and necrosis in response to I/R injury suggesting a protective role for one or both of these chaperones towards cardiomyocytes (Morrison et al. 2004). In αB-crystallin transgenic mice, Das and Dillmann have demonstrated ischemic cardioprotection (Ray et al. 2001).

11.2.4.3 Redox Modulation

Muscle contains high levels of mitochondria to comply with high energy requirements. The heart is made up of 55 % cardiomyocytes, with up to 35 % of the cardiomyocyte volume occupied by mitochondria and since mitochondria are the main source of reactive oxygen species (ROS), cells must adapt to deal with these harmful byproducts (Christians et al. 2012). Implications of αB-crystallin on redox state have mainly been explored through studies of the R120G mutant form. Transgenic mice overexpressing R120G αB-crystallin specifically in cardiomyocytes were shown to have a redox balance skewed towards reductive stress (Rajasekaran et al. 2007). In this model, R120G αB-crystallin-induced cardiomyopathy and aggregate formation were rescued by the intercross with mice having decreased levels of an antioxidative enzyme glucose-6-phosphate dehydrogenase, implicating an effect of the mutant form of αB-crystallin on redox state. αB-Crystallin cannot be thiolated due to its lack of cysteine residues, indicating that αB-crystallin does not undergo modification in response to extreme shifts of redox conditions.

11.3 Protective Effects of αB-Crystallin in Stressed or Pathological States

Many instances of protective roles for αB-crystallin have been described in the literature and correspond with its role as a stress response protein. Roles for αB-crystallin in I/R injury of the heart and exercise-induced injury in skeletal muscle will be discussed here.

11.3.1 Cardiac Muscle: Ischemia/Reperfusion

Neonatal and adult rat cardiomyocytes are protected from I/R injury by overexpression of αB-crystallin (Martin et al. 1997). Transgenic overexpression of αB-crystallin also protects the mouse heart in ex vivo I/R, decreasing the extent of infarction, lowering levels of oxidative stress, and decreasing apoptosis and necrosis (Ray et al. 2001). Conversely, DKO mouse hearts were more prone to I/R damage, and showed reduced contractile recovery, increased apoptosis and necrosis, and lower levels of the reduced form of glutathione, a protective molecule for oxidative damage (Morrison et al. 2004). To better understand the individual roles of αB-crystallin and HspB2 during I/R, Pinz and colleagues crossed DKO mice with mice transgenically overexpressing *CRYAB* and through I/R experiments, determined that αB-crystallin seems to be important for maintaining structure and diastolic function (Pinz et al. 2008). Through its mitochondrial interactions, as discussed previously, αB-crystallin may aid in maintaining mitochondrial integrity in response to I/R stress (Boelens 2014). In response to I/R, αB-crystallin also translocates to the mitochondria (Whittaker et al. 2009) and interacts with VDAC; this interaction may be critical for preventing apoptosis induced through the mitochondrial pathway (Mitra et al. 2013; Chis et al. 2012).

Sixty-minute cardiac ischemia in pigs resulted in overstretching of cardiomyocytes as observed by increased Z-disc spacing and increased distance between α-actinin and titin antibody staining (Golenhofen et al. 1999). In this study, αB-crystallin translocated from the cytoplasm to the I-band portion of titin in response to ischemia and following reperfusion, it remained bound to titin in cardiomyocytes that were no longer capable of contraction (Golenhofen et al. 1999). This result suggests that αB-crystallin is essential for mitigating the refolding of elastic titin domains in response to cardiomyocyte stretch as a result of ischemia, and αB-crystallin remains bound to the unfolded domains of titin in cardiomyocytes that are too damaged to recover contractile function.

11.3.2 Skeletal Muscle: Exercise

Eccentric contraction is the repeated stretching of actively contracting muscles and this can damage muscle fibers especially in untrained muscles, whereby damage occurs with concomitant inflammation (Koh 2002). Using ex vivo rabbit skeletal muscles, it was shown that desmin is lost in damaged fibers immediately after eccentric contraction and continues up to 3 days following injury (Friden and Lieber 2001). Dystrophin seems to be affected in a similar manner as a result of eccentric contraction in rat tibialis anterior muscles (Komulainen et al. 1998). An increase in free radical production in rabbit muscle occurred 24 h post-eccentric contraction, attributable to infiltrating inflammatory cells (Best et al. 1999).

Increases in the level of *CRYAB* were seen at 6 and 24 h after eccentric contraction (Kostek et al. 2007), which may be induced by mechanical and/or oxidative stress (Koh 2002). In a study involving human subjects using biopsies of the quadriceps muscle, αB-crystallin was upregulated 30 min following eccentric exercise and gradually decreased for 1 week after (Paulsen et al. 2007). Additionally, αB-crystallin responded immediately to eccentric exercise by binding to and fractioning with the cytoskeletal/myofibrillar proteins. Mice exposed to eccentric contractions showed immediate translocation of αB-crystallin from the soluble to insoluble fraction by western blot and cytosol to the Z-disc by immunostaining (Koh and Escobedo 2004). Phosphorylation of αB-crystallin was also observed during this time and may be important for its protective activity.

The temperature of muscles rises during exercise and may result in thermal injury. Chicken skeletal muscle myosin was unfolded by heat shock in vitro and in the presence of αB-crystallin, myosin retained enzymatic activity and aggregation was prevented (Melkani et al. 2006). αB-Crystallin was shown in vitro to prevent the acidic-induced aggregation of the N2A region of skeletal muscle titin, which contains Ig domains that are prone to unfolding during exercise (Kotter et al. 2014). It also is bound to the Ig domain regions of titin in skeletal muscle biopsies from patients with limb girdle muscular dystrophy type 2a; while in healthy human biopsies, αB-crystallin had cytosolic localization with faint Z-disc staining (Kotter et al. 2014). Ischemic rat skeletal muscle showed αB-crystallin bound along the length of the I-band, likely to the Ig domains of titin (Golenhofen et al. 2004). It is likely that this ability of αB-crystallin to protect titin Ig domains from aggregation upon unfolding due to stretch is also active during exercise (Kotter et al. 2014), especially eccentric contraction, where muscle is damaged from overstretching (Koh 2002).

11.4 αB-Crystallin-Opathies: Mutations and Human Pathologies

To date, 14 naturally occurring mutations in human *CRYAB* have been identified and published in the literature (Christians et al. 2012). Of these, ten are dominant mutations and four exhibit recessive inheritance patterns with disease pathologies ranging from skeletal myopathy, cardiomyopathy, cataracts, or some combination of the three. Eight of these mutations result in skeletal and/or cardiac myopathy (see Table 11.1) and can be classified as αB-crystallin-opathies (Selcen 2011; Sanbe et al. 2011; Christians et al. 2012). Variable penetrance and expressivity of the myopathy-causing

Table 11.1 Myopathy causing mutations in *CRYAB*

Genetic mutation	Resulting protein change	Type of myopathy	Dominant or recessive	Onset	Reference
c.60delC	S21Afs24X	Skeletal	Recessive	Infantile	Del Bigio et al. (2011)
350 G>C	D109H	Both	Dominant	Adult	Sacconi et al. (2012)
c.343delT	S115Pfs129X	Skeletal	Recessive	Infantile	Forrest et al. (2011)
358 A>G	R120G	Both	Dominant	Adult	Vicart et al. (1998)
451 C>T	Q151X	Skeletal	Dominant	Adult	Selcen and Engel 2003)
460 G>A	G154S	Skeletal or Cardiac	Dominant	Adult	Pilotto et al. (2006), Reilich et al. (2010)
c.464delCT	L155fs163X	Skeletal	Dominant	Adult	Selcen and Engel (2003)
495 G>A	R157H	Cardiac	Dominant	Adult	Inagaki et al. (2006)

mutations in *CRYAB* suggest genetic modifier and/or environmental factor contribution to the disease state (Christians et al. 2014). Additionally, for the dominant mutations, relative levels of the wildtype compared with the mutant form of αB-crystallin may also contribute to disease and could be modified by degradation, which may change with environmental stressors (Christians et al. 2014).

11.4.1 Point Mutations

First described in 1978 (Fardeau et al. 1978), the prototype and most well studied of the human mutations in αB-crystallin, arginine 120 mutated to a glycine (R120G) (Vicart et al. 1998), resulting in multisystem disorders and pathologies, was not identified until 1998 (Vicart et al. 1998). This disease affects a large, French pedigree and muscle biopsies characterized by electron microscopy revealed the buildup of dense, granulo-filamentous material within the sarcoplasm (Fardeau et al. 1978), which contained the intermediate filament protein desmin (Rappaport et al. 1988; Fardeau et al. 1978) as well as αB-crystallin, and therefore is classified as a desmin-related myopathy (DRM) (Vicart et al. 1998). Patient biopsies revealed a "rubbed-out" appearance when stained for myosin adenosine triphosphatase (ATPase) activity, potentially indicating a loss of myosin (Brady et al. 2001; Vicart et al. 1998; Fardeau et al. 1978). The R120G mutation (358 A>G) occurs in the α-crystallin domain and disrupts the structure as well as reduces the chaperone activity of αB-crystallin (Bova et al. 1999; Perng et al. 1999b). The binding affinity of αB-crystallin for desmin in vitro is increased by R120G, and results in the aberrant aggregation of desmin, preferentially during de novo desmin network synthesis rather than the aggregation of previously formed networks (Perng et al. 2004). This mutation shows closure of the hydrophobic groove present in αB-crystallin dimers, which may alter substrate binding as well as oligomerization (Clark et al. 2011).

In vitro work shows that the R120G mutant form of αB-crystallin has complete loss of chaperone activity towards the N2B region of cardiac titin (Zhu et al. 2009).

Transgenic overexpression of R120G in the mouse heart portrays pathology similar to patients, showing dense desmin and αB-crystallin positive aggregates, and disruption of the cytoskeletal structure and mitochondrial architecture, resulting in cardiomyopathy and death in early adulthood (Wang et al. 2001). Overexpression of wildtype *CRYAB*, on the other hand, was relatively unremarkable (Wang et al. 2001). A knockin mouse model of R120G exhibits early onset cataracts and skeletal myopathy, with increased severity depending on dose (heterozygous versus homozygous knockin), though mortality was not increased (Andley et al. 2011), as is the case in the cardiac-specific overexpression of R120G (Wang et al. 2001). Cardiac function was not reported on in the publication of the R120G knockin animals (Andley et al. 2011).

Aggregates formed as a result of the R120G mutation contain amyloid oligomers and are classified as aggresomes (Sanbe et al. 2004), characteristic of neurodegenerative diseases, which result from transport of the aggregates along microtubules to the perinuclear region (Garcia-Mata et al. 2002). The dysfunction of mitochondria is one of the first defects observed in R120G overexpressing mouse hearts (Maloyan et al. 2005). Cardiomyocytes exhibit reduced mitochondrial oxygen consumption, changes in the permeability transition pore, and poor inner membrane potential, thereby activating apoptotic pathways and adversely affecting the hearts of transgenic animals. Transgenic overexpression of R120G in the heart also yields reductive stress, due to altered activity of enzymes that generate redox intermediates, including glucose-6-phosphate dehydrogenase, glutathione reductase, and glutathione peroxidase (Rajasekaran et al. 2007). These animals have elevated levels of autophagic activity by the age of 2 months, and this response is thought to be cytoprotective, possibly minimizing aggregate presence (Tannous et al. 2008). A study involving treatment of R120G transgenic mouse hearts with oxypurinol to rescue mitochondrial defects, determined that contractility was not rescued indicating further mechanical defects due to disarray of sarcomeres and accumulation of aggregates (Maloyan et al. 2009). Additionally, it has been suggested that aggregate formation and disease onset may be repressed at a young age due to competition with wildtype αB-crystallin (Andley et al. 2011; Vicart et al. 1998; Perng et al. 2004), which may also be the case for other dominant mutations described below.

The less studied ability of αB-crystallin to localize to nuclear speckles is eliminated by the R120G mutation, indicating a possible impact of the mutation on transcript processing (Vandenijssel et al. 2003). Overexpression of certain interacting partners has been found to attenuate the detrimental effects of R120G.

Interaction of R120G with BAG3, a co-chaperone involved in chaperone-assisted selective autophagy, increased R120G solubility, prevented its aggregation, and inhibited cell death induced by R120G overexpression in cell culture (Hishiya et al. 2011). Overexpression of HspB8 can inhibit R120G-induced aggresome formation and block the progression of cardiomyopathy (Sanbe et al. 2007, 2009). Voluntary exercise (Maloyan et al. 2007) and BCL-2 overexpression (Maloyan et al. 2009) also delay the onset and reduce the unfavorable effects of the R120G mutation.

The R120G disease state is likely due to a combination of loss and gain of function effects of the mutation (Sanbe et al. 2011). The broad range of roles played by αB-crystallin invites the potential for many factors to contribute to the R120G disease state, including: aggregation with desmin, loss of/altered chaperone function, altered mitochondrial architecture and function, and reductive stress. Since the initial discovery of R120G, other mutations have been identified in patient populations and can also be described as αB-crystallin-opathies.

A family with members harboring the dominant point mutation, D109H, in αB-crystallin exhibits similar multisystem pathology to patients with the R120G mutation (Sacconi et al. 2012). Residues D109 and R120 interact with each other during dimerization of αB-crystallin, and therefore disruption of this interaction may be responsible for the similar disease characteristics among patients with these two mutations (Sacconi et al. 2012) including effects on the hydrophobic groove, oligomerization (Clark et al. 2011), and desmin interaction (Perng et al. 2004). Overexpression of YFP-tagged constructs in HeLa cells showed similar aggregation of D109H and R120G as well as a similar decrease in aggregation and levels of apoptosis with Hsp27 co-overexpression (Raju and Abraham 2013), supporting a similar mechanism of action of the two mutations; however, further analysis is necessary.

A dominant glycine 154 to serine (G154S) mutation in a conserved residue was identified in a 48 year old patient diagnosed with mild dilated cardiomyopathy and slightly elevated skeletal creatine phosphokinase levels, indicating potential involvement of the skeletal muscle as well (Pilotto et al. 2006). The turnover rate of the protein is predicted to increase with this mutation (Pilotto et al. 2006). A second patient with this same mutation, G154S, presented in his sixties with progressive distal leg weakness and atrophy affecting mostly limb muscles, as well as mild difficulty swallowing (Reilich et al. 2010). The patient's biopsy showed disrupted myofibrillar structure, vacuolization, and desmin/αB-crystallin positive aggregates. No cataracts, cardiac or respiratory involvement were noted in this patient. The G154S mutation appears to have affects in both cardiac and skeletal muscle, and the specific presentation of the patient's symptoms may rely upon other confounding genetic or environmental factors causing stress in the muscle that result in the apparent tissue-specific disease manifestation. A potential increased turnover rate of this mutant form of the protein (Pilotto et al. 2006) may help to suppress symptoms exhibited by patients into later adulthood.

A patient with familial dilated cardiomyopathy, presenting after age 40, was identified to have a dominant, mutation of arginine 157 to a histidine residue (R157H), which was shown to decrease binding to the cardiac-specific N2B region of titin (Inagaki et al. 2006). This mutation, unlike the others, did not result in aberrant localization of αB-crystallin or in aggregation, and therefore, there may be an alternate disease mechanism causing dilated cardiomyopathy in this patient compared with the desmin-related cardiomyopathy in other patients with αB-crystallin-opathies. Reduced ability of αB-crystallin to bind to titin leaves titin vulnerable to stress and may predispose patients to heart failure (Inagaki et al. 2006). An in vitro chaperone assay showed that R157H maintained some chaperone activity towards the N2B region of titin, albeit reduced compared to wildtype αB-crystallin (Zhu et al. 2009). Interestingly, this patient and family members did not exhibit skeletal muscle weakness or disease, though αB-crystallin does bind and chaperone the Ig domains in the N2A region of titin in skeletal muscle (Kotter et al. 2014; et al. 2004), in addition to the cardiac N2B region (Kotter et al. 2014), which may indicate that αB-crystallin binds the alternate regions of titin in a different way and the R157H mutation specifically disrupts the association of αB-crystallin with the N2B region. In vitro binding assays using the skeletal N2A region of titin with R157H αB-crystallin could verify this. Alternatively, absence of specific environmental stress to the skeletal muscle may be responsible for apparent lack of skeletal muscle involvement. The R157H mutation may also have other yet-unstudied implications that leave cardiac muscle more vulnerable than skeletal muscle.

11.4.2 Dominant C-Terminal Truncation Mutations

Other than the aforementioned missense mutations, the remaining mutations in αB-crystallin causing myopathy are due to frameshift or nonsense mutations resulting in varying degrees of c-terminal truncation, some with additional, novel peptide sequences present. In vitro work using site-directed mutagenesis to mutate the two terminal lysine residues in αB-crystallin to leucine or glycine greatly reduced its chaperone activity, suggesting that the lysines in the c-terminus may be critical for interacting with unfolded proteins through charge-charge interactions (Plater et al. 1996). Loss of this ability by c-terminal truncation of αB-crystallin may prevent stable substrate binding and therefore reduce chaperone activity. It has also been suggested that the flexible c-terminal region folds over, and through the I/V/L-X-I/V/L motif, binds to the β4/8 groove of the adjacent monomer, contributing to dimer formation, and potentially blocking the substrate binding face of αB-crystallin, thereby modulating substrate binding and chaperone activity (Ghosh et al. 2006; Delbecq et al. 2012). Expression of only the α-crystallin domain of αB-crystallin results in dimer formation with retained chaperone activity, but the lost ability to oligomerize (Feil et al. 2001). The high potential of c-terminal truncated mutants to aggregate, suggests a critical role for the c-terminus in preventing self-aggregation of αB-crystallin possibly through promotion of oligomerization, and expression of the mutant forms may induce co-aggregation with wildtype αB-crystallin inducing a dominant negative effect, thereby inhibiting the chaperone function of wildtype αB-crystallin (Hayes et al. 2008). The myopathy-causing, dominant, c-terminal truncation mutations will now be described.

A patient harboring the dominant 464delCT mutation in *CRYAB* presented at age 52 with respiratory trouble due to reduced diaphragmatic movement, as well as leg weakness and difficulty swallowing, and died at age 58 due to respiratory failure (Selcen and Engel 2003). This mutation generates a frameshift resulting in eight missense codons before a premature stop codon. A 53 year old patient with a dominant 451C→T transition generating a nonsense mutation (Q151X) suffered for 10 years with slowly progressive leg weakness and atrophy and exhibited elevated creatine kinase levels (Selcen and Engel 2003). Neither patient had cataracts or cardiomyopathy. Abnormal fiber regions in both patients show intense desmin and αB-crystallin staining, and in both patients, the wildtype allele was preferentially expressed compared with the mutant forms.

Q151X results in extreme loss of protein stability and prevents oligomerization, while increasing in vitro chaperone activity towards citrate synthase and desmin (Hayes et al. 2008). The increase in chaperone activity may be due to potential loss of capping of the chaperone site by interaction with the c-terminal region (Delbecq et al. 2012; Ghosh et al. 2006), suggesting the c-terminus may be responsible for reducing αB-crystallin substrate binding for some substrates under normal conditions (Hayes et al. 2008). The 464delCT mutation results in the introduction of a novel peptide into the c-terminus resulting in decreased in vitro chaperone function towards desmin, as well as aggregation and loss of solubility that can be partially recovered through mixture with wildtype αB-crystallin (Hayes et al. 2008). Overexpression of Hsp27 also rescues the solubility and prevents aggregation of 464delCT, mediated by the ubiquitin proteasome system in H9C2 cells (a rat embryonic cardiomyocyte cell line) (Zhang et al. 2010).

11.4.3 Recessive Mutations

With mutations in *CRYAB*, especially the recessive mutations, a major question is to what extent the pathology observed is due to loss of function effects of αB-crystallin. DKO mice show progressive skeletal muscle deterioration with age following a normal development, with no apparent impact on the heart (Brady et al. 2001), except under stressful conditions such as I/R or transverse aortic constriction (TAC) (Morrison et al. 2004; Kumarapeli et al. 2008). These mice grow normally until 40 weeks, after which they lose weight, due to the inability to eat properly and exhibit severe kyphosis, hunched posture, both due to loss of corresponding musculature, with the tongue, head, and axial muscles being most greatly affected. Fatty replacement, macrophage infiltration, fibrosis, and vacuolization were observed in deteriorating muscles. Amorphous, flocculent, electron-opaque material was also noted corresponding with loss of myofibrils. Increased staining for desmin was also observed in affected fibers. Attempted determination of fiber type based on fiber type-specific myosin failed due to lack of myosin detection in deteriorating fibers (Brady et al. 2001), as confirmed in additional studies showing reduced levels of myosin heavy chain in skeletal muscle of DKO mice by western blot (Neppl et al. 2014). This may be similar to muscles of patients with the R120G mutation who have loss of myosin in their muscle cells (Vicart et al. 1998; Fardeau et al. 1978). Potentially, oxidative, slow-twitch fibers, which have the highest levels of αB-crystallin and are enriched in large truncal muscles, are preferentially degraded in the DKO mice (Brady et al. 2001) and in patients with recessive *CRYAB* mutations (Forrest et al. 2011; Del Bigio et al. 2011). Neppl and colleagues showed that 1 year old DKO mice have a significantly reduced number of satellite cells in their skeletal muscle and a reduced ability to regenerate muscle following cardiotoxin-induced injury, shown by decreased cross-sectional myofiber area and increased fibrosis 2 weeks post-injury, indicating a potential role for αB-crystallin in skeletal muscle regeneration (Neppl et al. 2014).

This DKO mouse model is confounded due to the double knockout nature. HspB2, or myotonic dystrophy protein kinase binding protein (MKBP), activates and protects myotonic dystrophy protein kinase (DMPK) (Suzuki et al. 1998). *DMPK -/-* mice show minor decreases in size of head and neck muscles with age (Jansen et al. 1996) and late onset myopathy (Reddy et al. 1996). It is, therefore, possible that through its activation and protection of DMPK and other potential functions of HspB2, the loss of HspB2 also contributes to the pathogenesis seen in DKO animals. Subsequent studies have attempted to define the specific roles for loss of αB-crystallin and HspB2. A genetic study comparing the effects of I/R or inotropic stimulation in the hearts of wildtype, DKO, *CRYAB* transgenic, and DKO crossed with *CRYAB* transgenic (effectively expressing αB-crystallin without HspB2) concluded that αB-crystallin is responsible for structural remodeling and mechanical maintenance, while HspB2 is tasked with energetic balance maintenance in the stressed heart (Pinz et al. 2008).

CRYAB transgenic mouse hearts subjected to TAC showed reduced NFAT transactivation and attenuated hypertrophic response, while DKO hearts exhibit increased NFAT transactivation at baseline and develop cardiac insufficiencies in response to TAC, leading the authors to conclude that αB-crystallin prevents cardiac hypertrophic responses, possibly through inhibition of NFAT signaling (Kumarapeli et al. 2008). A single, cardiac specific knockout of *HSPB2* was also generated and shows under basal conditions that cardiac function, hypertrophic responses, and mitochondrial metabolism were unchanged; however, when animals were subjected to TAC, mitochondrial energetics were reduced (Ishiwata et al. 2012), in agreement with the previously suggested role for HspB2 in maintaining energetic balance (Pinz et al. 2008). To our knowledge, a whole body knockout of either *CRYAB* or *HSPB2* has yet to be published and would further distinguish the distinct roles for each in both skeletal and cardiac muscle. Additionally, identification of recessive mutants in *CRYAB* affecting muscle, discussed below, may assist in identifying loss of αB-crystallin function phenotypes, though these mutants may also have gain of function effects since the mutant proteins are expressed to some extent.

Fatal, infantile onset muscular dystrophy was identified in a cohort of Canadian aboriginals all harboring a homozygous c.60C deletion in the n-terminus of *CRYAB*, predicting a serine to alanine mutation at residue 21 and a premature stop codon after 23 missense residues, with unaffected parents both being heterozygous for this mutation (Del Bigio et al. 2011). Patients developed rigid muscles with elevated serum creatine kinase levels and died of respiratory insufficiency shortly after birth, except one child surviving to 4 years of age on mechanical ventilation. Axial muscles were more severely affected than appendicular muscles, Z-disc disarray, dense inclusions, vacuole presence, immune cell infiltration, and necrotic and regenerating fibers were noted. Deteriorating fibers stained strongly for myotilin, desmin, which concentrated at the periphery of inclusions, and αB-crystallin (using an N-terminal specific antibody), which was highly expressed in inclusions and dimly throughout other regions of the fibers. Hearts examined in these patients were normal. This mutant mRNA is likely a target for nonsense-mediated mRNA decay due to the premature stop codon occurring upstream of the final splice junction (Amrani et al. 2006), which may attenuate its expression, though mutant protein is still expressed (Del Bigio et al. 2011). The authors hypothesize that the lack of association of αB-crystallin with titin in this case may be responsible for the severe muscle stiffness observed in patients.

A second recessive mutation in *CRYAB*, c.343delT, results in a frameshift mutation generating a predicted 127 amino acid protein, instead of the wildtype 175 amino acids (Forrest et al. 2011). The patient identified with this mutation, born from unrelated, heterozygous parents, presented starting at age 4 months with progressive feeding difficulties and respiratory distress, requiring ventilation, severe muscle stiffness affecting mostly axial muscles with some limb involvement, and elevated creatine kinase levels. No cardiomyopathy or cataracts were noted. Muscle biopsy showed increased vacuolization, increased lipid content, globular inclusions, fibrosis, and disrupted myofibrillar structure. Abnormal fibers were intensely positive for myotilin, desmin, and αB-crystallin, which was also detected in a 15 kDa truncated form by western blot. Appearance of symptoms following normal development suggests a role for αB-crystallin in remodeling of myofibrillar structure following contraction, rather than a developmental role.

Similarities between patients with these two recessive mutations in *CRYAB* and DKO mice, including skeletal muscle deterioration with fatty deposits, fibrosis, inflammation, and vacuolization, indicate potentially that the effects of these recessive mutants result from loss of αB-crystallin function. Patients develop more severe symptoms as infants, whereas the DKO mice develop progressive symptoms into adulthood. Additionally, patients exhibit severe muscle stiffness, which is not reported in the DKO mice. The phenotypic differences could be due to multiple factors including: the confounding effects of loss of HspB2 in the DKO mice, species variability of requirement for αB-crystallin in skeletal muscle, or potential additional gain of toxic function effects of the mutant protein expression in patients with these recessive mutations. Further examination is necessary to resolve these issues.

11.5 Differential Functions of αB-Crystallin in Skeletal and Cardiac Muscle

Tissue specificity of pathology correlating with the various mutations brings about interesting questions with regards to the roles and requirements for αB-crystallin in cardiac and skeletal muscle. Though the muscles are both striated, the structure and function is very different. Potentially, various mutations in αB-crystallin could affect tissues differently. The late-onset of disease with the majority of these mutations also suggests age-related stress and/or the presence of genetic modifiers may be key factors in disease manifestation. With variable patient lifestyles and different stress encounters, it is difficult to determine in the relatively small number of patients whether, given time and/or the correct stress, patients may also develop clinical symptoms in the yet-unaffected tissue. Nevertheless, we discuss in this section some potential reasons for tissue-specific impacts of αB-crystallin.

Detailed analysis of the cellular localization of αB-crystallin using immunohistochemistry of rat organs indicates potential expression of αB-crystallin in highly oxidative cells, as suggested by a correlation between αB-crystallin expression and markers of oxidative activity (Iwaki et al. 1990). All type 1 aerobic fibers (slow-twitch) and about half of type 2 anaerobic fibers (fast-twitch), corresponding to mostly type 2A fibers, were positive for αB-crystallin, with more intense staining in type 1 fibers. Additional studies confirmed higher expression of αB-crystallin in slow- compared to fast-twitch fibers (Golenhofen et al. 2004; Atomi et al. 2000). Variable expression in different types of skeletal muscle fibers may account for specific fiber-type degradation, but this would not account for the lack of cardiac involvement, since αB-crystallin levels are comparably high in the heart (Golenhofen et al. 2004). Higher αB-crystallin expression in oxidative muscle (cardiac and type 1 skeletal fibers) may also be due to requirements of αB-crystallin for maintaining redox balance (Rajasekaran et al. 2007).

αB-Crystallin plays a role in maintaining the desmin network in muscle, which is crucial for proper sarcomere alignment as well as mitochondrial architecture (Perng et al. 1999a; Lockard and Bloom 1993; Wang et al. 2001; Maloyan et al. 2005). Differences between muscle types in mitochondrial architecture and requirements for appropriate coupling (Milner et al. 2000; Reipert et al. 1999; Rambourg and Segretain 1980; Rappaport et al. 1998) could underlie tissue specificity of disease as well as exemplify normal roles for αB-crystallin.

Sarcomeric localization of αB-crystallin in ischemic skeletal muscle is similar to cardiac muscle, with I-band and intermediate filament localization, though at the I-band, αB-crystallin binds to titin along the length of the I-band region in skeletal muscle, where in cardiac muscle it binds to a narrow region (Golenhofen et al. 2004), identified as the N2B element of titin (Bullard et al. 2004; Golenhofen et al. 2002). Skeletal muscle is susceptible to stretch injury on a day-to-day basis as a result of exercise, which can unfold titin Ig domains (Friden and Lieber 2001; Koh 2002), whereas in cardiac muscle, titin plays a role in the maintenance of passive stiffness during diastole, and, in patients with heart failure, isoforms of titin switch to more compliant forms, with stretching of titin isoforms occurring in cardiomyocytes due to I/R injury and hypertrophic response in heart failure (Linke 2008; Golenhofen et al. 1999). Stretching induces unfolding of the elastic domains of titin (Rief 1997; Minajeva et al. 2001), which may require αB-crystallin to prevent aggregation (Bullard et al. 2004; Golenhofen et al. 2002, 2004; Kotter et al. 2014). Potentially the different isoforms of titin expressed in cardiac and skeletal muscle (Labeit and Kolmerer 1995), as well the apparent variable binding of αB-crystallin to the titin domains may be a reason for tissue specificity of disease. Or more simply, tissue susceptibility to stretch may also explain this phenomenon.

Development of cardiac and skeletal muscle appear not to require αB-crystallin, since DKO mice (Brady et al. 2001), as well as all patients with mutations in *CRYAB*, develop symptoms following normal development (see Table 11.1). The major role of αB-crystallin seems to be in the maintenance of muscle tissue as well as protection during stress, as described above. Figure 11.1 discusses the potential impacts of alterations in levels of αB-crystallin or mutations in *CRYAB* on various stages of skeletal muscle myogenesis. Maintenance of skeletal muscle requires the activation of resident, quiescent progenitor cells, known as satellite cells, identified by expression of the transcription factor Pax7, to sequentially express the myogenic regulatory factors (MRFs), including MyoD and myogenin, which regulate muscle differentiation (Weintraub 1993; Le Grand and Rudnicki 2007). Proliferating myoblasts expressing MyoD must turn off MyoD expression, exit the cell cycle, turn on myogenin, and fuse to form multinucleated myotubes (Weintraub 1993;

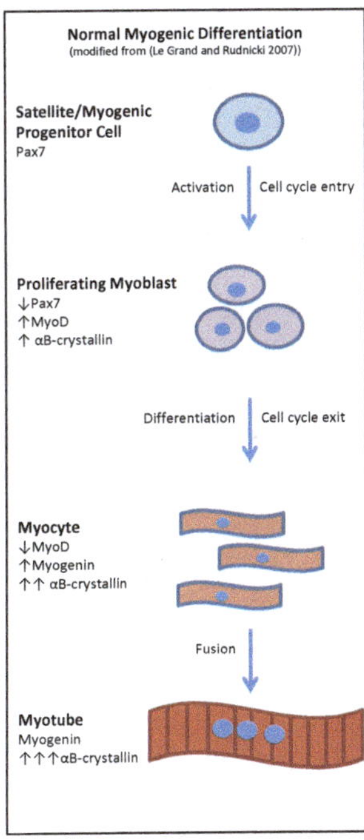

Overexpression of αB-crystallin in C2C12 Myoblasts	Loss of αB-crystallin in DKO animals	Recessive mutant αB-crystallin	Dominant mutant αB-crystallin
	Lower basal percentage of satellite cells (Nepplet al. 2014). Greater increase in percentage of satellite cells with injury (Neppl et al. 2014). Enhanced cell cycle entry?	Lower basal percentage of satellite cells?	Lower basal percentage of satellite cells?
Lower expression and increased degradation of MyoD (Singh et al. 2010) Defective cell cycle exit (Singh et al. 2010)	Increased proliferation (Neppl et al. 2014) Altered MyoD levels? Defective cell cycle exit? Defective differentiation?	Loss/gain of αB-crystallin function? Increased proliferation? Altered MyoD levels? Defective cell cycle exit? Defective differentiation?	Loss/gain of αB-crystallin function? Increased proliferation? Altered MyoD levels? Defective cell cycle exit? Defective differentiation?
	Electron dense material buildup (Brady et al. 2001) Desmin inclusions (Brady et al. 2001) Z-disc dissaray (Brady et al. 2001) Vacuolization (Brady et al. 2001) Smaller fiber diameter with injury-stimulated regeneration (Neppl et al. 2014) Decreased MHC (Brady et al. 2001)	Dense inclusions (Forrest et al. 2011; Del Bigio et al. 2011) Desmin inclusions (Forrest et al. 2011; Del Bigio et al. 2011) Z-disc dissaray (Forrest et al. 2011; Del Bigio et al. 2011) Vacuolization (Forrest et al. 2011; Del Bigio et al. 2011) Necrosis (Forrest et al. 2011; Del Bigio et al. 2011) Decreased MHC? Loss of chaperone function towards titin/desmin?	Desmin/αB-crystallin positive aggregates (Sacconi et al. 2012; Vicart et al. 1998; Selcen and Engel 2003; Reilich et al. 2010) Z-disc dissaray (Sacconi et al. 2012; Vicart et al. 1998; Selcen and Engel 2003; Reilich et al. 2010) Loss of chaperone function towards titin /desmin (Zhu et al. 2009; Perng et al. 1999b) Decreased MHC (Vicart et al. 1998)

Fig. 11.1 The schematic represents normal myogenic differentiation of satellite cells to myotubes, indicating relative expression levels of defining transcription factors throughout the process (Modified from Le Grand and Rudnicki 2007). Satellite cells/myogenic progenitor cells, expressing Pax7, are activated and enter the cell cycle to become proliferating myoblasts, with an increase in MyoD expression and concomitant decrease in Pax7 expression. Proliferating myoblasts must exit the cell cycle to fulfill differentiation to myocytes, which downregulate MyoD expression and upregulate myogenin. Myocytes then fuse into multi-nucleated myotubes, which maintain expression of myogenin. During myogenesis, αB-crystallin is expressed at low levels in proliferating myoblasts and is greatly upregulated upon differentiation to myotubes (Adhikari et al. 2004; Singh et al. 2010). The table to the right of the schematic indicates variations in myogenesis that occur as a result of overexpression of αB-crystallin, loss of αB-crystallin, recessive mutations in αB-crystallin, and dominant mutations in αB-crystallin. *Red text* indicates a testable hypothesis for that condition

Le Grand and Rudnicki 2007). Overexpression of αB-crystallin significantly delays this process in C2C12 myoblasts, with cells showing a defect in cell cycle exit and lower levels of MyoD (Singh et al. 2010). DKO mice exhibit reduced levels of satellite cells under basal conditions at 1 year of age, and unexpectedly, a cardiotoxin injury model of DKO mouse tibialis anterior muscles revealed a threefold increase in the percentage of satellite cells in response to injury; though the mice also showed decreased cross-sectional myofiber area and increased fibrosis (Neppl et al. 2014). This study suggests that loss of αB-crystallin results in an increase in proliferation, with lower levels of p21 and notch signaling molecules, possibly through modulation of Argonaute 2 activity, as shown by enrichment of miRNAs on Argonaute 2 in injured DKO skeletal muscles and co-immunoprecipitation of Argonaute 2 with αB-crystallin (Neppl et al. 2014). Argonaute 2 is an essential component of the RISC complex for miRNA-mediated silencing; the possibility of αB-crystallin as an allosteric regulator of the RISC complex may be a reason for its broad cellular impact (Neppl et al. 2014). The study by Neppl and colleagues suggests that with the loss of αB-crystallin, the regenerative response of satellite cells is skewed towards proliferation with inefficient differentiation to myotubes (Neppl et al. 2014). This is contradictory to data suggesting that overexpression of αB-crystallin in C2C12 myoblasts leads to increased proliferation through delayed exit from the cell cycle with a decrease in p21 expression (Singh et al. 2010). Both studies indicate a role for αB-crystallin in cell cycle regulation during skeletal muscle regeneration (Neppl et al. 2014; Singh et al. 2010). Additional experiments addressing the model system used (i.e. in vivo or cell culture, loss of αB-crystallin or overexpression) are necessary to reconcile the exact role for αB-crystallin in cell cycle regulation under basal and stressed conditions.

Due to withdrawal of growth factors, myoblasts must also resist apoptosis when differentiating to myotubes. αB-crystallin is induced in and contributes to apoptosis-resistance in C2C12 myoblasts through inhibition of caspase 3 activation, and overexpression of αB-crystallin prevents apoptosis occurring as a result of differentiation (Kamradt et al. 2001, 2002). These studies suggest that αB-crystallin may be important for modulating myoblast differentiation during tissue regeneration by multifaceted means including impact on proliferation and cell cycle exit, levels of regulatory MyoD, and inhibition of apoptosis. Cardiac muscle does not have the luxury to regenerate as skeletal muscle does (Mercola et al. 2011), implicating that loss of αB-crystallin function may manifest in skeletal muscle instead of cardiac muscle under healthy conditions due to requirements in regenerative myogenesis.

11.6 Stem Cells as a Model Systems for Studying αB-Crystallin in Cardiac and Skeletal Muscle

Many model systems have been employed in the study of αB-crystallin as discussed above, including rodents and larger mammalian models, some with whole body or tissue-specific overexpression/knockout, human tissue biopsies from healthy and diseased patients, cell culture with primary cells or cell lines, as well as in vitro biochemical analysis. To more fully understand the differential implications of αB-crystallin in cardiac and skeletal muscle, it is crucial to use models in which both cell types can be analyzed in parallel. Animal models with whole body genetic modifications, cell culture using both cardiac and skeletal muscle cell lines, and in vitro experiments analyzing interactions with cardiac- and skeletal-specific isoforms will be useful in accomplishing this feat. Stem cells offer additional potential for analyzing the impact of αB-crystallin on cardiac and skeletal muscle. Induced pluripotent stem cells (iPSCs), a technology first developed in the lab of 2012 Nobel Prize winner, Shinya Yamanaka (Takahashi and Yamanaka 2006), are an invaluable tool offering the ability to model cell-autonomous disease using a patient's own cells and the potential to screen drug compounds or provide cell therapy (Robinton and Daley 2012). Somatic cells are isolated from patients, which can be done now through minimally invasive procedures including blood draws and urine sample collection, and reprogrammed into induced pluripotent stem cells through a variety of techniques involving the overexpression of key transcription factors (Chou et al. 2011; Zhou et al. 2012). Once reprogrammed, these cells are pluripotent and can be directed to differentiate to either cardiac or skeletal muscle cells, which can be studied in culture (Hosoyama et al. 2014; Zhang et al. 2012; Lian et al. 2013; Darabi et al. 2012).

Our lab has previously generated and characterized mouse iPSC-derived cardiomyocytes from transgenic R120G *CRYAB* mice, and shown that cardiomyocytes derived from these cells exhibit αB-crystallin positive aggregates as well as activation of the hypertrophic response (Limphong et al. 2013), indicating that iPSCs have the potential to recapitulate at least some phenotypes observed in αB-crystallinopathies. Many published examples of the use of iPSCs for modeling cardiac diseases exist in the literature, including long QT (Itzhaki et al. 2011) and LEOPARD syndromes (Carvajal-Vergara et al. 2010), with the list rapidly growing. Genetic manipulation is also feasible in iPSCs (Yusa et al. 2011; An et al. 2012; Fong et al. 2013), which allows for the generation of gene-corrected control cells from patient cells harboring mutations, or for insertion of mutations of interest into wildtype iPSC lines for analysis. Caveats of the use of iPSCs as a model include purity of differentiated cultures, which has improved drastically over time and is likely to continue improving with advancement of differentiation methods, lack of support cell presence in the culture (i.e. fibroblasts, endothelial cells, etc.) that may play a role in disease pathogenesis, and variability between iPSC lines with differentiation, which can be minimized through the use of genetic manipulation for the generation of appropriate controls. The iPSC system could be very effective for looking at cell-type specific impacts of αB-crystallin in mutant and wildtype form, which is a focus of pursuit in our laboratory.

11.7 Conclusions and Future Directions

Overall, much work has been done to describe the function of αB-crystallin in cardiac and skeletal muscle both in healthy and diseased states. The use of model systems that allow for direct comparison of the two cell/tissue types will better define differential roles of αB-crystallin in each tissue and help to determine stressors or genetic modifiers that may contribute to disease susceptibility. The study of mutant forms of αB-crystallin existing in diseased patients will provide insights into not only how the mutation affects the structure and function of αB-crystallin, but also into the function and requirements of wildtype αB-crystallin during tissue maintenance and in disease compromised tissues.

The use of iPSCs to model protein misfolding disorders such as αB-crystallin-opathies, provides not only an in vitro disease model with the ability to analyze various impacted cell types and determine mechanistic details of the disease, but also the potential for small molecule screening and possibly regenerative cell therapies to treat patients. Small molecules that reduce the presence of toxic protein aggregates may be beneficial in protein misfolding diseases and can be screened for using the clinically relevant cell type generated from iPSCs (Ebert et al. 2012). The probable impact of loss of αB-crystallin on satellite cells suggests the potential for cell therapy in transplant of gene-corrected autologous or unaffected, nonautologous muscle progenitor cells to prevent skeletal myopathy in patients with recessive, early onset αB-crystallin-opathies. Engraftment of human iPSC-derived skeletal muscle progenitor cells has been shown to seed the satellite compartment in dystrophic mice suggesting the potential for long term benefits from this type of cell therapy (Darabi et al. 2012). The rapidly advancing field of iPSCs will likely lend itself to mechanistic disease modeling, small molecule screening, and regenerative therapies in protein misfolding disorders in the years to come.

Acknowledgments Lindsey Barber provided excellent editorial assistance during preparation of this manuscript.

References

Adhikari AS, Sridhar Rao K, Rangaraj N, Parnaik VK, Mohan Rao C (2004) Heat stress-induced localization of small heat shock proteins in mouse myoblasts: intranuclear lamin A/C speckles as target for alphaB-crystallin and Hsp25. Exp Cell Res 299(2):393–403. doi:10.1016/j.yexcr.2004.05.032

Ahmad MF, Raman B, Ramakrishna T, Rao Ch M (2008) Effect of phosphorylation on alpha B-crystallin: differences in stability, subunit exchange and chaperone activity of homo and mixed oligomers of alpha B-crystallin and its phosphorylation-mimicking mutant. J Mol Biol 375(4):1040–1051. doi:10.1016/j.jmb.2007.11.019

Amrani N, Sachs MS, Jacobson A (2006) Early nonsense: mRNA decay solves a translational problem. Nat Rev Mol Cell Biol 7(6):415–425. doi:10.1038/nrm1942

An MC, Zhang N, Scott G, Montoro D, Wittkop T, Mooney S, Melov S, Ellerby LM (2012) Genetic correction of Huntington's disease phenotypes in induced pluripotent stem cells. Cell Stem Cell 11(2):253–263. doi:10.1016/j.stem.2012.04.026

Andley UP, Hamilton PD, Ravi N, Weihl CC (2011) A knock-in mouse model for the R120G mutation of alphaB-crystallin recapitulates human hereditary myopathy and cataracts. PLoS one 6(3):e17671. doi:10.1371/journal.pone.0017671

Aquilina JA, Benesch JL, Ding LL, Yaron O, Horwitz J, Robinson CV (2004) Phosphorylation of alphaB-crystallin alters chaperone function through loss of dimeric substructure. J Biol Chem 279(27):28675–28680. doi:10.1074/jbc.M403348200

Atomi Y, Yamada S, Nishida T (1991) Early changes of αB-crystallin mRNA in rat skeletal muscle to mechanical tension and denervation. Biochem Biophys Res Commun 181(3):1323–1330
Atomi Y, Toro K, Masuda T, Hatta H (2000) Fiber-type-specific aB-crystallin distribution and its shifts with T3 and PTU treatments in rat hindlimb muscles. J Appl Physiol 88:1355–1364
Bagneris C, Bateman OA, Naylor CE, Cronin N, Boelens WC, Keep NH, Slingsby C (2009) Crystal structures of alpha-crystallin domain dimers of alphaB-crystallin and Hsp20. J Mol Biol 392(5):1242–1252. doi:10.1016/j.jmb.2009.07.069
Basha E, O'Neill H, Vierling E (2012) Small heat shock proteins and alpha-crystallins: dynamic proteins with flexible functions. Trends Biochem Sci 37(3):106–117. doi:10.1016/j.tibs.2011.11.005
Bennardini F, Wrzosek A, Chiesi M (1992) Alpha B-crystallin in cardiac tissue. Association with actin and desmin filaments. Circ Res 71(2):288–294. doi:10.1161/01.res.71.2.288
Best TM, Fiebig R, Corr DT, Brickson S, Ji L (1999) Free radical activity antioxidant enzyme and glutathione changes with muscle stretch injury in rabbits. J Appl Physiol 87:74–82
Bloemendal H (1981) The lens proteins. Molecular and cellular biology of the eye lens. Wiley, New York, pp 1–47
Boelens WC (2014) Cell biological roles of alphaB-crystallin. Prog Biophys Mol Biol. doi:10.1016/j.pbiomolbio.2014.02.005
Bova MP, Ding LL, Horwitz J, Fung BKK (1997) Subunit exchange of aA-crystallin. J Biol Chem 272(47):29511–29517
Bova MP, Yaron O, Huang Q, Ding L, Haley DA, Stewart PL, Horwitz J (1999) Mutation R120G in aB-crystallin, which is linked to a desmin-related myopathy, results in an irregular structure and defective chaperone-like function. Proc Natl Acad Sci U S A 96:6137–6142
Brady JP, Garland DL, Green DE, Tamm ET, Giblin FJ, Wawrousek EF (2001) aB-crystallin in lens development and muscle integrity: a gene knockout approach. IOVS 42(12):2924–2934
Bruinsma IB, de Jager M, Carrano A, Versleijen AA, Veerhuis R, Boelens W, Rozemuller AJ, de Waal RM, Verbeek MM (2011) Small heat shock proteins induce a cerebral inflammatory reaction. J Neurosci 31(33):11992–12000. doi:10.1523/JNEUROSCI. 0945-11.2011
Bullard B, Ferguson C, Minajeva A, Leake MC, Gautel M, Labeit D, Ding L, Labeit S, Horwitz J, Leonard KR, Linke WA (2004) Association of the chaperone alphaB-crystallin with titin in heart muscle. J Biol Chem 279(9):7917–7924. doi:10.1074/jbc.M307473200
Carra S, Seguin SJ, Lambert H, Landry J (2008) HspB8 chaperone activity toward poly(Q)-containing proteins depends on its association with Bag3, a stimulator of macroautophagy. J Biol Chem 283(3):1437–1444. doi:10.1074/jbc.M706304200
Carra S, Brunsting JF, Lambert H, Landry J, Kampinga HH (2009) HspB8 participates in protein quality control by a non-chaperone-like mechanism that requires eIF2{alpha} phosphorylation. J Biol Chem 284(9):5523–5532. doi:10.1074/jbc.M807440200
Carvajal-Vergara X, Sevilla A, D'Souza SL, Ang YS, Schaniel C, Lee DF, Yang L, Kaplan AD, Adler ED, Rozov R, Ge Y, Cohen N, Edelmann LJ, Chang B, Waghray A, Su J, Pardo S, Lichtenbelt KD, Tartaglia M, Gelb BD, Lemischka IR (2010) Patient-specific induced pluripotent stem-cell-derived models of LEOPARD syndrome. Nature 465(7299):808–812. doi:10.1038/nature09005
Carver JA, Guerreiro N, Nicholls KA, Truscott RJW (1995) On the interaction of a-crystallin with unfolded proteins. Biochem Biophys Acta 1252:251–260
Chis R, Sharma P, Bousette N, Miyake T, Wilson A, Backx PH, Gramolini AO (2012) Alpha-crystallin B prevents apoptosis after H2O2 exposure in mouse neonatal cardiomyocytes. Am J Physiol Heart Circ Physiol 303:H967–H978. doi:10.1152/ajpheart.00040.2012.-/-Crystallin
Chou BK, Mali P, Huang X, Ye Z, Dowey SN, Resar LM, Zou C, Zhang YA, Tong J, Cheng L (2011) Efficient human iPS cell derivation by a non-integrating plasmid from blood cells with unique epigenetic and gene expression signatures. Cell Res 21(3):518–529. doi:10.1038/cr.2011.12
Christians ES, Yan LJ, Benjamin IJ (2002) Heat shock factor 1 and heat shock proteins: critical partners in protection against acute cell injury. Crit Care Med 30(1):S43–S50
Christians ES, Ishiwata T, Benjamin IJ (2012) Small heat shock proteins in redox metabolism: implications for cardiovascular diseases. Int J Biochem Cell Biol 44(10):1632–1645. doi:10.1016/j.biocel.2012.06.006

Christians ES, Banerjee Mustafi S, Benjamin IJ (2014) Chaperones and cardiac misfolding protein diseases. Curr Protein Pept Sci 15(3):189–204

Clark AR, Naylor CE, Bagneris C, Keep NH, Slingsby C (2011) Crystal structure of R120G disease mutant of human alphaB-crystallin domain dimer shows closure of a groove. J Mol Biol 408(1):118–134. doi:10.1016/j.jmb.2011.02.020

Darabi R, Arpke RW, Irion S, Dimos JT, Grskovic M, Kyba M, Perlingeiro RC (2012) Human ES- and iPS-derived myogenic progenitors restore DYSTROPHIN and improve contractility upon transplantation in dystrophic mice. Cell Stem Cell 10(5):610–619. doi:10.1016/j.stem.2012.02.015

de Jong WW, Caspers GJ, Leunissen JAM (1998) Genealogy of the alpha-crystallin—small heatshock protein superfamily. Int J Biol Macromol 22:151–162

Del Bigio MR, Chudley AE, Sarnat HB, Campbell C, Goobie S, Chodirker BN, Selcen D (2011) Infantile muscular dystrophy in Canadian aboriginals is an alphaB-crystallinopathy. Ann Neurol 69(5):866–871. doi:10.1002/ana.22331

Delbecq SP, Jehle S, Klevit R (2012) Binding determinants of the small heat shock protein, alphaB-crystallin: recognition of the 'IxI' motif. EMBO J 31(24):4587–4594. doi:10.1038/emboj.2012.318

den Engelsman J, Keijsers V, de Jong WW, Boelens WC (2003) The small heat-shock protein alpha B-crystallin promotes FBX4-dependent ubiquitination. J Biol Chem 278(7):4699–4704. doi:10.1074/jbc.M211403200

den Engelsman J, Gerrits D, de Jong WW, Robbins J, Kato K, Boelens WC (2005) Nuclear import of {alpha}B-crystallin is phosphorylation-dependent and hampered by hyperphosphorylation of the myopathy-related mutant R120G. J Biol Chem 280(44):37139–37148. doi:10.1074/jbc.M504106200

Dubin RA, Wawrousek EF, Piatigorsky J (1989) Expression of the murine alpha B-crystallin gene is not restricted to the lens. Mol Cell Biol 9(3):1083–1091. doi:10.1128/mcb.9.3.1083

Ebert AD, Liang P, Wu JC (2012) Induced pluripotent stem cells as a disease modeling and drug screening platform. J Cardiovasc Pharmacol 60(4):408–416. doi:10.1097/FJC.0b013e318247f642

Ecroyd H, Meehan S, Horwitz J, Aquilina JA, Benesch JL, Robinson CV, Macphee CE, Carver JA (2007) Mimicking phosphorylation of alphaB-crystallin affects its chaperone activity. Biochem J 401(1):129–141. doi:10.1042/BJ20060981

Ellis RJ, van der Vies SM (1991) Molecular chaperones. Annu Rev Biochem 60(1):321–347. doi:10.1146/annurev.bi.60.070191.001541

Fardeau M, Godet-Guillain J, Tome FM, Collin H, Gaudeau S, Boffety C, Vernant P (1978) Une nouvelle affection musculaire familiale définie par l'accumulation intra-sarcoplasmique d'un matériel granulofilamentaire dense en microscopie electronique. [A new familial muscular disorder demonstrated by the intra-sarcoplasmic accumulation of a granulo-filamentous material which is dense on electron microscopy (author's transl)]. Rev Neurol (Paris) 134:411–425

Feil IK, Malfois M, Hendle J, van Der Zandt H, Svergun DI (2001) A novel quaternary structure of the dimeric alpha-crystallin domain with chaperone-like activity. J Biol Chem 276(15):12024–12029. doi:10.1074/jbc.M010856200

Fong H, Wang C, Knoferle J, Walker D, Balestra ME, Tong LM, Leung L, Ring KL, Seeley WW, Karydas A, Kshirsagar MA, Boxer AL, Kosik KS, Miller BL, Huang Y (2013) Genetic correction of tauopathy phenotypes in neurons derived from human induced pluripotent stem cells. Stem Cell Rep 1(3):226–234. doi:10.1016/j.stemcr.2013.08.001

Forrest KM, Al-Sarraj S, Sewry C, Buk S, Tan SV, Pitt M, Durward A, McDougall M, Irving M, Hanna MG, Matthews E, Sarkozy A, Hudson J, Barresi R, Bushby K, Jungbluth H, Wraige E (2011) Infantile onset myofibrillar myopathy due to recessive CRYAB mutations. Neuromuscul Disord 21(1):37–40. doi:10.1016/j.nmd.2010.11.003

Friden J, Lieber RL (2001) Eccentric exercise-induced injuries to contractile and cytoskeletal muscle fibre components. Acta Physiol Scand 171:321–326

Garcia-Mata R, Gao Y, Sztul E (2002) Hassles with taking out the garbage: aggravating aggresomes. Traffic 3:388–396
Ghosh JG, Estrada MR, Clark JI (2006) Structure-based analysis of the beta8 interactive sequence of human alphaB crystallin. Biochemistry 45:9878–9886
Ghosh JG, Houck SA, Clark JI (2007a) Interactive sequences in the stress protein and molecular chaperone human alphaB crystallin recognize and modulate the assembly of filaments. Int J Biochem Cell Biol 39(10):1804–1815. doi:10.1016/j.biocel.2007.04.027
Ghosh JG, Shenoy AK, Clark JI (2007b) Interactions between important regulatory proteins and human alphaB crystallin. Biochemistry 46:6308–6317
Golenhofen N, Htun P, Ness W, Koob R, Schaper W, Drenckhahn D (1999) Binding of stress protein cryab to cardiac myofibrils correlates with the degree of myocardial damage during I/R in vivo. J Mol Cell Cardiol 31:569–580
Golenhofen N, Arbeiter A, Koob R, Drenckhahn D (2002) Ischemia-induced association of the stress protein alpha B-crystallin with I-band portion of cardiac titin. J Mol Cell Cardiol 34(3):309–319. doi:10.1006/jmcc.2001.1513
Golenhofen N, Perng MD, Quinlan RA, Drenckhahn D (2004) Comparison of the small heat shock proteins alphaB-crystallin, MKBP, HSP25, HSP20, and cvHSP in heart and skeletal muscle. Histochem Cell Biol 122(5):415–425. doi:10.1007/s00418-004-0711-z
Golenhofen N, Redel A, Wawrousek EF, Drenckhahn D (2006) Ischemia-induced increase of stiffness of alphaB-crystallin/HSPB2-deficient myocardium. Pflugers Arch 451(4):518–525. doi:10.1007/s00424-005-1488-1
Gopal-Srivastava R, Piatigorsky J (1993) The murine alpha B-crystallin/small heat shock protein enhancer: identification of alpha BE-1, alpha BE-2, alpha BE-3 and MRF control elements. Mol Cell Biol 13(11):7144–7152
Gopal-Srivastava R, Haynes JI II, Piatigorsky J (1995) Regulation of the murine aB-crystallin/small heath shock protein gene in cardiac muscle. Mol Cell Biol 15(12):7081–7090
Hayes VH, Devlin G, Quinlan RA (2008) Truncation of alphaB-crystallin by the myopathy-causing Q151X mutation significantly destabilizes the protein leading to aggregate formation in transfected cells. J Biol Chem 283(16):10500–10512. doi:10.1074/jbc.M706453200
Hishiya A, Salman MN, Carra S, Kampinga HH, Takayama S (2011) BAG3 directly interacts with mutated alphaB-crystallin to suppress its aggregation and toxicity. PLoS ONE 6(3):e16828. doi:10.1371/journal.pone.0016828
Horwitz J (1992) a-Crystallin can function as a molecular chaperone. Proc Natl Acad Sci U S A 89:10449–10453
Hosoyama T, McGivern JV, Van Dyke JM, Ebert AD, Suzuki M (2014) Derivation of myogenic progenitors directly from human pluripotent stem cells using a sphere-based culture. Stem Cells Transl Med. doi:10.5966/sctm.2013-0143
Houck SA, Landsbury A, Clark JI, Quinlan RA (2011) Multiple sites in alphaB-crystallin modulate its interaction with desmin filaments assembled in vitro. PLoS ONE 6(11):e25859. doi:10.1371/journal.pone.0025859.g001
Hu WF, Gong L, Cao Z, Ma H, Ji W, Deng M, Liu M, Hu XH, Chen P, Yan Q, Chen HG, Liu J, Sun S, Zhang L, Liu JP, Wawrousek E, Li DW (2012) αA- and αB-crystallins interact with caspase-3 and Bax to guard mouse lens development. Curr Mol Med 12(2):177–187
Ignolia TD, Craig EA (1982) Four small Drosophila heat shock proteins are related to each other and to mammalian alpha crystallin. Proc Natl Acad Sci U S A 79:2360–2364
Inagaki N, Hayashi T, Arimura T, Koga Y, Takahashi M, Shibata H, Teraoka K, Chikamori T, Yamashina A, Kimura A (2006) Alpha B-crystallin mutation in dilated cardiomyopathy. Biochem Biophys Res Commun 342(2):379–386. doi:10.1016/j.bbrc.2006.01.154
Ishiwata T, Orosz A, Wang X, Mustafi SB, Pratt GW, Christians ES, Boudina S, Abel ED, Benjamin IJ (2012) HSPB2 is dispensable for the cardiac hypertrophic response but reduces mitochondrial energetics following pressure overload in mice. PLoS one 7(8):e42118. doi:10.1371/journal.pone.0042118

Ito H, Okamoto K, Nakayama H, Isobe T, Kato K (1997) Phosphorylation of alphaB-crystallin in response to various types of stress. J Biol Chem 272:29934–29941

Itzhaki I, Maizels L, Huber I, Zwi-Dantsis L, Caspi O, Winterstern A, Feldman O, Gepstein A, Arbel G, Hammerman H, Boulos M, Gepstein L (2011) Modelling the long QT syndrome with induced pluripotent stem cells. Nature 471(7337):225–229. doi:10.1038/nature09747

Iwaki T, Kume-Iwaki A, Goldman JE (1990) Cellular distribution of alpha B-crystallin in non-lenticular tissues. J Histochem Cytochem 38(1):31–39. doi:10.1177/38.1.2294148

Jansen G, Groenen PJTA, Bachner D, Jap PHK, Coerwinkel M, Oerlemans F, van den Broek W, Gohlsch B, Pette D, Plomp JJ, Molenaar PC, Nederhoff MGJ, van Echteld CJA, Dekker M, Berns A, Hameister H, Wieringa B (1996) Abnormal myotonic dystrophy protein kinase levels produce only mild myopathy in mice. Nat Genet 13:316–324

Jehle S, Rajagopal P, Bardiaux B, Markovic S, Kuhne R, Stout JR, Higman VA, Klevit RE, van Rossum BJ, Oschkinat H (2010) Solid-state NMR and SAXS studies provide a structural basis for the activation of alphaB-crystallin oligomers. Nat Struct Mol Biol 17(9):1037–1042. doi:10.1038/nsmb.1891

Kamradt MC, Chen F, Cryns VL (2001) The small heat shock protein alpha B-crystallin negatively regulates cytochrome c- and caspase-8-dependent activation of caspase-3 by inhibiting its autoproteolytic maturation. J Biol Chem 276(19):16059–16063. doi:10.1074/jbc.C100107200

Kamradt MC, Chen F, Sam S, Cryns VL (2002) The small heat shock protein alpha B-crystallin negatively regulates apoptosis during myogenic differentiation by inhibiting caspase-3 activation. J Biol Chem 277(41):38731–38736. doi:10.1074/jbc.M201770200

Kappé G, Leunissen JM, Jong W (2002) Evolution and diversity of prokaryotic small heat shock proteins. In: Arrigo A-P, Müller WEG (eds) Small stress proteins, vol 28, Progress in molecular and subcellular biology. Springer, Berlin/Heidelberg, pp 1–17. doi:10.1007/978-3-642-56348-5_1

Kato K, Ito H, Kamei K, Inaguma Y, Iwamoto I, Saga S (1998) Phosphorylation of αB-crystallin in mitotic cells and identification of enzymatic activities responsible for phosphorylation. J Biol Chem 273:28346–28354

Koh TJ (2002) Do small heat shock proteins protect skeletal muscle from injury. Exerc Sports Sci Rev 30(3):117–121

Koh TJ, Escobedo J (2004) Cytoskeletal disruption and small heat shock protein translocation immediately after lengthening contractions. Am J Physiol Cell Physiol 286(3):C713–C722. doi:10.1152/ajpcell.00341.2003

Komulainen J, Takala TES, Kuipers H, Hesselink MKC (1998) The disruption of myofibre structures in rat skeletal muscle after forced lengthening contractions. Eur J Physiol 436:735–741

Kostek MC, Chen YW, Cuthbertson DJ, Shi R, Fedele MJ, Esser KA, Rennie MJ (2007) Gene expression responses over 24 h to lengthening and shortening contractions in human muscle: major changes in CSRP3, MUSTN1, SIX1, and FBXO32. Physiol Genomics 31:42–52. doi:10.1152/physiolgenomics.00151.2006.-Resistance

Kotter S, Unger A, Hamdani N, Lang P, Vorgerd M, Nagel-Steger L, Linke WA (2014) Human myocytes are protected from titin aggregation-induced stiffening by small heat shock proteins. J Cell Biol 204(2):187–202. doi:10.1083/jcb.201306077

Krishnamoorthy V, Donofrio AJ, Martin JL (2013) O-GlcNAcylation of alphaB-crystallin regulates its stress-induced translocation and cytoprotection. Mol Cell Biochem 379(1–2):59–68. doi:10.1007/s11010-013-1627-5

Kumarapeli AR, Su H, Huang W, Tang M, Zheng H, Horak KM, Li M, Wang X (2008) Alpha B-crystallin suppresses pressure overload cardiac hypertrophy. Circ Res 103(12):1473–1482. doi:10.1161/CIRCRESAHA.108.180117

Labeit S, Kolmerer B (1995) Titins: giant proteins in charge of muscle ultrastructure and elasticity. Science 270:293–296

Laganowsky A, Benesch JL, Landau M, Ding L, Sawaya MR, Cascio D, Huang Q, Robinson CV, Horwitz J, Eisenberg D (2010) Crystal structures of truncated alphaA and alphaB crystallins reveal structural mechanisms of polydispersity important for eye lens function. Protein Sci 19(5):1031–1043. doi:10.1002/pro.380

Le Grand F, Rudnicki MA (2007) Skeletal muscle satellite cells and adult myogenesis. Curr Opin Cell Biol 19(6):628–633. doi:10.1016/j.ceb.2007.09.012

Li DW, Liu JP, Mao YW, Xiang H, Wang J, Ma WY, Dong Z, Pike HM, Brown RE, Reed JC (2005) Calcium-activated RAF/MEK/ERK signaling pathway mediates p53-dependent apoptosis and is abrogated by alpha B-crystallin through inhibition of RAS activation. Mol Biol Cell 16(9):4437–4453. doi:10.1091/mbc.E05-01-0010

Lian X, Zhang J, Azarin SM, Zhu K, Hazeltine LB, Bao X, Hsiao C, Kamp TJ, Palecek SP (2013) Directed cardiomyocyte differentiation from human pluripotent stem cells by modulating Wnt/beta-catenin signaling under fully defined conditions. Nat Protoc 8(1):162–175. doi:10.1038/nprot.2012.150

Liang P, MacRae TH (1997) Molecular chaperones and the cytoskeleton. J Cell Sci 110:1431–1440

Limphong P, Zhang H, Christians E, Liu Q, Riedel M, Ivey K, Cheng P, Mitzelfelt K, Taylor G, Winge D, Srivastava D, Benjamin I (2013) Modeling human protein aggregation cardiomyopathy using murine induced pluripotent stem cells. Stem Cells Transl Med 2(3):161–166. doi:10.5966/sctm. 2012-0073

Linke WA (2008) Sense and stretchability: the role of titin and titin-associated proteins in myocardial stress-sensing and mechanical dysfunction. Cardiovasc Res 77(4):637–648. doi:10.1016/j.cardiores.2007.03.029

Liu JP, Schlosser R, Ma WY, Dong Z, Feng H, Liu L, Huang XQ, Liu Y, Li DW (2004) Human alphaA- and alphaB-crystallins prevent UVA-induced apoptosis through regulation of PKCalpha, RAF/MEK/ERK and AKT signaling pathways. Exp Eye Res 79(3):393–403. doi:10.1016/j.exer.2004.06.015

Liu S, Li J, Tao Y, Xiao X (2007) Small heat shock protein alphaB-crystallin binds to p53 to sequester its translocation to mitochondria during hydrogen peroxide-induced apoptosis. Biochem Biophys Res Commun 354(1):109–114. doi:10.1016/j.bbrc.2006.12.152

Lockard VG, Bloom S (1993) Trans-cellular desmin-lamin B intermediate filament network in cardiac myocytes. J Mol Cell Cardiol 25:303–309

Lutsch G, Vetter R, Offhauss U, Wieske M, Gröne H-J, Klemenz R, Schimke I, Stahl J, Benndorf R (1997) Abundance and location of the small heat shock proteins HSP25 and αB-crystallin in rat and human heart. Circulation 96:3466–3476

Maloyan A, Sanbe A, Osinska H, Westfall M, Robinson D, Imahashi K, Murphy E, Robbins J (2005) Mitochondrial dysfunction and apoptosis underlie the pathogenic process in alpha-B-crystallin desmin-related cardiomyopathy. Circulation 112(22):3451–3461. doi:10.1161/CIRCULATIONAHA.105.572552

Maloyan A, Gulick J, Glabe CG, Kayed R, Robbins J (2007) Exercise reverses preamyloid oligomer and prolongs survival in alphaB-crystallin-based desmin-related cardiomyopathy. Proc Natl Acad Sci U S A 104(14):5995–6000. doi:10.1073/pnas.0609202104

Maloyan A, Osinska H, Lammerding J, Lee RT, Cingolani OH, Kass DA, Lorenz JN, Robbins J (2009) Biochemical and mechanical dysfunction in a mouse model of desmin-related myopathy. Circ Res 104(8):1021–1028. doi:10.1161/CIRCRESAHA.108.193516

Mao YW, Liu JP, Xiang H, Li DW (2004) Human alphaA- and alphaB-crystallins bind to Bax and Bcl-X(S) to sequester their translocation during staurosporine-induced apoptosis. Cell Death Differ 11(5):512–526. doi:10.1038/sj.cdd.4401384

Martin JL, Mestril R, Hilal-Dandan R, Brunton LL, Dillmann WH (1997) Small heat shock proteins and protection against ischemic injury in cardiac myocytes. Circulation 96:4343–4348

Melkani GC, Cammarato A, Bernstein SI (2006) AlphaB-crystallin maintains skeletal muscle myosin enzymatic activity and prevents its aggregation under heat-shock stress. J Mol Biol 358(3):635–645. doi:10.1016/j.jmb.2006.02.043

Mercola M, Ruiz-Lozano P, Schneider MD (2011) Cardiac muscle regeneration: lessons from development. Genes Dev 25(4):299–309. doi:10.1101/gad.2018411

Milner D, Mavroidis M, Weisleder N, Capetanaki Y (2000) Desmin cytoskeleton linked to muscle mitochondrial distribution and respiratory function. J Cell Biol 150(6):1283–1297

Minajeva A, Kulke M, Fernandez JM, Linke WA (2001) Unfolding of titin domains explains the viscoelastic behavior of skeletal myofibrils. Biophys J 80:1442–1451

Mitra A, Basak T, Datta K, Naskar S, Sengupta S, Sarkar S (2013) Role of alpha-crystallin B as a regulatory switch in modulating cardiomyocyte apoptosis by mitochondria or endoplasmic reticulum during cardiac hypertrophy and myocardial infarction. Cell Death Dis 4:e582. doi:10.1038/cddis.2013.114

Morimoto RI (1998) Regulation of the heat shock transcriptional response: cross talk between a family of heat shock factors, molecular chaperones, and negative regulators. Genes Dev 12(24):3788–3796. doi:10.1101/gad.12.24.3788

Morner CT (1894) Untersuchung der Proteinsubstanzen in den lichtbrechenden Medien des Auges (Examination of the protein-substances in the refractory media of the eye). Hoppe Seyler's Z Physiol Chem 18(61):106

Morrison LE, Hoover HE, Thuerauf DJ, Glembotski CC (2003) Mimicking phosphorylation of alphaB-crystallin on serine-59 is necessary and sufficient to provide maximal protection of cardiac myocytes from apoptosis. Circ Res 92(2):203–211. doi:10.1161/01.res.0000052989.83995.a5

Morrison LE, Whittaker RJ, Klepper RE, Wawrousek EF, Glembotski CC (2004) Roles for αB-crystallin and HSPB2 in protecting the myocardium from ischemia-reperfusion-induced damage in a KO mouse model. Am J Physiol Heart Circ Physiol 286:H847–H855

Mymrikov EV, Seit-Nebi AS, Gusev NB (2011) Large potentials of small heat shock proteins. Physiol Rev 91:1123–1159. doi:10.1152/physrev.00023.2010.-Modern

Neppl RL, Kataoka M, Wang DZ (2014) Crystallin-alphaB regulates skeletal muscle homeostasis via modulation of argonaute2 activity. J Biol Chem 289(24):17240–17248. doi:10.1074/jbc.M114.549584

Neufer PD, Benjamin IJ (1996) Differential expression of alpha B-crystallin and Hsp27 in skeletal muscle during continuous contractile activity. Relationship to myogenic regulatory factors. J Biol Chem 271(39):24089–24095. doi:10.1074/jbc.271.39.24089

Ogata T, Yamaksak Y (1997) Ultra-high-resolution scanning electron microscopy of mitochondria and sarcoplasmic reticulum arrangement in human red, white, and intermediate muscle fibers. Anat Rec 248:214–223

Paulsen G, Vissing K, Kalhovde JM, Ugelstad I, Bayer ML, Kadi F, Schjerling P, Hallen J, Raastad T (2007) Maximal eccentric exercise induces a rapid accumulation of small heat shock proteins on myofibrils and a delayed HSP70 response in humans. Am J Physiol Regul Integr Comp Physiol 293:R844–R853. doi:10.1152/ajpregu.00677.2006.-In

Perng MD, Cairns L, Vandenijssel P, Prescott A, Hutcheson AM, Quinlan RA (1999a) Intermediate filament interactions can be altered by HSP27 and alphaB-crystallin. J Cell Sci 112:2099–2112

Perng MD, Muchowski PJ, Vandenijssel P, Wu JS, Hutcheson AM, Clark JI, Quinlan RA (1999b) The cardiomyopathy and lens cataract mutation in alphaB-crystalln alters its protein structure, chaperone activity, and interaction with intermediate filaments in vitro. J Biol Chem 274:33235–33243

Perng MD, Wen SF, van den IP, Prescott AR, Quinlan RA (2004) Desmin aggregate formation by R120G alphaB-crystallin is caused by altered filament interactions and is dependent upon network status in cells. Mol Biol Cell 15(5):2335–2346. doi:10.1091/mbc.E03-12-0893

Peschek J, Braun N, Rohrberg J, Back KC, Kriehuber T, Kastenmüller A, Weinkauf S, Buchner J (2013) Regulated structural transitions unleash the chaperone activity of alphaB-crystallin. Proc Natl Acad Sci U S A 110(40):E3780–E3789

Pilotto A, Marziliano N, Pasotti M, Grasso M, Costante AM, Arbustini E (2006) AlphaB-crystallin mutation in dilated cardiomyopathies: low prevalence in a consecutive series of 200 unrelated probands. Biochem Biophys Res Commun 346(4):1115–1117. doi:10.1016/j.bbrc.2006.05.203

Pinz I, Robbins J, Rajasekaran NS, Benjamin IJ, Ingwall JS (2008) Unmasking different mechanical and energetic roles for the small heat shock proteins CryAB and HSPB2 using genetically modified mouse hearts. FASEB J 22(1):84–92. doi:10.1096/fj.07-8130com

Plater ML, Goode D, Crabbe MJC (1996) Effects of site-directed mutations on the chaperone-like activity of alphaB-crystallin. J Biol Chem 271(45):28558–28566

Rajaraman K, Raman B, Ramakrishna T, Rao Ch M (2001) Interaction of human recombinant alphaA- and alphaB-crystallins with early and late unfolding intermediates of citrate synthase on its thermal denaturation. FEBS Lett 497:118–123

Rajasekaran NS, Connell P, Christians ES, Yan LJ, Taylor RP, Orosz A, Zhang XQ, Stevenson TJ, Peshock RM, Leopold JA, Barry WH, Loscalzo J, Odelberg SJ, Benjamin IJ (2007) Human alpha B-crystallin mutation causes oxido-reductive stress and protein aggregation cardiomyopathy in mice. Cell 130(3):427–439. doi:10.1016/j.cell.2007.06.044

Raju I, Abraham EC (2013) Mutants of human alphaB-crystallin cause enhanced protein aggregation and apoptosis in mammalian cells: influence of co-expression of HspB1. Biochem Biophys Res Commun 430(1):107–112. doi:10.1016/j.bbrc.2012.11.051

Rambourg A, Segretain D (1980) Three-dimensional electron microscopy of mitochondria and endoplasmic reticulum in the red muscle fiber of the rat diaphragm. Anat Rec 197(1):33–48

Rappaport L, Contard F, Samuel JL, Delcayre C, Marotte F, Tome F, Fardeau M (1988) Storage of phosphorylated desmin in a familial myopathy. FEBS Lett 231(2):421–425

Rappaport L, Oliviero P, Samuel JL (1998) Cytoskeleton and mitochondrial morphology and function. Mol Cell Biochem 184:101–105

Ray PS, Martin JL, Swanson EA, Otani H, Dillmann WH, Das DK (2001) Transgene overexpression of aB crystallin confers simultaneous protection against cardiomyocyte apoptosis and necrosis during myocardial ischemia and reperfusion. FASEB J 15:393–402

Reddy S, Smith DBJ, Rich MM, Leferovich JM, Reilly P, Davis BM, Tran K, Rayburn H, Bronson R, Cros D, Balice-Gordon RJ, Housman D (1996) Mice lacking the myotonic dystrophy protein kinase develop a late onset progressive myopathy. Nat Genet 13:325–335

Reilich P, Schoser B, Schramm N, Krause S, Schessl J, Kress W, Muller-Hocker J, Walter MC, Lochmuller H (2010) The p.G154S mutation of the alpha-B crystallin gene (CRYAB) causes late-onset distal myopathy. Neuromuscul Disord 20(4):255–259. doi:10.1016/j.nmd.2010.01.012

Reipert S, Steinbock F, Fischer I, Bittner RE, Zeold A, Wiche G (1999) Association of mitochondria with plectin and desmin intermediate filaments in striated muscle. Exp Cell Res 252:479–491

Rief M (1997) Reversible unfolding of individual titin immunoglobulin domains by AFM. Science 276(5315):1109–1112. doi:10.1126/science.276.5315.1109

Robinton DA, Daley GQ (2012) The promise of induced pluripotent stem cells in research and therapy. Nature 481(7381):295–305. doi:10.1038/nature10761

Sacconi S, Feasson L, Antoine JC, Pecheux C, Bernard R, Cobo AM, Casarin A, Salviati L, Desnuelle C, Urtizberea A (2012) A novel CRYAB mutation resulting in multisystemic disease. Neuromuscul Dis 22(1):66–72. doi:10.1016/j.nmd.2011.07.004

Sanbe A, Osinska H, Saffitz JE, Glabe CG, Kayed R, Maloyan A, Robbins J (2004) Desmin-related cardiomyopathy in transgenic mice: a cardiac amyloidosis. Proc Natl Acad Sci U S A 101(27):10132–10136. doi:10.1073/pnas.0401900101

Sanbe A, Yamauchi J, Miyamoto Y, Fujiwara Y, Murabe M, Tanoue A (2007) Interruption of CryAB-amyloid oligomer formation by HSP22. J Biol Chem 282(1):555–563. doi:10.1074/jbc.M605481200

Sanbe A, Daicho T, Mizutani R, Endo T, Miyauchi N, Yamauchi J, Tanonaka K, Glabe C, Tanoue A (2009) Protective effect of geranylgeranylacetone via enhancement of HSPB8 induction in desmin-related cardiomyopathy. PLoS one 4(4):e5351. doi:10.1371/journal.pone.0005351

Sanbe A, Marunouchi T, Yamauchi J, Tanonaka K, Nishigori H, Tanoue A (2011) Cardioprotective effect of nicorandil, a mitochondrial ATP-sensitive potassium channel opener, prolongs survival in HSPB5 R120G transgenic mice. PLoS one 6(4):e18922. doi:10.1371/journal.pone.0018922

Selcen D (2011) Myofibrillar myopathies. Neuromuscular Dis NMD 21(3):161–171. doi:10.1016/j.nmd.2010.12.007

Selcen D, Engel AG (2003) Myofibrillar myopathy caused by novel dominant negative cryAB mutations. Ann Neurol 54(6):804–810

Singh BN, Rao KS, Rao Ch M (2010) Ubiquitin-proteasome-mediated degradation and synthesis of MyoD is modulated by alphaB-crystallin, a small heat shock protein, during muscle differentiation. Biochim Biophys Acta 1803(2):288–299. doi:10.1016/j.bbamcr.2009.11.009

Smirnova E, Chebotareva N, Gurvits B (2013) Transient transformation of oligomeric structure of alpha-crystallin during its chaperone action. Int J Biol Macromol 55:62–68. doi:10.1016/j.ijbiomac.2012.12.013

Spector DL, Lamond AI (2011) Nuclear speckles. Cold Spring Harb Perspect Biol 3(2):1–12. a000646. doi:10.1101/cshperspect.a000646

Suzuki A, Sugiyama Y, Hayashi Y, Nyu-i N, Yoshida M, Nonaka I, Ishiura S, Arahata K, Ohno S (1998) MKBP, a novel member of the small heat shock protein family, binds and activates the myotonic dystrophy protein kinase. J Cell Biol 140:1113–1124

Takahashi, Yamanaka (2006) Induction of pluripotent stem cells from mouse embryonic and adult fibroblast cultures by defined factors. Cell 126(4):663–676

Tannous P, Zhu H, Johnstone JL, Shelton JM, Rajasekaran NS, Benjamin IJ, Nguyen L, Gerard RD, Levine B, Rothermel BA, Hill JA (2008) Autophagy is an adaptive response in desmin-related cardiomyopathy. Proc Natl Acad Sci U S A 105(28):9745–9750. doi:10.1073/pnas.0706802105

Taylor RP, Benjamin IJ (2005) Small heat shock proteins: a new classification scheme in mammals. J Mol Cell Cardiol 38(3):433–444. doi:10.1016/j.yjmcc.2004.12.014

Thompson HS, Scordilis SP, Clarkson PM, Lohrer WA (2001) A single bout of eccentric exercise increases Hsp27 and HSC/HSP70 in human skeletal muscle. Acta Physiol Scand 171:187–193

Vandenijssel P, Wheelock R, Prescott A, Russell P, Quinlan R (2003) Nuclear speckle localisation of the small heat shock protein alphaB-crystallin and its inhibition by the R120G cardiomyopathy-linked mutation. Exp Cell Res 287(2):249–261. doi:10.1016/s0014-4827(03)00092-2

Vicart CA, Guicheney P, Li Z, Prévost MC, Faure A, Chateau D, Chapon F, Tomé F, Dupret JM, Paulin D, Fardeau M (1998) A missense mutation in the alphaB-crystallin chaperone gene causes a desmin-related myopathy. Nat Genet 20(1):92–95

Wang X, Osinska H, Klevitsky R, Gerdes AM, Nieman M, Lorenz J, Hewett T, Robbins J (2001) Expression of R120G- B-crystallin causes aberrant desmin and B-crystallin aggregation and cardiomyopathy in mice. Circ Res 89(1):84–91. doi:10.1161/hh1301.092688

Watanabe G, Kato S, Nakata H, Ishida T, Ohuchi N, Ishioka C (2009) alphaB-crystallin: a novel p53-target gene required for p53-dependent apoptosis. Cancer Sci 100(12):2368–2375. doi:10.1111/j.1349-7006.2009.01316.x

Weintraub H (1993) The MyoD family and myogenesis- redundancy, networks, and thresholds. Cell 75:1241–1244

Whittaker R, Gude GN, Sussman MA, Gottlieb RA, Glembotski CC (2009) Kinetics of the translocation and phosphorylation of alphaB-crystallin in mouse heart mitochondria during ex vivo ischemia. Am J Physiol Heart Circ Physiol 296:H1633–H1642. doi:10.1152/ajpheart.01227.2008.-/Bcrystallin

Yan LJ, Christians EC, Liu L, Xiao X, Sohal RS, Benjamin IJ (2002) Mouse heat shock transcription factor 1 deficiency alters cardiac redox homeostasis and increases mitochondrial oxidative damage. EMBO J 21(19):5164–5172

Yusa K, Rashid ST, Strick-Marchand H, Varela I, Liu PQ, Paschon DE, Miranda E, Ordonez A, Hannan NR, Rouhani FJ, Darche S, Alexander G, Marciniak SJ, Fusaki N, Hasegawa M, Holmes MC, Di Santo JP, Lomas DA, Bradley A, Vallier L (2011) Targeted gene correction of alpha1-antitrypsin deficiency in induced pluripotent stem cells. Nature 478(7369):391–394. doi:10.1038/nature10424

Zhang H, Rajasekaran NS, Orosz A, Xiao X, Rechsteiner M, Benjamin IJ (2010) Selective degradation of aggregate-prone CryAB mutants by HSPB1 is mediated by ubiquitin-proteasome pathways. J Mol Cell Cardiol 49(6):918–930. doi:10.1016/j.yjmcc.2010.09.004

Zhang J, Klos M, Wilson GF, Herman AM, Lian X, Raval KK, Barron MR, Hou L, Soerens AG, Yu J, Palecek SP, Lyons GE, Thomson JA, Herron TJ, Jalife J, Kamp TJ (2012)

Extracellular matrix promotes highly efficient cardiac differentiation of human pluripotent stem cells: the matrix sandwich method. Circ Res 111(9):1125–1136. doi:10.1161/CIRCRESAHA.112.273144

Zhou T, Benda C, Dunzinger S, Huang Y, Ho JC, Yang J, Wang Y, Zhang Y, Zhuang Q, Li Y, Bao X, Tse HF, Grillari J, Grillari-Voglauer R, Pei D, Esteban MA (2012) Generation of human induced pluripotent stem cells from urine samples. Nat Protoc 7(12):2080–2089. doi:10.1038/nprot.2012.115

Zhu Y, Bogomolovas J, Labeit S, Granzier H (2009) Single molecule force spectroscopy of the cardiac titin N2B element: effects of the molecular chaperone alphaB-crystallin with disease-causing mutations. J Biol Chem 284(20):13914–13923. doi:10.1074/jbc.M809743200

Chapter 12
Role of Small Heat Shock Protein HspB5 in Cancer

Wilbert C. Boelens

Abstract HspB5, also called αB-crystallin, is a ubiquitous small heat shock protein (sHSP) that is strongly induced by a variety of stresses, but that also functions constitutively in multiple cell types. Extensive research has demonstrated that HspB5 acts as an ATP-independent molecular chaperone by binding unfolding proteins and protecting cells from damage due to irreversible protein aggregation. As a result of its importance in protein homeostasis HspB5 is of significant interest to many areas of cell biology, including the development of cancer. However, the molecular understanding of HspB5's role in cancer is only beginning to emerge. In this chapter an overview is given of data that provide insight into the oncogenic role of HspB5 in human cancer.

Keywords HspB5 • Crystallin • Cancer • Expression • Cell survival

12.1 Introduction

HspB5 belongs to the family of stress proteins, grouped together based on stretches of sequence homology, of which the most conserved part is the α-crystallin domain (Kappe et al. 2010). The human family of sHSPs contains ten members (HspB1–HspB10), of which HspB5 is stress inducible and therefore belongs to the family of heat shock proteins (Lanneau et al. 2010). In general, sHSPs form large oligomers with dynamic quaternary structures, which diversification likely reflects an adaptation to tolerate all kinds of stresses and to cope with different client proteins (Hochberg and Benesch 2014). The abundance of the different human sHSPs varies enormously, depending on growth conditions, developmental states, differentiation and oncogenic status of the cell (Arrigo 2012). HspB5 is remarkably highly expressed in the eye lens, where it forms mixed complexes with HspB4 (αA-crystallin), and in

W.C. Boelens (✉)
Department of Biomolecular Chemistry, Institute for Molecules and Materials and Radboud Institute for Molecular Life Sciences, Radboud University, Nijmegen, The Netherlands
e-mail: w.boelens@ncmls.ru.nl

R.M. Tanguay, L.E. Hightower (eds.), *The Big Book on Small Heat Shock Proteins*, Heat Shock Proteins 8, DOI 10.1007/978-3-319-16077-1_12

heart and skeletal muscle, where it forms mixed complexes with HspB1 (Hsp27), HspB6 (Hsp20) and HspB8. HspB5 displays ATP-independent chaperone activity by suppressing the aggregation of a large set of client proteins, thereby contributing to the balance between cell survival and cell death (Boelens 2014; Acunzo et al. 2012). Its chaperone function is regulated by phosphorylation at three different phosphorylation sites, Ser19, Ser45, and Ser59, all located within the N-terminal domain. Phosphorylation can be stimulated by various kinds of stresses, such as hyperthermia, oxidative stress or the exposure to cytotoxic drugs (Kato et al. 1998; Ito et al. 1997). Phosphorylation of HspB5 has been shown to correlate with a reduction in average oligomer size and enhanced interaction with client proteins (Peschek et al. 2013). By binding different client proteins and thereby modifying the stability and/or activity of these proteins, HspB5 is able to affect all kinds of cellular activities (Arrigo 2013). Besides HspB5 other sHSPs have been shown to play crucial roles in human cancer pathologies by interacting with pro-oncogenic client proteins, recently reviewed by Arrigo and Gilbert (Arrigo and Gibert 2014). Here, I will focus on the expression of HspB5 in cancer and how this expression may promote cancer formation.

12.2 HspB5 Expression in Cancers

One of the first reports describing the association of HspB5 expression and cancer was a study of human brain tumors (Aoyama et al. 1993). Elevated levels of HspB5 were demonstrated to occur in the more aggressive stages of gliomas. Especially the more infiltrative glioblastoma cells showed higher levels of HspB5 (Goplen et al. 2010). In breast cancer patients HspB5 expression was closely associated with advanced tumor grade progression, lymph vesicular invasion, and mortality, a phenomenon that was particularly observed in triple-negative breast tumors (Chelouche-Lev et al. 2004; Kim et al. 2011; Moyano et al. 2006). Furthermore, breast cancer patients with brain metastasis formation show higher levels of HspB5 than breast cancer patients without brain metastasis (Malin et al. 2014). Overexpression of HspB5 can transform immortalized human mammary epithelial cells that can develop invasive mammary carcinomas in nude mice with basal-like breast tumor features (Moyano et al. 2006). Consequently HspB5 can be considered as a biomarker in the diagnosis of breast cancers, especially in those with advanced grade (Ivanov et al. 2008; Sitterding et al. 2008; Tsang et al. 2012). Similarly, high levels of HspB5 are associated with low survival rate in hepatocellular carcinoma (Huang et al. 2013; Tang et al. 2009), ovarian carcinoma (Volkmann et al. 2013), clear cell renal cell carcinoma (Ho et al. 2013) and head and neck squamous cell carcinoma (van de Schootbrugge et al. 2013). In non-small-cell lung cancer only the nuclear and not the cytoplasmic localized HspB5 correlated with poor survival (Cherneva et al. 2010).

HspB5 does not correlate with poor prognosis in all cancer types analyzed and negative effects of HspB5 on cancer progression have been described as well.

In nasopharyngeal carcinomas down regulation of HspB5 has been associated with tumor progression (Lung et al. 2008). Overexpression of HspB5 in cell lines of this type of cancer suppressed progression-associated phenotypes such as loss of cell adhesion, invasion and interaction with the tumor microenvironment (Huang et al. 2012). Similarly, patients with cutaneous squamous cell carcinoma having perineural invasion, which is associated with a poor prognosis, showed lower HspB5 expression compared to patients without perineural invasion (Solares et al. 2010). Since HspB5 expression can have opposite effects in different cancer types, it is likely that the molecular mechanisms by which HspB5 influences cancer development will be diverse.

12.3 Transcription Factors Involved in the Expression of HspB5

The molecular mechanisms that regulate HspB5 transcription in normal, stress or pathological conditions are extremely complex and rely on the combinatory effects of many transcription factors (Fig. 12.1) (de Thonel et al. 2012). Multiple pathways coordinate the spatial and temporal expression of the *HspB5* gene during lens development through the action of master genes, like members of the Maf protein family and Pax6. The MyoD family proteins (muscle regulatory-binding site), regulate HspB5 expression in muscle cells by binding the E-box (Gopal-Srivastava and Piatigorsky 1994), but additional elements are required, since one E-box is not sufficient for full transcriptional activity (Gopal-Srivastava et al. 2000). The induction of HspB5 transcription by various stresses proceeds via heat shock transcription factor (HSF)-dependent mechanisms, by which also concomitant induction of other stress proteins is observed.

The mechanisms underlying deregulated HspB5 expression in cancers are poorly understood. The overexpression of HspB5 in some cancers has been reported to be, at least partly, independent of HSF regulation (de Thonel et al. 2012). In basal-like breast cancer, in which the expression of HspB5 is correlated with a poor prognosis

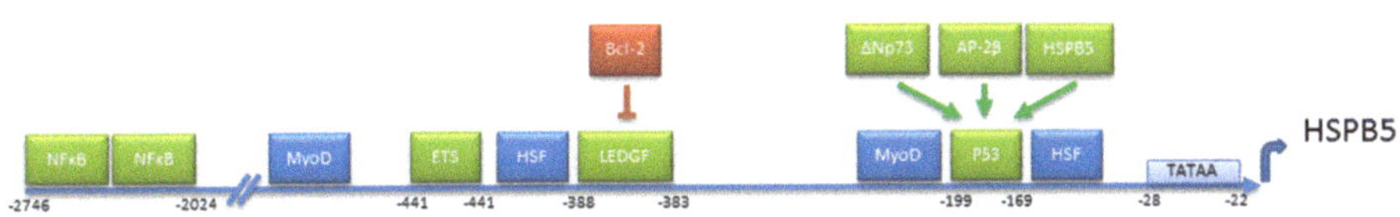

Fig. 12.1 Transcription factors able to interact with the promoter region of *HspB5*. The transcription factors known to influence the expression of HspB5 in tumors are indicated in *green*. Transcription factors that regulate the expression of HspB5 in normal tissues are indicated in *blue*. The factors above the transcription factors indirectly stimulate (*green*) or inhibit (*red*) HspB5 expression. The numbers indicate the location of transcription binding sites relative to the transcription start site. TATAA indicates the TATA box. The *HspB2* gene located upstream of *HspB5* is not shown

and may contribute to the aggressive phenotype, the proto-oncogene *Ets1*, a member of the ETS transcription factor family, appears to be involved in the upregulation of HspB5 expression. Silencing *Ets1* reduced both *HspB5* promoter activity and protein levels. Furthermore, Ets1 recognized a palindromic element corresponding to the ETS-DNA consensus motif in the human *HspB5* promoter, which is highly conserved among mammalian species (Bosman et al. 2010).

A substantial fraction of tumor cells may experience cycling hypoxia, characterized by transient episodes of hypoxia and reoxygenation (Koritzinsky and Wouters 2013). Such tumor cells are under a unique burden of stress, mediated by excessive production of reactive oxygen species (ROS). ROS has been shown to induce the expression of HspB5 (van de Schootbrugge et al. 2014). One of the transcription factors stimulated by ROS is the lens epithelial derived growth factor (LEDGF), which is able to bind to the stress response element of the *HspB5* gene, thereby stimulating its transcription (Shin et al. 2008). Remarkably, the anti-apoptotic family member Bcl-2, a pro-oncoprotein able to protect cells under various stress conditions, has been shown to reduce the activity of LEDGF in rabbit lens epithelial cells, thereby down-regulating the expression of HspB5 (Mao et al. 2001). The Bcl-2-induced down-regulation of HspB5 was shown to be responsible for a decreased protection against oxidative stress. However, the expression of another Bcl-2 family member, Bcl-2-Like 12 (Bcl2L12) showed a significant upregulation of HspB5 on mRNA and protein levels in cortical astrocytes (Stegh et al. 2008). This protein is only distantly related to the canonical Bcl-2 family members and part of its anti-apoptotic activity might be a consequence of its ability to increase the expression of HspB5. Whether Bcl2L12 enhances HspB5 expression by affecting the activity of LEDGF is not known.

The tumor suppressor protein p53 is a transcription factor that trans-activates various genes in response to DNA-damaging stress. By a search for new target genes using a cDNA microarray system, *HspB5* was identified as a gene that is trans-activated by p53 (Watanabe et al. 2009). Furthermore, a conserved response element for p53 was found in the *HspB5* promoter (Evans et al. 2010). Induction of *HspB5* by genotoxic stress could be inhibited by siRNAs targeting p53 and ectopic expression of p53-induced HspB5 mRNA and protein expression, showing that p53 is responsible for HspB5 induction in response to DNA-damaging stress. However, p53 does not seem to be sufficient for HspB5 induction. The N-terminal truncated form of p73, a p53-related transcription factor lacking a transactivation domain, the expression of which is also induced by p53, was needed as well (Evans et al. 2010). In addition, the transcription factor AP-2β was shown to upregulate the transcription of the *HspB5* gene by stimulating the activity of p53 (Hu et al. 2012). Notably, HspB5 is able to directly interact with the p53 protein and can affect the functioning of p53 (Watanabe et al. 2009; Jin et al. 2009; Liu et al. 2007). HspB5 has been shown to contribute to p53 stability, which is probably related to the chaperone activity of HspB5 (Peschek et al. 2013; Jin et al. 2009). Overexpression of HspB5 increased p53 protein and, in contrast, repression of HspB5 decreased p53 protein (Watanabe et al. 2009). These findings point to a link between HspB5 and genotoxic stress that is regulated by cooperative actions of p53.

Another transcription factor that plays a role in the oncogenic activity of HspB5 is NFκB, a key factor that regulates expression of a large number of genes involved in apoptosis, proliferation, tumorigenesis and metastasis. Loss-of-function mutations of the tumor suppressors tuberous sclerosis complex 1 or 2 (*TSC1/2*) have been shown to cause a rare multi-system genetic disease that causes benign tumors to grow in the brain and on other vital organs such as the kidneys, heart, eyes, lungs, and skin. Patients with this disease showed an increased expression of HspB5 in the afflicted organs (Wang et al. 2013). One of the transcription factors that was found to be activated in cells with inactivated *TSC1* or *TSC2* genes is NFκB. By computational analysis, two NFκB-binding sites were found within the promoter region on *HspB5* gene. Mutations in these putative NFκB-binding sites markedly attenuated the *HspB5* promoter activity in *TSC2-/-* MEFs, indicating that these sites were responsible for the upregulation of HspB5.

The studies described above have brought many insights on the action of control elements and transcription factors that operate on *HspB5* gene transcription, however further elucidation of the mechanisms of *HspB5* gene transcription is still needed to help to decipher the role of HspB5 in cancer development.

12.4 Involvement of HspB5 in Cell Survival

Several chemotherapeutic agents act through inducing cell death of neoplastic cells, which means that protective effects are disadvantageous for the outcome of the treatment. Accumulating evidence suggests that HspB5 can prevent cell death triggered by various stimuli, including hyperthermia, oxidative stress or the exposure to cytotoxic drugs (Acunzo et al. 2012). The cellular protection might be due to the prevention of intracellular damage induced by the stressors. HspB5 may prevent the aggregation or facilitate renaturation of proteins and protect cytoskeletal elements against ischemic injury or depolymerizing agents (Xi et al. 2006; Singh et al. 2007). Furthermore, HspB5 may prevent cell death by interacting with key pro-apoptotic proteins, thereby blocking apoptosis at different levels (Fig. 12.2). In human retinal pigment epithelial cell, HspB5 inhibits apoptosis induced by staurosporine by interacting with the pro-apoptotic proteins Bax and Bcl-Xs and inhibiting their translocation from the cytosol to the mitochondria (Mao et al. 2004). Also cell death induced by hydrogen peroxide and staurosporin might be inhibited this way (Mao et al. 2001; Hamann et al. 2013). Remarkably, phosphorylation at Ser59 may specifically stimulate the interaction with the anti-apoptotic protein Bcl-2 in MCF cells, thereby reducing the anti-apoptotic activity of this protein and making the cells more sensitive for the chemotherapeutic agent vinblastine (Launay et al. 2010). This effect is opposed to the phosphorylation-induced anti-apoptotic activity observed in cardiac myocytes and astrocyte cell lines (Morrison et al. 2003; Li and Reiser 2011). The pro- and anti-apoptotic activities of HspB5 indicate that the regulation of the protective mechanisms of phosphorylated HspB5 is complex and may depend on cellular background and stress conditions.

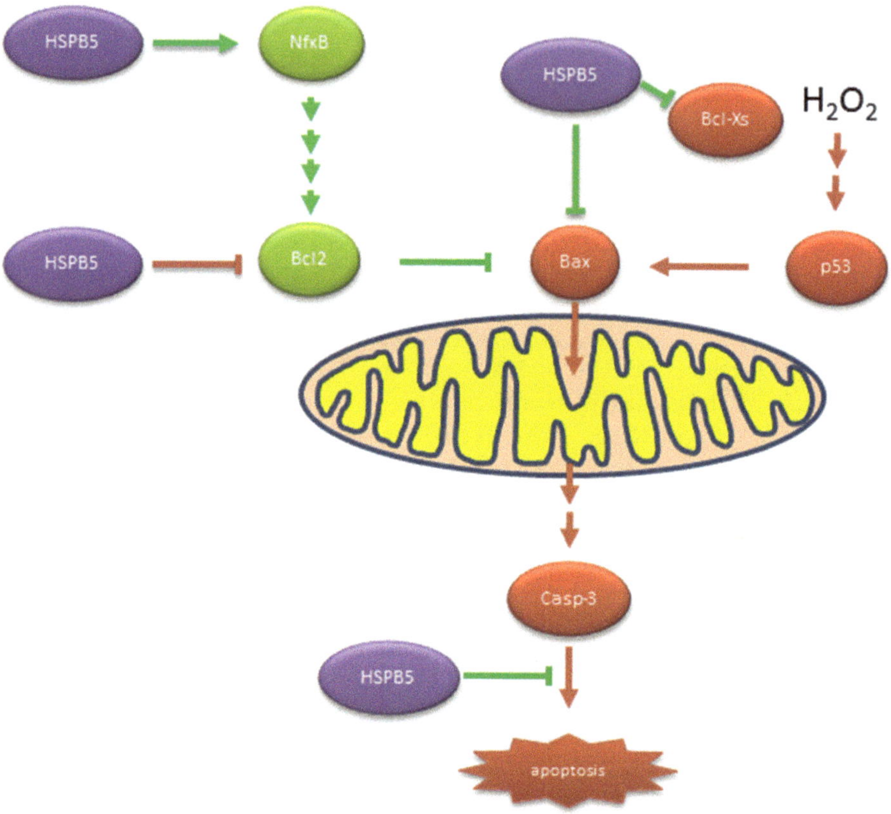

Fig. 12.2 The involvement of HspB5 in regulation of apoptosis. Schematic illustration showing the apoptotic and survival pathways controlled by HspB5. The pro-apoptotic signals are indicated in *red* and the anti-apoptotic signals in *green*

Downstream in the apoptotic pathway, HspB5 directly binds to partially processed caspase-3, thereby inhibiting the pro-apoptotic function of this protein (Kamradt et al. 2001). This interaction desensitizes cancer cells to chemotherapeutic treatments. Moreover, HspB5 may inhibit caspase-3 activation in cells that are primed for apoptosis by inactivation of the retinoblastoma tumor suppressor protein Rb. Rb plays an integral role in G1-S checkpoint control and consequently is a frequent target for inactivation in cancer. Prevention of Rb-induced apoptosis in cells may promote oncogenic transformation and thus in this context HspB5 may actually have an oncogenic activity (Petrovic et al. 2013).

In the mouse skeletal myoblast C2C12 cells HspB5 was observed to prevent apoptosis induced by hydrogen peroxide treatment (Liu et al. 2007). As described above, HspB5 is able to interact with p53 (Liu et al. 2007; Jin et al. 2009; Watanabe et al. 2009). Activated p53 can induce apoptosis by transactivating the expression of multiple pro-apoptotic genes, but also participates in apoptosis by acting directly at the mitochondrial membrane. At this location, p53 physically interacts with anti-apoptotic Bcl-2 family proteins, thereby stimulating the release of cytochrome c from the mitochondria, subsequently leading to apoptosis (Galluzzi et al. 2011). By interacting with p53, HspB5 may inhibit the oxidative-stress induced translocation of p53 from cytoplasm to mitochondria and in this way prevent induction of apoptosis (Liu et al. 2007).

In another study using C2C12 cells, treatment with TNF-α induced apoptosis, but also led to association of HspB5 with IKKβ. This interaction facilitated the degradation of phosphorylated IκBα, a prime step in NFκB activation. The ability to activate NFκB was dependent on the phosphorylation status of HspB5. This process may be stimulated by a feed-forward loop, because activated NFκB can enhance the expression of HspB5 (Wang et al. 2013). The HspB5-dependent NFκB activation protected myoblasts from TNF-α induced cytotoxicity by enhancing the expression of the anti-apoptotic protein Bcl-2.

Taken together, these studies show that HspB5 has several mechanisms to exert its anti-apoptotic activity, allowing protecting cancer cells in different ways, likely depending on cell type and stress conditions.

12.5 Involvement of HspB5 in Cell Cycle Progression

Cell cycle progression is determined by the balance of positive regulators, cyclin-dependent-protein kinases, relative to negative regulators, cyclin-dependent kinase inhibitors. The family of D-type cyclins (D1, D2, and D3) are the regulatory subunits that control the G1/S-phase transition. Of the three D-type cyclins, cyclin D1 is most frequently overexpressed in human cancer (Pestell 2013). Accumulation of cyclin D1 is tightly regulated through various mechanisms including transcription, protein localization and ubiquitin-dependent proteolysis. The inhibition of ubiquitin-dependent proteolysis of cyclin D1 is thought to be a primary mechanism of cyclin D1 overexpression in human tumors. For the proteolysis, cyclin D1 phosphorylation at Thr-286 by GSK3β is required. One of the ubiquitin ligases involved in the degradation of cyclin D1 is the Skp1-Cul1-F box (SCF) E3 ubiquitin ligase, which contains the F-box protein Fbx4 and HspB5 (Lin et al. 2006). HspB5 has been shown to directly interact with Fbx4, which was stimulated by mimicking the phosphorylation of HspB5 (den Engelsman et al. 2003). Both Fbx4 and HspB5 are responsible for SCF substrate specificity. Knockdown of either Fbx4 or HspB5 in cells reduced cyclin D1 ubiquitination and reduced cyclin D1 proteolytic turnover resulting in accelerated cell cycle progression (Lin et al. 2006). To assess the potential contribution of HspB5 and Fbx4 to cyclin D1 overexpression in human cancer, the expression of HspB5 and Fbx4 have been assessed in primary esophageal cancers where cyclin D1 is known to be overexpressed in nearly 40 % of the cases (Lin et al. 2006). Strikingly, approximately 20 % of esophageal carcinomas exhibited loss of either HspB5 or Fbx4 as determined by immunohistochemistry. Furthermore, in 15 % of the primary esophageal tumors inactivating mutations in the *Fbx4* gene were observed (Barbash and Diehl 2008). In melanoma cells the serine/threonine kinase B-RAF-MEK signaling is often hyperactivated to stimulate cell proliferation. This aberrant signaling has been shown to reduce the expression of HspB5 leading to a partial stabilization of cyclin D1 (Hu and Aplin 2010). These results show that inactivation of $SCF^{Fbx4/HspB5}$E3 ligase activity is not only restricted to esophageal cancers, a tumor where cyclin D1 is thought to be a driving force, but may also affect cell growth in other types of cancers.

12.6 Involvement of HspB5 in Modulating Tumor Neovascularization

HspB5 interacts with several important human growth factors, including vascular endothelial growth factor (VEGF) (Ghosh et al. 2007). The interaction of HspB5 with VEGF is dependent on the same region HspB5 that is involved in the binding of misfolded proteins, indicating that HspB5 may chaperone and stabilize misfolded VEGF. VEGF is an endothelial cell-specific mitogen that promotes vascular angiogenesis and is induced by hypoxic stress, a known stimulator of angiogenesis. As solid tumors grow, the inner mass become hypoxic, which will stimulate the production of VEGF to enhance neovascularization. In HspB5-deficient mice the tumor vasculature has been found to display high levels of apoptotic cells and decreased vessel formation, indicating that HspB5 is an angiogenic modulator (Dimberg et al. 2008). Furthermore, in HspB5 knock-out mice the expression of VEGF was found to remain low during retinal angiogenesis (Kase et al. 2010). A possible explanation for these observations is that increased VEGF expression results in both properly folded and misfolded VEGF in the endoplasmic reticulum (ER). The misfolded VEGF cannot be transported to the Golgi apparatus for secretion and will be exported to the cytoplasm for degradation by the ubiquitin-proteasome system. However, when HspB5 is present, it may bind and stimulate the refolding of misfolded VEGF, thereby enhancing the production of properly folded VEGF. Part of the rescued VEGF may not to be secreted, but used for the intracrine VEGF signaling, a process crucial for the vascular homeostasis (Ruan et al. 2011).

12.7 Involvement of HspB5 in Metastasis Formation

Invasion of cancer cells in other tissues is one of the main causes of death of cancer patients, but the molecular and cellular mechanisms underlying tumor metastasis are still not well understood. The spreading of cancer cells through the blood and lymph systems is a multistep process (Geiger and Peeper 2009). An important step in cell spreading is Epithelial-to-Mesenchymal Transition (EMT), a process that is characterized by the loss of polarity, the loss of epithelial markers, reorganization of the cytoskeleton and the acquisition of mesenchymal markers (Lamouille et al. 2014). EMT can be seen as the acquisition of extreme plasticity by epithelial cells. In cancer, EMT enables malignant cells to acquire a migratory phenotype and is thus associated with tumor invasiveness. In hepatocellular carcinoma (HCC) HspB5 expression has been shown to induce EMT through activation of the extracellular-regulated protein kinase (ERK) cascade (Huang et al. 2013). The up-regulation of ERK1/2 activity in this cell type may occur via the interaction of HspB5 with 14-3-3ζ protein. Both overexpression of HspB5 and 14-3-3ζ correlated with poor survival outcome of HCC, indicating that the HspB5-14-3-3ζ complex promotes HCC progression.

Another important step in cell spreading is that cells have to detach from their environment. Normally, detached cells undergo anoikis, a type of cell death induced by inappropriate loss of cell adhesion. Hence, anoikis suppression is required for tumor cells to be able to metastasize to distant sites (Tan et al. 2013). HspB5 may confer protection to cells by its protective activity (see above) and this way may help to avoid anoikis.

A next step in cell spreading concerns the invasion of neighboring tissue. To this end tumor cells must acquire the ability to migrate through the basal lamina that separates tumor mass from stroma, which often occurs at sites where the lamina is incomplete. Many proteins are involved in the process of cell migration and in metastasizing tumors mutations in genes coding for proteins that mediate the attachment between cells and their environment have been found (Guo and Giancotti 2004). Overexpression of HspB5 in human mammary epithelial cells caused loss of polarity and increased migration and invasion (Moyano et al. 2006). Furthermore, these cells formed invasive carcinomas in nude mice. The oncogenic changes were dependent on HspB5 overexpression, since knockdown of HspB5 expression suppressed the abnormal phenotype. The transformation appeared to be dependent on the phosphorylation state of HspB5, as a pseudophosphorylation mutant of HspB5, which mimics an irreversible form of stress-induced phosphorylation, did not confer neoplastic changes.

Cell motility is tightly coupled to the biochemical and mechanical properties of the actin cytoskeleton. Actin filaments are semi-flexible polymers, which in conjunction with the molecular motor myosin are able to exert or resist against force in a cellular environment. To modulate the mechanical properties, actin filaments can organize into a variety of architectures generating a diversity of cellular organizations including branched or crosslinked networks in lamellipodia, parallel bundles in filopodia, and antiparallel structures in contractile fibers. In migrating lens epithelial cells, HspB5 was found to localize to the lamellipodia (Maddala and Rao 2005). HspB5 exhibited a clear co-localization with the actin meshwork and regulatory proteins involved in actin dynamics and cell adhesion, suggesting a role for HspB5 in actin dynamics during cell migration. Localization of HspB5 to the lamellipodia appeared to depend on Ser59 phosphorylation, since inhibition of the p38 MAP kinase diminished the accumulation in lamellipodia.

Another important step in cell spreading entails intravasation into pre-existing and newly formed blood and lymph vessels. Normally, angiogenesis is rare and occurs mainly during wound healing and the female reproductive cycle. The growth of new vessels is controlled by a balance of angiogenic activators and angiogenic inhibitors (Bergers and Benjamin 2003). During the "angiogenic switch", upon which a tumor activates vascularisation, this balance is disturbed. A major player in angiogenesis is vascular endothelial growth factor (VEGF) (Roskoski 2007). Increased VEGF secretion is often correlated with metastasis formation (Hu et al. 2009) and worse outcome for the patient (Bremnes et al. 2006). HspB5 is able to enhance the production of VEGF (see above) and this may affect the intravasation of tumor cells, thereby promoting metastasis formation (van de Schootbrugge et al. 2013).

12.8 Conclusion and Future Perspective

The combined literature data described here indicates that HspB5 is involved in several vital cellular processes, finely regulating the balance between life and death of the cells by protecting cells under unfavorable conditions. Until recently, it was believed that HspB5, like most other sHsps, is a specialized chaperone whose activity was to attenuate the damages to cellular proteins by mediating their storage until they could be refolded. The recent findings show that HspB5 is involved in an impressive number of cancer-related processes. The multiple cellular functions of HspB5 might be a result of its fundamental property to interact with client proteins to control their folding. If the folding of client proteins is not adapted to cellular conditions, HspB5 can participate in their refolding, degradation or modulation of their enzyme activity. Hence, the proteins that are controlled by HspB5 could have crucial functions in the development of cancer. However, a major drawback to understand the functions of HspB5 in cancer is the fact that the cellular conditions that allow HspB5 to interact with specific client proteins are still not well characterized. Future work should be directed toward analysis and definition of HspB5 interactions with client proteins involved in crucial cellular functions, such as proteins of signaling pathways. These studies might indicate certain client proteins as useful targets for therapeutic manipulation, for example to induce cancer cell death or to sensitize them for current therapeutic approaches.

References

Acunzo J, Katsogiannou M, Rocchi P (2012) Small heat shock proteins HSP27 (HspB1), alphaB-crystallin (HspB5) and HSP22 (HspB8) as regulators of cell death. Int J Biochem Cell Biol 44(10):1622–1631. doi:10.1016/j.biocel.2012.04.002

Aoyama A, Steiger RH, Frohli E, Schafer R, von Deimling A, Wiestler OD, Klemenz R (1993) Expression of alpha B-crystallin in human brain tumors. Int J Cancer 55(5):760–764

Arrigo AP (2012) Pathology-dependent effects linked to small heat shock proteins expression: an update. Scientifica 2012:185641. doi:10.6064/2012/185641

Arrigo AP (2013) Human small heat shock proteins: protein interactomes of homo- and hetero-oligomeric complexes: an update. FEBS Lett 587(13):1959–1969. doi:10.1016/j.febslet.2013.05.011

Arrigo AP, Gibert B (2014) HspB1, HspB5 and HspB4 in human cancers: potent oncogenic role of some of their client proteins. Cancers (Basel) 6(1):333–365. doi:10.3390/cancers6010333

Barbash O, Diehl JA (2008) SCF(Fbx4/alphaB-crystallin) E3 ligase: when one is not enough. Cell Cycle 7(19):2983–2986

Bergers G, Benjamin LE (2003) Tumorigenesis and the angiogenic switch. Nat Rev Cancer 3(6):401–410. doi:10.1038/nrc1093

Boelens WC (2014) Cell biological roles of alphaB-crystallin. Prog Biophys Mol Biol. doi:10.1016/j.pbiomolbio.2014.02.005

Bosman JD, Yehiely F, Evans JR, Cryns VL (2010) Regulation of alphaB-crystallin gene expression by the transcription factor Ets1 in breast cancer. Breast Cancer Res Treat 119(1):63–70. doi:10.1007/s10549-009-0330-4

Bremnes RM, Camps C, Sirera R (2006) Angiogenesis in non-small cell lung cancer: the prognostic impact of neoangiogenesis and the cytokines VEGF and bFGF in tumours and blood. Lung Cancer 51(2):143–158. doi:10.1016/j.lungcan.2005.09.005

Chelouche-Lev D, Kluger HM, Berger AJ, Rimm DL, Price JE (2004) alphaB-crystallin as a marker of lymph node involvement in breast carcinoma. Cancer 100(12):2543–2548. doi:10.1002/cncr.20304

Cherneva R, Petrov D, Georgiev O, Slavova Y, Toncheva D, Stamenova M, Trifonova N (2010) Expression profile of the small heat-shock protein alpha-B-crystallin in operated-on non-small-cell lung cancer patients: clinical implication. Eur J Cardiothorac Surg 37(1):44–50. doi:10.1016/j.ejcts.2009.06.038

de Thonel A, Le Mouel A, Mezger V (2012) Transcriptional regulation of small HSP-HSF1 and beyond. Int J Biochem Cell Biol 44(10):1593–1612. doi:10.1016/j.biocel.2012.06.012

den Engelsman J, Keijsers V, de Jong WW, Boelens WC (2003) The small heat-shock protein alpha B-crystallin promotes FBX4-dependent ubiquitination. J Biol Chem 278(7):4699–4704. doi:10.1074/jbc.M211403200

Dimberg A, Rylova S, Dieterich LC, Olsson AK, Schiller P, Wikner C, Bohman S, Botling J, Lukinius A, Wawrousek EF, Claesson-Welsh L (2008) alphaB-crystallin promotes tumor angiogenesis by increasing vascular survival during tube morphogenesis. Blood 111(4):2015–2023. doi:10.1182/blood-2007-04-087841

Evans JR, Bosman JD, Brown-Endres L, Yehiely F, Cryns VL (2010) Induction of the small heat shock protein alphaB-crystallin by genotoxic stress is mediated by p53 and p73. Breast Cancer Res Treat 122(1):159–168. doi:10.1007/s10549-009-0542-7

Galluzzi L, Morselli E, Kepp O, Vitale I, Pinti M, Kroemer G (2011) Mitochondrial liaisons of p53. Antioxid Redox Signal 15(6):1691–1714. doi:10.1089/ars.2010.3504

Geiger TR, Peeper DS (2009) Metastasis mechanisms. Biochim Biophys Acta 1796(2):293–308. doi:10.1016/j.bbcan.2009.07.006

Ghosh JG, Shenoy AK Jr, Clark JI (2007) Interactions between important regulatory proteins and human alphaB crystallin. Biochemistry 46(21):6308–6317. doi:10.1021/bi700149h

Gopal-Srivastava R, Piatigorsky J (1994) Identification of a lens-specific regulatory region (LSR) of the murine alpha B-crystallin gene. Nucleic Acids Res 22(7):1281–1286

Gopal-Srivastava R, Kays WT, Piatigorsky J (2000) Enhancer-independent promoter activity of the mouse alphaB-crystallin/small heat shock protein gene in the lens and cornea of transgenic mice. Mech Dev 92(2):125–134

Goplen D, Bougnaud S, Rajcevic U, Boe SO, Skaftnesmo KO, Voges J, Enger PO, Wang J, Tysnes BB, Laerum OD, Niclou S, Bjerkvig R (2010) alphaB-crystallin is elevated in highly infiltrative apoptosis-resistant glioblastoma cells. Am J Pathol 177(4):1618–1628. doi:10.2353/ajpath.2010.090063

Guo W, Giancotti FG (2004) Integrin signalling during tumour progression. Nat Rev Mol Cell Biol 5(10):816–826. doi:10.1038/nrm1490

Hamann S, Metrailler S, Schorderet DF, Cottet S (2013) Analysis of the cytoprotective role of alpha-crystallins in cell survival and implication of the alphaA-crystallin C-terminal extension domain in preventing Bax-induced apoptosis. PLoS One 8(2):e55372. doi:10.1371/journal.pone.0055372

Ho PY, Chueh SC, Chiou SH, Wang SM, Lin WC, Lee IL, Yang HY, Peng HC, Lai MK (2013) AlphaB-crystallin in clear cell renal cell carcinoma: tumor progression and prognostic significance. Urol Oncol 31(7):1367–1377. doi:10.1016/j.urolonc.2012.01.015

Hochberg GK, Benesch JL (2014) Dynamical structure of alphaB-crystallin. Prog Biophys Mol Biol. doi:10.1016/j.pbiomolbio.2014.03.003

Hu R, Aplin AE (2010) alphaB-crystallin is mutant B-RAF regulated and contributes to cyclin D1 turnover in melanocytic cells. Pigment Cell Melanoma Res 23(2):201–209. doi:10.1111/j.1755-148X.2010.00668.x

Hu Z, Fan C, Livasy C, He X, Oh DS, Ewend MG, Carey LA, Subramanian S, West R, Ikpatt F, Olopade OI, van de Rijn M, Perou CM (2009) A compact VEGF signature associated with distant metastases and poor outcomes. BMC Med 7:9. doi:10.1186/1741-7015-7-9

Hu X, Wang L, Sun W, Xiao L, Wu Y, Zhuo Y, Nie D, Zhou J, Zhang J (2012) AP-2beta enhances p53-mediated transcription of the alphaB-crystallin gene through stabilizing p53. Mol Biol Rep 39(1):209–214. doi:10.1007/s11033-011-0727-0

Huang Z, Cheng Y, Chiu PM, Cheung FM, Nicholls JM, Kwong DL, Lee AW, Zabarovsky ER, Stanbridge EJ, Lung HL, Lung ML (2012) Tumor suppressor Alpha B-crystallin (CRYAB) associates with the cadherin/catenin adherens junction and impairs NPC progression-associated properties. Oncogene 31(32):3709–3720. doi:10.1038/onc.2011.529

Huang XY, Ke AW, Shi GM, Zhang X, Zhang C, Shi YH, Wang XY, Ding ZB, Xiao YS, Yan J, Qiu SJ, Fan J, Zhou J (2013) alphaB-crystallin complexes with 14-3-3zeta to induce epithelial-mesenchymal transition and resistance to sorafenib in hepatocellular carcinoma. Hepatology 57(6):2235–2247. doi:10.1002/hep.26255

Ito H, Okamoto K, Nakayama H, Isobe T, Kato K (1997) Phosphorylation of alphaB-crystallin in response to various types of stress. J Biol Chem 272(47):29934–29941

Ivanov O, Chen F, Wiley EL, Keswani A, Diaz LK, Memmel HC, Rademaker A, Gradishar WJ, Morrow M, Khan SA, Cryns VL (2008) alphaB-crystallin is a novel predictor of resistance to neoadjuvant chemotherapy in breast cancer. Breast Cancer Res Treat 111(3):411–417. doi:10.1007/s10549-007-9796-0

Jin X, Moskophidis D, Hu Y, Phillips A, Mivechi NF (2009) Heat shock factor 1 deficiency via its downstream target gene alphaB-crystallin (Hspb5) impairs p53 degradation. J Cell Biochem 107(3):504–515. doi:10.1002/jcb.22151

Kamradt MC, Chen F, Cryns VL (2001) The small heat shock protein alpha B-crystallin negatively regulates cytochrome c- and caspase-8-dependent activation of caspase-3 by inhibiting its autoproteolytic maturation. J Biol Chem 276(19):16059–16063. doi:10.1074/jbc.C100107200

Kappe G, Boelens WC, de Jong WW (2010) Why proteins without an alpha-crystallin domain should not be included in the human small heat shock protein family HSPB. Cell Stress Chaperones 15(4):457–461. doi:10.1007/s12192-009-0155-4

Kase S, He S, Sonoda S, Kitamura M, Spee C, Wawrousek E, Ryan SJ, Kannan R, Hinton DR (2010) alphaB-crystallin regulation of angiogenesis by modulation of VEGF. Blood 115(16):3398–3406. doi:10.1182/blood-2009-01-197095

Kato K, Ito H, Kamei K, Inaguma Y, Iwamoto I, Saga S (1998) Phosphorylation of alphaB-crystallin in mitotic cells and identification of enzymatic activities responsible for phosphorylation. J Biol Chem 273(43):28346–28354

Kim HS, Lee Y, Lim YA, Kang HJ, Kim LS (2011) alphaB-Crystallin is a novel oncoprotein associated with poor prognosis in breast cancer. J Breast Cancer 14(1):14–19. doi:10.4048/jbc.2011.14.1.14

Koritzinsky M, Wouters BG (2013) The roles of reactive oxygen species and autophagy in mediating the tolerance of tumor cells to cycling hypoxia. Semin Radiat Oncol 23(4):252–261. doi:10.1016/j.semradonc.2013.05.006

Lamouille S, Xu J, Derynck R (2014) Molecular mechanisms of epithelial-mesenchymal transition. Nat Rev Mol Cell Biol 15(3):178–196. doi:10.1038/nrm3758

Lanneau D, Wettstein G, Bonniaud P, Garrido C (2010) Heat shock proteins: cell protection through protein triage. Sci World J 10:1543–1552. doi:10.1100/tsw.2010.152

Launay N, Tarze A, Vicart P, Lilienbaum A (2010) Serine 59 phosphorylation of {alpha} B-crystallin down-regulates its anti-apoptotic function by binding and sequestering Bcl-2 in breast cancer cells. J Biol Chem 285(48):37324–37332. doi:10.1074/jbc.M110.124388

Li R, Reiser G (2011) Phosphorylation of Ser45 and Ser59 of alphaB-crystallin and p38/extracellular regulated kinase activity determine alphaB-crystallin-mediated protection of rat brain astrocytes from C2-ceramide- and staurosporine-induced cell death. J Neurochem 118(3):354–364. doi:10.1111/j.1471-4159.2011.07317.x

Lin DI, Barbash O, Kumar KG, Weber JD, Harper JW, Klein-Szanto AJ, Rustgi A, Fuchs SY, Diehl JA (2006) Phosphorylation-dependent ubiquitination of cyclin D1 by the SCF(FBX4-alphaB crystallin) complex. Mol Cell 24(3):355–366. doi:10.1016/j.molcel.2006.09.007

Liu S, Li J, Tao Y, Xiao X (2007) Small heat shock protein alphaB-crystallin binds to p53 to sequester its translocation to mitochondria during hydrogen peroxide-induced apoptosis. Biochem Biophys Res Commun 354(1):109–114. doi:10.1016/j.bbrc.2006.12.152

Lung HL, Lo CC, Wong CC, Cheung AK, Cheong KF, Wong N, Kwong FM, Chan KC, Law EW, Tsao SW, Chua D, Sham JS, Cheng Y, Stanbridge EJ, Robertson GP, Lung ML (2008) Identification of tumor suppressive activity by irradiation microcell-mediated chromosome transfer and involvement of alpha B-crystallin in nasopharyngeal carcinoma. Int J Cancer 122(6):1288–1296. doi:10.1002/ijc.23259

Maddala R, Rao VP (2005) alpha-Crystallin localizes to the leading edges of migrating lens epithelial cells. Exp Cell Res 306(1):203–215. doi:10.1016/j.yexcr.2005.01.026

Malin D, Strekalova E, Petrovic V, Deal AM, Al Ahmad A, Adamo B, Miller CR, Ugolkov A, Livasy C, Fritchie K, Hamilton E, Blackwell K, Geradts J, Ewend M, Carey L, Shusta EV, Anders CK, Cryns VL (2014) alphaB-crystallin: a novel regulator of breast cancer metastasis to the brain. Clin Cancer Res 20(1):56–67. doi:10.1158/1078-0432.CCR-13-1255

Mao YW, Xiang H, Wang J, Korsmeyer S, Reddan J, Li DW (2001) Human bcl-2 gene attenuates the ability of rabbit lens epithelial cells against H2O2-induced apoptosis through down-regulation of the alpha B-crystallin gene. J Biol Chem 276(46):43435–43445. doi:10.1074/jbc.M102195200

Mao YW, Liu JP, Xiang H, Li DW (2004) Human alphaA- and alphaB-crystallins bind to Bax and Bcl-X(S) to sequester their translocation during staurosporine-induced apoptosis. Cell Death Differ 11(5):512–526. doi:10.1038/sj.cdd.4401384

Morrison LE, Hoover HE, Thuerauf DJ, Glembotski CC (2003) Mimicking phosphorylation of alphaB-crystallin on serine-59 is necessary and sufficient to provide maximal protection of cardiac myocytes from apoptosis. Circ Res 92(2):203–211

Moyano JV, Evans JR, Chen F, Lu M, Werner ME, Yehiely F, Diaz LK, Turbin D, Karaca G, Wiley E, Nielsen TO, Perou CM, Cryns VL (2006) AlphaB-crystallin is a novel oncoprotein that predicts poor clinical outcome in breast cancer. J Clin Invest 116(1):261–270. doi:10.1172/JCI25888

Peschek J, Braun N, Rohrberg J, Back KC, Kriehuber T, Kastenmuller A, Weinkauf S, Buchner J (2013) Regulated structural transitions unleash the chaperone activity of alphaB-crystallin. Proc Natl Acad Sci U S A 110(40):E3780–E3789. doi:10.1073/pnas.1308898110

Pestell RG (2013) New roles of cyclin D1. Am J Pathol 183(1):3–9. doi:10.1016/j.ajpath.2013.03.001

Petrovic V, Malin D, Cryns VL (2013) alphaB-crystallin promotes oncogenic transformation and inhibits caspase activation in cells primed for apoptosis by Rb inactivation. Breast Cancer Res Treat 138(2):415–425. doi:10.1007/s10549-013-2465-6

Roskoski R Jr (2007) Vascular endothelial growth factor (VEGF) signaling in tumor progression. Crit Rev Oncol Hematol 62(3):179–213. doi:10.1016/j.critrevonc.2007.01.006

Ruan Q, Han S, Jiang WG, Boulton ME, Chen ZJ, Law BK, Cai J (2011) alphaB-crystallin, an effector of unfolded protein response, confers anti-VEGF resistance to breast cancer via maintenance of intracrine VEGF in endothelial cells. Mol Cancer Res 9(12):1632–1643. doi:10.1158/1541-7786.MCR-11-0327

Shin JH, Piao CS, Lim CM, Lee JK (2008) LEDGF binding to stress response element increases alphaB-crystallin expression in astrocytes with oxidative stress. Neurosci Lett 435(2):131–136. doi:10.1016/j.neulet.2008.02.029

Singh BN, Rao KS, Ramakrishna T, Rangaraj N, Rao Ch M (2007) Association of alphaB-crystallin, a small heat shock protein, with actin: role in modulating actin filament dynamics in vivo. J Mol Biol 366(3):756–767. doi:10.1016/j.jmb.2006.12.012

Sitterding SM, Wiseman WR, Schiller CL, Luan C, Chen F, Moyano JV, Watkin WG, Wiley EL, Cryns VL, Diaz LK (2008) AlphaB-crystallin: a novel marker of invasive basal-like and metaplastic breast carcinomas. Ann Diagn Pathol 12(1):33–40. doi:10.1016/j.anndiagpath.2007.02.004

Solares CA, Boyle GM, Brown I, Parsons PG, Panizza B (2010) Reduced alphaB-crystallin staining in perineural invasion of head and neck cutaneous squamous cell carcinoma. Otolaryngol Head Neck Surg 142(3 Suppl 1):S15–S19. doi:10.1016/j.otohns.2009.12.001

Stegh AH, Kesari S, Mahoney JE, Jenq HT, Forloney KL, Protopopov A, Louis DN, Chin L, DePinho RA (2008) Bcl2L12-mediated inhibition of effector caspase-3 and caspase-7 via distinct mechanisms in glioblastoma. Proc Natl Acad Sci U S A 105(31):10703–10708. doi:10.1073/pnas.0712034105

Tan K, Goldstein D, Crowe P, Yang JL (2013) Uncovering a key to the process of metastasis in human cancers: a review of critical regulators of anoikis. J Cancer Res Clin Oncol 139(11):1795–1805. doi:10.1007/s00432-013-1482-5

Tang Q, Liu YF, Zhu XJ, Li YH, Zhu J, Zhang JP, Feng ZQ, Guan XH (2009) Expression and prognostic significance of the alpha B-crystallin gene in human hepatocellular carcinoma. Hum Pathol 40(3):300–305. doi:10.1016/j.humpath.2008.09.002

Tsang JY, Lai MW, Wong KH, Chan SK, Lam CC, Tsang AK, Yu AM, Tan PH, Tse GM (2012) alphaB-crystallin is a useful marker for triple negative and basal breast cancers. Histopathology 61(3):378–386. doi:10.1111/j.1365-2559.2012.04234.x

van de Schootbrugge C, Bussink J, Span PN, Sweep FC, Grenman R, Stegeman H, Pruijn GJ, Kaanders JH, Boelens WC (2013) alphaB-crystallin stimulates VEGF secretion and tumor cell migration and correlates with enhanced distant metastasis in head and neck squamous cell carcinoma. BMC Cancer 13:128. doi:10.1186/1471-2407-13-128

van de Schootbrugge C, Schults EM, Bussink J, Span PN, Grenman R, Pruijn GJ, Kaanders JH, Boelens WC (2014) Effect of hypoxia on the expression of alphaB-crystallin in head and neck squamous cell carcinoma. BMC Cancer 14:252. doi:10.1186/1471-2407-14-252

Volkmann J, Reuning U, Rudelius M, Hafner N, Schuster T, Becker VRA, Weimer J, Hilpert F, Kiechle M, Durst M, Arnold N, Schmalfeldt B, Meindl A, Ramser J (2013) High expression of crystallin alphaB represents an independent molecular marker for unfavourable ovarian cancer patient outcome and impairs TRAIL- and cisplatin-induced apoptosis in human ovarian cancer cells. Int J Cancer J Int Cancer 132(12):2820–2832. doi:10.1002/ijc.27975

Wang F, Chen X, Li C, Sun Q, Chen Y, Wang Y, Peng H, Liu Z, Chen R, Liu K, Yan H, Ye BH, Kwiatkowski DJ, Zhang H (2013) Pivotal role of augmented alphaB-crystallin in tumor development induced by deficient TSC1/2 complex. Oncogene. doi:10.1038/onc.2013.401

Watanabe G, Kato S, Nakata H, Ishida T, Ohuchi N, Ishioka C (2009) alphaB-crystallin: a novel p53-target gene required for p53-dependent apoptosis. Cancer Sci 100(12):2368–2375. doi:10.1111/j.1349-7006.2009.01316.x

Xi JH, Bai F, McGaha R, Andley UP (2006) Alpha-crystallin expression affects microtubule assembly and prevents their aggregation. FASEB J 20(7):846–857. doi:10.1096/fj.05-5532com

Chapter 13
Small Heat Shock Proteins and Fibrosis

Pierre-Simon Bellaye, Olivier Burgy, Julien Colas, Sebastien Causse, Carmen Garrido, and Philippe Bonniaud

Abstract Small heat shock proteins (sHSP) are involved in many essential cellular mechanisms both in physiologic and pathologic conditions. HSP27 (HSPB1), αB-crystallin (HSPB5) and HSP20 (HSPB6), the most studied members, are stress-inducible chaperones with an anti-aggregation function. They have been shown to inhibit apoptosis by interacting with proteins involved in programmed cell death such as cytochrome c or caspases, to have anti-oxidant properties and/or to modulate protein homeostasis by participating in the proteasomal degradation of specific proteins under stress conditions. Heat shock proteins accumulate in cancer cells and this overexpression is needed for the cancer cells' survival. Accordingly, the inhibition of heat shock proteins such as HSP27 is an emerging strategy in cancer therapy (already is in phase II clinical trials). Fibrogenesis and cancer share several properties as both pathologies are characterized by genetic alterations, uncontrolled cell proliferation, altered cell interaction and communication and tissue invasion.

The current review will discuss the involvement of HSP27, αB-crystallin and HSP20 in the process of fibrogenesis and in established organ fibrosis. We will dissect the impact of their function inducing the recruitment of inflammatory cells, the

P.-S. Bellaye • O. Burgy • J. Colas • S. Causse
INSERM U866 équipe labellisée Ligue contre le cancer, 21000 Dijon, France

Faculty of Medicine and Pharmacy, University of Burgundy, Dijon 21033, France

C. Garrido (✉)
INSERM U866 équipe labellisée Ligue contre le cancer, 21000 Dijon, France

Faculty of Medicine and Pharmacy, University of Burgundy, Dijon 21033, France

Anticancer centre Georges François Leclerc, Dijon, France

Faculty of Medicine and Pharmacy, INSERM UMR866,
7, boulevard Jeanne d'Arc, 21079 Dijon, France
e-mail: cgarrido@u-bourgogne.fr

P. Bonniaud
INSERM U866 équipe labellisée Ligue contre le cancer, 21000 Dijon, France

Faculty of Medicine and Pharmacy, University of Burgundy, Dijon 21033, France

Service de Pneumologie, CHU Dijon, 21079 Dijon, France

R.M. Tanguay, L.E. Hightower (eds.), *The Big Book on Small Heat Shock Proteins*, Heat Shock Proteins 8, DOI 10.1007/978-3-319-16077-1_13

secretion of profibrotic cytokines, mainly Transforming Growth Factor-β1 (TGF-β1), and modulating oxidative stress, apoptosis and/or the proteasome system in the fibrotic process.

Keywords HSP27 • Alpha-B-crystallin • HSP20 • Fibrosis • Collagen • TGFbeta • EMT

13.1 Heat Shock Proteins

Cells are constantly exposed to several aggressions caused by many environmental factors (heat shock, oxidative stress, UV), exposure to pharmacologic toxic agents (heavy metals, alcohol, chemotherapy) or certain pathological conditions. Protective mechanisms are established in cells in order to maintain their functions. The "heat shock response" was first highlighted in 1964 in drosophila by Ritossa (1964; Tissieres et al. 1974) and is characterized by the expression of a particular protein superfamily, the Heat Shock Proteins (HSPs). HSPs are conserved cellular proteins present in many species such as yeast, bacteria, plants, animals and humans. See Kampinga et al. (2009) for HUGO-approved nomenclature of human HSPs.

In normal conditions HSPs are able to interact with mature and immature proteins in order to help the folding of newly synthesized proteins, stabilize them, help them to reach organelles such as the mitochondria or the endoplasmic reticulum, and participate in the protein turn-over (Gething and Sambrook 1992). During stress, proper folding of proteins can be affected, leading to a loss of protein function. The accumulation of misfolded proteins enhances the formation of aggregates, which disrupts the cellular machinery and can lead to cell death. The stress-inducible proteins ensures the correct folding of misfolded proteins or, if refolding is impaired, they direct them towards the proteasome system where the defective proteins are degraded (Burel et al. 1992; Lanneau et al. 2010). While HSPs already form up to 2–3 % of total cellular proteins at their baseline level, their expression is strongly induced upon stress. This induction is mediated by specific transcription factors, called heat shock factors (HSF), the best characterized of these being HSF-1, which has the ability to bind DNA at specific sites called Heat Shock Elements (HSE) found in the promoters of HSP genes (Arrigo 2005). In normal conditions, HSF-1 is sequestrated in the cytoplasm by HSP70, HSP90 and several co-chaperones (Voellmy 2006; Conde et al. 2009). In stress conditions, the HSPs release HSF-1 to bind misfolded proteins, allowing the phosphorylation, activation and trimerization of HSF-1, which migrates into the nucleus, interacts with HSE and finally induces the transcription of HSP genes.

In mammals, heat shock proteins are grouped into six major families according to their molecular weight: HSP100, HSP90, HSP70, HSP60, HSP40 and the small Heat Shock Protein family (sHSP, (Vos et al. 2008)). Although HSPs share many common properties, each family has specificities in regards to their cellular localization, their dependence on ATP, their substrate specificity and the type of diseases in which they may be involved.

sHSP are low molecular weight proteins (15–30 kDa) that possess a highly dynamic structure. sHSP can modulate their quaternary structure and act as monomers or dimers but they also can form, through interactions with themselves or other sHSPs, homo or hetero-oligomers containing up to 50 subunits (Benesch et al. 2008). Structurally, all sHSP share a C-terminal domain, referred to as alpha-crystallin domain, constituting an intermolecular interaction site (Hayes et al. 2009). HSP27, αB-crystallin and HSP20 are the most studied members of sHSP and have strong anti-aggregation chaperone activity, although their mode of action to prevent aggregation is distinct (McDonald et al. 2012).

sHSPs are involved in many essential cellular mechanisms both in physiologic and pathologic conditions. For instance, several studies have shown that sHSPs could inhibit apoptosis by interacting with proteins involved in programmed cell death such as cytochrome c, caspases or Apaf-1 (Apoptotic protease-activating factor 1 (Bruey et al. 2000; Didelot et al. 2006; Kamradt et al. 2002). sHSPs can also be involved in cytokine induction, inflammation (Tamura et al. 2012; Salinthone et al. 2008) and cytoskeleton related diseases (Wettstein et al. 2012).

sHSPs have been largely described as implicated in the pathogenesis of cancer (Jego et al. 2013). sHSPs take part in several molecular mechanisms of cancerous cells, including proliferation (Hayashi et al. 2012), invasion (Lemieux et al. 1997), angiogenesis (Thuringer et al. 2013) and metastasis formation (Pavan et al. 2014; van de Schootbrugge et al. 2013; Malin et al. 2014). In several reports, sHSPs appeared beneficial for cancerous cells and thereby deleterious for patients affected by a wide range of cancer types (Rappa et al. 2012; Ma et al. 2013; Ruan et al. 2011; Ivanov et al. 2008). sHSP inhibition thus appears to be of great interest in order to improve the efficiency of chemotherapy and disease outcome (Jakubowicz-Gil et al. 2013; Gibert et al. 2013). Fibrogenesis and cancer share several properties as both pathologies are characterized by genetic alterations, uncontrolled cell proliferation, altered cell interaction and communication and tissue invasion (Vancheri et al. 2010). Besides, processes in which sHSPs have been implicated such as the recruitment of inflammatory cells, abundant secretion of profibrotic cytokines, mainly Transforming Growth Factor-β1 (TGF-β1), increase in oxidative stress, apoptosis and degradation via the proteasome system, are all events that occur during fibrogenesis suggesting a role for sHSP.

The current review will discuss the involvement of the sHSP HSP27, HSP20 and αB-crystallin in the process of fibrogenesis and in established organ fibrosis.

13.2 The Fibrotic Processes

Fibrosis is an essential process of tissue healing, promoting wound repair, re-epithelialization and restoration of the normal function of the affected organs in cases of chronic injury. This protective mechanism can be highly deleterious when it is out of control, leading to organ failure and diseases such as liver, cardiac, renal or pulmonary fibrosis. Whereas the process of fibrogenesis can be slightly different depending on the affected organ, the major steps remain identical (Kisseleva and

Brenner 2008; Calabresi et al. 2007). Epithelial cell injury and apoptosis appears to be the triggering events of the process. Epithelial cell death results in the recruitment of inflammatory cells and an important secretion of pro-fibrotic cytokines such as TGF-β1 (Fig. 13.1). This pro-fibrotic environment leads to the activation of neighboring cells and contributes to the progression of fibrosis. One of the major events during fibrosis is the abnormal and massive increase in extracellular matrix (ECM) deposition with notably collagen accumulation (Fig. 13.1). Myofibroblasts are considered to be the key cells of the fibrotic process since they are essentially responsible for ECM synthesis and are associated with disease progression (Phan 2002). Several hypothesis have been proposed to explain the origin of these aggressive cells. Resident fibroblasts and circulating mesenchymal cells derived from bone marrow (fibrocytes) are thought to contribute to the pool of myofibroblasts and ECM deposition (Kisseleva and Brenner 2008; Bucala et al. 1994) under TGF-β1 stimulation. Epithelial cells are also major players in fibrogenesis since they can trans-differentiate and acquire a myofibroblastic phenotype, and thus contribute to the accumulation of ECM in the lung tissue. This transformation, called Epithelial-to-Mesenchymal Transition (EMT), is initiated by Transforming Growth Factor-β1 (TGF-β1), a growth factor essential to fibrosis (Klugman et al. 2012; Carew et al. 2012). While during normal wound healing myofibroblasts are necessary for wound contraction and closure, their persistent activation and apoptosis inhibition cause

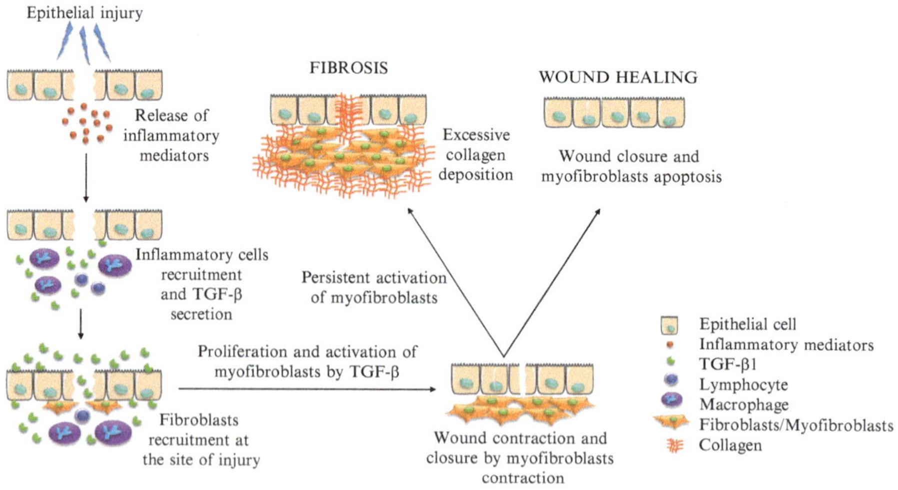

Fig. 13.1 Wound healing vs fibrosis. Injury of the epithelium causes the release of inflammatory mediators that induce the recruitment of immune cells such as lymphocytes and macrophages at the site of injury. These immune cells secrete several cytokines including TGF-β1 that promotes the activation of neighbouring cells which, in turn, will secrete pro-fibrotic mediators. This pro-fibrotic environment allows the recruitment and the differentiation of fibroblasts/myofibroblasts which enable the closure of the wound. Myofibroblasts are then eliminated by apoptosis. However, in case of chronic injury and persistent activation of myofibroblasts, the accumulation of extracellular matrix becomes deleterious and leads to fibrosis

abnormal and excessive wound healing (Fig. 13.1). The fibrotic process is thus maintained by both the apoptosis of epithelial cells which leads to the production of pro-fibrotic cytokines, mainly TGF-β1, but also by the inhibition of apoptosis in myofibroblasts, which produces an excess of ECM causing the disease. This "paradox" has been widely discussed in the literature and the key role of TGF-β1 has been confirmed (Thannickal and Horowitz 2006; Drakopanagiotakis et al. 2008). Although TGF-β1 inhibits cellular proliferation in several cell types, it stimulates mesenchymal cell proliferation, induces the secretion of ECM, inhibits apoptosis in myofibroblasts and, therefore, is strongly pro-fibrotic (Border and Noble 1994). In addition, the ability of TGF-β1 to induce epithelial cell apoptosis may be another mechanism whereby fibrosis is promoted, since impaired epithelial repair is increasingly recognized as an important mechanism in fibrogenesis (Barbas-Filho et al. 2001). Furthermore, TGF-β1 levels are increased in fibrotic lung tissue from patients affected with pulmonary fibrosis (Coker et al. 2001). The induction of severe prolonged pulmonary fibrosis in rodents by overexpression of active TGF-β1 in the lung is another confirmation of its central role in fibrogenesis (Sime et al. 1997).

TGF-β is a 25 kDa protein consisting of two identical 12.5 kDa subunits joined covalently by disulphide bonds. There are, at least, three mammalian isoforms of TGF-β, designated β1, β2 and β3 which all mediate signal through the same surface receptors. The TGF-β1 isoform is the most commonly secreted isoform, and is consistently found to be upregulated in fibrotic diseases (Khalil et al. 1996). TGF-β signaling occurs essentially through two main receptors: TGF-β receptors type I and II, TβRI and TβRII respectively. These receptors are serine/threonine kinases. TβRII is constitutively active and after binding of its ligand, activates TβRI via phosphorylation (Wrana et al. 1994). In turn, TβRI activates a large number of intracellular signaling pathways through kinase activation, including the Smad pathway essential for fibrogenesis (Massague and Wotton 2000; Wells 2000). Smad proteins are direct effectors of the activated TGF-β receptor complex and mediate signaling from cell surface receptors to the nucleus (Eickelberg 2001; Schiller et al. 2004). They can be grouped in three classes: Receptor-activated Smad (R-Smad: Smad1, 2, 3, 5 and 8); common-mediator Smad (Co-Smad: Smad4) and inhibitory Smad (I-Smad: Smad6 and 7). Transphosphorylation of TβRI activates its kinase domain which may then phosphorylate R-Smads (Fig. 13.2). Once activated, R-Smads bind to Smad4 and translocate to the nucleus in order to regulate the transcription of a large number of genes. The regulation of R-Smads activation is mediated in the cytoplasm by I-Smads that prevent binding and phosphorylation of R-Smad by activated TGF-β receptors. Protein levels of R-Smads are regulated in part by ubiquitin-proteasome-mediated degradation. Smurf proteins (Smurf 1 and 2) are E3-ubiquitin ligases able to ubiquitinate R-Smads and direct them to the proteasome system (Kavsak et al. 2000, Fig. 13.2). Smad3 appears to be a crucial element in the signal transduction pathways involved in wound healing and fibrosis (Roberts 2002; Roberts et al. 2001; Verrecchia et al. 2001). Indeed, Smad3 deficient mice are protected against radiation-induced skin fibrosis (Flanders et al. 2002) and against bleomycin-induced (Zhao et al. 2002) and TGF-β1-induced pulmonary fibrosis (Bonniaud et al. 2004a).

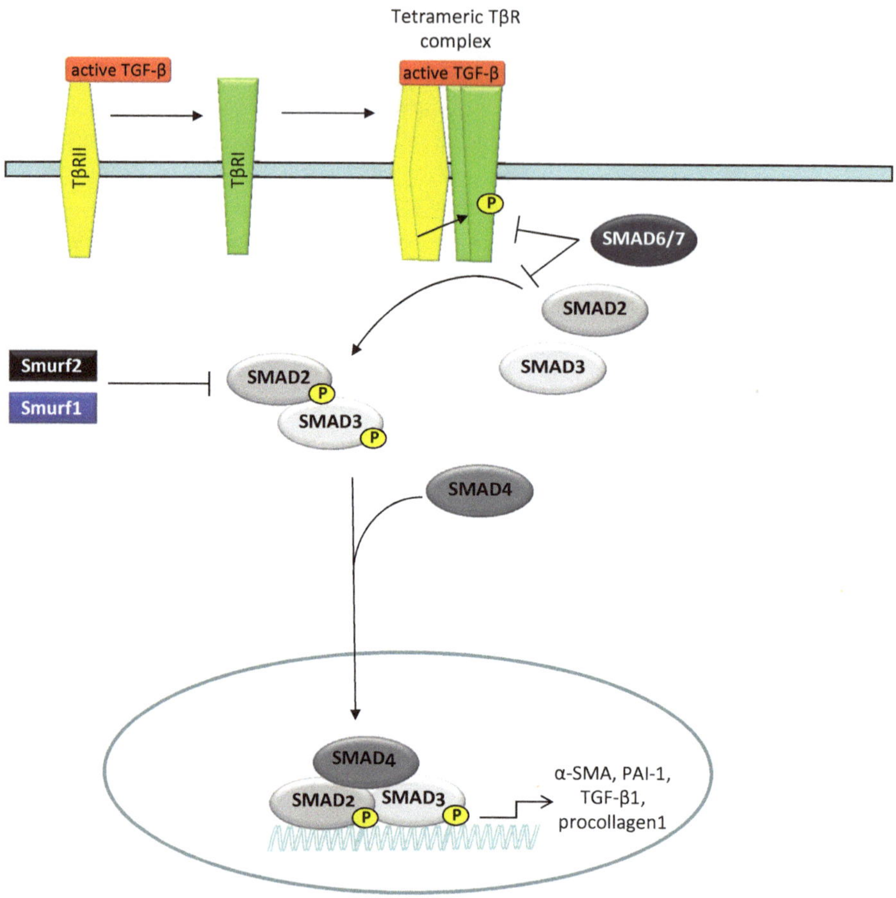

Fig. 13.2 Smad-dependent TGF-β1 pathway. Transphosphorylation of TβRI activates its kinase domain, which may then phosphorylate Smad2 and Smad3. Once activated, Smad2/3 bind to Smad4 and translocate to the nucleus in order to regulate the transcription of a large number of genes. The regulation of Smad2/3 activation takes place in the cytoplasm by Smad6 and Smad7 (Smad6/7) that prevent binding and phosphorylation of Smad2/3. Protein levels of Smad2/3 are regulated in part by ubiquitin-proteasome-mediated degradation. Smurf proteins (Smurf 1 and 2) are E3-ubiquitin ligases that ubiquitinate Smad2/3

13.3 HSP27: Favors Fibrosis

HSP27 is the most studied and the most expressed HSP at specific stages of development and cellular differentiation. Its basal expression is relatively low but is largely increased during heat shock and various cellular stresses. HSP27 is constitutively expressed in the cytoplasm (Lindquist and Craig 1988) but in case of heat shock it rapidly translocates into the nucleus (Arrigo et al. 1988). The N-terminal domain contains a WDPF motif required for the formation of oligomers whereas the

C-terminal domain is involved in the stabilization of those oligomers. In the cell, there is an equilibrium between the dimeric HSP27 and the larger oligomeric forms. Oligomerization is a dynamic process driven by the phosphorylation status of HSP27 (Kato et al. 1994b). Phosphorylated HSP27 forms dimers and the dephosphorylated status of HSP27 drives its oligomerization (Mehlen et al. 1997; Mehlen and Arrigo 1994). HSP27 can be phosphorylated mainly on three serines located at positions 15, 78 and 82. This phosphorylation is regulated by MK2 (MAPKAP Kinase 2) and p38MAPK (Stokoe et al. 1992).

HSP27 is known to be involved in cytoskeleton dynamics. Indeed, the dephosphorylated form of HSP27 is able to bind to the terminal end of actin filaments and inhibit their polymerization, whereas its phosphorylated form separates from actin and allows polymerization (Kostenko et al. 2009).

Besides its effect on actin, HSP27 has been shown to have anti-apoptotic properties (Bruey et al. 2000) and to promote cell survival. For instance, HSP27 induces the ubiquitination and degradation of the NF-κB inhibitor, I-κBα, thus allowing nuclear translocation of NF-κB and activation of target genes promoting cell survival such as Bcl-2, BcL-X_L and IAP (inhibitor of Apoptosis Protein, (Parcellier et al. 2003)).

HSP27 has also been demonstrated to be overexpressed in cancerous cells, protecting them from apoptosis and favoring their proliferation (Garrido et al. 2003). In addition, HSP27 seems to promote metastasis formation through its effect on cytoskeletal components and to favor resistance to chemotherapeutic agents (Hino et al. 2000; Katoh et al. 2000; Li et al. 2011). Choi et al. showed that HSP27 expression was directly responsible for colorectal cancer cell resistance to the chemotherapeutic agent irinotecan. By inhibiting HSP27 authors could decrease the resistance of those cells to the treatment (Choi et al. 2007). Moreover, Zhuang et al. demonstrated that HSP27 inhibition sensitized A549 cancerous lung cells to apoptosis induced by TRAIL (Tumor necrosis factor-alpha-Related Apoptosis-Inducing Ligand, (Zhuang et al. 2010)).

HSP27 has been reported as involved in corneas wound healing in mice. Song et al. demonstrated that the expression of the phosphorylated form of HSP27 was greater in wounded corneal epithelial cells than in unwounded controls and promoted epithelial cell migration thus favoring wound healing (Song et al. 2014). Further, during wound healing, it has been shown that the phosphorylation of HSP27 is enhanced after skin incision (Hirano et al. 2002). Dermal fibroblasts contraction is correlated with the level of HSP27. Therefore, fibroblasts expressing low levels of HSP27 possess less contractile properties (Hirano et al. 2004). The authors demonstrated that the effect of HSP27 on wound contraction was mediated by the activation of mitogen-activated protein kinase (MAPK)-activated protein kinase 2/3 (MAPKAPK 2/3) by phosphorylated HSP27. Moreover, Crowe et al. recently demonstrated *in vivo* that HSP27 depleted fibroblasts, reduced proliferation properties and that deficient mice for HSP27 had an impaired wound healing, characterized by a decrease in re-epithelialization and collagen deposition (Doshi et al. 2008; Crowe et al. 2013).

The number of studies investigating the role of HSP27 in fibrogenesis is quite limited. However, an up-regulation of HSP27 in lungs from patients affected by pulmonary fibrosis has been reported in cells located nearby fibroblastic foci, more specifically in regions where the ECM-producing cells are found during the disease (Korfei et al. 2011). Another study showed that HSP27 was up-regulated in epithelial cells located around fibroblastic foci. The authors suggested that epithelial cell proliferation and migration could play a role in the morphologic changes occurring in the lungs during pulmonary fibrosis (Chilosi et al. 2006). Two other articles showed that HSP27 was involved in the process of EMT. Vargha et al. showed in 2008 that HSP27 was upregulated *in vitro* during EMT in mesothelial cells from peritoneum whereas it was diminished *in vivo* during chronic EMT in dialysis induced peritoneal fibrosis (Vargha et al. 2008). The authors suggested that the discordant HSP27 expression between *in vivo* and *in vitro* EMT might be due to morphological changes that are more important during *in vitro* TGF-β1-induced EMT. Indeed, in their model, during *in vivo* EMT induced by peritoneal dialysis fluids, although cells underwent EMT only a subpopulation of cells presented morphological fibroblast-like changes. In vitro, TGF-β1 was associated with strong morphological changes and α-SMA up-regulation. The correlation between α-SMA and HSP27 up-regulation might reflect the alteration of the cytoskeleton induced by TGF-β1 during myofibroblastic transformation of mesothelial cells. *In vitro*, HSP27 up-regulation protected cells from apoptosis induced by peritoneal dialysis fluids (Vargha et al. 2008). Vidyasagar et al. suspected the involvement of HSP27 in renal fibrosis via its role on EMT. The authors showed that HSP27 was overexpressed during kidney fibrosis and that it co-localized with TGF-β1, α-SMA and E-cadherin. Overexpression of HSP27 in renal epithelial cells induced E-cadherin expression and repressed the expression of Snail, the major transcription factor promoting EMT. Authors concluded that HSP27 had a protective role against fibrosis and that its up-regulation could slow down fibrosis and EMT processes (Vidyasagar et al. 2008). The same authors have recently demonstrated using transgenic mice, that HSP27 upregulation in tubular epithelial cells lead to decreased fibrogenesis associated with a decline in phosphorylated p38MAPK, collagen III, and α-SMA, following induction of renal fibrosis. In their model, E-cadherin levels remained unchanged in tubular cells after fibrosis induction (Vidyasagar et al. 2013).

Another recent study has highlighted the effect of HSP27 overexpression during EMT in epithelial and mesothelial cells during pulmonary and pleural fibrosis. In this study, HSP27 was upregulated within fibroblastic foci *in vivo* in patients affected by pulmonary fibrosis (Wettstein et al. 2013). HSP27 induced phenotypic changes in cells, which acquired a myofibroblast-like shape along with α-SMA upregulation and E-cadherin downregulation. Furthermore, the inhibition of HSP27 *in vivo* using OGX-427, an antisense oligonucleotide already in clinical trials as a chemo-sensitizing agent in cancer therapy (Zoubeidi and Gleave 2012), protected rats from pleural fibrosis induced by TGF-β1 overexpression. In addition to the inhibition of fibrosis, repression of HSP27 also inhibited the migration of pleural "myofibroblasts" towards the lung parenchyma, an essential event in the *in vivo* model of

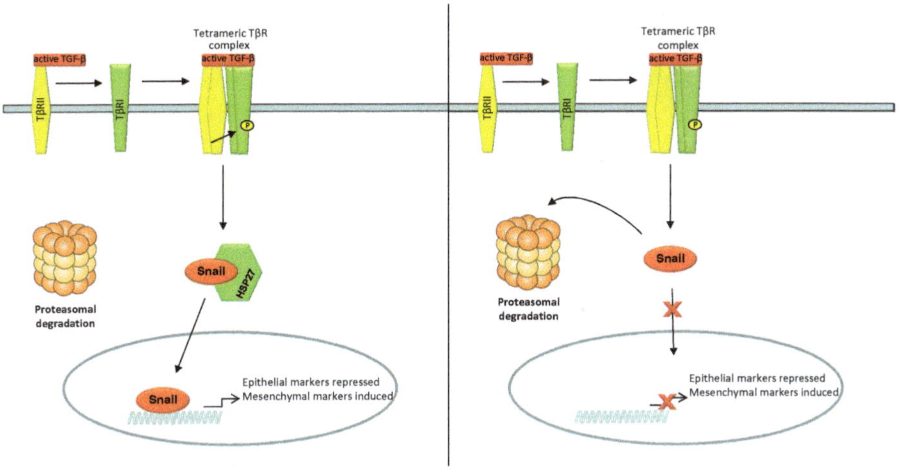

Fig. 13.3 HSP27 stabilizes the transcription factor Snail and favors EMT. *Left panel*: HSP27 is able to bind Snail and protect it against proteasomal degradation. After activation of the TGF-β signaling pathway, Snail translocates to the nucleus and induces transcription of EMT genes. *Right panel*: When HSP27 is inhibited, Snail degradation by the proteasome is enhanced, thus inhibiting the EMT process and preventing fibrosis

TGF-β1 induced pleural fibrosis (Wettstein et al. 2013). The therapeutic use of OGX-427 against pleuro-pulmonary fibrosis has been patented (Oncogenics).

HSP27 chaperones and stabilizes the essential transcription factor for the induction of EMT, Snail (Fig. 13.3). Indeed, HSP27 protects Snail from degradation by the proteasome, leading to its nuclear accumulation and activation of its EMT inducing target genes. By extension, inhibition of HSP27 *in vivo* induced degradation of Snail and stalled EMT confirming that HSP27 may be an interesting therapeutic target against pulmonary fibrosis (Wettstein et al. 2013).

Thus, HSP27 is a well-documented pro-fibrotic agent in pulmonary fibrosis. However, this function has yet to be confirmed in other organs since HSP27 has been reported to protect from renal fibrosis *in vitro* and *in vivo* (Vidyasagar et al. 2008, 2013).

13.4 αB-crystallin: A Pro-fibrotic Factor

Like HSP27, αB-crystallin is another ubiquitous sHSP, highly inducible under stress conditions in many organs such as the brain, heart, smooth muscles and lungs (Bhat and Nagineni 1989). HSP27 and αB-crystallin are two closely related proteins and it has been shown that they co-localized in many organs in normal and pathological conditions. Some publications even report a synergistic role for these two sHSP, which are able to interact with each other (MacIntyre et al. 2008). In 2001, a study showed that HSP27 prevented structural changes and aggregation of αB-crystallin induced by heat shock and thus may have a role in its stabilization (Fu and Liang 2003). Like HSP27, αB-crystallin has a chaperone activity and is able to bind hydrophobic areas on the surface of misfolded proteins, preventing their aggregation and protecting cellular integrity (Markov et al. 2008). The function of αB-crystallin is modulated by its oligomerization and its phosphorylation at three serine residues, the serines 19, 45 and 59. αB-crystallin has a role in many physiological and pathological processes such as cell growth and cell differentiation, apoptosis, tumorigenesis, signal transduction and the modulation of cytoskeletal proteins such as intermediate filaments, the latter being probably one of the most important functions of αB-crystallin (Nicholl and Quinlan 1994; Wisniewski and Goldman 1998; Launay et al. 2006). Indeed, αB-crystallin is known to interact with intermediate filaments and contributes to their homeostasis. The damages caused by mutations of αB-crystallin on the cytoskeleton architecture, which sometimes lead to severe diseases in humans, shows the importance of these interactions (Wettstein et al. 2012).

Brady et al. established the role of αB-crystallin in skeletal and cardiac muscle through the study of mice deficient for the gene encoding αB-crystallin (Brady et al. 2001). Surprisingly, the eye lens structure of αB-crystallin deficient mice was normal suggesting that the development and maintenance of lens transparency do not strictly require the presence of this chaperone. Although the lens does not seem to be affected by the absence of αB-crystallin, this mutation causes a severe phenotype in aged mice characterized by a hunched posture, a significant loss of body weight after 40 weeks, and muscle cell degeneration. αB-crystallin is also expressed at high levels during early embryonic development of the heart and also in the fully formed heart. However, the heart structure of αB-crystallin deficient mice appeared normal even in older mice (Brady et al. 2001).

AlphaB-crystallin is involved in many cellular processes, including apoptosis. Several studies have demonstrated that αB-crystallin overexpression has a protective effect against a wide range of apoptotic stimuli, whereas its inhibition significantly sensitizes cells to programmed cell death (Kamradt et al. 2001, 2002, 2005; Garrido et al. 2012). The anti-apoptotic effect of αB-crystallin has been shown to involve the inhibition of caspase-3, a key apoptosis effector protein (Kamradt et al. 2001, 2002; Mao et al. 2001; Morrison et al. 2003). It has been demonstrated that αB-crystallin interacts with procapase-3, and thereby inhibits its cleavage and activation. This effect of αB-crystallin is critical to its anti-apoptotic function. Morrison et al. demonstrated in cardiomyocytes that phosphorylation of αB-crystallin on serine 59 appeared as necessary and sufficient to ensure maximal protection against apoptosis (Morrison et al. 2003). AlphaB-crystallin was also able to bind to pro-apoptotic proteins such as Bax, Bcl-Xs (Mao et al. 2004) and p53 (Liu et al. 2007) and prevent their translocation into the mitochondria, thus inhibiting apoptosis.

Another recently identified function of αB-crystallin is its ability to facilitate the degradation of specific proteins by the ubiquitin/proteasome system (Garrido 2002). Indeed, αB-crystallin phosphorylated on serine 19 and 45 is able to interact with FBX4, a component of the SCF ubiquitin-ligase (SKP1/CUL1/F-box) which ubiquitinates client proteins to address them to the proteasome (den Engelsman et al. 2003). The αB-crystallin/FBX4 complex facilitates the degradation of various substrates.

αB-crystallin has been shown to be upregulated in several human diseases (Cherneva et al. 2010, 2012; Aoyama et al. 1993; Lo et al. 2007; Pinder et al. 1994) but its role on fibrogenesis remains elusive. In a model of liver fibrosis, it has been shown *in vitro* and *in vivo* that αB-crystallin was quickly upregulated after HSC (hepatic stellate cells) activation and acquired a nuclear localization in heat shock conditions. αB-crystallin was totally absent in non-activated HSCs suggesting a role in the activation of these cells, a key step during liver fibrosis (Lang et al. 2000; van de Bovenkamp et al. 2008). Rezzani et al. also showed that overexpression of αB-crystallin is induced by cyclosporine A (CsA) in a model of vascular fibrosis. CsA has a direct effect on the structure and the spatial arrangement of the cytoskeleton and especially on the expression of vimentin and desmin. The authors suggested that αB-crystallin expression is induced in order to protect cells from the toxic effects of the cytoskeletal remodeling, thereby offering protection from vascular fibrosis (Rezzani et al. 2005). Interestingly, different studies have shown that αB-crsytallin expression can be induced by TGF-β1 (Yu et al. 2007; Welge-Lussen et al. 1999) and an article recently published by Huang et al. showed a role for αB-crystallin in the EMT process in a model of hepatocellular carcinoma. The authors showed that αB-crystallin formed a complex with the protein 14-3-3ζ, protecting it from degradation by the proteasome (Fig. 13.4). This results in an increase of the "pool" of 14-3-3ζ in the cell, leading to activation of the ERK signaling pathway (Fig. 13.4). Therefore, αB-crystallin favors the activation of the ERK phosphorylation cascade leading to the activation of the transcription factor Slug, an

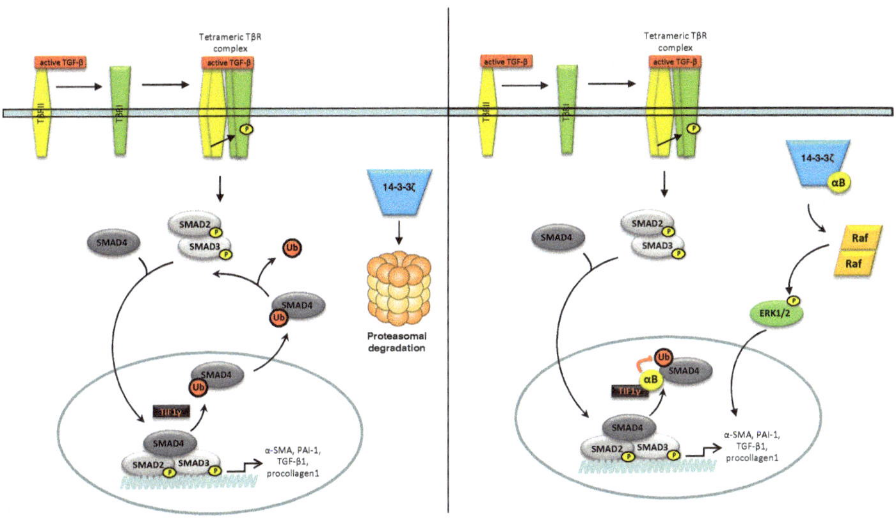

Fig. 13.4 αB-crystallin inhibition favors Smad4 nuclear export. *Left panel*: In the absence of αB-crystallin Smad4 is monoubiquitinated by TIF1γ enhancing Smad4's nuclear export. *Right*: αB-crystallin reduces the protein level of TIF1γ thus limiting Smad4 monoubiquitination and nuclear export. The TGF-β1 pathway is therefore inhibited. AlphaB-crystallin can also interact with the protein 14-3-3ζ allowing Raf dimerization and activation of the ERK phosphorylation cascade

inducer of EMT (Huang et al. 2013). This process highlighted in this article in a model of hepatic carcinoma may be particularly important in fibrogenesis since, as already mentioned, EMT is a process involved in fibrosis.

A study recently demonstrated a role for αB-crystallin in pulmonary fibrosis. For the first time, αB-crystallin has been shown to be overexpressed in hyperplastic alveolar epithelial cells and also within fibroblastic foci in lungs from patients affected by pulmonary fibrosis. Furthermore, αB-crystallin was upregulated during fibrogenesis in mouse models of pulmonary fibrosis induced by several agents including bleomycin, TGF-β1 and IL-1β. Mouse lacking αB-crystallin were not able to produce TGF-β1 in an efficient way and the authors identified a direct role for αB-crystallin on the TGF-β1 pathway. Indeed, αB-crystallin was able to modulate the localization of the protein Smad4, which is responsible for the translocation of Smad2 and Smad3 into the nucleus where they can activate TGF-β1 responsive genes (Fig. 13.4). The lack of αB-crystallin abrogated Smad4s' nuclear localization, thereby blocking the TGF-β1 pathway and subsequent fibrosis (Bellaye et al. 2014). The nuclear localization of Smad4 is driven by its monoubiquitination by TIF1γ. *In vitro*, the overexpression of αB-crystallin disrupted the interaction between TIF1γ and Smad4, mono-ubiquitination of Smad4 and its nuclear export (Fig. 13.4). As a consequence, Smad4 was sequestrated in the nucleus, favoring the TGF-β-induced fibrosis (Bellaye et al. 2014).

Those different studies highlight αB-crystallin as a new potential therapeutic target in fibrosis and the urgent need of specific αB-crystallin inhibitors.

13.5 HSP20: Anti-fibrotic Protein?

HSP20 belongs to the small heat shock protein family and has a molecular mass of 20 kD. The structure of HSP20 is highly homologous to that of αB-crystallin and HSP27. HSP20 can form oligomers along with HSP27 and αB-crystallin and isolated HSP20 forms low and high molecular mass complexes (Gusev et al. 2005). HSP20 possess several sites of phosphorylation which modulates its interaction with other proteins and its cellular functions. HSP20 exhibits a chaperone activity but molecular mechanisms involving HSP20 might be different from that of other sHSP. HSP20 is mainly expressed in skeletal muscle, diaphragm, heart, and smooth muscles (Kato et al. 1994a). It has been shown that HSP20 might have some interactions with the cytoskeleton and could promote smooth muscle relaxation. Indeed, HSP20 co-immunoprecipitates with actin and α-actinin and phosphorylation of HSP20 decreases this interaction. During contraction, non-phosphorylated HSP20 stabilizes actin filaments. HSP20 phosphorylation induces its dissociation from polymerized actin resulting in smooth muscle cell relaxation (Brophy et al. 1999; Tessier et al. 2003). As many sHSP, phosphorylated HSP20 also exhibits anti-apoptotic properties by blocking the proteolytic activation of caspase3 by interacting with pro-caspase3 (Fan et al. 2004).

During fibrosis, TGF-β1 stimulates the expression of connective tissue growth factor (CTGF) in fibroblasts which could mediate some of the profibrotic effects of TGF-β such as fibroblast differentiation and collagen deposition (Bonniaud et al. 2004b). It has been shown that CTGF expression is strongly modulated by the cytoskeleton, thus stabilization of the actin cytoskeleton increases CTGF expression and actin depolymerization decreases the expression of CTGF (Muehlich et al. 2007). Lopes et al. used a peptide analogue of the phosphorylated form of HSP20 (AZX100), which is known to induce the disruption of the actin cytoskeleton (Dreiza et al. 2005), in order to reduce the expression of CTGF and subsequent fibrosis (Lopes et al. 2009). The authors demonstrated in keloid fibroblasts that AZX100 could reduce CTGF expression induced by TGF-β and subsequent collagen I overproduction characterizing fibrosis. CTGF and collagen expression was also reduced by AZX100 after stimulation of keloid fibroblasts by other profibrotic agents such as lysophosphatidic acid (LPA), thrombin, and endothelin (ET). In addition, the HSP20 analogue hampered the formation of actin stress fibers occurring during the process of wound healing. Further, using an *in vivo* model of rodent dermal scarring, they showed that AZX100 significantly improves collagen fiber orientation, density, and maturity thus limiting excessive scarring (Lopes et al. 2009). However, the mechanism involving ATX100 was not elucidated in this paper. The authors demonstrated that the HSP20 analogue did not enhance Smad3 phosphorylation suggesting that the effect of AZX100 on the cytoskeleton and CTGF is independent of SMAD signaling.

It has been shown in a model of cardiac fibrosis that the upregulation of HSP20 protects mesenchymal stem cells (MSC) against hypoxia-induced apoptosis *in vitro*. Indeed, the survival of MSC overexpressing HSP20 was improved, compared to normal MSC, after transplantation into the infarcted heart. After transplantation of MSC overexpressing HSP20 the cardiac function of infarcted myocardium was improved in correlation with a decrease in cardiac fibrosis (Wang et al. 2009). The authors investigated the underlying mechanism and identified that the beneficial effects of HSP20 was due to the permanent activation of Akt signaling favoring MSC survival and an increased secretion of growth factors such as VEGF, FGF-2, and IGF-1, providing cardioprotective and pro-angiogenic effects (Wang et al. 2009).

13.6 Concluding Remarks

Small heat shock proteins are involved in several cellular processes and display a wide range of functions. Further investigation is needed to clearly determine their roles in fibrogenesis but recent publications demonstrate their potential interest in therapeutic approaches against fibrotic diseases. Interestingly, depending on the HSPs, these can have opposite effects on fibrotic processes. Few publications reveals that HSP20 could be anti-fibrotic whereas HSP27 and αB-crystallin have been demonstrated to favor fibrogenesis through their role on the enhancement of the TGF-β1 pathway by stabilizing either Snail (HSP27) or by favoring the nuclear localization of Smads (αB-crystallin). As the major role of the TGF-β pathway in fibrosis is now well documented, modulation of the expression and activity of sHSP might represent a new therapeutic approach for fibrotic disorders such as pulmonary fibrosis for which no curative treatment exist. Indeed, inhibitors of HSP27 or αB-crystallin and HSP20 analoges (AZX100) appear to be of interest as it prevents fibrogenesis in several animal models of tissue fibrosis. However, their efficiency has yet to be evaluated in clinical trials.

References

Aoyama A, Steiger RH, Frohli E, Schafer R, von Deimling A, Wiestler OD, Klemenz R (1993) Expression of alpha B-crystallin in human brain tumors. Int J Cancer 55(5):760–764

Arrigo AP (2005) Heat shock proteins as molecular chaperones. Med Sci 21(6–7):619–625. doi:10.1051/medsci/2005216-7619

Arrigo AP, Suhan JP, Welch WJ (1988) Dynamic changes in the structure and intracellular locale of the mammalian low-molecular-weight heat shock protein. Mol Cell Biol 8(12):5059–5071

Barbas-Filho JV, Ferreira MA, Sesso A, Kairalla RA, Carvalho CR, Capelozzi VL (2001) Evidence of type II pneumocyte apoptosis in the pathogenesis of idiopathic pulmonary fibrosis (IFP)/usual interstitial pneumonia (UIP). J Clin Pathol 54(2):132–138

Bellaye PS, Wettstein G, Burgy O, Besnard V, Joannes A, Colas J, Causse S, Marchal-Somme J, Fabre A, Crestani B, Kolb M, Gauldie J, Camus P, Garrido C, Bonniaud P (2014) The small heat-shock protein alphaB-crystallin is essential for the nuclear localization of Smad4: impact on pulmonary fibrosis. J Pathol 232(4):458–472. doi:10.1002/path.4314

Benesch JL, Ayoub M, Robinson CV, Aquilina JA (2008) Small heat shock protein activity is regulated by variable oligomeric substructure. J Biol Chem 283(42):28513–28517. doi:10.1074/jbc.M804729200

Bhat SP, Nagineni CN (1989) alpha B subunit of lens-specific protein alpha-crystallin is present in other ocular and non-ocular tissues. Biochem Biophys Res Commun 158(1):319–325

Bonniaud P, Kolb M, Galt T, Robertson J, Robbins C, Stampfli M, Lavery C, Margetts PJ, Roberts AB, Gauldie J (2004a) Smad3 null mice develop airspace enlargement and are resistant to TGF-beta-mediated pulmonary fibrosis. J Immunol 173(3):2099–2108

Bonniaud P, Martin G, Margetts PJ, Ask K, Robertson J, Gauldie J, Kolb M (2004b) Connective tissue growth factor is crucial to inducing a profibrotic environment in "fibrosis-resistant" BALB/c mouse lungs. Am J Respir Cell Mol Biol 31(5):510–516. doi:10.1165/rcmb.2004-0158OC

Border WA, Noble NA (1994) Transforming growth factor beta in tissue fibrosis. N Engl J Med 331(19):1286–1292. doi:10.1056/NEJM199411103311907

Brady JP, Garland DL, Green DE, Tamm ER, Giblin FJ, Wawrousek EF (2001) AlphaB-crystallin in lens development and muscle integrity: a gene knockout approach. Invest Ophthalmol Vis Sci 42(12):2924–2934

Brophy CM, Lamb S, Graham A (1999) The small heat shock-related protein-20 is an actin-associated protein. J Vasc Surg 29(2):326–333

Bruey JM, Ducasse C, Bonniaud P, Ravagnan L, Susin SA, Diaz-Latoud C, Gurbuxani S, Arrigo AP, Kroemer G, Solary E, Garrido C (2000) Hsp27 negatively regulates cell death by interacting with cytochrome c. Nat Cell Biol 2(9):645–652. doi:10.1038/35023595

Bucala R, Spiegel LA, Chesney J, Hogan M, Cerami A (1994) Circulating fibrocytes define a new leukocyte subpopulation that mediates tissue repair. Mol Med 1(1):71–81

Burel C, Mezger V, Pinto M, Rallu M, Trigon S, Morange M (1992) Mammalian heat shock protein families. Expression and functions. Experientia 48(7):629–634

Calabresi PA, Giovannoni G, Confavreux C, Galetta SL, Havrdova E, Hutchinson M, Kappos L, Miller DH, O'Connor PW, Phillips JT, Polman CH, Radue EW, Rudick RA, Stuart WH, Lublin FD, Wajgt A, Weinstock-Guttman B, Wynn DR, Lynn F, Panzara MA (2007) The incidence and significance of anti-natalizumab antibodies: results from AFFIRM and SENTINEL. Neurology 69(14):1391–1403. doi:10.1212/01.wnl.0000277457.17420.b5

Carew RM, Wang B, Kantharidis P (2012) The role of EMT in renal fibrosis. Cell Tissue Res 347(1):103–116. doi:10.1007/s00441-011-1227-1

Cherneva R, Petrov D, Georgiev O, Slavova Y, Toncheva D, Stamenova M, Trifonova N (2010) Expression profile of the small heat-shock protein alpha-B-crystallin in operated-on non-small-cell lung cancer patients: clinical implication. Eur J Cardiothorac Surg 37(1):44–50. doi:10.1016/j.ejcts.2009.06.038

Cherneva RV, Georgiev OB, Petrova DS, Trifonova NL, Stamenova M, Ivanova V, Vlasov VI (2012) The role of small heat-shock protein alphaB-crystalline (HspB5) in COPD pathogenesis. Int J Chron Obstruct Pulmon Dis 7:633–640. doi:10.2147/COPD.S34929

Chilosi M, Zamo A, Doglioni C, Reghellin D, Lestani M, Montagna L, Pedron S, Ennas MG, Cancellieri A, Murer B, Poletti V (2006) Migratory marker expression in fibroblast foci of idiopathic pulmonary fibrosis. Respir Res 7:95. doi:10.1186/1465-9921-7-95

Choi DH, Ha JS, Lee WH, Song JK, Kim GY, Park JH, Cha HJ, Lee BJ, Park JW (2007) Heat shock protein 27 is associated with irinotecan resistance in human colorectal cancer cells. FEBS Lett 581(8):1649–1656. doi:10.1016/j.febslet.2007.02.075

Coker RK, Laurent GJ, Jeffery PK, du Bois RM, Black CM, McAnulty RJ (2001) Localisation of transforming growth factor beta1 and beta3 mRNA transcripts in normal and fibrotic human lung. Thorax 56(7):549–556

Conde R, Belak ZR, Nair M, O'Carroll RF, Ovsenek N (2009) Modulation of Hsf1 activity by novobiocin and geldanamycin. Biochem Cell Biol 87(6):845–851. doi:10.1139/o09-049

Crowe J, Aubareda A, McNamee K, Przybycien PM, Lu X, Williams RO, Bou-Gharios G, Saklatvala J, Dean JL (2013) Heat shock protein B1-deficient mice display impaired wound healing. PLoS One 8(10):e77383. doi:10.1371/journal.pone.0077383

den Engelsman J, Keijsers V, de Jong WW, Boelens WC (2003) The small heat-shock protein alpha B-crystallin promotes FBX4-dependent ubiquitination. J Biol Chem 278(7):4699–4704. doi:10.1074/jbc.M211403200

Didelot C, Schmitt E, Brunet M, Maingret L, Parcellier A, Garrido C (2006) Heat shock proteins: endogenous modulators of apoptotic cell death. Handb Exp Pharmacol 172:171–198

Doshi B, Perdrizet GA, Hightower LE (2008) Wound healing from a cellular stress response perspective. Cell Stress Chaperones 13:393–399. doi:10.1007/s12192-008-0059-8

Drakopanagiotakis F, Xifteri A, Polychronopoulos V, Bouros D (2008) Apoptosis in lung injury and fibrosis. Eur Respir J 32(6):1631–1638. doi:10.1183/09031936.00176807

Dreiza CM, Brophy CM, Komalavilas P, Furnish EJ, Joshi L, Pallero MA, Murphy-Ullrich JE, von Rechenberg M, Ho YS, Richardson B, Xu N, Zhen Y, Peltier JM, Panitch A (2005) Transducible heat shock protein 20 (HSP20) phosphopeptide alters cytoskeletal dynamics. FASEB J 19(2):261–263. doi:10.1096/fj.04-2911fje

Eickelberg O (2001) Endless healing: TGF-beta, SMADs, and fibrosis. FEBS Lett 506(1):11–14

Fan GC, Chu G, Mitton B, Song Q, Yuan Q, Kranias EG (2004) Small heat-shock protein Hsp20 phosphorylation inhibits beta-agonist-induced cardiac apoptosis. Circ Res 94(11):1474–1482. doi:10.1161/01.RES.0000129179.66631.00

Flanders KC, Sullivan CD, Fujii M, Sowers A, Anzano MA, Arabshahi A, Major C, Deng C, Russo A, Mitchell JB, Roberts AB (2002) Mice lacking Smad3 are protected against cutaneous injury induced by ionizing radiation. Am J Pathol 160(3):1057–1068. doi:10.1016/S0002-9440(10)64926-7

Fu L, Liang JJ (2003) Enhanced stability of alpha B-crystallin in the presence of small heat shock protein Hsp27. Biochem Biophys Res Commun 302(4):710–714

Garrido C (2002) Size matters: of the small HSP27 and its large oligomers. Cell Death Differ 9(5):483–485. doi:10.1038/sj/cdd/4401005

Garrido C, Schmitt E, Cande C, Vahsen N, Parcellier A, Kroemer G (2003) HSP27 and HSP70: potentially oncogenic apoptosis inhibitors. Cell Cycle 2(6):579–584

Garrido C, Paul C, Seigneuric R, Kampinga HH (2012) The small heat shock proteins family: the long forgotten chaperones. Int J Biochem Cell Biol 44(10):1588–1592. doi:10.1016/j.biocel.2012.02.022

Gething MJ, Sambrook J (1992) Protein folding in the cell. Nature 355(6355):33–45. doi:10.1038/355033a0

Gibert B, Simon S, Dimitrova V, Diaz-Latoud C, Arrigo AP (2013) Peptide aptamers: tools to negatively or positively modulate HSPB1(27) function. Philos Trans R Soc Lond B Biol Sci 368(1617):20120075. doi:10.1098/rstb.2012.0075

Gusev NB, Bukach OV, Marston SB (2005) Structure, properties, and probable physiological role of small heat shock protein with molecular mass 20 kD (Hsp20, HspB6). Biochemistry (Mosc) 70(6):629–637

Hayashi N, Peacock JW, Beraldi E, Zoubeidi A, Gleave ME, Ong CJ (2012) Hsp27 silencing coordinately inhibits proliferation and promotes Fas-induced apoptosis by regulating the PEA-15 molecular switch. Cell Death Differ 19(6):990–1002. doi:10.1038/cdd.2011.184

Hayes D, Napoli V, Mazurkie A, Stafford WF, Graceffa P (2009) Phosphorylation dependence of hsp27 multimeric size and molecular chaperone function. J Biol Chem 284(28):18801–18807. doi:10.1074/jbc.M109.011353

Hino M, Kurogi K, Okubo MA, Murata-Hori M, Hosoya H (2000) Small heat shock protein 27 (HSP27) associates with tubulin/microtubules in HeLa cells. Biochem Biophys Res Commun 271(1):164–169. doi:10.1006/bbrc.2000.2553

Hirano S, Rees RS, Gilmont RR (2002) MAP kinase pathways involving hsp27 regulate fibroblast-mediated wound contraction. J Surg Res 102(2):77–84. doi:10.1006/jsre.2001.6315

Hirano S, Shelden EA, Gilmont RR (2004) HSP27 regulates fibroblast adhesion, motility, and matrix contraction. Cell Stress Chaperones 9(1):29–37

Huang XY, Ke AW, Shi GM, Zhang X, Zhang C, Shi YH, Wang XY, Ding ZB, Xiao YS, Yan J, Qiu SJ, Fan J, Zhou J (2013) alphaB-crystallin complexes with 14-3-3zeta to induce epithelial-mesenchymal transition and resistance to sorafenib in hepatocellular carcinoma. Hepatology. doi:10.1002/hep.26255

Ivanov O, Chen F, Wiley EL, Keswani A, Diaz LK, Memmel HC, Rademaker A, Gradishar WJ, Morrow M, Khan SA, Cryns VL (2008) alphaB-crystallin is a novel predictor of resistance to neoadjuvant chemotherapy in breast cancer. Breast Cancer Res Treat 111(3):411–417. doi:10.1007/s10549-007-9796-0

Jakubowicz-Gil J, Langner E, Badziul D, Wertel I, Rzeski W (2013) Silencing of Hsp27 and Hsp72 in glioma cells as a tool for programmed cell death induction upon temozolomide and quercetin treatment. Toxicol Appl Pharmacol 273(3):580–589. doi:10.1016/j.taap.2013.10.003

Jego G, Hazoume A, Seigneuric R, Garrido C (2013) Targeting heat shock proteins in cancer. Cancer Lett 332(2):275–285. doi:10.1016/j.canlet.2010.10.014

Kampinga HH, Hageman J, Vos MJ, Kubota H, Tanguay RM, Bruford EA, Cheetham ME, Chen B, Hightower LE (2009) Guidelines for the nomenclature of the human heat shock proteins. Cell Stress Chaperones 14:105–111. doi:10.1007/s12192-008-0068-7

Kamradt MC, Chen F, Cryns VL (2001) The small heat shock protein alpha B-crystallin negatively regulates cytochrome c- and caspase-8-dependent activation of caspase-3 by inhibiting its auto-proteolytic maturation. J Biol Chem 276(19):16059–16063. doi:10.1074/jbc.C100107200

Kamradt MC, Chen F, Sam S, Cryns VL (2002) The small heat shock protein alpha B-crystallin negatively regulates apoptosis during myogenic differentiation by inhibiting caspase-3 activation. J Biol Chem 277(41):38731–38736. doi:10.1074/jbc.M201770200

Kamradt MC, Lu M, Werner ME, Kwan T, Chen F, Strohecker A, Oshita S, Wilkinson JC, Yu C, Oliver PG, Duckett CS, Buchsbaum DJ, LoBuglio AF, Jordan VC, Cryns VL (2005) The small heat shock protein alpha B-crystallin is a novel inhibitor of TRAIL-induced apoptosis that suppresses the activation of caspase-3. J Biol Chem 280(12):11059–11066. doi:10.1074/jbc.M413382200

Kato K, Goto S, Inaguma Y, Hasegawa K, Morishita R, Asano T (1994a) Purification and characterization of a 20-kDa protein that is highly homologous to alpha B crystallin. J Biol Chem 269(21):15302–15309

Kato K, Hasegawa K, Goto S, Inaguma Y (1994b) Dissociation as a result of phosphorylation of an aggregated form of the small stress protein, hsp27. J Biol Chem 269(15):11274–11278

Katoh M, Koninkx J, Schumacher U (2000) Heat shock protein expression in human tumours grown in severe combined immunodeficient mice. Cancer Lett 161(1):113–120

Kavsak P, Rasmussen RK, Causing CG, Bonni S, Zhu H, Thomsen GH, Wrana JL (2000) Smad7 binds to Smurf2 to form an E3 ubiquitin ligase that targets the TGF beta receptor for degradation. Mol Cell 6(6):1365–1375

Khalil N, O'Connor RN, Flanders KC, Unruh H (1996) TGF-beta 1, but not TGF-beta 2 or TGF-beta 3, is differentially present in epithelial cells of advanced pulmonary fibrosis: an immunohistochemical study. Am J Respir Cell Mol Biol 14(2):131–138. doi:10.1165/ajrcmb.14.2.8630262

Kisseleva T, Brenner DA (2008) Mechanisms of fibrogenesis. Exp Biol Med (Maywood) 233(2):109–122. doi:10.3181/0707-MR-190

Klugman J, Lee JC, Nelson SL (2012) School co-ethnicity and Hispanic parental involvement. Soc Sci Res 41(5):1320–1337. doi:10.1016/j.ssresearch.2012.05.005

Korfei M, Schmitt S, Ruppert C, Henneke I, Markart P, Loeh B, Mahavadi P, Wygrecka M, Klepetko W, Fink L, Bonniaud P, Preissner KT, Lochnit G, Schaefer L, Seeger W, Guenther A (2011) Comparative proteomic analysis of lung tissue from patients with idiopathic pulmonary fibrosis (IPF) and lung transplant donor lungs. J Proteome Res 10(5):2185–2205. doi:10.1021/pr1009355

Kostenko S, Johannessen M, Moens U (2009) PKA-induced F-actin rearrangement requires phosphorylation of Hsp27 by the MAPKAP kinase MK5. Cell Signal 21(5):712–718. doi:10.1016/j.cellsig.2009.01.009

Lang A, Schrum LW, Schoonhoven R, Tuvia S, Solis-Herruzo JA, Tsukamoto H, Brenner DA, Rippe RA (2000) Expression of small heat shock protein alphaB-crystallin is induced after hepatic stellate cell activation. Am J Physiol Gastrointest Liver Physiol 279(6):G1333–G1342

Lanneau D, Wettstein G, Bonniaud P, Garrido C (2010) Heat shock proteins: cell protection through protein triage. Sci World J 10:1543–1552. doi:10.1100/tsw.2010.152

Launay N, Goudeau B, Kato K, Vicart P, Lilienbaum A (2006) Cell signaling pathways to alphaB-crystallin following stresses of the cytoskeleton. Exp Cell Res 312(18):3570–3584. doi:10.1016/j.yexcr.2006.07.025

Lemieux P, Oesterreich S, Lawrence JA, Steeg PS, Hilsenbeck SG, Harvey JM, Fuqua SA (1997) The small heat shock protein hsp27 increases invasiveness but decreases motility of breast cancer cells. Invasion Metastasis 17(3):113–123

Li GP, Wang H, Lai YK, Chen SC, Lin MC, Lu G, Zhang JF, He XG, Qian CN, Kung HF (2011) Proteomic profiling between CNE-2 and its strongly metastatic subclone S-18 and functional characterization of HSP27 in metastasis of nasopharyngeal carcinoma. Proteomics 11(14):2911–2920. doi:10.1002/pmic.201000483

Lindquist S, Craig EA (1988) The heat-shock proteins. Annu Rev Genet 22:631–677. doi:10.1146/annurev.ge.22.120188.003215

Liu S, Li J, Tao Y, Xiao X (2007) Small heat shock protein alphaB-crystallin binds to p53 to sequester its translocation to mitochondria during hydrogen peroxide-induced apoptosis. Biochem Biophys Res Commun 354(1):109–114. doi:10.1016/j.bbrc.2006.12.152

Lo WY, Tsai MH, Tsai Y, Hua CH, Tsai FJ, Huang SY, Tsai CH, Lai CC (2007) Identification of over-expressed proteins in oral squamous cell carcinoma (OSCC) patients by clinical proteomic analysis. Clin Chim Acta 376(1–2):101–107. doi:10.1016/j.cca.2006.06.030

Lopes LB, Furnish EJ, Komalavilas P, Flynn CR, Ashby P, Hansen A, Ly DP, Yang GP, Longaker MT, Panitch A, Brophy CM (2009) Cell permeant peptide analogues of the small heat shock

protein, HSP20, reduce TGF-beta1-induced CTGF expression in keloid fibroblasts. J Invest Dermatol 129(3):590–598. doi:10.1038/jid.2008.264
Ma W, Teng Y, Hua H, Hou J, Luo T, Jiang Y (2013) Upregulation of heat shock protein 27 confers resistance to actinomycin D-induced apoptosis in cancer cells. FEBS J 280(18):4612–4624. doi:10.1111/febs.12432
MacIntyre DA, Tyson EK, Read M, Smith R, Yeo G, Kwek K, Chan EC (2008) Contraction in human myometrium is associated with changes in small heat shock proteins. Endocrinology 149(1):245–252. doi:10.1210/en.2007-0662
Malin D, Strekalova E, Petrovic V, Deal AM, Al Ahmad A, Adamo B, Miller CR, Ugolkov A, Livasy C, Fritchie K, Hamilton E, Blackwell K, Geradts J, Ewend M, Carey L, Shusta EV, Anders CK, Cryns VL (2014) alphaB-crystallin: a novel regulator of breast cancer metastasis to the brain. Clin Cancer Res 20(1):56–67. doi:10.1158/1078-0432.CCR-13-1255
Mao YW, Xiang H, Wang J, Korsmeyer S, Reddan J, Li DW (2001) Human bcl-2 gene attenuates the ability of rabbit lens epithelial cells against H2O2-induced apoptosis through down-regulation of the alpha B-crystallin gene. J Biol Chem 276(46):43435–43445. doi:10.1074/jbc.M102195200
Mao YW, Liu JP, Xiang H, Li DW (2004) Human alphaA- and alphaB-crystallins bind to Bax and Bcl-X(S) to sequester their translocation during staurosporine-induced apoptosis. Cell Death Differ 11(5):512–526. doi:10.1038/sj.cdd.4401384
Markov DI, Pivovarova AV, Chernik IS, Gusev NB, Levitsky DI (2008) Small heat shock protein Hsp27 protects myosin S1 from heat-induced aggregation, but not from thermal denaturation and ATPase inactivation. FEBS Lett 582(10):1407–1412. doi:10.1016/j.febslet.2008.03.035
Massague J, Wotton D (2000) Transcriptional control by the TGF-beta/Smad signaling system. EMBO J 19(8):1745–1754. doi:10.1093/emboj/19.8.1745
McDonald ET, Bortolus M, Koteiche HA, McHaourab HS (2012) Sequence, structure, and dynamic determinants of Hsp27 (HspB1) equilibrium dissociation are encoded by the N-terminal domain. Biochemistry 51(6):1257–1268. doi:10.1021/bi2017624
Mehlen P, Arrigo AP (1994) The serum-induced phosphorylation of mammalian hsp27 correlates with changes in its intracellular localization and levels of oligomerization. Eur J Biochem 221(1):327–334
Mehlen P, Hickey E, Weber LA, Arrigo AP (1997) Large unphosphorylated aggregates as the active form of hsp27 which controls intracellular reactive oxygen species and glutathione levels and generates a protection against TNFalpha in NIH-3T3-ras cells. Biochem Biophys Res Commun 241(1):187–192. doi:10.1006/bbrc.1997.7635
Morrison LE, Hoover HE, Thuerauf DJ, Glembotski CC (2003) Mimicking phosphorylation of alphaB-crystallin on serine-59 is necessary and sufficient to provide maximal protection of cardiac myocytes from apoptosis. Circ Res 92(2):203–211
Muehlich S, Cicha I, Garlichs CD, Krueger B, Posern G, Goppelt-Struebe M (2007) Actin-dependent regulation of connective tissue growth factor. Am J Physiol Cell Physiol 292(5):C1732–C1738. doi:10.1152/ajpcell.00552.2006
Nicholl ID, Quinlan RA (1994) Chaperone activity of alpha-crystallins modulates intermediate filament assembly. EMBO J 13(4):945–953
Parcellier A, Schmitt E, Gurbuxani S, Seigneurin-Berny D, Pance A, Chantome A, Plenchette S, Khochbin S, Solary E, Garrido C (2003) HSP27 is a ubiquitin-binding protein involved in I-kappaBalpha proteasomal degradation. Mol Cell Biol 23(16):5790–5802
Pavan S, Musiani D, Torchiaro E, Migliardi G, Gai M, Di Cunto F, Erriquez J, Olivero M, Di Renzo MF (2014) HSP27 is required for invasion and metastasis triggered by hepatocyte growth factor. Int J Cancer 134(6):1289–1299. doi:10.1002/ijc.28464
Phan SH (2002) The myofibroblast in pulmonary fibrosis. Chest 122(6 Suppl):286S–289S
Pinder SE, Balsitis M, Ellis IO, Landon M, Mayer RJ, Lowe J (1994) The expression of alpha B-crystallin in epithelial tumours: a useful tumour marker? J Pathol 174(3):209–215. doi:10.1002/path.1711740310

Rappa F, Farina F, Zummo G, David S, Campanella C, Carini F, Tomasello G, Damiani P, Cappello F, DEM EC, Macario AJ (2012) HSP-molecular chaperones in cancer biogenesis and tumor therapy: an overview. Anticancer Res 32(12):5139–5150

Rezzani R, Rodella L, Buffoli B, Giugno L, Stacchiotti A, Bianchi R (2005) Cyclosporine a induces vascular fibrosis and heat shock protein expression in rat. Int Immunopharmacol 5(1):169–176. doi:10.1016/j.intimp.2004.07.018

Ritossa FM (1964) Experimental activation of specific loci in polytene chromosomes of drosophila. Exp Cell Res 35:601–607

Roberts AB (2002) The ever-increasing complexity of TGF-beta signaling. Cytokine Growth Factor Rev 13(1):3–5

Roberts AB, Piek E, Bottinger EP, Ashcroft G, Mitchell JB, Flanders KC (2001) Is Smad3 a major player in signal transduction pathways leading to fibrogenesis? Chest 120(1 Suppl):43S–47S

Ruan Q, Han S, Jiang WG, Boulton ME, Chen ZJ, Law BK, Cai J (2011) alphaB-crystallin, an effector of unfolded protein response, confers anti-VEGF resistance to breast cancer via maintenance of intracrine VEGF in endothelial cells. Mol Cancer Res 9(12):1632–1643. doi:10.1158/1541-7786.MCR-11-0327

Salinthone S, Tyagi M, Gerthoffer WT (2008) Small heat shock proteins in smooth muscle. Pharmacol Ther 119(1):44–54. doi:10.1016/j.pharmthera.2008.04.005

Schiller M, Javelaud D, Mauviel A (2004) TGF-beta-induced SMAD signaling and gene regulation: consequences for extracellular matrix remodeling and wound healing. J Dermatol Sci 35(2):83–92. doi:10.1016/j.jdermsci.2003.12.006

Sime PJ, Xing Z, Graham FL, Csaky KG, Gauldie J (1997) Adenovector-mediated gene transfer of active transforming growth factor-beta1 induces prolonged severe fibrosis in rat lung. J Clin Invest 100(4):768–776. doi:10.1172/JCI119590

Song IS, Kang SS, Kim ES, Park HM, Choi CY, Tchah H, Kim JY (2014) Heat shock protein 27 phosphorylation is involved in epithelial cell apoptosis as well as epithelial migration during corneal epithelial wound healing. Exp Eye Res 118:36–41. doi:10.1016/j.exer.2013.11.002

Stokoe D, Engel K, Campbell DG, Cohen P, Gaestel M (1992) Identification of MAPKAP kinase 2 as a major enzyme responsible for the phosphorylation of the small mammalian heat shock proteins. FEBS Lett 313(3):307–313

Tamura Y, Torigoe T, Kutomi G, Hirata K, Sato N (2012) New paradigm for intrinsic function of heat shock proteins as endogenous ligands in inflammation and innate immunity. Curr Mol Med 12(9):1198–1206

Tessier DJ, Komalavilas P, Panitch A, Joshi L, Brophy CM (2003) The small heat shock protein (HSP) 20 is dynamically associated with the actin cross-linking protein actinin. J Surg Res 111(1):152–157

Thannickal VJ, Horowitz JC (2006) Evolving concepts of apoptosis in idiopathic pulmonary fibrosis. Proc Am Thorac Soc 3(4):350–356. doi:10.1513/pats.200601-001TK

Thuringer D, Jego G, Wettstein G, Terrier O, Cronier L, Yousfi N, Hebrard S, Bouchot A, Hazoume A, Joly AL, Gleave M, Rosa-Calatrava M, Solary E, Garrido C (2013) Extracellular HSP27 mediates angiogenesis through Toll-like receptor 3. FASEB J. doi:10.1096/fj.12-226977

Tissieres A, Mitchell HK, Tracy UM (1974) Protein synthesis in salivary glands of Drosophila melanogaster: relation to chromosome puffs. J Mol Biol 84(3):389–398

van de Bovenkamp M, Groothuis GM, Meijer DK, Olinga P (2008) Liver slices as a model to study fibrogenesis and test the effects of anti-fibrotic drugs on fibrogenic cells in human liver. Toxicol Vitro 22(3):771–778. doi:10.1016/j.tiv.2007.11.019

Van De Schootbrugge C, Bussink J, Span PN, Sweep FC, Grenman R, Stegeman H, Pruijn GJ, Kaanders JH, Boelens WC (2013) alphaB-crystallin stimulates VEGF secretion and tumor cell migration and correlates with enhanced distant metastasis in head and neck squamous cell carcinoma. BMC Cancer 13:128. doi:10.1186/1471-2407-13-128

Vancheri C, Failla M, Crimi N, Raghu G (2010) Idiopathic pulmonary fibrosis: a disease with similarities and links to cancer biology. Eur Respir J 35(3):496–504. doi:10.1183/09031936.00077309

Vargha R, Bender TO, Riesenhuber A, Endemann M, Kratochwill K, Aufricht C (2008) Effects of epithelial-to-mesenchymal transition on acute stress response in human peritoneal mesothelial cells. Nephrol Dial Transplant 23(11):3494–3500. doi:10.1093/ndt/gfn353

Verrecchia F, Chu ML, Mauviel A (2001) Identification of novel TGF-beta /Smad gene targets in dermal fibroblasts using a combined cDNA microarray/promoter transactivation approach. J Biol Chem 276(20):17058–17062. doi:10.1074/jbc.M100754200

Vidyasagar A, Reese S, Acun Z, Hullett D, Djamali A (2008) HSP27 is involved in the pathogenesis of kidney tubulointerstitial fibrosis. Am J Physiol Renal Physiol 295(3):F707–F716. doi:10.1152/ajprenal.90240.2008

Vidyasagar A, Reese SR, Hafez O, Huang LJ, Swain WF, Jacobson LM, Torrealba JR, Chammas PE, Wilson NA, Djamali A (2013) Tubular expression of heat-shock protein 27 inhibits fibrogenesis in obstructive nephropathy. Kidney Int 83(1):84–92. doi:10.1038/ki.2012.336

Voellmy R (2006) Feedback regulation of the heat shock response. Handb Exp Pharmacol 172:43–68

Vos MJ, Hageman J, Carra S, Kampinga HH (2008) Structural and functional diversities between members of the human HSPB, HSPH, HSPA, and DNAJ chaperone families. Biochemistry 47(27):7001–7011. doi:10.1021/bi800639z

Wang X, Zhao T, Huang W, Wang T, Qian J, Xu M, Kranias EG, Wang Y, Fan GC (2009) Hsp20-engineered mesenchymal stem cells are resistant to oxidative stress via enhanced activation of Akt and increased secretion of growth factors. Stem Cells 27(12):3021–3031. doi:10.1002/stem.230

Welge-Lussen U, May CA, Eichhorn M, Bloemendal H, Lutjen-Drecoll E (1999) AlphaB-crystallin in the trabecular meshwork is inducible by transforming growth factor-beta. Invest Ophthalmol Vis Sci 40(10):2235–2241

Wells RG (2000) Fibrogenesis.V. TGF-beta signaling pathways. Am J Physiol Gastrointest Liver Physiol 279(5):G845–G850

Wettstein G, Bellaye PS, Micheau O, Bonniaud P (2012) Small heat shock proteins and the cytoskeleton: an essential interplay for cell integrity? Int J Biochem Cell Biol 44(10):1680–1686. doi:10.1016/j.biocel.2012.05.024

Wettstein G, Bellaye PS, Kolb M, Hammann A, Crestani B, Soler P, Marchal-Somme J, Hazoume A, Gauldie J, Gunther A, Micheau O, Gleave M, Camus P, Garrido C, Bonniaud P (2013) Inhibition of HSP27 blocks fibrosis development and EMT features by promoting Snail degradation. FASEB J 27(4):1549–1560. doi:10.1096/fj.12-220053

Wisniewski T, Goldman JE (1998) Alpha B-crystallin is associated with intermediate filaments in astrocytoma cells. Neurochem Res 23(3):385–392

Wrana JL, Attisano L, Wieser R, Ventura F, Massague J (1994) Mechanism of activation of the TGF-beta receptor. Nature 370(6488):341–347. doi:10.1038/370341a0

Yu AL, Fuchshofer R, Birke M, Priglinger SG, Eibl KH, Kampik A, Bloemendal H, Welge-Lussen U (2007) Hypoxia/reoxygenation and TGF-beta increase alphaB-crystallin expression in human optic nerve head astrocytes. Exp Eye Res 84(4):694–706. doi:10.1016/j.exer.2006.12.008

Zhao J, Shi W, Wang YL, Chen H, Bringas P Jr, Datto MB, Frederick JP, Wang XF, Warburton D (2002) Smad3 deficiency attenuates bleomycin-induced pulmonary fibrosis in mice. Am J Physiol Lung Cell Mol Physiol 282(3):L585–L593. doi:10.1152/ajplung.00151.2001

Zhuang H, Jiang W, Cheng W, Qian K, Dong W, Cao L, Huang Q, Li S, Dou F, Chiu JF, Fang XX, Lu M, Hua ZC (2010) Down-regulation of HSP27 sensitizes TRAIL-resistant tumor cell to TRAIL-induced apoptosis. Lung Cancer 68(1):27–38. doi:10.1016/j.lungcan.2009.05.014

Zoubeidi A, Gleave M (2012) Small heat shock proteins in cancer therapy and prognosis. Int J Biochem Cell Biol 44(10):1646–1656. doi:10.1016/j.biocel.2012.04.010

Chapter 14
Neurodegenerative Diseases, Sex Differences and the 27 kDa Heat Shock Protein in the Nervous System

Danielle A. Rioux, Kathleen Murphy, Michael J. Esser, and R. William Currie

Abstract There are sex differences in degenerative disease prevalence in humans. Most models of degenerative disease use male animals. Examining female and male responses to stress may give insight into disease prevalence. Heat shock proteins are chaperones linked to damaged proteins in degenerative diseases and may be expressed differentially in females and males. We have characterized the induced expression of Hsp27 and Hsp70 in the brain and the heart of female and male rats. Rats were heat shocked, brains were removed 24 h after, and western analyses were done to quantify the expression of these proteins. Immunofluorescence was used to localize Hsp70 and Hsp27 in the hippocampus. Overall, male rats have significantly greater induced expression of both Hsp27 and Hsp70 in the brain. In the hippocampus, Hsp27 was localized in astrocytes, and Hsp70 was localized in blood vessels and microglia, following heat shock. Therefore, the difference in expression of heat shock proteins seen in the brain of male and female rats following heat shock suggests that sex hormones may have an impact on degenerative disease prevalence.

Keywords Hsp27 (HSPB1) • Hippocampus • Sex differences • Heat shock • Immunofluorescence

14.1 Introduction

As a result of the increased life expectancy, chronic neurodegenerative diseases such as amyotrophic lateral sclerosis (ALS), Alzheimer's disease (AD), and Parkinson's disease (PD) are becoming more common (Hebert et al. 2003). These

D.A. Rioux • K. Murphy • R.W. Currie (✉)
Department of Medical Neuroscience, Dalhousie University, Halifax, NS, Canada
e-mail: wcurrie@dal.ca

M.J. Esser
Department of Pediatrics, Alberta Children's Hospital Research Institute, Calgary, AB, Canada

R.M. Tanguay, L.E. Hightower (eds.), *The Big Book on Small Heat Shock Proteins*, Heat Shock Proteins 8, DOI 10.1007/978-3-319-16077-1_14

neurodegenerative diseases have a strong prevalence for either males or females. However, most research into these diseases tends to be done in male animal models of disease. One common feature of all these diseases of aging is the accumulation of abnormal protein aggregates and the cellular response to proteotoxicity with altered expression levels of various heat shock proteins. In this chapter, we will summarize the localization and levels of Hsp27 in the brain of normal and heat shocked adult female and male rats. This characterization may provide insights into female and male differences in disease prevalence.

14.2 Expression of Heat Shock Proteins in the Diseased Brain

Amyotrophic lateral sclerosis, Parkinson's disease, and Alzheimer's disease are all neurodegenerative diseases that have an accumulation of intracellular and extracellular protein aggregates, protein misfolding, and neuronal loss in the central nervous system (Muchowski and Wacker 2005). In ALS, motor neurons of the spinal cord and motor cortex are selectively lost (Martin 1999). In PD dopaminergic neurons in the substantia nigra are lost (Martin 1999). In AD neuronal loss is prominent in the entorhinal cortex and hippocampus (Martin 1999; Bruijn et al. 2004; Schapira and Olanow 2004).

Neurons are post mitotic and are therefore unable to dilute the toxic effects of misfolded and damaged proteins through cell division. Thus neurons are especially vulnerable to aggregated and misfolded proteins and often die by apoptosis (Muchowski and Wacker 2005). Heat shock proteins (HSP) function as chaperones to prevent apoptosis, protein misfolding, and protein aggregation. HSP are potent suppressors of neurodegenerative diseases (Renkawek et al. 1994; Muchowski and Wacker 2005). Transgenic upregulation of Hsp27 and Hsp70 in animal models of neurodegenerative diseases seem to be protective by buffering the toxicity of aggregated or damaged proteins (Patel et al. 2005).

14.3 Disease Prevalence in the Brain

Amyotrophic lateral sclerosis is a neurodegenerative disease of upper and lower motoneurons. In North America and European countries, while peak incidence of ALS in men is between the ages of 70–74 and peak incidence in women is between the ages of 65–74, men are two times more likely than women to be diagnosed with ALS between the ages of 25–55. Between the ages of 55–75, men are 1.5 times more likely than post-menopausal women to be diagnosed with ALS (Huisman et al. 2011a). This suggests that estrogen in women may be a factor in the prevalence of ALS in females and males.

Alzheimer's disease is a degenerative condition and the most common form of dementia in people over the age of 60. AD is most often diagnosed between the ages of 65–69 years of age, and especially in women, where prevalence is approximately twice that in men (Moschetti et al. 2012). This female prevalence appears to be linked to the levels of estrogen in women, as this disease occurs approximately 10–20 years past the age at which most women begin menopause (Kawas et al. 1997; Gao et al. 1998; Hampel et al. 2011). This suggests the decline in estrogen in post-menopausal women may make women vulnerable to AD.

Parkinson's disease is a degenerative disease of the central nervous system resulting from death of dopaminergic neurons in the substantia nigra. PD is 1.5–2 times more common in men, compared to women, and the average age for diagnosis is usually between the ages of 50–60 years. Given that menopause affects women between the ages of 45–65 and the majority of cases of PD are diagnosed between the ages of 50–60, it appears that the age of menopausal women coincides with the age of greatest PD diagnoses (Van Den Eeden et al. 2003), which highlights estrogen levels as a possible factor contributing to PD onset (Kompoliti 2003).

14.4 Sex Bias in Animal Models of Disease

Diseases of the nervous system affect males and females disproportionately. While various research funding agencies mandate that women and minorities be included in research studies (Wald and Wu 2010; Beery and Zucker 2011), males remain the predominate sex used in animal studies of these diseases. As a result promising treatments identified in male animal models have failed to translate to clinical trials (van der Worp et al. 2010). The lack of research done in female animal models may be a major factor contributing to the failures seen in clinical trials related to therapeutics.

Of some 2,000 published animal studies examining human diseases, about 80 % had a strong male bias, particularly in neuroscience and pharmacology (Beery and Zucker 2011). Most researchers avoid female animals because of concerns with variations seen in data due to the fluctuations of estrogen levels during the estrous cycle. In order to take into consideration the varying levels of estrogen during the estrous cycle, compared to males, four times as many female animals, corresponding to the four stages of the estrous cycle, may be required in a study (Wald and Wu 2010). Taking estrogen levels into consideration is important in order to develop results that can be translatable to human populations for purposes of studying disease manifestation and therapeutics to treat and manage diseases.

One common feature of neurodegenerative diseases is the accumulation of abnormally folded protein within cells. These intracellular protein aggregates seem to be proteotoxic and the proteotoxicity induces expression of various heat shock proteins. Heat shock proteins are in all cells of the body and protect the cells against physical and chemical stressors, including proteotoxicity. HSP act as molecular chaperones (Benarroch 2011) and are involved in a variety of degenerative diseases, such as ALS, AD, and PD buffering the effects of the various stressors, such as reactive oxygen species (Renkawek et al. 1994; Xu and Wick 1996; Maatkamp et al. 2004; Muchowski and Wacker 2005; Huang et al. 2006; Sharp et al. 2007; reviewed by Xu et al. 2012).

14.5 Expression of Heat Shock Proteins

Heat shock proteins were discovered in 1962, and are a set of highly conserved chaperone proteins found in eukaryotes and prokaryotes (Ritossa 1962, 1996). Some HSP are constitutively expressed in cells and function to promote proper folding and assembly of polypeptides, and are involved in cell signaling and protein trafficking. Other HSP are rapidly induced in response to cellular stress, like hyperthermia and ischemic injury, and function to prevent the formation of protein aggregates (Benarroch 2011).

Mammalian HSP have been classified into five families according to their molecular mass (in kDa): HSP110 (HSPH), HSP90 (HSPC), HSP70 (HSPA), HSP40 (DNAJ) and small HSP (HSPB, 15–30 kDa) including Hsp27 (Lindquist and Craig 1988; Kampinga et al. 2009). Each family of HSP is composed of members that have constitutive and induced expression. For example the HSP70 family includes two isoforms, Hsc70 (HSPA8) and Hsp70 (HSPA1A). Under normal conditions Hsc70 is abundant while Hsp70 is at low or undetectable levels in most tissues. Following heat shock or ischemic injury, Hsp70 is synthesized at high levels and is abundant in tissues for days (Currie and White 1983; Karmazyn et al. 1990). Hsp27 (HSPB1) is even more interesting being abundant under normal conditions in muscle and even in some specific neurons of the brain stem and spinal cord, but being undetectable in cerebrum. After various injuries, Hsp27 is highly inducible in neurons and glial cells in the central nervous system (Plumier et al. 1997a, b; Krueger-Naug et al. 2000).

Hsp27 has several functions, such as: (1) a molecular chaperone (Diaz-Latoud et al. 2005) by inhibiting actin polymerization (Weiske et al. 2001), (2) an antioxidant by reducing the effect of tumor necrosis factor-α-mediated cell death, by phosphorylation (Mehlen et al. 1995), and (3) a suppressor of signaling events, leading to apoptosis, by blocking cytochrome c release therefore suppressing caspase-3 (Pandey et al. 2000). Importantly, the functional diversity of Hsp27 appears to be regulated by its phosphorylation state (Rogalla et al. 1996).

Under normal conditions in the cytosol, Hsp27 is in an oligomeric form and following a stress, the monomeric Hsp27 becomes phosphorylated by several mechanisms, including mitogen-activated-protein-kinases, which causes Hsp27 to either stay in a monomeric phosphorylated form or for it to dimerize. Monomeric phosphorylated Hsp27 can then bind to actin and prevent protein damage and apoptosis. In contrast, dimeric phosphorylated Hsp27 targets denatured proteins to the proteasome (Ferns et al. 2006). Normal levels of Hsp27 in the rat may provide neurons with a line of defense against protein misfolding and aggregation resulting from neurodegenerative diseases that impair normal neuronal functioning (Chen and Brown 2007).

14.6 Sex Differences in Expression of Heat Shock Proteins in the Brain

As noted above, most research on expression of HSP has been done in male animals, even though there are sex differences in neurological disease prevalence (Huisman et al. 2011b; Van Den Eeden et al. 2003) and HSP are strongly associated with various degenerative diseases of the brain.

Curiously, in a study that used various regions of the brain in both female and male rats to examine the constitutive levels of several Hsp, the results were reported for the females and males as a combined group, except in one region: the hypothalamus (Bodega et al. 2002). Hsp70 levels were greater in the hypothalamus of 3-week old male rats, compared to same age female rats. By 8 months of age, female rats had greater levels of Hsp70 in the hypothalamus, compare to male rats.

In the brain of adult male rats, Hsp70 is present at very low levels (Krueger et al. 1999). Following hyperthermic treatment, Hsp70 levels in male rats significantly increased in the hippocampus, specifically in glial cells, at 3, 6, 12 and 24 h following hyperthermic treatment, compared to controls that were not heat shocked. Similarly, using female rats, Hsp70 was significantly increased within 4 h following heat shock, and was detected in glial cells and vascular cells in the cerebral cortex and hippocampus (Krueger et al. 1999; Bechtold et al. 2000; Pavlik et al. 2003).

Most studies examining the Hsp27 levels in the brain are done in male animals. Hsp27 is undetectable in the cerebral cortex of unstressed male rats, and is present normally in the brain stem and spinal cord (Plumier et al. 1997b). After heat shock, by 2 h, Hsp27 is expressed at high levels in many cells (mostly glial) in the male rat brain including the hippocampus (Krueger-Naug et al. 2000; Bechtold and Brown 2003). Hsp27 levels peaked at 24 h following hyperthermia and returned to constitutive or basal levels only after more than 6 days (Krueger-Naug et al. 2000). Thus, most research on the induced expression of Hsp70 and Hsp27 has been done in male rats and there is little or no information on the expression levels of Hsp70 and Hsp27 in the female rat brain.

14.7 Sex Specific Gonadal Hormones

Sex specific disease prevalence may be related to gonadal hormones. The most obvious hormonal differences between females and males are the circulating levels of estrogen and testosterone, respectively. Additionally, in females levels of progesterone may also influence disease prevalence. Age-related hormonal changes increase the risk of neurodegenerative and cardiovascular diseases (Kalin and Zumoff 1990; Margo and Winn 2006; Barron and Pike 2012). Throughout a lifetime, the brain is a hormonally responsive organ and therefore affected by the fluctuating levels of sex hormones, especially with age (Kalin and Zumoff 1990; Margo and Winn 2006; Barron and Pike 2012). Here we will focus on estrogen in the brain and expression of heat shock proteins.

14.8 Estrogen in the Brain

There are three types of biologically active estrogens, estrone, estriole and 17-beta-estradiol (E2). E2 is produced by the ovaries and is the most concentrated in plasma during reproductive years and is the most potent in estrogenic effects. Plasma E2 concentrations fluctuate throughout the estrous cycle in females (Wojtys et al. 2002). For example E2 plasma level concentrations range from 499 to 573 pg/mL, in diestrus; 145 to 2,178 pg/mL, in proestrus; 161 pg/mL, in estrus; and 401 pg/mL, in metaestrus (Shaikh 1971).

Premenopausal women between 20 and 50 years of age have E2 plasma concentrations of approximately 81 pg/mL, whereas postmenopausal women between 51 and 65 years of age have E2 plasma concentrations of approximately 35 pg/mL. On average males have E2 plasma concentrations of approximately 18 pg/mL that is consistent between of 20 and 65 years of age (Massafra et al. 2002).

Besides its physiological role in the female reproductive system, E2 modulates the nervous system by activation of the estrogen receptors that can alter gene transcription in the nucleus or acutely activate kinase signaling in the cytosol. Estrogen receptor-alpha (ER-α) with a molecular mass of 67 kDa and estrogen receptor-beta (ER-β) with a molecular mass of 59 kDa are both highly homologous and expressed in the nervous system (reviewed by Kumar et al. 2011).

In the hippocampus, estrogen receptors are found in interneurons throughout the dentate gyrus and striatum radiatum with the number of positive cells and intensity of immunoreactivity is equivalent in female and male rats (Weiland et al. 1997). However, there may be sex differences in the localization of ER-α in the hippocampus. ER-α may be more abundant in dendritic spines of CA1 region of the female hippocampus compared to the male hippocampus (Romeo et al. 2005). In the female rat hippocampus, ER-α immunoreactivity is predominantly in the nuclei of interneurons, in the CA1 region and in the hilus of the dentate gyrus (Milner et al. 2001). Some ER-α immunoreactivity is localized in the perikarya, in dendritic spines, axons and axon terminals of pyramidal and granular cells within the CA1 region. Some extranuclear ER-α immunoreactivity is localized in astrocytes throughout the hippocampal formation (Milner et al. 2001).

In the male rat hippocampus, ER-α immunoreactivity is predominantly in the nuclei of pyramidal cells of the CA1 and CA3 regions (Sakuma et al. 2009).

In comparison, ER-β immunoreactivity is localized to similar hippocampal regions in both female and male rats, including the dentate gyrus, CA1, and CA3 regions, and most is non-nuclear. Furthermore, ER-β immunoreactivity is localized in the cytoplasm, dendritic spines, dendrites, axons and axon terminals of pyramidal and granule cells (Azcoitia et al. 1999; Milner et al. 2005; Sakuma et al. 2009).

In addition to estrogen, Hsp90 also binds to estrogen receptors (Milne and Noble 2008). The estrogen receptors are normally bound to Hsp90. When E2 binds to the receptor, Hsp90 is displaced and free to bind to the HSF1, leading to decreased trimerization of HSF1 and decreased transcription of HSP (Milne and Noble 2008). Estrogen and the estrogen receptor translocate to the nucleus and bind to the estrogen response element, increasing transcription of estrogen regulated genes (Milne and Noble 2008). However, if there is stress and damaged proteins present, Hsp90 chaperones the damaged proteins leaving HSF1 free to trimerize, translocate to the nucleus and increase the transcription of HSP (Milne and Noble 2008). This is the proposed mechanism for estrogen regulation of HSP expression.

14.9 The Effects of Estrogen on the Expression of Heat Shock Proteins in the Brain

Estrogen supplementation in males increases the levels of HSP in the hippocampus. Six hours after the administration of E2 in male rats, there is significantly more Hsp27 and Hsp70 immunoreactive cells in cerebral blood vessels, compared with the brains of control rats (Lu et al. 2002). Additionally, immunoreactive Hsp27 levels in astrocytes and in hippocampal neurons are increased in the brains of E2 treated animals (Lu et al. 2002). Also, in the brains of animals given E2, the heat shock transcription factor-1 levels are significantly greater, compared to animals that did not receive E2 supplementation (Lu et al. 2002).

Similarly, in the cortex, HSF1 is at a lower level after heat shock treatment in ovariectomized female rats treated with estrogen rather than with testosterone, and perhaps Hsp70 levels were also lower in the cortex of estrogen-treated rats, compared to testosterone-treated rats (Papasozomenos and Papasozomenos 2008). Hsp70 is linked with gonadal hormones and estrogen levels may influence the expression of heat shock protein in the brain.

It seems clear that in the brain these male and female differences in levels of Hsp27 and Hsp70 have not been explored. However, it seems reasonable to expect sex differences in levels of HSP because of the differences in prevalence of diseases of proteopathies in the brain. Due to differences in the prevalence of diseases associated with abnormal protein folding, it seems reasonable to consider that there may be sex differences in the baseline and/or induced levels of Hsp27 and Hsp70 in the brain of female and male rats.

We have now explored the levels of Hsp27 and Hsp70 in the brain of female and male control rats and in female and male rats 24 h after heat shock treatment, and to identify cell types expressing Hsp27 and Hsp70 in the hippocampus of female and male rats.

14.10 Western Analysis of Hsp70 and Hsp27 in the Hippocampus

Male and female Sprague-Dawley rats (Charles River, Quebec, Canada; 225–250 g) were cared for in accordance with the *Guide for Care and Use of Experimental Animals* of the Canadian Council on Animal Care. The experimental procedures used were approved by the Dalhousie University Committee on Laboratory Animals. The animals were weighed and given an intra-peritoneal injection of sodium pentobarbital (CEVA, McGill, 35 mg/kg). Once the animals were anesthetized, based on pedal reflex, a lubricated thermometer was inserted rectally. The control rats were constantly monitored but were not warmed. The heat shock rats were placed on an insulated heating pad (48 °C), covered, and constantly monitored. Once the core body temperature reached 42 °C, this temperature was maintained for 15 min, by adjusting the heating pad. After 15 min at 42 °C, the animals were removed from the pad and allowed to recover. Once the rats became ambulatory and drank and ate on their own, they were placed in a clean cage.

Twenty-four hours after the control and heat shock treatments, female and male rats were given an overdose of sodium pentobarbital and hippocampus harvested for analysis. Hsp70 and Hsp27 levels were detected by western analysis in the hippocampus of control and 24 h heat shock female and male rats. β-tubulin levels were detected and served as a loading control and for normalization of the levels of Hsp70 and Hsp27.

By western analysis, little or no Hsp70 was detected in the hippocampus of control female and control male rats (Fig. 14.1a). Hsp27 was easily detectable and at similar levels in the hippocampus of control female and male rats (Fig. 14.1a). Statistical analysis, using an unpaired two-tailed t-test, revealed no differences in levels of Hsp70 or Hsp27 in the hippocampus between control female ($n=5$) and male ($n=5$) rats ($p>0.05$).

Hsp70 and Hsp27 levels in the hippocampus were low in control female rats and appeared to be increased in female rats 24 h following heat shock treatment (Fig. 14.1b). Statistical analysis, using an unpaired two-tailed t-test, revealed significantly greater levels of Hsp70 ($p<0.05$; Fig. 14.1b) and Hsp27 ($p<0.05$;

Fig. 14.1 (continued) to be greater in the hippocampus 24 h following heat shock treatment in female rats compared to control female rats. Statistical analysis revealed significant differences in Hsp70 and Hsp27 levels in the hippocampus of control female ($n=5$) and 24 h after heat shock female ($n=6$) rats (*, $p<0.05$). (**c**) By Western analysis, Hsp70 and Hsp27 levels appeared to be greater in the hippocampus 24 h following heat shock treatment in male rats compared to control male rats. Statistical analyses revealed significant differences in Hsp70 and Hsp27 levels in the hippocampus of control male ($n=5$) and 24 h after heat shock male ($n=6$) rats (**, $p<0.01$). (**d**) By Western analysis, Hsp70 and Hsp27 levels appeared to be greater 24 h following heat shock treatment in the hippocampus of male rats compared to female rats. Statistical analyses revealed significant differences in Hsp70 and Hsp27 levels in the hippocampus of 24 h after heat shock female ($n=6$) and 24 h after heat shock male ($n=6$) rats (*, $p<0.05$). Data are presented as mean ± SEM

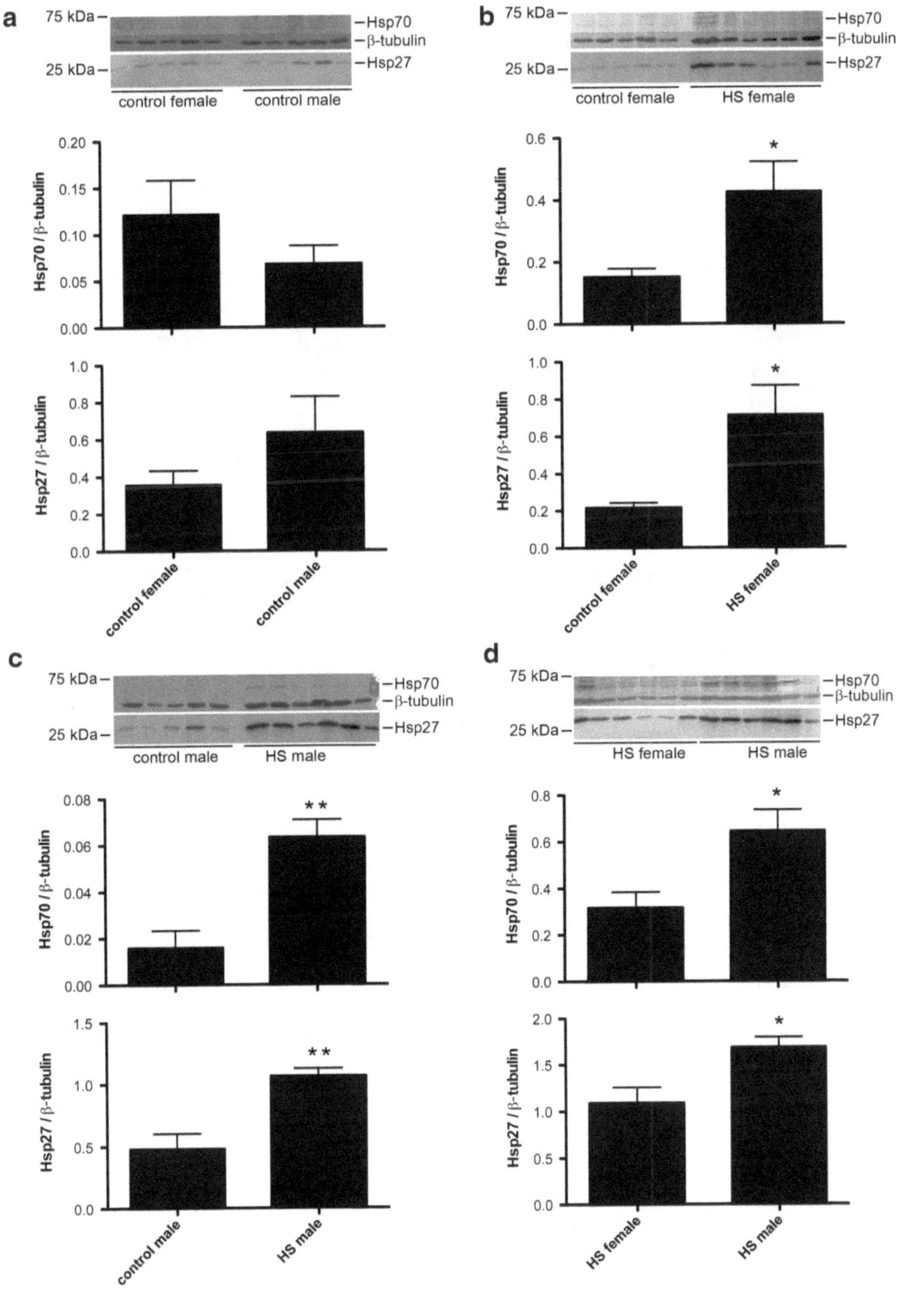

Fig. 14.1 Western analysis of Hsp70 and Hsp27 levels in the hippocampus of control and heat shocked female and male rats. (**a**) By Western analysis, little or no Hsp70 and low levels of Hsp27 were detected in the hippocampus of control female and male rats. Statistical analyses revealed no significant differences in the levels of Hsp70 and Hsp27 in the hippocampus of control female (n=5) and male (n=5) rats (p>0.05). (**b**) By Western analysis, Hsp70 and Hsp27 levels appeared

Fig. 14.1b) in the hippocampus of heat shock female (n = 6) rats, compared to control female (n = 5) rats.

Hsp70 and Hsp27 levels in the hippocampus were low in control male rats and appeared to be increased in male rats 24 h following heat shock treatment (Fig. 14.1c). Statistical analysis, using an unpaired two-tailed t-test, revealed significantly greater levels of Hsp70 ($p < 0.01$; Fig. 14.1c) and Hsp27 ($p < 0.01$; Fig. 14.1c) in the hippocampus of heat shock male (n = 6) rats, compared to control male (n = 5) rats.

Hsp70 and Hsp27 levels in the hippocampus were easily detectable 24 h following heat shock treatment in both female and male rats (Fig. 14.1d). Heat shock male rats appeared to have greater levels of Hsp70 and Hsp27 than heat shock female rats. Statistical analysis, using an unpaired two-tailed t-test, revealed significantly greater levels of Hsp70 ($p < 0.05$; Fig. 14.1d) and Hsp27 ($p < 0.05$; Fig. 14.1d) in the hippocampus of heat shock male (n = 6) rats, compared to heat shock female (n = 6) rats.

14.11 Localization of Hsp70 and Hsp27 in the Hippocampus by Immunofluorescence Microscopy

At 24 h following control or heat shock treatment, animals were deeply anesthetized with an intra-peritoneal injection of sodium pentobarbital (100 mg/kg) and then perfused transcardially with 0.9 % saline for approximately 1 min to remove blood. They were then perfused with 4 % paraformaldehyde in 0.1 M sodium phosphate buffer (pH = 7.4, 4 °C) for approximately 2 min. Brains were then harvested for immunofluorescence microscopy. Brains were then post-fixed in 4 % paraformaldehyde in 0.1 M sodium phosphate buffer (pH = 7.4) for approximately 48 h at 4 °C, followed by immersion in 30 % sucrose made up in 0.1 M phosphate buffer (pH = 7.4). Serial coronal 30 μm sections of the brain were cut on a freezing microtome (Leica) then stored in a refrigerator (4 °C) in Millonig's buffer (0.1 M sodium phosphate with 0.03 % sodium azide, pH = 7.4) until being processed for immunofluorescence microscopy.

14.12 Hsp70 Immunofluorescence

The free-floating 30 μm coronal brain sections were washed 3 × 10 min in phosphate buffered saline with Triton-X 100 (PBS-T), and sections were incubated in 1 % hydrogen peroxide made in PBS-T for 20 min. Sections were washed 3 × 10 min in PBS-T before incubating for 1 h in 2 % donkey serum (Millipore) in PBS-T. Hippocampal sections immunoreacted with antibodies against Hsp70, were also immunoreacted with antibodies against either glial fibrillary acidic protein (GFAP),

Table 14.1 Hsp70, GFAP, NeuN, Iba1, and factor VIII primary and secondary antibodies for immunofluorescence microscopy

Primary antibodies	Secondary antibodies
Hsp70 monoclonal mouse antibody at 1:500 (Enzo Scientific, ADI-SPA-810)	Donkey anti-mouse IgG antibody conjugated to Alexa Fluor 488 at 1:500 (Invitrogen, A21202)
	Donkey anti-mouse IgG antibody conjugated to Alexa Fluor 555 at 1:500 (Invitrogen, A31570)
GFAP polyclonal rabbit antibody at 1:500 (Dako, Z0334)	Donkey anti-rabbit IgG antibody conjugated to Alexa Fluor 488 at 1:500 (Invitrogen, A21206)
NeuN polyclonal rabbit antibody conjugated to Alexa Fluor 488 at 1:500 (Millipore, ABN78A4)	[a]
Iba1 polyclonal rabbit antibody at 1:500 (Wako, 019–19741)	Donkey anti-rabbit IgG antibody conjugated to Alexa Fluor 488 at 1:500 (Invitrogen, A21206)
Factor VIII polyclonal rabbit at 1:500 (Oncogene Research Products, PC313)	Donkey anti-rabbit IgG antibody conjugated to Alexa Fluor 555 at 1:500 (Invitrogen, A31570)

[a]The NeuN polyclonal rabbit antibody conjugated to Alexa Fluor 488 no primary control sections were incubated with donkey anti-rabbit IgG antibody conjugated to Alexa Fluor 488 at 1:500 (Invitrogen, A21206)

neuronal nuclei (NeuN), ionized calcium binding adaptor molecule 1 (Iba1), or Factor VIII. The sections were incubated for 1 h at room temperature and then put at 4 °C on an orbital shaker for 3 days (see Table 14.1 for antibody information).

After 3 days of incubating in the primary antibodies, sections were washed 3 × 10 min in PBS. Sections were then incubated on an orbital shaker at room temperature for 2 h in fluorescent secondary antibodies diluted in 2 % donkey serum (see Table 14.1 for secondary antibody information).

14.13 Hsp27 Immunofluorescence

Free-floating 30 μm coronal brain sections were washed 3 × 10 min in PBS-T, and then incubated in 1 % hydrogen peroxide made in PBS-T for 20 min. Sections were washed 3 × 10 min in PBS-T before incubating for 1 h in 2 % donkey serum (Millipore). Hippocampal sections immunoreacted with antibodies against Hsp27, were also immunoreacted with antibodies against either GFAP, NeuN, Iba1 or Hsp70. The sections were incubated for 1 h at room temperature and then put at 4 °C on a shaker for 3 days (see Table 14.2 for antibody information).

After 3 days of incubating in the primary antibodies, sections were washed 3 × 10 min in PBS. Sections were then incubated on a shaker at room temperature for 2 h in fluorescent secondary antibodies diluted in 2 % donkey serum (see Table 14.2 for secondary antibody information).

Table 14.2 Hsp27, GFAP, NeuN, Iba1, and Hsp70 primary and secondary antibodies for immunofluorescence microscopy

Primary antibodies	Secondary antibodies
Hsp27 polyclonal rabbit antibody at 1:500 (Enzo Scientific, ADI-SPA-801)	donkey anti-rabbit IgG antibody conjugated to Alexa Fluor 488 at 1:500 (Invitrogen, A21206)
GFAP monoclonal mouse antibody conjugated to Cy3 at 1:500 (Sigma, C92050)	[a]
NeuN monoclonal mouse antibody at 1:500 (Millipore, MAB377)	donkey anti-mouse IgG antibody conjugated to Alexa Fluor 555 at 1:500 (Invitrogen, A31570)
	donkey anti-mouse IgG antibody conjugated to Alexa Fluor 488 at 1:500 (Invitrogen, A21202)
Iba1 monoclonal mouse antibody at 1:500 (Abcam, ab15690)	donkey anti-mouse IgG antibody conjugated to Alexa Fluor 555 at 1:500 (Invitrogen, A31570)
Hsp70 monoclonal mouse antibody at 1:500 (Enzo Scientific, ADI-SPA-810)	donkey anti-mouse IgG antibody conjugated to Alexa Fluor 555 at 1:500 (Invitrogen, A31570)

[a]The GFAP monoclonal mouse antibody conjugated to Cy3 no primary control sections were incubated with donkey anti-mouse IgG antibody conjugated to Cy3 at 1:500 (Jackson Immuno Research, 715-165-151)

14.14 Immunoreactivity Specificity Controls

No Hsp70 immunoreactive cells were detected in the hippocampus with omission of the primary Hsp70 monoclonal mouse antibody or omission of the secondary donkey anti-mouse IgG antibody conjugated to Alexa Fluor 555 or the secondary donkey anti-mouse IgG antibody conjugated to Alexa Fluor 488 (data not shown).

No Hsp27 immunoreactive cells were detected with omission of the primary Hsp27 polyclonal rabbit antibody or omission of the secondary donkey anti-rabbit IgG antibody conjugated to Alexa Fluor 488 (data not shown).

No immunoreactivity was detected with omission of the primary GFAP polyclonal rabbit antibody or with omission of the secondary donkey anti-rabbit IgG antibody conjugated to Alexa Fluor 488. No immunoreactivity was detected with omission of the primary GFAP monoclonal mouse antibody conjugated to Cy3 or with incubation with inappropriate secondary donkey anti-mouse IgG antibody conjugated to Cy3, or with secondary donkey anti-rabbit IgG antibody conjugated to Alexa Fluor 488 (data not shown).

No immunoreactive neurons were detected with omission of the primary NeuN monoclonal mouse antibody and with the omission of the donkey anti-mouse IgG secondary antibody conjugated to Alexa Fluor 488. No immunoreactivity was detected with omission of the primary NeuN polyclonal rabbit antibody conjugated to Alexa Fluor 488 or with incubation with inappropriate secondary donkey anti-rabbit IgG antibody conjugated to Alexa Fluor 488, or with secondary donkey anti-mouse IgG antibody conjugated to Alexa Fluor 555 (data not shown).

No immunoreactive microglia were detected with omission of the primary Iba1 polyclonal rabbit antibody and with the omission of the secondary donkey anti-rabbit IgG antibody conjugated to Alexa Fluor 488. No Iba1 immunoreactivity was detected with omission of the primary Iba1 monoclonal mouse antibody or the secondary donkey anti-mouse IgG antibody conjugated to Alexa Fluor 555 (data not shown).

No immunoreactivity was detected with omission of the primary Factor VIII polyclonal rabbit antibody and with omission of the secondary donkey anti-rabbit IgG antibody conjugated to Alexa Fluor 555 (data not shown).

14.15 Immunofluorescence Double Labeling for Hsp70 and GFAP, NeuN, and Iba1

Hsp70 immunoreactivity was undetectable in the hippocampus of control female rats (Fig. 14.2A1, A2, A3) and was detectable in many cells and small blood vessels 24 h following heat shock in the hippocampus of female rats (Fig. 14.2B1, B2, B3). Hsp70 immunoreactivity was detectable in a few cells of the hippocampus of control male rats (Fig. 14.2C1, C2, C3) and in many cells and small blood vessels 24 h following heat shock in the hippocampus of male rats (Fig. 14.2D1, D2, D3).

GFAP immunoreactivity was detectable in many cells in the hippocampus in control female (Fig. 14.2A1), heat shock female (Fig. 14.2B1), control male (Fig. 14.2C1), and heat shock male (Fig. 14.2D1) rats. Micrographs show no colocalization of Hsp70 and GFAP.

NeuN immunoreactivity was detectable in many cells in the hippocampus of control female (Fig. 14.2A2), heat shock female (Fig. 14.2B2), control male (Fig. 14.2C2), and heat shock male (Fig. 14.2D2) rats. Micrographs show no colocalization of Hsp70 and NeuN.

Iba1 immunoreactivity was detectable in many cells in the hippocampus of control female (Fig. 14.2A3), heat shock female (Fig. 14.2B3), control male (Fig. 14.2C3), and heat shock male (Fig. 14.2D3) rats. Micrographs of Hsp70 and Iba1 immunoreactivity show some colocalization of Hsp70 and Iba1. After heat shock in the female, Hsp70 immunoreactivity (Fig. 14.2B3, arrows) was detected mostly in Iba1 negative cells and in some small blood vessels. After heat shock in the male, Hsp70 immunoreactivity (Fig. 14.2D3, arrows) was detected in some Iba1 immunoreactive cells and some small blood vessels.

14.16 Immunofluorescence Double Labeling for Hsp27 and GFAP, NeuN, and Iba1

Hsp27 immunoreactivity was undetectable in the hippocampus of control female rats (a few fluorescent red blood cells are evident) (Fig. 14.3A1, A2, A3) and was detectable in small blood vessels and many astrocytes in the hippocampus 24 h

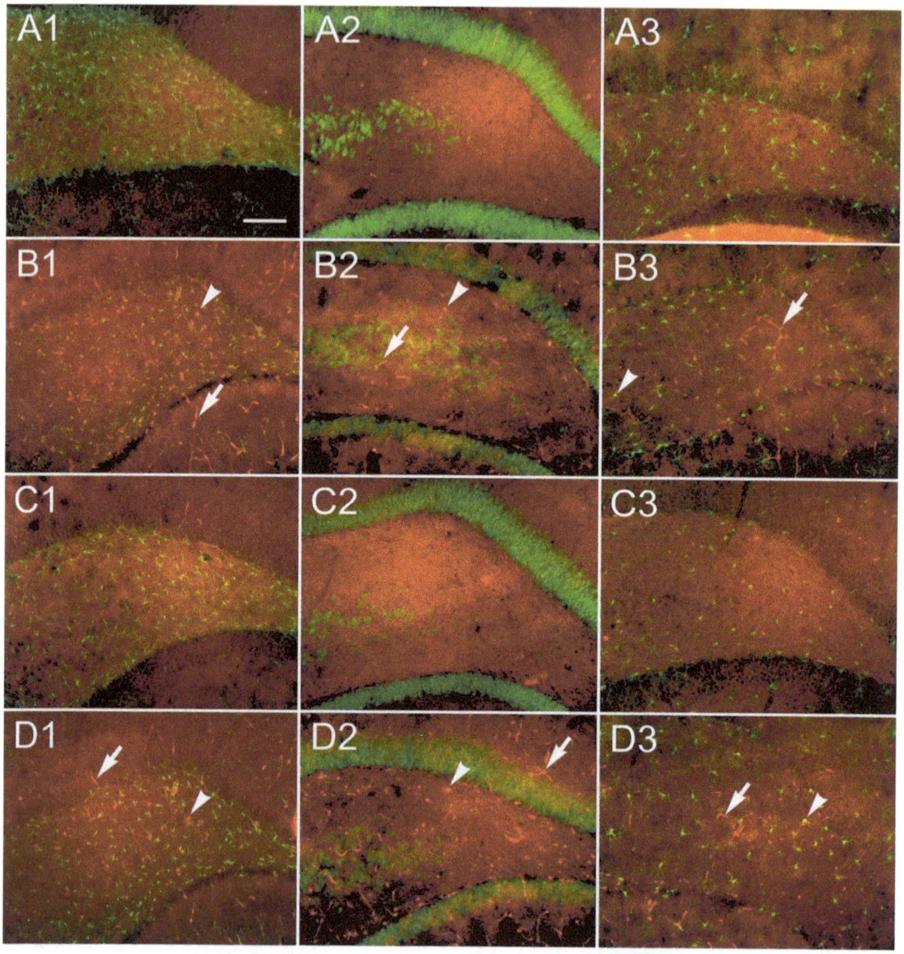

Fig. 14.2 Hsp70 and GFAP, NeuN and Iba1 immunoreactivity in the hippocampus of female and male rats. Hsp70 immunoreactivity (*red*) was undetectable in the hippocampus of control female rats (**A1**, **A2**, **A3**) and was detectable in many cells (*arrowheads*) and small blood vessels (*arrows*) 24 h following heat shock in the hippocampus of female rats (**B1**, **B2**, **B3**). Hsp70 immunoreactivity was detectable in a few cells of the hippocampus of control male rats (**C1**, **C2**, **C3**) and in many cells (*arrowheads*) and small blood vessels (*arrows*) 24 h following heat shock in the hippocampus of male rats (**D1**, **D2**, **D3**). GFAP immunoreactivity (*green*) was detectable in many cells in the hippocampus in control female (**A1**), heat shock female (**B1**), control male (**C1**), and heat shock male (**D1**) rats and was not colocalized with Hsp70 immunoreactivity. NeuN immunoreactivity (*green*) was detectable in many cells in the hippocampus of control female (**A2**), heat shock female (**B2**), control male (**C2**), and heat shock male (**D2**) rats and was not colocalized with Hsp70 immunoreactivity. Iba1 immunoreactivity (*green*) was detectable in many cells in the hippocampus of control female (**A3**), heat shock female (**B3**), control male (**C3**), and heat shock male (**D3**) rats and some colocalization with Hsp70 immunoreactivity (*red + green = yellow*) was detected (**B3**, **D3**, *arrowheads*). The scale bar in **A1** represents 100 microns in all micrographs

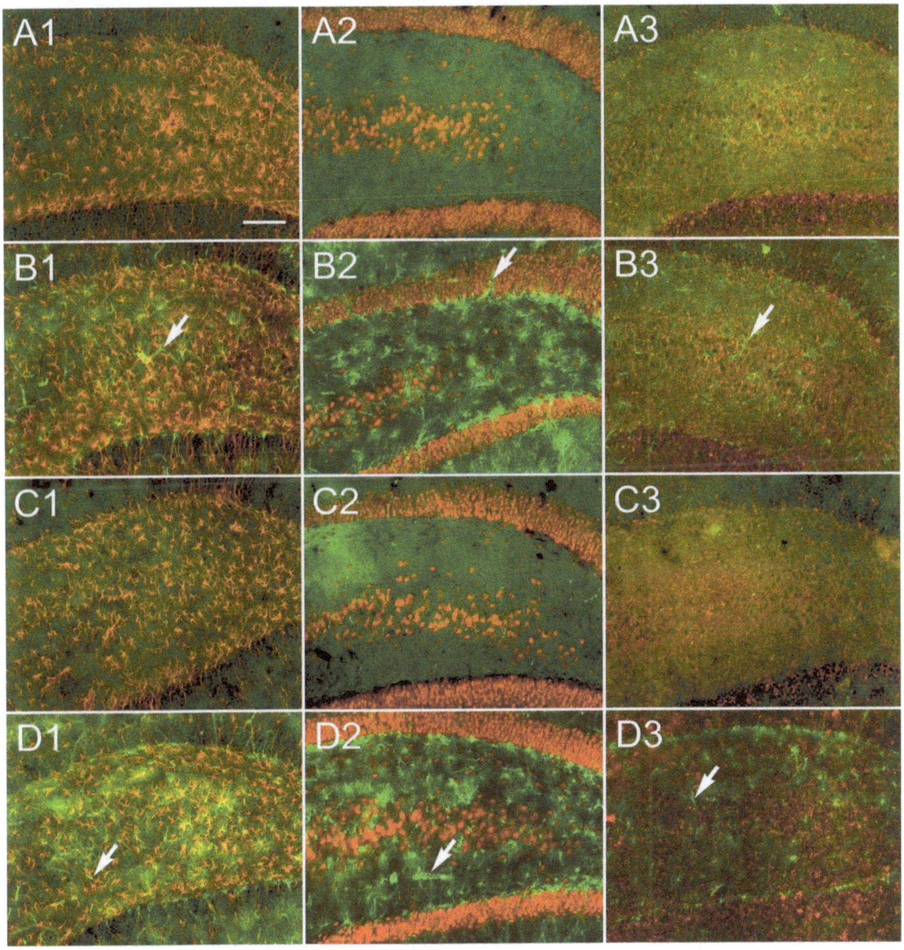

Fig. 14.3 Hsp27 and GFAP, NeuN and Iba1 immunoreactivity in the hippocampus of female and male rats. Hsp27 immunoreactivity (*green*) was undetectable in the hippocampus of control female rats (a few fluorescent red blood cells are evident) (**A1**, **A2**, **A3**) and was detectable in small blood vessels and many astrocytes in the hippocampus 24 h following heat shock in female rats (**B1**, **B2**, **B3**). Hsp27 immunoreactivity was detectable at a low level in some astrocytes in the hippocampus of control male rats (**C1**, **C2**, **C3**) and in some small blood vessels and many astrocytes 24 h following heat shock in the hippocampus of male rats (**D1**, **D2**, **D3**). GFAP immunoreactivity (*red*) was detectable in many astrocytes in the hippocampus of control female (**A1**), heat shock female (**B1**), control male (**C1**), and heat shock male (**D1**) rats and was colocalized with Hsp27 immunoreactivity (*arrows*). NeuN immunoreactivity (*red*) was detectable in many neurons in the hippocampus of control female (**A2**), heat shock female (**B2**), control male (**C2**), and heat shock male (**D2**) rats and was not colocalized with Hsp27 immunoreactivity (*arrows*). Iba1 immunoreactivity (*red*) was detectable in many cells in the hippocampus of control female (**A3**), heat shock female (**B3**), control male (**C3**), and heat shock male (**D3**) rats and was not colocalized with Hsp27 immunoreactivity (*arrows*). The scale bar in **A1** represents 100 microns in all micrographs

following heat shock in female rats (Fig. 14.3B1, B2, B3). Hsp27 immunoreactivity was detectable at a low level in some astrocytes in the hippocampus of control male rats (Fig. 14.3C1, C2, C3) and in some small blood vessels and many astrocytes 24 h following heat shock in the hippocampus of male rats (Fig. 14.3D1, D2, D3).

GFAP immunoreactivity was detectable in many astrocytes in the hippocampus of control female (Fig. 14.3A1), heat shock female (Fig. 14.3B1), control male (Fig. 14.3C1), and heat shock male (Fig. 14.3D1) rats. Micrographs of Hsp27 and GFAP immunoreactivity show some colocalization of Hsp27 and GFAP (arrows), particularly after heat shock.

NeuN immunoreactivity was detectable in many neurons in the hippocampus of control female (Fig. 14.3A2), heat shock female (Fig. 14.3B2), control male (Fig. 14.3C2), and heat shock male (Fig. 14.3D2) rats. Micrographs show no colocalization of Hsp27 and NeuN.

Iba1 immunoreactivity was detectable in many cells in the hippocampus of control female (Fig. 14.3A3), heat shock female (Fig. 14.3B3), control male (Fig. 14.3C3), and heat shock male (Fig. 14.3D3) rats. Micrographs show no colocalization of Hsp27 and Iba1.

14.17 Immunofluorescence Double Labeling for Hsp70 and Factor VIII, Hsp70 and Hsp27

Immunofluorescence double labeling for Hsp70 and Factor VIII (Fig. 14.4) revealed similar Hsp70 immunoreactivity as before. Hsp70 immunoreactivity was undetectable in the hippocampus of control female rats (Fig. 14.4A1) and was detectable in many cells and small blood vessels of the hippocampus 24 h following heat shock in female rats (Fig. 14.4B1). Hsp70 immunoreactivity was not detectable in the hippocampus of control male rats (Fig. 14.4C1) and was detected in many cells and small blood vessels of the hippocampus 24 h following heat shock in male rats (Fig. 14.4D1). Factor VIII immunoreactivity was detectable in many small blood vessels in the hippocampus of control female (Fig. 14.4A1), heat shock female (Fig. 14.4B1), control male (Fig. 14.4C1), and heat shock male (Fig. 14.4D1) rats. Micrographs of Hsp70 and Factor VIII immunoreactivity show colocalization of Hsp70 and Factor VIII, particularly in small blood vessels after heat shock

Fig. 14.4 (continued) (**D2**, **D3**). Hsp27 immunoreactivity (*green*, in **A2**, **B2**, **B3**, **C2**, **D2**, **D3**) was detectable in a few small blood vessels in the hippocampus of control female rats (**A2**) and was detectable in small blood vessels and many astrocytes in the hippocampus 24 h following heat shock in female rats (**B2**, **B3**). Hsp27 immunoreactivity was detectable in a few small blood vessels in the hippocampus of control male rats (**C2**) and was detected in some small blood vessels and many cells in the hippocampus 24 h following heat shock in male rats (**D2**, **D3**). *Arrows* on **B2** and **D2** are on the identical structures at high magnification in **B3** and **D3**, respectively. The scale bar in **A1** represents 100 microns in **A1**, **B1**, **C1**, **D1**, **A2**, **B2**, **C2**, and **D2**. The scale bar in **B3** represents 100 microns in **B3** and **D3**

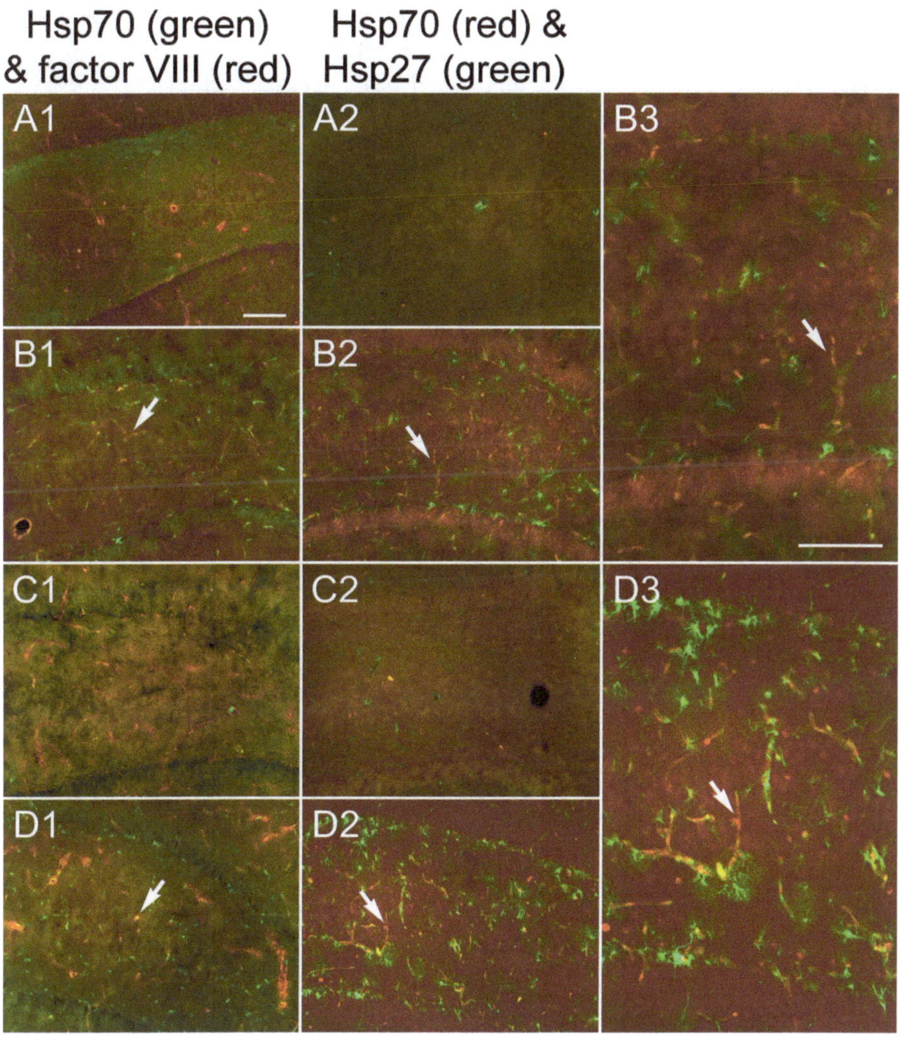

Fig. 14.4 Hsp70, Factor VIII and Hsp27 immunoreactivity in the hippocampus of female and male rats. Hsp70 immunoreactivity (*green*, in **A1**, **B1**, **C1**, **D1**) was undetectable in the hippocampus of control female rats (**A1**) and was detectable in many cells and small blood vessels of the hippocampus 24 h following heat shock in female rats (**B1**). Hsp70 immunoreactivity was undetectable in the hippocampus of control male rats (**C1**) and was detected in many cells and small blood vessels of the hippocampus 24 h following heat shock in male rats (**D1**). Factor VIII immunoreactivity (*red*, in **A1**, **B1**, **C1**, **D1**) was detectable in many small blood vessels in the hippocampus of control female (**A1**), heat shock female (**B1**), control male (**C1**), and heat shock male (**D1**) rats and showed some colocalization with Hsp70 immunoreactivity, particularly in small blood vessels (*arrows*, **B1**, **D1**). Hsp70 immunoreactivity (*red*, in **A2**, **B2**, **B3**, **C2**, **D2**, **D3**) was undetectable in the hippocampus of control female rats (**A2**) and was detectable in many cells and small blood vessels in the hippocampus 24 h following heat shock in female rats (**B2**, **B3**). Hsp70 immunoreactivity was undetectable in the hippocampus of control male rats (**C2**) and was detectable in small blood vessels and in many cells in the hippocampus 24 h following heat shock in male rats

(Fig. 14.4B1, D1). Hsp70 immunoreactive cells that were negative for Factor VIII immunoreactivity are also evident (Fig. 14.4B1, D1, arrows).

Immunofluorescence double labeling for Hsp27 and Hsp70 (Fig. 14.4A2, B2, C2, D2) revealed similar patterns of Hsp27 immunoreactivity and Hsp70 immunoreactivity as seen before. Hsp27 immunoreactivity was detectable in a few small blood vessels in the hippocampus of control female rats (Fig. 14.4A2) and was detectable in small blood vessels and many astrocytes in the hippocampus 24 h following heat shock in female rats (Fig. 14.4B2). Hsp27 immunoreactivity was detectable in a few small blood vessels in the hippocampus of control male rats (Fig. 14.4C2) and was detected in some small blood vessels and many cells in the hippocampus 24 h following heat shock in male rats (Fig. 14.4D2). Hsp70 immunoreactivity (red) was undetectable in the hippocampus of control female rats (Fig. 14.4A2) and was detectable in many cells and small blood vessels in the hippocampus 24 h following heat shock in female rats (Fig. 14.4B2). Hsp70 immunoreactivity was undetectable in the hippocampus of control male rats (Fig. 14.4C2) and was detectable in small blood vessels and in many cells in the hippocampus 24 h following heat shock in male rats (Fig. 14.4D2). Micrographs of Hsp27 and Hsp70 immunoreactivity show some colocalization of Hsp27 and Hsp70 in small blood vessels and in many cells, particularly after heat shock (Fig. 14.4B3, D3; arrows).

14.18 Discussion

The findings of this work are that: (1) In the hippocampus, Hsp70 and Hsp27 levels are similar in control female and control male rats. (2) In the hippocampus, Hsp70 and Hsp27 levels are significantly greater in male rats, compared to levels in female rats, 24 h after heat shock treatment. (3) In the hippocampus, Hsp70 is colocalized in microglia and blood vessels, Hsp27 is colocalized in astrocytes, and Hsp70 and Hsp27 are closely associated in blood vessels and astrocytes, respectively, in female and male rats 24 h after heat shock treatment. These results are important for understanding sex differences in neurodegenerative disease prevalence.

14.19 Hsp70 and Hsp27 Expression in the Hippocampi of Control Rats

In the absence of stress, Hsp70 and Hsp27 are expressed at low levels in the hippocampus. In this study, western analyses indicate that Hsp70 levels in the hippocampus are similar in control female and male rats (Fig. 14.1a). This finding is consistent with previous studies (Olazábal et al. 1992a; Thorp et al. 2007) and supports the notion that there are no sex differences in the baseline levels of Hsp70 in the hippocampus of female and male rats. The low level of Hsp70 observed in the hippocampus of control animals is also similar to that seen in other studies done exclusively with male rats (Currie and White 1983; David et al. 1994; Armstrong et al. 1996; Krueger et al. 1999).

In the present study, Hsp27 levels in the hippocampus were not different between control female rats and control male rats (Fig. 14.1a). This finding appears to be the first comparison of constitutive Hsp27 levels in the hippocampus of female and male rats. The low level of Hsp27 observed in the hippocampus of control animals is also similar to that seen in other studies done exclusively with male rats (Plumier et al. 1996, 1997a, b; Krueger-Naug et al. 2000). While Hsp27 is at low levels in the hippocampus, Hsp27 is abundant in neurons of the brain stem and spinal cord in control male rats (Plumier et al. 1997b).

14.20 Hsp70 and Hsp27 Expression in the Hippocampi of Heat Shock Rats

In the current study, 24 h after heat shock treatment hippocampal Hsp70 and Hsp27 levels are greater in both the female (Fig. 14.1b) and male (Fig. 14.1c) rats compared control rats. Most interestingly Hsp70 and Hsp27 levels are greater in male rats, compared to female rats, 24 h after heat shock treatment (Fig. 14.1d). Similarly, following an acute stressor, such as tail shock, Hsp70 levels are significantly greater in the hippocampus in male rats, compared to female rats. Although E2 plasma levels were not measured in our study for comparison with Hsp levels, the estrous cycle has no impact on induced levels of Hsp70 in the brain of female rats (Nickerson et al. 2006). The highly inducible nature of Hsp70 in the brain or other tissues has been examined for many years (Schlesinger et al. 1982; Currie and White 1983; David et al. 1994; Armstrong et al. 1996; Krueger et al. 1999; Bechtold et al. 2000). While these studies examined male rats, it is also clear that Hsp70 is inducible to high levels in the hippocampus of female rats (Pavlik et al. 2003).

Hsp70 hippocampal induced levels were significantly greater in male rats, compared to female rats. (Fig. 14.1d). While Hsp70 is at low levels in the rat brain, the constitutively expressed Hsc70 is abundant in the brain (Manzerra et al. 1997). Hsc70 has similar chaperone functions as Hsp70 and likely regulates HSF1. The lower levels of induced Hsp70 observed in the hippocampus of female rats may be related to higher levels of Hsc70 in the absence of stress (Paroo et al. 1999). Hsp70 and Hsc70 may be involved in the interaction between estrogen and its receptor, since both Hsp70 and Hsc70 have been found to bind E2 (Gacad and Adams 1998).

Hsp27 is expressed at high levels in the hippocampus after heat shock in both female (Fig. 14.1b) and male (Fig. 14.1c) rats and it is clear that Hsp27 expression is greater in males than in females after heat shock (Fig. 14.1d). Our finding of the highly inducible nature of Hsp27 after heat shock is similar to other studies done with male rats (Krueger-Naug et al. 2000; Bechtold and Brown 2003). Our finding of greater expression of Hsp27 in the hippocampus of male rats after heat shock appears to be the first comparison of inducible Hsp27 levels in female and male rats.

14.21 Antibody Specificity

Negative and positive Hsp70 immunofluorescence in the hippocampus of control male and heat shock male rats, respectively, suggests the immunoreactivity is specific for the inducible Hsp70. The negative immunofluorescence for the no primary antibody and no secondary antibody also suggest the specific nature of the immunoreaction for Hsp70.

Finding Hsp27 immunofluorescence in a few cells in the hippocampus of a control male rat and many cells after heat shock in a male rat suggests the immunoreactivity is specific for Hsp27. The negative immunofluorescence for the no primary antibody and no secondary antibody also suggest the specific nature of the immunoreaction for Hsp27.

GFAP immunofluorescence in the hippocampus of heat shock male rat is specific for the glial acidic fibrillary protein in astrocytes, as suggested by the no primary antibody and no secondary antibody negative immunoreaction.

NeuN immunofluorescence in the hippocampus of heat shock male rat is specific for the neuronal nuclei in neurons, as indicated by the no primary antibody and no secondary antibody negative immunoreaction.

Iba1 immunofluorescence in the hippocampus of heat shock male rat is specific for the ionized calcium binding adapter molecular protein in microglia, as indicated by the no primary antibody and no secondary antibody negative immunoreaction.

Factor VIII immunofluorescence in the hippocampus of heat shock male rats is specific for the factor VIII protein in and around blood vessels, as indicated by the no primary antibody and no secondary antibody negative immunoreaction.

14.22 Hsp70 and Hsp27 Immunofluorescence in the Hippocampus of Rats

The low level of Hsp70 immunofluorescence in the hippocampus of control female (Fig. 14.2A1, A2, A3) and control male (Fig. 14.2C1, C2, C3) rats is similar to that seen in previous studies (Armstrong et al. 1996; Pavlik et al. 2003). By immunofluorescence, Hsp70 is not detected in control female rats (Pavlik et al. 2003) and by immunohistochemistry Hsp70 is not detected in control male rats (Armstrong et al. 1996). Hsp70 immunofluorescence 24 h after heat shock treatment in female and male rat hippocampi was not colocalized with either GFAP (Fig. 14.2B1, D1) or NeuN (Fig. 14.2B2, D2) immunofluorescence and is similar to previous findings (Lee et al. 2002). However, Hsp70 immunofluorescence 24 h after heat shock treatment is detectable and is colocalized with Iba1 in both female and male rats (Fig. 14.2B3 and D3, respectively).

In this study, Hsp27 immunofluorescence in the hippocampus of control female (Fig. 14.3A1, A2, A3) and control male (Fig. 14.3C1, C2, C3) rats was at low levels. This finding is similar to previous work done with only male animals (Plumier et al.

1996, 1997a, b; Krueger-Naug et al. 2000). Following heat shock, Hsp27 immunofluorescence is detectable and is colocalized with GFAP immunofluorescence in both female and male rats (Fig. 14.3B1 and D1, respectively). Hsp27 immunofluorescence is not colocalized with NeuN or Iba1 immunofluorescence (Fig. 14.3B2, D2 or B3, D3, respectively). Colocalization of Hsp27 with GFAP is indicative of immunoreactivity in astrocytes (Plumier et al. 1997a; Lee et al. 2002; Bechtold and Brown 2003). Hsp27 appears not to be localized in neurons or microglia as indicated by the NeuN and Iba1 immunofluorescence, respectively (Lee et al. 2002; Maeda et al. 2008).

Our finding of Hsp27 immunoreactivity in astrocytes in both female and males is similar to previous studies that were done in male animals only (Krueger-Naug et al. 2000; Bechtold and Brown 2003). This study appears to the first showing localization of Hsp27 in astrocytes in the female hippocampus.

14.23 Factor VIII, Hsp70 and Hsp27 Immunofluorescence in the Hippocampus of Rats

Factor VIII immunofluorescence (Fig. 14.4A1, B1, C1, D1) is indicative of immunoreactivity in small blood vessels (Theilen and Kuschinsky 1992). After heat shock treatment some induced Hsp70 immunofluorescence is co-localized with Factor VIII immunofluorescence in female (Fig. 14.4B1) and male (Fig. 14.4D1) hippocampus. Our finding of Hsp70 localization in microglia (Fig. 14.2B3, D3) and blood vessels is similar to previous studies showing Hsp70 colocalized in microglia, oligodendrocytes and blood vessels, using OX-42 antibody, GSII lectin antibody, and histological morphology, respectively (Pavlik et al. 2003). Similarly, Hsp70 has also been localized in vimentin-positive glia (astrocytes) in the hippocampus of male rats after heat shock (Krueger et al. 1999).

Following heat shock, Hsp27 immunofluorescence is associated with some Hsp70 immunofluorescence in the hippocampus of female and male rats 24 h after heat shock (Fig. 14.4B2 and D2). The male rats have greater levels of both Hsp70 and Hsp27, compared to the female rats, 24 h after heat shock treatment (Fig. 14.4B2 and D2). As well, Hsp70 colocalized with microglia and in small blood vessels, Hsp27 is in astrocytes, and Hsp27-expressing astrocytes are in close proximity to Hsp70-expressing blood vessels, which is evident in the hippocampus of both females (Fig. 14.4B3) and males (Fig. 14.4D3) 24 h after heat shock treatment. The resulting cellular damage within the blood vessels 24 h following heat shock treatment likely induced the expression of Hsp70. Normally, astrocytes associate with endothelial cells in blood vessels in the brain to help protect the blood brain barrier, especially following stress (Bechtold and Brown 2003). Heat shock treatment likely caused increased permeability in the endothelial cells and cellular damage. As a result, Hsp27-expressing astrocytes in association with the Hsp70-expressing endothelial cells in the blood vessel wall may be facilitating restoration of homeostasis following heat stress (Plumier et al. 1996; Krueger-Naug et al. 2000; Pavlik et al. 2003).

Heat shock triggers thermal and osmotic stressors that induce protein instability and aggregation, leading to activation of the HSF1 in order to maintain protein homeostasis (Cotto and Morimoto 1999). For example, heat shock triggers hypo-osmotic stress in mammalian cells leading to activation of HSF1 and iso-osmotic conditions restore activated HSF1 to normal levels (Caruccio et al. 1997). Hsp27 levels are also induced following stress (heat shock, ischemia, etc.) and is a good marker of reactive gliosis (Plumier et al. 1997a). Elevated expression of Hsp27 after heat shock facilitates protein repair and actin filament reorganization and stabilization (Lavoie et al. 1995) and prevents apoptosis through inhibition of caspase activity (Garrido et al. 1999).

14.24 Effects of Estrogen on Expression of HSP

An obvious difference between females and males is the levels of circulating estrogen. Women have approximately 2–4 times greater plasma levels of E2, compared to men (Massafra et al. 2002) and plasma levels of E2 in female rats is also greater compared to male rats (Pratchayasakul et al. 2011). E2 functions by binding to ER-α or ER-β. In the hippocampus the overall density of ER-α and ER-β are not different between female and male rats. However, the distribution of ER-α in the female hippocampus is widespread, and is present in both nuclear and extranuclear sites in the pyramidal and granule cells and in their dendrites and axons (Milner et al. 2001, 2005; Romeo et al. 2005). In female rats, ER-α is also found in astrocytes throughout the hippocampus (Milner et al. 2005). In male rats, the ER-α is primarily located in pyramidal cells in the CA1 and CA2 regions of the hippocampus (Azcoitia et al. 1999; Milner et al. 2001, 2005; Sakuma et al. 2009).

The difference in ER-α distribution in the hippocampus may regulate the estrogen signaling pathways. Estrogen receptors in the nuclei act by genomic signaling and those bound to the plasma membrane, act by non-genomic signaling. The ER-α acts by both of these signaling pathways in the hippocampus, whereas the ER-β acts primarily by non-genomic signaling, as it is located in extranuclear sites in the hippocampus (Milner et al. 2001, 2005). The sex differences in plasma estrogen levels, signaling pathways and the hippocampal distribution of the ER-α, may contribute to some of the sex differences in expression of heat shock proteins in the hippocampus.

In unstressed conditions, estrogen may regulate levels of HSP differentially in females and males. Estrogen may also regulate the expression of HSP following stressful conditions. In this study, the levels of Hsp70 and Hsp27 are greater in hippocampus of heat shock male rats compared to heat shocked female rats. In other studies the levels of Hsp70 have been examined in hearts of female and male rats. Hsp70 levels are significantly greater in hearts of male rats compared to female rats after 2 weeks of treadmill exercise training (Paroo et al. 1999). E2 supplementation in male rats suppressed the exercise induced increase of Hsp70 in the hearts to the level seen in hearts of female rats (Paroo et al. 1999). Removal of E2 by ovariectomy significantly increased the level of Hsp70 in the hearts of exercised rats compared to intact female rats (Paroo et al. 2002). A similar increase in Hsp70 in the heart is seen in ovariectomized rats compared to intact female rats after heat shock treatment (Shinohara et al. 1993). Therefore, circulating estrogen levels may have an effect on the sex differences in expression of heat shock proteins in both the brain and the heart of rats.

Estrogen seems to be regulating the expression of HSP differentially between males and females. Interestingly, ovariectomized rats that were treated with E2 had increased levels of Hsp90 in the hypothalamus, compared to ovariectomized rats that were not given E2 (Olazábal et al. 1992b). This suggests that E2 in female rats may regulate higher Hsp90 levels in the cortex and hypothalamus, compared to that in male rats (Elaković et al. 2007). However in other tissues, including the heart, Hsp90 levels are higher in male rats (Voss et al. 2003). Interaction of Hsp90 with both the estrogen receptor and the HSF1 may account for these differences. The estrogen receptor is maintained in the cytoplasm bound to Hsp90 (Milne and Noble 2008). Estrogen freely penetrates cell membranes and binds with the estrogen receptor in the cytoplasm, releasing Hsp90. Freed Hsp90 can bind and inactivate any free HSF1 in the cytoplasm, thus suppressing the transcription of various heat shock genes regulated by HSF1. In normal conditions, Hsp90 and/or Hsc70 bind and maintain HSF1 in an inactive state in the cytoplasm (Zou et al. 1998; Guo et al. 2001; Anckar and Sistonen 2011). Metabolic stressors such as heat shock, intense exercise, or ischemia leads to the accumulation of damaged or denatured proteins within cells. Various HSP including Hsp90 and Hsc70 are recruited to damaged proteins, freeing HSF1. HSF1 trimerizes, becomes phosphorylated, and translocates to the nucleus and binds to the heat shock elements in the promoter region of heat shock genes and increases their transcription (Morimoto 1998; Anckar and Sistonen 2011). Thus E2 interaction with its receptors and Hsp90 may be regulating the differential expression of various HSP in females and males (Milne and Noble 2008).

In addition to interacting with HSF1, and estrogen receptors, Hsp90 (and Hsc70) also interacts with glucocorticoid receptors (Pratt 1993; Jones et al. 2004). These interactions of Hsp90 may also explain the differential glucocorticoid levels in females and males after various stresses. Following a stressor, female rats have less activation of HSF1 and less expression of HSP, and higher cortisol levels compared to male rats (Jezová et al. 1994; Beiko et al. 2004; Li and Sánchez 2005). High glucocorticoid levels binding to the glucocorticoid receptors free Hsp90 that binds and inactivates HSF1 (Li and Sánchez 2005). These sex differences in response to stress may be indicative of differential activation of the hypothalamic pituitary adrenal axis in that following an acute stress, levels of cortisol secretion are greater and maintained at higher levels in females, compared to males (Gallucci et al. 1993; Kant et al. 1983).

Estrogen may mediate protection through a mechanism that stabilizes cellular membranes and prevents protein denaturation, resulting in less oxidative stress and expression of heat shock proteins (Paroo et al. 2002). E2 dampens oxidative stress by increasing antioxidants following injury (Essig and Nosek 1997; Kher et al. 2005). Oxidative stress leads to activation of the p38 mitogen activated protein kinase, nuclear factor kappa B and accumulation of inflammatory cytokines, like tumor necrosis factor, that may lead to activation of the apoptotic pathways (reviewed by Kher et al. 2005). In a normal state, male rats have significantly greater phosphorylated (activated) p38 mitogen activated protein kinase, compared to females, indicating that normally females have lower levels of oxidative stress (Wang et al. 1998). Lower levels of oxidative stress in females may be a factor contributing to sex differences in degenerative disease prevalence, that disproportionately affect males, compared to females. As stated previously, this sex difference equalizes once women begin menopause, and E2 levels drop drastically. Low levels of E2 play a role in the development of degenerative diseases. In neurodegeneration, damaged proteins aggregate and lead to proteotoxicity and the release of reactive oxygen species (Muchowski and Wacker 2005; reviewed by Xu et al. 2012).

While a strong case for estrogen regulating expression of HSPs in females can be made, other related possibilities also come to mind. As mentioned above, estrogen may stabilize cellular membranes preventing protein denaturation, resulting in less oxidative stress and less expression of heat shock proteins (Paroo et al. 2002). In males, the increased amounts of Hsp70 and Hsp27 in the hippocampus may be due to more thermolabile cell membranes that lead to greater proteotoxicity and activation of the heat shock response after hyperthermia. Alternatively, brain cells of male rats may contain relatively higher amounts of thermolabile proteins. High levels of thermolabile proteins in males would be expected to cause the production of higher levels of HSPs in response to hyperthermia. Estrogen or lack of estrogen may be paramount in the regulation of HSPs in females and males, respectively.

14.25 Conclusions

HSP are differentially regulated in female and male rats. Overall, male rats have greater induced levels of Hsp70 and Hsp27 in the hippocampus compared to levels in female rats, 24 h after heat shock treatment. In the hippocampus 24 h after heat shock, Hsp70 is expressed in microglia and small blood vessels, and Hsp27 is expressed in astrocytes and astrocytic processes around small blood vessels. Sex differences in E2 regulation of the expression of HSP in the brain after stress may help to explain neurodegenerative disease prevalence. The age-related accumulation of damaged proteins in neurodegenerative disease induces a sex-specific heat shock response that appears to be regulated at the level of E2 interacting with the estrogen receptor and Hsp90. This interaction may explain sex differences in disease prevalence.

References

Anckar J, Sistonen L (2011) Regulation of HSF1 function in the heat stress response: implications in aging and disease. Annu Rev Biochem 80:1089–1115

Armstrong JN, Plumier JC, Robertson HA, Currie RW (1996) The inducible 70,000 molecular/weight heat shock protein is expressed in the degenerating dentate hilus and piriform cortex after systemic administration of kainic acid in the rat. Neuroscience 74:685–693

Azcoitia I, Sierra A, Garcia-Segura LM (1999) Localization of estrogen receptor beta-immunoreactivity in astrocytes of the adult rat brain. Glia 26:260–267

Barron AM, Pike CJ (2012) Sex hormones, aging, and Alzheimer's disease. Front Biosci 4:976–997

Bechtold DA, Brown IR (2003) Induction of Hsp27 and Hsp32 stress proteins and vimentin in glial cells of the rat hippocampus following hyperthermia. Neurochem Res 28:1163–1173

Bechtold DA, Rush SJ, Brown IR (2000) Localization of the heat-shock protein Hsp70 to the synapse following hyperthermic stress in the brain. J Neurochem 74:641–646

Beery AK, Zucker I (2011) Males still dominate animal studies. Nature 465:690

Beiko J, Lander R, Hampson E, Boon F, Cain DP (2004) Contribution of sex differences in the acute stress response to sex differences in water maze performance in the rat. Behav Brain Res 151:239–253

Benarroch E (2011) Heat shock proteins: multiple neuroprotective functions and implications for neurologic disease. Neurology 76:660–667

Bodega G, Hernández C, Suárez I, Martin M, Fernández B (2002) HSP70 constitutive expression in rat central nervous system from postnatal development to maturity. J Histochem Cytochem 50:1161–1168

Bruijn LI, Miller TM, Cleveland DW (2004) Unraveling the mechanisms involved in motor neuron degeneration in ALS. Annu Rev Neurosci 27:723–749

Caruccio L, Bae S, Liu AY, Chen KY (1997) The heat-shock transcription factor HSF1 is rapidly activated by either hyper- or hypo-osmotic stress in mammalian cells. Biochem J 327:341–347

Chen S, Brown IR (2007) Neuronal expression of constitutive heat shock proteins: implications for neurodegenerative diseases. Cell Stress Chaperones 12:51–58

Cotto JJ, Morimoto RI (1999) Stress-induced activation of the heat-shock response: cell and molecular biology of heat-shock factors. Biochem Soc Symp 64:105–118

Currie RW, White FP (1983) Characterization of the synthesis and accumulation of a 71-kilodalton protein induced in rat tissues after hyperthermia. Can J Biochem Cell Biol 61:438–446

David JC, Currie RW, Robertson HA (1994) Expression and distribution of hsp71 and hsc73 messenger RNAs in rat brain following heat shock: effect of dizocilpine maleate. Neuroscience 62:945–954

Diaz-Latoud C, Buache E, Javouhey E, Arrigo AP (2005) Substitution of the unique cysteine residue of murine Hsp25 interferes with the protective activity of this stress protein through inhibition of dimer formation. Antioxid Redox Signal 7:436–445

Elaković I, Brkljacić J, Matić G (2007) Long-term imipramine treatment affects rat brain and pituitary corticosteroid receptors and heat shock proteins levels in a gender-specific manner. J Neural Transm 114:1069–1080

Essig DA, Nosek TM (1997) Muscle fatigue and induction of stress protein genes: a dual function of reactive oxygen species? Can J Appl Physiol 22:409–428

Ferns G, Shams S, Shafi S (2006) Heat shock protein 27: its potential role in vascular disease. Int J Exp Pathol 87:253–274

Gacad MA, Adams JS (1998) Proteins in the heat shock-70 family specifically bind 25-hydroxyvitamin D3 and 17beta-estradiol. J Clin Endocrinol Metabol 83:1264–1267

Gallucci WT, Baum A, Laue L, Rabin DS, Chrousos GP, Gold PW, Kling MA (1993) Sex differences in sensitivity of the hypothalamic-pituitary-adrenal axis. Health Psychol 12:420–425

Gao S, Hendrie HC, Hall KS, Hui S (1998) The relationships between age, sex, and the incidence of dementia and Alzheimer disease: a meta-analysis. Arch Gen Psychiatry 55:809–815

Garrido C, Bruey JM, Fromentin A, Hammann A, Arrigo AP, Solary E (1999) HSP27 inhibits cytochrome c-dependent activation of procaspase-9. FASEB J 13:2061–2070

Guo Y, Guettouche T, Fenna M, Boellmann F, Pratt WB, Toft DO, Smith DF, Voellmy R (2001) Evidence for a mechanism of repression of heat shock factor 1 transcriptional activity by a multichaperone complex. J Biol Chem 276:45791–45799

Hampel H, Prvulovic D, Teipel S, Jessen F, Luckhaus C, Frölich L, Riepe MW, Dodel R, Leyhe T, Bertram L, Hoffmann W, Faltraco F (2011) The future of Alzheimer's disease: the next 10 years. Prog Neurobiol 95:718–728

Hebert LE, Scherr PA, Bienias JL, Bennett DA, Evans DA (2003) Alzheimer disease in the US population prevalence estimates using the 2000 census. JAMA 60:1119–1122

Huang C, Cheng H, Hao S, Zhou H, Zhang X, Gao J, Sun QH, Hu H, Wang CC (2006) Heat shock protein 70 inhibits alpha-synuclein fibril formation via interactions with diverse intermediates. J Mol Biol 364:323–336

Huisman MH, de Jong SW, van Doormaal PT, Weinreich SS, Schelhaas HJ, van der Kooi AJ, de Visser M, Veldink JH, van den Berg LH (2011a) Population based epidemiology of amyotrophic lateral sclerosis using capture-recapture methodology. J Neurol Neurosurg Psychiatry 82:1165–1170

Huisman MH, de Jong SW, Verwijs MC, Schelhaas HJ, van der Kooi AJ, de Visser M, Veldink JH, van den Berg LH (2011b) Family history of neurodegenerative and vascular diseases in ALS: a population-based study. Neurology 4:1363–1369

Jezová D, Kvetnanský R, Vigas M (1994) Sex differences in endocrine response to hyperthermia in sauna. Acta Physiol Scand 150:293–298

Jones TJ, Li D, Wolf IM, Wadekar SA, Periyasamy S, Sánchez ER (2004) Enhancement of glucocorticoid receptor-mediated gene expression by constitutively active heat shock factor 1. Mol Endocrinol 18:509–520

Kalin MF, Zumoff B (1990) Sex hormones and coronary disease: a review of the clinical studies. Steroids 8:330–352

Kampinga HH, Hageman J, Vos MJ, Kubota H, Tanguay RM, Bruford EA, Cheetham ME, Chen B, Hightower LE (2009) Guidelines for the nomenclature of the human heat shock proteins. Cell Stress Chaperones 14:105–111

Kant GJ, Lenox RH, Bunnell BN, Mougey EH, Pennington LL, Meyerhoff JL (1983) Comparison of stress response in male and female rats: pituitary cyclic AMP and plasma prolactin, growth hormone and corticosterone. Psychoneuroendocrinology 4:421–428

Karmazyn M, Mailer K, Currie RW (1990) Acquisition and decay of heat-shock-enhanced postischemic ventricular recovery. Am J Physiol 259:424–431

Kawas C, Resnick S, Morrison A, Brookmeyer R, Corrada M, Zonderman A, Bacal C, Lingle DD, Metter E (1997) A prospective study of estrogen replacement therapy and the risk of developing Alzheimer's disease: the Baltimore longitudinal study of aging. Neurology 4:1517–1521

Kher A, Wang M, Tsai BM, Pitcher JM, Greenbaum ES, Nagy RD, Patel KM, Wairiuko GM, Markel TA, Meldrum DR (2005) Sex differences in the myocardial inflammatory response to acute injury. Shock 23:1–10

Kompoliti K (2003) Estrogen and Parkinson's disease. Front Biosci 8:391–400

Krueger AM, Armstrong JN, Plumier JC, Robertson HA, Currie RW (1999) Cell specific expression of Hsp70 in neurons and glia of the rat hippocampus after hyperthermia and kainic acid-induced seizure activity. Mol Brain Res 71:265–278

Krueger-Naug AM, Hopkins DA, Armstrong JN, Plumier JC, Currie RW (2000) Hyperthermic induction of the 27-kDa heat shock protein (Hsp27) in neuroglia and neurons of the rat central nervous system. J Comp Neurol 428:495–510

Kumar R, Zakharov MN, Khan SH, Miki R, Jang H, Toraldo G, Singh R, Bhasin S, Jasuja R (2011) The dynamic structure of the estrogen receptor. J Amino Acids 2011:ID 812540, 1–7

Lavoie JN, Lambert H, Hickey E, Weber LA, Landry J (1995) Modulation of cellular thermoresistance and actin filament stability accompanies phosphorylation-induced changes in the oligomeric structure of heat shock protein 27. Mol Cell Biol 15:505–516

Lee MY, Choi YS, Choi JS, Min DS, Chun MH, Kim ON, Lee SB, Kim SY (2002) An immunohistochemical study of APG-2 protein in the rat hippocampus after transient forebrain ischemia. Brain Res 924:237–241
Li D, Sánchez ER (2005) Glucocorticoid receptor and heat shock factor 1: novel mechanism of reciprocal regulation. Vitam Horm 71:239–262
Lindquist S, Craig EA (1988) The heat-shock proteins. Annu Rev Genet 22:631–677
Lu A, Ran RQ, Clark J, Reilly M, Nee A, Sharp FR (2002) 17-beta-estradiol induces heat shock proteins in brain arteries and potentiates ischemic heat shock protein induction in glia and neurons. J Cereb Blood Flow Metab 22:183–195
Maatkamp A, Vlug A, Haasdijk E, Troost D, French PJ, Jaarsma D (2004) Decrease of Hsp25 protein expression precedes degeneration of motoneurons in ALS-SOD1 mice. Eur J Neurosci 20:14–28
Maeda T, Nagata K, Yoshida Y (2008) Exogenous l-DOPA induce no dopamine immuno-reactivity in striatal astroglias and microglias of adult rats with extensive nigro-striatal dopaminergic denervation. Neurosci Lett 433:255–258
Manzerra P, Rush SJ, Brown IR (1997) Tissue-specific differences in heat shock protein hsc70 and hsp70 in the control and hyperthermic rabbit. J Cell Physiol 170:130–137
Margo K, Winn R (2006) Testosterone treatments: why, when, and how? Am Fam Physician 73:1591–1598
Martin JB (1999) Molecular basis of the neurodegenerative disorders. N Engl J Med 340:1970–1980
Massafra C, Gioia D, De Felice C, Muscettola M, Longini M, Buonocore G (2002) Gender-related differences in erythrocyte glutathione peroxidase activity in healthy subjects. Clin Endocrinol (Oxf) 57:663–667
Mehlen P, Kretz-Remy C, Briolay J, Fostan P, Mirault ME, Arrigo AP (1995) Intracellular reactive oxygen species as apparent modulators of heat-shock protein 27 (hsp27) structural organization and phosphorylation in basal and tumour necrosis factor alpha-treated T47D human carcinoma cells. Biochem J 312:367–375
Milne KJ, Noble EG (2008) Response of the myocardium to exercise: sex-specific regulation of Hsp70. Med Sci Sports Exerc 40:655–663
Milner TA, McEwen BS, Hayashi S, Li CJ, Reagan LP, Alves SE (2001) Ultrastructural evidence that hippocampal alpha estrogen receptors are located at extranuclear sites. J Comp Neurol 429:355–371
Milner TA, Ayoola K, Drake CT, Herrick SP, Tabori NE, McEwen BS, Warrier S, Alves SE (2005) Ultrastructural localization of estrogen receptor beta immunoreactivity in the rat hippocampal formation. J Comp Neurol 491:81–95
Morimoto RI (1998) Regulation of the heat shock transcriptional response: cross talk between a family of heat shock factors, molecular chaperones, and negative regulators. Genes Dev 12:3788–3796
Moschetti K, Cummings PL, Sorvillo F, Kuo T (2012) Burden of Alzheimer's disease-related mortality in the United States, 1999–2008. J Am Geriatr Soc 60:1509–1514
Muchowski PJ, Wacker JL (2005) Modulation of neurodegeneration by molecular chaperones. Nat Rev Neurosci 6:11–22
Nickerson M, Kennedy SL, Johnson JD, Fleshner M (2006) Sexual dimorphism of the intracellular heat shock protein 72 response. J Appl Physiol 101:566–575
Olazábal UE, Pfaff DW, Mobbs CV (1992a) Sex differences in the regulation of heat shock protein 70 kDa and 90 kDa in the rat ventromedial hypothalamus by estrogen. Brain Res 596:311–314
Olazábal UE, Pfaff DW, Mobbs CV (1992b) Estrogenic regulation of heat shock protein 90 kDa in the rat ventromedial hypothalamus and uterus. Mol Cell Endocrinol 84:175–183
Pandey P, Farber R, Nakazawa A, Kumar S, Bharti A, Nalin C, Weichselbaum R, Kufe D, Kharbanda S (2000) Hsp27 functions as a negative regulator of cytochrome c-dependent activation of procaspase-3. Oncogene 19:1975–1981

Papasozomenos SC, Papasozomenos H (2008) The activation of heat shock transcription factor 1 is differentially regulated in the brain of estrogen- and testosterone-treated heat-shocked rats. J Alzheimers Dis 15:375–390

Paroo Z, Tiidus PM, Noble EG (1999) Estrogen attenuates HSP 72 expression in acutely exercised male rodents. Eur J Appl Physiol Occup Physiol 80:180–184

Paroo Z, Haist JV, Karmazyn M, Noble EG (2002) Exercise improves postischemic cardiac function in males but not females: consequences of a novel sex-specific heat shock protein 70 response. Circ Res 90:911–917

Patel YJ, Payne Smith MD, de Belleroche J, Latchman DS (2005) Hsp27 and Hsp70 administered in combination have a potent protective effect against FALS-associated SOD1-mutant-induced cell death in mammalian neuronal cells. Brain Res Mol Brain Res 134:256–274

Pavlik A, Aneja IS, Lexa J, Al-Zoabi BA (2003) Identification of cerebral neurons and glial cell types inducing heat shock protein Hsp70 following heat stress in the rat. Brain Res 973:179–189

Plumier JC, Armstrong JN, Landry J, Babity JM, Robertson HA, Currie RW (1996) Expression of the 27,000 mol. wt heat shock protein following kainic acid-induced status epilepticus in the rat. Neuroscience 75:849–856

Plumier JC, David J-C, Robertson HA, Currie RW (1997a) Cortical application of potassium chloride induces the low-molecular weight heat shock protein (Hsp27) in astrocytes. J Cereb Blood Flow Metab 17:781–790

Plumier JC, Hopkins DA, Robertson HA, Currie RW (1997b) Constitutive expression of the 27-kDa heat shock protein (Hsp27) in sensory and motor neurons of the rat nervous system. J Comp Neurol 384:409–428

Pratchayasakul W, Chattipakorn N, Chattipakorn SC (2011) Effects of estrogen in preventing neuronal insulin resistance in hippocampus of obese rats are different between genders. Life Sci 89:702–707

Pratt WB (1993) The role of heat shock proteins in regulating the function, folding, and trafficking of the glucocorticoid receptor. J Biol Chem 268:21455–21458

Renkawek K, Bosman GJ, de Jong WW (1994) Expression of small heat-shock protein hsp 27 in reactive gliosis in Alzheimer disease and other types of dementia. Acta Neuropathol 87:511–519

Ritossa F (1962) A new puffing pattern induced by temperature shock and DNP in Drosphila. Experientia 18:571–573

Ritossa F (1996) Discovery of heat shock proteins. Cell Stress Chaperones 1:97–98

Rogalla P, Rohen C, Bonk U, Bullerdiek J (1996) Telomeric repeat fragment lengths are not correlated to histological grading in 85 breast cancers. Cancer Lett 106:155–161

Romeo RD, McCarthy JB, Wang A, Milner TA, McEwen BS (2005) Sex differences in hippocampal estradiol-induced N-methyl-D-aspartic acid binding and ultrastructural localization of estrogen receptor-alpha. Neuroendocrinology 81:391–399

Sakuma S, Tokuhara D, Hattori H, Matsuoka O, Yamano T (2009) Expression of estrogen receptor alpha and beta in reactive astrocytes at the male rat hippocampus after status epilepticus. Neuropathology 29:55–62

Schapira AH, Olanow CW (2004) Neuroprotection in Parkinson disease: mysteries, myths, and misconceptions. JAMA 291:358–364

Schlesinger MJ, Ashburner M, Tissieres A (1982) Heat shock from bacteria to man. Cold Spring Harbor Laboratory, Cold Spring Harbor

Shaikh AA (1971) Estrone and estradiol levels in the ovarian venous blood from rats during the estrous cycle and pregnancy. Biol Reprod 5:297–307

Sharp PS, Akbar MT, Bouri S, Senda A, Joshi K, Chen HJ, Latchman DS, Wells DJ, de Belleroche J (2007) Protective effects of heat shock protein 27 in a model of ALS occur in the early stages of disease progression. Neurobiol Dis 30:42–55

Shinohara H, Inaguma Y, Goto S, Inagaki T, Kato K (1993) Alpha B crystallin and HSP28 are enhanced in the cerebral cortex of patients with Alzheimer's disease. J Neurol Sci 119:203–208

Theilen H, Kuschinsky W (1992) Fluorescence labeling of the capillary network in rat brains. J Cereb Blood Flow Metab 12:347–350

Thorp DB, Haist JV, Leppard J, Milne KJ, Karmazyn M, Noble EG (2007) Exercise training improves myocardial tolerance to ischemia in male but not in female rats. Am J Physiol Regul Integr Comp Physiol 293:363–371

Van Den Eeden SK, Tanner CM, Bernstein AL, Fross RD, Leimpeter A, Bloch DA, Nelson LM (2003) Incidence of Parkinson's disease: variation by age, gender, and race/ethnicity. Am J Epidemiol 157:1015–1022

van der Worp HB, Howells DW, Sena ES, Porritt MJ, Rewell S, O'Collins V, Macleod MR (2010) Can animal models of disease reliably inform human studies? PLoS Med 7:e1000245

Voss MR, Stallone JN, Li M, Cornelussen RN, Knuefermann P, Knowlton AA (2003) Gender differences in the expression of heat shock proteins: the effect of estrogen. Am J Physiol Heart Circ Physiol 285:687–692

Wald C, Wu C (2010) Biomedical research. Of mice and women: the bias in animal models. Science 327:1571–1572

Wang X, Martindale JL, Liu Y, Holbrook NJ (1998) The cellular response to oxidative stress: influences of mitogen-activated protein kinase signalling pathways on cell survival. Biochem J 333:291–300

Weiland NG, Orikasa C, Hayashi S, McEwen BS (1997) Distribution and hormone regulation of estrogen receptor immunoreactive cells in the hippocampus of male and female rats. J Comp Neurol 388:603–612

Weiske J, Schöneberg T, Schröder W, Hatzfeld M, Tauber R, Huber O (2001) The fate of desmosomal proteins in apoptotic cells. J Biol Chem 276:41175–41181

Wojtys EM, Huston LJ, Boynton MD, Spindler KP, Lindenfeld TN (2002) The effect of the menstrual cycle on anterior cruciate ligament injuries in women as determined by hormone levels. Am J Sports Med 30:182–188

Xu Q, Wick G (1996) The role of heat shock proteins in protection and pathophysiology of the arterial wall. Mol Med Today 2:372–379

Xu Q, Metzler B, Jahangiri M, Mandal K (2012) Molecular chaperones and heat shock proteins in atherosclerosis. Am J Physiol Heart Circ Physiol 302:506–514

Zou J, Guo Y, Guettouche T, Smith DF, Voellmy R (1998) Repression of heat shock transcription factor HSF1 activation by HSP90 (HSP90 complex) that forms a stress-sensitive complex with HSF1. Cell 94:471–480

Chapter 15
HspB5/αB-Crystallin in the Brain

Nikola Golenhofen and Britta Bartelt-Kirbach

Abstract Many organs including the brain exhibit powerful endogenous cytoprotective mechanisms to survive recurrent cellular stress events, i.e. the development of stress tolerance. Investigation of the molecular mechanisms underlying this neuroprotective phenomenon is of special interest since it may provide the basis to develop new therapeutic strategies for the treatment of neurological diseases. One important mechanism is the upregulation of heat shock proteins. Here, we will review the neuroprotective potential of HspB5/αB-crystallin. HspB5 is expressed in glia as well as in neurons and upregulated in certain cell types or subset of cells at pathophysiological conditions. HspB5 is found to be associated with the disease-causing pathological protein aggregates, such as amyloid plaques in Alzheimer's disease or Rosenthal fibers in Alexander disease. One possible function of HspB5 is to counteract the aggregation process leading to increased cell survival. However, HspB5 may act additionally via its non-chaperone functions, such as anti-inflammatory, anti-apoptotic properties or association with cytoskeletal proteins influencing filament assembly. The cytoprotective activity of HspB5 is regulated by phosphorylation. Interestingly, in neurons HspB5 is recruited to axons and dendrites by phosphorylation, however, to this end little is known about the molecular targets of phosphoHspB5 in neurons. Identifying the impact of phosphorylation of HspB5 in glia and neurons and the targets of HspB5 may be useful to develop new therapeutic strategies for neurological diseases.

Keywords HspB5 • Neuroprotection • Neurodegenerative diseases • Stress response • Phosphorylation

N. Golenhofen (✉) • B. Bartelt-Kirbach
Institute of Anatomy and Cell Biology, University of Ulm, 89081 Ulm, Germany
e-mail: nikola.golenhofen@uni-ulm.de

R.M. Tanguay, L.E. Hightower (eds.), *The Big Book on Small Heat Shock Proteins*, Heat Shock Proteins 8, DOI 10.1007/978-3-319-16077-1_15

15.1 Introduction

The brain as well as other organs exhibit powerful endogenous cytoprotective mechanisms to survive recurrent cellular stress events. It is able to develop stress tolerance which has been shown in animal models and in retrospective case-control studies in humans (Kitagawa 2012; Kitagawa et al. 1990; Moncayo et al. 2000; Obrenovitch 2008). Unravelling the molecular mechanisms underlying this neuroprotective phenomenon is of special interest since it may provide the basis to develop new therapeutic strategies for the treatment of neurological diseases. It has become clear that induction of stress tolerance is a multifactorial event which includes triggering the stress response resulting in upregulation of heat shock proteins. The following review will focus on one member of the small heat shock protein family, i.e. HspB5 also called αB-crystallin.

15.2 Expression of HspB5 in the Brain

HspB5 belongs to the family of small heat shock proteins which comprises 11 members, named HspB1 –HspB11 in accordance with the guidelines of the HUGO Gene Nomenclature Committee (Bellyei et al. 2007; Kappe et al. 2003). There exist alternative names for many of the small Hsps that are related to their discovery or their molecular weight. Thus, HspB5 was formerly called αB-crystallin referring to its high abundance in the eye lens responsible for the refractive properties and lens transparency (Bloemendal 1981; Wistow and Piatigorsky 1988). It is still debated if the HspB protein family consists of 11 or 10 members. Hsp16.2/HspB11 has been recently proposed as the 11th member due to its high sequence similarities with some other small Hsps (sHsps). However, it contains no typical α-crystallin-domain which is the characteristic hallmark of sHsps and, thus, should not be included into the sHsp family (Bellyei et al. 2007; Kappe et al. 2010).

Despite the existence of many reports investigating individual sHsps in the brain there are only two systematic comparative studies showing the expression pattern of all sHsps in the brain. Quraishe et al. (2008) found in mouse brain HspB1, HspB5, HspB6, HspB7 and HspB8 expressed using classic RT-PCR (HspB11 was not investigated) and Bartelt-Kirbach et al. (Kirbach and Golenhofen 2011) showed expression of HspB1, HspB2, HspB3, HspB5, HspB6, HspB8 and HspB11 in rat brain by quantitative real-time RT-PCR. The differences in both studies might be due to the different species or more likely due to the different methods used. HspB2 and HspB3 were expressed at a very low level and not in all brain regions. Quantitative real-time RT-PCR displays a higher sensitivity compared to classic RT-PCR and, therefore, HspB2 and HspB3 were probably not detected by classic RT-PCR. HspB5 (followed by HspB6) had the highest expression level in brain among all sHsps but expression in brain is still much lower than in heart muscle, namely only 1–2 % compared to heart (Kirbach and Golenhofen 2011). This fits quite well to the quantification of HspB5 protein content performed by Kato et al. (1991) earlier who found that HspB5 protein in brain amounted to 0.3–2 % of heart tissue depending on the brain region investigated.

Now, 25 years after the discovery of HspB5 in the brain and other non-lenticular tissues (Bhat and Nagineni 1989; Iwaki et al. 1990) the function of HspB5 in the brain is still mainly unclear. Initial studies focused on HspB5 in glia, later also in neurons themselves. There is only little information about the neuronal function at physiological conditions, but it has become clear that HspB5 may play an important role in various neurological diseases even if its functions and molecular targets are only partially identified.

15.2.1 Cellular Distribution of HspB5: Glia versus Neurons

One key question is in which cell type of the brain HspB5 is expressed to exert its function. Nervous tissue consists of neurons and glial cells. Neurons have the unique property to be electrically excitable important for processing and transmission of information. Glial cells have various supportive functions, such as regulation of the internal environment of the tissue, building up the myelin sheath or guidance of neurons during embryogenesis.

The first immunohistochemical studies of rat brain showed localization of HspB5 in oligodendrocytes most abundant in the white matter and in some astrocytes but not in neurons (Iwaki et al. 1990). This confinement of HspB5 to glial cells was confirmed in normal canine and human brain in a more detailed follow-up study (Iwaki et al. 1992). However, in brains of patients with neurodegenerative disorders HspB5 was found in addition in some single neurons as well as in groups of astrocytes of affected brain areas (Iwaki et al. 1992). This suggested that HspB5 expression might be part of a reactive response in various pathological conditions.

Indeed HspB5 is inducible in cultured primary astrocytes and glioma cell lines by various types of cellular stresses (Head et al. 1994, 1996; Iwaki et al. 1993; Kato et al. 1993). By overexpression or knockdown experiments it has become clear that elevated HspB5 levels in astrocytes are of protective value leading to thermoresistance, increased cell survival or inhibition of apoptosis after cellular stress (Iwaki et al. 1994, 1995; Li and Reiser 2011; Shin et al. 2009). HspB5 is also induced in vivo in astrocytes after stress, such as kainic acid treatment (Che et al. 2001), middle cerebral artery occlusion (Piao et al. 2005), intracerebral hemorrhage (Ke et al. 2013) or in several neurodegenerative diseases and animal models thereof (see below).

Stress-responsiveness of HspB5 was also shown in cultured oligodendrocytes (Goldbaum and Richter-Landsberg 2001) and, thus, this process might also be important at pathophysiological conditions and might play an especially important role in demyelinating diseases such as multiple sclerosis (see below). Kim et al. suggested that additionally HspB5 of oligodendrocytes might be important during brain development (Kim et al. 2012).

Whereas HspB5 has been investigated in detail in glia at normal or pathophysiological conditions, little attention has been paid to neuronally derived HspB5. This is probably attributable to the fact that HspB5 expression initially was not detected in neurons and later only in single "ballooned" neurons under pathological conditions (Iwaki et al. 1992; Kato et al. 1992a, b; Lowe et al. 1992). It was also found in neurons after cerebral infarction and olivary hypertrophy (Fukushima et al. 2006; Minami et al. 2003). Recently, using cultured hippocampal neurons it could be confirmed that HspB5 is responsive to several stress conditions, such as heat shock, oxidative, sodium arsenite or hyperosmotic stress in neurons themselves (Bartelt-Kirbach and Golenhofen 2014; Kirbach and Golenhofen 2011).

15.3 HspB5 in Neurological Diseases

HspB5 as well as other Hsps plays an important role in neurological diseases. Most neurodegenerative disorders share a common pathology, namely an accumulation of misfolded proteins leading to pathological protein aggregates and finally to cellular dysfunction and cell death. Hsps acting mostly as chaperones are especially important in these diseases since they are in general able to refold proteins or may pass irreversibly misfolded proteins to the proteasome for degradation. For an overview of expression of HspB5 in neurological diseases (see Table 15.1).

15.3.1 Alexander Disease

Alexander disease is a rare but severe disease of the central nervous system first described in 1949 (Alexander 1949). The major clinical symptoms are macrocephaly, leukodystrophy, developmental retardation or cognitive delays. It has become clear that Alexander disease is a genetic disorder with most patients having mutations of the intermediate filament protein, glial fibrillary acidic protein (GFAP) (Brenner et al. 2001; Li et al. 2005). The pathological hallmark is the presence of Rosenthal fibers in astrocytes which are intracellular protein aggregates consisting mainly of GFAP, HspB1 and HspB5 (Head et al. 1993; Iwaki et al. 1989, 1993). However, little is known both about the mechanisms by which GFAP mutations cause the disease and the functional role of HspB5 in Rosenthal fibres. HspB5 binds to GFAP and regulates filament assembly in vitro (Nicholl and Quinlan 1994). Mutated GFAP leads to formation of GFAP containing cytoplasmic aggregates which then associate with HspB5 (Der Perng et al. 2006). In a mouse model of Alexander disease elevated HspB5 levels were beneficial. HspB5 reduced GFAP accumulation, reduced Rosenthal fiber formation and rescued animals from terminal seizures (Hagemann et al. 2009). Recent work has shown that GFAP inhibits the proteasome and that this effect is increased in mutated GFAP leading to less proteasomal degradation of GFAP. Interestingly, HspB5 reverses this inhibitory effect of mutated GFAP on the proteasome which might be one mechanism of how HspB5 exerts its cytoprotective effect in Alexander disease (Tang et al. 2010).

Table 15.1 Overview about expression of HspB5 in neurological diseases

			Cellular distribution/upregulation		
Disease	Species	Brain region	Glia	Neurons	Others
Alexander disease	Human	Affected brain areas	Rosenthal fibers of astrocytes (Head et al. 1993; Iwaki et al. 1989, 1993)		
	In vitro				Binds to GFAP and regulates filament assembly (Nicholl and Quinlan 1994)
Alzheimer's disease	Human	Affected brain areas	Reactive astrocytes, microglia, oligodendrocytes (Bjorkdahl et al. 2008; Renkawek et al. 1994; Wilhelmus et al. 2006b)		
	Human	Areas especially rich in extracellular neurofibrillary tangles		Subset of neurons (Mao et al. 2001)	
	In vitro				Binds to β-amyloid and inhibits fibril formation (Shammas et al. 2011; Stege et al. 1999; Wilhelmus et al. 2006a)
Other tauopathies	Human	Affected brain areas		Ballooned neurons (Kato et al. 1992a, b; Lowe et al. 1992)	
	Human	Affected neocortex and basal ganglia	Reactive astrocytes, oligodendrocytes (Dabir et al. 2004)	Ballooned neurons (Dabir et al. 2004)	
	Human	Whole brain homogenate (Iwaki et al. 1992)			

(continued)

Table 15.1 (continued)

Disease	Species	Brain region	Cellular distribution/upregulation		
			Glia	Neurons	Others
Parkinson's disease	Human	Cortex, substantia nigra, hippocampus	Reactive astrocytes and microglia (Renkawek et al. 1999)		
	Human	Substantia nigra	Reactive astrocytes and microglia (Jellinger 2000)	Neurons with axonal spheroids + "Lewy" neurites (Jellinger 2000)	
	Human	Cerebral cortex, amygdala, and ventral claustrum (severe cases); anterior temporal and insular mesocortex (mild cases)		Subset of neurons (Braak et al. 2001)	
	Human	Whole brain homogenate (Iwaki et al. 1992)			
	Mouse	Affected brain regions (alpha-SynA53T transgenic mice)	Subgroups of glia (Wang et al. 2008)		
	Drosophila				Suppression of alpha-synuclein aggregation (Tue et al. 2012)
	In vitro				Inhibits growth of alpha-synuclein fibrils (Rekas et al. 2004; Waudby et al. 2010)
Poly-Q-diseases	Mouse	Reduction in whole brain homogenates (Zabel et al. 2002)			
	Mouse	Ocular lens			Enhanced aggregation of ectopic expressed mutant huntingtin in HspB5 –/– mice (Muchowski et al. 2008)

	Drosophila				Suppression of poly-Q aggregation (Tue et al. 2012)
					Suppression of ataxin-3 toxicity (Bilen and Bonini 2007)
	In vitro				Inhibition of first stage of ataxin-3 aggregation (Robertson et al. 2010)
Amyotrophic lateral sclerosis	Human	Whole brain homogenate (Iwaki et al. 1992)			
	Mouse	Spinal cord	Subgroups of glia (Wang et al. 2008), reactive glia (Vleminckx et al. 2002)		
	Mouse				HspB5 cofractionates with insoluble mutant SOD1 (Wang et al. 2003)
	In vitro				Inhibition of SOD1 aggregate growth (Yerbury et al. 2013)
Multiple sclerosis	Human	Preactive lesions	Oligodendrocytes (van Noort et al. 1995)		
	Human				Myelin (van Noort et al. 1995)
Prion diseases	Human	Whole brain homogenate (Iwaki et al. 1992)			
	Human	Affected brain regions		Ballooned neurons (Kato et al. 1992a, b)	
	Human	Affected brain regions, spongiotic tissue	Astrocytes (Renkawek et al. 1992)		
	Hamster	Whole brain	Astrocytes (Wang et al. 2013)		
	In vitro				HspB5 effectively shears prion protein fibrils (Sun et al. 2008)

15.3.2 Alzheimer's Disease

Alzheimer's disease (AD) is a prevalent age-related neurodegenerative disease characterized by its major symptom, dementia (Selkoe 2001). The pathological hallmarks of AD are neurofibrillary tangles and senile plaques. Senile plaques are extracellular amyloid-rich plaques consisting mainly of a 40–42 amino acid long peptide amyloid-β. In contrast, neurofibrillary tangles are formed by hyperphosphorylation of the microtubule associated protein tau and neurofilaments. HspB5 is upregulated in AD brains and localized in all type of glia cells in the areas rich in senile plaques (Bjorkdahl et al. 2008; Renkawek et al. 1994; Wilhelmus et al. 2006b). It was also found in neurons in the limbic system of AD brains especially in late stages of AD and areas rich in neurofibrillary tangles (Mao et al. 2001).

HspB5 binds to amyloid-β in vitro and prevents fibril formation (Ecroyd and Carver 2009; Shammas et al. 2011; Stege et al. 1999; Wilhelmus et al. 2006a). However, there are contradictory reports about the impact of this inhibition on amyloid toxicity. By the increase of non-fibrillar forms of amyloid-β even more toxic forms may be produced actually resulting in increased cell death (Stege et al. 1999). However, under other experimental conditions amyloid-β toxicity was reduced by binding to HspB5 (Dehle et al. 2010; Wilhelmus et al. 2006a). Further studies are needed to clarify if HspB5-amyloid-β interaction is an important process in Alzheimer's disease or if intracellular (upregulated) HspB5 in reactive glia or in affected neurons may exert its effect by other molecular targets.

15.3.3 Other Tauopathies

Apart from Alzheimer's disease also other neurodegenerative diseases result from the abnormal aggregation of tau protein (Lee et al. 2001). These are summarized under frontotemporal dementias (FTD, also known as Pick's disease) and include corticobasal degeneration (CBD), progressive supranuclear palsy (PSP) and familial tauopathy with parkinsonism linked to chromosome 17 (FTDP-17) (Kertesz et al. 2003; Tolnay and Probst 2003). Higher immunoreactivity of HspB5 was seen in brains of eight PSP patients compared to controls (Iwaki et al. 1992). HspB5 has especially been found upregulated in ballooned neurons of such diseases, including AD (Kato et al. 1992a, b; Lowe et al. 1992). Ballooned neurons are reported to contain phosphorylated neurofilament protein but seldom tangles of tau protein (Dickson et al. 1986; Mori and Oda 1997). In addition to neurons, immunoreactivity of HspB5 was also seen in glial inclusions in CBD, PSP and FTDP-17 patients (Dabir et al. 2004). HspB5 signals were not found in neuronal tau inclusions but the staining of ballooned neurons in affected areas could be confirmed. In summary, this points to an association of HspB5 with abnormal neurofilament fibrils rather than with tau fibrils, at least in neurons. It is therefore still unclear if HspB5 associates at all with abnormal tau protein aggregates and what role it plays in the pathogenesis of these diseases.

15.3.4 Parkinson's Disease

Parkinson's disease (PD) is characterized by a pathological accumulation of α-synuclein protein in so-called Lewy bodies in neurons primarily of the substantia nigra. The loss of dopamin-producing neurons in this area results in the typical symptoms of tremor, rigidity, brady- or akinesia and postural instability (Jankovic 2008).

HspB5 is upregulated in PD brains (Iwaki et al. 1992) in different cell types. Strong HspB5 immunoreactivity was found in reactive astrocytes and microglia (Jellinger 2000), although this is probably only true for PD cases with additional AD dementia (Renkawek et al. 1999). Ballooned neurons, which have been described for several other neurodegenerative diseases, were also detected in PD, their appearance correlating with disease severity and sites of disease specific appearance of Lewy bodies (Braak et al. 2001). Interestingly, however, the telencephalic HspB5-immunopositive neurons mostly did not develop Lewy bodies and neurons with Lewy bodies seldom were HspB5-immunopositive (Braak et al. 2001). This is in contrast to another group, which found the Lewy bodies/neurites and axonal spheroids immunopositive for HspB5 (Jellinger 2000). Therefore, it is still not clear if HspB5 associates directly with the α-synuclein aggregates in Lewy bodies. However, in vitro as well as drosophila studies show not only an interaction of both proteins but also that HspB5 is a potent inhibitor of α-synuclein fibril formation (Rekas et al. 2004; Tue et al. 2012; Waudby et al. 2010). It would be of interest if HspB5 overexpression would have neuroprotective effects in a parkinson mouse model.

15.3.5 Poly-Q-Diseases

This group of diseases is characterized by autosomal dominant inherited mutations in genes for huntingtin or ataxins, leading to expanded poly-glutamine (poly-Q) repeats in the resulting protein. These poly-Q proteins form toxic amyloid-like aggregates and ultimately result in either Morbus Huntington or spinocerebellar ataxias. In Morbus Huntington a degeneration of GABAergic neurons in the striatum is seen, leading to the typical clinical symptoms, such as hyperkinesia and later-on dementia. Death occurs on average 20 years after disease onset. Spinocerebellar ataxias are characterized by progressive loss of the fine coordination of movement while the mental capacity is usually not impaired. HspB5 expression has been found to decrease in affected brains of a Huntington mouse model over the time course of disease (Zabel et al. 2002). On the other hand, there is evidence that HspB5 has a protective effect from expression of mutant huntingtin in a HspB5-knockout mouse model (Muchowski et al. 2008) and from the fact that it can inhibit protein aggregation in drosophila and in vitro (Bilen and Bonini 2007; Robertson et al. 2010; Tue et al. 2012). It remains to be investigated if HspB5 is upregulated in specific cell types in affected areas of patients and whether it plays a functional role in this disease.

15.3.6 Amyotrophic Lateral Sclerosis

Amyotrophic lateral sclerosis (ALS) is a progressive lethal disease due to the degeneration of motor neurons in brainstem, spinal cord and motor cortex. Most ALS cases are sporadic, but 5–10 % are hereditary cases, of which approximately 20 % carry mutations in the superoxide dismutase gene (SOD1) (Deng et al. 1993; Rosen et al. 1993). HspB5 is upregulated in ALS brains and in the spinal cord in single neurons in the anterior horn (Iwaki et al. 1992). ALS is also characterized by the so-called "ballooned" neurons seen in several neurodegenerative diseases. However, unlike in Pick's disease or Creutzfelt-Jakob disease, these neurons do not show HspB5 expression in ALS patients (Kato et al. 1992a). Only a single ALS case has been reported where eosinophilic fibrillary neuronal inclusions in the cerebral cortex were also HspB5-positive (Arima et al. 1998). In transgenic mice expressing mutant SOD1 HspB5 was found upregulated in subgroups of glial cells (oligodendrocytes, reactive astroglia) in the spinal cord (Vleminckx et al. 2002; Wang et al. 2008). HspB5, which was shown before to co-fractionate with mutated insoluble SOD1 (Wang et al. 2003), was recently reported to bind to SOD1 aggregates and inhibit aggregate growth in vitro (Yerbury et al. 2013). It therefore seems likely that HspB5 has a protective effect in ALS at least in cases caused by mutated SOD1.

15.3.7 Multiple Sclerosis

Multiple sclerosis (MS) is an immune-mediated demyelinating disease of the central nervous system. However, endogeneous neurodegenerative processes seem to be involved additionally in the pathogenesis of the disease. The major clinical symptoms are motor and sensory deficits, visual problems, cognitive dysfunction and tiredness. HspB5 was first linked to MS in 1995 when it was considered as a possible autoantigen because HspB5 derived from myelin of MS patients was a strong activator of human T-cells (van Noort et al. 1995). However, later it was shown that HspB5 reactive T-cells do not induce autoimmune encephalomyelitis (EAE) in mice which does not fit to the hypothesis of HspB5 as an autoantigen (Wang et al. 2006). In the meantime it has become clear that the role of HspB5 in MS is much more complex and that it mainly fulfils beneficial effects via its anti-inflammatory activity. HspB5-deficient mice show more severe inflammation and demyelination after EAE, an experimental model of MS. Moreover, intravenous administration of recombinant HspB5 ameliorated clinical symptoms and neuroinflammation (Ousman et al. 2007). Administration of HspB5 is also advantageous after experimental stroke resulting in reduced stroke volume and inflammatory cytokines associated with stroke pathology (Arac et al. 2011). This makes HspB5 an attractive therapeutic target in neuroinflammatory diseases and future research should be directed to investigating applications of HspB5 in humans. Indeed one recent study investigated the influence of HspB5 on the cytokine production of $CD4^+$ T-cells isolated from MS patients in vitro and found that HspB5 was able to suppress T-cell activation derived from a subset of MS patients (Quach et al. 2013). In addition to this anti-inflammatory function of HspB5 in MS it plays a role in preactive MS lesions before infiltration of the tissue with lymphocytes takes place. HspB5 is selectively upregulated in preactive MS lesions where it activates surrounding microglia. This process is thought to be part of an anti-inflammatory repair response aiming to restore tissue homeostasis (Bsibsi et al. 2013; van Noort et al. 2010, 2012).

15.3.8 Prion Diseases

Prions are endogenous proteins that become infectious when misfolded. They then cause transmissible spongiform encephalopathies as Creutzfeldt-Jakob (CJD) in human, bovine spongiform encephalopathy (BSE) in cattle or scrapie in sheep (DeArmond and Prusiner 2003; Prusiner 1998). HspB5 is upregulated in CJD brains compared to controls and many neurons in the deeper layers of the neocortex in these patients were HspB5 positive (Iwaki et al. 1992). Ballooned neurons are also seen in this disease and most of them express HspB5 (Kato et al. 1992a, b). HspB5 was also found to be upregulated in astrocytes, most intensively in the spongiotic tissue (Renkawek et al. 1992). Recently, a study of scrapie infected hamsters revealed an increased expression of HspB5 at the terminal disease stages, mainly in astrocytes (Wang et al. 2013). This group did not find a colocalization of HspB5 with prion protein deposits in the infected animals. Prion proteins form toxic fibrils which have to be fragmented in order to replicate. In vitro, HspB5 was able to shear prion fibrils fast and effectively, therefore it might even contribute to disease progression (Sun et al. 2008).

15.4 Concluding Remarks/Perspectives

Taken together, HspB5 seems to play many different roles in the brain and especially in neurological diseases. It is expressed in glia as well as in neurons and upregulated in certain cell types or subset of cells at pathophysiological conditions. Often it is found associated with pathological protein aggregates characteristic of the respective disease. As described above it can be concluded that in most conditions HspB5 counteracts the aggregation process and thus will exert a neuroprotective effect. However, it has become clear that this typical chaperone-like function of HspB5 is not the only way of action of HspB5 in the brain. HspB5 displays additionally other non-chaperone functions, such as anti-inflammatory, anti-apoptotic functions or association with cytoskeletal proteins influencing filament assembly. The importance of the anti-inflammatory effect of HspB5 has been highlighted above in the section about MS.

HspB5 function is known to be regulated by phosphorylation in several cell types and tissues (Ito et al. 1997; Morrison et al. 2003), but unfortunately little attention has been drawn to this aspect in the brain. We could recently show that in cultured hippocampal neurons phosphorylated HspB5 displays different subcellular distributions compared to the unphosphorylated form and that the localization was dependent on which of the three phosphorylation sites was actually phosphorylated (Schmidt et al. 2012). This suggests that phosphorylation of HspB5 leads to interaction with different specific molecular targets within one cell. The distribution pattern of phosphoHspB5 hinted at filamentous structures within dendrites, axons and components of synapses as possible targets. This led us to the hypothesis that stress-induced phosphorylation of HspB5 with subsequent translocation to synaptic structures and neuronal processes may provide one yet unknown mechanism of how HspB5 exerts a neuroprotective effect under pathophysiological conditions. It has been shown indeed that HspB5 is phosphorylated at serine 45 and 59 in several neurological diseases, such as Alexander disease, Alzheimer's disease and in Down syndrome (Kato et al. 2001; Palminiello et al. 2009). However, Palminiello et al. (2009) showed at least in Down-Syndrome that phosphorylation of HspB5 was restricted to glial cells and was not detected in neurons. In glial cells phosphorylation of HspB5 is important for the anti-apoptotic effect of HspB5 (Li and Reiser 2011).

More experiments are needed to investigate the phosphorylation of HspB5 in neurons themselves in neurological diseases. There might be important effects of phosphoHspB5 on neurons in addition to the well described effect on pathological protein aggregates and HspB5 in glial cells. Identifying the role of phosphorylation of HspB5 in neurons and the molecular targets of phosphoHspB5 in neurons may help to develop new therapeutic strategies for neurological diseases.

References

Alexander WS (1949) Progressive fibrinoid degeneration of fibrillary astrocytes associated with mental retardation in a hydrocephalic infant. Brain 72:373–381, 3 pl

Arac A, Brownell SE, Rothbard JB, Chen C, Ko RM, Pereira MP, Albers GW, Steinman L, Steinberg GK (2011) Systemic augmentation of alphaB-crystallin provides therapeutic benefit twelve hours post-stroke onset via immune modulation. Proc Natl Acad Sci U S A 108:13287–13292

Arima K, Ogawa M, Sunohara N, Nishio T, Shimomura Y, Hirai S, Eto K (1998) Immunohistochemical and ultrastructural characterization of ubiquitinated eosinophilic fibrillary neuronal inclusions in sporadic amyotrophic lateral sclerosis. Acta Neuropathol 96:75–85

Bartelt-Kirbach B, Golenhofen N (2014) Reaction of small heat shock proteins to different kinds of cellular stress in cultured rat hippocampal neurons. Cell Stress Chaperones 19:145–153

Bellyei S, Szigeti A, Pozsgai E, Boronkai A, Gomori E, Hocsak E, Farkas R, Sumegi B, Gallyas F Jr (2007) Preventing apoptotic cell death by a novel small heat shock protein. Eur J Cell Biol 86:161–171

Bhat SP, Nagineni CN (1989) alpha B subunit of lens-specific protein alpha-crystallin is present in other ocular and non-ocular tissues. Biochem Biophys Res Commun 158:319–325

Bilen J, Bonini NM (2007) Genome-wide screen for modifiers of ataxin-3 neurodegeneration in Drosophila. PLoS Genet 3:1950–1964

Bjorkdahl C, Sjogren MJ, Zhou X, Concha H, Avila J, Winblad B, Pei JJ (2008) Small heat shock proteins Hsp27 or alphaB-crystallin and the protein components of neurofibrillary tangles: tau and neurofilaments. J Neurosci Res 86:1343–1352

Bloemendal H (1981) Molecular and cellular biology of the eye lens. Wiley, New York

Braak H, Del Tredici K, Sandmann-Kiel D, Rub U, Schultz C (2001) Nerve cells expressing heat-shock proteins in Parkinson's disease. Acta Neuropathol 102:449–454

Brenner M, Johnson AB, Boespflug-Tanguy O, Rodriguez D, Goldman JE, Messing A (2001) Mutations in GFAP, encoding glial fibrillary acidic protein, are associated with Alexander disease. Nat Genet 27:117–120

Bsibsi M, Holtman IR, Gerritsen WH, Eggen BJ, Boddeke E, van der Valk P, van Noort JM, Amor S (2013) Alpha-B-crystallin induces an immune-regulatory and antiviral microglial response in preactive multiple sclerosis lesions. J Neuropathol Exp Neurol 72:970–979

Che Y, Piao CS, Han PL, Lee JK (2001) Delayed induction of alpha B-crystallin in activated glia cells of hippocampus in kainic acid-treated mouse brain. J Neurosci Res 65:425–431

Dabir DV, Trojanowski JQ, Richter-Landsberg C, Lee VM, Forman MS (2004) Expression of the small heat-shock protein alphaB-crystallin in tauopathies with glial pathology. Am J Pathol 164:155–166

DeArmond SJ, Prusiner SB (2003) Perspectives on prion biology, prion disease pathogenesis, and pharmacologic approaches to treatment. Clin Lab Med 23:1–41

Dehle FC, Ecroyd H, Musgrave IF, Carver JA (2010) alphaB-Crystallin inhibits the cell toxicity associated with amyloid fibril formation by kappa-casein and the amyloid-beta peptide. Cell Stress Chaperones 15:1013–1026

Deng HX, Hentati A, Tainer JA, Iqbal Z, Cayabyab A, Hung WY, Getzoff ED, Hu P, Herzfeldt B, Roos RP et al (1993) Amyotrophic lateral sclerosis and structural defects in Cu, Zn superoxide dismutase. Science 261:1047–1051

Der Perng M, Su M, Wen SF, Li R, Gibbon T, Prescott AR, Brenner M, Quinlan RA (2006) The Alexander disease-causing glial fibrillary acidic protein mutant, R416W, accumulates into Rosenthal fibers by a pathway that involves filament aggregation and the association of alpha B-crystallin and HSP27. Am J Hum Genet 79:197–213

Dickson DW, Yen SH, Suzuki KI, Davies P, Garcia JH, Hirano A (1986) Ballooned neurons in select neurodegenerative diseases contain phosphorylated neurofilament epitopes. Acta Neuropathol 71:216–223

Ecroyd H, Carver JA (2009) Crystallin proteins and amyloid fibrils. Cell Mol Life Sci 66:62–81

Fukushima K, Mizuno Y, Takatama M, Okamoto K (2006) Increased neuronal expression of alpha B-crystallin in human olivary hypertrophy. Neuropathology 26:196–200

Goldbaum O, Richter-Landsberg C (2001) Stress proteins in oligodendrocytes: differential effects of heat shock and oxidative stress. J Neurochem 78:1233–1242

Hagemann TL, Boelens WC, Wawrousek EF, Messing A (2009) Suppression of GFAP toxicity by alphaB-crystallin in mouse models of Alexander disease. Hum Mol Genet 18:1190–1199

Head MW, Corbin E, Goldman JE (1993) Overexpression and abnormal modification of the stress proteins alpha B-crystallin and HSP27 in Alexander disease. Am J Pathol 143:1743–1753

Head MW, Corbin E, Goldman JE (1994) Coordinate and independent regulation of alpha B-crystallin and hsp27 expression in response to physiological stress. J Cell Physiol 159:41–50

Head MW, Hurwitz L, Goldman JE (1996) Transcription regulation of alpha B-crystallin in astrocytes: analysis of HSF and AP1 activation by different types of physiological stress. J Cell Sci 109(Pt 5):1029–1039

Ito H, Okamoto K, Nakayama H, Isobe T, Kato K (1997) Phosphorylation of alphaB-crystallin in response to various types of stress. J Biol Chem 272:29934–29941

Iwaki T, Kume-Iwaki A, Liem RK, Goldman JE (1989) Alpha B-crystallin is expressed in non-lenticular tissues and accumulates in Alexander's disease brain. Cell 57:71–78

Iwaki T, Kume-Iwaki A, Goldman JE (1990) Cellular distribution of alpha B-crystallin in non-lenticular tissues. J Histochem Cytochem 38:31–39

Iwaki T, Wisniewski T, Iwaki A, Corbin E, Tomokane N, Tateishi J, Goldman JE (1992) Accumulation of alpha B-crystallin in central nervous system glia and neurons in pathologic conditions. Am J Pathol 140:345–356

Iwaki T, Iwaki A, Tateishi J, Sakaki Y, Goldman JE (1993) Alpha B-crystallin and 27-kd heat shock protein are regulated by stress conditions in the central nervous system and accumulate in Rosenthal fibers. Am J Pathol 143:487–495

Iwaki T, Iwaki A, Tateishi J, Goldman JE (1994) Sense and antisense modification of glial alpha B-crystallin production results in alterations of stress fiber formation and thermoresistance. J Cell Biol 125:1385–1393

Iwaki T, Iwaki A, Fukumaki Y, Tateishi J (1995) Alpha B-crystallin in C6 glioma cells supports their survival in elevated extracellular K+: the implication of a protective role of alpha B-crystallin accumulation in reactive glia. Brain Res 673:47–52

Jankovic J (2008) Parkinson's disease: clinical features and diagnosis. J Neurol Neurosurg Psychiatry 79:368–376

Jellinger KA (2000) Cell death mechanisms in Parkinson's disease. J Neural Transm 107:1–29

Kappe G, Franck E, Verschuure P, Boelens WC, Leunissen JA, de Jong WW (2003) The human genome encodes 10 alpha-crystallin-related small heat shock proteins: HspB1-10. Cell Stress Chaperones 8:53–61

Kappe G, Boelens WC, de Jong WW (2010) Why proteins without an alpha-crystallin domain should not be included in the human small heat shock protein family HSPB. Cell Stress Chaperones 15:457–461

Kato K, Shinohara H, Kurobe N, Inaguma Y, Shimizu K, Ohshima K (1991) Tissue distribution and developmental profiles of immunoreactive alpha B crystallin in the rat determined with a sensitive immunoassay system. Biochim Biophys Acta 1074:201–208

Kato S, Hirano A, Umahara T, Kato M, Herz F, Ohama E (1992a) Comparative immunohistochemical study on the expression of alpha B crystallin, ubiquitin and stress-response protein 27 in ballooned neurons in various disorders. Neuropathol Appl Neurobiol 18:335–340

Kato S, Hirano A, Umahara T, Llena JF, Herz F, Ohama E (1992b) Ultrastructural and immunohistochemical studies on ballooned cortical neurons in Creutzfeldt-Jakob disease: expression of alpha B-crystallin, ubiquitin and stress-response protein 27. Acta Neuropathol 84:443–448

Kato K, Goto S, Hasegawa K, Inaguma Y (1993) Coinduction of two low-molecular-weight stress proteins, alpha B crystallin and HSP28, by heat or arsenite stress in human glioma cells. J Biochem 114:640–647

Kato K, Inaguma Y, Ito H, Iida K, Iwamoto I, Kamei K, Ochi N, Ohta H, Kishikawa M (2001) Ser-59 is the major phosphorylation site in alphaB-crystallin accumulated in the brains of patients with Alexander's disease. J Neurochem 76:730–736

Ke K, Li L, Rui Y, Zheng H, Tan X, Xu W, Cao J, Xu J, Cui G, Xu G, Cao M (2013) Increased expression of small heat shock protein alphaB-crystallin after intracerebral hemorrhage in adult rats. J Mol Neurosci 51:159–169

Kertesz A, Hillis A, Munoz DG (2003) Frontotemporal degeneration, pick's disease, pick complex, and ravel. Ann Neurol 54(Suppl 5):S1–S2

Kim JY, Song SH, Kim HN, Kim DW, Sohn HJ, Lee EY, Cho SS, Seo JH (2012) alpha b-crystallin is expressed in myelinating oligodendrocytes of the developing and adult avian retina. Neurochem Res 37:2135–2142

Kirbach BB, Golenhofen N (2011) Differential expression and induction of small heat shock proteins in rat brain and cultured hippocampal neurons. J Neurosci Res 89:162–175

Kitagawa K (2012) Ischemic tolerance in the brain: endogenous adaptive machinery against ischemic stress. J Neurosci Res 90:1043–1054

Kitagawa K, Matsumoto M, Tagaya M, Hata R, Ueda H, Niinobe M, Handa N, Fukunaga R, Kimura K, Mikoshiba K et al (1990) 'Ischemic tolerance' phenomenon found in the brain. Brain Res 528:21–24

Lee VM, Goedert M, Trojanowski JQ (2001) Neurodegenerative tauopathies. Annu Rev Neurosci 24:1121–1159

Li R, Reiser G (2011) Phosphorylation of Ser45 and Ser59 of alphaB-crystallin and p38/extracellular regulated kinase activity determine alphaB-crystallin-mediated protection of rat brain astrocytes from C2-ceramide- and staurosporine-induced cell death. J Neurochem 118:354–364

Li R, Johnson AB, Salomons G, Goldman JE, Naidu S, Quinlan R, Cree B, Ruyle SZ, Banwell B, D'Hooghe M, Siebert JR, Rolf CM, Cox H, Reddy A, Gutierrez-Solana LG, Collins A, Weller RO, Messing A, van der Knaap MS, Brenner M (2005) Glial fibrillary acidic protein mutations in infantile, juvenile, and adult forms of Alexander disease. Ann Neurol 57:310–326

Lowe J, Errington DR, Lennox G, Pike I, Spendlove I, Landon M, Mayer RJ (1992) Ballooned neurons in several neurodegenerative diseases and stroke contain alpha B crystallin. Neuropathol Appl Neurobiol 18:341–350

Mao JJ, Katayama S, Watanabe C, Harada Y, Noda K, Yamamura Y, Nakamura S (2001) The relationship between alphaB-crystallin and neurofibrillary tangles in Alzheimer's disease. Neuropathol Appl Neurobiol 27:180–188

Minami M, Mizutani T, Kawanishi R, Suzuki Y, Mori H (2003) Neuronal expression of alphaB crystallin in cerebral infarction. Acta Neuropathol 105:549–554

Moncayo J, de Freitas GR, Bogousslavsky J, Altieri M, van Melle G (2000) Do transient ischemic attacks have a neuroprotective effect? Neurology 54:2089–2094

Mori H, Oda M (1997) Ballooned neurons in cortocobasal degeneneration and progressive supranuclear palsy. Neuropathology 17:248–252

Morrison LE, Hoover HE, Thuerauf DJ, Glembotski CC (2003) Mimicking phosphorylation of alphaB-crystallin on serine-59 is necessary and sufficient to provide maximal protection of cardiac myocytes from apoptosis. Circ Res 92:203–211

Muchowski PJ, Ramsden R, Nguyen Q, Arnett EE, Greiling TM, Anderson SK, Clark JI (2008) Noninvasive measurement of protein aggregation by mutant huntingtin fragments or alpha-synuclein in the lens. J Biol Chem 283:6330–6336

Nicholl ID, Quinlan RA (1994) Chaperone activity of alpha-crystallins modulates intermediate filament assembly. EMBO J 13:945–953

Obrenovitch TP (2008) Molecular physiology of preconditioning-induced brain tolerance to ischemia. Physiol Rev 88:211–247

Ousman SS, Tomooka BH, van Noort JM, Wawrousek EF, O'Connor KC, Hafler DA, Sobel RA, Robinson WH, Steinman L (2007) Protective and therapeutic role for alphaB-crystallin in autoimmune demyelination. Nature 448:474–479

Palminiello S, Jarzabek K, Kaur K, Walus M, Rabe A, Albertini G, Golabek AA, Kida E (2009) Upregulation of phosphorylated alphaB-crystallin in the brain of children and young adults with Down syndrome. Brain Res 1268:162–173

Piao CS, Kim SW, Kim JB, Lee JK (2005) Co-induction of alphaB-crystallin and MAPKAPK-2 in astrocytes in the penumbra after transient focal cerebral ischemia. Exp Brain Res 163:421–429

Prusiner SB (1998) Prions. Proc Natl Acad Sci U S A 95:13363–13383

Quach QL, Metz LM, Thomas JC, Rothbard JB, Steinman L, Ousman SS (2013) CRYAB modulates the activation of CD4+ T cells from relapsing-remitting multiple sclerosis patients. Mult Scler 19:1867–1877

Quraishe S, Asuni A, Boelens WC, O'Connor V, Wyttenbach A (2008) Expression of the small heat shock protein family in the mouse CNS: differential anatomical and biochemical compartmentalization. Neuroscience 153:483–491

Rekas A, Adda CG, Andrew Aquilina J, Barnham KJ, Sunde M, Galatis D, Williamson NA, Masters CL, Anders RF, Robinson CV, Cappai R, Carver JA (2004) Interaction of the molecular chaperone alphaB-crystallin with alpha-synuclein: effects on amyloid fibril formation and chaperone activity. J Mol Biol 340:1167–1183

Renkawek K, de Jong WW, Merck KB, Frenken CW, van Workum FP, Bosman GJ (1992) alpha B-crystallin is present in reactive glia in Creutzfeldt-Jakob disease. Acta Neuropathol 83:324–327

Renkawek K, Voorter CE, Bosman GJ, van Workum FP, de Jong WW (1994) Expression of alpha B-crystallin in Alzheimer's disease. Acta Neuropathol 87:155–160

Renkawek K, Stege GJ, Bosman GJ (1999) Dementia, gliosis and expression of the small heat shock proteins hsp27 and alpha B-crystallin in Parkinson's disease. Neuroreport 10:2273–2276

Robertson AL, Headey SJ, Saunders HM, Ecroyd H, Scanlon MJ, Carver JA, Bottomley SP (2010) Small heat-shock proteins interact with a flanking domain to suppress polyglutamine aggregation. Proc Natl Acad Sci U S A 107:10424–10429

Rosen DR, Siddique T, Patterson D, Figlewicz DA, Sapp P, Hentati A, Donaldson D, Goto J, O'Regan JP, Deng HX et al (1993) Mutations in Cu/Zn superoxide dismutase gene are associated with familial amyotrophic lateral sclerosis. Nature 362:59–62

Schmidt T, Bartelt-Kirbach B, Golenhofen N (2012) Phosphorylation-dependent subcellular localization of the small heat shock proteins HspB1/Hsp25 and HspB5/alphaB-crystallin in cultured hippocampal neurons. Histochem Cell Biol 138:407–418

Selkoe DJ (2001) Alzheimer's disease: genes, proteins, and therapy. Physiol Rev 81:741–766

Shammas SL, Waudby CA, Wang S, Buell AK, Knowles TP, Ecroyd H, Welland ME, Carver JA, Dobson CM, Meehan S (2011) Binding of the molecular chaperone alphaB-crystallin to Abeta amyloid fibrils inhibits fibril elongation. Biophys J 101:1681–1689

Shin JH, Kim SW, Lim CM, Jeong JY, Piao CS, Lee JK (2009) alphaB-crystallin suppresses oxidative stress-induced astrocyte apoptosis by inhibiting caspase-3 activation. Neurosci Res 64:355–361

Stege GJ, Renkawek K, Overkamp PS, Verschuure P, van Rijk AF, Reijnen-Aalbers A, Boelens WC, Bosman GJ, de Jong WW (1999) The molecular chaperone alphaB-crystallin enhances amyloid beta neurotoxicity. Biochem Biophys Res Commun 262:152–156

Sun Y, Makarava N, Lee CI, Laksanalamai P, Robb FT, Baskakov IV (2008) Conformational stability of PrP amyloid fibrils controls their smallest possible fragment size. J Mol Biol 376:1155–1167

Tang G, Perng MD, Wilk S, Quinlan R, Goldman JE (2010) Oligomers of mutant glial fibrillary acidic protein (GFAP) Inhibit the proteasome system in Alexander disease astrocytes, and the small heat shock protein alphaB-crystallin reverses the inhibition. J Biol Chem 285:10527–10537

Tolnay M, Probst A (2003) The neuropathological spectrum of neurodegenerative tauopathies. IUBMB Life 55:299–305

Tue NT, Shimaji K, Tanaka N, Yamaguchi M (2012) Effect of alphaB-crystallin on protein aggregation in Drosophila. J Biomed Biotechnol 2012:252049

van Noort JM, van Sechel AC, Bajramovic JJ, el Ouagmiri M, Polman CH, Lassmann H, Ravid R (1995) The small heat-shock protein alpha B-crystallin as candidate autoantigen in multiple sclerosis. Nature 375:798–801

van Noort JM, Bsibsi M, Gerritsen WH, van der Valk P, Bajramovic JJ, Steinman L, Amor S (2010) Alphab-crystallin is a target for adaptive immune responses and a trigger of innate responses in preactive multiple sclerosis lesions. J Neuropathol Exp Neurol 69:694–703

van Noort JM, Bsibsi M, Nacken P, Gerritsen WH, Amor S (2012) The link between small heat shock proteins and the immune system. Int J Biochem Cell Biol 44:1670–1679

Vleminckx V, Van Damme P, Goffin K, Delye H, Van Den Bosch L, Robberecht W (2002) Upregulation of HSP27 in a transgenic model of ALS. J Neuropathol Exp Neurol 61:968–974

Wang J, Slunt H, Gonzales V, Fromholt D, Coonfield M, Copeland NG, Jenkins NA, Borchelt DR (2003) Copper-binding-site-null SOD1 causes ALS in transgenic mice: aggregates of non-native SOD1 delineate a common feature. Hum Mol Genet 12:2753–2764

Wang C, Chou YK, Rich CM, Link JM, Afentoulis ME, van Noort JM, Wawrousek EF, Offner H, Vandenbark AA (2006) AlphaB-crystallin-reactive T cells from knockout mice are not encephalitogenic. J Neuroimmunol 176:51–62

Wang J, Martin E, Gonzales V, Borchelt DR, Lee MK (2008) Differential regulation of small heat shock proteins in transgenic mouse models of neurodegenerative diseases. Neurobiol Aging 29:586–597

Wang K, Ren K, Yan YE, Wang H, Zhng BY, Liu Y, Dong XP, Zhang J (2013) Analysis of the expressions of alphaB-crystallin in the brain tissues of agent 263K-infected hamsters at terminal stage. Bing Du Xue Bao 29:192–196

Waudby CA, Knowles TP, Devlin GL, Skepper JN, Ecroyd H, Carver JA, Welland ME, Christodoulou J, Dobson CM, Meehan S (2010) The interaction of alphaB-crystallin with mature alpha-synuclein amyloid fibrils inhibits their elongation. Biophys J 98:843–851

Wilhelmus MM, Boelens WC, Otte-Holler I, Kamps B, de Waal RM, Verbeek MM (2006a) Small heat shock proteins inhibit amyloid-beta protein aggregation and cerebrovascular amyloid-beta protein toxicity. Brain Res 1089:67–78

Wilhelmus MM, Otte-Holler I, Wesseling P, de Waal RM, Boelens WC, Verbeek MM (2006b) Specific association of small heat shock proteins with the pathological hallmarks of Alzheimer's disease brains. Neuropathol Appl Neurobiol 32:119–130

Wistow GJ, Piatigorsky J (1988) Lens crystallins: the evolution and expression of proteins for a highly specialized tissue. Annu Rev Biochem 57:479–504

Yerbury JJ, Gower D, Vanags L, Roberts K, Lee JA, Ecroyd H (2013) The small heat shock proteins alphaB-crystallin and Hsp27 suppress SOD1 aggregation in vitro. Cell Stress Chaperones 18:251–257

Zabel C, Chamrad DC, Priller J, Woodman B, Meyer HE, Bates GP, Klose J (2002) Alterations in the mouse and human proteome caused by Huntington's disease. Mol Cell Proteomics 1:366–375

Chapter 16
Small HSP Variants and Human Diseases

Huan Guo and Tangchun Wu

Abstract Small heat shock proteins (sHSPs) are key members of the HSP superfamily that are ubiquitous among all organisms. The amino acid sequences of the sHSPs are not as highly conserved as the larger HSPs but it comprises a characteristic α-crystallin domain (ACD). The mammalian sHSP family, now referred to as the HSPB family, includes 11 members: HSP27/HSPB1, MKBP (myotonic dystrophy protein kinase-binding protein)/HSPB2, HSPB3, alphaA-crystallin / HSPB4, alphaB-crystallin/HSPB5, HSP20/HSPB6, cvHSP (cardiovascular heat shock protein)/HSPB7, HSP22/HSPB8, HSPB9, ODF1 (outer dense fiber protein)/HSPB10, and HSPB11. They have been reported to have a wide range of cellular functions, including endowing cells with thermotolerance and acting as molecular chaperones. Because of these functions, sHSPs can participate in a large number of fundamental cellular processes such as controlling protein folding, F-actin-dependent processes, cytoprotection/anti-apoptosis, differentiation, cell proliferation, and gene expression, and thereby are involved in many pathological diseases, such as neurodegenerative diseases, cancer, and cardiovascular diseases. Genetic mutation or variations of sHSPs genes may change their expression levels and affect protein functions, thus contributing to cell malfunction especially during stress. Here we examined the current reports regarding the mutations or variations of sHSPs genes, and we analyzed their associations with the development, progression, and prognosis of some human diseases. We also addressed the biological functions for the potential causal sHSPs variations and discussed their possible implications in human diseases.

Keywords Small heat shock protein • Molecular chaperones • Genetic variations • Mutation • Polymorphism

---

H. Guo • T. Wu, M.D., Ph.D. (✉)
Department of Occupational and Environmental Health, State Key Laboratory of Environmental Health (Incubating), School of Public Health, Tongji Medical College, Huazhong University of Science and Technology, 13 Hangkong Rd, Wuhan 430030, China
e-mail: wut@mails.tjmu.edu.cn

R.M. Tanguay, L.E. Hightower (eds.), *The Big Book on Small Heat Shock Proteins*, Heat Shock Proteins 8, DOI 10.1007/978-3-319-16077-1_16

16.1 Introduction

Small heat shock proteins (sHSPs) are a family of low-molecular-weight chaperones (12–43 kDa) that form oligomeric structures ranging from 9 to 50 subunits (~1 MDa). The human genome encodes for 11 different small HSP members, HSPB1-11 (Table 16.1), with the exception of HSPB11, are characterized by a conserved C-terminal sequence of 80–100 amino acid residues termed the α-crystallin domain, flanked by a variable, structurally disordered N-terminal arm and a short, flexible C-terminal extension (Kappe et al. 2003; Kim et al. 1998; Carver and Lindner 1998; Kampinga et al. 2009). The sHSPs share complex oligomerization and phosphorylation properties allowing them to interact and modulate the activity of many client proteins. Functional study of sHSPs is expanding quickly as the applications of genomics and proteomics have revealed the characteristics and medical importance of sHSPs.

Expression of sHsps can be induced in response to various kinds of stress including heat shock, oxidative stress, osmostress and ischemia, but some sHsps are expressed constitutively under physiological conditions (Arrigo 2013). The sHSPs are expressed during embryonic development and may function as chaperones to ensure protein homeostasis. Functioning as molecular chaperones, sHSPs can prevent stress-induced aggregation of denatured proteins and promote their folding to native conformations when favorable conditions return. Intracellular quantities and cellular localizations of sHSPs change in response to development, physiological stressors such as anoxia/hypoxia, heat and oxidation, and in relation to pathological status (Ciocca et al. 1993; Merendino et al. 2002). They regulate a large number of fundamental cellular processes and are associated with the pathology of a variety of human diseases.

Table 16.1 Members of small heat shock protein family

Gene ID	Alias	Description	Chromosome location
HSPB1	*HSP27*	Heat shock 27 kDa protein 1	Chr7: 75,769,859..75,771,548
HSPB2	*MKBP*	Myotonic dystrophy protein kinase-binding protein	Chr 11: 111,288,709..111,290,027
HSPB3		Heat shock 27 kDa protein 3	Chr5: 53,787,202..53,787,964
CRYAA	*HSPB4*	alpha A-crystallin,	Chr21: 43,462,210..43,465,982
CRYAB	*HSPB5*	alpha B-crystallin	Chr11: 111,284,560..111,287,683
HSPB6	*HSP20*	Heat shock protein, alpha-crystallin-related, B6	Chr19: 40,937,310..40,939,770
HSPB7	*cvHSP*	Cardiovascular heat shock protein	Chr1: 16,213,112..16,217,538
HSPB8	*Hsp22*	Heat shock 22 kDa protein 8	Chr12: 118,100,978..118,116,933
HSPB9		Heat shock protein, alpha-crystallin-related, B9	Chr17: 37,528,395..37,528,872
ODF1	*HSPB10*	Outer dense fiber protein	Chr8: 103,633,036..103,642,422

Through direct sequencing of tissue or blood DNA samples, mutations and single nucleotide polymorphisms (SNPs) have been found in human sHSP genes. Epidemiological and biological studies have also been carried out on the relationships and functional testing for these sHSP variants in human diseases. This chapter reviews current research into the implications of sHSP variations on several human diseases (Table 16.2).

16.2 Small HSPs Variants and Human Disease

16.2.1 HSPB1 *Variants in Lung Cancer*

Heat shock protein 27 (Hsp27: HSPB1) is the most widely studied key member of the human small heat shock protein family. Tumors of various human cancers, including lung cancer, have been reported to exhibit over-expression of Hsp27 (Kim et al. 2007). The tumorigenic potential of Hsp27 has been demonstrated in both experimental animal models (Garrido et al. 1998) and cell biological studies (Kamada et al. 2007). It can mediate tumorigenesis and progression through inhibition of programmed cell death and senescence, two essential traits of cancer cells (Mosser and Morimoto 2004; Ciocca and Calderwood 2005; Calderwood et al. 2006). Hsp27 has also been considered as an independent prognosis marker for various cancers, since overexpression of Hsp27 was reported to be involved in chemo- or radiotherapy resistance of tumor cells and was thus considered to be associated with poor prognosis in cancers of stomach, liver, prostate, and osteosarcoma (Ciocca and Calderwood 2005; Lee et al. 2005; Zhang and Shen 2007). Elevated Hsp27 can suppress p53 activation and inhibit the p53/p21 mediated senescence pathway to facilitate survival of tumor cells (O'Callaghan-Sunol et al. 2007).

The functional single nucleotide polymorphism (SNP) −1271G>C (rs2868371), located in the promoter region of *HSPB1* genes, was reported to be associated with lung cancer risk, prognosis, and adverse toxicity of radio(chemo)therapy for non-small cell lung cancer (NSCLC). The epidemiological study carried out by Guo et al. showed that the −1271C allele was associated with a significantly increased lung cancer risk in two independent Chinese case-control studies (odd ratio = 1.26 and 1.49, 1152 pairs and 500 pairs, respectively), but it conferred a favorable survival for patients with advanced NSCLC in two independent cohorts (248 and 335 advanced NSCLC patients, adjusted hazard ratio, 0.66 and 0.75, respectively) (Guo et al. 2010). The biological function of this variation was further investigated by a genotype-phenotype correlation analysis, reporter assay, and tissue examination. They found that in subjects occupationally exposed to PAHs, the −1271C allele carriers exhibited significantly higher DNA damage levels than −1271G allele carriers. The in vitro reporter assay showed that the −1271C allele-driven promoter decreased the expression of Hsp27 by 27–53 % more than −1271G allele-driven promoter in both normal bronchial epithelial and malignant cancer cells. In addition, the Western

Table 16.2 sHSP variants and human diseases

Diseases	Traits	sHSP genes	Variations	Biological functions for variations	Ref
Lung cancer	Risk, prognosis, DNA damage levels	*HSPB1*	rs2868371 (−1271 G>C)	Compared with the -1271G allele, −1271C allele decreased the Hsp27 expression	O'Callaghan-Sunol et al. (2007)
	Radiation-induced esophageal toxicity	*HSPB1*	rs2868371 (−1271 G>C)		Lopez Guerra et al. (2011)
	Radiation pneumonitis	*HSPB1*	rs2868371 (−1271 G>C)		Pang et al. (2013)
Cataract	Development	*CRYAA*	R116C	R116C substitution diminished the protection ability, protein structure, and chaperon activity of αA-crystallin	Bloemendal et al. (2004), Litt et al. (1998), Andley et al. (2002), Shroff et al. (2000), Cobb and Petrash (2000)
	Development	*CRYAA*	R89C	R49C mutant protein was abnormally localized to the nucleus and failed to protect from staurosporine-induced apoptotic cell death.	Bera and Abraham (2002)
	Development	*CRYAB*	R120G	R120G modified its protein structure, defected chaperone activity, and changed interaction with intermediate filaments for αB-crystallin.	Horwitz (2000), Bova et al. (1999)
Neurological disorders					
amyotrophic lateral sclerosis, ALS	Development	*HSPB1*	−217 T>C	−217C mutant allele impaired the HSF-HSE binding, decreased to 50% promoter activity as compared to the wild-type -217T allele in neuronal and non-neuronal cells in both normal and heat shock status.	Batulan et al. (2006)

hereditary distal motor neuropathy, HMN	Development	*HSPB1*	R127W, S135F, T151I, P182L	S135F mutation reduced viability of neuronal cells *in vitro* and affected the neurofilament assembly.	Der Perng and Quinlan (2004)
hereditary distal motor neuropathy, HMN	Development	*HSPB8*	K141N, K141E	K141N and K141E mutations showed greater binding to *HSPB1*.	Evgrafov et al. (2004)
Charcot-Marie-Tooth disease, CMT	Development	*HSPB8*	S135F, R136W	S135F mutation reduced viability of neuronal cells *in vitro* and affected the neurofilament assembly.	Der Perng and Quinlan (2004)
desmin-related myopathies, DRM	Development	*CRYAB*	R120G	R120G modified its protein structure, defected chaperone activity, and changed interaction with intermediate filaments for αB-crystallin.	Irobi et al. (2004), Horwitz (2000)
Dilated cardiomyopathy, DCM	Risk	*HSPB7*	rs1739843 C>T	None	Maron et al. (2006)
Heart failure	Risk	*HSPB7*	rs1739843 C>T	None	Stark et al. (2010)
Multiple sclerosis, MS	Risk	*CRYAB*	rs2234702 (−650 C>G)	None	Laaksonen et al. (2003)
	Risk	*CRYAB*	rs14133 (−249 C>G)	None	van Veen et al. (2003)
Type 1 diabetes, T1D	Risk, development of IAA-positivity	*CRYAB*	rs2234702 (−650 C>G)	None	Handel et al. (2009)

blot analysis also demonstrated that −1271CC carriers showed a lower Hsp27 expression in normal lung tissue than −1271GG carriers. Taken together, the *HSPB1* promoter −1271C allele may contribute to an increased lung cancer risk but a favorable survival in advanced NSCLC patients, possibly by reducing the Hsp27 protein expression levels that appeared to impair the host DNA repair capacity (Guo et al. 2010). The reduced DNA repair capacity driven by the −1271C allele may increase the lung cancer risk and sensitivity of lung cancer cells to radio(chemo)therapy, thus conferring a favorable survival.

The role of Hsp27 in clinical resistance to radiation has been described previously (Kassem et al. 2002). The Hsp27 level prior to treatment was suggested as a prognostic biomarker for radiation tolerance. NSCLC patients with the greatest interleukocyte Hsp27 levels before treatment suffered more severe acute radiation morbidity (Guisasola et al. 2006). Lopez Guerra et al. investigated the association of the *HSPB1* rs2868371 polymorphism with radiation-induced esophageal toxicity in radio(chemo)therapy treated NSCLC patients. Their study demonstrated that patients with the *HSPB1* −1271CC genotype had a significantly higher risk of radiation-induced esophageal toxicity after radiotherapy for NSCLC than patients with other genotypes (Lopez Guerra et al. 2011). Furthermore, another study carried out by Pang et al. investigated the association of this polymorphism with risk of radiation pneumonitis (RP) for NSCLC patients after chemoradiation, and they reported that the −1271CC genotype of *HSPB1* was associated with severe RP after chemoradiation for NSCLC (Pang et al. 2013).

*16.2.2 CRYAA (*HSPB4*) and CRYAB (*HSPB5*) Mutations with Cataract*

sHSPs can prevent cataract in the mammalian lens and guard against ischemic and reperfusion injury due to heart attack and stroke. Lens tissue derives from cells containing large amounts of densely packed proteins known as α-, β- and γ-crystallins, which function for the lifespan of an organism and are essential for vision (Sun and MacRae 2005). α-crystallins are crucial structural elements and molecular chaperones that can maintain lens transparency. As the chaperoning capability of α-crystallin declines, lens proteins are more likely to aggregate, a characteristic linking cataract to other protein folding diseases (Bloemendal et al. 2004).

The R116C mutation in αA-crystallin gene (*CRYAA*: *HSPB4*) was responsible for autosomal dominant congenital cataract, a common cause of infant blindness (Litt et al. 1998). Functional studies of this missense mutation suggested that the R116C substitution diminished the protective ability of αA-crystallin against stress-induced lens epithelial cell apoptosis (Andley et al. 2002), changed the protein structure (Shroff et al. 2000; Cobb and Petrash 2000), and destroyed the chaperone activity of αA-crystallin (Shroff et al. 2000); however, the impaired chaperoning activity may not completely explain cataract development (Bera and Abraham 2002).

Sequencing of *CRYAA* from cataract patients identified another C to T transition in exon 1 of *CRYAA* that was predicted to result in the nonconservative substitution of cysteine for arginine at codon 49 (R49C). Transfection studies of lens epithelial cells revealed that, unlike wild-type *CRYAA*, the R49C mutant protein was abnormally localized to the nucleus and failed to protect from staurosporine-induced apoptotic cell death. This study revealed the first dominant cataract mutation in *CRYAA* located outside the phylogenetically conserved 'α-crystallin core domain' of the sHSP family (Mackay et al. 2003).

αB-crystallin (HSPB5) is also a main component of the lens where it maintains a clear amorphous composition by preventing denatured proteins from aggregating to form opaque inclusions (Horwitz 2000). The R120G mutation of *CRYAB* has also been reported to be associated with cataract (Bova et al. 1999). Like *CRYAA* R120, the *CRYAB* R120 was also located in the α-crystallin conserved region as arginine. By using α-lactalbumin, alcohol dehydrogenase, and insulin as target proteins, in vitro assays indicated that the *CRYAB* R120G had a reduced or completely lost chaperone-like function for αB-crystallin (Bova et al. 1999). This mutation was also found to modify its protein structure, chaperone activity, and interaction with intermediate filaments in vitro, although binding of the modified protein to filaments increases in comparison to wild-type aB-crystallin (Perng et al. 1999). These altered functions caused by *CRYAB* R120G substitution may point to its role in development of cataract.

16.2.3 sHSP Mutations in Neurological Disorders

16.2.3.1 *HSPB1* in Amyotrophic Lateral Sclerosis

Amyotrophic lateral sclerosis (ALS) is a devastating adult-onset neurodegenerative disease characterized by the progressive and selective loss of upper motor neurons in the motor cortex and lower motor neurons in the brain stem and spinal cord (Pasinelli and Brown 2006). Hsp27 upregulation and phosphorylation is required for injured sensory and motor neuron survival (Benn et al. 2002). The overexpression of HSP27 in motor neurons can facilitate neuroprotective capacities by protecting against copper-zinc superoxide dismutase 1 (SOD1)-mutant induced cell death in mammalian neuronal cells (Batulan et al. 2006). Through sequencing, Dierick et al. detected three nucleotide variants of *HSPB1* in ALS patients, one of which, −217T>C, is located in the consensus sequence of a functionally characterized heat shock element (HSE) essential for stress-induced transcription of *HSPB1* (Dierick et al. 2007). This variant can impair the HSP27 stress response as determined by reporter assay and electrophoretic mobility shift assays (EMSA). The −217C mutant allele decreased to 50 % promoter activity as compared to the wild-type −217T allele in neuronal and non-neuronal cells. Following heat shock, the HSE variant attenuated significantly the stress related increase in transcription of *HSPB1*.

Compared to the −217T allele, the −217C mutant allele dramatically reduced HSF binding to the conserved HSE region (−222 to −214) (Dierick et al. 2007).

16.2.3.2 *HSPB1* and *HSPB8* Mutations in Distal Neuropathies

Two published studies have identified genetic links between *HSPB1, HSPB8* and the distal neuropathies, Charcot-Marie-Tooth disease (CMT), and hereditary distal motor neuropathy (HMN), revealing important roles for these specific sHSPs in motor neurons. It has been demonstrated that mutations in *HSPB1* and *HSPB8* can be observed in specific families with CMT and HMN caused by premature axonal loss, possibly due to neuronal death and subsequent degeneration (Der Perng and Quinlan 2004). For *HSPB1* gene, the mutations R127W, S135F, T151I, and P182L were found in families with distal HMN, while the mutations S135F and R136W were found in families with CMT (Evgrafov et al. 2004). Function analysis of the S135F mutation showed that the mutant HSP27 reduced viability of neuronal cells in vitro and affected neurofilament assembly, suggesting that it is responsible for premature axonal degeneration, the direct cause of both CMT2 and HMN. For *HSPB8* gene, the mutations K141N or K141E were found in families with distal HMN, while the mutation K141N was found in families with CMT (Irobi et al. 2004). BothK141NandK141Emutationstargetthesameaminoacid.Co-immunoprecipitation experiments showed greater binding of both *HSPB8* mutants to the interacting partner *HSPB1*, while the expression of mutant *HSPB8* in cultured cells promoted formation of intracellular aggregates. These results suggested that mutations in sHSPs have an important role in neurodegenerative disorders.

16.2.3.3 *CRYAB* (*HSPB5*) Mutation and Desmin-Related Myopathies

Desmin-related myopathies (DRM) are a set of inherited neuromuscular disorders, characterized by adult onset and delayed accumulation of aggregates of desmin, a protein belonging to the type III intermediate filament family, in the sarcoplasma of skeletal and cardiac muscles. aB-crystallin (HSPB5) is abundantly expressed in nonocular tissues such as skeletal and cardiac muscle, and it can aggregate with desmin. One pedigree analysis of a French DRM family located the causal region of DRM to chromosome 11q21-23 (Vicart et al. 1998), which contains the αB-crystallin gene. The R120G missense mutation in αB-crystallin (*CRYAB*) was found to co-segregates with the disease phenotype in DRM family. Further examination of muscle cell lines transfected with the mutant *CRYAB* cDNA showed intracellular aggregates that contain both desmin and αB-crystallin, which was also observed in muscle fibers from DRM patients (Vicart et al. 1998). These results are the first to identify a defect in a molecular chaperone as a cause for an inherited human muscle disorder. It was also mentioned above that by doing biochemical analysis, it was found that the αB-crystallin R120G missense mutation can disrupt aB-crystallin structure, chaperone activity, and intermediate filament interaction, demonstrating the functional importance of residue R120 (Bova et al. 1999).

16.2.4 **HSPB7** *SNP in Dilated Cardiomyopathy and Heart Failure*

HSPB7 is also called cardiovascular HSP because of its selective expression in cardiovascular tissues (Kappe et al. 2003). Dilated cardiomyopathy (DCM) is a severe disease of the heart muscle and often leads to chronic heart failure. It is characterized by systolic dysfunction as well as dilation and impaired contraction of the ventricles, often leading to chronic heart failure and eventually requiring cardiac transplantation (Maron et al. 2006). A genome-wide association study was carried out in a case-control association study of 664 idiopathic DCM cases and 1,874 controls from Germany, followed by replication in one Germany (564 cases and 981 controls) and in two French populations (433 cases and 395 controls, 249 cases and 380 controls, respectively), in order to explore genetic variations for susceptibility to idiopathic DCM (Stark et al. 2010). The analysis revealed a significantly protective role for the intronic *HSPB7* SNP rs1739843 T allele in risk of idiopathic DCM in each study panel and in all Europeans [OR(95 % CI)=0.72(0.65–0.78)] ($P=5.28\times10^{-13}$). This association was further validated by Cappola et al. who reported a protective effect of rs1739843 T allele in both ischemic and non-ischemic heart failure for Caucasians (Cappola et al. 2010), which is in conformity with the results of DCM.

Li et al. further investigated the association of *HSPB7* SNPs (rs1739843, rs7523558, and rs6660685) with risk of DCM among a Han Chinese population (Li et al. 2013). However, they did not find any differences in genotypes or allele frequencies of *HSPB7* variations between DCM patients and control subjects (Li et al. 2013). These results suggested that many factors, including different races, the different enrollment criteria for DCM patients, and different patient status, may account for the difference between the results of Caucasian and Han Chinese populations. Further population-level fine mapping analysis and biological studies are needed to locate the causal *HSPB7* SNPs and their biological functions.

16.2.5 **CRYAB** *(***HSPB5***) SNP and Multiple Sclerosis*

Alpha B-crystallin is one of the strongest T cell immune response targets in multiple sclerosis (MS) patients (Bajramovic et al. 2000). The transcription level of alpha B-crystallin was elevated in brain lesions of multiple sclerosis (MS) patients (Chabas et al. 2001). Alpha B-crystallin is also a strong autoantibody target in the cerebrospinal fluid of MS patients (Ousman et al. 2007). Due to its abundant expression in the central nervous system and its auto antigenic nature, alpha B-crystallin was thought to be directly involved in MS and to enhance MS pathogenesis.

The genetic association between *CRYAB* gene and multiple sclerosis (MS) was identified by a genome-wide scan using microsatellite marker (Ebers et al. 1996; Laaksonen et al. 2003). The −650C allele of *CRYAB* −650C>G (rs2234702) polymorphism was reported to be associated with MS susceptibility and a non-inflammatory, neurodegenerative phenotype in a Dutch population (van Veen et al. 2003), while the −249 C>G (rs14133) was reported to be associated with MS and subtypes of MS in a Danish population (Stoevring et al. 2007). However, functional studies for these variations have not been explored yet.

16.2.6 CRYAB(HSPB5) *SNP and Type 1 Diabetes*

Type 1 diabetes (T1D) has a multi-factorial etiology involving genetic susceptibility and poorly defined environmental factors (Atkinson and Eisenbarth 2001). T1D and MS share similarities in disease epidemiology, influence from environmental factors, genetic risk factors and pathogenic mechanisms (Handel et al. 2009). Sun et al. investigated the associations of *CRYAB* variants with risk of T1D in a Swedish population (Sun et al. 2012). They found that the *CRYAB* −650 C>G (rs2234702) polymorphism, which had been reported to be associated with the development of MS in a Dutch population, was also found to confer a susceptibility relationship with the development of T1D in the Swedish cohort (OR = 1.48, P = 0.03). The *CRYAB* −650C was also associated with the development of IAA-positivity in T1D patients, especially in those carrying T1D high-risk HLA haplotypes (Sun et al. 2012).

16.3 Perspectives

In this chapter, we mainly focus on describing research regarding the associations and biological functions of sHSPs variations with human diseases. Considering the ubiquitous expression and the important role of small HSPs in the development of human tissues such as lens, neuron, cardiac, skeletal muscle, and other tissues, sHSPs were reported to be key regulators associated with the development of cataract, cancer, neurodegenerative and cardiovascular diseases. Although studies have revealed many new insights into the importance of sHSPs for human diseases, many questions remain unclear in terms of their mechanism of action, and of sHSPs variations. Functional exploration of sHSP variations is of importance as it may offer specific targets for risk, diagnostic tests, drug development, treatment and prevention implications.

sHSPs are a typical group of responders to environmental stresses, such as heat shock, environmental chemicals, ionizing radiations, among others (Fig. 16.1). All these environmental stresses may cause early harmful effects such as oxidative and

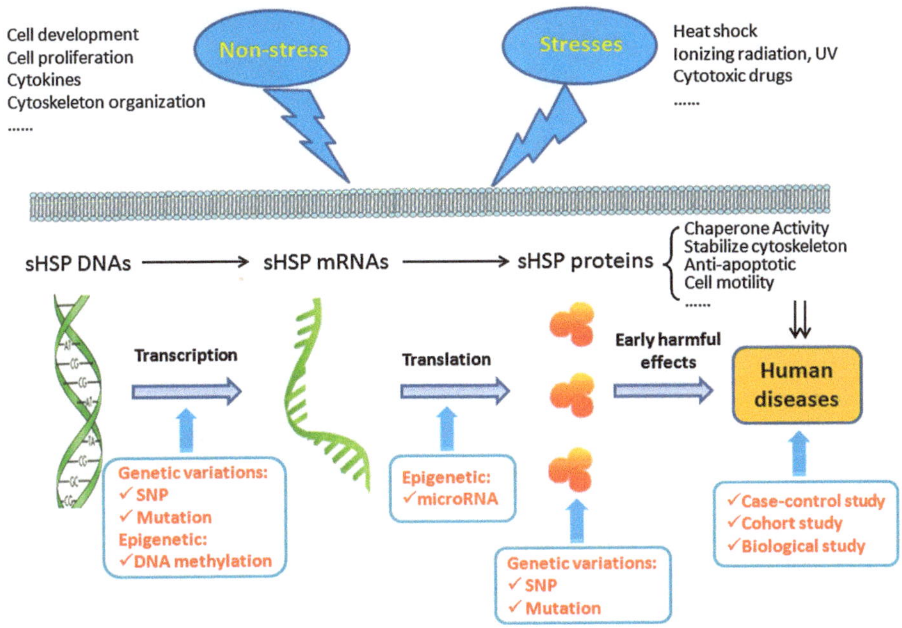

Fig. 16.1 Small Heat shock proteins (sHSPs) have multifaceted functions during health and disease. sHSPs are modulated under normal and stress conditions, and the expression of sHSPs can be affected by both genetic variants, epigenetic changes and protein levels. Large sample sized case-control and cohort studies, complemented by well-designed biological experiments are urgently needed to investigate the association and function of genetic factors and gene-environment interactions

genetic damage before the final development of diseases. Over the past two decades, studies exploring the association of sHSP gene polymorphisms and risk of human diseases were mainly performed by tissue examination or case-control studies. Few studies have investigated the associations of these variations with early biological effects of human diseases. Because of the limited power of case-control studies in the identification of etiological factors, future epidemiological studies including follow-up design, especially large sample sized cohort studies are urgently needed to verify the causal variations and effects of gene-environmental interaction. Moreover, the biological functions of positive variations, such as *HSPB7* rs1739843, *CRYAB* (*HSPB5*) rs2234702 and rs14133, need to be further explored by well designed biological experiments.

The genetic variations located at the regulatory region, 5′UTR, promoter, 3′UTR, may affect the expressions of sHSPs at the mRNA level. In addition, the epigenetic modification of DNA methylation may also affect the mRNA expression. In the stage of post-transcriptional regulation, epigenetic factors such as microRNA profiles have emerged as crucial effectors. As a consequence, researchers should pay more attention to epigenetic changes and epigenetic-genetic interaction when exploring the mechanism/pathways that control the expression and function of sHSPs in human diseases.

As vital members of the sHSP family, Hsp27 and αB-crystallin act to cause diverse effects depending on the cell type or signal. For example, in normal cells, their expressions are associated with decreased proliferation and cellular differentiation whereas in cancer cells, elevated Hsp27 and αB-crystallin reduces proliferation but confers resistance to treatment (chemotherapy or radiotherapy) and increases migratory potential and invasiveness in many cancers (Calderwood et al. 2007). Approaches aimed at modulating expression and/or activities of Hsp27 are currently emerging for cancer therapy (Arrigo et al. 2007). For this kind of review, identification and verification of the causal variations and key regulators may help to uncover predictive biomarkers and provide personalized cancer therapies in the future. Large sample size case-control and cohort studies, accompanied by well-designed biological experiments are urgently needed to investigate the association and function of genetic factors and gene-environment interactions.

References

Andley UP, Patel HC, Xi JH (2002) The R116C mutation in alpha A-crystallin diminishes its protective ability against stress-induced lens epithelial cell apoptosis. J Biol Chem 277(12):10178–10186. doi:10.1074/jbc.M109211200

Arrigo AP (2013) Human small heat shock proteins: protein interactomes of homo- and hetero-oligomeric complexes: an update. FEBS Lett 587(13):1959–1969. doi:10.1016/j.febslet.2013.05.011

Arrigo AP, Simon S, Gibert B, Kretz-Remy C, Nivon M, Czekalla A, Guillet D, Moulin M, Diaz-Latoud C, Vicart P (2007) Hsp27 (HspB1) and alphaB-crystallin (HspB5) as therapeutic targets. FEBS Lett 581(19):3665–3674. doi:10.1016/j.febslet.2007.04.033

Atkinson MA, Eisenbarth GS (2001) Type 1 diabetes: new perspectives on disease pathogenesis and treatment. Lancet 358(9277):221–229. doi:10.1016/S0140-6736(01)05415-0

Bajramovic JJ, Plomp AC, Goes A, Koevoets C, Newcombe J, Cuzner ML, van Noort JM (2000) Presentation of alpha B-crystallin to T cells in active multiple sclerosis lesions: an early event following inflammatory demyelination. J Immunol 164(8):4359–4366

Batulan Z, Taylor DM, Aarons RJ, Minotti S, Doroudchi MM, Nalbantoglu J, Durham HD (2006) Induction of multiple heat shock proteins and neuroprotection in a primary culture model of familial amyotrophic lateral sclerosis. Neurobiol Dis 24(2):213–225. doi:10.1016/j.nbd.2006.06.017

Benn SC, Perrelet D, Kato AC, Scholz J, Decosterd I, Mannion RJ, Bakowska JC, Woolf CJ (2002) Hsp27 upregulation and phosphorylation is required for injured sensory and motor neuron survival. Neuron 36(1):45–56

Bera S, Abraham EC (2002) The alphaA-crystallin R116C mutant has a higher affinity for forming heteroaggregates with alphaB-crystallin. Biochemistry 41(1):297–305

Bloemendal H, de Jong W, Jaenicke R, Lubsen NH, Slingsby C, Tardieu A (2004) Ageing and vision: structure, stability and function of lens crystallins. Prog Biophys Mol Biol 86(3):407–485. doi:10.1016/j.pbiomolbio.2003.11.012

Bova MP, Yaron O, Huang Q, Ding L, Haley DA, Stewart PL, Horwitz J (1999) Mutation R120G in alphaB-crystallin, which is linked to a desmin-related myopathy, results in an irregular structure and defective chaperone-like function. Proc Natl Acad Sci U S A 96(11):6137–6142

Calderwood SK, Khaleque MA, Sawyer DB, Ciocca DR (2006) Heat shock proteins in cancer: chaperones of tumorigenesis. Trends Biochem Sci 31(3):164–172

Calderwood SK, Sherman MY, Ciocca DR (2007) Heat shock proteins volume 2: heat shock proteins in cancer. Springer, Netherlands, pp 73–92

Cappola TP, Li M, He J, Ky B, Gilmore J, Qu L, Keating B, Reilly M, Kim CE, Glessner J, Frackelton E, Hakonarson H, Syed F, Hindes A, Matkovich SJ, Cresci S, Dorn GW 2nd (2010) Common variants in HSPB7 and FRMD4B associated with advanced heart failure. Circ Cardiovasc Genet 3(2):147–154. doi:10.1161/CIRCGENETICS.109.898395

Carver JA, Lindner RA (1998) NMR spectroscopy of alpha-crystallin. Insights into the structure, interactions and chaperone action of small heat-shock proteins. Int J Biol Macromol 22(3–4):197–209

Chabas D, Baranzini SE, Mitchell D, Bernard CC, Rittling SR, Denhardt DT, Sobel RA, Lock C, Karpuj M, Pedotti R, Heller R, Oksenberg JR, Steinman L (2001) The influence of the proinflammatory cytokine, osteopontin, on autoimmune demyelinating disease. Science 294(5547):1731–1735. doi:10.1126/science.1062960

Ciocca DR, Calderwood SK (2005) Heat shock proteins in cancer: diagnostic, prognostic, predictive, and treatment implications. Cell Stress Chaperones 10(2):86–103

Ciocca DR, Oesterreich S, Chamness GC, McGuire WL, Fuqua SA (1993) Biological and clinical implications of heat shock protein 27,000 (Hsp27): a review. J Natl Cancer Inst 85(19):1558–1570

Cobb BA, Petrash JM (2000) Structural and functional changes in the alpha A-crystallin R116C mutant in hereditary cataracts. Biochemistry 39(51):15791–15798

Der Perng M, Quinlan RA (2004) Neuronal diseases: small heat shock proteins calm your nerves. Curr Biol 14(15):R625–R626. doi:10.1016/j.cub.2004.07.047

Dierick I, Irobi J, Janssens S, Theuns J, Lemmens R, Jacobs A, Corsmit E, Hersmus N, Van Den Bosch L, Robberecht W, De Jonghe P, Van Broeckhoven C, Timmerman V (2007) Genetic variant in the HSPB1 promoter region impairs the HSP27 stress response. Hum Mutat 28(8):830–840

Ebers GC, Kukay K, Bulman DE, Sadovnick AD, Rice G, Anderson C, Armstrong H, Cousin K, Bell RB, Hader W, Paty DW, Hashimoto S, Oger J, Duquette P, Warren S, Gray T, O'Connor P, Nath A, Auty A, Metz L, Francis G, Paulseth JE, Murray TJ, Pryse-Phillips W, Nelson R, Freedman M, Brunet D, Bouchard JP, Hinds D, Risch N (1996) A full genome search in multiple sclerosis. Nat Genet 13(4):472–476. doi:10.1038/ng0896-472

Evgrafov OV, Mersiyanova I, Irobi J, Van Den Bosch L, Dierick I, Leung CL, Schagina O, Verpoorten N, Van Impe K, Fedotov V, Dadali E, Auer-Grumbach M, Windpassinger C, Wagner K, Mitrovic Z, Hilton-Jones D, Talbot K, Martin JJ, Vasserman N, Tverskaya S, Polyakov A, Liem RK, Gettemans J, Robberecht W, De Jonghe P, Timmerman V (2004) Mutant small heat-shock protein 27 causes axonal Charcot-Marie-Tooth disease and distal hereditary motor neuropathy. Nat Genet 36(6):602–606

Garrido C, Fromentin A, Bonnotte B, Favre N, Moutet M, Arrigo AP, Mehlen P, Solary E (1998) Heat shock protein 27 enhances the tumorigenicity of immunogenic rat colon carcinoma cell clones. Cancer Res 58(23):5495–5499

Guisasola MC, Calvo F, Marcos P, Simon I, Villanueva FJ, Andres E, Suarez A, Garcia-Barreno P (2006) Peripheral leukocyte response to oncological radiotherapy: expression of heat shock proteins. Int J Radiat Biol 82(3):171–179. doi:10.1080/09553000600643516

Guo H, Bai Y, Xu P, Hu Z, Liu L, Wang F, Jin G, Deng Q, Tu Y, Feng M, Lu D, Shen H, Wu T (2010) Functional promoter −1271G>C variant of HSPB1 predicts lung cancer risk and survival. J Clin Oncol 28(11):1928–1935. doi:10.1200/JCO.2009.24.4954

Handel AE, Handunnetthi L, Ebers GC, Ramagopalan SV (2009) Type 1 diabetes mellitus and multiple sclerosis: common etiological features. Nat Rev Endocrinol 5(12):655–664. doi:10.1038/nrendo.2009.216

Horwitz J (2000) The function of alpha-crystallin in vision. Semin Cell Dev Biol 11(1):53–60. doi:10.1006/scdb.1999.0351

Irobi J, Van Impe K, Seeman P, Jordanova A, Dierick I, Verpoorten N, Michalik A, De Vriendt E, Jacobs A, Van Gerwen V, Vennekens K, Mazanec R, Tournev I, Hilton-Jones D, Talbot K, Kremensky I, Van Den Bosch L, Robberecht W, Van Vandekerckhove J, Van Broeckhoven C, Gettemans J, De Jonghe P, Timmerman V (2004) Hot-spot residue in small heat-shock protein 22 causes distal motor neuropathy. Nat Genet 36(6):597–601. doi:10.1038/ng1328

Kamada M, So A, Muramaki M, Rocchi P, Beraldi E, Gleave M (2007) Hsp27 knockdown using nucleotide-based therapies inhibit tumor growth and enhance chemotherapy in human bladder cancer cells. Mol Cancer Ther 6(1):299–308

Kampinga HH, Hageman J, Vos MJ, Kubota H, Tanguay RM, Bruford EA, Cheetham ME, Chen B, Hightower LE (2009) Guidelines for the nomenclature of the human heat shock proteins. Cell Stress Chaperones 14(1):105–111. doi:10.1007/s12192-008-0068-7

Kappe G, Franck E, Verschuure P, Boelens WC, Leunissen JA, de Jong WW (2003) The human genome encodes 10 alpha-crystallin-related small heat shock proteins: HspB1-10. Cell Stress Chaperones 8(1):53–61

Kassem H, Sangar V, Cowan R, Clarke N, Margison GP (2002) A potential role of heat shock proteins and nicotinamide N-methyl transferase in predicting response to radiation in bladder cancer. Int J Cancer 101(5):454–460. doi:10.1002/ijc.10631

Kim KK, Kim R, Kim SH (1998) Crystal structure of a small heat-shock protein. Nature 394(6693):595–599. doi:10.1038/29106

Kim EH, Lee HJ, Lee DH, Bae S, Soh JW, Jeoung D, Kim J, Cho CK, Lee YJ, Lee YS (2007) Inhibition of heat shock protein 27-mediated resistance to DNA damaging agents by a novel PKC delta-V5 heptapeptide. Cancer Res 67(13):6333–6341

Laaksonen M, Jonasdottir A, Fossdal R, Ruutiainen J, Sawcer S, Compston A, Benediktsson K, Thorlacius T, Gulcher J, Ilonen J (2003) A whole genome association study in Finnish multiple sclerosis patients with 3669 markers. J Neuroimmunol 143(1–2):70–73

Lee YJ, Lee DH, Cho CK, Chung HY, Bae S, Jhon GJ, Soh JW, Jeoung DI, Lee SJ, Lee YS (2005) HSP25 inhibits radiation-induced apoptosis through reduction of PKC delta-mediated ROS production. Oncogene 24(23):3715–3725

Li X, Luo R, Mo X, Jiang R, Kong H, Hua W, Wu X (2013) Polymorphism of ZBTB17 gene is associated with idiopathic dilated cardiomyopathy: a case control study in a Han Chinese population. Eur J Med Res 18:10. doi:10.1186/2047-783X-18-10

Litt M, Kramer P, LaMorticella DM, Murphey W, Lovrien EW, Weleber RG (1998) Autosomal dominant congenital cataract associated with a missense mutation in the human alpha crystallin gene CRYAA. Hum Mol Genet 7(3):471–474

Lopez Guerra JL, Wei Q, Yuan X, Gomez D, Liu Z, Zhuang Y, Yin M, Li M, Wang LE, Cox JD, Liao Z (2011) Functional promoter rs2868371 variant of HSPB1 associates with radiation-induced esophageal toxicity in patients with non-small-cell lung cancer treated with radio(chemo)therapy. Radiother Oncol 101(2):271–277. doi:10.1016/j.radonc.2011.08.039

Mackay DS, Andley UP, Shiels A (2003) Cell death triggered by a novel mutation in the alphaA-crystallin gene underlies autosomal dominant cataract linked to chromosome 21q. Eur J Hum Genet 11(10):784–793. doi:10.1038/sj.ejhg.5201046

Maron BJ, Towbin JA, Thiene G, Antzelevitch C, Corrado D, Arnett D, Moss AJ, Seidman CE, Young JB (2006) Contemporary definitions and classification of the cardiomyopathies: an American Heart Association Scientific Statement from the Council on Clinical Cardiology, Heart Failure and Transplantation Committee; Quality of Care and Outcomes Research and Functional Genomics and Translational Biology Interdisciplinary Working Groups; and Council on Epidemiology and Prevention. Circulation 113(14):1807–1816. doi:10.1161/CIRCULATIONAHA.106.174287

Merendino AM, Paul C, Vignola AM, Costa MA, Melis M, Chiappara G, Izzo V, Bousquet J, Arrigo AP (2002) Heat shock protein-27 protects human bronchial epithelial cells against oxidative stress-mediated apoptosis: possible implication in asthma. Cell Stress Chaperones 7(3):269–280

Mosser DD, Morimoto RI (2004) Molecular chaperones and the stress of oncogenesis. Oncogene 23(16):2907–2918

O'Callaghan-Sunol C, Gabai VL, Sherman MY (2007) Hsp27 modulates p53 signaling and suppresses cellular senescence. Cancer Res 67(24):11779–11788

Ousman SS, Tomooka BH, van Noort JM, Wawrousek EF, O'Connor KC, Hafler DA, Sobel RA, Robinson WH, Steinman L (2007) Protective and therapeutic role for alphaB-crystallin in autoimmune demyelination. Nature 448(7152):474–479. doi:10.1038/nature05935

Pang Q, Wei Q, Xu T, Yuan X, Lopez Guerra JL, Levy LB, Liu Z, Gomez DR, Zhuang Y, Wang LE, Mohan R, Komaki R, Liao Z (2013) Functional promoter variant rs2868371 of HSPB1 is associated with risk of radiation pneumonitis after chemoradiation for non-small cell lung cancer. Int J Radiat Oncol Biol Phys 85(5):1332–1339. doi:10.1016/j.ijrobp.2012.10.011

Pasinelli P, Brown RH (2006) Molecular biology of amyotrophic lateral sclerosis: insights from genetics. Nat Rev Neurosci 7(9):710–723. doi:10.1038/nrn1971

Perng MD, Muchowski PJ, van Den IP, Wu GJ, Hutcheson AM, Clark JI, Quinlan RA (1999) The cardiomyopathy and lens cataract mutation in alphaB-crystallin alters its protein structure, chaperone activity, and interaction with intermediate filaments in vitro. J Biol Chem 274(47):33235–33243

Shroff NP, Cherian-Shaw M, Bera S, Abraham EC (2000) Mutation of R116C results in highly oligomerized alpha A-crystallin with modified structure and defective chaperone-like function. Biochemistry 39(6):1420–1426

Stark K, Esslinger UB, Reinhard W, Petrov G, Winkler T, Komajda M, Isnard R, Charron P, Villard E, Cambien F, Tiret L, Aumont MC, Dubourg O, Trochu JN, Fauchier L, Degroote P, Richter A, Maisch B, Wichter T, Zollbrecht C, Grassl M, Schunkert H, Linsel-Nitschke P, Erdmann J, Baumert J, Illig T, Klopp N, Wichmann HE, Meisinger C, Koenig W, Lichtner P, Meitinger T, Schillert A, Konig IR, Hetzer R, Heid IM, Regitz-Zagrosek V, Hengstenberg C (2010) Genetic association study identifies HSPB7 as a risk gene for idiopathic dilated cardiomyopathy. PLoS Genet 6(10):e1001167. doi:10.1371/journal.pgen.1001167

Stoevring B, Frederiksen JL, Christiansen M (2007) CRYAB promoter polymorphisms: influence on multiple sclerosis susceptibility and clinical presentation. Clin Chim Acta 375(1–2):57–62. doi:10.1016/j.cca.2006.06.017

Sun Y, MacRae TH (2005) The small heat shock proteins and their role in human disease. FEBS J 272(11):2613–2627. doi:10.1111/j.1742-4658.2005.04708.x

Sun C, Sedimbi SK, Ashok AK, Sanjeevi CB (2012) CRYAB-650 C>G (rs2234702) affects susceptibility to Type 1 diabetes and IAA-positivity in Swedish population. Hum Immunol 73(7):759–766. doi:10.1016/j.humimm.2012.04.004

van Veen T, van Winsen L, Crusius JB, Kalkers NF, Barkhof F, Pena AS, Polman CH, Uitdehaag BM (2003) [Alpha]B-crystallin genotype has impact on the multiple sclerosis phenotype. Neurology 61(9):1245–1249

Vicart P, Caron A, Guicheney P, Li Z, Prevost MC, Faure A, Chateau D, Chapon F, Tome F, Dupret JM, Paulin D, Fardeau M (1998) A missense mutation in the alphaB-crystallin chaperone gene causes a desmin-related myopathy. Nat Genet 20(1):92–95. doi:10.1038/1765

Zhang Y, Shen X (2007) Heat shock protein 27 protects L929 cells from cisplatin-induced apoptosis by enhancing Akt activation and abating suppression of thioredoxin reductase activity. Clin Cancer Res 13(10):2855–2864

Part IV
sHsps Cellular and Molecular Mechanisms of Action

Chapter 17
The Dynamic Duo of Small Heat Proteins and IFs Maintain Cell Homeostasis, Resist Cellular Stress and Enable Evolution in Cells and Tissues

Ming Der Perng and Roy A. Quinlan

Abstract As with many proteins, their initial christening with a name corrals our thinking on their functions and roles. IFs were not named as heat shock proteins, but along with heat shock proteins, they continue to be expressed in response to heat shock. In this review we outline the case for exposing the identity of the "dynamic duo" as small heat shock proteins (sHSPs) and their IF (IF) partners in their battle to assuage stress. Together they maintain the homeostasis of and integrate key cellular processes within cells that allow potential evolutionary opportunity to be seized (Quinlan RA, Ellis RJ, Philos Trans R Soc Lond B Biol Sci 368:20130091, 2013). Both sHSPs and IFs form dynamic polymeric structures – nano-particles and intermediate (nano) filaments respectively. Both have a wide range of interaction (client) proteins. IFs provide a convenient matrix to host small heat shock proteins and potentiate their role as chaperones, whilst IF function is sHSP-dependent. Together, it is their emerging capacity as integrators of both cell homeostasis and the stress response within and between tissues, however, which is the most exciting. This is because the dynamic duo co-operate on the two central chaperone activities – assisting protein assemblies and dealing with the consequences of protein damage. We propose that the sHSP-IF partnership is key to understanding the stress response, not least because it emphasizes the role of these protein chaperones in assisting the building, regulation and turnover of the IF protein assemblies as well as in diseases caused by their malfunction.

Keywords CRYAB • aB-crystallin • Desmin • Vimentin • GFAP • Autophagy • Cardiomyopathy • Proteostasis • Proteinopathy

---

M. Der Perng (✉)
Institute of Molecular Medicine, College of Life Sciences,
National Tsing Hua University, Hsinchu 300, Taiwan
e-mail: mdperng@life.nthu.edu.tw

R.A. Quinlan
Biophysical Sciences Institute, University of Durham, South Road,
Durham DH1 3LE, UK
e-mail: r.a.quinlan@dur.ac.uk

R.M. Tanguay, L.E. Hightower (eds.), *The Big Book on Small Heat Shock Proteins*, Heat Shock Proteins 8, DOI 10.1007/978-3-319-16077-1_17

Abbreviations

CRYAB	αB-crystallin
DRM	Desmin related myopathy
GFAP	Glial fibrillary acidic protein
HSP	Heat shock protein
HSR	Heat shock response
IF	Intermediate filament
NF-L	Neurofilament light chain

17.1 Where It All Began – The Identification of the Dynamic Duo of sHSPs and IFs as Stress Proteins

Heat shock helped identify the protein chaperones (e.g. Kelley and Schlesinger 1978), but taking the chicken embryo fibroblast as an example, vimentin was also robustly labeled with ^{35}S-methionine along with the resident small heat shock protein (Collier and Schlesinger 1986). Similar results were obtained in other cells expressing both protein chaperones and IF proteins (Vilaboa et al. 1997; Slater et al. 1981; Shyy et al. 1989). Indeed, an interesting observation from these early studies was that preconditioning stresses helped protect the IF cytoskeleton against heat-stress induced collapse (e.g. (Welch and Mizzen 1988)) and leading to the suggestion that sHSPs were instrumental in the recovery of the IF cytoskeleton after heat shock (Collier and Schlesinger 1986), highlighting the interdependence of IF cytoskeleton and protein chaperones. We (Landsbury et al. 2010) and others (Toivola et al. 2010) have made the case for IFs to be included as stress proteins, but it is the partnership between sHSPs and IFs (Landsbury et al. 2010; Quinlan 2002) that brings synergy to their individual stress protein roles and which is discussed here.

Now two decades on from the first isolation of a specific, soluble IF-sHSP complex (Nicholl and Quinlan 1994), there are three strong lines of evidence pointing to the importance of the functional partnership between sHSPs and IFs and their combatting of stress in cells and tissues. The first, and by far the strongest, is the phenocopying by mutations in both sHSP and IF proteins (e.g. cardiomyopathy caused by mutations in αB-crystallin (Vicart et al. 1998)) and desmin; (Goldfarb et al. 1998); cataract caused by mutations in both αA-crystallin (Su et al. 2012), αB-crystallin (Liu et al. 2006c), vimentin (Müller et al. 2009), BFSP1 (Wang et al. 2013; Ramachandran et al. 2007) and BFSP2 (Zhang et al. 2004; Ma et al. 2008; Zhang et al. 2006; Liu et al. 2014); neuropathies caused by mutations in HSP27 (Evgrafov et al. 2004), HSP22 (Irobi et al. 2004) and neurofilament light chain, NFL (Jordanova et al. 2003; Mersiyanova et al. 2000). Recent reviews have elaborated on the role for keratins (Toivola et al. 2010; Pan et al. 2013) and nuclear lamins (Shimi and Goldman 2014) in combatting different stresses. Table 17.1 gives a listing of the currently known mutations in sHSPs and their associated cytoskeletal proteins that cause human diseases, an obvious stress endpoint, but one where the consequences

Table 17.1 Mutations in sHSPs and their interacting cytoskeletal proteins that phenocopy each other in human diseases or where there are proven biochemical interactions

(A): sHSPs genes and mutations linked to human disease

sHSP	Gene symbol; gene coordinates;-(synonyms)	Chromosomal location (GRCh37)	Mutation details	Phenotypes (MIM)	Reference
HSP27	HSPB1; (CMT2F; DKFZp586P1322; HMN2B; HS.76067; Hsp25; HSP27; HSP28; SRP27)	7q11.23 (7:75,931,874 – 75,933,613)	G34R	Charcot-Marie-Tooth disease, axonal, type 2 F (606595)	Capponi et al. (2011)
			P39L	Neuropathy, distal hereditary motor, type IIB (608634)	Houlden et al. (2008)
			G41K		Capponi et al. (2011)
			G84R		Houlden et al. (2008), and James et al. (2008)
			L99M (recessive)		Houlden et al. (2008)
			R127W		Evgrafov et al. (2004), Dierick et al. (2008), and Tang et al. (2005a)
			S135F		Evgrafov et al. (2004)
			S135M		Houlden et al. (2008)
			R136W		Evgrafov et al. (2004)
			R136L		Capponi et al. (2011)
			R140G		Ackerley et al. (2006)
			K141Q		Ikeda et al. (2009)
			T151I		Evgrafov et al. (2004)
			S158fs		Capponi et al. (2011)
			T164A	Amyotrophic lateral sclerosis – genetic modifier?	Lin et al. (2011)

(continued)

Table 17.1 (continued)

(A): sHSPs genes and mutations linked to human disease					
sHSP	Gene symbol; gene coordinates;-(synonyms)	Chromosomal location (GRCh37)	Mutation details	Phenotypes (MIM)	Reference
			E175X		Rossor et al. (2012)
			T180I		Capponi et al. (2011)
			T180L		Luigetti et al. (2010)
			P182L		Evgrafov et al. (2004) and Ackerley et al. (2006)
			P182S		Kijima et al. (2005)
			R188W		Capponi et al. (2011)
			(c.-217 T>C)		Dierick et al. (2007)
HSPB2/MKBP	HSPB2 (Hs.78846; LOH11CR1K; MGC133245; MKBP)	11q22-q23.1 (11:111,783,459 – 111,784,816)	R7S	Neuronopathy, distal hereditary motor, type IIC (613376)	Kolb et al. (2010)
HSPB3/HSPL27	HSPB3 (HSPL27)	5q11.2 (5:53,751,430 – 53,752,213)			
αA-crystallin	CRYAA (HSPB4, CRYA1)	21q22.3 (21:44,589,140 – 44,592,914)	W9X: recessive	Cataract 9, multiple types (123580)	
HSPB4				Cataract with microcornea	Pras et al. (2000)
			R12C		Hansen et al. (2007)
			R21L		Devi et al. (2008)
			R21Q	Cataract and microophthalmia	Laurie et al. (2013)
			R21W		Graw et al. (2006)
			R49C	Cataract with microcornea	Hansen et al. (2007)
			R54C		Devi et al. (2008)
			R54L	Cataract, Y sutures	Mackay et al. (2003), Devi et al. (2008) and Khan et al. (2007)

			R54P	Age related cataract	Yang et al. (2013)
			R71L		Su et al. (2012)
			G98R	Cataract sometimes associated with micro-ophthalmia and microcornea	Bhagyalaxmi et al. (2009)
			R116C		Santhiya et al. (2006)
			R116H		Litt et al. (1998), Beby et al. (2007) and Vanita et al. (2006)
			R117,HY118del		Richter et al. (2008)
					Gu et al. (2008)
					Sun et al. (2011)
αB-crystallin	CRYAB (CRYA2; CTPP2; HSPB5)	11q22.3-q23.1 (11:111,779,349 – 111,782,472)	R11H	Cardiomyopathy, dilated, 1II (615184)	Chen et al. (2009)
HSPB5			P20S	Cataract 16, multiple types (613763)	Liu et al. (2006b)
			Ser21AlafsX24 (homozygous)	Myopathy, myofibrillar, 2 (608810)	Del Bigio et al. (2011)
			R56W (recessive)	Myopathy, myofibrillar, fatal infantile hypertrophy, alpha-B crystallin-related (613869)	Safieh et al. (2009)
			R69C		Sun et al. (2011)
			D109H		Sacconi et al. (2012)
			Ser115ProfsX14 (recessive)		Forrest et al. (2011)
			R120G		Vicart et al. (1998)
			D140N		Liu et al. (2006c)
			del450A		Berry et al. (2001)

(continued)

Table 17.1 (continued)

(A): sHSPs genes and mutations linked to human disease					
sHSP	Gene symbol; gene coordinates;-(synonyms)	Chromosomal location (GRCh37)	Mutation details	Phenotypes (MIM)	Reference
			del464CT		Selcen and Engel (2003)
			Q151X		Selcen and Engel (2003)
			G154S		Reilich et al. (2010)
			R157H		Inagaki et al. (2006)
			A171T		Devi et al. (2008)
HSP20	HSPB6 (AA387366; Gm479; Hsp20; MGC107687)	19q13.13 (19:36,245,466 – 36,247,929)	P20L	Diminished cardioprotection – genetic modifier to cardiomyopathy	Nicolaou et al. (2008)
cvHSP	HSPB7 (CVHSP)	1p36.23-p34.3 (1:16,340,522 – 16,346,083)		Charcot Marie Tooth Disease type IIF (608673)	
HSP22	HSPB8	12q24 (12:119,616,594 – 119,632,550)	K141E	Neuropathy, distal hereditary motor, type IIA (158590)	Tang et al. (2005b), Irobi et al. (2004) and Nakhro et al. (2013)
			K141N		
			K141T		
HSPB9	HSPB9 (CT51; FLJ27437)	17q21.2 (17: 40,274,756-40,275,371)		Link to male fertility?	Yang et al. (2012)
ODF1	HSP10 (HSPB10; MGC129928; MGC129929; ODF; ODF2; ODF27; ODFP; ODFPG; ODFPGA; ODFPGB; RT7; SODF)	8q22 (8:103,563,847 – 103,573,244)			Fontaine et al. (2003)

(B): Disease Causing Mutations in Cytoskeletal Proteins that interact with sHSPs

Protein	Gene symbol (synonyms)	Chromosomal location	Mutation details (MIM)	Phenotypes	Reference
Actin (cardiac)	ACTC1	15q14 (15:35,080,296 – 35,087,926)	For current listing of disease causing mutations see http://www.hgmd.cf.ac.uk/ac/gene.php?gene=ACTC1	Atrial septal defect 5 (612794)	
				Cardiomyopathy, dilated, 1R (613424)	
				Cardiomyopathy, familial hypertrophic, 11 (612098)	
				Left ventricular noncompaction 4 (613424)	
CP49	BFSP2 (CP47; CP49; LIFL-L; MGC142078; MGC142080; PHAKOSIN)	3q21-25 (3:133,118,789 – 133,194,055)	E223del;	Cataract 12, multiple types (611597)	Jakobs et al. (2000), Zhang et al. (2004) and Cui et al. (2007)
				Cataract and myopia	
			R287W	Cataract	Conley et al. (2000)
			R339H	Cataract	Ma et al. (2008)
Desmin	DES (CMD1I; CSM1; CSM2; FLJ12025; FLJ39719; FLJ41013; FLJ41793)	2q35 (2:220,283,098 – 220,291,460)	See http://www.interfil.org/details.php?id=NM_001927 for current listing of desmin mutations	Muscular dystrophy, limb-girdle, type 2R (615325)	Brodehl et al. (2013)
			A120D	Cardiomyopathy, dilated, 1I (604765)	
			1289-2A>G	Myopathy, myofibrillar, 1 (601419)	
				Scapuloperoneal syndrome, neurogenic, Kaeser type (181400)	
				Arrhythmogenic right ventricular cardiomyopathy	
				Autosomal recessive limb girdle muscular dystrophy (LGMD2)	Cetin et al. (2013)

(continued)

Table 17.1 (continued)

(B): Disease Causing Mutations in Cytoskeletal Proteins that interact with sHSPs

Protein	Gene symbol (synonyms)	Chromosomal location	Mutation details (MIM)	Phenotypes	Reference
Filensin	BFSP1 (CP115; CP94; LIFL-H)	20p12.1-p11.23 (20:17,474,549 – 17,549,864)	Deletion encompassing exon 6, c.736-1384_c.957-66 del – frameshift and loss of exon 6; recessive	Cataract 33 (611391)	Ramachandran et al. (2007)
			D348N		Wang et al. (2013)
NFL	NEFL (CMT1F; CMT2E; FLJ53642; NF-L; NF68; NFL)	8p21 (8:24,808,467 – 24,814,382)	See http://www.interfil.org/details.php?id=NM_006158 for current listing of NFL mutations	Charcot Marie Tooth Disease type IIE (607684)	
				Charcot Marie Tooth Disease type IF (607734)	
Titin	TTN (connectin)	2q31.2 (2:179,390,715 – 179,672,149)	CRYAB mutations decrease binding to N2B (R120G, D157N) and I26/27 (R120G) regions of TTN	Cardiomyopathy, dilated, 1G (604145)	For listing of current 127 TTN mutations see Chauveau et al. (2014).
			See Chauveau et al. (2014) for current listing	Cardiomyopathy, familial hypertrophic, 9 (613765)	
				Muscular dystrophy, limb-girdle, type 2 J (608807)	
				Myopathy, early-onset, with fatal cardiomyopathy (611705)	
				Myopathy, proximal, with early respiratory muscle involvement (603689)	
				Tibial muscular dystrophy, tardive (600334)	
Vimentin	VIM (FLJ36605)	10p13 (10:17,269,933 – 17,279,591)	E151K	Cataract 30, pulverulent (116300)	Müller et al. (2009)

Data extracted from OMIM (http://omim.org/entry/188840), HGMD (http://www.hgmd.cf.ac.uk/ac/index.php) and/or LOVD (http://www.dmd.nl/nmdb/home.php?select_db=TTN) and the Intermediate Filament Database (http://www.interfil.org/proteins.php) databases and updated from the primary literature of protein damage have overwhelmed the affected cells. The fact that only four (HSP27 (HSPB1), αA-crystallin (HSPB4), αB-crystallin (HSPB5) HSP22 (HSPB8)) of the ten sHSPs (Kappe et al. 2003) that have so far been found with disease causing mutations is interesting. We suggest that this reflects the fact that IF aggregates in these diseases are the easiest and most prominent histopathological feature facilitating their discovery (Quinlan and van den IJssel 1999; Selcen 2011; Carra et al. 2013; Strnad et al. 2012; Lowe et al. 1988). Our focus is therefore on these four sHSPs (HSP27 (HSPB1), αA-crystallin (HSPB4) and αB-crystallin (HSPB5), and to a lesser extent HSP22 (HSPB8)) that also have the potential to all form mixed oligomers (Skouri-Panet et al. 2012) justifying an overlap in their client proteins. The consequences of damaging either sHSPs or IFs by mutation therefore compromises the role of the dynamic duo in supporting the formation of protein assemblies, but also further compromises the cell by undermining the chaperone machinery to exacerbate the problem of accumulating misfolded proteins.

The second line of evidence is that compromising the IF network either by mutation or by the removal of individual proteins increases the stress susceptibility of the cells and tissue in which they are usually expressed (keratin, e.g.(Bar et al. 2014; Caulin et al. 2000; Ku et al. 2007; Alam et al. 2013; Weerasinghe et al. 2014); vimentin, e.g.(Eckes et al. 1998; Da et al. 2014; Perlson et al. 2005); GFAP, e.g.(Hagemann et al. 2006; Li et al. 2008; Lundkvist et al. 2004; Nawashiro et al. 1998); desmin, e.g.(Li et al. 1997), lamins e.g.(Sullivan et al. 1999; Butin-Israeli et al. 2013; Malhas et al. 2009; Saha et al. 2013; Seco-Cervera et al. 2014; Shimi et al. 2011; Shimi and Goldman 2014; Verstraeten et al. 2009); neurofilaments e.g.(Yum et al. 2009; Zhai et al. 2007) and lastly the orphan proteins expressed in the lens (Fudge et al. 2011)). The targeted knockout of sHSPs can also compromise the stress response (HSP27, (O'Shaughnessy et al. 2007; Sur et al. 2008; Crowe et al. 2013); αA-crystallin,(Yaung et al. 2008); αB-crystallin, (Andley et al. 2001; Wawrousek and Brady 1998; Jiao et al. 2014; Kumarapeli et al. 2008; Shao et al. 2013; Yaung et al. 2008; Morrison et al. 2004; Brady et al. 1997); HSP20, (Ishiwata et al. 2012)). The fact that compromising either the sHSP or the IF protein changes the stress response is the key observation.

The last line of evidence supporting the functional interaction of sHSPs and IFs is the synergy of the partnership between sHSPs and IFs. Here we evidence the important sHSP function of facilitating protein dynamic assemblies as complexes, polymers and scaffolds. The latter all involve protein-protein interactions and the folding energy landscape for protein-protein interactions has traps that need to be avoided (Zheng et al. 2012; Kozakov et al. 2014), adding an additional layer of complexity upon the folding landscape of individual proteins (Hartl and Hayer-Hartl 2009), but opening opportunity for sHSPs to facilitate successful protein-protein interactions in protein assemblies. The efficiency of sHSP chaperones can be dramatically increased when tethered to a support (Garvey et al. 2011), as seen for other chaperone classes (Altamirano et al. 1997; Antonio-Perez et al. 2012). There is no doubt that some of the sHSP complement in cells resides on IFs (Fig. 17.1) and this is increased during stressful events. Therefore, we suggest this is a reasonable expectation of the association with IFs. For example, the association of HSP70 with keratin filaments is required for maintaining aPKC activity

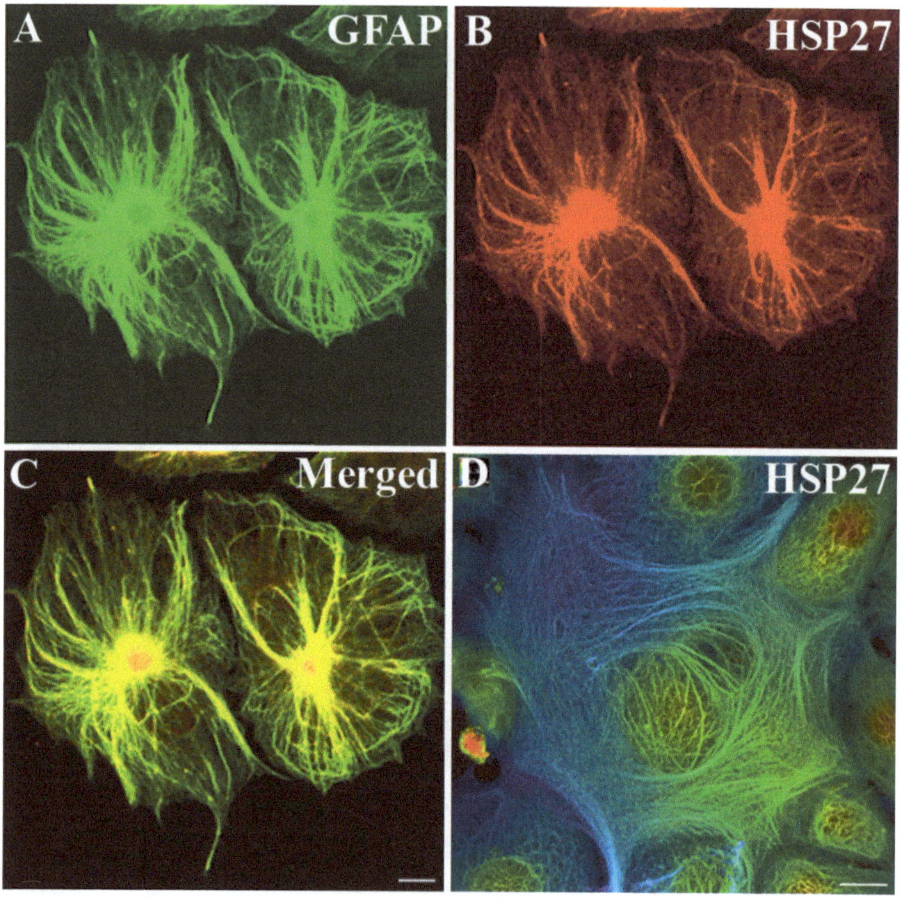

Fig. 17.1 The sHSP-IF partnership – seeing is believing
The codistribution of HSP27 with GFAP was visualized by double label immunofluorescence microscopy using both the anti-HSP27 mouse monoclonal antibody (**a**) and the rabbit polyclonal anti-GFAP antibodies (**b**). The immunofluorescence for GFAP is in the *green* channel (**a**), whereas the staining for HSP27 is in the *red* channel (**b**). Merged images show the superimposition of both the *green* and *red* signals, and overlapping areas appearing *yellow* (**c**). Notice that HSP27 associated with the full range of filament arrangements from bundles to the very fine filament distributions. (**d**) MCF7 cells stained with a monoclonal anti-HSP27 antibody, which decorated with the keratin-containing IF networks. The signal has been color-coded for depth, with blue representing the signal at an optical plane just above the cell support and *red* being furthest away. Bars, 10 μm

(Mashukova et al. 2009, 2014) and HSP27 regulates the activity of cdk5 and the subsequent phosphorylation of neurofilaments (Holmgren et al. 2013). In the epidermis, the regulation of filaggrin expression by HSP27 involves phosphorylation and transition to the keratin filament network (Jonak et al. 2011; Robitaille et al. 2010). These examples evidence how the IF scaffold facilitates sHSP function within the context of the unstressed cell environment and illustrate how these protein chaperones help integrate the different cellular signaling pathways (Fig. 17.2). It also evidences that sHSP chaperone function includes assisting protein complex formation and not just preventing protein unfolding and aggregation. Then there is the potential for the sHSP association with IFs to regulate subunit exchange. This is essential, for example, to αB-crystallin function (Delbecq and Klevit 2013; Stengel et al. 2010). The R120G mutation is thought to decrease subunit dynamics (Clark et al. 2011) and it also increases its binding to both desmin (Perng et al. 2004) and vimentin (Song et al. 2008), which in turn induces IF aggregation (Vicart et al. 1998; Perng et al. 1999b), compromising both the chaperone function of αB-crystallin and the function of IF networks.

From an IF standpoint, sHSPs assist both IF dynamics (Nicholl and Quinlan 1994) and filament network formation and maintenance require sHSPs (Perng et al. 1999a, b, 2004; Kayser et al. 2013; Zhai et al. 2007; Koyama and Goldman 1999). The overexpression of αB-crystallin can, for instance, ameliorate the disease causing consequences of mutations in both GFAP (Hagemann et al. 2009) and desmin (Wang et al. 2003) and the overexpression of either HSP27 or αB-crystallin can compensate for the loss of desmin (Soumaka et al. 2008). Many of the diseases caused by mutations in IF proteins (Table 17.1) are typified by protein aggregates that usually contain sHSPs (Quinlan and van den IJssel 1999; Selcen 2011; Dabir et al. 2004; Bjorkdahl et al. 2008) and the sequestration of these chaperones by the IF aggregates is proposed to be part of the disease mechanism. sHSPs shift from the soluble to IF fraction of the cell is a consistent observation as part of the HSR e.g.(Djabali et al. 1997; Perng et al. 1999a; Bennardini et al. 1992) and binding *in vitro* is temperature dependent (Elliott et al. 2013; Perng et al. 2004), a detail that allowed the Km to measured for αB-crystallin binding to desmin (Perng et al. 2004). The interaction between αB-crystallin and desmin, influences the caspase-mediated cleavage of desmin (Chen et al. 2003) to propagate caspase-mediated cell death as in the case of the αB-crystallin mutant R120G (Elliott et al. 2013). This appears to be specific to the partnership of αB-crystallin with desmin (Elliott et al. 2013), suggesting a potential hierarchy in the IF-sHSP partnerships where the partnership is essential.

There are also two very important additional factors to consider for sHSPs and IFs. They impact the appreciation of their role in ameliorating stress, but also in understanding the role of sHSPs in the assembly of protein complexes. Both protein families exhibit a complex pattern of tissue and cell expression (Kato et al. 1991a, b; Pan et al. 2013; Kampinga and Garrido 2012) and the interactome for sHSPs is rich and diverse (Arrigo and Gibert 2013). Taking αB-crystallin (HSPB5) as an example, every key pathway is impacted by this protein (Fig. 17.2), from the cell cycle to genome stability, from apoptosis to proteostasis, from the cytoskeleton to metabolic enzymes. This explains their potential to integrate (Quinlan and Ellis 2013) and influence the proliferation, differentiation of individual cells and their integration into tissues and animal physiology. Equally, IF proteins are widely expressed and mutations in IF proteins are the cause of a broad range of diseases (Omary 2009). They perform roles that now are recognized to far exceed the mechanical roles first attributed, but instead include signaling, (Kim and Coulombe

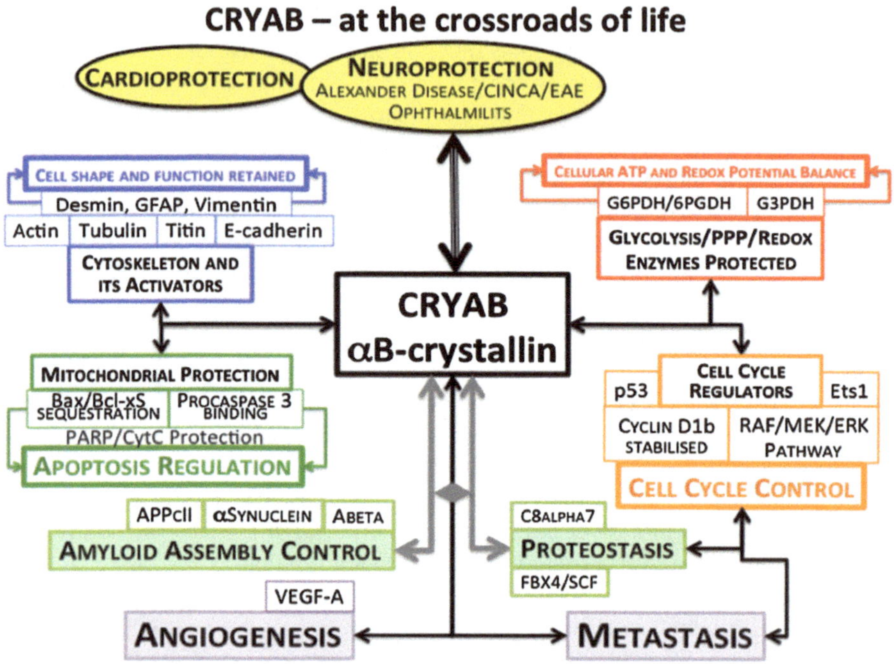

Fig. 17.2 αB-crystallin at the crossroads of life and its role as an evolutionary capacitor
The post-stroke injection of αB-crystallin protein reduced both stroke volume and inflammatory cytokines endogenous anti-inflammatory and neuroprotectant molecule produced after stroke (Arac et al. 2011), but also can assist in the recovery after spinal cord injury (Klopstein et al. 2012). This follows initial studies that used lens α-crystallin to counter neuroinflammation induced by silver nitrate injection (Masilamoni et al. 2006). Recently it has been shown that astrocytic DRD2 activation suppresses inflammation in the central nervous system and this is a αB-crystallin -dependent mechanism (Shao et al. 2012). αB-crystallin overexpression also protects against autoimmune demyelination in an EAE model (Ousman et al. 2007). In the retina, the protection of RPE cells is mediated by αB-crystallin secretion in exosomes (Sreekumar et al. 2010). Cardiomyocytes are also protected by αB-crystallin against oxidative stress induced by hydrogen peroxide (Chis et al. 2012) and by ischaemia (Morrison et al. 2004; Martin et al. 1997). Its overexpression can also protect against short-term overload cardiac hypertrophy (Kumarapeli et al. 2008). Its overexpression can reduce the aggregation of a desmin mutant that causes cardiomyopathy (Wang et al. 2003) and also reverse the formation of protein inclusions caused by the overexpression of GFAP (Koyama and Goldman 1999) in tissue culture models. Its overexpression in a model of Alexander disease caused by the aggregation of GFAP, significantly reduces morbidity (Hagemann et al. 2009)
Amyloid Assembly Control and Influence of Genetic Traits
Amyloid fibril assembly of Aβ (Narayan et al. 2012), κ-casein (Dehle et al. 2010), α-synuclein (Waudby et al. 2010) and also for apolipoprotein CII (Binger et al. 2012) can be inhibited by αB-crystallin to prevent fibril formation. αB-crystallin can suppress amyloid-mediated genetic traits In Drosophila (Tue et al. 2012) and can both prevent and cure the [PSI+] phenotype induced by Sup35 prions in S. cerevisiae (Duennwald et al. 2012). In mice the absence of CRYAB in combination with another sHSP, HSPB2, exacerbates the behavioral changes in an Alzheimer model (Ojha et al. 2011). In contrast the overexpression of a HSP70 class chaperone, Hspa13, accelerated prion-mediated disease in mice (Grizenkova et al. 2012). These studies support the argument that αB-crystallin, like HSP90, has capacitor-like activity with respect to genetic traits
Proteostasis and Cell Cycle Control
The overexpression of αB-crystallin can reverse the morbidity in mice caused by GFAP (Hagemann et al. 2009) and in cellular models this is likely due to increased proteasomal activity

Fig. 17.2 (continued) (Li et al. 2011). Similarly in a cardiomyopathy model, the overexpression of αB-crystallin also reduced desmin aggregation (Wang et al. 2003) and can dissolve GFAP aggregates (Koyama and Goldman 1999). Whilst these represent disease scenarios, proteastasis in unstressed cells is essential for regulating the cell cycle. αB-crystallin can bind the C8/alpha7 proteasome subunit (Boelens et al. 2001) and it also binds BAG3 (Hishiya et al. 2011), a regulator of chaperone-based aggresome targeting (Gamerdinger et al. 2011). It also interacts directly with the FBX4/SCF complex in a phosphorylation dependent manner (den Engelsman et al. 2003) to regulate the proteasomal degradation of SCF targets, such as cyclin D1, TP53 (Barbash et al. 2009; Jin et al. 2009; Lin et al. 2006) and MyoD (Singh et al. 2010; Tintignac et al. 2005). It binds directly to TP53 (Watanabe et al. 2009; Liu et al. 2007). In the canonical wnt-signalling pathway, αB-crystallin binds to the E-cadherin-β-catenin complex to maintain a membrane location for the complex (Huang et al. 2012) Recent data suggest that αB-crystallin could regulate systemic expansion of IMCs through a cell-intrinsic mechanism (Dieterich et al. 2013)

Metastasis

The direct involvement of αB-crystallin in the regulation of the cell cycle combined with the fact that the overexpression of αB-crystallin itself can complete the transformation process in immortalized human mammary epithelial cells (Moyano et al. 2006) CRYAB is a marker of poor prognosis in cancer (Moyano et al. 2006; Ivanov et al. 2008) and its ability to prevent apoptosis contributes to the aggressiveness of these and other tumours (Goplen et al. 2010; Volkmann et al. 2012). CRYAB expression in breast cancer is regulated by both TP53 and Ets1 (Bosman et al. 2010). αB-crystallin is associated significantly with ovarian (Volkmann et al. 2012), glioblastoma (Goplen et al. 2010) and hepatocellular carcinomas (Tang et al. 2009). αB-crystallin in combination with HSPB2 promote tumour vascularization (Dimberg et al. 2008) via a VEGF-A based mechanism (Kase et al. 2010). Also αB-crystallin regulates expansion of CD11b+Gr-1+ immature myeloid cells during tumor progression (Dieterich et al. 2013). αB-crystallin has been identified as a potential target in multiple myeloma because its knockdown also prevents tubular morphogenesis as an anti-angiogenic strategy (Berardi et al. 2012)

Mitochondrial Protection and Apoptosis Inhibition

In RPE cells loss of αB-crystallin induces mitochondrial dysfunction (Dou et al. 2012). αB-crystallin binds to and regulates Bax and caspase 3 during lens development (Hu et al. 2012). Its binding to mitochondria is stress induced (Chis et al. 2012; McGreal et al. 2012) and it is essential to the calcineurin A protection against apoptosis in a cardiac hypertrophy murine model (Bousette et al. 2010). αB-crystallin binds directly to Bax and Bcl-Xs (Mao et al. 2004), procaspase 3 (Kamradt et al. 2001; Mao et al. 2001) to prevent the degradation of PARP and the degradation (Kamradt et al. 2001) and oxidation (McGreal et al. 2012) of cytochrome C. The targeted expression of R120G αB-crystallin to mouse cardiomyocytes induces mitochondrial dysfunction and increased apoptosis (Maloyan et al. 2005)

Metabolic flux – ATP levels and Redox Potential

Levels of reduced glutathione in isolated hepatocytes are restored in the livers of mice lacking αB-crystallin and HSPB2 (Masilamoni et al. 2005). ATP levels are also αB-crystallin /HSPB2 dependent in mice (Benjamin et al. 2007). The R120G mutation in αB-crystallin causes oxido-redox stress by altering the levels of NADPH (Rajasekaran et al. 2007)

Cytoskeleton – a prime fascia example of an assembly chaperone role for CRYAB

Actin filament assembly can be regulated by CRYAB, which also protects the filaments against cytochalasin induced disassembly and also against (Singh et al. 2007) which also colocalises with actin in the leading edge of motile cells (Maddala and Rao 2005). The association with actin is phosphorylation dependent (Singh et al. 2007) Maintains myosin function and protects it during heat shock (Melkani et al. 2006). Titin (Golenhofen et al. 2002; Bullard et al. 2004; Kotter et al. 2014). Interaction with cadherin molecules (Thedieck et al. 2008). Tubulin assembly can be inhibited (Ohto-Fujita et al. 2007) but depending on the ratio CRYAB can also promote tubulin assembly (Houck and Clark 2010). It is however, the phenol-copying of the human disease cardiomyopathy by mutations in both CRYAB (Vicart et al. 1998) and desmin (Goldfarb et al. 1998), which confirmed CRYAB as assembly chaperone for intermediate filaments (Ngai et al. 1990; Nicholl and Quinlan 1994)

2007), cell growth, apoptosis and cell motility, processes that are regulated by a complex pattern of posttranslational modifications (Snider and Omary 2014).

17.2 The Disease Mechanisms of Desmin Related Myopathy and the Consequences for the Dynamic Duo

Identification of pathogenic mutations in desmin and αB-crystallin, analysis of their disease phenotypes, and successful modeling of underlying conditions in transgenic mice have shaped our understanding of the critical pathogenic events leading to desmin related myopathies (DRM). It is also important to emphasize here that DRM is an example of how mutations drive the aggregation of proteins via misfolding and thus energetically trapping the protein in an unfavourable conformation, and protein chaperones like sHSPs are not exempt from such an outcome. A transgenic mouse model expressing R120G αB-crystallin specifically in the heart has been generated, which recapitulated the key features of DRM phenotype observed in humans, with the formation of perinuclear aggregates and progression to heart failure (Wang et al. 2001b). Analysis of aggresome composition in R120G αB-crystallin-expressing cardiomyocytes also suggests that they form an amyloid (Sanbe et al. 2004) as (Meehan et al. 2007). Indeed subsequent studies showed that the expression of wild type αB-crystallin in R120G αB-crystallin-expressing cardiomyocytes led to a significant decrease in aggresomal accumulation, enhanced amyloid oligomer levels, and increased cytotoxicity (Sanbe et al. 2005). These observations suggest that amyloid oligomers could be toxic, whereas aggresome accumulation may be cytoprotective by effectively lowering the level of amyloid oligomers. However, experimental evidence also suggests that toxicity could be attributed to both misfolded oligomers and aggregated proteins (Tsunemi et al. 2012; Fatouros et al. 2012; Douglas et al. 2009). Although the toxicity of aggregated proteins remains a matter of controversy (Ross and Poirier 2005), it could be that misfolded proteins interact and recruit other cellular components during the aggregation process, thus compromising further cellular homeostasis. Another possibility is that amyloid-like aggregates cause aberrant protein turnover by inhibiting the ubiquitin-proteasome system or altering autophagy (Rubinsztein 2006).

17.3 Mitochondrial Dysfunction and Cell Death in DRM

DRM-causing mutations resulted in early perturbations in mitochondrial structure and function (Maloyan et al. 2005), which may be related to the disease phenotype. Desmin forms a transcytoplasmic IF network in association with the sarcomeric Z-discs in muscle and it also interacts with mitochondria to ensure their proper positioning so that the ATP produced meets the energy requirements of the contractile apparatus (Capetanaki et al. 2007). Indeed, studies in desmin knockout mice first revealed the importance of desmin IFs in mitochondrial positioning, shape and function in both skeletal and cardiac muscles (Milner et al. 2000). Overexpression of anti-apoptotic protein Bcl-2 in these desmin-null mice corrected the mitochondrial defects and ameliorated the cardiomyopathy (Weisleder et al. 2004), despite the fact that sarcomere organization was not completely restored. These data strongly suggest that mitochondria contribute to the IF disease pathology. This is supported by observations in patients diagnosed with DRM, where the focal clustering of morphologically and functionally compromised mitochondria was reported e.g.(Schroder et al. 2003). It is also by studies in mice expressing either mutant desmin or R120G αB-crystallin also demonstrated that mitochondrial dysfunction plays a very early and critical role in the developing DRM pathology (Wang et al. 2001a, b; Sanbe et al. 2005). The overexpression of Bcl-2 using a cardiac specific promoter also ameliorated αB-crystallin R120G pathology and decreased morbidity (Maloyan et al. 2010). This was also associated with decreased mitochondrial abnormalities and mutant protein aggregation, restoration of cardiac function, prevention of cardiac hypertrophy, and attenuation of apoptosis. However, inhibition of apoptosis resulted in the upregulation of autophagy and alternative death pathways, with the net result of increased necrosis. The mechanisms in which how the multiple death pathways talk to one another in cardiomyocytes are still being investigated.

17.4 The Role of the Protein Degradation Machinery in DRM

Aggregates are a consistent histopathological feature of DRM. The ubiquitin-proteasome system (UPS) and autophagy are the two primary proteolytic routes for removing misfolded and aggregated proteins from the cell. Deficits in the protein degradation mechanisms can lead to an imbalance in protein homeostasis and increased steady-state levels of misfolded proteins that are prone to aggregation (Su and Wang 2010). The buildup of this material leads to the formation of aggresomes and protein inclusions (Kopito 2000; Willis et al. 2010).

The UPS is responsible for degrading most cellular proteins through a series of steps in which a protein is first ubiquitinated and then shuttled to the proteasome for degradation (Willis et al. 2010). It is becoming increasingly apparent that compromised proteasomal activity is an important contributing pathology to cardiovascular disease (Schlossarek and Carrier 2011) and that deficits in the UPS degradation pathway may play a role in the inclusion body phenotype of DRM (Wang and Robbins 2006). It has been shown that protein aggregation directly impaired the function of the ubiquitin-proteasome system (Bence et al. 2001), a finding also observed in cultured cardiomyocytes (Dong et al. 2004). In the R120G αB-crystallin mouse model, it was shown that R120G αB-crystallin causes an apparent impairment of proteasome function, with a concomitant increase in 20S proteasome proteins and a decrease in components of the 19S proteosomal subunit (Chen et al. 2005). Similar observations were seen in the mouse model of DRM featuring protein aggregation caused by cardiac-expression of mutant desmin (Liu et al. 2006a), suggesting that the impairment in proteasomal degradation may be attributed to insufficient delivery of substrate to the proteasome. However, the mechanism of this inhibition of UPS remains unknown. Measurement of the proteasomal degradation as a function of aggregate levels in neonatal rat ventricular myocytes confirmed that inhibition occurred in a dose-dependent manner (Liu et al. 2006a). Coexpression of either αB-crystallin or Hsp70 reduced the amount of misfolded desmin and increased proteasomal degradation, effects that were replicated using Congo red, a dye that inhibits amyloid-like aggregate accumulation in cardiomyocytes (Sanbe et al. 2004). Stimulating the UPS side of the proteostais response by the overexpression of UBC9, a sumo E2-ligase, also helped reduce the aggregate load in R120G αB-crystallin expressing transgenic animals (Gupta et al. 2014). These findings prove that protein aggregates formed in DRM directly impaired proteasomal degradation. Importantly, these studies also suggest that UPS impairment may represent an important pathogenic event underlying cardiac proteinopathies featuring abnormal protein aggregation.

Autophagy plays a key role in the clearance of misfolded proteins and defective organelles. The importance of autophagy in maintaining physiologic homeostasis is illustrated by the fact that impaired autophagic function is a potential cause of misfolded protein accumulations, cytoplasmic aggregate load, and a number of human diseases, including DRM (Tannous et al. 2008). There is an early increase in autophagic response in the transgenic mouse cardiomyocytes expressing R120G αB-crystallin (Tannous et al. 2008). Inhibiting autophagy by crossing these R120G αB-crystallin mice with animals harboring an inactive allele of *beclin 1*, a gene required for autophagy, increased the rate of intracellular aggregate accumulation, accelerated ventricular dysfunction, and early mortality. Consistent with these observations, beclin-1 is downregulated in R120G αB-crystallin mice (Maloyan et al. 2010). These findings point to an increase in autophagy as an adaptive response to misfolded protein generation in this DRM model (Tannous et al. 2008). Importantly, these data also suggest increasing autophagy can be protective in the R120G αB-crystallin -mediated DRM model. In fact, overexpressing Atg7, a key component of autophagosome synthesis, increased autophagic flux and led to a reduction in amyloid oligomer and aggregate contents in R120G αB-crystallin-expressing cardiomyocytes (Pattison et al. 2011). Conversely, loss of function mediated by small interfering RNA-mediated knockdown of Atg7 in the R120G αB-crystallin background significantly decreased autophagy and increased R120G aggregate content and cytotoxicity. Increased expression of ATG7 in R120G αB-crystallin mice enhanced autophagy and reduced intracellular aggregates, which ameliorated ventricular dysfunction, decreased cardiac hypertrophy and prolonged survival in this DRM model (Bhuiyan et al. 2013). Interestingly, increased autophagic flux in DRM mouse hearts also correlates with an increase in the autophagy-related protein p62 (Zheng and Wang 2010). The depletion of p62 impaired aggresome and autophagosome formation, leading to decreased cell viability in cultured neonatal rat ventricular myocytes expressing mutant desmin or αB-crystallin. In the R120G αB-crystallin knock-in mice, altered p62 expression and impaired autophagy were also observed in lens epithelial cells (Wignes et al. 2013). These data suggest that increasing autophagy could potentially be protective in DRM, and that autophagy modulation may provide a basis for designing therapeutic strategies to treat cardiac proteinopathies such as DRM.

The discussion thus far has considered the role that the proteostasis pathways play in combatting aggregates formed from mutant sHSPs and IF proteins, but there is also the impact that the loss of these two components have on proteostasis itself. For instance αB-crystallin is part of the FBX4/SCF complex needed for the turnover of important cell cycle regulators (Fig. 17.2). The IF protein, glial fibrillary acidic protein (GFAP), is part of the lysosomal translocation complex required for chaperone mediated autophagy (Bandyopadhyay et al. 2010). Inhibition of the proteasome system feeds back to represses GFAP expression (Middeldorp et al. 2009), but the effect upon autophagy was not investigated. IFs, along with actin microfilaments are reportedly docking sites for the 26S proteasome (Arcangeletti et al. 1992). Individually therefore, both sHSPs and IF proteins are directly involved in proteostasis, but their individual and combined impact needs to be investigated further (Fig. 17.3).

17.5 The Extracellular Dimension to sHSPs and IFs

So far it has been the intracellular consequences of mutant sHSP and IFs in cardiomyopathy that have been considered. It interesting though that levels of αB-crystallin were reported to be raised in the serum of a hamster model of cardiomyopathy (Kozawa et al. 2001) and now it is evident that there is a release of αB-crystallin from muscle after exercise (Kozawa et al. 2001) and as a result of injury (Jorgensen et al. 2013) and disease such as cataract (Ranjan et al. 2008) and multiple sclerosis (Rothbard et al. 2012). HSP27 is released after global ischaemia of the human myocardium HSP20, released after injury and like αB-crystallin was also elevated in serum taken from the hamster cardiomyopathy model (Kozawa et al. 2002). This release will likely involve the exosome pathway as reported for αB-crystallin (Sreekumar et al. 2010; Gangalum et al. 2011; Clayton et al. 2005). Our current understanding of the exosome pathway is limited, but as exosomes are thought to be derived from multivesicular endosomes (Raposo and Stoorvogel 2013), there is a clear overlap with autophagy. Not only that but the intermediate filament protein GFAP is an integral part of the lysosomal membrane translocation complex needed to support chaperone mediated autophagy (Bandyopadhyay et al. 2010). IF proteins are also released from diseased tissues and there are serum signatures that are exploited diagnostically for instance for desmin (Ma et al. 2009), GFAP (Jany et al. 2013) and NF-L (Hjalmarsson et al. 2014) the oldest of these being the TROMA1 antigen, which is derived from keratins (Magin et al. 1986). Uptake of exosomes into cells appears to be an HSP27-ERK1/2 dependent process (Svensson et al.

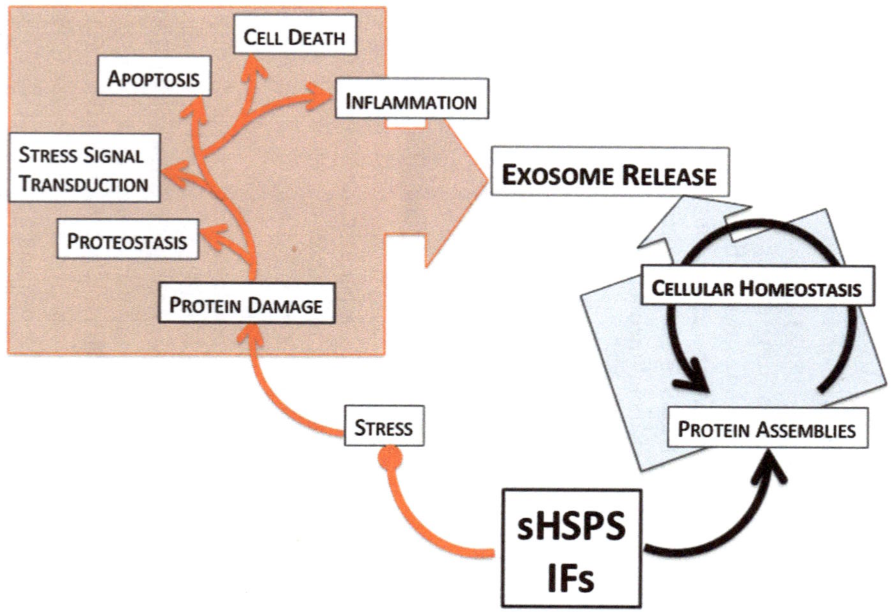

Fig. 17.3 The Dynamic Duo dealing with Protein Assemblies and Protein Damage
The two arenas for the sHSP-IF dynamic duo are during homeostasis (in the blue corner) and when there are stressful events (in the red corner) caused by disease, mutation or environment. IN both cases, the individual cell is broadcasting to its neighbours and to the rest of other tissues the stress status via exosomes, a mechanism that can be used to support organismal proteostasis

2013). The exosome pathway is a mechanism to integrate physiological responses in an organism (Pegtel et al. 2014) and this led some to consider that increasing the extracellular concentrations of sHSPs or minichaperones derived from them (Nahomi et al. 2013) could be beneficial.

There are now examples in the literature to support this hypothesis. The post-stroke injection of αB-crystallin, an endogenous anti-inflammatory and neuroprotectant molecule, not only reduced both stroke volume and inflammatory cytokines endogenous anti-inflammatory and neuroprotectant molecule produced after stroke (Arac et al. 2011), but also can assist in the recovery after spinal cord injury (Klopstein et al. 2012). This follows initial studies that injected lens α-crystallin (αB-crystallin and αA-crystallin purified from the eye lens) intra-peritoneally to counter neuroinflammation induced by silver nitrate injection (Masilamoni et al. 2006). In a MPTP induced mouse model of Parkinson's disease, it has been shown that astrocytic DRD2 activation suppresses inflammation in the central nervous system in a CRYAB-dependent mechanism (Shao et al. 2012). In the retina, the protection of RPE cells is mediated by CRYAB secretion in exosomes (Sreekumar et al. 2010). The relevance of these observations to the recently proposed mechanism observed in *C. elegans* for organismal proteostasis (van Oosten-Hawle and Morimoto 2014) are evident and clearly the exosome provides a mechanism for such signaling to occur, one that is key to the integrated physiological response needed for organisms to survive and to evolve. For example, the foetal response to maternal infection is to elevate sHSP gene expression in brain tissue (Garbett et al. 2012). The sHSPs are clearly an important component in this response.

The mechanism of this protection by αB-crystallin is varied and complex, but there are some indications. For instance αB-crystallin can bind preferentially proinflammatory molecules in human plasma (Rothbard et al. 2012), and it binds to blood platelets and prevents the ADP-stimulated release of platelet granules (Enomoto et al. 2009). Blood platelets are also regulated by HSP20 and HSP20 is acknowledged as a cardiokine promoting myocardial angiogenesis (Zhang et al. 2012), modulates myometrial relaxation during pregnancy (Tyson et al. 2008). sHSP involvement in propagating intracellular signaling, as well as inflammatory responses (Sur et al. 2008; Jin et al. 2014) resulting in decreased wound healing (Crowe et al. 2013) as well as influencing VEGF signalling in the promotion of angiogenesis in a tumour model (Lee et al. 2012). These examples illustrate how sHSPs like αB-crystallin affect whole-animal physiology, something that is also beginning to emerge for IFs (Alam et al. 2013; Leterrier et al. 1996). This review makes the case for the sHSPs and IFs being considered as functional partners in cell homeostasis and in the cellular response to stresses, but given the role of IFs in integrating individual cells within the context of tissues (Snider and Omary 2014) there is the possibility that this partnership has broader physiological significance given their synergistic interaction with sHSPs.

Acknowledgements The financial support of the National Science Council to MDP is gratefully acknowledged. RAQ thanks the financial support of the Leverhulme Trust (RPG-2012-554).

References

Ackerley S, James PA, Kalli A, French S, Davies KE, Talbot K (2006) A mutation in the small heat-shock protein HSPB1 leading to distal hereditary motor neuronopathy disrupts neurofilament assembly and the axonal transport of specific cellular cargoes. Hum Mol Genet 15:347–354

Alam CM, Silvander JS, Daniel EN, Tao GZ, Kvarnstrom SM, Alam P, Omary MB, Hanninen A, Toivola DM (2013) Keratin 8 modulates beta-cell stress responses and normoglycaemia. J Cell Sci 126:5635–5644

Altamirano MM, Golbik R, Zahn R, Buckle AM, Fersht AR (1997) Refolding chromatography with immobilized mini-chaperones. Proc Natl Acad Sci U S A 94:3576–3578

Andley UP, Song Z, Wawrousek EF, Brady JP, Bassnett S, Fleming TP (2001) Lens epithelial cells derived from alphaB-crystallin knockout mice demonstrate hyperproliferation and genomic instability. FASEB J 15:221–229

Antonio-Perez A, Ramon-Luing LA, Ortega-Lopez J (2012) Chromatographic refolding of rhodanese and lysozyme assisted by the GroEL apical domain, DsbA and DsbC immobilized on cellulose. J Chromatogr A 1248:122–129

Arac A, Brownell SE, Rothbard JB, Chen C, Ko RM, Pereira MP, Albers GW, Steinman L, Steinberg GK (2011) Systemic augmentation of alphaB-crystallin provides therapeutic benefit twelve hours post-stroke onset via immune modulation. Proc Natl Acad Sci U S A 108:13287–13292

Arcangeletti C, Olink CM, Minisini R, Huesca M, Chezzi C, Scherrer K (1992) Patterns of cytodistribution of prosomal antigens on the vimentin and cytokeratin networks of monkey kidney cells. Eur J Cell Biol 59:464–476

Arrigo AP, Gibert B (2013) Protein interactomes of three stress inducible small heat shock proteins: HspB1, HspB5 and HspB8. Int J Hyperthermia 29:409–422

Bandyopadhyay U, Sridhar S, Kaushik S, Kiffin R, Cuervo AM (2010) Identification of regulators of chaperone-mediated autophagy. Mol Cell 39:535–547

Bar J, Kumar V, Roth W, Schwarz N, Richter M, Leube RE, Magin TM (2014) Skin fragility and impaired desmosomal adhesion in mice lacking all keratins. J Invest Dermatol 134:1012–1022

Barbash O, Egan E, Pontano LL, Kosak J, Diehl JA (2009) Lysine 269 is essential for cyclin D1 ubiquitylation by the SCF(Fbx4/alphaB-crystallin) ligase and subsequent proteasome-dependent degradation. Oncogene 28:4317–4325

Beby F, Commeaux C, Bozon M, Denis P, Edery P, Morle L (2007) New phenotype associated with an Arg116Cys mutation in the CRYAA gene: nuclear cataract, iris coloboma, and microphthalmia. Arch Ophthalmol 125:213–216

Bence NF, Sampat RM, Kopito RR (2001) Impairment of the ubiquitin-proteasome system by protein aggregation. Science 292:1552–1555

Benjamin IJ, Guo Y, Srinivasan S, Boudina S, Taylor RP, Rajasekaran NS, Gottlieb R, Wawrousek EF, Abel ED, Bolli R (2007) CRYAB and HSPB2 deficiency alters cardiac metabolism and paradoxically confers protection against myocardial ischemia in aging mice. Am J Physiol Heart Circ Physiol 293:H3201–H3209

Bennardini F, Wrzosek A, Chiesi M (1992) Alpha B-crystallin in cardiac tissue. Association with actin and desmin filaments. Circ Res 71:288–294

Berardi S, Caivano A, Ria R, Nico B, Savino R, Terracciano R, De Tullio G, Ferrucci A, De Luisi A, Moschetta M, Mangialardi G, Catacchio I, Basile A, Guarini A, Zito A, Ditonno P, Musto P, Dammacco F, Ribatti D, Vacca A (2012) Four proteins governing overangiogenic endothelial cell phenotype in patients with multiple myeloma are plausible therapeutic targets. Oncogene 31:2258–2269

Berry V, Francis P, Reddy MA, Collyer D, Vithana E, Mackay I, Dawson G, Carey AH, Moore A, Bhattacharya SS, Quinlan RA (2001) Alpha-B crystallin gene (CRYAB) mutation causes dominant congenital posterior polar cataract in humans. Am J Hum Genet 69:1141–1145

Bhagyalaxmi SG, Srinivas P, Barton KA, Kumar KR, Vidyavathi M, Petrash JM, Bhanuprakash Reddy G, Padma T (2009) A novel mutation (F71L) in alphaA-crystallin with defective chaperone-like function associated with age-related cataract. Biochim Biophys Acta 1792:974–981

Bhuiyan MS, Pattison JS, Osinska H, James J, Gulick J, Mclendon PM, Hill JA, Sadoshima J, Robbins J (2013) Enhanced autophagy ameliorates cardiac proteinopathy. J Clin Invest 123:5284–5297

Binger KJ, Ecroyd H, Yang S, Carver JA, Howlett GJ, Griffin MD (2012) Avoiding the oligomeric state: alphaB-crystallin inhibits fragmentation and induces dissociation of apolipoprotein C-II amyloid fibrils. FASEB J 27:1214

Bjorkdahl C, Sjogren MJ, Zhou X, Concha H, Avila J, Winblad B, Pei JJ (2008) Small heat shock proteins Hsp27 or alphaB-crystallin and the protein components of neurofibrillary tangles: tau and neurofilaments. J Neurosci Res 86:1343–1352

Boelens WC, Croes Y, De Jong WW (2001) Interaction between alphaB-crystallin and the human 20S proteasomal subunit C8/alpha7. Biochim Biophys Acta 1544:311–319

Bosman JD, Yehiely F, Evans JR, Cryns VL (2010) Regulation of alphaB-crystallin gene expression by the transcription factor Ets1 in breast cancer. Breast Cancer Res Treat 119:63–70

Bousette N, Chugh S, Fong V, Isserlin R, Kim KH, Volchuk A, Backx PH, Liu P, Kislinger T, Maclennan DH, Emili A, Gramolini AO (2010) Constitutively active calcineurin induces cardiac endoplasmic reticulum stress and protects against apoptosis that is mediated by alpha-crystallin-B. Proc Natl Acad Sci U S A 107:18481–18486

Brady JP, Garland D, Duglastabor Y, Robison WG, Groome A, Wawrousek EF (1997) Targeted disruption of the mouse alpha A-crystallin gene induces cataract and cytoplasmic inclusion bodies containing the small heat shock protein alpha B-crystallin. Proc Natl Acad Sci U S A 94:884–889

Brodehl A, Dieding M, Klauke B, Dec E, Madaan S, Huang T, Gargus J, Fatima A, Saric T, Cakar H, Walhorn V, Tonsing K, Skrzipczyk T, Cebulla R, Gerdes D, Schulz U, Gummert J, Svendsen JH, Olesen MS, Anselmetti D, Christensen AH, Kimonis V, Milting H (2013) The novel desmin mutant p.A120D impairs filament formation, prevents intercalated disk localization, and causes sudden cardiac death. Circ Cardiovasc Genet 6:615–623

Bullard B, Ferguson C, Minajeva A, Leake MC, Gautel M, Labeit D, Ding L, Labeit S, Horwitz J, Leonard KR, Linke WA (2004) Association of the chaperone alphaB-crystallin with titin in heart muscle. J Biol Chem 279:7917–7924

Butin-Israeli V, Adam SA, Goldman RD (2013) Regulation of nucleotide excision repair by nuclear lamin b1. PLoS One 8:e69169

Capetanaki Y, Bloch RJ, Kouloumenta A, Mavroidis M, Psarras S (2007) Muscle intermediate filaments and their links to membranes and membranous organelles. Exp Cell Res 313:2063–2076

Capponi S, Geroldi A, Fossa P, Grandis M, Ciotti P, Gulli R, Schenone A, Mandich P, Bellone E (2011) HSPB1 and HSPB8 in inherited neuropathies: study of an Italian cohort of dHMN and CMT2 patients. J Peripher Nerv Syst 16:287–294

Carra S, Rusmini P, Crippa V, Giorgetti E, Boncoraglio A, Cristofani R, Naujock M, Meister M, Minoia M, Kampinga HH, Poletti A (2013) Different anti-aggregation and pro-degradative functions of the members of the mammalian sHSP family in neurological disorders. Philos Trans R Soc Lond B Biol Sci 368:20110409

Caulin C, Ware CF, Magin TM, Oshima RG (2000) Keratin-dependent, epithelial resistance to tumor necrosis factor-induced apoptosis. J Cell Biol 149:17–22

Cetin N, Balci-Hayta B, Gundesli H, Korkusuz P, Purali N, Talim B, Tan E, Selcen D, Erdem-Ozdamar S, Dincer P (2013) A novel desmin mutation leading to autosomal recessive limb-girdle muscular dystrophy: distinct histopathological outcomes compared with desminopathies. J Med Genet 50:437–443

Chauveau C, Rowell J, Ferreiro A (2014) A rising titan: TTN review and mutation update. Hum Mutat 35:1046

Chen F, Chang R, Trivedi M, Capetanaki Y, Cryns VL (2003) Caspase proteolysis of desmin produces a dominant-negative inhibitor of intermediate filaments and promotes apoptosis. J Biol Chem 278:6848–6853

Chen Q, Liu JB, Horak KM, Zheng H, Kumarapeli AR, Li J, Li F, Gerdes AM, Wawrousek EF, Wang X (2005) Intrasarcoplasmic amyloidosis impairs proteolytic function of proteasomes in cardiomyocytes by compromising substrate uptake. Circ Res 97:1018–1026

Chen Q, Ma J, Yan M, Mothobi ME, Liu Y, Zheng F (2009) A novel mutation in CRYAB associated with autosomal dominant congenital nuclear cataract in a Chinese family. Mol Vis 15:1359–1365

Chis R, Sharma P, Bousette N, Miyake T, Wilson A, Backx PH, Gramolini AO (2012) alpha-crystallin B prevents apoptosis after H2O2 exposure in mouse neonatal cardiomyocytes. Am J Physiol Heart Circ Physiol 303:H967–H978

Clark AR, Naylor CE, Bagneris C, Keep NH, Slingsby C (2011) Crystal structure of R120G disease mutant of human alphaB-crystallin domain dimer shows closure of a groove. J Mol Biol 408:118–134

Clayton A, Turkes A, Navabi H, Mason MD, Tabi Z (2005) Induction of heat shock proteins in B-cell exosomes. J Cell Sci 118:3631–3638

Collier NC, Schlesinger MJ (1986) Induction of heat-shock proteins in the embryonic chicken lens. Exp Eye Res 43:103–117

Conley YP, Erturk D, Keverline A, Mah TS, Keravala A, Barnes LR, Bruchis A, Hess JF, Fitzgerald PG, Weeks DE, Ferrell RE, Gorin MB (2000) A juvenile-onset, progressive cataract locus on chromosome 3q21-q22 is associated with a missense mutation in the beaded filament structural protein-2. Am J Hum Genet 66:1426–1431

Crowe J, Aubareda A, Mcnamee K, Przybycien PM, Lu X, Williams RO, Bou-Gharios G, Saklatvala J, Dean JL (2013) Heat shock protein B1-deficient mice display impaired wound healing. PLoS One 8:e77383

Cui X, Gao L, Jin Y, Zhang Y, Bai J, Feng G, Gao W, Liu P, He L, Fu S (2007) The E233del mutation in BFSP2 causes a progressive autosomal dominant congenital cataract in a Chinese family. Mol Vis 13:2023–2029

Da Q, Behymer M, Correa JI, Vijayan KV, Cruz MA (2014) Platelet adhesion involves a novel interaction between vimentin and von Willebrand factor under high shear stress. Blood 123:2715–2721

Dabir DV, Trojanowski JQ, Richter-Landsberg C, Lee VM, Forman MS (2004) Expression of the small heat-shock protein alphaB-crystallin in tauopathies with glial pathology. Am J Pathol 164:155–166

Dehle FC, Ecroyd H, Musgrave IF, Carver JA (2010) alphaB-crystallin inhibits the cell toxicity associated with amyloid fibril formation by kappa-casein and the amyloid-beta peptide. Cell Stress Chaperones 15:1013–1026

Del Bigio MR, Chudley AE, Sarnat HB, Campbell C, Goobie S, Chodirker BN, Selcen D (2011) Infantile muscular dystrophy in Canadian aboriginals is an alphaB-crystallinopathy. Ann Neurol 69:866–871

Delbecq SP, Klevit RE (2013) One size does not fit all: the oligomeric states of alphaB crystallin. FEBS Lett 587:1073–1080

Den Engelsman J, Keijsers V, De Jong WW, Boelens WC (2003) The small heat-shock protein alpha B-crystallin promotes FBX4-dependent ubiquitination. J Biol Chem 278:4699–4704

Devi RR, Yao W, Vijayalakshmi P, Sergeev YV, Sundaresan P, Hejtmancik JF (2008) Crystallin gene mutations in Indian families with inherited pediatric cataract. Mol Vis 14:1157–1170

Dierick I, Irobi J, Janssens S, Theuns J, Lemmens R, Jacobs A, Corsmit E, Hersmus N, Van Den Bosch L, Robberecht W, De Jonghe P, Van Broeckhoven C, Timmerman V (2007) Genetic variant in the HSPB1 promoter region impairs the HSP27 stress response. Hum Mutat 28:830

Dierick I, Baets J, Irobi J, Jacobs A, De Vriendt E, Deconinck T, Merlini L, Van Den Bergh P, Rasic VM, Robberecht W, Fischer D, Morales RJ, Mitrovic Z, Seeman P, Mazanec R, Kochanski A, Jordanova A, Auer-Grumbach M, Helderman-Van Den Enden AT, Wokke JH, Nelis E, De

Jonghe P, Timmerman V (2008) Relative contribution of mutations in genes for autosomal dominant distal hereditary motor neuropathies: a genotype-phenotype correlation study. Brain 131:1217–1227

Dieterich LC, Schiller P, Huang H, Wawrousek EF, Loskog A, Wanders A, Moons L, Dimberg A (2013) alphaB-Crystallin regulates expansion of CD11b+Gr-1+ immature myeloid cells during tumor progression. FASEB J 27:151–162

Dimberg A, Rylova S, Dieterich LC, Olsson AK, Schiller P, Wikner C, Bohman S, Botling J, Lukinius A, Wawrousek EF, Claesson-Welsh L (2008) alphaB-crystallin promotes tumor angiogenesis by increasing vascular survival during tube morphogenesis. Blood 111:2015–2023

Djabali K, Denechaud B, Landon F, Portier MM (1997) alpha B-crystallin interacts with intermediate filaments in response to stress. J Cell Sci 110:2759–2769

Dong X, Liu J, Zheng H, Glasford JW, Huang W, Chen QH, Harden NR, Li F, Gerdes AM, Wang X (2004) In situ dynamically monitoring the proteolytic function of the ubiquitin-proteasome system in cultured cardiac myocytes. Am J Physiol Heart Circ Physiol 287:H1417–H1425

Dou G, Sreekumar PG, Spee C, He S, Ryan SJ, Kannan R, Hinton DR (2012) Deficiency of alphaB crystallin augments ER stress-induced apoptosis by enhancing mitochondrial dysfunction. Free Radic Biol Med 53:1111–1122

Douglas PM, Summers DW, Cyr DM (2009) Molecular chaperones antagonize proteotoxicity by differentially modulating protein aggregation pathways. Prion 3:51–58

Duennwald ML, Echeverria A, Shorter J (2012) Small heat shock proteins potentiate amyloid dissolution by protein disaggregases from yeast and humans. PLoS Biol 10:e1001346

Eckes B, Dogic D, Colucci-Guyon E, Wang N, Maniotis A, Ingber D, Merckling A, Langa F, Aumailley M, Delouvee A, Koteliansky V, Babinet C, Krieg T (1998) Impaired mechanical stability, migration and contractile capacity in vimentin-deficient fibroblasts. J Cell Sci 111:1897–1907

Elliott JL, Der Perng M, Prescott AR, Jansen KA, Koenderink GH, Quinlan RA (2013) The specificity of the interaction between alphaB-crystallin and desmin filaments and its impact on filament aggregation and cell viability. Philos Trans R Soc Lond B Biol Sci 368:20120375

Enomoto Y, Adachi S, Matsushima-Nishiwaki R, Niwa M, Tokuda H, Akamatsu S, Doi T, Kato H, Yoshimura S, Ogura S, Iwama T, Kozawa O (2009) alphaB-crystallin extracellularly suppresses ADP-induced granule secretion from human platelets. FEBS Lett 583:2464–2468

Evgrafov OV, Mersiyanova I, Irobi J, Van Den Bosch L, Dierick I, Leung CL, Schagina O, Verpoorten N, Van Impe K, Fedotov V, Dadali E, Auer-Grumbach M, Windpassinger C, Wagner K, Mitrovic Z, Hilton-Jones D, Talbot K, Martin JJ, Vasserman N, Tverskaya S, Polyakov A, Liem RK, Gettemans J, Robberecht W, De Jonghe P, Timmerman V (2004) Mutant small heat-shock protein 27 causes axonal Charcot-Marie-Tooth disease and distal hereditary motor neuropathy. Nat Genet 36:602–606

Fatouros C, Pir GJ, Biernat J, Koushika SP, Mandelkow E, Mandelkow EM, Schmidt E, Baumeister R (2012) Inhibition of tau aggregation in a novel Caenorhabditis elegans model of tauopathy mitigates proteotoxicity. Hum Mol Genet 21:3587–3603

Fontaine JM, Rest JS, Welsh MJ, Benndorf R (2003) The sperm outer dense fiber protein is the 10th member of the superfamily of mammalian small stress proteins. Cell Stress Chaperones 8:62–69

Forrest KM, Al-Sarraj S, Sewry C, Buk S, Tan SV, Pitt M, Durward A, Mcdougall M, Irving M, Hanna MG, Matthews E, Sarkozy A, Hudson J, Barresi R, Bushby K, Jungbluth H, Wraige E (2011) Infantile onset myofibrillar myopathy due to recessive CRYAB mutations. Neuromuscul Disord 21:37–40

Fudge DS, Mccuaig JV, Van Stralen S, Hess JF, Wang H, Mathias RT, Fitzgerald PG (2011) Intermediate filaments regulate tissue size and stiffness in the murine lens. Invest Ophthalmol Vis Sci 52:3860–3867

Gamerdinger M, Kaya AM, Wolfrum U, Clement AM, Behl C (2011) BAG3 mediates chaperone-based aggresome-targeting and selective autophagy of misfolded proteins. EMBO Rep 12:149–156

Gangalum RK, Atanasov IC, Zhou ZH, Bhat SP (2011) AlphaB-crystallin is found in detergent-resistant membrane microdomains and is secreted via exosomes from human retinal pigment epithelial cells. J Biol Chem 286:3261–3269

Garbett KA, Hsiao EY, Kalman S, Patterson PH, Mirnics K (2012) Effects of maternal immune activation on gene expression patterns in the fetal brain. Transcult Psychiatry 2:e98

Garvey M, Griesser SS, Griesser HJ, Thierry B, Nussio MR, Shapter JG, Ecroyd H, Giorgetti S, Bellotti V, Gerrard JA, Carver JA (2011) Enhanced molecular chaperone activity of the small heat-shock protein alphaB-cystallin following covalent immobilization onto a solid-phase support. Biopolymers 95:376–389

Goldfarb LG, Park KY, Cervenakova L, Gorokhova S, Lee HS, Vasconcelos O, Nagle JW, Semino-Mora C, Sivakumar K, Dalakas MC (1998) Missense mutations in desmin associated with familial cardiac and skeletal myopathy. Nat Genet 19:402–403

Golenhofen N, Arbeiter A, Koob R, Drenckhahn D (2002) Ischemia-induced association of the stress protein alpha B-crystallin with I-band portion of cardiac titin. J Mol Cell Cardiol 34:309–319

Goplen D, Bougnaud S, Rajcevic U, Boe SO, Skaftnesmo KO, Voges J, Enger PO, Wang J, Tysnes BB, Laerum OD, Niclou S, Bjerkvig R (2010) alphaB-crystallin is elevated in highly infiltrative apoptosis-resistant glioblastoma cells. Am J Pathol 177:1618–1628

Graw J, Klopp N, Illig T, Preising MN, Lorenz B (2006) Congenital cataract and macular hypoplasia in humans associated with a de novo mutation in CRYAA and compound heterozygous mutations in P. Graefes Arch Clin Exp Ophthalmol 244:912–919

Grizenkova J, Akhtar S, Hummerich H, Tomlinson A, Asante EA, Wenborn A, Fizet J, Poulter M, Wiseman FK, Fisher EM, Tybulewicz VL, Brandner S, Collinge J, Lloyd SE (2012) Overexpression of the Hspa13 (Stch) gene reduces prion disease incubation time in mice. Proc Natl Acad Sci U S A 109:13722–13727

Gu F, Luo W, Li X, Wang Z, Lu S, Zhang M, Zhao B, Zhu S, Feng S, Yan YB, Huang S, Ma X (2008) A novel mutation in AlphaA-crystallin (CRYAA) caused autosomal dominant congenital cataract in a large Chinese family. Hum Mutat 29:769

Gupta MK, Gulick J, Liu R, Wang X, Molkentin JD, Robbins J (2014) Sumo E2 ligase UBC9 is required for efficient protein quality control in cardiomyocytes. Circ Res 115:721

Hagemann TL, Connor JX, Messing A (2006) Alexander disease-associated glial fibrillary acidic protein mutations in mice induce Rosenthal fiber formation and a white matter stress response. J Neurosci 26:11162–11173

Hagemann TL, Boelens WC, Wawrousek EF, Messing A (2009) Suppression of GFAP toxicity by alphaB-crystallin in mouse models of Alexander disease. Hum Mol Genet 18:1190–1199

Hansen L, Yao W, Eiberg H, Kjaer KW, Baggesen K, Hejtmancik JF, Rosenberg T (2007) Genetic heterogeneity in microcornea-cataract: five novel mutations in CRYAA, CRYGD, and GJA8. Invest Ophthalmol Vis Sci 48:3937–3944

Hartl FU, Hayer-Hartl M (2009) Converging concepts of protein folding in vitro and in vivo. Nat Struct Mol Biol 16:574–581

Hishiya A, Salman MN, Carra S, Kampinga HH, Takayama S (2011) BAG3 directly interacts with mutated alphaB-crystallin to suppress its aggregation and toxicity. PLoS One 6:e16828

Hjalmarsson C, Bjerke M, Andersson B, Blennow K, Zetterberg H, Aberg ND, Olsson B, Eckerstrom C, Bokemark L, Wallin A (2014) Neuronal and glia-related biomarkers in cerebrospinal fluid of patients with acute ischemic stroke. J Cent Nerv Syst Dis 6:51–58

Holmgren A, Bouhy D, De Winter V, Asselbergh B, Timmermans JP, Irobi J, Timmerman V (2013) Charcot-Marie-Tooth causing HSPB1 mutations increase Cdk5-mediated phosphorylation of neurofilaments. Acta Neuropathol 126:93–108

Houck SA, Clark JI (2010) Dynamic subunit exchange and the regulation of microtubule assembly by the stress response protein human alphaB crystallin. PLoS One 5:e11795

Houlden H, Laura M, Wavrant-De Vrieze F, Blake J, Wood N, Reilly MM (2008) Mutations in the HSP27 (HSPB1) gene cause dominant, recessive, and sporadic distal HMN/CMT type 2. Neurology 71:1660–1668

Hu WF, Gong L, Cao Z, Ma H, Ji W, Deng M, Liu M, Hu XH, Chen P, Yan Q, Chen HG, Liu J, Sun S, Zhang L, Liu JP, Wawrousek E, Li DW (2012) alphaA- and alphaB-crystallins interact with caspase-3 and Bax to guard mouse lens development. Curr Mol Med 12:177–187

Huang Z, Cheng Y, Chiu PM, Cheung FM, Nicholls JM, Kwong DL, Lee AW, Zabarovsky ER, Stanbridge EJ, Lung HL, Lung ML (2012) Tumor suppressor Alpha B-crystallin (CRYAB) associates with the cadherin/catenin adherens junction and impairs NPC progression-associated properties. Oncogene 31:3709–3720

Ikeda Y, Abe A, Ishida C, Takahashi K, Hayasaka K, Yamada M (2009) A clinical phenotype of distal hereditary motor neuronopathy type II with a novel HSPB1 mutation. J Neurol Sci 277:9–12

Inagaki N, Hayashi T, Arimura T, Koga Y, Takahashi M, Shibata H, Teraoka K, Chikamori T, Yamashina A, Kimura A (2006) Alpha B-crystallin mutation in dilated cardiomyopathy. Biochem Biophys Res Commun 342:379–386

Irobi J, Van Impe K, Seeman P, Jordanova A, Dierick I, Verpoorten N, Michalik A, De Vriendt E, Jacobs A, Van Gerwen V, Vennekens K, Mazanec R, Tournev I, Hilton-Jones D, Talbot K, Kremensky I, Van Den Bosch L, Robberecht W, Van Vandekerckhove J, Broeckhoven C, Gettemans J, De Jonghe P, Timmerman V (2004) Hot-spot residue in small heat-shock protein 22 causes distal motor neuropathy. Nat Genet 36:597–601

Ishiwata T, Orosz A, Wang X, Mustafi SB, Pratt GW, Christians ES, Boudina S, Abel ED, Benjamin IJ (2012) HSPB2 is dispensable for the cardiac hypertrophic response but reduces mitochondrial energetics following pressure overload in mice. PLoS One 7:e42118

Ivanov O, Chen F, Wiley EL, Keswani A, Diaz LK, Memmel HC, Rademaker A, Gradishar WJ, Morrow M, Khan SA, Cryns VL (2008) alphaB-crystallin is a novel predictor of resistance to neoadjuvant chemotherapy in breast cancer. Breast Cancer Res Treat 111:411–417

Jakobs PM, Hess JF, Fitzgerald PG, Kramer P, Weleber RG, Litt M (2000) Autosomal-dominant congenital cataract associated with a deletion mutation in the human beaded filament protein gene BFSP2. Am J Hum Genet 66:1432–1436

James PA, Rankin J, Talbot K (2008) Asymmetrical late onset motor neuropathy associated with a novel mutation in the small heat shock protein HSPB1 (HSP27). J Neurol Neurosurg Psychiatry 79:461–463

Jany PL, Hagemann TL, Messing A (2013) GFAP expression as an indicator of disease severity in mouse models of Alexander disease. ASN Neuro 5:e00109

Jiao Q, Sanbe A, Zhang X, Liu JP, Minamisawa S (2014) alphaB-Crystallin R120G variant causes cardiac arrhythmias and alterations in the expression of Ca(2+) -handling proteins and endoplasmic reticulum stress in mice. Clin Exp Pharmacol Physiol 41:589–599

Jin X, Moskophidis D, Hu Y, Phillips A, Mivechi NF (2009) Heat shock factor 1 deficiency via its downstream target gene alphaB-crystallin (Hspb5) impairs p53 degradation. J Cell Biochem 107:504–515

Jin C, Cleveland JC, Ao L, Li J, Zeng Q, Fullerton DA, Meng X (2014) Human myocardium releases heat shock protein 27 (HSP27) after global ischemia: the proinflammatory effect of extracellular HSP27 through toll-like receptor (TLR)-2 and TLR4. Mol Med 20:280–289

Jonak C, Mildner M, Klosner G, Paulitschke V, Kunstfeld R, Pehamberger H, Tschachler E, Trautinger F (2011) The hsp27kD heat shock protein and p38-MAPK signaling are required for regular epidermal differentiation. J Dermatol Sci 61:32–37

Jordanova A, De Jonghe P, Boerkoel CF, Takashima H, De Vriendt E, Ceuterick C, Martin JJ, Butler IJ, Mancias P, Papasozomenos S, Terespolsky D, Potocki L, Brown CW, Shy M, Rita DA, Tournev I, Kremensky I, Lupski JR, Timmerman V (2003) Mutations in the neurofilament light chain gene (NEFL) cause early onset severe Charcot-Marie-Tooth disease. Brain 126:590–597

Jorgensen A, Foster PP, Brubakk AO, Eftedal I (2013) Effects of hyperbaric oxygen preconditioning on cardiac stress markers after simulated diving. Physiol Rep 1:e00169

Kampinga HH, Garrido C (2012) HSPBs: small proteins with big implications in human disease. Int J Biochem Cell Biol 44:1706–1710

Kamradt MC, Chen F, Cryns VL (2001) The small heat shock protein alpha B-crystallin negatively regulates cytochrome c- and caspase-8-dependent activation of caspase-3 by inhibiting its autoproteolytic maturation. J Biol Chem 276:16059–16063
Kappe G, Franck E, Verschuure P, Boelens WC, Leunissen JA, De Jong WW (2003) The human genome encodes 10 alpha-crystallin-related small heat shock proteins: HspB1-10. Cell Stress Chaperones 8:53–61
Kase S, He S, Sonoda S, Kitamura M, Spee C, Wawrousek E, Ryan SJ, Kannan R, Hinton DR (2010) alphaB-crystallin regulation of angiogenesis by modulation of VEGF. Blood 115:3398–3406
Kato K, Shinohara H, Kurobe N, Goto S, Inaguma Y, Ohshima K (1991a) Immunoreactive alpha A crystallin in rat non-lenticular tissues detected with a sensitive immunoassay method. Biochim Biophys Acta 1080:173–180
Kato K, Shinohara H, Kurobe N, Inaguma Y, Shimizu K, Ohshima K (1991b) Tissue distribution and developmental profiles of immunoreactive aB crystallin in the rat non-lenticular tissues determined with a sensitive immunoassay system. Biochim Biophys Acta 1074:201–208
Kayser J, Haslbeck M, Dempfle L, Krause M, Grashoff C, Buchner J, Herrmann H, Bausch AR (2013) The small heat shock protein Hsp27 affects assembly dynamics and structure of keratin intermediate filament networks. Biophys J 105:1778–1785
Kelley PM, Schlesinger MJ (1978) The effect of amino acid analogues and heat shock on gene expression in chicken embryo fibroblasts. Cell 15:1277–1286
Khan AO, Aldahmesh MA, Meyer B (2007) Recessive congenital total cataract with microcornea and heterozygote carrier signs caused by a novel missense CRYAA mutation (R54C). Am J Ophthalmol 144:949–952
Kijima K, Numakura C, Goto T, Takahashi T, Otagiri T, Umetsu K, Hayasaka K (2005) Small heat shock protein 27 mutation in a Japanese patient with distal hereditary motor neuropathy. J Hum Genet 50:473–476
Kim S, Coulombe PA (2007) Intermediate filament scaffolds fulfill mechanical, organizational, and signaling functions in the cytoplasm. Genes Dev 21:1581–1597
Klopstein A, Santos-Nogueira E, Francos-Quijorna I, Redensek A, David S, Navarro X, Lopez-Vales R (2012) Beneficial effects of alphaB-crystallin in spinal cord contusion injury. J Neurosci 32:14478–14488
Kolb SJ, Snyder PJ, Poi EJ, Renard EA, Bartlett A, Gu S, Sutton S, Arnold WD, Freimer ML, Lawson VH, Kissel JT, Prior TW (2010) Mutant small heat shock protein B3 causes motor neuropathy: utility of a candidate gene approach. Neurology 74:502–506
Kopito RR (2000) Aggresomes, inclusion bodies and protein aggregation. Trends Cell Biol 10:524–530
Kotter S, Unger A, Hamdani N, Lang P, Vorgerd M, Nagel-Steger L, Linke WA (2014) Human myocytes are protected from titin aggregation-induced stiffening by small heat shock proteins. J Cell Biol 204:187–202
Koyama Y, Goldman JE (1999) Formation of GFAP cytoplasmic inclusions in astrocytes and their disaggregation by alphaB-crystallin. Am J Pathol 154:1563–1572
Kozakov D, Li K, Hall DR, Beglov D, Zheng J, Vakili P, Schueler-Furman O, Paschalidis I, Clore GM, Vajda S (2014) Encounter complexes and dimensionality reduction in protein-protein association. Elife 3:e01370
Kozawa O, Matsuno H, Niwa M, Hatakeyama D, Kato K, Uematsu T (2001) AlphaB-crystallin, a low-molecular-weight heat shock protein, acts as a regulator of platelet function. Cell Stress Chaperones 6:21–28
Kozawa O, Matsuno H, Niwa M, Hatakeyama D, Oiso Y, Kato K, Uematsu T (2002) HSP20, low-molecular-weight heat shock-related protein, acts extracellularly as a regulator of platelet functions: a novel defense mechanism. Life Sci 72:113–124
Ku NO, Strnad P, Zhong BH, Tao GZ, Omary MB (2007) Keratins let liver live: mutations predispose to liver disease and crosslinking generates Mallory-Denk bodies. Hepatology 46:1639–1649

Kumarapeli AR, Su H, Huang W, Tang M, Zheng H, Horak KM, Li M, Wang X (2008) Alpha B-crystallin suppresses pressure overload cardiac hypertrophy. Circ Res 103:1473–1482

Landsbury A, Perng MD, Pohl E, Quinlan RA (2010) Chapter 1.2: Functional symbiosis between the intermediate filament cytoskeleton and small heat shock proteins. In: Simon S, Arrigo AP (eds) Small stress proteins and human diseases. Nova Science, Hauppauge

Laurie KJ, Dave A, Straga T, Souzeau E, Chataway T, Sykes MJ, Casey T, Teo T, Pater J, Craig JE, Sharma S, Burdon KP (2013) Identification of a novel oligomerization disrupting mutation in CRYAlphaA associated with congenital cataract in a South Australian family. Hum Mutat 34:435–438

Lee YJ, Lee HJ, Choi SH, Jin YB, An HJ, Kang JH, Yoon SS, Lee YS (2012) Soluble HSPB1 regulates VEGF-mediated angiogenesis through their direct interaction. Angiogenesis 15:229–242

Leterrier JF, Kas J, Hartwig J, Vegners R, Janmey PA (1996) Mechanical effects of neurofilament cross-bridges. Modulation by phosphorylation, lipids, and interactions with F-actin. J Biol Chem 271:15687–15694

Li ZL, Mericskay M, Agbulut O, Butlerbrowne G, Carlsson L, Thornell LE, Babinet C, Paulin D (1997) Desmin is essential for the tensile strength and integrity of myofibrils but not for myogenic commitment, differentiation, and fusion of skeletal muscle. J Cell Biol 139:129–144

Li L, Lundkvist A, Andersson D, Wilhelmsson U, Nagai N, Pardo AC, Nodin C, Stahlberg A, Aprico K, Larsson K, Yabe T, Moons L, Fotheringham A, Davies I, Carmeliet P, Schwartz JP, Pekna M, Kubista M, Blomstrand F, Maragakis N, Nilsson M, Pekny M (2008) Protective role of reactive astrocytes in brain ischemia. J Cereb Blood Flow Metab 28:468–481

Li J, Horak KM, Su H, Sanbe A, Robbins J, Wang X (2011) Enhancement of proteasomal function protects against cardiac proteinopathy and ischemia/reperfusion injury in mice. J Clin Invest 121:3689–3700

Lin DI, Barbash O, Kumar KG, Weber JD, Harper JW, Klein-Szanto AJ, Rustgi A, Fuchs SY, Diehl JA (2006) Phosphorylation-dependent ubiquitination of cyclin D1 by the SCF(FBX4-alphaB crystallin) complex. Mol Cell 24:355–366

Lin KP, Soong BW, Yang CC, Huang LW, Chang MH, Lee IH, Antonellis A, Lee YC (2011) The mutational spectrum in a cohort of Charcot-Marie-Tooth disease type 2 among the Han Chinese in Taiwan. PLoS One 6:e29393

Litt M, Kramer P, Lamorticella DM, Murphey W, Lovrien EW, Weleber RG (1998) Autosomal dominant congenital cataract associated with a missense mutation in the human alpha crystallin gene CRYAA. Hum Mol Genet 7:471–474

Liu J, Tang M, Mestril R, Wang X (2006a) Aberrant protein aggregation is essential for a mutant desmin to impair the proteolytic function of the ubiquitin-proteasome system in cardiomyocytes. J Mol Cell Cardiol 40:451–454

Liu M, Ke T, Wang Z, Yang Q, Chang W, Jiang F, Tang Z, Li H, Ren X, Wang X, Wang T, Li Q, Yang J, Liu J, Wang QK (2006b) Identification of a CRYAB mutation associated with autosomal dominant posterior polar cataract in a Chinese family. Invest Ophthalmol Vis Sci 47:3461–3466

Liu Y, Zhang X, Luo L, Wu M, Zeng R, Cheng G, Hu B, Liu B, Liang JJ, Shang F (2006c) A novel alphaB-crystallin mutation associated with autosomal dominant congenital lamellar cataract. Invest Ophthalmol Vis Sci 47:1069–1075

Liu S, Li J, Tao Y, Xiao X (2007) Small heat shock protein alphaB-crystallin binds to p53 to sequester its translocation to mitochondria during hydrogen peroxide-induced apoptosis. Biochem Biophys Res Commun 354:109–114

Liu Q, Wang KJ, Zhu SQ (2014) A novel p.G112E mutation in BFSP2 associated with autosomal dominant pulverulent cataract with sutural opacities. Curr Eye Res 39:1013–1019

Lowe J, Blanchard A, Morrell K, Lennox G, Reynolds L, Billet M, Landon M, Mayer RJ (1988) Ubiquitin is a common factor in intermediate filament inclusion bodies of diverse type in man, including those of Parkinson's disease, Pick's disease and Alzheimer's disease, as well as rosenthal fibres in cerebellar astrocytomas, cytoplasmic bodies in muscles, and mallory bodies in alcoholic liver disease. J Pathol 155:9–15

Luigetti M, Fabrizi GM, Madia F, Ferrarini M, Conte A, Del Grande A, Tasca G, Tonali PA, Sabatelli M (2010) A novel HSPB1 mutation in an Italian patient with CMT2/dHMN phenotype. J Neurol Sci 298:114–117

Lundkvist A, Reichenbach A, Betsholtz C, Carmeliet P, Wolburg H, Pekny M (2004) Under stress, the absence of intermediate filaments from Muller cells in the retina has structural and functional consequences. J Cell Sci 117:3481–3488

Ma X, Li FF, Wang SZ, Gao C, Zhang M, Zhu SQ (2008) A new mutation in BFSP2 (G1091A) causes autosomal dominant congenital lamellar cataracts. Mol Vis 14:1906–1911

Ma Y, Peng J, Liu W, Zhang P, Huang L, Gao B, Shen T, Zhou Y, Chen H, Chu Z, Zhang M, Qin H (2009) Proteomics identification of desmin as a potential oncofetal diagnostic and prognostic biomarker in colorectal cancer. Mol Cell Proteomics 8:1878–1890

Mackay DS, Andley UP, Shiels A (2003) Cell death triggered by a novel mutation in the alphaA-crystallin gene underlies autosomal dominant cataract linked to chromosome 21q. Eur J Hum Genet 11:784–793

Maddala R, Rao VP (2005) alpha-Crystallin localizes to the leading edges of migrating lens epithelial cells. Exp Cell Res 306:203–215

Magin TM, Jorcano JL, Franke WW (1986) Cytokeratin expression in simple epithelia. II cDNA cloning and sequence characteristics of bovine cytokeratin A (no.8). Differentiation 30:254–264

Malhas AN, Lee CF, Vaux DJ (2009) Lamin B1 controls oxidative stress responses via Oct-1. J Cell Biol 184:45–55

Maloyan A, Sanbe A, Osinska H, Westfall M, Robinson D, Imahashi K, Murphy E, Robbins J (2005) Mitochondrial dysfunction and apoptosis underlie the pathogenic process in alpha-B-crystallin desmin-related cardiomyopathy. Circulation 112:3451–3461

Maloyan A, Sayegh J, Osinska H, Chua BH, Robbins J (2010) Manipulation of death pathways in desmin-related cardiomyopathy. Circ Res 106:1524–1532

Mao YW, Xiang H, Wang J, Korsmeyer S, Reddan J, Li DW (2001) Human bcl-2 gene attenuates the ability of rabbit lens epithelial cells against H2O2-induced apoptosis through down-regulation of the alpha B-crystallin gene. J Biol Chem 276:43435–43445

Mao YW, Liu JP, Xiang H, Li DW (2004) Human alphaA- and alphaB-crystallins bind to Bax and Bcl-X(S) to sequester their translocation during staurosporine-induced apoptosis. Cell Death Differ 11:512–526

Martin JL, Mestril R, Hilal-Dandan R, Brunton LL, Dillmann WH (1997) Small heat shock proteins and protection against ischemic injury in cardiac myocytes. Circulation 96:4343–4348

Mashukova A, Oriolo AS, Wald FA, Casanova ML, Kroger C, Magin TM, Omary MB, Salas PJ (2009) Rescue of atypical protein kinase C in epithelia by the cytoskeleton and Hsp70 family chaperones. J Cell Sci 122:2491–2503

Mashukova A, Kozhekbaeva Z, Forteza R, Dulam V, Figueroa Y, Warren R, Salas PJ (2014) The BAG-1 isoform BAG-1M regulates keratin-associated Hsp70 chaperoning of aPKC in intestinal cells during activation of inflammatory signaling. J Cell Sci 127:3568–3577

Masilamoni JG, Jesudason EP, Bharathi SN, Jayakumar R (2005) The protective effect of alpha-crystallin against acute inflammation in mice. Biochim Biophys Acta 1740:411–420

Masilamoni JG, Jesudason EP, Baben B, Jebaraj CE, Dhandayuthapani S, Jayakumar R (2006) Molecular chaperone alpha-crystallin prevents detrimental effects of neuroinflammation. Biochim Biophys Acta 1762:284–293

Mcgreal RS, Kantorow WL, Chauss DC, Wei J, Brennan LA, Kantorow M (2012) alphaB-crystallin/sHSP protects cytochrome c and mitochondrial function against oxidative stress in lens and retinal cells. Biochim Biophys Acta 1820:921–930

Meehan S, Knowles TP, Baldwin AJ, Smith JF, Squires AM, Clements P, Treweek TM, Ecroyd H, Tartaglia GG, Vendruscolo M, Macphee CE, Dobson CM, Carver JA (2007) Characterisation of amyloid fibril formation by small heat-shock chaperone proteins human alphaA-, alphaB- and R120G alphaB-crystallins. J Mol Biol 372:470–484

Melkani GC, Cammarato A, Bernstein SI (2006) alphaB-crystallin maintains skeletal muscle myosin enzymatic activity and prevents its aggregation under heat-shock stress. J Mol Biol 358:635–645

Mersiyanova IV, Perepelov AV, Polyakov AV, Sitnikov VF, Dadali EL, Oparin RB, Petrin AN, Evgrafov OV (2000) A new variant of Charcot-Marie-Tooth disease type 2 is probably the result of a mutation in the neurofilament-light gene. Am J Hum Genet 67:37–46

Middeldorp J, Kamphuis W, Sluijs JA, Achoui D, Leenaars CH, Feenstra MG, Van Tijn P, Fischer DF, Berkers C, Ovaa H, Quinlan RA, Hol EM (2009) Intermediate filament transcription in astrocytes is repressed by proteasome inhibition. FASEB J 23:2710–2726

Milner DJ, Mavroidis M, Weisleder N, Capetanaki Y (2000) Desmin cytoskeleton linked to muscle mitochondrial distribution and respiratory function. J Cell Biol 150:1283–1298

Morrison LE, Whittaker RJ, Klepper RE, Wawrousek EF, Glembotski CC (2004) Roles for alphaB-crystallin and HSPB2 in protecting the myocardium from ischemia-reperfusion-induced damage in a KO mouse model. Am J Physiol Heart Circ Physiol 286:H847–H855

Moyano JV, Evans JR, Chen F, Lu M, Werner ME, Yehiely F, Diaz LK, Turbin D, Karaca G, Wiley E, Nielsen TO, Perou CM, Cryns VL (2006) AlphaB-crystallin is a novel oncoprotein that predicts poor clinical outcome in breast cancer. J Clin Invest 116:261–270

Müller M, Bhattacharya SS, Moore T, Prescott Q, Wedig T, Herrmann H, Magin TM (2009) Dominant cataract formation in association with a vimentin assembly disrupting mutation. Hum Mol Genet 18:1052–1057

Nahomi RB, Wang B, Raghavan CT, Voss O, Doseff AI, Santhoshkumar P, Nagaraj RH (2013) Chaperone peptides of alpha-crystallin inhibit epithelial cell apoptosis, protein insolubilization, and opacification in experimental cataracts. J Biol Chem 288:13022–13035

Nakhro K, Park JM, Kim YJ, Yoon BR, Yoo JH, Koo H, Choi BO, Chung KW (2013) A novel Lys141Thr mutation in small heat shock protein 22 (HSPB8) gene in Charcot-Marie-Tooth disease type 2L. Neuromuscul Disord 23:656–663

Narayan P, Meehan S, Carver JA, Wilson MR, Dobson CM, Klenerman D (2012) Amyloid-beta oligomers are sequestered by both intracellular and extracellular chaperones. Biochemistry 51:9270–9276

Nawashiro H, Messing A, Azzam N, Brenner M (1998) Mice lacking GFAP are hypersensitive to traumatic cerebrospinal injury. Neuroreport 9:1691–1696

Ngai J, Coleman TR, Lazarides E (1990) Localization of newly synthesized vimentin subunits reveals a novel mechanism of intermediate filament assembly. Cell 60:415–427

Nicholl ID, Quinlan RA (1994) Chaperone activity of a-crystallins modulates intermediate filament assembly. EMBO J 13:945–953

Nicolaou P, Knoll R, Haghighi K, Fan GC, Dorn GW 2nd, Hasenfub G, Kranias EG (2008) Human mutation in the anti-apoptotic heat shock protein 20 abrogates its cardioprotective effects. J Biol Chem 283:33465–33471

O'shaughnessy RF, Welti JC, Cooke JC, Avilion AA, Monks B, Birnbaum MJ, Byrne C (2007) AKT-dependent HspB1 (Hsp27) activity in epidermal differentiation. J Biol Chem 282:17297–17305

Ohto-Fujita E, Fujita Y, Atomi Y (2007) Analysis of the alphaB-crystallin domain responsible for inhibiting tubulin aggregation. Cell Stress Chaperones 12:163–171

Ojha J, Karmegam RV, Masilamoni JG, Terry AV, Cashikar AG (2011) Behavioral defects in chaperone-deficient Alzheimer's disease model mice. PLoS One 6:e16550

Omary MB (2009) "IF-pathies": a broad spectrum of intermediate filament-associated diseases. J Clin Invest 119:1756–1762

Ousman SS, Tomooka BH, Van Noort JM, Wawrousek EF, O'Connor KC, Hafler DA, Sobel RA, Robinson WH, Steinman L (2007) Protective and therapeutic role for alphaB-crystallin in autoimmune demyelination. Nature 448:474–479

Pan X, Hobbs RP, Coulombe PA (2013) The expanding significance of keratin intermediate filaments in normal and diseased epithelia. Curr Opin Cell Biol 25:47–56

Pattison JS, Osinska H, Robbins J (2011) Atg7 induces basal autophagy and rescues autophagic deficiency in CryABR120G cardiomyocytes. Circ Res 109:151–160

Pegtel DM, Peferoen L, Amor S (2014) Extracellular vesicles as modulators of cell-to-cell communication in the healthy and diseased brain. Philos Trans R Soc Lond B Biol Sci 369:20130516

Perlson E, Hanz S, Ben-Yaakov K, Segal-Ruder Y, Seger R, Fainzilber M (2005) Vimentin-dependent spatial translocation of an activated MAP kinase in injured nerve. Neuron 45:715–726

Perng MD, Cairns L, Van Den IJssel P, Prescott A, Hutcheson AM, Quinlan RA (1999a) Intermediate filament interactions can be altered by HSP27 and alphaB-crystallin. J Cell Sci 112(Pt 13):2099–2112

Perng MD, Muchowski PJ, Van Den IJssel P, Wu GJS, Clark JI, Quinlan RA (1999b) The cardiomyopathy and lens cataract mutation in aB-crystallin compromises secondary, tertiary and quaternary protein structure and reduces in vitro chaperone activity. J Biol Chem 274:33235–33243

Perng MD, Wen SF, Van Den IJssel P, Prescott AR, Quinlan RA (2004) Desmin aggregate formation by R120G alphaB-crystallin is caused by altered filament interactions and is dependent upon network status in cells. Mol Biol Cell 15:2335–2346

Pras E, Frydman M, Levy-Nissenbaum E, Bakhan T, Raz J, Assia EI, Goldman B (2000) A nonsense mutation (W9X) in CRYAA causes autosomal recessive cataract in an inbred Jewish Persian family. Invest Ophthalmol Vis Sci 41:3511–3515

Quinlan R (2002) Cytoskeletal competence requires protein chaperones. Prog Mol Subcell Biol 28:219–233

Quinlan RA, Ellis RJ (2013) Chaperones: needed for both the good times and the bad times. Philos Trans R Soc Lond B Biol Sci 368:20130091

Quinlan R, Van Den IJSSEL P (1999) Fatal attraction: when chaperone turns harlot [news]. Nat Med 5:25–26

Rajasekaran NS, Connell P, Christians ES, Yan LJ, Taylor RP, Orosz A, Zhang XQ, Stevenson TJ, Peshock RM, Leopold JA, Barry WH, Loscalzo J, Odelberg SJ, Benjamin IJ (2007) Human alpha B-crystallin mutation causes oxido-reductive stress and protein aggregation cardiomyopathy in mice. Cell 130:427–439

Ramachandran RD, Perumalsamy V, Hejtmancik JF (2007) Autosomal recessive juvenile onset cataract associated with mutation in BFSP1. Hum Genet 121:475–482

Ranjan M, Nayak S, Kosuri T, Rao BS (2008) Immunochemical detection of glycated lens crystallins and their circulating autoantibodies in human serum during aging. Mol Vis 14:2056–2066

Raposo G, Stoorvogel W (2013) Extracellular vesicles: exosomes, microvesicles, and friends. J Cell Biol 200:373–383

Reilich P, Schoser B, Schramm N, Krause S, Schessl J, Kress W, Muller-Hocker J, Walter MC, Lochmuller H (2010) The p.G154S mutation of the alpha-B crystallin gene (CRYAB) causes late-onset distal myopathy. Neuromuscul Disord 20:255–259

Richter L, Flodman P, Barria Von-Bischhoffshausen F, Burch D, Brown S, Nguyen L, Turner J, Spence MA, Bateman JB (2008) Clinical variability of autosomal dominant cataract, microcornea and corneal opacity and novel mutation in the alpha A crystallin gene (CRYAA). Am J Med Genet A 146:833–842

Robitaille H, Simard-Bisson C, Larouche D, Tanguay RM, Blouin R, Germain L (2010) The small heat-shock protein Hsp27 undergoes ERK-dependent phosphorylation and redistribution to the cytoskeleton in response to dual leucine zipper-bearing kinase expression. J Invest Dermatol 130:74–85

Ross CA, Poirier MA (2005) Opinion: what is the role of protein aggregation in neurodegeneration? Nat Rev Mol Cell Biol 6:891–898

Rossor AM, Davidson GL, Blake J, Polke JM, Murphy SM, Houlden H, Innes A, Kalmar B, Greensmith L, Reilly MM (2012) A novel p.Glu175X premature stop mutation in the C-terminal end of HSP27 is a cause of CMT2. J Peripher Nerv Syst 17:201–205
Rothbard JB, Kurnellas MP, Brownell S, Adams CM, Su L, Axtell RC, Chen R, Fathman CG, Robinson WH, Steinman L (2012) Therapeutic effects of systemic administration of chaperone alphaB-crystallin associated with binding proinflammatory plasma proteins. J Biol Chem 287:9708–9721
Rubinsztein DC (2006) The roles of intracellular protein-degradation pathways in neurodegeneration. Nature 443:780–786
Sacconi S, Feasson L, Antoine JC, Pecheux C, Bernard R, Cobo AM, Casarin A, Salviati L, Desnuelle C, Urtizberea A (2012) A novel CRYAB mutation resulting in multisystemic disease. Neuromuscul Disord 22:66–72
Safieh LA, Khan AO, Alkuraya FS (2009) Identification of a novel CRYAB mutation associated with autosomal recessive juvenile cataract in a Saudi family. Mol Vis 15:980–984
Saha B, Zitnik G, Johnson S, Nguyen Q, Risques RA, Martin GM, Oshima J (2013) DNA damage accumulation and TRF2 degradation in atypical Werner syndrome fibroblasts with LMNA mutations. Front Genet 4:129
Sanbe A, Osinska H, Saffitz JE, Glabe CG, Kayed R, Maloyan A, Robbins J (2004) Desmin-related cardiomyopathy in transgenic mice: a cardiac amyloidosis. Proc Natl Acad Sci U S A 101:10132–10136
Sanbe A, Osinska H, Villa C, Gulick J, Klevitsky R, Glabe CG, Kayed R, Robbins J (2005) Reversal of amyloid-induced heart disease in desmin-related cardiomyopathy. Proc Natl Acad Sci U S A 102:13592–13597
Santhiya ST, Soker T, Klopp N, Illig T, Prakash MV, Selvaraj B, Gopinath PM, Graw J (2006) Identification of a novel, putative cataract-causing allele in CRYAA (G98R) in an Indian family. Mol Vis 12:768–773
Schlossarek S, Carrier L (2011) The ubiquitin-proteasome system in cardiomyopathies. Curr Opin Cardiol 26:190–195
Schroder R, Goudeau B, Simon MC, Fischer D, Eggermann T, Clemen CS, Li Z, Reimann J, Xue Z, Rudnik-Schoneborn S, Zerres K, Van Der Ven PF, Furst DO, Kunz WS, Vicart P (2003) On noxious desmin: functional effects of a novel heterozygous desmin insertion mutation on the extrasarcomeric desmin cytoskeleton and mitochondria. Hum Mol Genet 12:657–669
Seco-Cervera M, Spis M, Garcia-Gimenez JL, Ibanez-Cabellos JS, Velazquez-Ledesma A, Esmoris I, Banuls S, Perez-Machado G, Pallardo FV (2014) Oxidative stress and antioxidant response in fibroblasts from Werner and atypical Werner syndromes. Aging (Albany NY) 6:231–245
Selcen D (2011) Myofibrillar myopathies. Neuromuscul Disord 21:161–171
Selcen D, Engel AG (2003) Myofibrillar myopathy caused by novel dominant negative alpha B-crystallin mutations. Ann Neurol 54:804–810
Shao W, Zhang SZ, Tang M, Zhang XH, Zhou Z, Yin YQ, Zhou, QB, Huang YY, Liu YJ, Wawrousek E, Chen T, Li SB, Xu M, Zhou JN, Hu G, Zhou JW (2012) Suppression of neuroinflammation by astrocytic dopamine D2 receptors via alphaB-crystallin. Nature 494:90–94
Shao W, Zhang SZ, Tang M, Zhang XH, Zhou Z, Yin YQ, Zhou QB, Huang YY, Liu YJ, Wawrousek E, Chen T, Li SB, Xu M, Zhou JN, Hu G, Zhou JW (2013) Suppression of neuroinflammation by astrocytic dopamine D2 receptors via alphaB-crystallin. Nature 494:90–94
Shimi T, Goldman RD (2014) Nuclear lamins and oxidative stress in cell proliferation and longevity. Adv Exp Med Biol 773:415–430
Shimi T, Butin-Israeli V, Adam SA, Hamanaka RB, Goldman AE, Lucas CA, Shumaker DK, Kosak ST, Chandel NS, Goldman RD (2011) The role of nuclear lamin B1 in cell proliferation and senescence. Genes Dev 25:2579–2593
Shyy T-T, Asch BB, Asch HL (1989) Concurrent collapse of keratin filaments, aggregation of organelles, and inhibition of protein synthesis during the heat shock response in mammary epithelial cells. J Cell Biol 108:997–1008

Singh BN, Rao KS, Ramakrishna T, Rangaraj N, Rao CHM (2007) Association of alphaB-crystallin, a small heat shock protein, with actin: role in modulating actin filament dynamics in vivo. J Mol Biol 366:756–767

Singh BN, Rao KS, Rao CHM (2010) Ubiquitin-proteasome-mediated degradation and synthesis of MyoD is modulated by alphaB-crystallin, a small heat shock protein, during muscle differentiation. Biochim Biophys Acta 1803:288–299

Skouri-Panet F, Michiel M, Ferard C, Duprat E, Finet S (2012) Structural and functional specificity of small heat shock protein HspB1 and HspB4, two cellular partners of HspB5: role of the in vitro hetero-complex formation in chaperone activity. Biochimie 94:975–984

Slater A, Cato AC, Sillar GM, Kioussis J, Burdon RH (1981) The pattern of protein synthesis induced by heat shock of HeLa cells. Eur J Biochem 117:341–346

Snider NT, Omary MB (2014) Post-translational modifications of intermediate filament proteins: mechanisms and functions. Nat Rev Mol Cell Biol 15:163–177

Song S, Hanson MJ, Liu BF, Chylack LT, Liang JJ (2008) Protein-protein interactions between lens vimentin and alphaB-crystallin using FRET acceptor photobleaching. Mol Vis 14:1282–1287

Soumaka E, Rapti K, Diokmetzidou A, Milner DJ, Mavroidis M, Capetanaki Y (2008) alpha B-Crystallin and HSP25 mediate protection of the desmin null myocardium. J Mol Cell Cardiol 44:793

Sreekumar PG, Kannan R, Kitamura M, Spee C, Barron E, Ryan SJ, Hinton DR (2010) alphaB crystallin is apically secreted within exosomes by polarized human retinal pigment epithelium and provides neuroprotection to adjacent cells. PLoS One 5:e12578

Stengel F, Baldwin AJ, Painter AJ, Jaya N, Basha E, Kay LE, Vierling E, Robinson CV, Benesch JL (2010) Quaternary dynamics and plasticity underlie small heat shock protein chaperone function. Proc Natl Acad Sci U S A 107:2007–2012

Strnad P, Paschke S, Jang KH, Ku NO (2012) Keratins: markers and modulators of liver disease. Curr Opin Gastroenterol 28:209–216

Su H, Wang X (2010) The ubiquitin-proteasome system in cardiac proteinopathy: a quality control perspective. Cardiovasc Res 85:253–262

Su D, Guo Y, Li Q, Guan L, Zhu S, Ma X (2012) A novel mutation in CRYAA is associated with autosomal dominant suture cataracts in a Chinese family. Mol Vis 18:3057–3063

Sullivan T, Escalante-Alcalde D, Bhatt H, Anver M, Bhat N, Nagashima K, Stewart CL, Burke B (1999) Loss of A-type lamin expression compromises nuclear envelope integrity leading to muscular dystrophy. J Cell Biol 147:913–920

Sun W, Xiao X, Li S, Guo X, Zhang Q (2011) Mutation analysis of 12 genes in Chinese families with congenital cataracts. Mol Vis 17:2197–2206

Sur R, Lyte PA, Southall MD (2008) Hsp27 regulates pro-inflammatory mediator release in keratinocytes by modulating NF-kappaB signaling. J Invest Dermatol 128:1116–1122

Svensson KJ, Christianson HC, Wittrup A, Bourseau-Guilmain E, Lindqvist E, Svensson LM, Morgelin M, Belting M (2013) Exosome uptake depends on ERK1/2-heat shock protein 27 signaling and lipid Raft-mediated endocytosis negatively regulated by caveolin-1. J Biol Chem 288:17713–17724

Tang B, Liu X, Zhao G, Luo W, Xia K, Pan Q, Cai F, Hu Z, Zhang C, Chen B, Zhang F, Shen L, Zhang R, Jiang H (2005a) Mutation analysis of the small heat shock protein 27 gene in chinese patients with Charcot-Marie-Tooth disease. Arch Neurol 62:1201–1207

Tang BS, Zhao GH, Luo W, Xia K, Cai F, Pan Q, Zhang RX, Zhang FF, Liu XM, Chen B, Zhang C, Shen L, Jiang H, Long ZG, Dai HP (2005b) Small heat-shock protein 22 mutated in autosomal dominant Charcot-Marie-Tooth disease type 2L. Hum Genet 116:222–224

Tang Q, Liu YF, Zhu XJ, Li YH, Zhu J, Zhang JP, Feng ZQ, Guan XH (2009) Expression and prognostic significance of the alpha B-crystallin gene in human hepatocellular carcinoma. Hum Pathol 40:300–305

Tannous P, Zhu H, Nemchenko A, Berry JM, Johnstone JL, Shelton JM, Miller FJ Jr, Rothermel BA, Hill JA (2008) Intracellular protein aggregation is a proximal trigger of cardiomyocyte autophagy. Circulation 117:3070–3078

Thedieck C, Kalbacher H, Kratzer U, Lammers R, Stevanovic S, Klein G (2008) alpha B-crystallin is a cytoplasmic interaction partner of the kidney-specific cadherin-16. J Mol Biol 378:145–153
Tintignac LA, Lagirand J, Batonnet S, Sirri V, Leibovitch MP, Leibovitch SA (2005) Degradation of MyoD mediated by the SCF (MAFbx) ubiquitin ligase. J Biol Chem 280:2847–2856
Toivola DM, Strnad P, Habtezion A, Omary MB (2010) Intermediate filaments take the heat as stress proteins. Trends Cell Biol 20:79–91
Tsunemi T, Ashe TD, Morrison BE, Soriano KR, Au J, Roque RA, Lazarowski ER, Damian VA, Masliah E, La Spada AR (2012) PGC-1alpha rescues Huntington's disease proteotoxicity by preventing oxidative stress and promoting TFEB function. Sci Transl Med 4:142ra97
Tue NT, Shimaji K, Tanaka N, Yamaguchi M (2012) Effect of alphaB-crystallin on protein aggregation in Drosophila. J Biomed Biotechnol 2012:252049
Tyson EK, Macintyre DA, Smith R, Chan EC, Read M (2008) Evidence that a protein kinase A substrate, small heat-shock protein 20, modulates myometrial relaxation in human pregnancy. Endocrinology 149:6157–6165
Van Oosten-Hawle P, Morimoto RI (2014) Organismal proteostasis: role of cell-nonautonomous regulation and transcellular chaperone signaling. Genes Dev 28:1533–1543
Vanita V, Singh JR, Hejtmancik JF, Nuernberg P, Hennies HC, Singh D, Sperling K (2006) A novel fan-shaped cataract-microcornea syndrome caused by a mutation of CRYAA in an Indian family. Mol Vis 12:518–522
Verstraeten VL, Caputo S, Van Steensel MA, Duband-Goulet I, Zinn-Justin S, Kamps M, Kuijpers HJ, Ostlund C, Worman HJ, Briede JJ, Le Dour C, Marcelis CL, Van Geel M, Steijlen PM, Van Den Wijngaard A, Ramaekers FC, Broers JL (2009) The R439C mutation in LMNA causes lamin oligomerization and susceptibility to oxidative stress. J Cell Mol Med 13:959–971
Vicart P, Caron A, Guicheney P, Li Z, Prevost MC, Faure A, Chateau D, Chapon F, Tome F, Dupret JM, Paulin D, Fardeau M (1998) A missense mutation in the alphaB-crystallin chaperone gene causes a desmin-related myopathy. Nat Genet 20:92–95
Vilaboa NE, Garcia-Bermejo L, Perez C, De Blas E, Calle C, Aller P (1997) Heat-shock and cadmium chloride increase the vimentin mRNA and protein levels in U-937 human promonocytic cells. J Cell Sci 110(Pt 2):201–207
Volkmann J, Reuning U, Rudelius M, Hafner N, Schuster T, Becker V Ros A, Weimer J, Hilpert F, Kiechle M, Durst M, Arnold N, Schmalfeldt B, Meindl A, Ramser J (2012) High expression of crystallin alphaB represents an independent molecular marker for unfavourable ovarian cancer patient outcome and impairs TRAIL- and cisplatin-induced apoptosis in human ovarian cancer cells. Int J Cancer 132:2820–2832
Wang X, Robbins J (2006) Heart failure and protein quality control. Circ Res 99:1315–1328
Wang X, Osinska H, Dorn GW 2nd, Nieman M, Lorenz JN, Gerdes AM, Witt S, Kimball T, Gulick J, Robbins J (2001a) Mouse model of desmin-related cardiomyopathy. Circulation 103:2402–2407
Wang X, Osinska H, Klevitsky R, Gerdes AM, Nieman M, Lorenz J, Hewett T, Robbins J (2001b) Expression of R120G-alphaB-crystallin causes aberrant desmin and alphaB-crystallin aggregation and cardiomyopathy in mice. Circ Res 89:84–91
Wang X, Klevitsky R, Huang W, Glasford J, Li F, Robbins J (2003) AlphaB-crystallin modulates protein aggregation of abnormal desmin. Circ Res 93:998–1005
Wang H, Zhang T, Wu D, Zhang J (2013) A novel beaded filament structural protein 1 (BFSP1) gene mutation associated with autosomal dominant congenital cataract in a Chinese family. Mol Vis 19:2590–2595
Watanabe G, Kato S, Nakata H, Ishida T, Ohuchi N, Ishioka C (2009) alphaB-crystallin: a novel p53-target gene required for p53-dependent apoptosis. Cancer Sci 100:2368–2375
Waudby CA, Knowles TP, Devlin GL, Skepper JN, Ecroyd H, Carver JA, Welland ME, Christodoulou J, Dobson CM, Meehan S (2010) The interaction of alphaB-crystallin with mature alpha-synuclein amyloid fibrils inhibits their elongation. Biophys J 98:843–851
Wawrousek EF, Brady JP (1998) aB-crystallin gene knockout mice develop a severe fatal phenotype late in life. Invest Ophthalmol Vis Sci 39:2400

Weerasinghe SV, Ku NO, Altshuler PJ, Kwan R, Omary MB (2014) Mutation of caspase-digestion sites in keratin 18 interferes with filament reorganization, and predisposes to hepatocyte necrosis and loss of membrane integrity. J Cell Sci 127:1464–1475

Weisleder N, Taffet GE, Capetanaki Y (2004) Bcl-2 overexpression corrects mitochondrial defects and ameliorates inherited desmin null cardiomyopathy. Proc Natl Acad Sci U S A 101:769–774

Welch WJ, Mizzen LA (1988) Characterization of the thermotolerant cell. II. Effects on the intracellular distribution of heat-shock protein 70, intermediate filaments, and small nuclear ribonucleoprotein complexes. J Cell Biol 106:974–981

Wignes JA, Goldman JW, Weihl CC, Bartley MG, Andley UP (2013) p62 expression and autophagy in alphaB-crystallin R120G mutant knock-in mouse model of hereditary cataract. Exp Eye Res 115:263–273

Willis MS, Townley-Tilson WH, Kang EY, Homeister JW, Patterson C (2010) Sent to destroy: the ubiquitin proteasome system regulates cell signaling and protein quality control in cardiovascular development and disease. Circ Res 106:463–478

Yang K, Meinhardt A, Zhang B, Grzmil P, Adham IM, Hoyer-Fender S (2012) The small heat shock protein ODF1/HSPB10 is essential for tight linkage of sperm head to tail and male fertility in mice. Mol Cell Biol 32:216–225

Yang Z, Su D, Li Q, Ma Z, Yang F, Zhu S, Ma X (2013) A R54L mutation of CRYAA associated with autosomal dominant nuclear cataracts in a Chinese family. Curr Eye Res 38:1221–1228

Yaung J, Kannan R, Wawrousek EF, Spee C, Sreekumar PG, Hinton DR (2008) Exacerbation of retinal degeneration in the absence of alpha crystallins in an in vivo model of chemically induced hypoxia. Exp Eye Res 86:355–365

Yum SW, Zhang J, Mo K, Li J, Scherer SS (2009) A novel recessive Nefl mutation causes a severe, early-onset axonal neuropathy. Ann Neurol 66:759–770

Zhai J, Lin H, Julien JP, Schlaepfer WW (2007) Disruption of neurofilament network with aggregation of light neurofilament protein: a common pathway leading to motor neuron degeneration due to Charcot-Marie-Tooth disease-linked mutations in NFL and HSPB1. Hum Mol Genet 16:3103–3116

Zhang Q, Guo X, Xiao X, Yi J, Jia X, Hejtmancik JF (2004) Clinical description and genome wide linkage study of Y-sutural cataract and myopia in a Chinese family. Mol Vis 10:890–900

Zhang L, Gao L, Li Z, Qin W, Gao W, Cui X, Feng G, Fu S, He L, Liu P (2006) Progressive sutural cataract associated with a BFSP2 mutation in a Chinese family. Mol Vis 12:1626–1631

Zhang X, Wang X, Zhu H, Kranias EG, Tang Y, Peng T, Chang J, Fan GC (2012) Hsp20 functions as a novel cardiokine in promoting angiogenesis via activation of VEGFR2. PLoS One 7:e32765

Zheng Q, Wang X (2010) Autophagy and the ubiquitin-proteasome system in cardiac dysfunction. Panminerva Med 52:9–25

Zheng W, Schafer NP, Davtyan A, Papoian GA, Wolynes PG (2012) Predictive energy landscapes for protein-protein association. Proc Natl Acad Sci U S A 109:19244–19249

Chapter 18
Regulation of Actin-Based Structure Dynamics by HspB Proteins and Partners

Solenn M. Guilbert, Alice-Anaïs Varlet, Margit Fuchs, Herman Lambert, Jacques Landry, and Josée N. Lavoie

Abstract Small heat shock proteins (HspB proteins) form a diverse family of proteins that have evolved distinct modes of action to protect cells from proteotoxic stress. Studies conducted within the last 25 years have revealed specialized roles for some HspB, in particular for HspB1, HspB8, and its cochaperone Bag3, in the modulation of actin-based cytoskeletal dynamics under physiological and stress conditions, which might be related to their linkage to human cancer. Little is known, however, on whether and how such biological activities on signaling are connected to pathways within the quality control network. In this chapter, we examine functional relationships between HspB proteins, the cochaperone Bag3 and actin dynamics and describe the mechanisms known so far that are responsible for their modulation of actin architecture. We further discuss on how such activities might be connected to quality control network. While some pieces of the puzzle might need to be inserted differently, we hope that this review will stimulate further studies to elucidate the mechanistic behind chaperone-mediated actin remodeling by HspB proteins and their partners.

Keywords HspB • Bag3 • Actin • Selective autophagy • Cytoskeletal dynamics

---

Solenn M. Guilbert, Alice-Anaïs Varlet are co-first authors.

S.M. Guilbert • J.N. Lavoie (✉) • A.-A. Varlet • M. Fuchs • H. Lambert • J. Landry
Centre de recherche sur le cancer de l'Université Laval, Quebec, QC, Canada

Centre de recherche du Centre Hospitalier Universitaire de Québec, Axe Oncologie, Quebec, QC, Canada

Groupe St-Patrick de recherche en Oncologie Fondamentale, Quebec, QC, Canada
e-mail: josee.lavoie@crhdq.ulaval.ca

R.M. Tanguay, L.E. Hightower (eds.), *The Big Book on Small Heat Shock Proteins*, Heat Shock Proteins 8, DOI 10.1007/978-3-319-16077-1_18

18.1 Introduction

Small heat shock proteins (sHsp or HspB) were discovered because of their strong induction after heat shock in *Drosophilia melanogaster* (Tissieres et al. 1974). Thereafter, they were viewed as proteins able to confer protection on cells against the deleterious effect of several stresses, including heat shock. In mammalian cells, HspB cytoprotective effects were first described in a breakthrough paper by Landry et al. in 1989 (Landry et al. 1989). Such protective functions were subsequently associated to a chaperone activity of HspB proteins (Jakob et al. 1993). Many of them were found to act in an ATP-independent manner, to bind non-native proteins to limit their aggregation and facilitate their subsequent refolding by ATP-dependent chaperone systems (reviewed in McHaourab et al. 2009). Accordingly, they are predicted to stand at the front line in cellular defense mechanisms against proteotoxic stress. Though, exactly how they function in vivo to recognize damaged or labile cellular components is not completely understood. There is currently no single model of structure and mechanism of action for this diverse family of proteins.

HspB proteins are typified by their small molecular weight (15–30 kDa) and a highly conserved C-terminal α-crystallin domain of approximately 90 amino acids. An N-terminal region that varies in sequence and length, and a C-terminal extension flank this α-crystallin domain, and confer specificity to each family member. In humans, there are ten HspB family members (HspB1–10) and a related protein (HSPC034/Hsp16.2) with no α-crystallin-like domain, which appear to share chaperone functions (Garrido et al. 2012; Kappe et al. 2003, 2010). Some are expressed ubiquitously e.g. HspB1 and HspB8, but others are tissue specific e.g. HspB9 (Morrow and Tanguay 2012; Vos et al. 2008). Most, but not all HspBs, are organized into dynamic oligomers with dimers as building blocks (reviewed in (Basha et al. 2012). Stress-induced phosphorylation of vertebrate HspBs is believed to promote exchange of subunits, altering the oligomeric architecture (Arrigo and Gibert 2012; Landry and Huot 1999; Welsh and Gaestel 1998). Yet, how phosphorylation affects the chaperone activity is still an open issue. Clearly, HspBs form a diverse family of proteins exhibiting considerable differences in sequence and tissue localization. This suggests that they have evolved distinct modes of action and makes it unlikely that they would function only as "general chaperones" in quality control mechanisms (Basha et al. 2012). Besides, there is evidence to support more specialized roles for HspB proteins in the regulation of actin dynamics.

In this chapter, we will examine functional connections between HspB proteins and actin dynamics that might be relevant to several physiological and pathological processes. We will first outline the initial work that led to the discovery of a role for HspB proteins in the regulation of actin using HspB1 as paradigm. We will feature more recent, unappreciated work that has clarified the mechanistic underpinnings HspB1-mediated regulation of actin dynamics, suggesting a function in actin monomer sequestration. We will then focus on novel findings involving the chaperone complex formed by HspB8-Bag3 in a selective form of autophagy called chaperone-assisted selective autophagy (CASA). This process was involved in the clearance and replenishment of components of actin-based structure, hence maintaining the homeostasis of mechanically strained tissues. Finally, based on data supporting a role for Bag3 in cell adhesion and motility and novel developments, we will discuss the possibility for a more widespread function for the HspB8-Bag3 chaperone complex in selective recognition, sequestration/or degradation of cytoskeletal proteins during dynamic remodeling of actin cytoskeleton in unstressed cells, and its potential relevance in cancer.

18.2 HspB1: A Bona Fide Regulator of Actin Dynamics Hijacked by Pathogenic Bacteria

HspB1 is the archetype and the most studied member among vertebrate HspB family of proteins. The pertinence of heat-shock-induced transcriptional activation of HspB1 for cellular thermoresistance was established by a series of studies in the early 1990s. It was demonstrated that cells which ectopically express high levels of HspB1 become resistant to heat shock and to a variety of other stress (Jaattela and Wissing 1993; Knauf et al. 1992; Landry et al. 1989; Lavoie et al. 1993a; Mehlen et al. 1995; Rollet et al. 1992). HspB1 was also the first member of this family of proteins to be related to the regulation of actin architecture in response to stress, but also during physiological stimulations (Lavoie et al. 1993a, b).

18.2.1 HspB1 and Actin Dynamics: A Relationship Relevant for Both Stressed and Unstressed Cells

Miron et al. first reported that HspB1 could regulate actin polymerization in cell-free systems. In their assays, purified HspB1 inhibited actin polymerization and even more, promoted rapid depolymerization of F-actin (Miron et al. 1988, 1991). It was concluded that HspB1 behaves like an F-actin barbed-end capping protein with little effect on actin nucleation. Shortly after, Lavoie et al. reported that HspB1 could promote cell survival in the presence of the actin polymerization inhibitor cytochalasin D, in a phosphorylation-dependent manner (Lavoie et al. 1993a). This protective effect was correlated with changes in the organization of the actin cytoskeleton upon overexpression of HspB1, providing further support for a direct activity on actin (Lavoie et al. 1995). Furthermore, Lavoie et al. found that phosphorylation of HspB1 not only act to prevent severe alterations of the actin cytoskeleton architecture in response to various stress, but also to modulate growth factor-induced actin remodeling and membrane dynamics (Lavoie et al. 1993b; 1995). For instance, wild type HspB1 was found to enhance growth factor-induced F-actin accumulation, whereas the HspB1 phosphorylation mutant exerted a dominant negative effect and inhibited actin reconfiguration in response to mitogens.

During that period, Benndorf et al. (1994) provided additional evidence reinforcing the role of HspB1 phosphorylation and its direct action on actin polymerization. Non-phosphorylated and phosphorylated Hsp25/HspB1 monomers as well as multimeric particles were isolated from Ehrlich ascites tumor cells and assessed for their activity on actin polymerization in a test tube. It was found that the actin polymerization-inhibiting activity of Hsp25/HspB1 is dependent on the degree of its phosphorylation and on its structural organization (Benndorf et al. 1994). Indeed, non-phosphorylated monomers inhibited actin polymerization while phosphorylated monomers and non-phosphorylated multimeric particles did not. Based on this work and others (Mounier and Arrigo 2002), it was proposed that unphosphorylated HspB1 inhibits actin assembly by capping the fast-growing end of actin filaments (F-actin), a model that was later disputed by During et al. in 2007 (as described in the next section).

In the meantime, the nature of HspB1 signal transduction pathway was elucidated. It was established that the phosphorylation-dependent effects of HspB1 on actin dynamics were triggered in a similar manner by toxic stress and physiological signals, through activation of the p38MAPK-MAPKAPK2 signaling axis (Ahlers et al. 1994; Ben-Levy et al. 1995; Guay et al. 1997; Huot et al. 1995; McLaughlin et al. 1996; Rouse et al. 1994; Stokoe et al. 1992; Zhou et al. 1993). Based on these initial studies that were performed mostly in fibroblasts overexpressing HspB1, it was proposed that the contribution of HspB1 as a modulator of actin remodeling in response to various signals would depend upon the cellular context, i.e. the concentration of HspB1 relative to other actin regulators (Landry and Huot 1995). In cells where HspB1 is highly expressed, e.g. during development, or after stress, activation of the p38MAPK signaling axis would be a key determinant of actin remodeling via phosphorylation of HspB1. In support of such model, the physiological importance of the p38MAPK-HspB1 pathway in modulating actin dynamics was clearly demonstrated in human umbilical vein endothelial cells that constitutively express high levels of HspB1 (Huot et al. 1997; Rousseau et al. 1997). The exact nature of the phosphorylation-regulated activity of HspB1 remained enigmatic for some time, but it was hypothesized that the ability of non-phosphorylated HspB1 to act as a F-actin capping protein in a test tube would be inhibited by its phosphorylation in cells. Release of phosphorylated HspB1 could promote F-actin polymerization and remodeling by uncapping the fast-growing end. Yet, direct mechanistic evidence for such dynamic interaction between HspB1-actin in vivo has been missing.

During the next decade, activation of the p38MAPK-MAPKAPK2-HspB1 signaling axis was extended to several contexts, including rat kidney mesenchymal cells, human keratinocytes, HeLa cells or aortic smooth muscle cells, and in animal models (Garmyn et al. 2001; Hirade et al. 2002; Ito et al. 2005; Kato et al. 1999; Muller et al. 1999; Vertii et al. 2006; Weber et al. 2005). Further, a role for HspB1 in the regulation of actin architecture in unstressed cells was confirmed. This is how HspB1 was involved in several processes involving extensive remodeling of the cell, for instance, in the regulation of cell migration, cell contraction, cell invasion and podocyte and lamellipodium formation (McMullen et al. 2005; Muller et al. 1999; Pichon et al. 2004; Shin et al. 2005; Smoyer and Ransom 2002). Similarly, other members of the HspB family of proteins were linked to the regulation of actin dynamics, suggesting that actin structures represent major cellular components interacting with these proteins. In particular, overexpression of HspB5 was found to protect cells against cytochalasin D, like HspB1, and interaction with actin appears to be regulated also by its phosphorylation status (Wang and Spector 1996). As HspB5 exhibits a distinct localization after heat shock and is organized in fibers that colocalize with F-actin, it is likely that its mode of action differs (Singh et al. 2007). Likewise, HspB6 was reported to bind actin and to regulate actin dynamics depending on its phosphorylation status (Brophy et al. 1999a, b). In a more recent study, it was further found that among the ten members of HspB family, HspB1, HspB6 and HspB7 show independent cardioprotective activity to prevent tachycardia remodeling during atrial fibrillation by reducing the formation of F-actin stress fibers (Ke et al. 2011). Intriguingly, HspB8 was shown to function by inhibiting the activity of RhoA GTPase upstream of F-actin assembly, pointing to a distinctive activity on the actin-regulatory machinery.

Notwithstanding a large body of evidence, the role of HspBs as mediators of actin-regulatory protein function has been largely ignored in most reviews on signaling pathways for actin assembly, perhaps, because of a misunderstanding of their mode of action. It is only recently that some clarifications came from studying the pathogenic effect of Bacillus anthracis on actin-based motility. The next section is dedicated to these findings, as they have received little attention from the sHsp community.

18.2.2 Toward a Mechanistic Understanding of HspB1-Actin Relationship

Many pathogens, including bacteria, have evolved gene products to subvert and reconfigure the actin cytoskeleton to promote infection and pathogenesis. Studying the molecular biology of interactions between pathogenic products and host cells has provided key information about the role of actin and essential mediators of actin-regulatory protein function, in particular, about the Rho GTPase signaling system (Lavoie et al. 2010; Taylor et al. 2011; Welch and Way 2013). Similarly, the phosphorylation-dependent function of HspB1 in actin dynamics has been identified as a key target of the anthrax lethal toxin that mediates its paralyzing effect on neutrophil actin-based motility, supporting its major role as a actin-regulatory protein (During et al. 2005, 2007).

Lethal toxin is thought to be the major virulence factor responsible for impaired immunity and death upon inhalation of anthrax, which causes fatal bacteremia (Moayeri and Leppla 2004). In order to identify the pathways underlying inhibition of actin-based motility by lethal toxin, During et al. have searched for proteins exhibiting changes in phosphorylation in response to lethal toxin, as it was known to affect MEK kinases. They identified HspB1 as a major protein showing decreased phosphorylation in response to lethal toxin (During et al. 2007). To explore the role of HspB1 as a mediator of lethal toxin-dependent inhibition of actin-based motility, the authors elegantly exploited *Listeria* as a probe to assess actin polymerization in cells. *Listeria* is an intracellular bacterium that stimulates actin assembly in host cells, to form actin-rich comet tails propelling the bacterium through the cytoplasm (Lambrechts et al. 2008). Remarkably, they found that lethal toxin dramatically impaired the assembly of *Listeria* actin tails. Such effect was paralleled by inhibition of HspB1 phosphorylation and could be mimicked by incubating cells with a p38MAPK inhibitor. Strong evidence was provided that HspB1 phosphorylation is a prime mediator of lethal toxin inhibition of actin assembly, using depletion-rescue approaches. Inhibition of *Listeria* actin-based motility was impaired by HspB1-specific siRNAs, but restored after microinjection of wild type or non-phosphorylated HspB1 (During et al. 2007). This was consistent with original findings by Lavoie et al. showing that phosphorylation of HspB1 promotes new actin assembly, while non-phosphorylatable HspB1 acts as dominant-negative and decreases F-actin assembly (Lavoie et al. 1995).

To get more mechanistic insights into the mode of action of HspB1, the authors examined the effects of recombinant HspB1 (wild type, non-phosphorylatable or pseudophosphorylated) on purified pyrenyl actin assembly and disassembly (Kouyama and Mihashi 1981). Cell-free assays were performed to assess its potential as a barbed-end capping protein, its activity on actin monomer sequestration and on disassembly of barbed filament end. It was found that increasing amounts of HspB1 had no effect on the rate of actin nucleation using as template spectrin-4.1 actin nuclei that contain actin filaments with free barbed ends. However, HspB1 inhibited actin assembly when pre-incubated with the monomeric actin solution, suggesting that it might function by sequestering actin monomers rather than by capping actin filaments. In further support, non-phosphorylatable and wild type HspB1 could inhibit actin assembly from gelsolin-actin nuclei, which are actin filaments already capped at their barbed end. In such assays, any slowdown of actin assembly could only be explained by actin monomer sequestration. Based on these detailed assays and immunolocalization analyses, a new model was proposed for the phosphorylation-dependent promotion of actin assembly by HspB1. Unphosphorylated HspB1 octamers, which were found to localize mainly at the leading edge of polarized cells, would provide a "reservoir" of actin monomers that would be mobilized via signal-induced phosphorylation (Fig. 18.1). Phosphorylation of HspB1, which promotes dimer formation, would release a pool of actin monomers for local promotion of actin assembly. Hence agonist stimulation would serve to uncap actin filaments through the activation of phosphoinositides and GTPase signaling systems (Soderling 2009), while freeing sequestered actin monomers through phosphorylation of HspB1 that would act in concert with other pathways to promote actin assembly (During et al. 2007). Phosphorylated HspB1 would move towards the cell center where it would be dephosphorylated, again bind actin monomers, and then shuttle them back to the leading edge, ready to be mobilized upon cell activation.

Through this study, the phosphorylation of HspB1 was firmly established as a key mechanism to regulate motility in nonmuscle cells by showing its subversion by pathogenic bacteria, 20 years after the initial discovery of its actin-modulatory

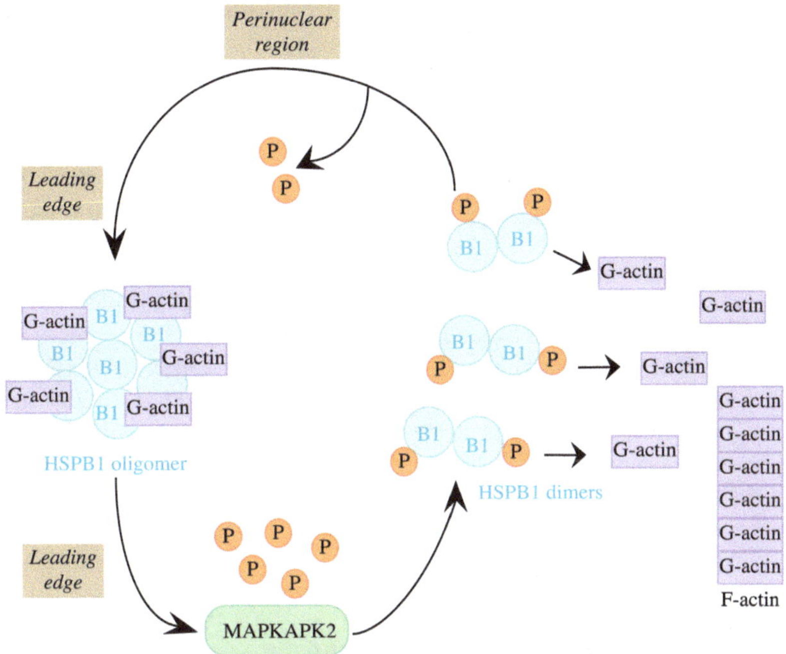

Fig. 18.1 Model describing how HspB1 would facilitate actin-based motility, through a phosphorylation cycle that could serve to shuttle actin monomers to sites of actin assembly (Based on During et al. 2007; *G-actin* monomeric actin, *F-actin* actin filament, *P* phosphate. See text for details)

activity. It remains to be determined whether other HspB proteins can exhibit similar activity, and if so, in which specific contexts. HspB8, however, has recently emerged as a peculiar member of this family with a clearly distinctive mode of action for the regulation of actin-based cellular structures. Studies that have uncovered functional relationships with clearance of proteins by selective autophagy will be outlined below.

18.3 HspB8-Bag3: How Clearance of Proteins by Selective Autophagy Might Interface with Chaperoning of Higher Order Actin-Based Structures

The *Hspb8* gene was first identified in melanoma cells as a gene encoding a protein kinase named H11, and in breast cancer cells by a screen of estrogen-induction effects on gene expression, where it was named E2IG1 (Charpentier et al. 2000; Smith et al. 2000). The existence of intrinsic protein kinase activity of HspB8 has been questioned owing to low homology to protein kinases, low level of autophosphorylation, lack of specific substrate, and use of HspB8 immunocomplex in autophosphorylation assays, raising the possibility for contaminations by associated protein kinases (Depre et al. 2002; Kim et al. 2004a, b). In contrast, there are many strong data in support of the chaperone-like activity of HspB8. Indeed, Benndorf et al. rediscovered and identified HspB8 as a typical small HspB family member with similarity to HspB1, mainly in the α-crystallin domain (Benndorf et al. 2001). Thereafter, HspB8 chaperone activity was demonstrated in vitro and in vivo (Carra et al. 2005; Chavez Zobel et al. 2003; Chowdary et al. 2004; Kim et al. 2004b, 2006). Moreover, our group and others have described HspB8 as a key contributor to the macroautophagy pathway that leads to degradation of misfolded protein (Carra et al. 2008a, b, 2010; Crippa et al. 2010; Kwok et al. 2011).

A functional interaction between HspB8 and the BAG-family member Bag3 was discovered by our group and reported in a paper by Carra et al. in 2008 (Carra et al. 2008a). Bag3 was identified as an obligate partner of HspB8 regulating its stability. In several cell lines, all HspB8 appear to be associated with Bag3, the depletion of which leads to a rapid destabilization of HspB8. In marked contrast to canonical HspBs that form oligomeric complexes (mainly homotypic) in cells, HspB8 was found to exhibit a distinctive structural organization: it mainly forms dimers, but it is associated in a stoichiometric complex with Bag3 (HspB8-Bag3 [2:1]) (Carra et al. 2008a). Fuchs et al. have elucidated the molecular determinants of such interaction, highlighting a key role for Bag3 IPV motifs (Ile-Pro-Val) that binds to the hydrophobic groove of HspB8 (Fuchs et al. 2010). IPV-like motifs are usually found in the C-terminus of HspBs and have been linked to stabilization of dimer-dimer conformation (Basha et al. 2012). Intriguingly, neither HspB8 nor HspB6 contain such motif, which might explain their low-oligomeric states and a possible functional requirement for binding to Bag3. Actually among the other ubiquitously expressed HspB proteins HspB1, HspB5 and HspB6, only HspB6 has been found to associate with Bag3, like HspB8; the latter, however, was established as the preferred partner of Bag3 (Fuchs et al. 2010). Such interaction between HspB8-Bag3 could be significant for cell physiology, given that mutation of one Bag3 IPV motifs (P209L) was linked to severe myopathy in human (Selcen et al. 2009). Together, these findings support a more widespread function for HspB IPV motifs in the formation of multi-proteins complexes. They also point to a role for HspB8 in some pathway regulating protein degradation, as BAG-family of cochaperones are known to regulate triage of substrates between refolding and degradation (Alberti et al. 2003; Arndt et al. 2005; Dai et al. 2005; Doong et al. 2002, 2003).

18.3.1 HspB8-Bag3 in Protein Quality Control: When the Aggresome Pathway Meets the Autophagic Machinery

There are currently six BAG-family of cochaperones characterized by a C-terminal BAG domain that binds the Hsp70 ATPase domain to regulate the fate of Hsp70 substrates (Kabbage and Dickman 2008). Yet Bag3 is known to inhibit rather than to stimulate proteasomal degradation of proteins (Doong et al. 2003; Gamerdinger et al. 2009). In their initial study, Carra et al. first involved the HspB8-Bag3 complex in the clearance of aggregated polyQ proteins by macroautophagy (Carra et al. 2008a). Macroautophagy, hereafter named autophagy, is an essential physiological process for the bulk degradation of cellular constituents during starvation, thereby recycling nutrients in the internal pool that can be used for survival means (Yang and Klionsky 2010). Thought to be unselective, autophagy now appears as a selective process for the degradation of damaged organelles and insoluble protein aggregates otherwise indigestible by the proteasome (Shaid et al. 2013). Selective targeting of substrate proteins is driven by a subset of autophagy receptors/adaptor proteins that includes a protein partner of Bag3: the p62/SQSTM1 adaptor protein, which sequesters cargo into forming autophagosome (Gamerdinger et al. 2009; Rogov et al. 2014). Autophagic receptors recognize substrate modifications like ubiquitination on the one hand, and mediates targeting to the autophagic machinery through interacting with LC3 anchored at the autophagosome membrane, on the other hand. In their search for a mechanism whereby overexpressed HspB8-Bag3 could promote the clearance of polyQ proteins, Carra et al. found that the accumulation of Htt43Q was impaired in the presence of inhibitors of autophagy. HspB8 and/or Bag3 were shown to promote the redistribution of GFP-LC3, a marker of autophagosomes, from a diffuse cytoplasmic pattern to punctate structures. They were also found to cause accumulation of lipidated LC3 and its association with membranes. Accordingly, it was proposed that HspB8-Bag3 function to accelerate the disposal of polyQ proteins by mediating selective autophagy.

At the time of these findings, there was little known on the specific role of Bag3 in protein quality control and on the mechanisms of selective autophagy. Thereafter, the relevance of Bag3 in autophagy was expanded by Gamerdinger et al. In a first paper in 2009, it was found that Bag3, owing to its capacity to recruit the autophagy machinery, is an essential regulator of protein quality control during aging (Gamerdinger et al. 2009). The increased autophagic flux regulated by Bag3 was linked to its association with the autophagic adaptor protein p62/SQSTM1. In a second paper in 2011, the authors reported a role for Bag3 in chaperone-based aggresome-targeting and selective autophagy of misfolded proteins. They found that Bag3 binds to dynein and controls the transport of microaggregates into aggresomes upon inhibition of the proteasome with MG132 (Gamerdinger et al. 2011). A role for Bag3 in the aggresome pathway was concomitantly corroborated by Zhang and Qian in 2011, who reported the contribution of CHIP, a Hsp70/Hsp90 ubiquitin ligase, providing a link with the protein degradation machinery (Zhang and Qian 2011). The mechanisms involved were further expanded in a study from Xu et al. in 2013, who found that phosphorylation of Bag3 directs the recruitment of dynein via binding to 14-3-3 protein, which would function as a molecular adaptor in Bag3-mediated aggresomal targeting of misfolded proteins (Xu et al. 2013). The aggresome pathway is believed to protect cells against the toxicity of small aggregates of misfolded proteins when proteasome-dependent degradation fails, by promoting their targeting and congression/sequestration into juxtanuclear structures that are also believed to facilitate their clearance by autophagy (Garcia-Mata et al.

2002; Kopito 2000; Rodriguez-Gonzalez et al. 2008). Aggresomal targeting of misfolded proteins is an active process whereby damaged proteins are loaded onto the dynein-dynactin motor complex for retrograde transport along microtubules (Olzmann and Chin 2008; Ravikumar et al. 2005).

Thus Bag3, owing to its multi-domains structure and its ability to function as a large scaffold, seems to provide a signaling platform to recruit the molecular machineries for recognition (chaperones), transport (dynein) and seclusion/clearance of misfolded proteins (Fig. 18.2). How Bag3 might coordinate these processes remains to be determined. Besides, it is unclear how its activity is related to key molecular adaptors like HDAC6 that were described to play a similar role in aggresome-targeting of misfolded proteins (Olzmann and Chin 2008). In fact, molecular mechanisms underlying its associations with other components of this pathway, like the autophagic adaptor p62/SQSTM1, are yet to be defined. Importantly, while Hsp70 has been firmly established as a key partner of Bag3 in the aggresome pathway, the exact role of HspB8, if any, has not been addressed. Given the involvement of HspBs in regulation of actin dynamics, and based on recent findings in muscle cells, it is conceivable that HspB8 could recruit the Bag3-dependent machinery for seclusion and/or degradation of a selective subset of cellular proteins regulating the dynamics and integrity of complex actin-based structures. Evidence to support such a model will be outlined in the following section.

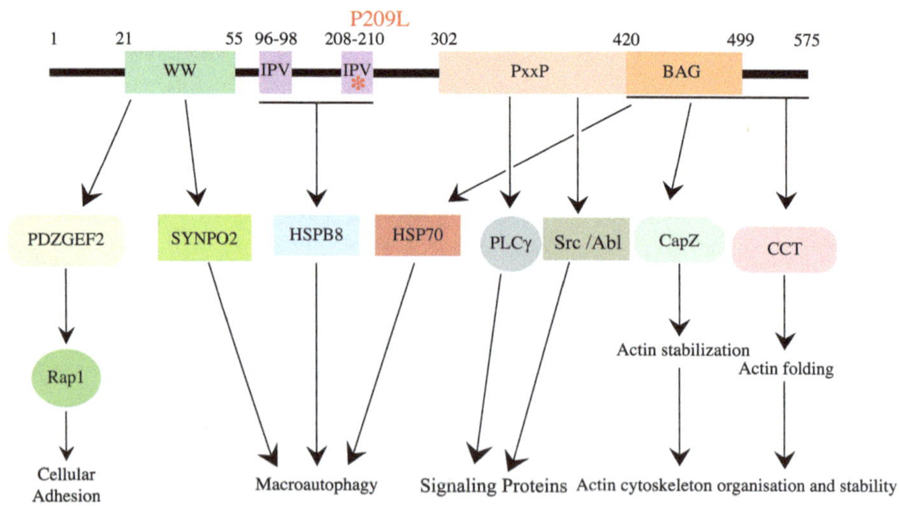

Fig. 18.2 Scheme of the modular structure of the cochaperone Bag3, emphasizing direct interactions with partners that have been involved in cytoskeletal proteostasis (CASA), or those associated with the effects of Bag3 on cell adhesion and motility; numbers represent the positions of amino acids within the Bag3 sequence; P209L, mutation linked to severe myopathy in human (see text for details)

18.3.2 Beyond Actin Assembly-Disassembly: Regulation of Cytoskeleton Protein Degradation Through Chaperone-Assisted Selective Autophagy (CASA)

Genetic evidence implicating Bag3 in the regulation of complex actin-based structures came from studies by the group of Takayama, who produced a mouse model by genetic ablation of *Bag3*. These mice were found to develop fulminant and severe myopathy (Homma et al. 2006). This phenotype was associated with disruption of Z-disk—the actin-contractile structure of myofibrils—followed by myofibrillar degeneration, involving Bag3 in homeostasis of cytoskeleton elements. These original findings were confirmed in a cellular model using shRNA-mediated knockdown of Bag3 in rat neonatal cardiomyocytes subjected to mechanical stretch. In this work, a new partner of Bag3-Hsc70 was identified which functions in F-actin stabilization: the F-actin-capping protein CapZ (Hishiya et al. 2010). It was shown in vitro that the Bag3-Hsc70 complex contributes to proper localization of CapZ to Z-disc to maintain functional structures. Together, these studies established a role for the Bag3-Hsp70 complex in the maintenance of the Z-disc, an actin-based structure exposed to mechanical stress during contraction of muscle cells.

Mechanistic insights into that process were provided by the group of Höhfeld, who described a HspB8/Bag3/Hsp70 complex in *Drosophila* that would be responsible for the maintenance of the Z-disk (Arndt et al. 2010). The complex was shown to facilitate the release of damaged proteins such as filamin (FLNA)—a mechanosensor in cells—and their disposal by a selective autophagy process. They proposed that HspB8 recognizes damaged filamin whereas processing of the cargo is dependent on Hsp70-CHIP mediated ubiquitylation, which leads to the recruitment of p62/SQSTM1 and stimulates autophagy. This is how the process called CASA—chaperone-assisted selective autophagy—was described as a mechanism essential for the maintenance of muscle cells integrity. In 2013, the same group further defined the CASA machinery as a tension-induced autophagy pathway essential for mechanotransduction not only in muscle cells, but also more generally, for adaptation of mammalian cells to mechanical strain (Ulbricht et al. 2013). It was found that Bag3, HspB8 and Hsp70 bind in a cooperative manner to a mechanosensitive region of filamin being exposed upon increased intracellular tension (Razinia et al. 2012). Synaptopodine-2 (SYNPO2)—a cytoskeleton adaptor protein and a FLNA interactor (Linnemann et al. 2010)—was identified as a new protein partner interacting with the WW domain of Bag3.

Based on a series of elegant experiments, they proposed a model in which Bag3 would hold together the machineries for the recognition, degradation, and transcriptional regulation of a prototype actin-binding protein denatured upon mechanical stress such as those experienced during actin-based contraction in both muscle and nonmuscle cells. In their model, SYNPO2 would link the CASA complex to the autophagic machinery by recruiting trafficking factors as VPS18, VPS16 and Atg7, while CHIP and p62/SQSTM1 would promote filamin ubiquitination and its autophagic clearance. Furthermore, Bag3, through its WW domain, also provides binding sites for proteins of the Hippo pathway—LATS1/2 and AMOTL1/2—that regulate the activity of YAP1/TAZ transcription regulators (Harvey et al. 2013). Therefore in mechanically strained cells, increased Bag3 expression and its binding to LATS1/2 and AMOTL1/2 would abrogate the cytoplasmic retention of YAP1 and promote transcriptional upregulation of cytoskeletal targets, including filamin (Dupont et al. 2011). Thus, the Bag3 complex has emerged as a crucial regulator of the homeostasis of actin-based structures by coordinating removal and disposal of damaged cytoskeleton components, while promoting a transcriptional response to compensate disposal (Ulbricht and Hohfeld 2013). Indeed, CASA activity and Bag3 expression were found to be adjusted to the extent of tension within the cytoskeleton; increased tension, just like heat stress, appears to activate HSF1 to stimulate compensatory expression of CASA-chaperones (Ulbricht et al. 2013).

Such a role for Bag3 could explain the progressive myopathy and Z-disk disruption in mice KO for Bag3 and the severe myopathy observed in patients with the P209L mutation, which, as aforementioned, is in the HspB8-Bag3 interaction motif (Fuchs et al. 2010). This provides support to the idea that HspB8 would function with Bag3 to preserve the integrity of this actin-based structure. Such groundbreaking findings also support a more general function for the CASA machinery in maintaining the integrity of actin-based structures in response to various stress in nonmuscle cells, and perhaps, in unstressed cells to allow rapid remodeling of higher order cytoskeletal structures upon cell activation. In support of this hypothesis, the CASA machinery was shown to localize along stress fibers that form when tension is generated inside cells during adhesion and migration in nonmuscle cells (Ulbricht and Hohfeld 2013). Further evidence linking Bag3 to the regulation of actin remodeling and cell adhesion, and potential relevance to pathological contexts such as cancer, will be discussed below.

18.4 Is HspB8-Bag3 Playing a More Widespread Function in the Regulation of the Dynamics of Actin-Based Structures in Unstressed Cells?

18.4.1 Bag3 in Cell Adhesion and Motility: Sustaining the Cancer Phenotype?

As mentioned before, Bag3 has a modular structure that provides multifaceted interactions enabling the protein to act as a molecular scaffold and as such, to modulate several biological processes including apoptosis, development and autophagy (Rosati et al. 2011). Bag3 is unique among the Bag proteins for the presence of a WW domain, which plays a critical role in selective autophagy as described above, and a proline-rich region (PXXP) that provides SH3-binding motifs (Fig. 18.2). These protein domains are generally found on proteins acting within signal transduction networks to connect plasma membrane receptors to submembranous cytoskeleton components, and to promote actin remodeling (Ilsley et al. 2002; Salah et al. 2012). Besides, the involvement of Bag3 in cytoskeleton organization, cell adhesion and motility is emerging as a key property of the cochaperone, which could be responsible for a large part of its linkage to human diseases such as cancer. In humans, expression of Bag3 is constitutive in a few normal cell types, but is found in several primary tumors and tumor cell lines of various origins (leukemia, lymphomas, neuroblastoma, pancreas, thyroid, breast and prostate carcinoma, melanoma, glioblastoma, kidney, colon and ovary cancers), where it is believed to support tumor cell motility (see for a review Rosati et al. 2011). Whether such function is related to the CASA activity described above, and whether molecular chaperones, in particular HspB8, function together with Bag3 to regulate adhesion and motility is an important, unresolved issue.

The finding that some cancers contain high levels of Bag3 raises the possibility that overexpression contributes to the malignant phenotype typified by tumor cell invasion and metastasis. In support of such hypothesis, Bag3 was involved in the modulation of epithelial-to-mesenchymal transition (EMT) in tissue culture cells, in human hepatocellular carcinoma cells (HCC)—a highly invasive tumor with high expression of Bag3 (Xiao et al. 2014). Li et al. also reported a phosphorylation-dependent effect of Bag3 on epithelial markers and cell shape in thyroid cancer cells (Li et al. 2013). EMT is a physiological process regulating cell migration during development. In cancer, however, it is associated with acquisition of mesenchymal characteristics, loss of epithelial polarity and cell-cell contacts, leading to increased invasiveness of tumor cells (Lamouille et al. 2014). EMT is driven by extensive remodeling of actin-based structures that support cell adhesion and polarity. Hence the involvement of Bag3 in EMT suggests that the cochaperone exerts crucial motility-promoting activity, perhaps, through modulation of actin structures. This is supported by the work of Iwasaki et al. who first explored a role for Bag3 in cell motility and adhesion using *Bag3*-deficient mouse embryonic fibroblasts, and by manipulating Bag3 protein levels in several cancer cell lines (Iwasaki et al. 2007). They found that Bag3 overexpression increases, while reduction of Bag3 decreases, cell motility in tissue cultures. Bag3 was localized to the cell leading edge, where actin remodeling and polymerization occur. Besides, a reduction of Bag3 protein level was associated with a decrease in Rac1 activity, supporting a role in actin polymerization at the leading edge of migrating cells. Moreover, knockdown of Bag3 protein level in a human tumor xenograft was shown to suppress invasion and metastasis in mice. This provides support to the possibility that overexpression of Bag3 in epithelial cancers could confer increased invasive properties to cells experiencing nutrient deprivation, a driving force for spreading of cancer cells. Actually, Bag3-promoting effects on cell motility were mostly observed when cells were cultured in low serum (Iwasaki et al. 2007).

Instead of the usual culprits, the chaperones, novel partners of Bag3 were linked to its effects on cell adhesion and motility. Through a yeast two-hybrid screen, Iwasaki et al. identified the PDZ domain containing guanine nucleotide exchange factor 2 (PDZGEF2)—a Rap1 guanine exchange factor—as another protein interacting with the WW domain of Bag3 (Iwasaki et al. 2010). Rap1 is a small

GTPase that has been involved in the control of establishment of cell polarity, activation of integrin-mediated cell adhesion, the regulation of cell-cell contacts and secretion (reviewed in Frische and Zwartkruis 2010). Consistently, Bag3 was shown to promote cell adhesion in Cos7 cells, in a PDZGEF2-dependent manner (Iwasaki et al. 2010). It is interesting to note that Rap1 was linked to the regulation of elements of the Hippo pathway, like Bag3 and the CASA machinery (Avruch et al. 2012; Ulbricht and Hohfeld 2013). Additionally, Bag3 seems to regulate, in a PXXP-dependent manner, the expression of CNN—a matricellular signaling protein promoting cell adhesion through integrins and proteoglycans, yet through unknown mechanisms (Kassis et al. 2009). Finally, the group of Tosco reported that Bag3 associates with the cytosolic chaperonin CCT to regulate actin folding (Fontanella et al. 2010), further highlighting a major role for Bag3 in actin-based dynamics. That Bag3 protein levels influence cell motility and adhesion was corroborated in several studies (reviewed in Rosati et al. 2011). In most of these studies, however, there have been little attempts to connect the effects of Bag3 on signaling cell adhesion and invasion, to the effects of molecular chaperones on these pathways.

18.4.2 Chaperone-Mediated Degradation of Proteins to Regulate Large-Scale Remodeling of Actin?

While attempting to fill these gaps, Colvin et al. recently demonstrated that Bag3 mediates the effects of Hsp70 on regulation of cancer-related signaling networks (Colvin et al. 2014). In particular, they found that Bag3, via its PXXP motif, interacts with the SH3 domain of the oncogenic tyrosine kinase Src, thereby mediating the effects of Hsp70 on Src signaling which is a crucial signaling hub controlling cell adhesion and motility (Brunton and Frame 2008; Colvin et al. 2014). Accordingly, the activity of Bag3 on cancer cell motility and adhesion could largely rely on its ability to integrate chaperone systems with a broad network of signaling pathways. Along the same line, it is conceivable that some of the effects of Bag3 on cancer-related processes involving large-scale remodeling of actin could rely on HspB8-dependent effects on selective targeting and clearance of cytoskeletal proteins (described in Sect. 18.3). Accumulating evidence indicates that vital processes for cancer cells like cell migration and cell division depend upon timely degradation of proteins to favor rapid changes in actin network architecture (Huang 2010). For instance, the progression through mitosis involves the most dramatic and spectacular changes in the overall structure of a cell, changes that dictates proper segregation of chromosomes and which are typically completed within an hour in highly dividing cells like cancer cells (Almonacid et al. 2014; Kunda and Baum 2009). While little is known on how these changes are regulated in time and space, selective protein sequestration and/or degradation are emerging as key mechanisms (Belaid et al. 2013, 2014; Werner et al. 2013).

Of particular interest here, it was shown that active RhoA—the master regulator of actomyosin dynamics at the cleavage furrow—needs to be sequestered into autophagosomes in late cytokinesis to allow disassembly of the contractile actin ring and final abscission of sister cells. Impaired autophagy was associated with cytokinesis failure, aneuploidy and motility defects, three processes that influence tumor progression. It was concluded that the autophagic degradation of active RhoA was required to restrict its activation in time and space and ensure fine-tuning of the Rho pathway (Belaid et al. 2013). How timely and specific recognition of protein substrates is controlled, however, is unknown. Based on recent developments, we infer that specialized chaperone systems, like the HspB8-Bag3 complex, could be mobilized during cell division to assist in the assembly-disassembly of macromolecular complex driving actin-based structures in highly dividing cells. In support of such idea, we recently uncovered a role for Bag3, in association with HspB8 in actin dynamics, spindle positioning and proper chromosome segregation during mitosis (Fuchs et al. unpublished data). Furthermore, the autophagic adaptor p62/SQSTM1—a protein partner of Bag3—was independently linked to mitotic progression and chromosome segregation (Linares et al. 2011). Nonetheless, functional relationships with quality control of mitotic organelles are yet to be discovered.

18.5 Concluding Remarks

Studies conducted over 25 years have unraveled specialized roles played by the small molecular chaperones HspB proteins in cytoskeletal dynamics and proteostasis. In this review, we have examined functional evidence supporting such functions for HspB1, HspB8 and its partner Bag3 that might be relevant to their effects in cancer cell survival and motility. Through this work, "unconventional chaperone activities" have been uncovered in protein sequestration (HspB1 and actin), shuttling/transport and autophagic degradation (the Bag3-HspB8 complex). Clearly, much more work is needed to clarify how individual HspB, and their partners, function to modulate actin remodeling in various contexts, and if the activities of Bag3 on signaling actin remodeling depend upon chaperones. How their emerging roles in quality control and dynamics of actin-based cellular structures are integrated within networks of signaling pathways in stressed and unstressed cells is yet to be dissected.

Although still speculative, we feel that there is sufficient evidence to propose a model that could serve as framework for future studies. In this model, HspB proteins and their partners could form a specific quality control network that would assist in the assembly and disassembly of macromolecular complexes controlling the dynamic architecture of actin in response to various signaling cascades. Some chaperones would act to sequester actin monomers and promote their release on cell activation, hence serving as large "reservoirs of polymerization-competent G-actin" that would support the assembly of actin structures (e.g. HspB1). Others would promote the autophagic clearance of cytoskeletal components to facilitate disassembly-assembly of actin structures, and limit signaling in a spatially and temporally restricted manner (e.g. HspB8-Bag3). The intimate relationship between HspB proteins and actin and its impact on selective autophagic degradation could even extend to HspB7 based on work from Vos et al. (2010, 2011), although the mechanism remains to be elucidated. Perturbations of actin architecture are among the common, primary features observed in response to a variety of insults, including oncogene activation, which also trigger the phosphorylation of HspB proteins. Then the phosphorylation-regulated function of HspB proteins could integrate multiple signals and coordinate actin-based cellular functions to support the cancer phenotype.

Considering the ever expanding dynamic processes relying on actin remodeling, including protein and organelle trafficking, cell polarity, directional cell migration, cell division and cellular traction/contraction, it should not be surprising to find elaborate connections between signaling pathways controlling actin assembly-disassembly and protein quality control network. Deregulation of many aspects linked to actin dynamics in cancer highlights the need for a better understanding of the complexities involved. Hence, follow up work should provide fascinating twists to the story of chaperone-mediated actin remodeling by HspB proteins and their partners.

Acknowledgments This work was supported by the Canadian Institutes of Health Research (Grant number 7088 to JNL and JL).

References

Ahlers A, Belka C, Gaestel M, Lamping N, Sott C, Herrmann F, Brach MA (1994) Interleukin-1-induced intracellular signaling pathways converge in the activation of mitogen-activated protein kinase and mitogen-activated protein kinase-activated protein kinase 2 and the subsequent phosphorylation of the 27-kilodalton heat shock protein in monocytic cells. Mol Pharmacol 46:1077–1083

Alberti S, Esser C, Hohfeld J (2003) BAG-1–a nucleotide exchange factor of Hsc70 with multiple cellular functions. Cell Stress Chaperones 8:225–231

Almonacid M, Terret ME, Verlhac MH (2014) Actin-based spindle positioning: new insights from female gametes. J Cell Sci 127:477–483

Arndt V, Daniel C, Nastainczyk W, Alberti S, Hohfeld J (2005) BAG-2 acts as an inhibitor of the chaperone-associated ubiquitin ligase CHIP. Mol Biol Cell 16:5891–5900

Arndt V, Dick N, Tawo R, Dreiseidler M, Wenzel D, Hesse M, Furst DO, Saftig P, Saint R, Fleischmann BK, Hoch M, Hohfeld J (2010) Chaperone-assisted selective autophagy is essential for muscle maintenance. Curr Biol 20:143–148

Arrigo AP, Gibert B (2012) HspB1 dynamic phospho-oligomeric structure dependent interactome as cancer therapeutic target. Curr Mol Med 12:1151–1163

Avruch J, Zhou D, Fitamant J, Bardeesy N, Mou F, Barrufet LR (2012) Protein kinases of the hippo pathway: regulation and substrates. Semin Cell Dev Biol 23:770–784

Basha E, O'Neill H, Vierling E (2012) Small heat shock proteins and alpha-crystallins: dynamic proteins with flexible functions. Trends Biochem Sci 37:106–117

Belaid A, Cerezo M, Chargui A, Corcelle-Termeau E, Pedeutour F, Giuliano S, Ilie M, Rubera I, Tauc M, Barale S, Bertolotto C, Brest P, Vouret-Craviari V, Klionsky DJ, Carle GF, Hofman P, Mograbi B (2013) Autophagy plays a critical role in the degradation of active RHOA, the control of cell cytokinesis, and genomic stability. Cancer Res 73:4311–4322

Belaid A, Ndiaye PD, Cerezo M, Cailleteau L, Brest P, Klionsky DJ, Carle GF, Hofman P, Mograbi B (2014) Autophagy and SQSTM1 on the RHOA(d) again: emerging roles of autophagy in the degradation of signaling proteins. Autophagy 10:201–208

Ben-Levy R, Leighton IA, Doza YN, Attwood P, Morrice N, Marshall CJ, Cohen P (1995) Identification of novel phosphorylation sites required for activation of MAPKAP kinase-2. EMBO J 14:5920–5930

Benndorf R, Hayess K, Ryazantsev S, Wieske M, Behlke J, Lutsch G (1994) Phosphorylation and supramolecular organization of murine small heat shock protein HSP25 abolish its actin polymerization-inhibiting activity. J Biol Chem 269:20780–20784

Benndorf R, Sun X, Gilmont RR, Biederman KJ, Molloy MP, Goodmurphy CW, Cheng H, Andrews PC, Welsh MJ (2001) HSP22, a new member of the small heat shock protein superfamily, interacts with mimic of phosphorylated HSP27 ((3D)HSP27). J Biol Chem 276:26753–26761

Brophy CM, Dickinson M, Woodrum D (1999a) Phosphorylation of the small heat shock-related protein, HSP20, in vascular smooth muscles is associated with changes in the macromolecular associations of HSP20. J Biol Chem 274:6324–6329

Brophy CM, Lamb S, Graham A (1999b) The small heat shock-related protein-20 is an actin-associated protein. J Vasc Surg 29:326–333

Brunton VG, Frame MC (2008) Src and focal adhesion kinase as therapeutic targets in cancer. Curr Opin Pharmacol 8:427–432

Carra S, Sivilotti M, Chavez Zobel AT, Lambert H, Landry J (2005) HspB8, a small heat shock protein mutated in human neuromuscular disorders, has in vivo chaperone activity in cultured cells. Hum Mol Genet 14:1659–1669

Carra S, Seguin SJ, Lambert H, Landry J (2008a) HspB8 chaperone activity toward poly(Q)-containing proteins depends on its association with Bag3, a stimulator of macroautophagy. J Biol Chem 283:1437–1444

Carra S, Seguin SJ, Landry J (2008b) HspB8 and Bag3: a new chaperone complex targeting misfolded proteins to macroautophagy. Autophagy 4:237–239

Carra S, Boncoraglio A, Kanon B, Brunsting JF, Minoia M, Rana A, Vos MJ, Seidel K, Sibon OC, Kampinga HH (2010) Identification of the Drosophila ortholog of HSPB8: implication of HSPB8 loss of function in protein folding diseases. J Biol Chem 285:37811–37822

Charpentier AH, Bednarek AK, Daniel RL, Hawkins KA, Laflin KJ, Gaddis S, MacLeod MC, Aldaz CM (2000) Effects of estrogen on global gene expression: identification of novel targets of estrogen action. Cancer Res 60:5977–5983

Chavez Zobel AT, Loranger A, Marceau N, Theriault JR, Lambert H, Landry J (2003) Distinct chaperone mechanisms can delay the formation of aggresomes by the myopathy-causing R120G alphaB-crystallin mutant. Hum Mol Genet 12:1609–1620

Chowdary TK, Raman B, Ramakrishna T, Rao CM (2004) Mammalian Hsp22 is a heat-inducible small heat-shock protein with chaperone-like activity. Biochem J 381:379–387

Colvin TA, Gabai VL, Gong J, Calderwood SK, Li H, Gummuluru S, Matchuk ON, Smirnova SG, Orlova NV, Zamulaeva IA, Garcia-Marcos M, Li X, Young Z, Rauch JN, Gestwicki JE, Takayama S, Sherman MY (2014) Hsp70-Bag3 interactions regulate cancer-related signaling networks. Cancer Res 74:4731–4740

Crippa V, Sau D, Rusmini P, Boncoraglio A, Onesto E, Bolzoni E, Galbiati M, Fontana E, Marino M, Carra S, Bendotti C, De Biasi S, Poletti A (2010) The small heat shock protein B8 (HspB8) promotes autophagic removal of misfolded proteins involved in amyotrophic lateral sclerosis (ALS). Hum Mol Genet 19:3440–3456

Dai Q, Qian SB, Li HH, McDonough H, Borchers C, Huang D, Takayama S, Younger JM, Ren HY, Cyr DM, Patterson C (2005) Regulation of the cytoplasmic quality control protein degradation pathway by BAG2. J Biol Chem 280:38673–38681

Depre C, Hase M, Gaussin V, Zajac A, Wang L, Hittinger L, Ghaleh B, Yu X, Kudej RK, Wagner T, Sadoshima J, Vatner SF (2002) H11 kinase is a novel mediator of myocardial hypertrophy in vivo. Circ Res 91:1007–1014

Doong H, Vrailas A, Kohn EC (2002) What's in the 'BAG'?–A functional domain analysis of the BAG-family proteins. Cancer Lett 188:25–32

Doong H, Rizzo K, Fang S, Kulpa V, Weissman AM, Kohn EC (2003) CAIR-1/BAG-3 abrogates heat shock protein-70 chaperone complex-mediated protein degradation: accumulation of poly-ubiquitinated Hsp90 client proteins. J Biol Chem 278:28490–28500

Dupont S, Morsut L, Aragona M, Enzo E, Giulitti S, Cordenonsi M, Zanconato F, Le Digabel J, Forcato M, Bicciato S, Elvassore N, Piccolo S (2011) Role of YAP/TAZ in mechanotransduction. Nature 474:179–183

During RL, Li W, Hao B, Koenig JM, Stephens DS, Quinn CP, Southwick FS (2005) Anthrax lethal toxin paralyzes neutrophil actin-based motility. J Infect Dis 192:837–845

During RL, Gibson BG, Li W, Bishai EA, Sidhu GS, Landry J, Southwick FS (2007) Anthrax lethal toxin paralyzes actin-based motility by blocking Hsp27 phosphorylation. EMBO J 26:2240–2250

Fontanella B, Birolo L, Infusini G, Cirulli C, Marzullo L, Pucci P, Turco MC, Tosco A (2010) The cochaperone BAG3 interacts with the cytosolic chaperonin CCT: new hints for actin folding. Int J Biochem Cell Biol 42:641–650

Frische EW, Zwartkruis FJ (2010) Rap1, a mercenary among the Ras-like GTPases. Dev Biol 340:1–9

Fuchs M, Poirier DJ, Seguin SJ, Lambert H, Carra S, Charette SJ, Landry J (2010) Identification of the key structural motifs involved in HspB8/HspB6-Bag3 interaction. Biochem J 425:245–255

Gamerdinger M, Hajieva P, Kaya AM, Wolfrum U, Hartl FU, Behl C (2009) Protein quality control during aging involves recruitment of the macroautophagy pathway by BAG3. Embo J 28:889–901

Gamerdinger M, Kaya AM, Wolfrum U, Clement AM, Behl C (2011) BAG3 mediates chaperone-based aggresome-targeting and selective autophagy of misfolded proteins. EMBO Rep 12:149–156

Garcia-Mata R, Gao YS, Sztul E (2002) Hassles with taking out the garbage: aggravating aggresomes. Traffic 3:388–396

Garmyn M, Mammone T, Pupe A, Gan D, Declercq L, Maes D (2001) Human keratinocytes respond to osmotic stress by p38 map kinase regulated induction of HSP70 and HSP27. J Invest Dermatol 117:1290–1295

Garrido C, Paul C, Seigneuric R, Kampinga HH (2012) The small heat shock proteins family: the long forgotten chaperones. Int J Biochem Cell Biol 44:1588–1592

Guay J, Lambert H, Gingras-Breton G, Lavoie JN, Huot J, Landry J (1997) Regulation of actin filament dynamics by p38 map kinase-mediated phosphorylation of heat shock protein 27. J Cell Sci 110:357–368

Harvey KF, Zhang X, Thomas DM (2013) The Hippo pathway and human cancer. Nat Rev Cancer 13:246–257

Hirade K, Kozawa O, Tanabe K, Niwa M, Matsuno H, Oiso Y, Akamatsu S, Ito H, Kato K, Katagiri Y, Uematsu T (2002) Thrombin stimulates dissociation and induction of HSP27 via p38 MAPK in vascular smooth muscle cells. Am J Physiol Heart Circ Physiol 283:H941–H948

Hishiya A, Kitazawa T, Takayama S (2010) BAG3 and Hsc70 interact with actin capping protein CapZ to maintain myofibrillar integrity under mechanical stress. Circ Res 107:1220–1231

Homma S, Iwasaki M, Shelton GD, Engvall E, Reed JC, Takayama S (2006) BAG3 deficiency results in fulminant myopathy and early lethality. Am J Pathol 169:761–773

Huang C (2010) Roles of E3 ubiquitin ligases in cell adhesion and migration. Cell Adh Migr 4:10–18

Huot J, Lambert H, Lavoie JN, Guimond A, Houle F, Landry J (1995) Characterization of 45-kDa/54-kDa HSP27 kinase, a stress-sensitive kinase which may activate the phosphorylation-dependent protective function of mammalian 27-kDa heat-shock protein HSP27. Eur J Biochem FEBS 227:416–427

Huot J, Houle F, Marceau F, Landry J (1997) Oxidative stress-induced actin reorganization mediated by the p38 mitogen-activated protein kinase/heat shock protein 27 pathway in vascular endothelial cells. Circ Res 80:383–392

Ilsley JL, Sudol M, Winder SJ (2002) The WW domain: linking cell signalling to the membrane cytoskeleton. Cell Signal 14:183–189

Ito H, Iwamoto I, Inaguma Y, Takizawa T, Nagata K, Asano T, Kato K (2005) Endoplasmic reticulum stress induces the phosphorylation of small heat shock protein, Hsp27. J Cell Biochem 95:932–941

Iwasaki M, Homma S, Hishiya A, Dolezal SJ, Reed JC, Takayama S (2007) BAG3 regulates motility and adhesion of epithelial cancer cells. Cancer Res 67:10252–10259

Iwasaki M, Tanaka R, Hishiya A, Homma S, Reed JC, Takayama S (2010) BAG3 directly associates with guanine nucleotide exchange factor of Rap1, PDZGEF2, and regulates cell adhesion. Biochem Biophys Res Commun 400:413–418

Jaattela M, Wissing D (1993) Heat-shock proteins protect cells from monocyte cytotoxicity: possible mechanism of self-protection. J Exp Med 177:231–236

Jakob U, Gaestel M, Engel K, Buchner J (1993) Small heat shock proteins are molecular chaperones. J Biol Chem 268:1517–1520

Kabbage M, Dickman MB (2008) The BAG proteins: a ubiquitous family of chaperone regulators. Cell Mol Life Sci 65:1390–1402

Kappe G, Franck E, Verschuure P, Boelens WC, Leunissen JA, de Jong WW (2003) The human genome encodes 10 alpha-crystallin-related small heat shock proteins: HspB1–10. Cell Stress Chaperones 8:53–61

Kappe G, Boelens WC, de Jong WW (2010) Why proteins without an alpha-crystallin domain should not be included in the human small heat shock protein family HSPB. Cell Stress Chaperones 15:457–461

Kassis JN, Virador VM, Guancial EA, Kimm D, Ho AS, Mishra M, Chuang EY, Cook J, Gius D, Kohn EC (2009) Genomic and phenotypic analysis reveals a key role for CCN1 (CYR61) in BAG3-modulated adhesion and invasion. J Pathol 218:495–504

Kato K, Ito H, Kamei K, Iwamoto I (1999) Selective stimulation of Hsp27 and alphaB-crystallin but not Hsp70 expression by p38 MAP kinase activation. Cell Stress Chaperones 4:94–101

Ke L, Meijering RA, Hoogstra-Berends F, Mackovicova K, Vos MJ, Van Gelder IC, Henning RH, Kampinga HH, Brundel BJ (2011) HSPB1, HSPB6, HSPB7 and HSPB8 protect against RhoA GTPase-induced remodeling in tachypaced atrial myocytes. PLoS One 6:e20395

Kim MV, Seit-Nebi AS, Gusev NB (2004a) The problem of protein kinase activity of small heat shock protein Hsp22 (H11 or HspB8). Biochem Biophys Res Commun 325:649–652

Kim MV, Seit-Nebi AS, Marston SB, Gusev NB (2004b) Some properties of human small heat shock protein Hsp22 (H11 or HspB8). Biochem Biophys Res Commun 315:796–801

Kim MV, Kasakov AS, Seit-Nebi AS, Marston SB, Gusev NB (2006) Structure and properties of K141E mutant of small heat shock protein HSP22 (HspB8, H11) that is expressed in human neuromuscular disorders. Arch Biochem Biophys 454:32–41

Knauf U, Bielka H, Gaestel M (1992) Over-expression of the small heat-shock protein, hsp25, inhibits growth of Ehrlich ascites tumor cells. FEBS Lett 309:297–302

Kopito RR (2000) Aggresomes, inclusion bodies and protein aggregation. Trends Cell Biol 10:524–530

Kouyama T, Mihashi K (1981) Fluorimetry study of N-(1-pyrenyl) iodoacetamide-labelled F-actin. Local structural change of actin protomer both on polymerization and on binding of heavy meromyosin. Eur J Biochem 114:33–38

Kunda P, Baum B (2009) The actin cytoskeleton in spindle assembly and positioning. Trends Cell Biol 19:174–179

Kwok AS, Phadwal K, Turner BJ, Oliver PL, Raw A, Simon AK, Talbot K, Agashe VR (2011) HspB8 mutation causing hereditary distal motor neuropathy impairs lysosomal delivery of autophagosomes. J Neurochem 119:1155–1161

Lambrechts A, Gevaert K, Cossart P, Vandekerckhove J, Van Troys M (2008) Listeria comet tails: the actin-based motility machinery at work. Trends Cell Biol 18:220–227

Lamouille S, Xu J, Derynck R (2014) Molecular mechanisms of epithelial-mesenchymal transition. Nat Rev Mol Cell Biol 15:178–196

Landry J, Huot J (1995) Modulation of actin dynamics during stress and physiological stimulation by a signaling pathway involving p38 MAP kinase and heat-shock protein 27. Biochem Cell Biol Biochim Biol Cell 73:703–707

Landry J, Huot J (1999) Regulation of actin dynamics by stress-activated protein kinase 2 (SAPK2)-dependent phosphorylation of heat-shock protein of 27 kDa (Hsp27). Biochem Soc Symp 64:79–89

Landry J, Chretien P, Lambert H, Hickey E, Weber LA (1989) Heat shock resistance conferred by expression of the human HSP27 gene in rodent cells. J Cell Biol 109:7–15

Lavoie JN, Gingras-Breton G, Tanguay RM, Landry J (1993a) Induction of Chinese hamster HSP27 gene expression in mouse cells confers resistance to heat shock. HSP27 stabilization of the microfilament organization. J Biol Chem 268:3420–3429

Lavoie JN, Hickey E, Weber LA, Landry J (1993b) Modulation of actin microfilament dynamics and fluid phase pinocytosis by phosphorylation of heat shock protein 27. J Biol Chem 268:24210–24214

Lavoie JN, Lambert H, Hickey E, Weber LA, Landry J (1995) Modulation of cellular thermoresistance and actin filament stability accompanies phosphorylation-induced changes in the oligomeric structure of heat shock protein 27. Mol Cell Biol 15:505–516

Lavoie JN, Landry MC, Faure RL, Champagne C (2010) Src-family kinase signaling, actin-mediated membrane trafficking and organellar dynamics in the control of cell fate: lessons to be learned from the adenovirus E4orf4 death factor. Cell Signal 22:1604–1614

Li N, Du ZX, Zong ZH, Liu BQ, Li C, Zhang Q, Wang HQ (2013) PKCdelta-mediated phosphorylation of BAG3 at Ser187 site induces epithelial-mesenchymal transition and enhances invasiveness in thyroid cancer FRO cells. Oncogene 32:4539–4548

Linares JF, Amanchy R, Greis K, Diaz-Meco MT, Moscat J (2011) Phosphorylation of p62 by cdk1 controls the timely transit of cells through mitosis and tumor cell proliferation. Mol Cell Biol 31:105–117

Linnemann A, van der Ven PF, Vakeel P, Albinus B, Simonis D, Bendas G, Schenk JA, Micheel B, Kley RA, Furst DO (2010) The sarcomeric Z-disc component myopodin is a multiadapter protein that interacts with filamin and alpha-actinin. Eur J Cell Biol 89:681–692

McHaourab HS, Godar JA, Stewart PL (2009) Structure and mechanism of protein stability sensors: chaperone activity of small heat shock proteins. Biochemistry 48:3828–3837

McLaughlin MM, Kumar S, McDonnell PC, Van Horn S, Lee JC, Livi GP, Young PR (1996) Identification of mitogen-activated protein (MAP) kinase-activated protein kinase-3, a novel substrate of CSBP p38 MAP kinase. J Biol Chem 271:8488–8492

McMullen ME, Bryant PW, Glembotski CC, Vincent PA, Pumiglia KM (2005) Activation of p38 has opposing effects on the proliferation and migration of endothelial cells. J Biol Chem 280:20995–21003

Mehlen P, Preville X, Chareyron P, Briolay J, Klemenz R, Arrigo AP (1995) Constitutive expression of human hsp27, Drosophila hsp27, or human alpha B-crystallin confers resistance to TNF- and oxidative stress-induced cytotoxicity in stably transfected murine L929 fibroblasts. J Immunol 154:363–374

Miron T, Wilchek M, Geiger B (1988) Characterization of an inhibitor of actin polymerization in vinculin-rich fraction of turkey gizzard smooth muscle. Eur J Biochem FEBS 178:543–553

Miron T, Vancompernolle K, Vandekerckhove J, Wilchek M, Geiger B (1991) A 25-kD inhibitor of actin polymerization is a low molecular mass heat shock protein. J Cell Biol 114:255–261

Moayeri M, Leppla SH (2004) The roles of anthrax toxin in pathogenesis. Curr Opin Microbiol 7:19–24

Morrow G, Tanguay RM (2012) Small heat shock protein expression and functions during development. Int J Biochem Cell Biol 44:1613–1621

Mounier N, Arrigo AP (2002) Actin cytoskeleton and small heat shock proteins: how do they interact? Cell Stress Chaperones 7:167–176

Muller E, Burger-Kentischer A, Neuhofer W, Fraek ML, Marz J, Thurau K, Beck FX (1999) Possible involvement of heat shock protein 25 in the angiotensin II-induced glomerular mesangial cell contraction via p38 MAP kinase. J Cell Physiol 181:462–469

Olzmann JA, Chin LS (2008) Parkin-mediated K63-linked polyubiquitination: a signal for targeting misfolded proteins to the aggresome-autophagy pathway. Autophagy 4:85–87

Pichon S, Bryckaert M, Berrou E (2004) Control of actin dynamics by p38 MAP kinase – Hsp27 distribution in the lamellipodium of smooth muscle cells. J Cell Sci 117:2569–2577

Ravikumar B, Acevedo-Arozena A, Imarisio S, Berger Z, Vacher C, O'Kane CJ, Brown SD, Rubinsztein DC (2005) Dynein mutations impair autophagic clearance of aggregate-prone proteins. Nat Genet 37:771–776

Razinia Z, Makela T, Ylanne J, Calderwood DA (2012) Filamins in mechanosensing and signaling. Annu Rev Biophys 41:227–246

Rodriguez-Gonzalez A, Lin T, Ikeda AK, Simms-Waldrip T, Fu C, Sakamoto KM (2008) Role of the aggresome pathway in cancer: targeting histone deacetylase 6-dependent protein degradation. Cancer Res 68:2557–2560

Rogov V, Dotsch V, Johansen T, Kirkin V (2014) Interactions between autophagy receptors and ubiquitin-like proteins form the molecular basis for selective autophagy. Mol Cell 53:167–178

Rollet E, Lavoie JN, Landry J, Tanguay RM (1992) Expression of Drosophila's 27 kDa heat shock protein into rodent cells confers thermal resistance. Biochem Biophys Res Commun 185:116–120

Rosati A, Graziano V, De Laurenzi V, Pascale M, Turco MC (2011) BAG3: a multifaceted protein that regulates major cell pathways. Cell Death Dis 2:e141

Rouse J, Cohen P, Trigon S, Morange M, Alonso-Llamazares A, Zamanillo D, Hunt T, Nebreda AR (1994) A novel kinase cascade triggered by stress and heat shock that stimulates MAPKAP kinase-2 and phosphorylation of the small heat shock proteins. Cell 78:1027–1037

Rousseau S, Houle F, Landry J, Huot J (1997) p38 MAP kinase activation by vascular endothelial growth factor mediates actin reorganization and cell migration in human endothelial cells. Oncogene 15:2169–2177

Salah Z, Alian A, Aqeilan RI (2012) WW domain-containing proteins: retrospectives and the future. Front Biosci 17:331–348

Selcen D, Muntoni F, Burton BK, Pegoraro E, Sewry C, Bite AV, Engel AG (2009) Mutation in BAG3 causes severe dominant childhood muscular dystrophy. Ann Neurol 65:83–89

Shaid S, Brandts CH, Serve H, Dikic I (2013) Ubiquitination and selective autophagy. Cell Death Differ 20:21–30

Shin KD, Lee MY, Shin DS, Lee S, Son KH, Koh S, Paik YK, Kwon BM, Han DC (2005) Blocking tumor cell migration and invasion with biphenyl isoxazole derivative KRIBB3, a synthetic molecule that inhibits Hsp27 phosphorylation. J Biol Chem 280:41439–41448

Singh BN, Rao KS, Ramakrishna T, Rangaraj N, Rao Ch M (2007) Association of alphaB-crystallin, a small heat shock protein, with actin: role in modulating actin filament dynamics in vivo. J Mol Biol 366:756–767

Smith CC, Yu YX, Kulka M, Aurelian L (2000) A novel human gene similar to the protein kinase (PK) coding domain of the large subunit of herpes simplex virus type 2 ribonucleotide reductase (ICP10) codes for a serine-threonine PK and is expressed in melanoma cells. J Biol Chem 275:25690–25699

Smoyer WE, Ransom RF (2002) Hsp27 regulates podocyte cytoskeletal changes in an in vitro model of podocyte process retraction. FASEB J 16:315–326

Soderling SH (2009) Grab your partner with both hands: cytoskeletal remodeling by Arp2/3 signaling. Sci Signal 2:pe5

Stokoe D, Engel K, Campbell DG, Cohen P, Gaestel M (1992) Identification of MAPKAP kinase 2 as a major enzyme responsible for the phosphorylation of the small mammalian heat shock proteins. FEBS Lett 313:307–313

Taylor MP, Koyuncu OO, Enquist LW (2011) Subversion of the actin cytoskeleton during viral infection. Nat Rev Microbiol 9:427–439

Tissieres A, Mitchell HK, Tracy UM (1974) Protein synthesis in salivary glands of Drosophila melanogaster: relation to chromosome puffs. J Mol Biol 84:389–398

Ulbricht A, Hohfeld J (2013) Tension-induced autophagy: may the chaperone be with you. Autophagy 9:920–922

Ulbricht A, Arndt V, Hohfeld J (2013) Chaperone-assisted proteostasis is essential for mechanotransduction in mammalian cells. Commun Integr Biol 6:e24925

Vertii A, Hakim C, Kotlyarov A, Gaestel M (2006) Analysis of properties of small heat shock protein Hsp25 in MAPK-activated protein kinase 2 (MK2)-deficient cells: MK2-dependent insolubilization of Hsp25 oligomers correlates with susceptibility to stress. J Biol Chem 281:26966–26975

Vos MJ, Hageman J, Carra S, Kampinga HH (2008) Structural and functional diversities between members of the human HSPB, HSPH, HSPA, and DNAJ chaperone families. Biochemistry 47:7001–7011

Vos MJ, Zijlstra MP, Kanon B, van Waarde-Verhagen MA, Brunt ER, Oosterveld-Hut HM, Carra S, Sibon OC, Kampinga HH (2010) HSPB7 is the most potent polyQ aggregation suppressor within the HSPB family of molecular chaperones. Hum Mol Genet 19:4677–4693

Vos MJ, Zijlstra MP, Carra S, Sibon OC, Kampinga HH (2011) Small heat shock proteins, protein degradation and protein aggregation diseases. Autophagy 7:101–103

Wang K, Spector A (1996) alpha-crystallin stabilizes actin filaments and prevents cytochalasin-induced depolymerization in a phosphorylation-dependent manner. Eur J Biochem FEBS 242:56–66

Weber NC, Toma O, Wolter JI, Wirthle NM, Schlack W, Preckel B (2005) Mechanisms of xenon- and isoflurane-induced preconditioning – a potential link to the cytoskeleton via the MAPKAPK-2/HSP27 pathway. Br J Pharmacol 146:445–455

Welch MD, Way M (2013) Arp2/3-mediated actin-based motility: a tail of pathogen abuse. Cell Host Microbe 14:242–255

Welsh MJ, Gaestel M (1998) Small heat-shock protein family: function in health and disease. Ann N Y Acad Sci 851:28–35

Werner A, Disanza A, Reifenberger N, Habeck G, Becker J, Calabrese M, Urlaub H, Lorenz H, Schulman B, Scita G, Melchior F (2013) SCFFbxw5 mediates transient degradation of actin remodeller Eps8 to allow proper mitotic progression. Nat Cell Biol 15:179–188

Xiao H, Cheng S, Tong R, Lv Z, Ding C, Du C, Xie H, Zhou L, Wu J, Zheng S (2014) BAG3 regulates epithelial-mesenchymal transition and angiogenesis in human hepatocellular carcinoma. Lab Invest J Tech Methods Pathol 94:252–261

Xu Z, Graham K, Foote M, Liang F, Rizkallah R, Hurt M, Wang Y, Wu Y, Zhou Y (2013) 14-3-3 protein targets misfolded chaperone-associated proteins to aggresomes. J Cell Sci 126:4173–4186

Yang Z, Klionsky DJ (2010) Eaten alive: a history of macroautophagy. Nat Cell Biol 12:814–822

Zhang X, Qian SB (2011) Chaperone-mediated hierarchical control in targeting misfolded proteins to aggresomes. Mol Biol Cell 22:3277–3288

Zhou M, Lambert H, Landry J (1993) Transient activation of a distinct serine protein kinase is responsible for 27-kDa heat shock protein phosphorylation in mitogen-stimulated and heat-shocked cells. J Biol Chem 268:35–43

Chapter 19
Heat Shock Alters Keratocyte Movement and Morphology: Exploring a Role for HSP27 (HSPB1)

Bindi M. Doshi, Lawrence E. Hightower, and Juliet Lee

Abstract HSP27 is essential for mammalian cell movement. To further explore the effects of heat shock and the mechanistic role of HSP27, we have initiated a study using a well-established model of rapidly moving cells, the fish keratocyte. Here we report that heat shock causes a decrease in cell speed. Since changes in cell morphology can drastically affect cell movement, we also monitored changes in cell morphology. Heat shock caused a decrease in the number of polar cells and an increase in those with one stuck adherent edge, indicating the occurrence of both cytoskeletal re-organization and increased adhesion to substrata. Analyses of HSP27 levels using Western blots showed they were relatively high in keratocytes prior to heat shock and remained high afterward. In contrast, Western blot analysis of HSP70 showed that it was induced strongly by heat shock, indicating that fish keratocytes mounted a robust heat shock response. Surprisingly, given the propensity of HSP27 to localize in nuclear/perinuclear regions following heat shock, the location of HSP27 in fish keratocytes was unchanged as shown by indirect immunostaining with anti-HSP27 antibodies. Fluorescence intensities of immunostained images of cells before and after heat shock were quantified using Image J software. The results of this analysis showed that fluorescence intensity decreased following heat shock, suggesting changes in HSP27 that affected antibody recognition. Possible roles for HSP27 in regulating actin filament dynamics, cell speed and morphology are discussed.

Keywords HSP27 • HSPB1 • Cell motility • Keratocytes • Heat shock • Cell morphology • Actin

---

B.M. Doshi • L.E. Hightower (✉) • J. Lee
Department of Molecular and Cell Biology, University of Connecticut, Storrs, CT 06269-3125, USA
e-mail: lawrence.hightower@uconn.edu

R.M. Tanguay, L.E. Hightower (eds.), *The Big Book on Small Heat Shock Proteins*, Heat Shock Proteins 8, DOI 10.1007/978-3-319-16077-1_19

19.1 Introduction

When we began our studies of human HSP27 (HSPB1), the most convincing evidence linking this protein to actin filament dynamics was its inhibition of actin polymerization in vitro (Miron et al. 1991; Benndorf et al. 1994). Our goal was to demonstrate this link in cells by studying the role of HSP27 in cell motility. We used a human colon cancer cell line and nontumorogenic colonocytes to show that HSP27 is critical for wound healing (Doshi et al. 2009). HSP27 levels were reduced using siRNA, resulting in loss of the cells' ability to close wounds in a cell culture-based assay. Immunoprecipitation pull-down assays were used to show that HSP27 and actin are in the same complex both before and after heat shock. Indirect immunostaining of cells was also done to evaluate co-localization of these two proteins. They showed co-localization before heat shock, little association by 3 h post heat shock, and increased association by 24 h after heat shock. Cells were still capable of movement when cytoprotection was established 3 h after heat shock.

Having established that HSP27 plays a vital role in regulating wound healing, we wished to study a motile cell type in which the effects of heat shock on cell speed and morphology can be measured readily. We chose fish epithelial keratocytes, because they are a well-studied model for fast moving cell types that the Lee laboratory has investigated in detail. In addition, we also have experience in analyses of the induction of HSP27 by heat in a fish hepatoma cell line (PLHC-1). HSPs 70, 30 and 27 have been characterized in these cells derived from the subtropical desert minnow species, *Peociliopsis lucida* (Norris et al. 1997). Using radioisotopic labeling, the heat shock temperature and the HSP27 induction time frame were determined. These results provided a starting point for determination of the heat shock response of the keratocytes derived from the tropical and subtropical species of Black Molly used for the experiments described herein.

The question remains why is HSP27 found with actin? It may regulate actin filament dynamics, and/or it may function as an actin binding protein providing structural support for the cytoskeleton. We sought to further investigate the function of HSP27 and actin interaction in keratocytes.

Keratocytes were obtained from Black Molly, *Molliniesia sphenops*, scales. Fish keratocytes are known for their fan-shaped lamella and tightly coordinated movement. As the leading lamella extends at the front, it pulls the cell body forward, and when the rear of the cell detaches, the entire cell can advance. Keratocyte movement is regulated by stretch-activated calcium channels (SACs) that promote rapid movement by allowing cells to sense and respond to increases in cytoskeletal tension (Lee et al. 1999). For example, when the cell rear is unable to detach, while the front is extending, cytoskeletal tension builds. When tension reaches a critical threshold level, SACs are activated, resulting in an increase in intracellular calcium and consequently release of the cell rear. After retraction at the rear of the cell, cytoskeletal tension drops and the cell is able to continue moving. Thus protrusion and retraction are tightly coordinated between the front and rear of the cell. Actin polymerization occurs at the front of the cell where various proteins are involved in regulating filament dynamics. Their functions can include stabilizing filaments, capping or uncapping filaments, and severing filaments. It is not known exactly how actin filaments are regulated by HSP27 or how heat shock affects filament dynamics. It is possible that HSP27 may regulate filament dynamics, as suggested by previous findings of HSP27 association with actin (Clarke et al. 2013; Doshi et al. 2009; Graceffa 2011; Miron et al. 1991; Mounier and Arrigo 2002).

To begin characterizing the heat shock response in keratocytes, changes in speed and morphology were observed before and after heat shock using phase contrast microscopy. Indirect immunostaining was used to determine HSP27 localization before and after heat shock. Western blot analysis was used to measure changes in HSP70 and HSP27 protein expression levels following heat shock.

19.2 Results

19.2.1 Heat Shock Decreased Cell Speed

To establish the effect of heat shock on cell movement, keratocytes were imaged in a heated chamber using differential interference (DIC) microscopy before, during, and after a 1-h heat shock at temperatures ranging from 37 to 41 °C. The observations after heat shock were made 1 h and 24 h after the termination of the heat shock. The speed of individual cells was obtained from time-lapse movies using the MTrackJ tool in Image J software. The back edge of each cell in the field of view was tracked throughout the image series to obtain the x, y coordinates of the cell position over time, which were used to calculate speed. We found that cell speed decreased in cells receiving heat shock at either 37 °C or 39 °C (Fig. 19.1). At 37 °C, average cell speed decreased significantly after a 1-h recovery period following heat shock, and decreased further by 24 h post heat shock. At 39 °C, average cell speed was also significantly decreased by 1 h and 24 h after heat shock. Heat shock temperatures of 40 and 41 °C for 1 h resulted in cell death (data not shown).

Changes in cell morphology can drastically influence cell speed. For example, keratocytes move most rapidly when they exhibit a fan shape that is indicative of a high degree of polarity. However, cell speed is reduced if polarity is lost such as when one edge becomes stuck, or the cell develops two lamellae. Many keratocytes showed a change in cell morphology as a result of heat shock (Fig. 19.2). To determine how the decrease in cell speed is related to changes in cell morphology, heat shocked cells were sorted into categories; polar, bipolar, one stuck edge, apolar, and rounded cells (Fig. 19.2a). Apolar cells were identified as cells having a lamella surrounding the cell body whereas rounded cells entirely lacked a lamella. The number of polar cells decreased after either a 37 or 39 °C heat shock. After a 37 °C heat shock for 1 h, keratocytes showed a doubling in the percentage of bipolar cells, no change in the percentage of cells with one stuck edge, apolar and rounded cells, consistent with substantial cytoskeletal rearrangement (Fig. 19.2b). After a 39 °C

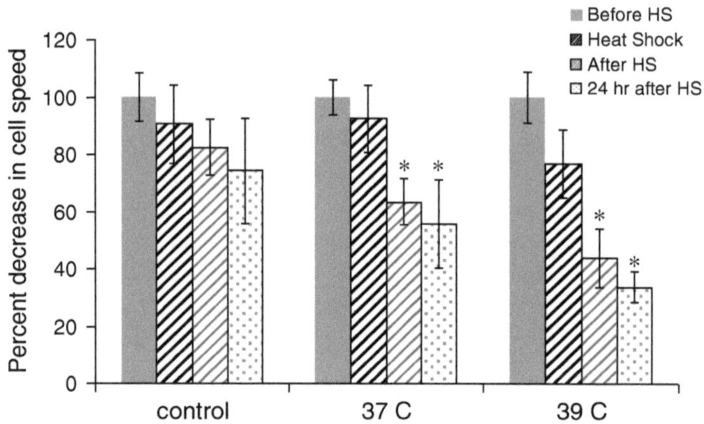

Fig. 19.1 Cell speed decreases in response to heat shock. Keratocyte average cell speed was measured before heat shock, during heat shock, 1 h after heat shock and 24 h after heat shock. Cells received a 1-h heat treatment at either 37 or 39 °C. Cells were maintained at room temperature at all other times. Control cells were monitored under the same conditions with the exclusion of a heat treatment. Individual cells were monitored and the results were averaged: Before HS (before heat shock) n = 19 for the 37 °C experiment and n = 34 for the 39 °C experiment; Heat Shock (during heat shock) n = 12 for 37 °C and n = 17 for 39 °C; After HS (1 h after heat shock) n = 12 at 37 °C and n = 11 at 39 °C; 24 h after HS (24 h after heat shock) n = 15 at 37 °C and n = 9 at 39 °C. Error bars denote standard error of the means. *Statistically significant at $P < 0.05$ (Student *t*-Test)

heat shock, the keratocyte population showed a greater percentage change in the same categories of cell morphology including a doubling of the number of cells with one stuck edge, indicating the occurrence of both cytoskeletal re-organization and increased adhesion strength (Fig. 19.2c).

19.2.2 HSP27 Localization Is Unaltered After Heat Shock

HSP27 is known for its translocation from the cytoplasmic region to the perinuclear/nuclear region after cells experience stress (Arrigo et al. 1988; Borrelli et al. 2002). As we observed previously (Doshi et al. 2009), SW480 colon cancer cells display HSP27 translocation to the perinuclear region as a result of heat shock. To establish whether HSP27 translocated to the perinuclear region in keratocytes following heat shock, indirect immunostaining was used to visualize HSP27 localization. Indirect immunostaining was done before and after heat shock, using primary HSP27 antibodies (Shelden laboratory) that were detected with a fluorescent secondary antibody. To monitor changes in cytoskeletal organization, fluorescent phalloidin was used to visualize actin (Fig. 19.3a). Immunofluorescence staining was performed on keratocytes prior to heat shock (Before 39 °C HS), immediately after receiving a 1-h heat shock (After 39 °C HS), and either 3 or 24 h after heat shock. Wide-field fluorescence microscopy was used to detect

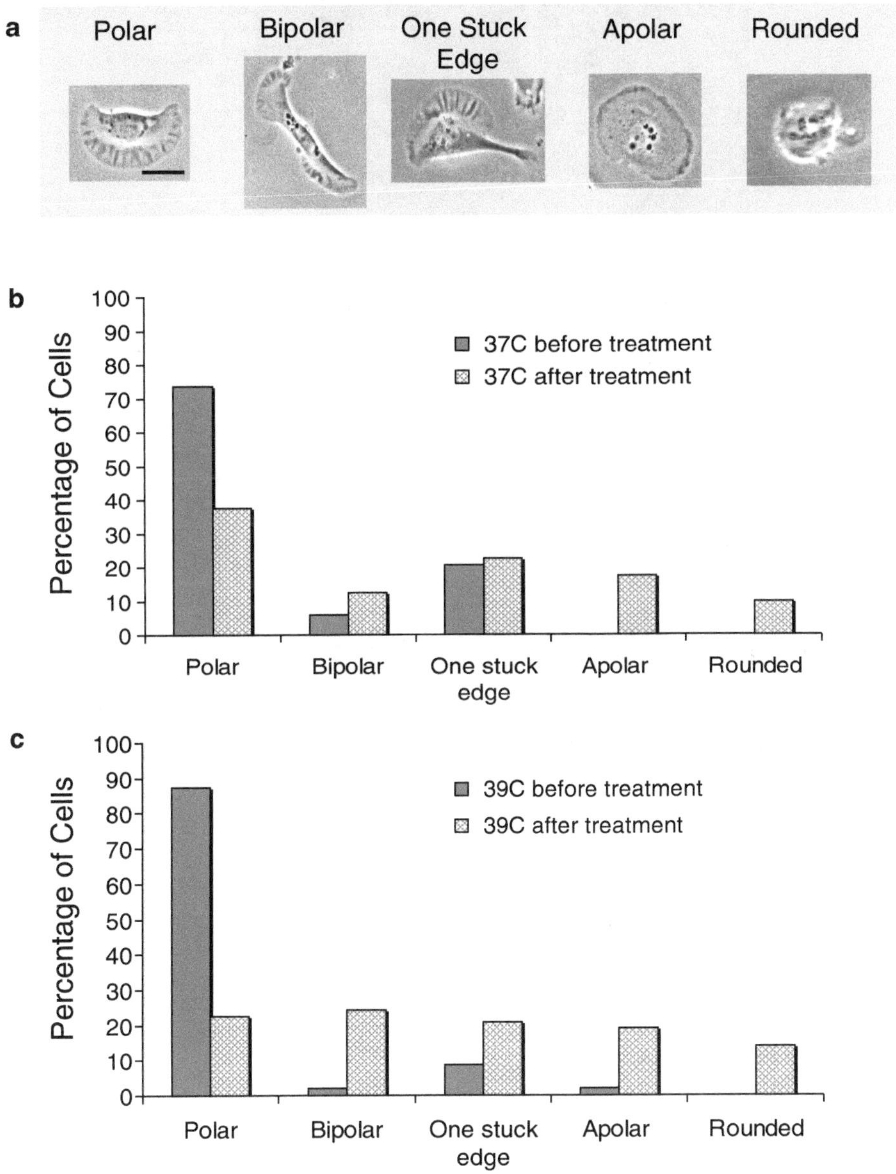

Fig. 19.2 Changes in cell morphology in response to heat shock. Individual keratocytes were classified into one of six different morphological categories (**a**). (**b**) Cell morphology before/after a 37 °C heat shock. Cell morphology was determined before (*dark grey bars*) and after (*light grey bars*) at 37 °C heat shock. Thirty-four cells were assessed before heat shock and 40 cells were assessed after a 39 °C heat shock. (**c**) Cell morphology before/after a 39 °C heat shock. Cell morphology was determined before (*dark grey bars*) and after (*light grey bars*) at 39 °C heat shock. Forty-seven cells were assessed before heat shock and 58 cells were assessed after a 39 °C heat shock

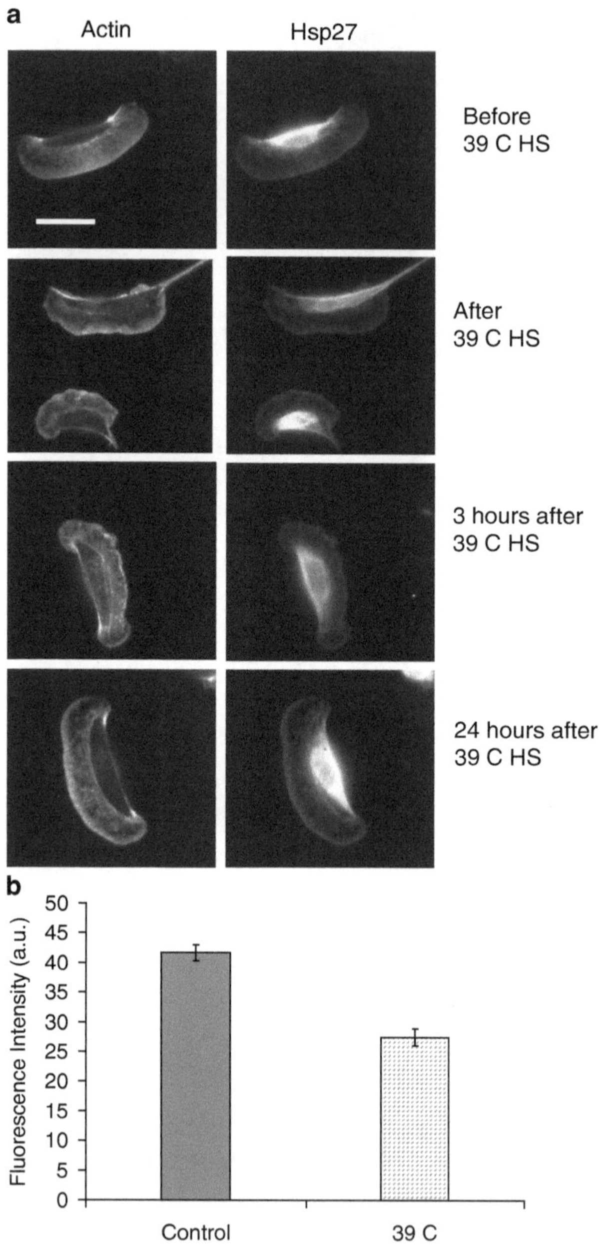

Fig. 19.3 Cellular location of HSP27 before and after heat shock. Keratocytes were fixed and stained for endogenous HSP27 using anti-HSP27 primary antibodies obtained from the Shelden laboratory and actin was detected using rhodamine-labeled phalloidin. Heat was applied at 39 °C for 1 h. (**a**) Images of representative unheated control cells fixed and subjected to indirect immunostaining (Before 39 °C HS, n = 33), cells stained immediately after heat shock (After 39 °C HS, n = 23), cells stained 3 h after heat shock (3 h after 39 °C HS, n = 31) and cells stained 24 h after heat shock (24 h after 39 °C HS, n = 20). Scale bar, 10 uM. (**b**) Fluorescence intensity levels derived from Image J software analysis of HSP27-stained cells shown in previous panel. The bar marked Control shows data obtained from cells without heating and the bar marked 39 °C shows data obtained from cells after a 1-h heat shock. The standard error of the means are shown

localization of HSP27. No change in HSP27 localization was observed between non-heat shocked and heat shocked cells. This result was confirmed using a primary HSP27 antibody from the Tanguay laboratory.

To quantify the amount of HSP27 immunostaining inside keratocytes, fluorescence intensity was measured using Image J Software. Fluorescence intensity of areas in cells stained with anti HSP-27 antibodies decreased after cells received heat shock (Fig. 19.3b).

19.2.3 HSP27 Expression Levels Were Unchanged Following Heat Shock

To detect changes in HSP70 and HSP27 expression levels before and after heat shock, protein levels were detected using Western blot analysis. HSP70 induction is a common characteristic of a heat shock response by virtually all organisms including fish (Norris et al. 1997). HSP70 protein expression levels were probed on immunoblots of SDS-PAGE gels of keratocyte lysates before and after heat shock (Fig. 19.4a). HSP70 levels increased compared to actin controls by 3 h after a 39 °C heat shock for 1 h, confirming that fish keratocytes could mount a heat shock response under our experimental conditions.

To determine whether HSP27 expression levels are affected by heat shock, immunoblots of cell lysates were made before and after heat shock and visualized using anti-HSP27 antibodies from either the Shelden or the Tanguay laboratory. No significant increases in HSP27 protein levels were detected 24 h after either a 37 or

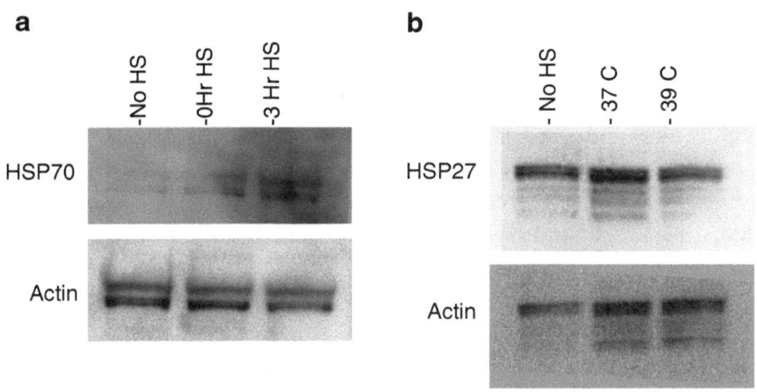

Fig. 19.4 Immunoblot analysis of HSP70 and HSP27 levels in keratocytes before and after heat shock. (**a**) Cell lysates were made from unheated control cells (No HS) and immediately after (0 h HS) or 3 h (3 h HS) after a 39 °C heat shock for 1 h. Proteins were separated by SDS-PAGE and the gels were processed by Western blotting using anti-actin and anti-HSP70 primary antibody. (**b**) Cell lysates were made from unheated control cells (No HS) and from cells 3 h after heating for 1 h at either 37 or 39 °C. The lysates were processed by Western blotting using anti-actin and anti-HSP27 (Shelden antibody)

39 °C heat treatment compared to actin controls (Fig. 19.4b). To assess whether HSP27 induction was occurring at shorter times post heat shock, protein extracts were taken before 1 h of heat shock at both temperatures, immediately after a 1-h heat shock, 3, 5 and 9 h post heat shock (data not shown). None of the lysates showed increased levels of HSP27 relative to actin controls regardless of which primary anti-HSP27 antibody was used. Immunoblots made from lysates of SW480 colon cancer cells were probed for HSP27 to verify that the antibodies we used could recognize keratocyte HSP27. Western blots showed that HSP27 levels increased after heat shock in SW480 cells, confirming that the antibody recognized HSP27 (data not shown). Since the amount of keratocyte HSP27 was high prior to heat shock and unchanged afterward, these results support the idea that keratocytes may express maximum levels of HSP27 at normal culture temperatures of 22 °C.

19.3 Discussion

HSP27 associates with actin in human cells in culture, indicating that it may be involved in actin filament dynamics (Doshi et al. 2009). It is often stated that HSP27 (HSPB1) is an actin filament barbed-end capping protein in vitro (Guay et al. 1997; Miron et al. 1991). This early work used the murine and avian HSP25 and numerous models of in vivo interactions between human HSP27 and actin filaments often have been based on studies of HSP25. More recent studies by Graceffa and colleagues using HSP27 (Graceffa 2011) and employing analytical ultracentrifugation, fluorescence spectroscopy and electron microscopy (EM) support the conclusion that HSP27 is a weak actin filament (F-actin) side-binding protein in vitro. Additional evidence from this study indicates that HSP27 binds F-actin as a monomer. Other investigators studying the effect of HSP27 on actin polymerization have concluded that it is probably not a barbed-end capping protein (During et al. 2007). They considered HSP27 to be an actin monomer sequestering protein. Graceffa stated in his paper that his work did not rule out the possibility that HSP27 is a barbed-end capping protein but rather it added support to the idea that HSP27 is also an F-actin side-binding protein. In this regard, we found the EM study reported by Graceffa quite interesting in that addition of HSP27 to F-actin did not result in dissociation of the actin filaments, which prompted him to conclude that HSP27 is not a strong sequesterer of actin monomers. It is important to note that the fluorescently tagged forms of HSP27 used in the Graceffa studies did not involve modification of the termini of HSP27, and so these are unlikely to interfere with actin binding. This possibility is supported by the recent finding that the attachment of fluorescent proteins to either the N-terminal or C-terminal ends of HSP27 and several other small HSPs alters their oligomeric structures, their chaperone-like activities and their ability to form hetero-oligomers (Datskevich and Gusev 2013).

A recent cell-based study (Clarke et al. 2013) may be more pertinent to our experiments with keratocytes. PC12 cells, a neuroendocrine rat pheochromocytoma line that expresses HSPB1 endogenously, was used. Association between

HSPB1 and F-actin was investigated using pull-down assays in conjunction with immunocytochemistry, confocal microscopy and Western blotting. The authors showed that F-actin is complexed with both phosphorylated and non-phosphorylated HSPB1 and that this association increases upon heat shock. Even though the details of this response are not known, it may account for the effects of heat shock on keratocytes.

The effects of heat shock on keratocyte movement were investigated to determine whether alterations in motility are related to HSP27 expression (Fig. 19.5a). HSP27 protein expression levels remained unchanged after heat shock, indicating that keratocyte HSP27 did not show the characteristic heat shock response common to many other cell lines. Cell speed decreased in response to heat shock and cell morphology was also affected. Keratocyte morphology changed from a polarized fan shape before heat shock, to either bipolar, apolar, rounded or having a stuck edge, after heat shock. Unlike many other cell types, fish keratocyte HSP27 did not translocate to either the nucleus or perinuclear region after heat shock. However, immunofluorescence staining for HSP27 decreased. This, together with the fact that HSP27 expression levels are unchanged by heat shock, indicates that HSP27 was less accessible to the primary antibody due to either a change in conformation or association state. This possibility is supported by evidence that HSP27-actin binding strength increases in response to heat shock. Given the abundance of actin filament networks in keratocytes, it is possible that HSP27 becomes tightly associated with these networks and unavailable for translocation to the nucleus as well as less accessible to the primary antibody.

Heat shock caused a significant decrease in keratocyte cell speed, which gave a strong indication that actin filament dynamics was altered. This may be due to the interaction of HSP27 with actin filaments. For example, if HSP27 acts as a plus end actin capping protein in vivo, it might promote a high rate of movement by maintaining a large pool of actin monomers in an unstressed cell. However, significant uncapping of HSP27 would decrease cell speed due to the depletion of actin monomers, as might be expected in a cell experiencing heat shock. HSP27 may also stabilize actin filaments during stress by binding to their sides. For example, fibroblasts over-expressing HSP27 were able to adhere better to collagen matrix than control or under-expressing HSP27 cells (Hirano et al. 2004). While its exact function in actin filament dynamics is unclear, numerous studies have shown that HSP27 is essential for cell migration. This has been demonstrated by studies where RNAi mediated reduction of HSP27 levels decreased migration regardless of cell type (Doshi et al. 2009; Garcia-Arguinzonis et al. 2010; Golembieski et al. 2008; Nomura et al. 2007).

The changes in keratocyte morphology and speed that occurred after heat shock suggest a change in actin filament dynamics. For example, some keratocytes appeared to become more firmly attached to the substratum, as indicated by cells possessing one stuck edge that is unable to retract. In addition, many cells lost polarity, becoming bipolar or apolar. Evidence suggests that the activity of mitogen-activated protein kinase (MAPK) might be involved in these changes, because MAPK phosphorylates HSP27 in response to stress (Stokoe et al. 1992), which is a major determinant of HSP27 function (Fig. 19.5b). Un-phosphorylated HSP27

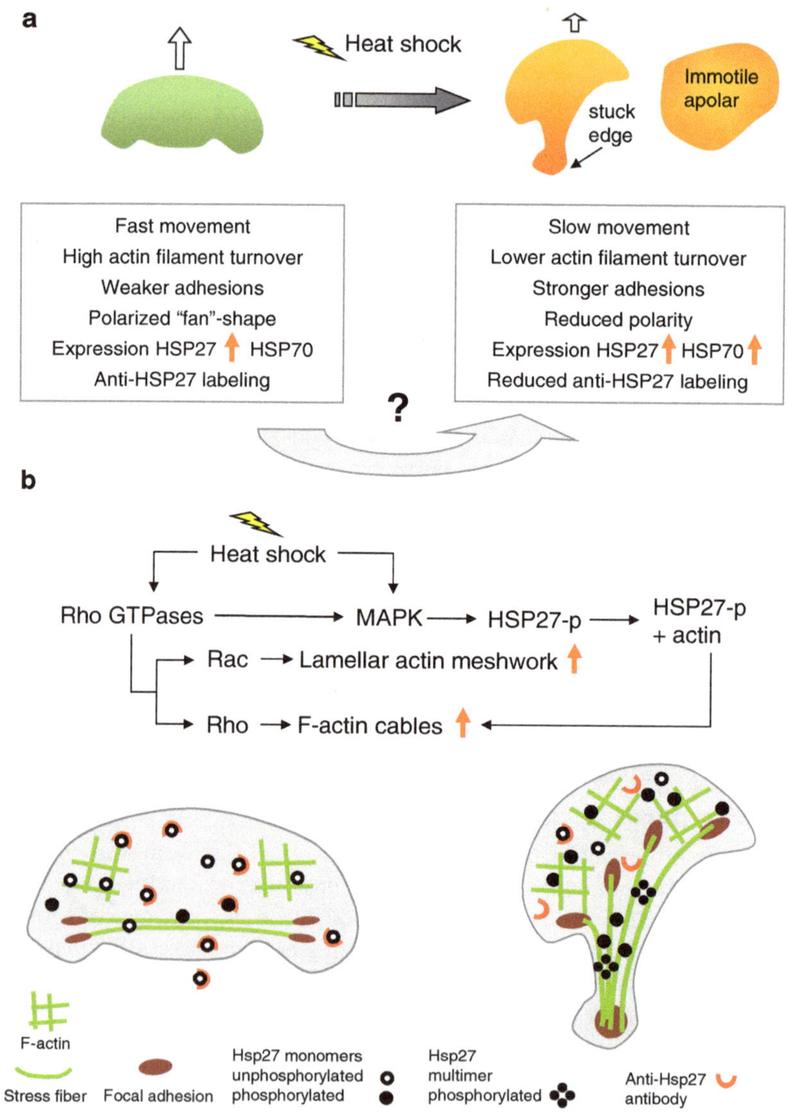

Fig. 19.5 Summary of effects of heat shock on keratocyte morphology and movement. (**a**) Diagram based on a summary of experimental results. (**b**) Diagram based on hypotheses derived from our experiments and current literature

cannot initiate new filament growth, and it can inhibit actin polymerization by behaving as an actin plus end capping protein (Benndorf et al. 1994; Miron et al. 1991). However, phosphorylated monomers and non-phosphorylated multimers permit actin polymerization in vitro (Benndorf et al. 1994).

A recent interesting paper (Stöhr and Hüttelmaier 2012) reported that IGF2 mRNA binding protein 1 (IGF2BP1), an oncofetal RNA-binding protein, is an oncogenic factor that regulates adhesion, migration and ultimately invasiveness of tumor cells through intracellular signaling pathways. Evidence was presented that IGF2BP1 interferes with phosphorylation of HSP27 directed by MAPK-activated protein kinase 5 (MK5). This protein kinase is also known as P38-regulated/activated protein kinase. This occurs indirectly through the facilitation of inhibition of MAPK4 mRNA translation by IGF2BP1. It is proposed that these alterations limit the ability of phosphorylated HSP27 to sequester G-actin, which in turn enhances cell adhesion and increases the speed of tumor cell migration.

Activation of the Rho GTPases has been shown to increase MAP kinase activity in response to stress (Du et al. 2010; Lee et al. 2001) and may also induce changes in keratocyte cell morphology. This is believed to occur through the activation of PAK (p21-activated kinase) which can activate the p38MAP kinase pathway together with Cdc42 and Rac. Increased Rac activity is consistent with our observation of apolar cells with large, spread lamellae, whereas the activation of Rho might explain why some keratocytes appear stuck at one edge, since this GTPase will increase stress fiber and adhesion formation. Increased levels of phosphorylated HSP27 might also contribute to this phenotype by stabilizing actin filaments, and indirectly decreasing adhesion turnover, which could reduce cell speed.

Translocation of HSP27 to the nucleus or to perinuclear regions after a heat shock has become a hallmark of the heat shock response in many different cell types (Arrigo et al. 1988; Ehrnsperger et al. 1999). Often, cytoskeletal networks collapse into the perinuclear region as well (Mounier and Arrigo 2002). We observed that in fish keratocytes the cytoskeletal structure remained stable after heat shock as observed by visualization of actin stress fibers. Under these circumstances, it is possible that cytoplasmic HSP27 becomes fully associated with actin stress fibers to stabilize them during heat shock. Consequently, these HSP27 molecules would not have been available to translocate, and no perinuclear localization of HSP27 due to a collapse of the filament network would have occurred. The decrease in fluorescence intensity of cytoplasmic HSP27 that is observed in keratocytes following heat shock may have been another manifestation of this increased association with the actin cytoskeleton. This is because the antibody binding site(s) on HSP27 may have been shielded by interaction with the actin filament network or conformational changes in HSP27 bound to these filaments may have reduced its affinity for the primary antibodies. Another possibility is that oligomerization of HSP27 following heat shock may have reduced antibody affinity due to conformational changes and shielding of antibody binding sites.

We observed that keratocytes have an elevated level of HSP27 without heat shock, when compared with the minimal levels of HSP27 expression seen in PLHC (*P. lucida* liver cells) without heat shock. Possibly since HSP27 is already upregulated in keratocytes prior to heat shock, HSP27 expression levels may not be able to increase further in response to heat shock. HSP27 present in high amounts in keratocytes and low levels in PLHC cells further exemplifies how regulation of this protein varies among cell types.

Fish have evolved a particularly rapid wound healing response that allows them to quickly repair skin lesions. According to our hypothesis, fish keratocytes would not require several hours of stress-induced gene expression in order to produce elevated levels of Hsp27 and to achieve maximum changes in cell morphology. This would allow rapid movement of keratocytes into damaged areas of skin to achieve wound closure. Fish liver cells presumably do not require such a rapid response and consequently maintain relatively low basal levels of HSP27. Our analysis of the heat shock response in fish keratocytes indicates that it develops over a period of several hours, during which keratocytes reduce their speed and become more tightly adhered to the substratum. In addition, the dramatic changes in cell morphology are consistent with a change in actin filament dynamics, perhaps involving stabilization of the cytoskeleton. We hypothesize that similar changes may occur during a fish skin wound response after the keratocytes have moved rapidly into the damaged area to initiate wound repair.

References

Arrigo AP, Suhan JP, Welch WJ (1988) Dynamic changes in the structure and intracellular locale of the mammalian low-molecular-weight heat shock protein. Mol Cell Biol 8(12):5059–5071

Benndorf R, Hayess K, Ryazantsev S, Wieske M, Behlke J, Lutsch G (1994) Phosphorylation and supramolecular organization of murine small heat shock protein HSP25 abolish its actin polymerization-inhibiting activity. J Biol Chem 269(32):20780–20784

Borrelli MJ, Bernock LJ, Landry J, Spitz DR, Weber LA, Hickey E, Freeman ML, Corry PM (2002) Stress protection by a fluorescent Hsp27 chimera that is independent of nuclear translocation or multimeric dissociation. Cell Stress Chaperones 7(3):281–296

Clarke JP, Mearow KM, Qin D, Tan L, You Q, Liu X (2013) Cell stress promotes the association of phosphorylated HspB1 with F-actin, expression of heat shock protein 27 and proliferating cell nuclear antigen in human retinoblastoma. PLoS One 8(7):e68978

Datskevich PN, Gusev NB (2013) Structure and properties of chimeric small heat shock proteins containing yellow fluorescent protein attached to their C-terminal ends. Cell Stress Chaperones. doi:10.1007/s12192-013-0477-0

Doshi BM, Hightower LE, Lee J (2009) The role of Hsp27 and actin in the regulation of movement in human cancer cells responding to heat shock. Cell Stress Chaperones 14:445–457. doi:10.1007/s12192-008-0098-1

Du J, Zhang L, Yang Y, Li W, Chen L, Ge Y, Sun C, Zhu Y, Gu L (2010) ATP depletion-induced actin rearrangement reduces cell adhesion via p38 MAPK-HSP27 signaling in renal proximal tubule cells. Cell Physiol Biochem 25(4–5):501–510

During RL, Gibson BG, Li W, Bishai EA, Sidhu GS, Landry J, Southwick FS (2007) Anthrax lethal toxin paralyzes actin-based motility by blocking Hsp27 phosphorylation. EMBO J 26(9):2240–2250. doi:10.1038/sj.emboj.7601687, 7601687 [pii]

Ehrnsperger M, Lilie H, Gaestel M, Buchner J (1999) The dynamics of Hsp25 quaternary structure. Structure and function of different oligomeric species. J Biol Chem 274(21):14867–14874. doi:10.1074/jbc.274.21.14867

Garcia-Arguinzonis M, Padro T, Lugano R, Llorente-Cortes V, Badimon L (2010) Low-density lipoproteins induce heat shock protein 27 dephosphorylation, oligomerization, and subcellular relocalization in human vascular smooth muscle cells. Arterioscler Thromb Vasc Biol. doi:10.1161/atvbaha.109.198440, ATVBAHA.109.198440

Golembieski WA, Thomas SL, Schultz CR, Yunker CK, McClung HM, Lemke N, Cazacu S, Barker T, Sage EH, Brodie C, Rempel SA (2008) HSP27 mediates SPARC-induced changes in glioma morphology, migration, and invasion. Glia 56(10):1061–1075. doi:10.1002/glia.20679

Graceffa P (2011) Hsp27-actin interaction. Biochem Res Int 2011:901572

Guay J, Lambert H, Gingras-Breton G, Lavoie JN, Huot J, Landry J (1997) Regulation of actin filament dynamics by p38 map kinase-mediated phosphorylation of heat shock protein 27. J Cell Sci 110(Pt 3):357–368

Hirano S, Shelden EA, Gilmont RR (2004) HSP27 regulates fibroblast adhesion, motility, and matrix contraction. Cell Stress Chaperones 9(1):29–37

Lee J, Ishihara A et al (1999) Regulation of cell movement is mediated by stretch-activated calcium channels. Nature 400(6742):382–386

Lee SH, Eom M, Lee SJ, Kim S, Park HJ, Park D (2001) BetaPix-enhanced p38 activation by Cdc42/Rac/PAK/MKK3/6-mediated pathway. Implication in the regulation of membrane ruffling. J Biol Chem 276(27):25066–25072. doi:10.1074/jbc.M010892200

Miron T, Vancompernolle K, Vandekerckhove J, Wilchek M, Geiger B (1991) A 25-kD inhibitor of actin polymerization is a low molecular mass heat shock protein. J Cell Biol 114(2):255–261. doi:10.1083/jcb.114.2.255

Mounier N, Arrigo AP (2002) Actin cytoskeleton and small heat shock proteins: how do they interact? Cell Stress Chaperones 7(2):167–176

Nomura N, Nomura M, Sugiyama K, Hamada J-I (2007) Phorbol 12-myristate 13-acetate (PMA)-induced migration of glioblastoma cells is mediated via p38MAPK/Hsp27 pathway. Biochem Pharmacol 74(5):690–701

Norris CE, Brown MA, Hickey E, Weber LA, Hightower LE (1997) Low-molecular-weight heat shock proteins in a desert fish (Poeciliopsis lucida): homologs of human Hsp27 and Xenopus Hsp30. Mol Biol Evol 14(10):1050–1061

Stöhr N, Hüttelmaier S (2012) IGF2BP1: a post-transcriptional “driver” of tumor cell migration. Cell Adhes Migr 6(4):312–318

Stokoe D, Engel K, Campbell DG, Cohen P, Gaestel M (1992) Identification of MAPKAP kinase 2 as a major enzyme responsible for the phosphorylation of the small mammalian heat shock proteins. FEBS Lett 313(3):307–313

Chapter 20
Reconsidering Old Data: Non-canonical HspB1 Species and the Enigma of the Cytoskeletal Function of HspB1

Rainer Benndorf and Peter R. Jungblut

Abstract At the cell biological level, numerous studies have provided evidence for the regulatory role of HspB1 (Hsp27, Hsp25) in the organization and stabilization of the actin cytoskeleton. Unfortunately, the underlying molecular mechanisms for this important function of HspB1 remain obscure. One factor that impedes a better understanding of this function of HspB1 is the conflicting accounts of its ability to inhibit the polymerization of actin.

Although various modifications of HspB1 have been reported, studies have been largely limited to the phosphorylation by the protein kinases MK2 and MK3. Phosphorylation by these MKs typically results in the formation of two or three HspB1 spots that can be visualized by two-dimensional gel electrophoresis. Together with the HspB1 spot of the non-phosphorylated HspB1, these spots contain the canonical HspB1 species that have been the focus of many studies in the past.

In contrast, a few studies report a large number (60 or more) of additional HspB1 species. The existence of these non-canonical HspB1 species cannot be explained by simple phosphorylation by the MKs. Instead, other modifications can be expected to contribute to these complex protein species patterns. Such complex HspB1 species patterns have been found in the human heart, extracts of mouse Ehrlich ascites tumor cells, and in Hela cells. It can be assumed that these non-canonical HspB1 species occur more widely, however, their detection requires suitable analytical methods. These complex modification patterns of HspB1 and of many other proteins have led to the protein speciation discourse and to the concept of the protein code that serves to better understand the functions of proteins.

Re-evaluating published data, we hypothesize that one of the non-canonical HspB1 species is involved in the inhibition of actin polymerization. This non-

R. Benndorf (✉)
Center for Clinical and Translational Research, The Research Institute at Nationwide Children's Hospital, Columbus, OH, USA

Department of Pediatrics, The Ohio State University, Columbus, OH, USA
e-mail: rbenndo@gmx.net

P.R. Jungblut
Core Facility Protein Analysis, Max-Planck-Institute for Infection Biology, Berlin, Germany
e-mail: jungblut@mpiib-berlin.mpg.de

R.M. Tanguay, L.E. Hightower (eds.), *The Big Book on Small Heat Shock Proteins*, Heat Shock Proteins 8, DOI 10.1007/978-3-319-16077-1_20

canonical protein species was present in native 'active' HspB1 preparations which inhibited the polymerization of actin, but was absent or nearly absent in preparations of 'inactive' recombinant and native phosphorylated HspB1, respectively. This non-canonical HspB1 species may have an additional post-translational modification or may adopt a conformation that differs from that of the canonical protein species, e.g. by exposing sequence segments to the surface that were shown to inhibit the polymerization of actin.

We propose that testing this hypothesis could help to clarify the role of HspB1 in the organization of the actin cytoskeleton. In this context we also propose that efforts should be made to unravel the chemical nature of a number of the non-canonical HspB1 species and their possible biological function(s).

Keywords HspB1 modification • Phosphorylation • Actin polymerization • Protein speciation discourse • Hypothesis

20.1 Introduction

In recent decades, numerous studies reported a regulatory role of HspB1 (Hsp27, Hsp25) in the organization and stabilization of the actin cytoskeleton at the cell biological level (Lavoie et al. 1993a, b, 1995; Mounier and Arrigo 2002; Koh and Escobedo 2004; Hirano et al. 2005; Robinson et al. 2010). With this evidence, there is little doubt about this aspect of the cellular function of HspB1. Unfortunately, at the molecular level the underlying mechanisms are not understood (see below). Moreover, for more than two decades, mammalian HspB1 has been known to be a prominent phosphoprotein which is phosphorylated by the protein kinases MK2 and MK3, or related enzymes (Arrigo and Welch 1987; Benndorf et al. 1988; Gaestel et al. 1991; Landry et al. 1992; Ronkina et al. 2008; Shiryaev et al. 2011). In addition, a number of other modifications of HspB1 have been reported, including phosphorylation by PKG (Butt et al. 2001) or other unknown protein kinases (see the UniProt database at http://www.uniprot.org/), acylation by various fatty acids (Oesterreich et al. 1991), sumoylation (Becker et al. 2013), methyl-glyoxyal modification (Schalkwijk et al 2006; van Heijst et al. 2006), O-GlcNAC glycosylation (Guo et al. 2012), oxidative dimerization (Zavialov et al. 1998a), glutathionylation (Zavialov et al. 1998b), ubiquitination (Sun et al. 2011), and cross-linking by transglutaminase (Merck et al. 1993), although the physiological implications of these modifications are mostly unknown.

In spite of this diversity, phosphorylation by MK2/MK3 is the most obvious and by far the most studied modification of HspB1. Human (and other primate) HspB1 contains three MK2/MK3 phosphorylation sites (Ser15, Ser78, Ser82) (Landry et al. 1992), whereas mouse (and other rodent) HspB1 contains two such sites (Ser15, Ser86) (Gaestel et al. 1991). Consistent with this, the overwhelming majority of studies demonstrated that human and mouse tissues typically contain HspB1 species that form four and three major HspB1 spots, respectively, as detected on two-dimensional isoelectric focusing-SDS polyacrylamide gel electrophoresis (2-DE) gels. These 'canonical HspB1 spots' contain non-phosphorylated, onefold phosphorylated, twofold phosphorylated and threefold phosphorylated HspB1 species.[1] However, a few studies report much more complex HspB1 patterns, with additional or 'non-canonical HspB1 species' being observed as they appeared in 'non-canonical HspB1 spots' on 2-DE gels. The existence of these additional HspB1 species cannot be explained by phosphorylation just by MK2/MK3 and related enzymes. Instead, additional modifications or other processes seem to play a role. This raises questions with regard to the chemical nature of these additional HspB1 species, and also to possible functional consequences including their biological importance.

Here we re-evaluate earlier reports on complex 2-DE HspB1 spot patterns with numerous additional or non-canonical HspB1 species in human hearts, in the mouse Ehrlich ascites tumor (EAT), and in Hela cells. Given this observed complexity of HspB1 patterns, it can be assumed that additional modifications other than phosphorylation by MK2 or MK3 contribute to the formation of this complexity, although the presence of certain HspB1 species may be restricted to specific tissues or to a distinct physiological situation. On this basis, we hypothesize that some of these non-canonical HspB1 species account for properties of HspB1 reported earlier, notably for its role in inhibition of actin polymerization. Studying these non-canonical HspB1 species may help to resolve the conflicting accounts on the role of HspB1 in the organization of the actin filaments.

[1] Traditionally, these protein spots were referred to as non-phosphorylated, onefold phosphorylated, twofold phosphorylated and threefold phosphorylated HspB1 isoforms. In this report, however, we do not use the term 'isoform' in connection with phosphorylation or other post-translational modifications, since the usage of this term should be restricted to genetic variants as defined by the IUPAC-IUBMB Joint Commission on Biochemical Nomenclature [Nomenclature of multiple forms of enzymes. In: Biochemical Nomenclature and Related Documents, 2nd edition. Edited by: Liébecq C. Colchester; Portland Press; 1992]. The more recent discourse on 'protein species' accommodates the situation that single 'spots' on 2-DE gels actually may represent more than one protein species (Jungblut et al. 2008). For example, the 'spot' representing onefold phosphorylated human HspB1 may contain HspB1 species phosphorylated at Ser15, or Ser78, or Ser82, or at one other amino acid residue, or mixtures thereof. In this chapter, we use the term 'HspB1 species' to refer to single or specific HspB1 forms, be they defined or not. This contrasts to the term 'canonical HspB1 spot' which designates those HspB1 forms (notably the non-phosphorylated, onefold, twofold and threefold phosphorylated HspB1 forms) as they appeared in most studies on 2-DE gels, and also to the term 'non-canonical HspB1 spot' which designates the numerous additional HspB1 forms as observed on 2-DE gels in a few studies on HspB1. Both the canonical and non-canonical HspB1 spots are formed by one or more of the canonical and non-canonical HspB1 species.

20.2 Puzzling Old and New Findings: Complex HspB1 Spot and Protein Species Patterns

20.2.1 Amazing Diversity of Non-canonical HspB1 Spots in Human Hearts

In analysis of protein patterns of healthy and diseased human heart tissues by the highly sensitive high resolution 2-DE in combination with western blotting and mass spectrometry, HspB1 was identified in atleast 59 spots (Otto et al. 1994; Thiede et al. 1996; Scheler et al. 1997, 1999). The proteins in these spots differ from each other in their molecular mass and/or their isolelectric point. A representative sample of immunostained HspB1-containing spots of a healthy human heart (left ventricle) is shown in Fig. 20.1 (modified from Scheler et al. 1999). This multitude of speciation is puzzling as the nature of the underlying protein modifications is only partially known. At least, protein species with apparent molecular masses of less than ~25 kDa can be assumed to result from partial proteolysis.

In Fig. 20.1, most of the HspB1 spots in the ~26–28 kDa range are grouped into two series with isoelectric points in the range of ~4.9–6.2. Among them are the main spots (h1, h2, h6, h7, h10, h12, h22, h17, h14) belonging either to the upper series with molecular masses of ~28 kDa, or to the lower series with molecular

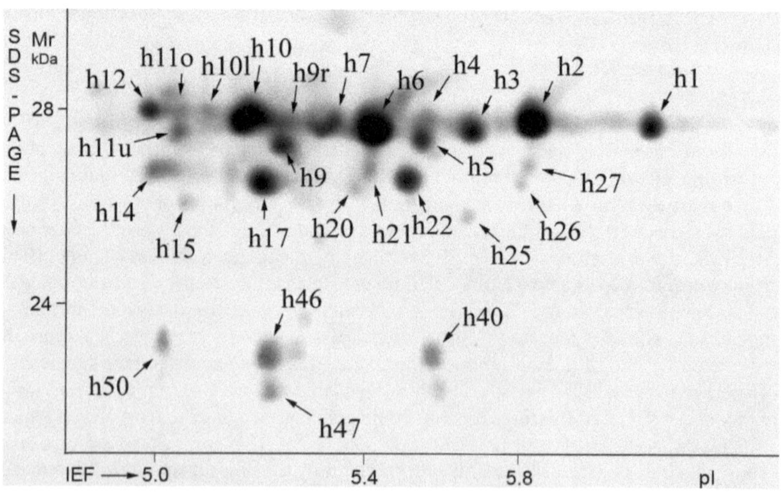

Fig. 20.1 Complex HspB1 species pattern in a healthy human myocardial left ventricle shown by high-resolution 2-DE followed by immunoblotting. For immunodetection, an HspB1-specific antibody was used. The identity of a number of these HspB1 species was also verified by mass spectrometry. Although the canonical HspB1 spots were not identified in this study, the candidates are spots h1, h2, h6, and h10 (The image was modified from Scheler et al. (1999) where more experimental details are given. Used with permission by the Wiley-VCH Verlag GmbH, Weinheim, Germany)

masses in the 26–27 kDa range. Some of these HspB1 spots can be assumed to represent the canonical protein species which differ by their degree of phosphorylation, although they were not exactly assigned in that study. The chemical nature of the difference of the molecular masses of ~1 kDa between both series is not known. Similarly, the nature of other HspB1 species, e.g. in the most acidic spots h12 and h14 or in the many spots with intermediate isoelectric points between the main spots, both in the upper and lower series, remains unknown. Simple phosphorylation steps by MK2/MK3 do not provide an adequate explanation for the existence of these non-canonical HspB1 spots. Interestingly, the intensity of some of these spots (e.g. h1, h4, h5, h9r, h15, h17, h20, h21, h25, h47, h50) on HspB1-immunoblots differed in control hearts and hearts from patients with dilated or ischemic cardiomyopathy (Scheler et al. 1999).

In a separate study, a few of these non-canonical HspB1 species with intermediate isoelectric points were also detected in human fetal and adult explanted heart samples, using the less sensitive standard 2-DE method (Lutsch et al. 1997).

In summary, human HspB1 can exist in a great number of protein species which includes the non-canonical species. On 2-DE gels, these HspB1 species form complex spot patterns including both canonical and non-canonical spots. It can be concluded that modifications other than phosphorylation at the three MK2/MK3 phosphorylation sites, or other processes, occur, resulting in the formation of such a complex protein species pattern. The occurrence of these non-canonical HspB1 species may not be limited to the heart. However, their detection requires both suitable samples and sufficiently sensitive analytical methods.

20.2.2 Phosphorylated and Non-phosphorylated Non-canonical HspB1 Species in Extracts of Ehrlich Ascites Tumor Cells

A similarly complex, yet informative, pattern of non-canonical HspB1 spots was observed on 2-DE gels from extracts of mouse EAT cells, as opposed to live EAT cells that contain HspB1 species which present as the three canonical spots (spots 1, 2, and 3 in Fig. 20.2a, panel a), of which two (spots 2 and 3) are phosphorylated (Fig. 20.2a, panel b) (Benndorf et al. 1988, 1992).

Surprisingly, a simple protein enrichment procedure (ammonium sulfate precipitation from the cell extract, yielding a 0.45S fraction) produced an apparent basic shift of spot 2, resulting in the formation of the new spot 1a, and thus presumably to a new HspB1 species (Fig. 20.2b, panel a), which typically was not observed in live cells. Incubation of this 0.45S fraction in the presence of ATP apparently reversed this shift, resulting in the restoration of spot 2 (panel b).

Since this 0.45S fraction contained active MK2 (Hayess and Benndorf 1997), incubation with [γ-^{32}P]ATP was used for identification of the phosphoproteins among the HspB1 species in the various spots (Fig. 20.2b, panel c). As expected,

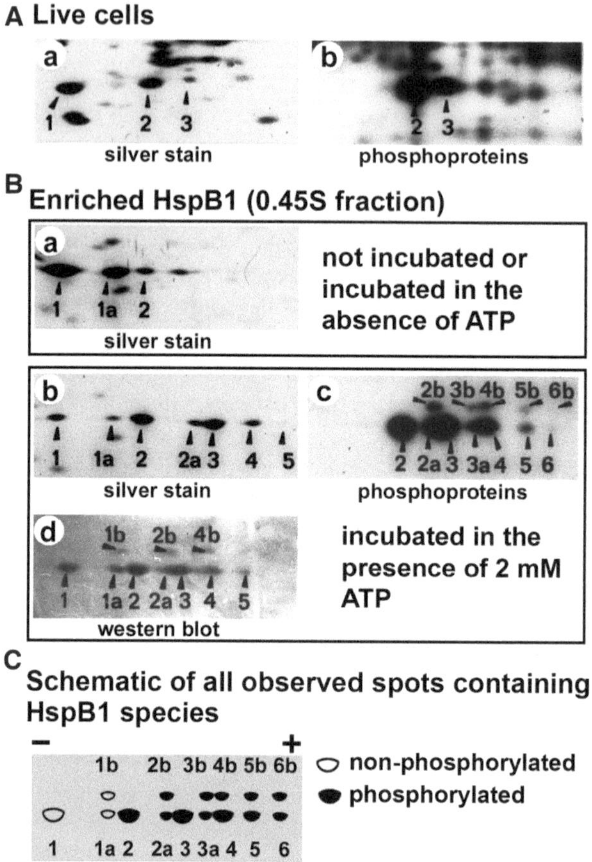

Fig. 20.2 Canonical and non-canonical spots of HspB1 in mouse EAT cells and fractionated cell extracts as separated by 2-DE. (**a**) Total (panel **a**) and phosphorylated (panel **b**) protein species in HspB1 spots in live EAT cells. In panel (**a**), the separated cell proteins were silver stained, whereas in panel (**b**) the cell proteins were labeled in vivo by incorporation of [^{32}P]phosphoric acid and visualized by autoradiography. Live cells contain only the canonical HspB1 spots 1 (containing non-phosphorylated HspB1 species), 2, and 3 (both containing phosphorylated HspB1 species). (**b**) Total (panels **a**, **b**, **d**) and phosphorylated (panel **c**) protein species in HspB1 spots in the 0.45S fraction of EAT cell extracts. Panels(**a**) and (**b**) show the spot distribution in the 0.45S fraction incubated in the absence or presence of ATP, respectively, as visualized by silver staining. In panel (**c**), the HspB1 species were labeled in vitro by incubation in the presence of [^{32}P]ATP and visualized by autoradiography. Panel (**d**) shows a western blot of all detectable HspB1 spots in the 0.45S fraction after incubation in the presence of ATP, using an antibody raised against the HspB1 isolated from spot 1. The non-incubated 0.45S fraction, or when incubated in the absence of ATP, contains the non-canonical HspB1 species in spot 1a, in addition to the canonical species in spots 1 and 2. The HspB1 species in spot 1a seems to have formed essentially from the species in spot 2. When incubated in the presence of ATP, the 0.45S fraction produces a number of non-canonical HspB1 spots, in addition to the canonical spots 2 and 3. These non-canonical spots were visualized by silver-staining (panel **b**), incorporation of ^{32}P into the phosphor-HspB1 species (panel **c**), and western blotting (panel **d**). (**c**) Schematic drawing of all HspB1 spots observed. Spots containing non-phosphorylated HspB1 species are indicated by open symbols, and spots containing phosphorylated HspB1 species by closed symbols. Numbers refer to HspB1 spots as used in this report (The images were taken from Benndorf et al. (1992), where more experimental details are given. Used with permission by Elsevier B.B., Amsterdam, The Netherlands (license number 3402661050555))

spots 2 and 3 were labeled (among others, see below), providing evidence for their phosphoprotein status also in this in vitro system. Unexpectedly, spot 1a was not labeled, like spot 1. This proves that a non-phosphorylated 'intermediate' HspB1 species exists, on 2-DE gels positioned between the non-phosphorylated canonical spot 1 and the onefold phosphorylated canonical spot 2.

The experiment shown in Fig. 20.2b also revealed the formation of a number of additional, non-canonical HspB1 species following the incubation in the presence of ATP, as shown by silver staining (panel b), incorporation of [γ-^{32}P]ATP (panel c), and western blotting (panel d). Apparently, the protein species in spots 1 and 1a were converted resulting in the formation of spots 2, 2a, 3, 3a, 4, 5, and 6 which all represent phosphoproteins. Among them, spots 2a, 3a, 4, 5, and 6 are non-canonical spots, containing non-canonical HspB1 species.

The highly sensitive radiolabeling method also detected the HspB1 species in spots 2b, 3b, 4b, 5b, and 6b, all of which are phosphoproteins (Fig. 20.2b, panel c). This b series of HspB1 species is distinguished by an increased molecular mass by ~1 kDa, compared to the masses of the protein species in spots 1 through 6. These HspB1 species may correspond to the high molecular mass forms that were observed in the human heart (cf. Fig. 20.1).

The identity of most of these HspB1 species in the various spots was verified by western blotting, using the incubated 0.45S fraction (Fig. 20.2b, panel d). Western blotting also detected an HspB1 species in spot 1b that was not phosphorylated. A schematic of all observed HspB1 spots observed in EAT cell extracts is shown in Fig. 20.2c.

Taken together, mouse HspB1 can form a number of non-canonical protein species in addition to the species typically contained in the three canonical spots, although they were detected only in cell extract fractions. The biochemical basis of the HspB1 species diversity as observed in this experiment is not known, other than phosphorylation at the two MK2/MK3 sites. In particular, the chemical basis of protein species 1a (contained in HspB1 spot 1a) is of interest as this species was present in the purified HspB1 fraction that was reported to inhibit the polymerization of actin in a cell-free system (see below). The protein species 1a was also detected in human hearts, suggesting that it fulfills a physiological function (Lutsch et al. 1997).

20.2.3 A Biological Discourse on Protein Speciation – Starting with HspB1

For the interpretation of proteomic data it is commonly assumed that the intensity of a spot on 2-DE gels corresponds to the gene expression or occurrence of the protein of interest. However, this approach frequently leads to misinterpretations, e.g. when the correlation of the spot intensity and cellular functions is to be studied. Proteomic analyses revealed that many proteins appear in multiple spots on 2-DE gels, suggesting that they exist in multiple and distinguishable forms. HspB1 was one of the early examples in which phosphorylation was identified as one reason for this protein species diversification (Arrigo and Welch 1987; Benndorf et al. 1988; Gaestel et al. 1991; Landry et al. 1992). Initially, the notion was that every single phosphorylation would result in a separate spot on these gels. This view, however, turned out to be incorrect as any phosphorylation leads to the same shift of the isoelectric point, independent of which position in the protein sequence is phosphorylated. Thus, the diversity of protein forms can be assumed to be much greater than the spot pattern indicates. Together with similar observations for other proteins, these considerations have led to the protein speciation discourse (Jungblut et al. 2008; Schlüter et al. 2009). After the synthesis, proteins can be modified in many ways with a great number of possible combinations, resulting in multiple protein species for each protein. This multiplicity of different protein species is the basis for the concept of the protein code which was initially described as histone code (Pannel et al. 2000), and later, in a more generalized form, as the protein code (Sims and Reinberg 2008). According to this protein speciation discourse, a given protein spot on 2-DE gels does not represent the expression of the respective gene. Instead, only the complete set of protein species of a given protein reflects the expression of its gene, as long as protein degradation and export can be ignored.

To complicate things, a recent publication by Thiede et al. (2013) demonstrated that ~50 % of 2-DE protein spots contained more than one identified protein or protein species, using the most sensitive analytical methods available at present that included stable isotopic labeling with amino acids in cell culture (SILAC) and 2-DE-Orbitrap-LC mass spectrometry. This approach allowed the quantification of proteins at the protein species level. High-resolution 2-DE resolved protein speciation was also demonstrated for a number of other proteomes, established from proteasomes (Schmidt et al. 2011), the eye lens (Hoehenwarter et al. 2006), *Mycobacterium tuberculosis* (Jungblut et al. 1999), *Helicobacter pylori* (Jungblut et al. 2000), and other bacteria or eukaryotic cells and tissues (cf. the 2D-PAGE database at http://www.mpiib-berlin.mpg.de/2D-PAGE). Similarly, protein species-specific regulation could be demonstrated in several experimental systems, including in an infection model of AGS cells with *Helicobacter pylori* (Holland et al. 2011).

The HspB1 example illustrates the magnitude of challenge for any analytical approach: In the case of human HspB1 with its 8 predicted phosphorylation sites (Uniprot database, Version 178, June 11, 2014), 9 protein spots can be expected for the 256 potential protein species, 1 spot without any phosphorylation, 1 spot with 1 phosphorylation at any phosphorylation site representing 8 protein species, 1 spot with 2 phosphorylations representing 28 protein species with different combinations of 2 phosphorylation sites, and so on. Further HspB1 modifications add more complexity. This great number of potential protein species explains the difficulty of their identification from such 2-DE spots, as several of the HspB1 species may exist only in small proportions.

20.2.4 HspB1 Species-Specific Regulation in Hela Cells

The above outlined isotopic labeling/2-DE-Orbitrap-LC mass spectrometry method provided evidence for the existence of multiple HspB1 species, and, most importantly, also for their specific regulation during apoptosis in HeLa cells (Thiede et al. 2013). By this approach ~1,200 proteins were identified, together with ~2,700 protein species. Thirteen of these protein species belonged to HspB1 (Fig. 20.3). Among them, five protein species were down-regulated during apoptosis by a factor <0.5, whereas eight protein species were up-regulated by a factor >2.0.

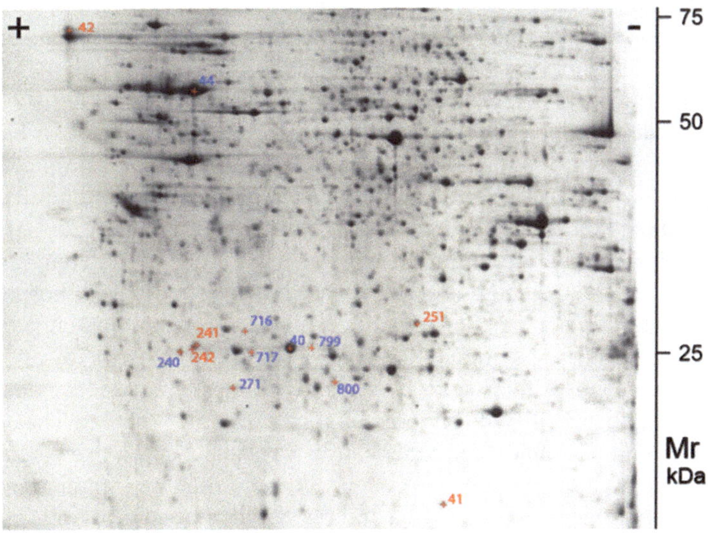

Fig. 20.3 HspB1 species-specific regulation during apoptosis in Hela cells. Non-labeled apoptotic Hela cells were mixed 1:1 with metabolically labeled SILAC control Hela cells. Subsequently, the cell proteins were separated by 2-DE in a pH gradient from 2 to 11 and an Mr range from 3 to 150 kDa with 23×30 cm large gels. After staining of the 2-DE gels with CBB-G250, all visible spots were excised and analyzed by LC-Orbitrap mass spectrometry (Thiede et al. 2013). The sector of the gel containing 13 spots of HspB1 species is shown; some of them were down-regulated (*red numbers*) by a factor (non-labeled/labeled) <0.5, whereas others were up-regulated (*blue numbers*) by a factor >2 during by apoptosis. The raw data are available in the 2D-PAGE database system at http://www.mpiib-berlin.mpg.de/2D-PAGE, enabling a more detailed analysis of the HspB1 species

20.3 The Unresolved Molecular Role of HspB1 in the Organization and Stabilization of the Actin Cytoskeleton

Given the widely accepted involvement of HspB1 in the organization and stabilization of the actin cytoskeleton at the cell biological level (Lavoie et al. 1993a, b, 1995; Mounier and Arrigo 2002; Koh and Escobedo 2004; Hirano et al. 2005; Robinson et al. 2010), the poor understanding of the underlying mechanism is frustrating. In a recent commentary on this subject, four candidate molecular mechanism involving HspB1 were proposed by Seit-Nebi et al. (2013): (i) binding of HspB1 to partially or completely denatured actin resulting in a stabilization of the microfilaments; (ii) interaction of HspB1 with a number of actin-binding proteins which may protect the microfilaments from disruption; (iii) indirect effects of HspB1 on the microfilaments through its interaction with the intermediate filaments and the microtubules, and (iv) indirect effects through the ability of HspB1 to control the cellular redox homeostasis. Other studies proposed that small oligomeric complexes of phosphorylated HspB1 coat the microfilaments, resulting in their contraction and protection (Mounier and Arrigo 2002; Hirano et al. 2005).

Another line of experimentation suggested a role for HspB1 as a barbed end actin capping protein. Originally this concept was proposed by Miron et al. (1988, 1991) who identified HspB1 in extracts of turkey gizzard smooth muscles, notably as a 'contaminant' of a vinculin preparation. Similarly, Benndorf et al. (1994) found an actin polymerization inhibiting activity in preparations of native mouse HspB1 purified from EAT cells, and this activity was abolished following phosphorylation by MK2. Since this actin polymerization inhibiting activity was eluted from gel filtration columns at the position of the HspB1 monomers/dimers, it was concluded that the non-phosphorylated HspB1 monomers or dimers are the active components, as opposed to phosphorylated and large oligomeric HspB1 species. The concept of HspB1 as an inhibitor of actin polymerization was further supported when two peptides derived from the mouse HspB1 sequence were found to inhibit the polymerization of actin (Wieske et al. 2001). In contrast to the native HspB1 and to the HspB1-derived peptides, recombinant HspB1 did not exhibit any actin polymerization inhibitory activity in these experiments (Benndorf et al. 1994; Butt et al. 2001).

A subsequent study did not confirm the inhibition of actin polymerization activity of HspB1 (Panasenko et al. 2003). Additionally, non-phosphorylated HspB1 typically forms large oligomers which dissociate upon phosphorylation into smaller oligomers or dimers (Rogalla et al. 1999; Hirano et al. 2005; Lelj-Garolla and Mauk 2005; Hayes et al. 2009). This was used as an argument against the existence of appreciable quantities of non-phosphorylated monomers in cells (Seit-Nebi et al. 2013).

Based on these reports, the concept of HspB1 as a pure barbed end actin capping protein with an actin polymerization inhibition activity has become obsolete (Graceffa 2011; Seit-Nebi et al. 2013). Indeed, there is the real possibility that the preparations of turkey gizzard and mouse EAT HspB1 that inhibited the polymerization of actin contained a highly active 'contaminant' with a similar molecular mass that remained undetected, misleadingly indicating that HspB1 was the inhibitor. However, this scenario does not explain the ability of the HspB1-derived peptides to inhibit the polymerization of actin, since these synthetic peptides did not contain any possible biological inhibitor of the actin polymerization (Wieske et al. 2001).

Re-evaluating these early experiments, it is striking that the 'active' non-phosphorylated low molecular mass EAT HspB1 contained notable amounts of the non-canonical protein species of spot 1a, as opposed to the 'inactive' HspB1 preparations (including both phosphorylated low molecular mass material and recombinant HspB1) which contained only minor amounts of protein species of this spot, or none (Gaestel et al. 1991; Benndorf et al. 1994). Thus, the non-canonical HspB1 protein species in spot 1a may have contained the 'active' component that inhibited the polymerization of actin in these experiments. The fact that the 'inactive' high molecular mass material also contained the HspB1 species from spot 1a does not impair this view as this species may have been buried inside the oligomeric complexes, thus preventing its interaction with actin. Taken together, there is the possibility that one of the non-canonical HspB1 spots (1a in Fig. 20.2) contains an HspB1 species that inhibits the polymerization of actin.

20.4 Hypothesis

Both the unsatisfying understanding of the role of HspB1 in the organization of the actin networks and the conflicting reports on its ability to inhibit the polymerization of actin require a new and unbiased approach to resolve this situation. One point of interest is the possible role of the non-canonical HspB1 species, in particular the protein species 1a (Fig. 20.2). Based on the published data (Benndorf et al. 1992, 1994), it can be hypothesized that the non-canonical spot 1a contains the 'active' HspB1 species that inhibits the polymerization of actin. Verification or falsification of this hypothesis seems to be feasible by suitable in vitro experimental approaches.

The assumption of the HspB1 species in spot 1a being the active component would explain why recombinant HspB1 (which does not seem to contain this protein species) did not show any actin polymerization inhibiting activity (Benndorf et al. 1994; Butt et al. 2001; Panasenko et al. 2003). Similarly, actin polymerization inhibition experiments using native HspB1 preparations may have produced negative results if the active protein species corresponding to spot 1a was not enriched in the used procedure (Panasenko et al. 2003).

Figure 20.2 can be interpreted such that the difference between the non-canonical HspB1 species 1a and the canonical species 2 is caused by just one phosphate group. If correct, this would imply that another yet unknown modification causes the pI difference between the protein species in spot 1 and that in spot 1a. Alternatively, the protein species in spot 1 and 1a may represent different conformational states of non-phosphorylated HspB1 that differ in their isoelectric points, be they induced by chemical modifications or energetically. If so, in the HspB1 species of spot 1a the segments of the protein sequence that inhibit actin polymerization in vitro (Wieske et al. 2001) may become exposed to the surface and thus become accessible to actin, enabling the inhibition of actin polymerization by HspB1. This view also implies that these sequence segments are buried inside the molecule in the protein species of HspB1 spots 1, 2 and 3. If correct, the above considerations may resolve the conflicting results on the inhibition of actin polymerization by HspB1 reported earlier.

20.5 Further Questions to Be Addressed

If verified, the existence of the non-canonical HspB1 species will raise a number of questions that need to be addressed: Do these HspB1 species (in particular those in spot 1a) have a physiological importance, or are they irrelevant for the function of HspB1? Is the differential occurrence of many protein species only a problem of detection sensitivity or is it functionally relevant? Given the list of reported HspB1 modifications (cf. Introduction), which of them actually contributes to the complex HspB1 species patterns, other than phosphorylation by MK2 and MK3? Do these additional protein species result from modifications and/or conformational changes that are difficult to detect in live cells, and how could such potentially short-lived transitional states be detected and experimentally studied? What are the characteristics of these non-canonical HspB1 species in terms of their conformational stability, interacting proteins, homo-oligomeric complex formation, hetero-oligomeric complex formation with other small heat shock proteins, and their chaperone-like activities? Is the complex HspB1 species pattern a suitable model for deciphering the protein code that may serve as example for other proteins?

20.6 Conclusions

In the past, the overwhelming majority of studies on HspB1 focused on the protein species contained in the three or four canonical spots as they are seen on 2-DE gels. These spots include one spot with non-phosphorylated HspB1 species and two or three spots containing MK2/MK3-phosphorylated protein species, depending on the studied animal species. In single tissues and conditions, however, the existence of a great number of additional HspB1 species has been reported, using suitable analytical methods. The chemical basis for this protein species diversity is not understood as long as the identification procedure relies on partial sequence information, although the involvement of other known HspB1 modifications can be assumed. It is possible that these non-canonical HspB1 species have distinguished features, e.g. in terms of their conformation, complex formation, interacting proteins, chaperone-like activity, etc. However, at present the physiological relevance of this HspB1 species complexity is not known. These complex modification patterns of HspB1, like those of many other proteins, have led to the protein speciation discourse which confirms the concept of the protein code that aims to better understand the functions of proteins.

One of the controversial functions of HspB1 is its role in inhibition of the polymerization of actin. Here we provide arguments in support of such a role for at least one of the protein species represented in one HspB1 spot. Re-evaluating published data, we hypothesize that one of the non-canonical HspB1 species inhibits the polymerization of actin, as opposed to the canonical protein species of various sources. If verified, this hypothesis has the potency to resolve the conflicting accounts on inhibition of the actin polymerization by HspB1.

References

Arrigo AP, Welch WJ (1987) Characterization and purification of the small 28,000-dalton mammalian heat shock protein. J Biol Chem 262:15359–15369

Becker J, Barysch SV, Karaca S, Dittner C, Hsiao HH, Diaz MB, Herzig S, Urlaub H, Melchior F (2013) Detecting endogenous SUMO targets in mammalian cells and tissues. Nat Struct Mol Biol 20:525–531

Benndorf R, Kraft R, Otto A, Stahl J, Böhm H, Bielka H (1988) Purification of the growth-related protein p25 of the Ehrlich ascites tumor and analysis of its isoforms. Biochem Int 17:225–234

Benndorf R, Hayess K, Stahl J, Bielka H (1992) Cell-free phosphorylation of the murine small heat-shock protein hsp25 by an endogenous kinase from Ehrlich ascites tumor cells. Biochim Biophys Acta 1136:203–207

Benndorf R, Hayess K, Ryazantsev S, Wieske M, Behlke J, Lutsch G (1994) Phosphorylation and supramolecular organization of murine small heat shock protein HSP25 abolish its actin polymerization-inhibiting activity. J Biol Chem 269:20780–20784

Butt E, Immler D, Meyer HE, Kotlyarov A, Laass K, Gaestel M (2001) Heat shock protein 27 is a substrate of cGMP-dependent protein kinase in intact human platelets: phosphorylation-induced actin polymerization caused by HSP27 mutants. J Biol Chem 276:7108–7113

Gaestel M, Schröder W, Benndorf R, Lippmann C, Buchner K, Hucho F, Erdmann VA, Bielka H (1991) Identification of the phosphorylation sites of the murine small heat shock protein hsp25. J Biol Chem 266:14721–14724

Graceffa P (2011) Hsp27-actin interaction. Biochem Res Int 2011:901572

Guo K, Gan L, Zhang S, Cui FJ, Cun W, Li Y, Kang NX, Gao MD, Liu KY (2012) Translocation of HSP27 into liver cancer cell nucleus may be associated with phosphorylation and O-GlcNAc glycosylation. Oncol Rep 28:494–500

Hayes D, Napoli V, Mazurkie A, Stafford WF, Graceffa P (2009) Phosphorylation dependence of hsp27 multimeric size and molecular chaperone function. J Biol Chem 284:18801–18807

Hayess K, Benndorf R (1997) Effect of protein kinase inhibitors on activity of mammalian small heat-shock protein (HSP25) kinase. Biochem Pharmacol 53:1239–1247

Hirano S, Sun X, DeGuzman CA, Ransom RF, McLeish KR, Smoyer WE, Shelden EA, Welsh MJ, Benndorf R (2005) p38 MAPK/HSP25 signaling mediates cadmium-induced contraction of mesangial cells and renal glomeruli. Am J Physiol Renal Physiol 288:F1133–F1143

Hoehenwarter W, Ackermann R, Zimny-Arndt U, Kumar NM, Jungblut PR (2006) The necessity of functional proteomics: protein species and molecular function elucidation exemplified by in vivo alpha A crystallin N-terminal truncation. Amino Acids 31:317–323

Holland C, Schmid M, Zimny-Arndt U, Rohloff J, Stein R, Jungblut PR, Meyer TF (2011) Quantitative phosphoproteomics reveals link between Helicobacter pylori infection and RNA splicing modulation in host cells. Proteomics 11:2798–2811

Jungblut PR, Schaible UE, Mollenkopf H-J, Zimny-Arndt U, Raupach B, Mattow J, Halada P, Lamer S, Kaufmann SHE (1999) Comparative proteome analysis of *Mycobacterium tubercu-*

losis and *Mycobacterium bovis* BCG strains: towards functional genomics of microbial pathogens. Mol Microbiol 33:1103–1117

Jungblut PR, Bumann D, Haas G, Zimny-Arndt U, Holland P, Lamer S, Siejak F, Aebischer A, Meyer TF (2000) Comparative proteome analysis of *H. pylori*. Mol Microbiol 36:710–725

Jungblut PR, Holzhütter HG, Apweiler R, Schlüter H (2008) The speciation of the proteome. Chem Cent J 2:16

Koh TJ, Escobedo J (2004) Cytoskeletal disruption and small heat shock protein translocation immediately after lengthening contractions. Am J Physiol Cell Physiol 286:C713–C722

Landry J, Lambert H, Zhou M, Lavoie JN, Hickey E, Weber LA, Anderson CW (1992) Human HSP27 is phosphorylated at serines 78 and 82 by heat shock and mitogen-activated kinases that recognize the same amino acid motif as S6 kinase II. J Biol Chem 267:794–803

Lavoie JN, Hickey E, Weber LA, Landry J (1993a) Modulation of actin microfilament dynamics and fluid phase pinocytosis by phosphorylation of heat shock protein 27. J Biol Chem 268:24210–24214

Lavoie JN, Gingras-Breton G, Tanguay RM, Landry J (1993b) Induction of Chinese hamster HSP27 gene expression in mouse cells confers resistance to heat shock. HSP27 stabilization of the microfilament organization. J Biol Chem 268:3420–3429

Lavoie JN, Lambert H, Hickey E, Weber LA, Landry J (1995) Modulation of cellular thermoresistance and actin filament stability accompanies phosphorylation-induced changes in the oligomeric structure of heat shock protein 27. Mol Cell Biol 15:505–516

Lelj-Garolla B, Mauk AG (2005) Self-association of a small heat shock protein. J Mol Biol 345:631–642

Lutsch G, Vetter R, Offhauss U, Wieske M, Gröne HJ, Klemenz R, Schimke I, Stahl J, Benndorf R (1997) Abundance and location of the small heat shock proteins HSP25 and alphaB-crystallin in rat and human heart. Circulation 96:3466–3476

Merck KB, Groenen PJ, Voorter CE, de Haard-Hoekman WA, Horwitz J, Bloemendal H, de Jong WW (1993) Structural and functional similarities of bovine alpha-crystallin and mouse small heat-shock protein. A family of chaperones. J Biol Chem 268:1046–1052

Miron T, Wilchek M, Geiger B (1988) Characterization of an inhibitor of actin polymerization in vinculin-rich fraction of turkey gizzard smooth muscle. Eur J Biochem 178:543–553

Miron T, Vancompernolle K, Vandekerckhove J, Wilchek M, Geiger B (1991) A 25-kD inhibitor of actin polymerization is a low molecular mass heat shock protein. J Cell Biol 114:255–2561

Mounier N, Arrigo AP (2002) Actin cytoskeleton and small heat shock proteins: how do they interact? Cell Stress Chaperones 7:167–176

Oesterreich S, Benndorf R, Reichmann G, Bielka H (1991) Phosphorylation and acylation of the growth-related murine small stress protein p25. In: Heilmeyer LMG (ed) Cellular regulation by protein phosphorylation, vol H56, NATO ASI series. Springer, Berlin/Heidelberg, pp 489–493

Otto A, Benndorf R, Wittmann-Liebold B, Jungblut P (1994) Identification of proteins on two-dimensional gels for the construction of a human heart 2-DE database. J Prot Chem 13:478–480

Panasenko OO, Kim MV, Marston SB, Gusev NB (2003) Interaction of the small heat shock protein with molecular mass 25 kDa (hsp25) with actin. Eur J Biochem 270:892–901

Pannell D, Osborne CS, Yao S, Sukonnik T, Pasceri P, Karaiskakis A, Okano M, Li E, Lipshitz HD, Ellis J (2000) Retrovirus vector silencing is de novo methylase independent and marked by a repressive histone code. EMBO J 19:5884–5894

Robinson AA, Dunn MJ, McCormack A, dos Remedios C, Rose ML (2010) Protective effect of phosphorylated Hsp27 in coronary arteries through actin stabilization. J Mol Cell Cardiol 49:370–379

Rogalla T, Ehrnsperger M, Preville X, Kotlyarov A, Lutsch G, Ducasse C, Paul C, Wieske M, Arrigo AP, Buchner J, Gaestel M (1999) Regulation of Hsp27 oligomerization, chaperone function, and protective activity against oxidative stress/tumor necrosis factor alpha by phosphorylation. J Biol Chem 274:18947–18956

Ronkina N, Kotlyarov A, Gaestel M (2008) MK2 and MK3–a pair of isoenzymes? Front Biosci 13:5511–5521

Schalkwijk CG, van Bezu J, van der Schors RC, Uchida K, Stehouwer CD, van Hinsbergh VW (2006) Heat-shock protein 27 is a major methylglyoxal-modified protein in endothelial cells. FEBS Lett 580:1565–1570

Scheler C, Müller EC, Stahl J, Müller-Werdan U, Salnikow J, Jungblut P (1997) Identification and characterization of heat shock protein 27 protein species in human myocardial two-dimensional electrophoresis patterns. Electrophoresis 18:2823–2831

Scheler C, Li XP, Salnikow J, Dunn MJ, Jungblut PR (1999) Comparison of two-dimensional electrophoresis patterns of heat shock protein Hsp27 species in normal and cardiomyopathic hearts. Electrophoresis 20:3623–3628

Schlüter H, Apweiler R, Holzhütter HG, Jungblut PR (2009) Finding one's way in proteomics: a protein species nomenclature. Chem Cent J 3:11

Schmidt F, Dahlmann B, Hustoft HK, Koehler CJ, Strozynski M, Kloss A, Zimny-Arndt U, Jungblut PR, Thiede B (2011) Quantitative proteome analysis of the 20S proteasome of apoptotic Jurkat T cells. Amino Acids 41:351–361

Seit-Nebi AS, Datskevich P, Gusev NB (2013) Commentary on paper: small heat shock proteins and the cytoskeleton: an essential interplay for cell integrity? (Wettstein et al.). Int J Biochem Cell Biol 45:344–346

Shiryaev A, Dumitriu G, Moens U (2011) Distinct roles of MK2 and MK5 in cAMP/PKA- and stress/p38MAPK-induced heat shock protein 27 phosphorylation. J Mol Signal 6:4

Sims RJ III, Reinberg D (2008) Is there a code embedded in proteins that is based on post-translational modifications? Nat Rev 9:1–6

Sun Y, Zhou M, Fu D, Xu B, Fang T, Ma Y, Chen J, Zhang J (2011) Ubiquitination of heat shock protein 27 is mediated by its interaction with Smad ubiquitination regulatory factor 2 in A549 cells. Exp Lung Res 37:568–573

Thiede B, Otto A, Zimny-Arndt U, Müller EC, Jungblut P (1996) Identification of human myocardial proteins separated by two-dimensional electrophoresis with matrix-assisted laser desorption/ionization mass spectrometry. Electrophoresis 17:588–599

Thiede B, Koehler GJ, Strozynski M, Treumann A, Stein R, Zimny-Arndt U, Schmid M, Jungblut PR (2013) High resolution quantitative proteomics of HeLa cells protein species using stable isotope labeling with amino acids in cell culture ?(SILAC), two-dimensional gel electrophoresis (2DE) and nano-liquid chromatography coupled to an LTQ-OrbitrapMass spectrometer. Mol Cell Proteomics 12:529–538

van Heijst JW, Niessen HW, Musters RJ, van Hinsbergh VW, Hoekman K, Schalkwijk CG (2006) Argpyrimidine-modified Heat shock protein 27 in human non-small cell lung cancer: a possible mechanism for evasion of apoptosis. Cancer Lett 241:309–319

Wieske M, Benndorf R, Behlke J, Dölling R, Grelle G, Bielka H, Lutsch G (2001) Defined sequence segments of the small heat shock proteins HSP25 and alphaB-crystallin inhibit actin polymerization. Eur J Biochem 268:2083–2090

Zavialov A, Benndorf R, Ehrnsperger M, Zav'yalov V, Dudich I, Buchner J, Gaestel M (1998a) The effect of the intersubunit disulfide bond on the structural and functional properties of the small heat shock protein Hsp25. Int J Biol Macromol 22:163–173

Zavialov A, Gaestel M, Korpela T, Zav'yalov V (1998b) Thiol/disulfide exchange between small heat shock protein 25 and glutathione. Biochim Biophys Acta 1388:123–132

Chapter 21
Role of HSPB8 in the Proteostasis Network: From Protein Synthesis to Protein Degradation and Beyond

Angelo Poletti and Serena Carra

Abstract Proper protein folding is crucial for protein stability and function; when folding fails, due to stress or genetic mutations, proteins may become toxic. Cells have evolved a complex protein quality control (PQC) system to protect against the toxicity exerted by aberrantly folded proteins, that may aggregate accumulating in various cellular compartments perturbing essential cellular activities, ultimately leading to cell and neuron death. The PQC comprises molecular chaperones, degradative systems (proteasome and autophagy) and components of the unfolded protein response. Prevention of protein aggregation, clearance of misfolded substrates and attenuation of translation, which decreases the amount of misfolding clients to levels manageable by the molecular chaperones, are all key steps for the maintenance of proteostasis and cell survival. In parallel, alterations of proteostasis may also (indirectly) influence RNA homeostasis; in fact, RNA-containing aggregates, known as stress granules, accumulate in cells with impaired PQC and autophagy colocalizing with proteinaceous aggregates in several neurodegenerative diseases. Among the different molecular chaperones, here we will focus on the small heat shock protein HSPB8, which is expressed in neurons in basal conditions and upregulated in response to misfolded protein accumulation. HSPB8 exerts protective functions in several models of protein conformation neurodegenerative diseases. The putative sites of action of HSPB8 that confer HSPB8 pro-survival and anti-aggregation functions are discussed, as well as its potential role at the crossroad between proteostasis and ribostasis.

A. Poletti
Dipartimento di Scienze Farmacologiche e Biomolecolari (DiSFeB), Universita' degli Studi di Milano, via Balzaretti 9, 20133 Milano, Italy

S. Carra (✉)
Dipartimento di Scienze Biomediche, Metaboliche e Neuroscienze, Universita' di Modena e Reggio Emilia, via G. Campi 287, 41125 Modena, Italy
e-mail: serena.carra@unimore.it

R.M. Tanguay, L.E. Hightower (eds.), *The Big Book on Small Heat Shock Proteins*, Heat Shock Proteins 8, DOI 10.1007/978-3-319-16077-1_21

Keywords Protein misfolding • Protein aggregation • Translation attenuation • HSPB8-BAG3 • Autophagy • Neurodegeneration • Polyglutamine diseases • Amyotrophic lateral sclerosis

Abbreviations

AD	Alzheimer's disease
ALS	Amyotrophic lateral sclerosis
ATX2	Ataxin 2
eIF2α	Eukaryotic initiation factor 2 on Ser51 of the α subunit
FTLD-U	Frontotemporal lobar degeneration with ubiquitin-positive inclusions
FUS	Fused in sarcoma
HD	Huntington's disease
hnRNPA1	Heterogeneous nuclear ribonucleoprotein A1
HSP	Heat shock protein
IBMPFD	Inclusion body myopathy with early-onset Paget disease and frontotemporal dementia
JNK	c-Jun N-terminal kinase
KD	Kennedy's disease
LAMP2A	Lysosome-associated membrane protein 2A
LC3	Microtubule-associated protein 1A/1B-light chain 3
MKK7	Mitogen-activated Protein Kinase Kinase 7
MSP	Multisystem proteinopathy
NBR1	Neighbor of BRCA1 gene 1
PD	Parkinson's disease
PERK	Protein kinase RNA-like endoplasmic reticulum kinase
polyQ	Polyglutamine
PQC	Protein quality control
RACK1	Receptor for Activated C Kinase 1
rhoA	Ras homolog gene family member A
RNPs	Ribonucleic proteins
ROCK1	Rho-associated coiled-coil containing protein kinase 1
SCA3	Spinocerebellar ataxia 3
SG	Stress granule
SOD1	Superoxide dismutase 1
SQSTM1	Sequestosome 1
TDP-43	TAR DNA-binding protein 43
TIA-1	T-cell intracytoplasmic antigen
TRAF2	TNF receptor-associated factor 2
UPS	Ubiquitin proteasome system
VCP	Valosin containing protein

21.1 Protein Quality Control: Guardian of Cellular Health

21.1.1 The Physiological Activation of the Protein Quality Control System

Protein homeostasis refers to the ability of cells to maintain an appropriate balance among protein synthesis, folding, assembly, translocation and clearance (Morimoto and Cuervo 2014). Protein homeostasis is ensured and modulated by the protein quality control (PQC) system and is essential for the long-term viability and health of the cells. Malfunction and deregulation of the PQC system are severe risk factors for the development of protein conformation diseases characterized by the accumulation of aggregated proteins (Bence et al. 2001; Douglas and Dillin 2010).

The PQC system includes molecular chaperones, degradative systems and stress-inducible pathways. Molecular chaperones (typically heat shock proteins, HSPs) are expressed in multiple cell compartments and survey protein quality by either assisting protein folding (both co- and post-translational folding) or directing aberrant proteins to degradation and thereby protecting against misfolding (Hartl 1996). Molecular chaperones also assist assembly and disassembly of macromolecular complexes, as well as protein translocation (Hartl and Hayer-Hartl 2002; Deuerling and Bukau 2004; Ron and Walter 2007). Unfolded substrates are repeatedly bound and held by molecular chaperones to avoid their irreversible aggregation (Hartl et al. 2011). There are several classes of molecular chaperones, which may also work with the assistance of co-chaperones and/or interactors of different nature; some chaperones are ATP-independent (e.g. small heat shock proteins, sHSPs/HSPBs) (Gobbo et al. 2011), while other are ATP-dependent (Hsc70/HSPA8 and Hsp70/HSPA1A) (Hartl et al. 2011). Both ATP-independent and ATP-dependent chaperones can bind to unfolded substrates. Instead, folding of the bound client to its native state is regulated by Hsp70 and ATP hydrolysis (Hartl et al. 2011). Nucleotide binding to Hsp70 and hydrolysis coupled to the release of the folded substrate is further regulated by co-chaperones including proteins of the Bag family (BAG1-BAG6), DNAJ/Hsp40 and Hip (Hartl et al. 2011), which also provide the specificity for interaction of chaperones with client proteins. When misfolding cannot be prevented (e.g. due to protein damage, denaturation, oxidation or a genetic mutation), proteins cannot mature into their fully active state, lose their function and become aggregation-prone, eventually acquiring toxic functions. Under these conditions, as well as when refolding fails, molecular chaperones assist the degradation of the bound client acting in concert with two major degradative systems: the ubiquitin-proteasome system (UPS) and lysosome-based degradation. The latter includes macroautophagy (here referred to as autophagy) and chaperone-mediated autophagy (CMA) (Cuervo and Wong 2014; Parzych and Klionsky 2014).

The UPS is a low capacity, but highly specific and selective proteolytic system that degrades short-lived proteins labeled with ubiquitin moieties (Ciechanover 2005; Ciechanover and Brundin 2003). The proteasome has a barrel shape, so that to be degraded by this system proteins must first be ubiquitinated and unfolded, in order to acquire the capability to enter its narrow central cavity; globular or irreversibly aggregated proteins cannot be processed by the proteasome. Among the molecular chaperones and co-chaperones that assist the proteasomal mediated degradation of ubiquitinated substrates are HSPAs/DNAJs and HSPA/BAG1 complexes (Alberti et al. 2002; Demand et al. 2001; Kampinga and Craig 2010).

Autophagy is a high capacity and non-specific multi-step process in which cytosolic material is sequestered in a double-membrane vesicle, the phagophore/autophagosome; the latter then fuses with the lysosome delivering its inner content for degradation by enzymatic hydrolysis, greatly facilitated by the acidic environment of the resulting intracellular compartment (Parzych and Klionsky 2014). Although autophagy has been initially described as a bulk nonspecific degradation process that mainly clears long-lived proteins and organelles, recent evidence has highlighted that autophagy plays a crucial role in the selective elimination of unwanted components such as dysfunctional organelles, pathogens and also aberrant protein aggregates (Klionsky and Emr 2000; Hara et al. 2006; Komatsu et al. 2006). In particular, it has emerged that autophagy can also selectively degrade ubiquitinated proteins (Bjorkoy et al. 2005). The recognition of ubiquitinated substrates is done by autophagy adaptors including SQSTM1 (also known as p62), NBR1 and valosin containing protein (VCP), which bind both to ubiquitin and the autophagosome-specific proteins members of the LC3/GABARAP/Gate16 family (Johansen and Lamark 2011). Specific chaperones and co-chaperones can participate in the delivery of bound clients to autophagosomes, including BAG3 (Carra et al. 2008a). Another example of the tight cooperation between chaperones and lysosome-based degradation is CMA, which removes a specific subset of proteins containing the pentapeptide lysosome-targeting motif (KFERQ). These substrates are directly translocated into the lysosome after docking to the LAMP2A and unfolding by a chaperone complex containing HSPA8 and the co-chaperones BAG1, HSPA8-interacting protein (Hip), Hsp-organising protein (Hop) and HSP40/DNAJB1 (Cuervo and Wong 2014).

21.1.2 Protein Damage and the PQC System

When cells are exposed to different types of acute or chronic stress, protein homeostasis is challenged and the protein-folding equilibrium is altered. Cells respond by increasing the expression of genes that protect against proteotoxic stress, including HSPs and do so by instantaneously stimulating the heat-shock response (HSR) (Anckar and Sistonen 2011; Morimoto 2008). The HSR is often mediated by specific sensors that activate transcription factors, such as the heat shock factor 1 (HSF-1), which controls several genes of the HSR. Existing and newly synthesized molecular chaperones capture the folding intermediates to prevent misfolding and aggregation and to facilitate refolding or degradation, and the entire PCQ system machinery is potentiated under these conditions. In parallel, cells temporarily attenuate translation allowing a decrease in the total intracellular levels of misfolding proteins to amounts manageable by the molecular chaperones (Tsaytler et al. 2011). Translation attenuation is triggered by the accumulation of unfolded proteins within the endoplasmic reticulum (ER) (Ron and Walter 2007). This, in fact, leads to PERK-mediated phosphorylation of eukaryotic initiation factor 2 on Ser51 of the α subunit (eIF2α), thereby resulting in the temporary sequestration of mRNAs encoding for "housekeeping" functions in cytoplasmic foci called stress granules (SGs) and inhibiting their translation (Kedersha and Anderson 2002). Sequestration of mRNAs encoding for "housekeeping" functions within SGs also prioritizes the synthesis of chaperones and enzymes needed for the stress response, while protecting and storing mRNAs during stress (Kedersha and Anderson 2002). Potential cell damage associated with the response to increased levels of misfolded proteins can be attenuated by the sequestration into SGs of scaffold proteins (e.g. JNK, MKK7, rhoA) and pro-apoptotic proteins (e.g. RACK1, ROCK1, TRAF2), thus. These proteins respectively modulate signaling cascades and inhibiting apoptosis during stress, generating a well-integrated stress response system (Buchan and Parker 2009; Kedersha et al. 2000; Takahashi et al. 2012). Once the stress is relieved, SGs, which are highly dynamic, disassemble, restoring proper translation within cells. Persistent SGs will be removed by autophagy, with the assistance of the ubiquitin chaperone VCP (Buchan et al. 2013), thereby also contributing to the recovery of protein homeostasis.

21.2 Protein Aggregation and Proteostasis Imbalance in Neurodegenerative Diseases

Misfolding and aggregation are common molecular events that affect several organs and tissues and are responsible for a large number of human diseases, including neurodegenerative and neuromuscular diseases. All the diseases that are characterized by the presence of proteinaceous aggregates have been called conformational diseases (Chiti and Dobson 2006). Conformational diseases affecting the brain include repeat expansion diseases, such as CAG-repeat/polyglutamine (polyQ) related diseases (e.g. Huntington's disease (HD), spinal and bulbar muscular atrophy (SBMA), and spinocerebellar ataxias (SCAs)), C9ORF72/GGGGCC related diseases (e.g. the most frequent familial forms of amyotrophic lateral sclerosis (ALS), frontotemporal dementia (FTD), etc.) as well as non-repeat expansion diseases including Parkinson's disease (PD), amyotrophic lateral sclerosis (ALS), prion disease, and Alzheimer's disease (AD), which may appear both in sporadic or familial forms. In the latter cases, the inheritance is associated to mutations of a specific gene whose product cannot fold properly. Little is known on the sporadic forms, but often alteration of components of the PCQ system is present, supporting a direct link between imbalance in proteostasis and disease onset and/or progression-severity. Depending on the type of disease (and type of disease-related protein) aggregates/fibrils accumulate in different neuronal populations/brain areas, which correlate with different symptoms and are always associated with a progressive (fatal) clinical course

(Cummings and Zoghbi 2000). However, it is important to underline that not all aggregated species are toxic, as neuronal death does not always correlate with the presence of aggregates. Protein aggregation is a multi-step process in which pre-fibrillar detergent soluble species assemble into large fibrillar non-detergent soluble species (Chiti and Dobson 2006). The neurotoxicity of the different forms of these aggregates/inclusions is still largely debated and in general large macromolecular aggregates are nowadays believed to be protective, while intermediate species would exert high toxicity. Macroscopic aggregates (inclusion bodies) would exert a protective role by trapping the early neurotoxic species into a specific subcellular compartment, waiting for their clearance from the cells. Microaggregates/intermediate species can exert toxicity acting at different steps, such as sequestration of cellular components, particularly transcription factors which are essential for neuronal survival (McCampbell et al. 2000; Kikis et al. 2010), alteration of the intracellular trafficking of molecules and organelles, possibly affecting axonal transport (Sau et al. 2011; Lee et al. 2004). Independent of the type of aggregate and/or of the precise step at which misfolded proteins exert their toxicity, the overall reduction or the prevention of protein misfolding and/or aggregation in neurons, as well as the enhanced clearance of the aggregating/aggregated species are considered to be neuroprotective. Decreased aggregation of misfolded proteins and increased clearance can be achieved a) by potentiating the molecular chaperones, which act in concert with the degradation systems UPS and/or autophagy and b) by directly stimulating autophagy with pharmacologic agents. Indeed, the protective role of specific molecular chaperones and co-chaperones (e.g. members of the HSPA, DNAJ and HSPB families) has been well established using both cellular and animal models of protein conformational neurodegenerative disorders or by potentiating autophagy-mediated clearance (Carra et al. 2008a; Crippa et al. 2010; Hageman et al. 2010; Ravikumar et al. 2004; Rubinsztein 2006; Vos et al. 2010).

21.3 Ribostasis, Stress Granules and Protein Conformation Diseases

Recent observations demonstrated that several neurodegenerative diseases are characterized by the accumulation of nuclear or cytoplasmic RNA-protein (RNP) aggregates. These RNP aggregates often contain stress granule markers and can colocalize with proteinaceous fibrillar aggregates. Most notably, in sporadic and familial forms of ALS, frontotemporal lobar degeneration with ubiquitin-positive inclusions (FTLD-U), multisystem proteinopathy (MSP), HD, AD (Schwab et al. 2008; Wilson et al. 2011) the nuclear RNA-binding protein, TDP-43 redistributes from the nucleus to cytoplasm, where it colocalizes with proteinaceous inclusions. These observations strongly suggest that a tight connection/cross-talk between protein and RNA homeostasis exists and that altered RNP/SG assembly, as well as impaired SG clearance may contribute to protein conformation diseases. This is further suggested by the fact that mutations in TDP-43 and in a number of other mRNA-binding proteins, including (e.g.: FUS, hnRNPA1, ATX2) are associated with neurodegenerative diseases, namely ALS, MSP, spinocerebellar ataxia type 2 (SCA2) or constitute a risk factor in other diseases (e.g. ATX2 is a risk factor for ALS (Laffita-Mesa et al. 2013)). Intriguingly, when mutated these mRNA-binding proteins aberrantly assemble into SGs upon stress, pointing to altered SG dynamics (assembly and disassembly) as important pathomechanism (Acosta et al. 2014). In fact, SGs play a crucial role in protecting and storing mRNAs during stress, as well as indirectly modulating signaling pathways (e.g. by sequestration of specific factors); altered SG dynamic may in turn affect RNA homeostasis. Thus, with this mechanism SG aggregates could alter the population of mRNAs available for translation, which may also contribute to disease.

What molecular events lead to persistence of SGs and colocalization of SG components with proteinaceous inclusions in protein conformation diseases is still largely unknown. Curiously, many of the mRNA-binding proteins involved in SG assembly (or recruited to SGs) contain prion-like domains (e.g. TIA-1, TDP-43, hnRNPA1), which allow them to self-aggregate/polymerize, and trigger SG assembly (Gilks et al. 2004; Li et al. 2013). However, unlike prionogenic fibrillar aggregates, which are irreversible, during normal SGs metabolism, the prion-like domains present in the mRNA-binding proteins reversibly assemble and disassemble ensuring their dynamic nature. The presence of deregulated expression or of mutations in RNA-binding proteins may confer the tendency to form stable amyloid structures mediated by the prion-like domains, which would cause a defective SG disassembly. Accumulation of persistent altered SGs may not only alter RNA metabolism and homeostasis, but also favor protein aggregation; in fact, improperly disassembled SGs or partly disassembled SGs may act as seed for aggregation, due to the presence of proteins with highly aggregation-prone prion-like domains. Alternatively, but not mutually exclusively, defects in clearance mechanisms, deregulated autophagy and accumulation of other aggregate-prone proteins, as it occurs in protein conformation diseases, could contribute to an increase in the frequency of amyloid initiation at SGs. This may change SG assembly dynamics and composition, favoring SG accumulation and/or coalescence with aggregates/inclusion bodies (which would participate in disease progression). In both scenarios, cells rely on an efficient PQC system and autophagy flux to properly clear unassembled SGs, thereby maintaining both RNA and protein homeostasis. In line with this hypothesis, mutations of VCP are associated with ALS and MSP (Inclusion body myopathy with early-onset Paget disease and frontotemporal dementia/IBMPFD), characterized by the accumulation of ubiquitinated inclusions that colocalize with the SG marker TDP-43 (Ju et al. 2008; Ju and Weihl 2010). Intriguingly, VCP is a ubiquitin chaperone involved in the autophagy-mediated clearance of SGs (Buchan et al. 2013), as well as in the modulation of autophagosome formation and autophagy-mediated ubiquitinated client degradation (Ju et al. 2009). This further points to a tight connection between altered proteostasis and RNA processing in disease and suggests that boosting molecular chaperones and degradative systems may be beneficial in preventing RNA aggregation and dysfunction (besides decreasing protein aggregation and aggregate-mediated toxicity).

21.4 eIF2α: Master Regulator of Protein Synthesis, SGs and Autophagy

As previously mentioned, upon stress cells activate a sophisticated integrated response aimed to conserve energy and divert cellular resources toward survival. This includes the temporary shut-down of translation (to decrease the load of unfolding chains for the existing molecular chaperones), upregulation of molecular chaperones (which are essential to avoid irreversible protein aggregation, as well as protein-RNA aberrant aggregation) and activation of autophagy (to clear the misfolded accumulating proteins) (Ron and Walter 2007). Key to these events is the phosphorylation of eIF2α. In fact, upon stress phosphorylation of eIF2α on one hand induces the conversion of LC3-I into LC3-II, which corresponds to the lipidated autophagosome-anchored form, and on the other hand induces the expression of key autophagy genes (e.g. ATG5) (Kouroku et al. 2007). Impairment in the activation of the eIF2α stress response renders the cells more vulnerable to the toxicity mediated by aggregate-prone proteins such as polyQ proteins, leading to accumulation of proteinaceous aggregates and activation of apoptosis (Kouroku et al. 2007). In parallel, phosphorylation of eIF2α promotes SG assembly (although SG formation can occur also independently of eIF2α phosphorylation) (Anderson and Kedersha 2002; Mazroui et al. 2007). SG assembly is triggered after translation shut-down and polyribosome disassembly and is not required for translation attenuation (Anderson and Kedersha 2008). While the beneficial role of autophagy stimulation and translation attenuation upon stress has been well documented, it has not yet been elucidated whether the SG response induced by phospho-eIF2α contributes directly, as an early event, in protein homeostasis and to what extent cross-talk between SG response and PQC exists. Interestingly, the ubiquitin chaperone VCP participates in SG clearance and several players of the PQC system are components of SGs or even modulate their assembly (e.g. ubiquitin and HDAC6, respectively) (Buchan et al. 2013; Kwon et al. 2007). This strongly points to a close interplay between the sensors of proteotoxicity (unfolded protein response, UPR), specific molecular chaperones, eIF2α phosphorylation, SG response and autophagy-mediated clearance. Thus, alteration in the activation of the eIF2α pathway or dysfunction of key players upstream or downstream of this pathway could also alter SG response (and indirectly RNA metabolism and/or stability upon stress), favouring protein overload and protein misfolding/accumulation. In parallel, due to the inefficient activation of autophagy, persistent SGs would accumulate (Buchan et al. 2013). In combination, these observations suggest that modulation of target proteins that can induce phospho-eIF2α, thereby promoting or facilitating temporary translation shut-down, autophagy and, eventually also SG response, may be beneficial in restoring and/or maintaining cell health. Interestingly, specific molecular chaperones such as HSPB8 can modulate (indirectly) via phospho-eIF2α both events, thereby exerting a protective function under proteotoxic stress conditions (see later) (Carra et al. 2009). To what extent molecular chaperones participate in the modulation of SG response and dynamic and whether/how this correlates with their protective/pro-survival role is still largely unknown and is certainly a field to be investigated in the future.

21.5 HSPBs: Implication in Neurodegenerative and Neuromuscular Diseases

Small HSPs (or HSPB) belong to the superfamily of ATP-independent chaperones. The mammalian HSPB protein family comprises ten members (HSPB1-10) (Fontaine et al. 2003; Kappe et al. 2003). From a functional point of view, two major functions have been attributed to HSPBs. First, some HSPB proteins can stabilize the cytoskeleton (actin based microfilaments and intermediate filaments), especially under stress conditions (see chapter by J. Lavoie and J. Landry) (Lavoie et al. 1995). Second, HSPB proteins participate in the maintenance of protein homeostasis by assisting the refolding (when possible) of misfolded aggregate-prone proteins, preventing their irreversible aggregation, and/or participating in their clearance (Carra et al. 2005; Vos et al. 2010). Due to their role in PQC, upregulation of some HSPBs has been implicated, indirectly or directly, in several neurodegenerative and neuromuscular disorders. Indeed, immunohistochemical studies done on post-mortem human tissues from patients suffering from e.g. PD, AD, HD show upregulation of HSPB1, HSPB5, HSPB8 in the areas characterized by neuronal damage/death and by reactive gliosis (Carra and Landry 2006; Seidel et al. 2012). Reactive gliosis is a reaction of the astrocytes to brain injury, resulting in morphological and functional changes of astrocytes and aimed at protecting the surrounding neuronal population from toxic insults and maintaining neuronal homeostasis. Interestingly, the highest expression levels of HSPB proteins are often observed in reactive astrocytes, which provide essential activities that preserve neuronal function. The pathological significance of HSPBs up-regulation in areas characterized by reactive gliosis and neurodegeneration is still largely not understood (Carra and Landry 2006); it is however assumed that it could be part of the stress response to neuronal damage to prevent/decrease the toxicity mediated by the aggregated mutated proteins (e.g. mutated polyQ proteins, mutated SOD1) (Carra et al. 2005; Crippa et al. 2010). This is further suggested by the findings that the transient overexpression of several HSPBs, in vitro, in mammalian cells and in Drosophila melanogaster models of protein conformation diseases, attenuates the aggregation of mutated proteins and protects against their mediated cytotoxicity (see also later: *HSPB8, autophagy and protein degradation*) (Carra et al. 2010; Vos et al. 2010; Gregory et al. 2012). Furthermore, transgenic flies overexpressing members of the Drosophila melanogaster (Dm) HSPBs are protected against mutated polyglutamine-induced neurodegeneration, as well as aging-related decline in locomotor behavior (extension of life span and protection against oxidative stress have been found in Dm-Hsp22 transgenic flies) (Morrow et al. 2004a, b). In combination these data suggest that several members of the HSPB family display protective functions and are important player for maintaining neuronal and muscular cell viability. This is further supported by the fact that mutations of several HSPB proteins (HSPB1, HSPB3, HSPB4, HSPB5, HSPB8) are associated with muscular and neurological disorders, including hereditary sensory and/or motor neuropathy (e.g. HSPB1, HSPB3, HSPB8), myofibrillar myopathy (HSPB5) and congenital cataract (HSPB4) (Boncoraglio et al. 2012; Irobi et al. 2004; Evgrafov et al. 2004; Kolb et al. 2010; Litt et al. 1998; Vicart et al. 1998).

In this chapter we will focus on the potentially protective function exerted by HSPB8 in neurodegenerative and neuromuscular diseases and we will highlight how HSPB8 may act at the crossroad of both protein synthesis and protein degradation, thereby participating in the maintenance of proteostasis.

21.6 HSPB8: At the Crossroad Between Protein Synthesis and Protein Aggregation

Differently from other members of the HSPB family that mainly exist as homo and/or hetero-oligomers containing other HSPB partners (e.g. HSPB1, HSPB5), in cells, HSPB8 forms a stable and stoichiometric complex with the HSPA8 co-chaperone Bcl-2 associated athanogene BAG3 that contains a well-defined 2:1 HSPB8:BAG3 ratio (Carra et al. 2008a). The different behavior of HSPB8, as compared to other "classical" HSPBs, was recently confirmed in vitro, using fluorescently labeled HSPBs and monitoring the type of hetero-oligomeric complexes formed. In particular, while HSPB1, HSPB5 and HSPB6 formed heterogeneous high molecular weight complexes and exchanged subunits, HSPB8 did not form stable complexes with either HSPB1 or HSPB5 (Datskevich et al. 2012). In mammalian cells, stability of HSPB8 is enhanced by its association with BAG3 (Carra et al. 2008a); this was confirmed in vitro studies, that also demonstrated that this interaction of HSPB8 and BAG3 leads both to an increase of thermal stability and an increased resistance to limited chymotrypsinolysis of HSPB8 (Shemetov and Gusev 2011). This makes HSPB8 an "atypical" member of the HSPB family, at least from the structural point of view. Although interaction of other members of the HSPB family with BAG3 has been reported (e.g. HSPB5 and HSPB6, but not HSPB1, can interact with BAG3) (Fuchs et al. 2010; Hishiya et al. 2010), the strength of these interactions was weaker as compared to HSPB8 affinity for BAG3 (Shemetov and Gusev 2011).

21.6.1 HSPB8, Autophagy and Protein Degradation

Concerning the functions of HSPB8, results from our laboratories, using cellular and Drosophila disease models of polyglutamine diseases, show that HSPB8, together with BAG3, reduced the aggregation of mutated polyglutamine proteins such as huntingtin, androgen receptor (AR) and ataxin 3, which are associated with HD, KD and SCA3 (Carra et al. 2005, 2008a, 2010). Similarly, the HSPB8-BAG3 complex inhibited the formation of insoluble species generated by misfolded mutated SOD1 and by a truncated form of TDP-43, associated/involved in familial or sporadic ALS (Crippa et al. 2010). Both the levels of high molecular weight oligomeric species and the insoluble aggregated species generated by mutated SOD1 and truncated TDP-43 were decreased by HSPB8 overexpression, suggesting that HSPB8 exerts its function on these misfolded proteins independently on their oligomeric state (Crippa et al. 2010). The reduced aggregation of misfolded proteins in cells overexpressing HSPB8 is due to the facilitation of the autophagic process (Carra et al. 2008a). In fact, in autophagy deficient cells (ATG5−/−) or upon inhibition of autophagy with 3-methyladenine and wortmannin, HSPB8 and BAG3 were no longer able to inhibit the aggregation/accumulation of mutated polyQ proteins, SOD1 and TDP-43 (Carra et al. 2005; Crippa et al. 2010). From the mechanistic point of view, such anti-aggregation and pro-degradative function of HSPB8 depends on its association with the partner BAG3, as knockdown of the latter diminishes HSPB8 ability to inhibit mutated protein aggregation (Carra et al. 2008b). Moreover, the work of several independent laboratories demonstrated that the HSPB8-BAG3-HSPA8 complex not only facilitates autophagy flux, but is involved also in client binding and targeting to autophagosomes for degradation (Arndt et al. 2010; Carra et al. 2008a; Gamerdinger et al. 2011). In particular, BAG3 possesses a dynein binding domain, which allows binding of the HSPB8-BAG3 complex to dynein and transport of the bound cargo to the microtubule organization center (MTOC) (Gamerdinger et al. 2011). Here, at the MTOC, autophagosomes are mainly assembled and highly concentrated, favouring cargo engulfing and degradation within autophagosomes. If not efficiently degraded, the bound cargo is targeted to the aggresome. Indeed, the aggresome is a dynamic structure that forms, at the MTOC, in response to an overload of improperly folded proteins (Kopito 2000). Thus, when autophagy flux is insufficient or dysregulated, as it occurs in neurodegenerative diseases, the dynein-mediated retrograde transport may not be fully paralleled by the rate of autophagosome formation; this would lead to the typical accumulation of aggresomes found in protein conformation diseases.

Concerning the stimulation of autophagy, overexpression of HSPB8 and/or BAG3 in mammalian cells (HeLa, HEK293T, but not in motoneuronal NSC34 cells) induces the LC3-II (autophagosome-anchored)/LC3-I ratio, which corresponds to an increased formation of autophagosomes (Carra et al. 2008b, 2009). The complex also increases the fusion of autophagosomes with lysosomes, as measured using specific inhibitors of the fusion step, thereby favoring autophagy-mediated degradation of different substrates (Carra et al. 2008b). In contrast, knocking down the HSPB8-BAG3 complex leads to a decreased activation of autophagy in basal conditions particularly under proteotoxic stress, thus rendering the cells more vulnerable to proteotoxicity (Carra et al. 2008b; Rapino et al. 2013); this is accompanied by a large increase in the accumulation of insoluble proteins within the cells. The activation of autophagy by HSPB8-BAG3 observed in the cell lines tested (e.g. HEK293T, HeLa) is a consequence of the induction of the phosphorylation of eIF2α. In fact, on one hand co-transfection of HSPB8 (and BAG3) with GADD34, which promotes the dephosphorylation of eIF2α, abrogates HSPB8 (and BAG3) mediated induction of autophagy (Carra et al. 2009). On the other hand, overexpression of HSPB8 (and BAG3) induces the phosphorylation of eIF2α, both in cells and in vitro. Such induction of phospho-eIF2α upon HSPB8-BAG3 overexpression is generally observed prior to obvious effects on the LC3-II/LC3-I ratio. In particular, while we observed induced phospho-eIF2α typically between 16 and 24 h after transfection of the chaperone complex, the maximal effects on autophagy were observed between 24 and 48 h post-transfection in cells overexpressing HSPB8-BAG3 ((Carra et al. 2009) and Carra, unpublished). As mentioned above phospho-eIF2α induces the expression of key autophagy genes, including ATG5, thus explaining the delay between these two HSPB8-BAG3-mediated effects.

Concerning client binding and targeting, the HSPB8-BAG3-HSPA8 complex interacts with the E3 ligase CHIP (Arndt et al. 2010; Crippa et al. 2010) and with the autophagy receptor protein p62/SQSTM1 (sequestosome1) (Gamerdinger et al. 2009). CHIP would ubiquitinate the HSPA8-bound substrates, while SQSTM1 is a multi-adaptor protein that simultaneously binds to ubiquitin and the autophagosome-associated protein LC3 (Bjorkoy et al. 2005; Pankiv et al. 2007), thereby linking polyubiquitinated proteins to the autophagic machinery. Actually, recent findings from our group show that BAG3 interacts, via HSPA8, with (poly)ubiquitinated proteins; whether all these clients are ubiquitinated via CHIP is however still unknown (Minoia et al. 2014). Once bound by the HSPB8-BAG3-HSPA8 complex, these clients are sequestered into cytoplasmic puncta that are labeled with the autophagic adapter/linker proteins SQSTM1, but also WIPI-1 and LC3. The sequestration of the ubiquitinated clients into SQSTM1-positive cytoplasmic puncta would avoid their proteasome-mediated degradation, thereby favoring their re-routing towards autophagy (Minoia et al. 2014). It is of note that while HSPB8 can colocalize with BAG3 and HSPA8 in these ubiquitin-SQSTM1-containing cytoplasmic puncta, it is not strictly required for BAG3 binding to ubiquitinated proteins as well as for BAG3-induced sequestration into cytoplasmic puncta, which depends entirely on BAG3 interaction with HSPA8. However, we observed that HSPB8 can also pull-down ubiquitinated proteins (Carra, unpublished), similarly to HSPA8; this opens the possibility that HSPB8 may exert an additive role to the one of HSPA8 and participate in the targeting of specific clients to degradation. Alternatively, since we found that HSPB8 colocalizes with BAG3 in ubiquitin-positive cytoplasmic puncta especially after prolonged inhibition of the proteasome, our data suggest that HSPB8 may cooperate with HSPA8 and BAG3 in client re-routing especially under severe stress conditions (Carra, unpublished).

21.6.2 HSPB8 and Protein Synthesis

As previously mentioned, we found that overexpression of HSPB8 in mammalian cells leads to phosphorylation of eIF2α (Carra et al. 2009), which acts as a dominant inhibitor of the guanine nucleotide exchange factor eIF2B and prevents the recycling of eIF2 between successive rounds of protein synthesis. As a consequence, translation is attenuated. A number of conditions result in eIF2α phosphorylation including heat shock, viral infection, nutrient deprivation, iron deficiency, and accumulation of unfolded or denatured proteins (Kimball 1999). In the case of cells overexpressing HSPB8, increased levels of phospho-eIF2α is accompanied by the induction of key transcription factors (e.g. ATF4) and of autophagy (Carra et al. 2009). Both attenuation of translation and induction of autophagy participate in decreasing the accumulation and, subsequently, the aggregation of mutated misfolded proteins mediated by HSPB8 (Carra et al. 2009). The ability of HSPB8 to inhibit protein synthesis has been further demonstrated in vitro, using recombinant HSPB8 and cDNAs encoding for different substrates, thereby allowing a generalization of its effect (Carra et al. 2009). From the mechanistic point of view, it is still largely unknown how HSPB8 can trigger the phosphorylation of eIF2α. HSPB8 shows sequence similarity to the protein kinase coding domain of the large subunit of herpes simplex virus type 2 ribonucleotide reductase (ICP10) and displays a Mn^{2+}-dependent protein kinase activity (Smith et al. 2000). This autokinase activity of HSPB8 depends on the lysine 113 and permits it to phosphorylate specific substrates such as e.g. myelin basic protein (Chowdary et al. 2004; Depre et al. 2002). Its putative autokinase, both wildtype HSPB8 and the kinase-dead K113G mutant forms of HSPB8 could induce the phosphorylation of eIF2α, thereby excluding any direct role of HSPB8 as kinase at the level of eIF2α (Carra et al. 2009).

In parallel to the eIF2α signaling pathway, the phosphoinositide-3 kinase (PI3K) transduction pathway is also activated in response to a wide range of stresses (Bang et al. 2000). Recently it has been reported that a cross-talk between these pathways exists and that the eIF2α kinase PKR, which leads to phosphorylation of eIF2α, also acts upstream of PI3K and turns on the Akt/PKB-FRAP/mTOR pathway. From the mechanistic point of view, the activation of the PI3K pathway is indirect and requires the inhibition of protein synthesis by phospho-eIF2α. As a result, the apoptotic and protein synthesis inhibitory effects exerted by phospho-eIF2α are counterbalanced/antagonized by the Akt/PKB-FRAP/mTOR pathway, which promotes cell proliferation and survival. Curiously, HSPB8 increases the phosphorylation of Akt and acts as survival and differentiation factor in hippocampal neurons (Ramirez-Rodriguez et al. 2013). Whether the induction of eIF2α phosphorylation is upstream of and required for the activation of Akt mediated by HSPB8 is currently unknown. Understanding how upregulation of HSPB8 leads to phosphorylation of eIF2α and whether this is linked to its role in cell survival (Akt activation) will unravel new targets whose modulation helps maintain proteostasis and boosts cell survival in stress and disease.

An inhibitory role in protein synthesis has been suggested also for another member of the HSPB family, HSPB1 (Cuesta et al. 2000). However, this seems mechanistically unrelated to phosphorylation of eIF2α. During heat shock, HSPB1 binds to the eukaryotic initiation factor eIF4G, promotes eIF4G insolubilization and prevents translation (Cuesta et al. 2000). Following heat shock, HSPB1 accumulates together with eIF4G in heat shock granules; similarly to HSPB1, we observed accumulation of HSPB8 in heat-shock granules (Carra, unpublished). What is the functional role of HSPB8 at the level of heat-shock granules and whether, similarly to HSPB1, HSPB8 can interact with eIF4G is not known.

21.6.3 HSPB8, eIF2α Phosphorylation and Stress Granules

As previously mentioned, upon stress, phosphorylation of eIF2α promotes SG assembly; however, it is of note that SGs can also be triggered in an eIF2α-independent manner, e.g. following inhibition of the proteasome or robust heat shock (Grousl et al. 2009; Mazroui et al. 2007). Formation of SGs is part of the integrated stress response aimed at cell survival and alteration in the dynamics of SGs has been associated with a number of pathological conditions. In particular, in protein conformation diseases, such as CAG/polyQ diseases or in ALS, colocalization of SG components with proteinaceous inclusions has been shown (Wolozin 2014; Ramaswami et al. 2013). This may result from the presence of mutated proteins that by itself can trigger SGs with altered dynamics (e.g. persistent SGs can be triggered by ALS-associated mutated FUS/TLS) as well as by the aberrant clearance of SGs, which, in turn, can derive from impairment of the PQC system (inhibition of autophagy and VCP) (Ramaswami et al. 2013; Buchan et al. 2013); the two processes are not mutually exclusive, and may coexist. Whether HSPB8 also influences the SG response, due to its action on phospho-eIF2α and whether, with its pro-autophagic activity, it participates in the clearance of persistent SGs, is currently unknown. However, considering that SG assembly is triggered by the self-aggregation of proteins that contain prion-like domains, it is plausible that specific chaperones are recruited into SGs to avoid irreversible aggregation of their components, thereby maintaining their dynamic nature. We found that HSPB8 is recruited into SGs following several stress conditions, including heat shock, treatment with arsenite and inhibitors of the proteasome (Fig. 21.1a and Carra, unpublished). HSPB8 also colocalizes with mutated ALS-associated FUS R518K in SGs (Fig. 21.1b). Combined, these observations open the possibility that HSPB8 may participate in the PQC at the level of SGs. HSPB8 may contribute to avoid irreversible protein aggregation within SGs. When irreversible aggregation occurs and/or in the presence of mutated or damaged components of SGs, HSPB8 may cooperate with BAG3 to target such components (or damaged and persistent SGs) to autophagy for degradation (Fig. 21.2). Indeed, SGs can be degraded by autophagy with the assistance of the ubiquitin chaperone VCP (Buchan et al. 2013).

Alternatively, but not mutually exclusive, HSPB8 may be recruited into SGs to serve specific functions, such as to regulate the activity of specific mRNA-binding proteins; for example HSPB8 interacts with two RNA-binding proteins: DEAD box protein Ddx20 (gemin3, DP103) and Src associated in mitosis of 68 kDa (Sam68) (Sun et al. 2010; Badri et al. 2006). The DEAD box proteins form a subgroup of the

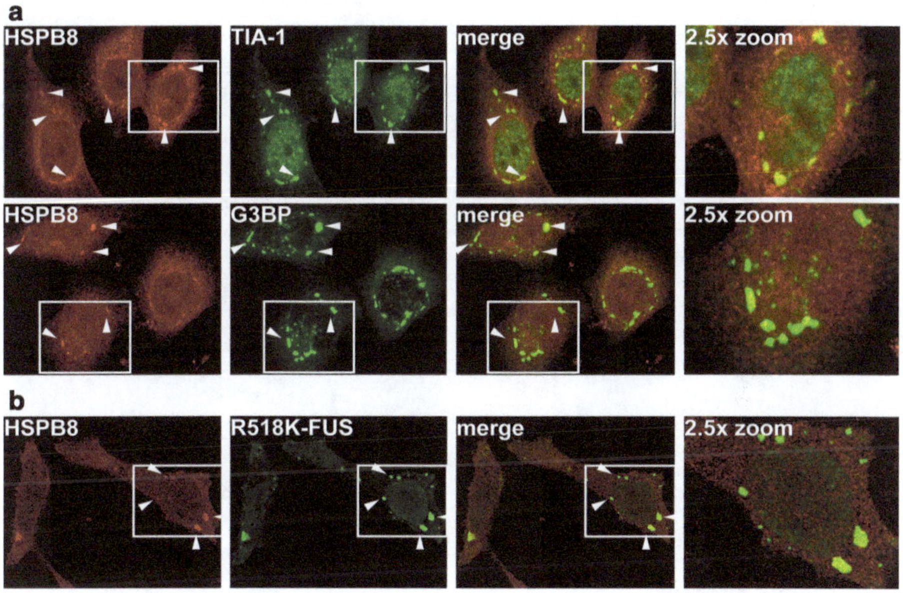

Fig. 21.1 HSPB8 is recruited into stress granules. (**a**) HeLa cells were heated at 43.5 °C for 45 min, fixed in 4 % formaldehyde for 10 min at room temperature, followed by permeabilization with cold acetone for 5 min. Cells were processed for immunofluorescence with anti-HSPB8, anti-TIA-1, anti-G3BP and DAPI. TIA-1 and G3BP were used as markers of stress granules. (**b**) HeLa cells were transfected by calcium phosphate with a cDNA encoding for HA-tagged R518K-FUS (kindly provided by Dr. U. Pandey). Twenty four hours post-transfection cells were fixed as described in A and processed for immunofluorescence with anti-HSPB8, anti-HA and DAPI. (**a**, **b**) A 2.5× magnification of the selected area is shown

DExD/H box family of helicases and have an ATP-dependent RNA unwinding (helicase) activity (Rocak and Linder 2004), and are involved in pre-mRNA processing, RNA turnover, RNA transcription, RNA export. In particular, Ddx20 associates with a protein, the survival of motor neuron (SMN) protein, that when mutated is responsible for spinal muscular atrophy (SMA). Also SMN localizes to SGs and its overexpression induces SGs (Charroux et al. 1999; Hua and Zhou 2004). SMN complexes are also involved in assembly and processing of diverse ribonucleoparticles (RNPs), including snRNPs (spliceosomes), snoRNPs, hnRNPs, transcriptosomes, and miRNPs (Pellizzoni et al. 2002). Sam68, associates with T-cell intracellular antigen-1 (TIA-1), a core component of SGs and localizes to SGs following oxidative stress (Henao-Mejia and He 2009; Henao-Mejia et al. 2009). At present, the functional significance of HSPB8 interaction with both Ddx20 and Sam68 is still largely unknown and may reflect a yet unraveled role of HSPB8 in modulating the function of these specific RNA-binding proteins, rather than their recruitment to SGs.

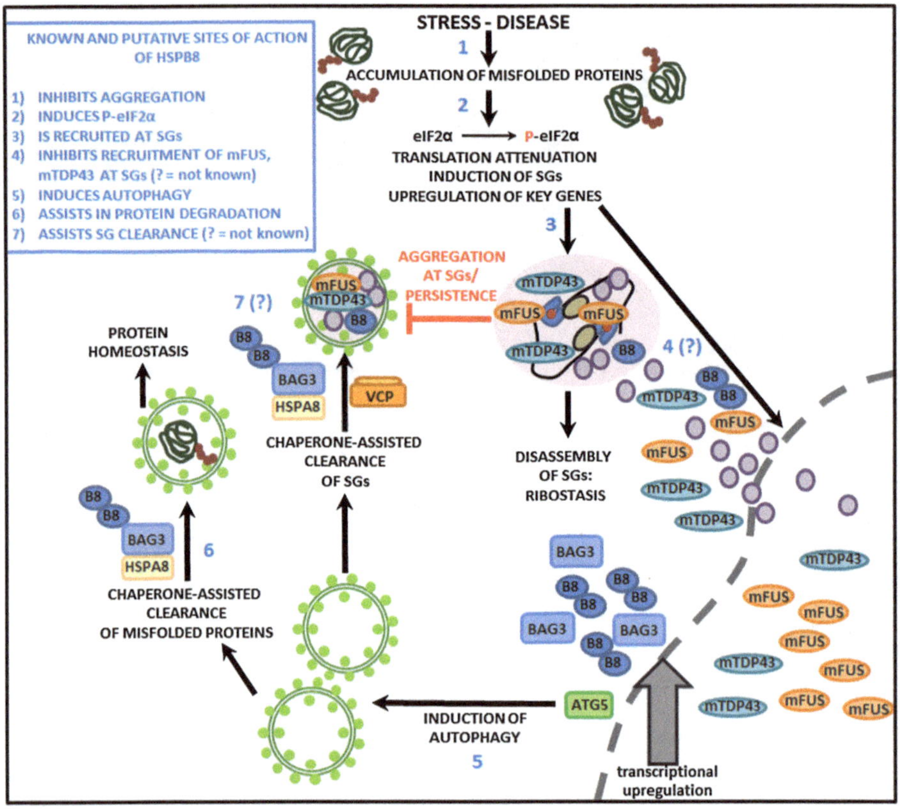

Fig. 21.2 Schematic model of the known and putative sites of action of HSPB8. (*1*) Upon proteotoxic stress conditions, due to either external insult, ageing or genetic mutations, the amount of aberrantly folded substrates exceeds the capacity of the cells to properly assist their folding or clearance; as a result aggregation-prone proteins accumulate, leading to the activation of the protein quality control system. (*2*) Phosphorylation of eIF2α represents an early event and has several consequences: it allows to temporarily attenuate translation; it induces the assembly of SGs (*3*) and it induces the expression of specific genes, including essential autophagy genes (ATG5). SG assembly is triggered by the self-aggregation of RNA-binding proteins that contains a prion-like domain, such as TIA-1, which translocates from the nucleus to the cytoplasm upon stress. Translocation into the cytoplasm and redistribution into SGs can also occur as a consequence of disease-associated mutations; in fact, TDP-43 and mutated FUS "aberrantly" redistribute into SGs. The molecular chaperone HSPB8 is also recruited at SGs but its function at this level is still largely unknown (*4*). (*5*) HSPB8, together with BAG3, is amongst the genes upregulated following proteotoxic stress (e.g. inhibition of the proteasome). HSPB8 and BAG3 participate in the stimulation of the autophagic flux and assist the targeting of damaged clients to autophagy (*6*), thereby allowing to restore proteostasis. HSPB8, which is recruited into SGs and where it colocalizes with mutated FUS, may also participate in the maintenance of SG dynamics. HSPB8 may prevent aberrant protein aggregation at SGs or may target altered and persistent SGs or components thereof to autophagy for clearance (*7*)

21.6.4 Mutated HSPB8 Is Associated with Motor Neuron Diseases

At present, three mutations (K141E, K141N and K141T) of HSPB8 have been associated with hereditary motor neuropathy (HMN) or with Charcot-Marie-Tooth type 2L (CMT-2L) disease and specifically target motor neurons (Irobi et al. 2004; Nakhro et al. 2013). The specific vulnerability of motor neurons to mutated HSPB8 has been confirmed by overexpression studies in primary neuronal motor neuron cultures, where K141E and K141N HSPB8 caused neurite degeneration (the K141T mutation has been discovered recently and little experimental information is available concerning its properties) (Irobi et al. 2010). Instead, no significant toxicity was observed in primary sensory neurons, cortical neurons or glial cells overexpressing mutated HSPB8 (Irobi et al. 2010). Although it is still unclear why motor neurons are particularly sensitive to mutated HSPB8, recent data obtained by our group suggest that K141E and K141N are mainly characterized by loss of function (LOF) rather than by a toxic gain of function (GOF). In fact, the anti-aggregation and pro-degradative activities exerted by wildtype HSPB8 on misfolded mutated proteins such as mutated AR, huntingtin, ataxin-3, SOD1, TDP-43 and the mutated form P182L of HSPB1, which is also associated with HMN, were significantly decreased by its mutations (Carra et al. 2005, 2010). The observation that mutated HSPB8 is characterized by a loss of function in PQC was further confirmed in vivo, using a Drosophila model of the CAG/polyQ disease SCA3 (Carra et al. 2010). The reduced ability of mutated HSPB8 to block the accumulation of aggregate-prone substrates may be due to a (partial) loss of HSPB8 function at the level of protein synthesis regulation (phosphorylation of eIF2α), autophagy induction and/or targeting of clients to autolysosomes for degradation. The latter is supported by experimental finding showing that overexpression of mutated HSPB8 caused the accumulation of autophagosomes that colocalise with protein aggregates, but fail to fuse with the lysosomes (Kwok et al. 2011). Instead, overexpression of wildtype HSPB8 increased the autophagic flux and colocalisation of autophagosomes with lysosomes (Kwok et al. 2011). Binding affinity of mutated HSPB8 to BAG3 was found to be decreased both in cells and in vitro, using recombinant proteins (Carra et al. 2010; Shemetov and Gusev 2011). In parallel, our recent data demonstrate that within the HSPB8-BAG3-HSPA8 complex, BAG3 seems to play a crucial role in client binding and re-routing towards autophagy as well as in stimulation of autophagic flux, while HSPB8 seems to act as helper (Minoia et al. 2014). In light of these observations, it is possible that the decreased ability of mutated HSPB8 to clear misfolded proteins and induce autophagic flux is a consequence of its reduced interaction/cooperation with BAG3 and of decreased HSPB8 stability (mutated HSPB8 by itself tends to aggregate) (Irobi et al. 2004). This would contribute to motor neuropathy. In fact, defective autophagosome maturation and autophagic vacuole accumulation have been shown in a number of protein conformational diseases as well as in HMNs/CMT disease. Moreover, genes directly involved in the endolysosomal and lysosomal pathway are mutated in HMNs (e.g. mutations in Rab7, in lipopolysaccharide-induced tumor necrosis factor-alpha factor/small integral membrane protein of lysosome/late endosome, LITAF/SIMPLE) (Verhoeven et al. 2003; Street et al. 2003; Saifi et al. 2005), further supporting that deregulated autophagy contributes to motor neuron diseases and knockout of key autophagy genes causes neurodegeneration with accumulation of ubiquitin-positive inclusions (Komatsu et al. 2006). Whether a decreased autophagy induction also occurs in cells expressing mutated HSPB8 as a consequence of impaired modulation of phospho-eIF2α and whether such mutant forms of HSPB8 may also indirectly affect SG response/dynamics under stress is still unknown. However, it is of note that interplay between autophagy and SG clearance exists (Buchan et al. 2013) and that both deregulated autophagy and accumulation of SG components are hallmarks of motor neuron diseases, including ALS, IBMPFD (Wolozin 2014; Ramaswami et al. 2013). Thus, altered clearance of SGs as a consequence of decreased autophagic flux may also occur and contribute to HSPB8-associated motor neuropathy. Future studies are needed to understand the exact mechanisms responsible for motor neuronal death in HSPB8-associated motor neuropathy and to explain what makes specifically motor neurons vulnerable to mutated HSPB8.

21.7 Conclusions and Perspectives

Accumulation of aggregated proteins is a hallmark of many neurodegenerative and muscular diseases, confirming that imbalance of protein homeostasis is deleterious for cell survival. To survive proteotoxic stress, due to external insults or genetic mutations, cells have evolved a well-orchestrated system, the protein quality (PQC) system. The PQC system avoids or limits irreversible protein aggregation, thereby maintaining normal protein homeostasis under many different intracellular or extracellular insults that affect protein stability and function. Key players of the PQC system are molecular chaperones, including both the heat shock proteins (HSPs) and the degradative pathways, mainly the proteasome and autophagy systems. Upregulation of chaperones that can boost protein degradation, is normally beneficial in protein conformation models, but the protective functions exerted by specific molecular chaperones may be linked to their action not only at the level of proteins, but also in ribonucleoprotein complexes. In fact, in several neurodegenerative diseases, RNA-containing SGs colocalize with proteinaceous aggregates; thus, imbalance of ribonucleoprotein homoestasis may, concomitantly with proteostasis imbalance, contribute to disease and cell death. Intriguingly, the formation of these SGs is triggered, upon proteotoxic stress, by the self-reversible aggregation of RNA-binding proteins that contain prion-like domains; this suggests that chaperones may also assist SG dynamics, avoiding irreversible RNA/protein aggregation at SGs (and SG persistence, which can be observed in neurodegenerative diseases). Alternatively, but not mutually exclusive, chaperones can recognize aberrant/damaged SGs and target them (or components thereof, such as mutated FUS, TDP-43) to autophagy for degradation. How and to what extent molecular chaperones act at the level of misfolded proteins, ribonucleoproteins and SGs, as well as on translation attenuation (which precedes SG formation) has attracted the attention of researchers. HSPB8 is an example of a chaperone, with multifaceted roles, which may act at different sites to help maintaining both protein and RNA homeostasis. In fact, HSPB8 promotes the autophagy-mediated clearance of aggregated proteins, induces the phosphorylation of eIF2α, which in turns triggers SGs and is itself recruited within SGs. Future research should investigate in more detail the link between protein and RNA homeostasis as well as the specific sites of action of chaperones, including HPB8. This will help unravel to what extent HSPB8 protective functions in protein conformation diseases are linked to its role in translation control, whether it assists SG dynamics or targets SG components to autophagy, which would also contribute to cell protection, or rather whether HSPB8 modulation of eIF2α phosphorylation and recruitment at SGs represent different, unrelated, activities. Therefore, a detailed characterization of the different partners and/or interactors recruited by HSPB8 to trigger one or all pathways acting synergistically to maintain protein and RNA balance, as well as the identification of small molecules capable of inducing HSPB8 expression in a cell-specific manner will be fundamental to finding innovative approaches to counteract human diseases associated with proteotoxicity.

Acknowledgments Serena Carra is grateful to MIUR (Rita Levi Montalcini Principal Research Fellowship). Serena Carra and Angelo Poletti are grateful to The AriSLA Foundation, Cariplo Foundation, AFM-Telethon France for support. Angelo Poletti is also graeful to Telethon, Italy, Regione Lombardia and the University of Milan for support.

References

Acosta JR, Goldsbury C, Winnick C, Badrock AP, Fraser ST, Laird AS, Hall TE, Don EK, Fifita JA, Blair IP, Nicholson GA, Cole NJ (2014) Mutant human FUS is ubiquitously mislocalized and generates persistent stress granules in primary cultured transgenic Zebrafish cells. PLoS One 9(6):e90572

Alberti S, Demand J, Esser C, Emmerich N, Schild H, Hohfeld J (2002) Ubiquitylation of BAG-1 suggests a novel regulatory mechanism during the sorting of chaperone substrates to the proteasome. J Biol Chem 277(48):45920–45927

Anckar J, Sistonen L (2011) Regulation of HSF1 function in the heat stress response: implications in aging and disease. Annu Rev Biochem 80:1089–1115

Anderson P, Kedersha N (2002) Stressful initiations. J Cell Sci 115(Pt 16):3227–3234

Anderson P, Kedersha N (2008) Stress granules: the Tao of RNA triage. Trends Biochem Sci 33(3):141–150

Arndt V, Dick N, Tawo R, Dreiseidler M, Wenzel D, Hesse M, Furst DO, Saftig P, Saint R, Fleischmann BK, Hoch M, Hohfeld J (2010) Chaperone-assisted selective autophagy is essential for muscle maintenance. Curr Biol 20(2):143–148

Badri KR, Modem S, Gerard HC, Khan I, Bagchi M, Hudson AP, Reddy TR (2006) Regulation of Sam68 activity by small heat shock protein 22. J Cell Biochem 99(5):1353–1362

Bang OS, Ha BG, Park EK, Kang SS (2000) Activation of Akt is induced by heat shock and involved in suppression of heat-shock-induced apoptosis of NIH3T3 cells. Biochem Biophys Res Commun 278(2):306–311

Bence NF, Sampat RM, Kopito RR (2001) Impairment of the ubiquitin-proteasome system by protein aggregation. Science 292(5521):1552–1555
Bjorkoy G, Lamark T, Brech A, Outzen H, Perander M, Overvatn A, Stenmark H, Johansen T (2005) p62/SQSTM1 forms protein aggregates degraded by autophagy and has a protective effect on huntingtin-induced cell death. J Cell Biol 171(4):603–614
Boncoraglio A, Minoia M, Carra S (2012) The family of mammalian small heat shock proteins (HSPBs): implications in protein deposit diseases and motor neuropathies. Int J Biochem Cell Biol 44(10):1657–1669
Buchan JR, Parker R (2009) Eukaryotic stress granules: the ins and outs of translation. Mol Cell 36(6):932–941
Buchan JR, Kolaitis RM, Taylor JP, Parker R (2013) Eukaryotic stress granules are cleared by autophagy and Cdc48/VCP function. Cell 153(7):1461–1474
Carra S, Landry J (2006) Small heat shock proteins in neurodegenerative diseases. In: Radons J, Multhoff G (eds) Heat shock proteins in biology and medicine, chapter 18:331–351
Carra S, Sivilotti M, Chavez Zobel AT, Lambert H, Landry J (2005) HspB8, a small heat shock protein mutated in human neuromuscular disorders, has in vivo chaperone activity in cultured cells. Hum Mol Genet 14(12):1659–1669
Carra S, Seguin SJ, Lambert H, Landry J (2008a) HspB8 chaperone activity toward poly(Q)-containing proteins depends on its association with Bag3, a stimulator of macroautophagy. J Biol Chem 283(3):1437–1444
Carra S, Seguin SJ, Landry J (2008b) HspB8 and Bag3: a new chaperone complex targeting misfolded proteins to macroautophagy. Autophagy 4(2):237–239
Carra S, Brunsting JF, Lambert H, Landry J, Kampinga HH (2009) HspB8 participates in protein quality control by a non-chaperone-like mechanism that requires eIF2{alpha} phosphorylation. J Biol Chem 284(9):5523–5532
Carra S, Boncoraglio A, Kanon B, Brunsting JF, Minoia M, Rana A, Vos MJ, Seidel K, Sibon OC, Kampinga HH (2010) Identification of the Drosophila ortholog of HSPB8: implication of HSPB8 loss of function in protein folding diseases. J Biol Chem 285(48):37811–37822
Charroux B, Pellizzoni L, Perkinson RA, Shevchenko A, Mann M, Dreyfuss G (1999) Gemin3: a novel DEAD box protein that interacts with SMN, the spinal muscular atrophy gene product, and is a component of gems. J Cell Biol 147(6):1181–1194
Chiti F, Dobson CM (2006) Protein misfolding, functional amyloid, and human disease. Annu Rev Biochem 75:333–366
Chowdary TK, Raman B, Ramakrishna T, Rao CM (2004) Mammalian Hsp22 is a heat-inducible small heat-shock protein with chaperone-like activity. Biochem J 381(Pt 2):379–387
Ciechanover A (2005) Proteolysis: from the lysosome to ubiquitin and the proteasome. Nat Rev 6(1):79–87
Ciechanover A, Brundin P (2003) The ubiquitin proteasome system in neurodegenerative diseases: sometimes the chicken, sometimes the egg. Neuron 40(2):427–446
Crippa V, Sau D, Rusmini P, Boncoraglio A, Onesto E, Bolzoni E, Galbiati M, Fontana E, Marino M, Carra S, Bendotti C, De Biasi S, Poletti A (2010) The small heat shock protein B8 (HspB8) promotes autophagic removal of misfolded proteins involved in amyotrophic lateral sclerosis (ALS). Hum Mol Genet 19(17):3440–3456
Cuervo AM, Wong E (2014) Chaperone-mediated autophagy: roles in disease and aging. Cell Res 24(1):92–104
Cuesta R, Laroia G, Schneider RJ (2000) Chaperone hsp27 inhibits translation during heat shock by binding eIF4G and facilitating dissociation of cap-initiation complexes. Genes Dev 14(12):1460–1470
Cummings CJ, Zoghbi HY (2000) Trinucleotide repeats: mechanisms and pathophysiology. Annu Rev Genomics Hum Genet 1:281–328
Datskevich PN, Mymrikov EV, Gusev NB (2012) Utilization of fluorescent chimeras for investigation of heterooligomeric complexes formed by human small heat shock proteins. Biochimie 94(8):1794–1804
Demand J, Alberti S, Patterson C, Hohfeld J (2001) Cooperation of a ubiquitin domain protein and an E3 ubiquitin ligase during chaperone/proteasome coupling. Curr Biol 11(20):1569–1577

Depre C, Hase M, Gaussin V, Zajac A, Wang L, Hittinger L, Ghaleh B, Yu X, Kudej RK, Wagner T, Sadoshima J, Vatner SF (2002) H11 kinase is a novel mediator of myocardial hypertrophy in vivo. Circ Res 91(11):1007–1014

Deuerling E, Bukau B (2004) Chaperone-assisted folding of newly synthesized proteins in the cytosol. Crit Rev Biochem Mol Biol 39(5–6):261–277

Douglas PM, Dillin A (2010) Protein homeostasis and aging in neurodegeneration. J Cell Biol 190(5):719–729

Evgrafov OV, Mersiyanova I, Irobi J, Van Den Bosch L, Dierick I, Leung CL, Schagina O, Verpoorten N, Van Impe K, Fedotov V, Dadali E, Auer-Grumbach M, Windpassinger C, Wagner K, Mitrovic Z, Hilton-Jones D, Talbot K, Martin JJ, Vasserman N, Tverskaya S, Polyakov A, Liem RK, Gettemans J, Robberecht W, De Jonghe P, Timmerman V (2004) Mutant small heat-shock protein 27 causes axonal Charcot-Marie-Tooth disease and distal hereditary motor neuropathy. Nat Genet 36(6):602–606

Fontaine JM, Rest JS, Welsh MJ, Benndorf R (2003) The sperm outer dense fiber protein is the 10th member of the super family of mammalian small stress proteins. Cell Stress Chaperones 8(1):62–69

Fuchs M, Poirier DJ, Seguin SJ, Lambert H, Carra S, Charette SJ, Landry J (2010) Identification of the key structural motifs involved in HspB8/HspB6-Bag3 interaction. Biochem J 425(1):245–255

Gamerdinger M, Hajieva P, Kaya AM, Wolfrum U, Hartl FU, Behl C (2009) Protein quality control during aging involves recruitment of the macroautophagy pathway by BAG3. EMBO J 28(7):889–901

Gamerdinger M, Kaya AM, Wolfrum U, Clement AM, Behl C (2011) BAG3 mediates chaperone-based aggresome-targeting and selective autophagy of misfolded proteins. EMBO Rep 12(2):149–156

Gilks N, Kedersha N, Ayodele M, Shen L, Stoecklin G, Dember LM, Anderson P (2004) Stress granule assembly is mediated by prion-like aggregation of TIA-1. Mol Biol Cell 15(12):5383–5398

Gobbo J, Gaucher-Di-Stasio C, Weidmann S, Guzzo J, Garrido C (2011) Quantification of HSP27 and HSP70 molecular chaperone activities. Methods Mol Biol 787:137–143

Gregory JM, Barros TP, Meehan S, Dobson CM, Luheshi LM (2012) The aggregation and neurotoxicity of TDP-43 and its ALS-associated 25 kDa fragment are differentially affected by molecular chaperones in Drosophila. PLoS One 7(2):e31899

Grousl T, Ivanov P, Frydlova I, Vasicova P, Janda F, Vojtova J, Malinska K, Malcova I, Novakova L, Janoskova D, Valasek L, Hasek J (2009) Robust heat shock induces eIF2alpha-phosphorylation-independent assembly of stress granules containing eIF3 and 40S ribosomal subunits in budding yeast, Saccharomyces cerevisiae. J Cell Sci 122(Pt 12):2078–2088

Hageman J, Rujano MA, van Waarde MA, Kakkar V, Dirks RP, Govorukhina N, Oosterveld-Hut HM, Lubsen NH, Kampinga HH (2010) A DNAJB chaperone subfamily with HDAC-dependent activities suppresses toxic protein aggregation. Mol Cell 37(3):355–369

Hara T, Nakamura K, Matsui M, Yamamoto A, Nakahara Y, Suzuki-Migishima R, Yokoyama M, Mishima K, Saito I, Okano H, Mizushima N (2006) Suppression of basal autophagy in neural cells causes neurodegenerative disease in mice. Nature 441:885–889

Hartl FU (1996) Molecular chaperones in cellular protein folding. Nature 381(6583):571–579

Hartl FU, Hayer-Hartl M (2002) Molecular chaperones in the cytosol: from nascent chain to folded protein. Science 295(5561):1852–1858

Hartl FU, Bracher A, Hayer-Hartl M (2011) Molecular chaperones in protein folding and proteostasis. Nature 475(7356):324–332

Henao-Mejia J, He JJ (2009) Sam68 relocalization into stress granules in response to oxidative stress through complexing with TIA-1. Exp Cell Res 315(19):3381–3395

Henao-Mejia J, Liu Y, Park IW, Zhang J, Sanford J, He JJ (2009) Suppression of HIV-1 Nef translation by Sam68 mutant-induced stress granules and nef mRNA sequestration. Mol Cell 33(1):87–96

Hishiya A, Salman MN, Carra S, Kampinga HH, Takayama S (2010) BAG3 directly interacts with mutated alphaB-crystallin to suppress its aggregation and toxicity. PLoS One 6(3):e16828

Hua Y, Zhou J (2004) Survival motor neuron protein facilitates assembly of stress granules. FEBS Lett 572(1–3):69–74

Irobi J, Van Impe K, Seeman P, Jordanova A, Dierick I, Verpoorten N, Michalik A, De Vriendt E, Jacobs A, Van Gerwen V, Vennekens K, Mazanec R, Tournev I, Hilton-Jones D, Talbot K, Kremensky I, Van Den Bosch L, Robberecht W, Van Vandekerckhove J, Broeckhoven C, Gettemans J, De Jonghe P, Timmerman V (2004) Hot-spot residue in small heat-shock protein 22 causes distal motor neuropathy. Nat Genet 36(6):597–601

Irobi J, Almeida-Souza L, Asselbergh B, De Winter V, Goethals S, Dierick I, Krishnan J, Timmermans JP, Robberecht W, De Jonghe P, Van Den Bosch L, Janssens S, Timmerman V (2010) Mutant HSPB8 causes motor neuron-specific neurite degeneration. Hum Mol Genet 19(16):3254–3265

Johansen T, Lamark T (2011) Selective autophagy mediated by autophagic adapter proteins. Autophagy 7(3):279–296

Ju JS, Weihl CC (2010) Inclusion body myopathy, Paget's disease of the bone and fronto-temporal dementia: a disorder of autophagy. Hum Mol Genet 19(R1):R38–R45

Ju JS, Miller SE, Hanson PI, Weihl CC (2008) Impaired protein aggregate handling and clearance underlie the pathogenesis of p97/VCP-associated disease. J Biol Chem 283(44):30289–30299

Ju JS, Fuentealba RA, Miller SE, Jackson E, Piwnica-Worms D, Baloh RH, Weihl CC (2009) Valosin-containing protein (VCP) is required for autophagy and is disrupted in VCP disease. J Cell Biol 187(6):875–888

Kampinga HH, Craig EA (2010) The HSP70 chaperone machinery: J proteins as drivers of functional specificity. Nat Rev 11(8):579–592

Kappe G, Franck E, Verschuure P, Boelens WC, Leunissen JA, de Jong WW (2003) The human genome encodes 10 alpha-crystallin-related small heat shock proteins: HspB1–10. Cell Stress Chaperones 8(1):53–61

Kedersha N, Anderson P (2002) Stress granules: sites of mRNA triage that regulate mRNA stability and translatability. Biochem Soc Trans 30(Pt 6):963–969

Kedersha N, Cho MR, Li W, Yacono PW, Chen S, Gilks N, Golan DE, Anderson P (2000) Dynamic shuttling of TIA-1 accompanies the recruitment of mRNA to mammalian stress granules. J Cell Biol 151(6):1257–1268

Kikis EA, Gidalevitz T, Morimoto RI (2010) Protein homeostasis in models of aging and age-related conformational disease. Adv Exp Med Biol 694:138–159

Kimball SR (1999) Eukaryotic initiation factor eIF2. Int J Biochem Cell Biol 31(1):25–29

Klionsky DJ, Emr SD (2000) Autophagy as a regulated pathway of cellular degradation. Science 290(5497):1717–1721

Kolb SJ, Snyder PJ, Poi EJ, Renard EA, Bartlett A, Gu S, Sutton S, Arnold WD, Freimer ML, Lawson VH, Kissel JT, Prior TW (2010) Mutant small heat shock protein B3 causes motor neuropathy: utility of a candidate gene approach. Neurology 74(6):502–506

Komatsu M, Waguri S, Chiba T, Murata S, Iwata J, Tanida I, Ueno T, Koike M, Uchiyama Y, Kominami E, Tanaka K (2006) Loss of autophagy in the central nervous system causes neurodegeneration in mice. Nature 441(7095):880–884

Kopito RR (2000) Aggresomes, inclusion bodies and protein aggregation. Trends Cell Biol 10(12):524–530

Kouroku Y, Fujita E, Tanida I, Ueno T, Isoai A, Kumagai H, Ogawa S, Kaufman RJ, Kominami E, Momoi T (2007) ER stress (PERK/eIF2alpha phosphorylation) mediates the polyglutamine-induced LC3 conversion, an essential step for autophagy formation. Cell Death Differ 14(2):230–239

Kwok AS, Phadwal K, Turner BJ, Oliver PL, Raw A, Simon AK, Talbot K, Agashe VR (2011) HspB8 mutation causing hereditary distal motor neuropathy impairs lysosomal delivery of autophagosomes. J Neurochem 119:1155–1161

Kwon S, Zhang Y, Matthias P (2007) The deacetylase HDAC6 is a novel critical component of stress granules involved in the stress response. Genes Dev 21(24):3381–3394

Laffita-Mesa JM, Rodriguez Pupo JM, Moreno Sera R, Vazquez Mojena Y, Kouri V, Laguna-Salvia L, Martinez-Godales M, Valdevila Figueira JA, Bauer PO, Rodriguez-Labrada R, Gonzalez Zaldivar Y, Paucar M, Svenningsson P, Velazquez Perez L (2013) De novo mutations in ataxin-2 gene and ALS risk. PLoS One 8(8):e70560

Lavoie JN, Lambert H, Hickey E, Weber LA, Landry J (1995) Modulation of cellular thermoresistance and actin filament stability accompanies phosphorylation-induced changes in the oligomeric structure of heat shock protein 27. Mol Cell Biol 15(1):505–516

Lee WC, Yoshihara M, Littleton JT (2004) Cytoplasmic aggregates trap polyglutamine-containing proteins and block axonal transport in a Drosophila model of Huntington's disease. Proc Natl Acad Sci U S A 101(9):3224–3229

Li YR, King OD, Shorter J, Gitler AD (2013) Stress granules as crucibles of ALS pathogenesis. J Cell Biol 201(3):361–372

Litt M, Kramer P, LaMorticella DM, Murphey W, Lovrien EW, Weleber RG (1998) Autosomal dominant congenital cataract associated with a missense mutation in the human alpha crystallin gene CRYAA. Hum Mol Genet 7(3):471–474

Mazroui R, Di Marco S, Kaufman RJ, Gallouzi IE (2007) Inhibition of the ubiquitin-proteasome system induces stress granule formation. Mol Biol Cell 18(7):2603–2618

McCampbell A, Taylor JP, Taye AA, Robitschek J, Li M, Walcott J, Merry D, Chai Y, Paulson H, Sobue G, Fischbeck KH (2000) CREB-binding protein sequestration by expanded polyglutamine. Hum Mol Genet 9(14):2197–2202

Minoia M, Boncoraglio A, Vinet J, Morelli FF, Brunsting JF, Poletti A, Krom S, Reits E, Kampinga HH, Carra S (2014) BAG3 induces the sequestration of proteasomal clients into cytoplasmic puncta: implications for a proteasome-to-autophagy switch. Autophagy 10(9):1603–1621

Morimoto RI (2008) Proteotoxic stress and inducible chaperone networks in neurodegenerative disease and aging. Genes Dev 22(11):1427–1438

Morimoto RI, Cuervo AM (2014) Proteostasis and the aging proteome in health and disease. J Gerontol 69(Suppl 1):S33–S38

Morrow G, Battistini S, Zhang P, Tanguay RM (2004a) Decreased lifespan in the absence of expression of the mitochondrial small heat shock protein Hsp22 in Drosophila. J Biol Chem 279(42):43382–43385

Morrow G, Samson M, Michaud S, Tanguay RM (2004b) Overexpression of the small mitochondrial Hsp22 extends Drosophila life span and increases resistance to oxidative stress. FASEB J 18(3):598–599

Nakhro K, Park JM, Kim YJ, Yoon BR, Yoo JH, Koo H, Choi BO, Chung KW (2013) A novel Lys141Thr mutation in small heat shock protein 22 (HSPB8) gene in Charcot-Marie-Tooth disease type 2L. Neuromuscul Disord 23(8):656–663

Pankiv S, Clausen TH, Lamark T, Brech A, Bruun JA, Outzen H, Overvatn A, Bjorkoy G, Johansen T (2007) p62/SQSTM1 binds directly to Atg8/LC3 to facilitate degradation of ubiquitinated protein aggregates by autophagy. J Biol Chem 282(33):24131–24145

Parzych KR, Klionsky DJ (2014) An overview of autophagy: morphology, mechanism, and regulation. Antioxid Redox Signal 20(3):460–473

Pellizzoni L, Yong J, Dreyfuss G (2002) Essential role for the SMN complex in the specificity of snRNP assembly. Science 298(5599):1775–1779

Ramaswami M, Taylor JP, Parker R (2013) Altered ribostasis: RNA-protein granules in degenerative disorders. Cell 154(4):727–736

Ramirez-Rodriguez G, Babu H, Klempin F, Krylyshkina O, Baekelandt V, Gijsbers R, Debyser Z, Overall RW, Nicola Z, Fabel K, Kempermann G (2013) The alpha crystallin domain of small heat shock protein b8 (Hspb8) acts as survival and differentiation factor in adult hippocampal neurogenesis. J Neurosci 33(13):5785–5796

Rapino F, Jung M, Fulda S (2013) BAG3 induction is required to mitigate proteotoxicity via selective autophagy following inhibition of constitutive protein degradation pathways. Oncogene 33:1713–1724

Ravikumar B, Vacher C, Berger Z, Davies JE, Luo S, Oroz LG, Scaravilli F, Easton DF, Duden R, O'Kane CJ, Rubinsztein DC (2004) Inhibition of mTOR induces autophagy and reduces toxicity of polyglutamine expansions in fly and mouse models of Huntington disease. Nat Genet 36(6):585–595

Rocak S, Linder P (2004) DEAD-box proteins: the driving forces behind RNA metabolism. Nat Rev 5(3):232–241

Ron D, Walter P (2007) Signal integration in the endoplasmic reticulum unfolded protein response. Nat Rev 8(7):519–529

Rubinsztein DC (2006) The roles of intracellular protein-degradation pathways in neurodegeneration. Nature 443(7113):780–786

Saifi GM, Szigeti K, Wiszniewski W, Shy ME, Krajewski K, Hausmanowa-Petrusewicz I, Kochanski A, Reeser S, Mancias P, Butler I, Lupski JR (2005) SIMPLE mutations in Charcot-Marie-Tooth disease and the potential role of its protein product in protein degradation. Hum Mutat 25(4):372–383

Sau D, Rusmini P, Crippa V, Onesto E, Bolzoni E, Ratti A, Poletti A (2011) Dysregulation of axonal transport and motorneuron diseases. Biol Cell 103(2):87–107

Schwab C, Arai T, Hasegawa M, Yu S, McGeer PL (2008) Colocalization of transactivation-responsive DNA-binding protein 43 and huntingtin in inclusions of Huntington disease. J Neuropathol Exp Neurol 67(12):1159–1165

Seidel K, Vinet J, Dunnen WF, Brunt ER, Meister M, Boncoraglio A, Zijlstra MP, Boddeke HW, Rüb U, Kampinga HH, Carra S (2012) The HSPB8-BAG3 chaperone complex is upregulated in astrocytes in the human brain affected by protein aggregation diseases. Neuropathol Appl Neurobiol 38(1):39–53

Shemetov AA, Gusev NB (2011) Biochemical characterization of small heat shock protein HspB8 (Hsp22)-Bag3 interaction. Arch Biochem Biophys 513(1):1–9

Smith CC, Yu YX, Kulka M, Aurelian L (2000) A novel human gene similar to the protein kinase (PK) coding domain of the large subunit of herpes simplex virus type 2 ribonucleotide reductase (ICP10) codes for a serine-threonine PK and is expressed in melanoma cells. J Biol Chem 275(33):25690–25699

Street VA, Bennett CL, Goldy JD, Shirk AJ, Kleopa KA, Tempel BL, Lipe HP, Scherer SS, Bird TD, Chance PF (2003) Mutation of a putative protein degradation gene LITAF/SIMPLE in Charcot-Marie-Tooth disease 1C. Neurology 60(1):22–26

Sun X, Fontaine JM, Hoppe AD, Carra S, DeGuzman C, Martin JL, Simon S, Vicart P, Welsh MJ, Landry J, Benndorf R (2010) Abnormal interaction of motor neuropathy-associated mutant HspB8 (Hsp22) forms with the RNA helicase Ddx20 (gemin3). Cell Stress Chaperones 15(5):567–582

Takahashi M, Higuchi M, Matsuki H, Yoshita M, Ohsawa T, Oie M, Fujii M (2012) Stress granules inhibit apoptosis by reducing reactive oxygen species production. Mol Cell Biol 33(4):815–829

Tsaytler P, Harding HP, Ron D, Bertolotti A (2011) Selective inhibition of a regulatory subunit of protein phosphatase 1 restores proteostasis. Science 332(6025):91–94

Verhoeven K, De Jonghe P, Coen K, Verpoorten N, Auer-Grumbach M, Kwon JM, FitzPatrick D, Schmedding E, De Vriendt E, Jacobs A, Van Gerwen V, Wagner K, Hartung HP, Timmerman V (2003) Mutations in the small GTP-ase late endosomal protein RAB7 cause Charcot-Marie-Tooth type 2B neuropathy. Am J Hum Genet 72(3):722–727

Vicart P, Caron A, Guicheney P, Li Z, Prevost MC, Faure A, Chateau D, Chapon F, Tome F, Dupret JM, Paulin D, Fardeau M (1998) A missense mutation in the alphaB-crystallin chaperone gene causes a desmin-related myopathy. Nat Genet 20(1):92–95

Vos MJ, Zijlstra MP, Kanon B, van Waarde-Verhagen MA, Brunt ER, Oosterveld-Hut HM, Carra S, Sibon OC, Kampinga HH (2010) HSPB7 is the most potent polyQ aggregation suppressor within the HSPB family of molecular chaperones. Hum Mol Genet 19:4677–4693

Wilson AC, Dugger BN, Dickson DW, Wang DS (2011) TDP-43 in aging and Alzheimer's disease – a review. Int J Clin Exp Pathol 4(2):147–155

Wolozin B (2014) Physiological protein aggregation run amuck: stress granules and the genesis of neurodegenerative disease. Discov Med 17(91):47–52

Chapter 22
Understanding What Small Heat Shock Proteins Do for Bacterial Cells

Zengyi Chang

Abstract Being initially discovered due to the increased transcription of their encoding genes under heat shock conditions, heat shock proteins (HSPs, interchangeably named as 'stress proteins' or 'molecular chaperones') are known for their dramatic increase in amount when an organism is exposed to a variety of stress conditions. The major known activity for HSPs is to prevent the aggregation of nascent polypeptide chains that are yet-folded or mature proteins that are denatured under stress conditions. By transiently binding to and subsequently releasing the client proteins, HSPs function to promote their folding and/or assembly in living organisms. Small heat shock proteins (sHSPs), being relatively small among the HSPs in terms of the molecular weight of a single polypeptide chain, was found to present in bacteria only years after they were first identified in animals and plants. In this chapter, I will provide a historical perspective on what we have learned about the structure, function and regulation of sHSPs in bacteria. Main aspects covered in this chapter include the following. sHSPs exist as large dynamic homo-oligomers and regulate their activities by modulating their oligomeric status in a stress-responsive manner; sHSPs exhibit effective chaperone-like activities under *in vitro* and *in vivo* conditions; The monomeric small heat shock proteins possess an immunoglobulin-like folding pattern; sHSPs associate with and affect the physical state of cellular membranes; sHSPs play a potential role for bacterial cells to enter the non-growing dormant state; The biological functions of sHSPs are explored via gene knockout studies. In the end, I will also briefly discuss some of the unresolved issues regarding the structure, function and regulation of sHSPs.

Keywords Cell membrane • Chaperone • Client proteins • Dormancy • Homo-oligomer

---

Z. Chang (✉)
State Key Laboratory of Protein and Plant Gene Research, Center for Protein Sciences, Center for History and Philosophy of Science, School of Life Sciences, Peking University, Beijing 100871, China
e-mail: changzy@pku.edu.cn

R.M. Tanguay, L.E. Hightower (eds.), *The Big Book on Small Heat Shock Proteins*, Heat Shock Proteins 8, DOI 10.1007/978-3-319-16077-1_22

22.1 A Brief History on the Discovery of Heat Shock Proteins

Each known organism lives and grows most effectively over an optimal temperature range. In this regard, temperature is one key factor that affects the life of all living organisms. Heat shock proteins as a group were discovered by Ferruccio Ritossa in the early 1960s, during which a new puffing pattern of the salivary gland polytene chromosomes in the embryos of the fruit fly *Drosophila* was found to be induced upon an unplanned temperature (heat) shock treatment at 37 °C (from their normal growing temperature of 25 °C). This was demonstrated to be due to rapid (within 2–3 min) new RNA synthesis (transcription) at the puff sites on the chromosomes (Ritossa 1962). The manuscript reporting this discovery was initially rejected by a highly reputable journal because the editor considered this finding irrelevant to the scientific community (Ritossa 1962, 1996).

A unique set of newly synthesized proteins, thus incorporating the radioisotope labeled amino acid [^{35}S]methionine, of fixed molecular masses between 70 and 20 kD were repeatedly detected in the salivary glands and other tissues of the fruit fly upon such heat shock treatment, whereas the synthesis of most other cellular proteins was found to be significantly suppressed (Tissieres et al. 1974; Lewis et al. 1975). These unique heat shock induced proteins of *Drosophila* were later termed 'heat shock' polypeptides or proteins, and commonly abbreviated as HSPs (Koninkx 1976; Ashburner and Bonner 1979). As in many cases, 'heat shock protein' is certainly not an ideal name for these proteins, because they were later found to be also expressed under normal temperature conditions (Cheney and Shearn 1983) and can also be induced by stress conditions other than heat shock (Lewis et al. 1975).

Although this selective induction of proteins by heat shock was initially thought to be unique to the flies, analogous heat shock or temperature shift-up response was later observed in avian and mammalian tissue culture cells (Kelley and Schlesinger 1978) and in bacterial cells (Yamamori et al. 1978) since the later 1970s. Further studies then showed that all organisms, from archaebacteria to eubacteria and from plants to animals, respond to heat or other stress conditions by producing the heat shock proteins (Craig and Lindquist 1988).

For years, heat shock genes were chosen as preferable and extremely successful model system for studying gene structure and regulation. Although the proteins encoded by the heat shock genes were found to be among the most highly conserved proteins across the species, the search for functions of the heat shock proteins turned out to be far more difficult and far less successful for many years. A function for heat shock proteins was gradually realized when some of them, often given various names to reflect the way that they were initially identified, were found to be intimately involved in protecting unfolded nascent proteins (during their maturation) or unfolded mature proteins resulted from denaturing conditions, often in an ATP-dependent manner. For example, high level of Hsp70 was found to accelerate the recovery of nucleolar morphology of mammalian cells after heat shock, with the Hsp70 rapidly and specifically dissociating from the insoluble nucleolar particles in the presence of ATP (Lewis and Pelham 1985); Additionally, the Hsp70-like protein was found to be identical to BiP ('immunoglobulin heavy-chain binding protein') that had been known to bind to the nascent heavy chain of IgG in the endoplasmic reticulum of B cells, as well as GRP78 ('glucose regulated protein 78') whose synthesis is induced by glucose starvation (Munro and Pelham 1986).

In light of these observations, it was speculated that one function of Hsp 70 (likely other heat shock proteins as well) is to tightly bind unfolded substrate (client) proteins under stress conditions but to release them after returning to the non-stressful conditions, using energy from ATP hydrolysis (Pelham 1986). Such roles of heat shock proteins to disrupt 'improper' interactions between other proteins by interacting with them in a transient manner led to their designation in general as 'molecular chaperones' (Ellis 1987).

22.2 A Brief History on the Discovery of the Small Heat Shock Proteins

Small heat shock proteins (often abbreviated as sHSPs) were initially identified as ones having a relatively small subunit sizes in the fruit fly *Drosophila*, in comparison with those 'larger' ones such as Hsp60, Hsp70, Hsp90 or Hsp100 family members. The small heat shock proteins were actually also detected in the initial attempts to identify newly synthesized proteins upon heat shock treatment of tissues or individuals of the fruit fly *Drosophila* (Tissieres et al. 1974; Lewis et al. 1975; Koninkx 1976). Four different forms of small heat shock proteins of 27, 26, 23, and 22 kD were then clearly distinguished from *Drosophila* tissues by fingerprinting analysis of trypsin-cleaved peptides (Mirault et al. 1978). When the DNA fragment encoding these four *hsp* genes, all located at *Drosophila* chromosomal subdivision 67B, were isolated and sequenced, it was revealed that their predicted amino acid sequences are not only homologous to each other, but also to that of the mammalian α-crystallin proteins known to be richly present in the eye lens and to form large oligomers (Ingolia and Craig 1982).

Proteins homologous to the fruit fly small heat shock proteins and mammalian α-crystallins were subsequently identified from such species as higher plant (Key et al. 1981; Nagao et al. 1985), animal (Russnak et al. 1983), and bacteria (Booth et al. 1988; Nerland et al. 1988; Verbon et al. 1992; Lee et al. 1992; Allen et al. 1992). In contrast to the extreme high level of conservation in amino acid sequences commonly found among the Hsp70s and Hsp60s, that for small heat shock proteins was found to be far less pronounced (Sauer and Dürre 1993; de Jong et al. 1993). Although the small heat shock proteins from different sources have been commonly found to be present as large insoluble heat shock granules that are resistant to detergent extraction in the nucleus under heat shock conditions, they were found to be soluble and perinuclear-located before or after the heat shock treatment (Arrigo et al. 1988; Collier et al. 1988). They were found to form large dynamic oligomers (Arrigo and Welch 1987; Behlke et al. 1991) and exhibit chaperone-like activities to suppress the aggregation of unfolded client proteins under *in vitro* conditions (Horwitz 1992; Jakob et al. 1993; Chang et al. 1996). Nevertheless, the actual function of this family of proteins, as for many proteins in living organisms, has been elusive.

22.3 The Small Heat Shock Proteins Were Found to Actually Possess the Largest Size in Their Native Assemblies Among the Heat Shock Proteins

Small heat shock proteins was initially defined on the basis of their small molecular weight as detected by the denaturing SDS-PAGE analysis (in combination with autoradiography) in the fruit fly *Drosophila* upon heat shock treatment (Ingolia and Craig 1982). Small heat shock proteins in other species were then identified as a result of their sharing amino acid sequence similarity with these *Drosophila* proteins, most pronounced in the so-called α-crystallin domain (de Jong et al. 1993). The name 'small heat shock protein' designates proteins possessing such a conserved domain, not including other low molecular weight heat shock proteins such as Hsp33, GroES, GrpE and Hsp15 (Narberhaus 2002). Human small heat shock proteins are named according to a HUGO-approved nomenclature (Kampinga et al. 2009). One of these proteins HSPB11 does not have the α-crystallin signature sequence but was included based on functional similarities. There is ongoing discussion about whether or not it should be included (Kappé et al. 2010). When small heat shock proteins were subjected to analysis by electron microscopy (Arrigo et al. 1988; Behlke et al. 1991; Chang et al. 1996; Shearstone and Baneyx 1999), or size exclusion chromatography (Hockertz et al. 1991; Shearstone and Baneyx 1999; Jiao et al. 2005), they were found to exist as large and dynamic oligomers, with an average native size between 400 and 800 kDa, being far larger than members of many other heat shock protein families.

22.4 Small Heat Shock Proteins from Bacteria Were Initially Identified by an Amino Acid Sequence Similarity to Small Heat Shock Proteins of the Eukaryotic Cells

Upon a shift-up of growth temperature from the normal 37 °C to the heat shock 42 °C for the *E. coli* bacterial cells, the level of five proteins having a size between 87 and 61 kD were found to be dramatically increased, whereas the low molecular weight heat shock proteins similar to those clearly detected in animal or plant cells were undetectable (Cooper and Ruettinger 1975; Yamamori et al. 1978). Due to this, for many years, it was not clear whether small heat shock proteins were present in bacteria.

A bacterial small heat shock protein was often revealed by its sharing an amino acid sequence similarity with the eukaryotic ones when the cDNA of a certain protein, named for other properties, was isolated and sequenced. These proteins were initially identified as antigens (Booth et al. 1988; Nerland et al. 1988; Verbon et al. 1992) or major membrane protein (Lee et al. 1992) of mycobacteria, inclusion body binding protein (IbpA and IbpB) of Gram-negative *E. coli* cells (Allen et al. 1992), a protein encoded by a gene whose mutation prevented the Gram-positive bacteria *Clostridium acetobutylicum* from producing acetone and butanol (Sauer and Dürre 1993). They were found to share sequence homology with the animal or plant small heat shock proteins only after their amino acid sequences were deduced from their encoding genes (Booth et al. 1988; Nerland et al. 1988; Verbon et al. 1992; Lee et al. 1992; Allen et al. 1992; Sauer and Dürre 1993). In retrospect, heat shock treatment of *Mycobacteria tuberculosis* at 48 °C (but not at 42 °C) significantly induced the production of a 15 kDa protein band (Young and Garbe 1991), which might be identical to the '14 kDa antigen' that was later found to be a member of the small heat shock protein family (Verbon et al. 1992). The IbpA and IbpB proteins were found to be induced by heat shock treatment, but to a level that was hardly visible by Coomassie blue staining and were detectable only by immunoblotting (Allen et al. 1992).

Small heat shock proteins have also been found in thermophilic cyanobacteria *Synechococcus vulcanus* (HspA) that normally grow at 50 °C (Roy and Nakamoto 1998), hyperthemophilic archaeon *Methanococcus jannaschii* (Hsp16.5) that normally grow at 85 °C (Kim et al. 1998a, b), as well as many other thermophilic bacteria (Laksanalamai and Robb 2004).

22.5 Bacterial Small Heat Shock Proteins Exist as Homo-oligomers with or Without a Fixed Number of Subunits (Monodispersed or Polydispersed)

All bacterial small heat shock proteins have been characterized as homo-oligomers under *in vitro* conditions. Some of them seem to be able to form oligomers of a fixed size, i.e., monodispersed, while others of a variety of sizes, i.e., polydispersed. Bacterial small heat shock proteins forming homo-oligomers of a fixed size include, for example, Hsp16.3 (or Acr1) from *Mycobacterial tuberculosis* (Chang et al. 1996; Kennaway et al. 2005), Hsp16.5 from the hyperthermophilic archaeon *Methanococcus jannaschii* (Kim et al. 1998a, b). Bacterial small heat shock proteins forming homo-oligomers of a diversity of sizes include, for example, the IbpB (Veinger et al. 1998; Kitagawa et al. 2002; Jiao et al. 2005) and IbpA (Matuszewska et al. 2005; Ratajczak et al. 2010) of *E. coli* cells. Under stress conditions, bacterial sHSPs have been observed to form supermolecular assemblies in the form of fibrils (Ratajczak et al. 2010; Shi et al. 2011), which was the form that bound unfolded client proteins. Given that small heat shock proteins have been observed to form large granules in animal cells under heat shock conditions (Arrigo et al. 1988; Collier et al. 1988), it would be interesting to find out whether such large assemblies are the functional forms of small heat shock proteins in bacterial cells. In particular, it needs to be demonstrated whether the thickened wall observed in the dormant state of mycobacteria (Cunningham and Spreadbury 1998) is related to such large forms of small heat shock proteins.

Another common feature for the quaternary structure of small heat shock proteins of bacteria and other species is the dynamic nature of the subunit assembly (as illustrated in Fig. 22.1). This allows a constant dissociation and reassociation of the oligomeric subunits of those of fixed oligomeric sizes (Gu et al. 2002), polydispersed sizes (Shearstone and Baneyx 1999; Jiao et al. 2005), or for those forming hetero-oligomers among the different isoforms (Studer and Narberhaus 2000). Such dynamic subunit interaction is apparently responsive to the physical and chemical ambient conditions, particularly such stress conditions as heat shock, allowing small heat shock proteins to effectively modulate their biological activities and to provide protection of their clients in the cells, correlating to the level of harshness of the stress conditions (Young et al. 1999; Mao et al. 2001; Gu et al. 2002; Bova et al. 2002; Fu et al. 2003, 2004; Fu and Chang 2004; Lentze et al. 2004; Jiao et al. 2005, 2008).

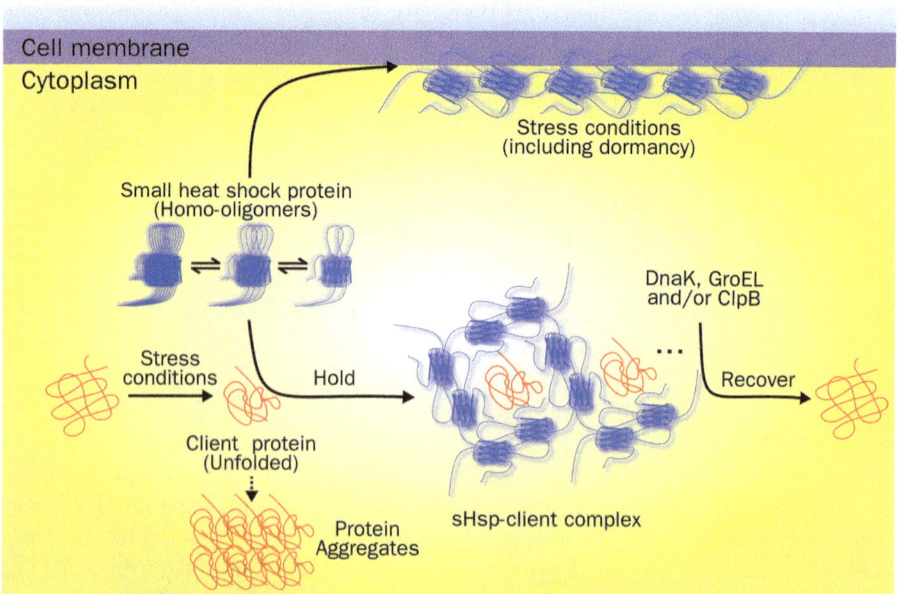

Fig. 22.1 Current understanding on the two major types of physiological roles of small heat shock proteins in bacterial cells. They may function as holdase-type of molecular chaperones to protect cellular client proteins under stress conditions. They may also function to protect the cell membrane when the bacterial cells enter the dormant state

22.6 Functioning as a Chaperone to Hold Unfolded Client Proteins in a Folding-Competent Structure

Other than possessing a dynamic oligomeric structure, another common feature of bacterial and other small heat shock proteins is to efficiently bind unfolded/denatured client (substrate) proteins under *in vitro* (Horwitz 1992; Jakob et al. 1993; Chang et al. 1996; Veinger et al. 1998; Kitagawa et al. 2002; Jiao et al. 2005; Matuszewska et al. 2005) or *in vivo* conditions (Laskowska et al. 1996; Veinger et al. 1998; Fu et al. 2013a, b), thus serving as a reservoir for unfolded client proteins. This will effectively prevent the client proteins from forming aggregates and meanwhile keeping them in a structure that is competent for subsequence refolding. Unlike members of the other families of heat shock proteins (e.g., Hsp60, Hsp70, Hsp90 or Hsp100), the lack of ATPase activity prevents small heat shock proteins from releasing the client proteins by themselves. Under *in vitro* conditions, the presence of Hsp70 (DnaK) and Hsp60 (GroEL) systems would help such client release to a small degree in an ATP-dependent manner (Veinger et al. 1998). Further studies demonstrated that when ClpB acts together with the DnaK chaperone system, the release and refolding of the substrates became far more effective both under *in vitro* and *in vivo* (Mogk et al. 2003a, b) conditions, suggesting that a functional chaperone triade of ClpB, DnaK and IbpA/B may act together to reverse protein aggregates in cells, as schematically illustrated in Fig. 22.1.

22.7 Monomeric Small Heat Shock Proteins Possess an Immunoglobulin Folding Pattern

The determination of the crystal structure of Hsp16.5, the small heat shock protein from the hyperthermophilic archaeon *Methanococcus jannaschii* (Kim et al. 1998a, b), as well as that of Hsp16.9 from plant wheat (van Montfort et al. 2001), revealed that the folding pattern of its monomer is similar to that of the immunoglobulin domain of IgG, despite of the lack of sequence similarity between them (Amzel and Poljak 1979; Kim et al. 1998a, b; Sun and MacRae 2005), i.e., consisting of a sandwich of two layers of β-sheets. In this sense, the small heat shock protein might be categorized into the immunoglobulin fold superfamily (Williams and Barclay 1988). Given that both small heat shock proteins and immunoglobulin superfamily members function to defend individuals or cells by binding to protein targets, this may implicate that they may have evolved under certain similar evolutionary selection forces. In this regard, accumulating evidence indicates that small heat shock proteins are secreted into the extracellular space and signal the activation of innate immunity or even the adaptive immune response in mammals (van Noort et al. 2012). Whether small heat shock proteins function as an antigen presentation protein or immune signaling molecule (as many other members of the immunoglobulin superfamily do) needs to be further investigated.

22.8 Protecting Cell Membrane Components

Many bacterial small heat shock proteins have been found to interact with the cell membranes. As a matter of fact, the small heat shock protein from *Mycobacterial tuberculosis*, Hsp16.3, was initially isolated as a major membrane protein and identified as a member of the small heat shock protein only after its amino acid sequence was made available (Verbon et al. 1992; Lee et al. 1992). This Hsp16.3 protein, being significantly induced in stationary phase mycobacterial cells exposed to low oxygen tension, was found to be largely located at the cell envelope, causing a thickening of the cell wall (Cunningham and Spreadbury 1998). Its association with the mycobacterial plasma membrane was found to depend on the dissociation of the oligomeric structure of Hsp16.3 (Chen et al. 2003; Zhang et al. 2005).

The small heat shock protein from the Gram-positive acidophilic bacterium *Leuconostoc oenos* was also found to be associated with the cell membrane (Jobin et al. 1997). The small heat shock proteins from the photosynthetic cyanobacterium *Synechocystis*, HSP17, was also found to be associated with the thylakoid membrane (Horvath et al. 1998) or cell membrane (Török et al. 2001) and to be able to regulate membrane physical properties *in vitro*, likely regulating membrane fluidity *in vivo*, and preserving membrane structure and integrity during the early stages of stress conditions (Tsvetkova et al. 2002). The small heat shock proteins from lactic acid bacterium *Oenococcus oeni* (Coucheney et al. 2005; Maitre et al. 2012, 2014) and *Lactobacillus plantarum* strain WCFS1 (Capozzi et al. 2011) were also found to interact with the cell membrane and to modulate the lipid physical state under stress (heat shock or ethanol) conditions. The small heat shock proteins from *E. coli* cells, IbpA and IbpB, were also detected in the membrane fractions (Miyake et al. 1993; Laskowska et al. 1996). In a recent study, we observed that heterologous expression of the small heat shock protein HSP17 from the nematode animal *C. elegans* enabled the Gram-negative bacterial *E. coli* cells to grow at the lethal temperature of 50 °C, conceivably by maintaining the integrity of the cell envelope (Ezemaduka et al. 2014).

Apparently, association with the cell membranes is a phenomenon commonly observed not only for bacterial small heat shock proteins, but also for eukaryotic ones (Nakamoto and Vígh 2007; van Noort et al. 2012). Protecting the membrane components is undoubtedly essential for cells to survive under stress conditions. In this regard, small heat shock proteins may not only protect the membrane proteins, but also the lipid bilayer such that the fluidity of the cell membranes is maintained under stress conditions (Török et al. 2001; Tsvetkova et al. 2002; Nakamoto and Vígh 2007; Horváth et al. 2008), as illustrated in Fig. 22.1.

22.9 Small Heat Shock Protein May Function During Bacterial Dormancy

It is a common phenomenon for bacterial cells and other organisms to enter dormancy, a non-growing state, through which individuals may survive for an extended period of time under severe stress conditions (Lewis 2007; Wang et al.

2009). Interestingly, small heat shock proteins have been found to accumulate in dormant bacterial cells. The small heat shock protein Hsp16.3 was identified as the most predominantly up-regulated protein in *M. tuberculosis* during the stationary phase or under reduced oxygen tension, conditions under which cells enter the dormant state (Yuan et al. 1996; Cunningham and Spreadbury 1998). Similarly, the small heat shock protein Lo18 was also found to be dramatically induced during the stationary growth phase of the lactic acid bacterium *Leuconostoc oenos* (Guzzo et al. 1997). The small heat shock proteins produced in the stationary phase seem to be extremely stable and apparently allow the cells to live for a long period of time (Guzzo et al. 1997). In dormant cells, small heat shock proteins may function at both the cell membrane and among cytoplasmic components to enable the cells to survive for a certain length of time under harsh environmental conditions (Fig. 22.1).

22.10 Understanding the Function of Bacterial Small Heat Shock Proteins by Gene-Knockout Studies

One way to understand the physiological function of a protein is to observe phenotypic changes when the encoding gene is removed from the genome of a particular organism. Gene knockout studies revealed that the small heat shock proteins are apparently dispensable for bacterial cells to grow at high temperatures, as demonstrated for the Gram-negative *E. coli* IbpA and IbpB (Thomas and Baneyx 1998) and Gram-positive *Streptomyces albus* Hsp18 (Servant and Mazodier 1995) or *Mycobacterial tuberculosis* Hsp16.3 (Yuan et al. 1998). On the other hand, the absence of Hsp16.3 was found to significantly impair the growth of the *M. tuberculosis* cells in macrophages (Yuan et al. 1998), the common host cells of this pathogen. The disruption of the gene CotM encoding the small heat shock protein of *Bacillus subtilis*, a Gram-positive soil bacterium, although it did not cause any growth defect at high temperatures, it resulted in abnormal spore (a dormant state) outer coat assembly (Henriques et al. 1997).

Gene knockout techniques are not yet feasible on the thermophilic bacterial cells (Laksanalamai and Robb 2004). The function of small heat shock proteins in such bacteria is therefore even less explored. Gene knockout studies will be effective only when the modified organisms exhibit certain phenotypic changes, and this relies much on the experimental condition under which the function of that particular protein becomes important. Finding such a condition and a phenotypic change is often, if not always, challenging. In contrast to structure studies, learning about the function of a protein in living cells is far more important and also far more difficult. In this regard, small heat shock proteins are certainly not exceptions.

22.11 Summary and Future Prospects

As a family of proteins ubiquitously present in all forms of life (including thermophilic bacteria that normally grow at a temperature as high as 85 °C), small heat shock proteins apparently have been working in cells since ancient times. The major biological properties characteristic of bacterial small heat shock proteins include the following: First, they form dynamic oligomers (Chang et al. 1996; Shearstone and Baneyx 1999; Jiao et al. 2005; Kennaway et al. 2005; Ratajczak et al. 2010; Shi et al. 2011), a feature essential for their functioning (Gu et al. 2002; Zhang et al. 2005). Second, they exhibit effective chaperone-like activities to suppress the aggregation of non-native client proteins under *in vitro* (Veinger et al. 1998; Kitagawa et al. 2002; Jiao et al. 2005; Matuszewska et al. 2005) or *in vivo* conditions (Veinger et al. 1998; Mogk et al. 2003b; Fu et al. 2013a, b). Third, they bind to and affect the physical state of cellular membranes (Verbon et al. 1992; Lee et al. 1992; Jobin et al. 1997; Horvath et al. 1998; Török et al. 2001; Tsvetkova et al. 2002; Chen et al. 2003; Zhang et al. 2005; Coucheney et al. 2005; Capozzi et al. 2011; Maitre et al. 2012, 2014), likely another key aspect of small heat shock protein functioning in cells. Fourth, they accumulate to a predominant level during the non-growing (or slow-growing) dormant phase (Yuan et al. 1996; Guzzo et al. 1997; Cunningham and Spreadbury 1998), where they are likely to function in maintaining the unique dormant state of bacterial cells, allowing them to survive under harsh conditions.

It is undoubtedly a great challenge to find out what small heat shock proteins do in bacterial cells. What we need to do is at least the following: First, we need to explore in detail the relationship between the oligomeric state of small heat shock proteins and their binding capacities toward the client proteins, to find out whether such binding absolutely relies on the oligomeric state, or if the monomeric form, after being dissociated from the oligomer, can also bind. Second, we have to identify the substrate (client) proteins of the small heat shock proteins in living bacterial cells, upon exposure to different stress conditions to find out whether there is any specificity in choosing the substrates. Thirdly, we need to unveil the fate of the bound client proteins upon returning to the non-stress conditions. In particular, what are the functional partners (likely other molecular chaperones or proteases) the small heat shock proteins rely on in handling the bound client proteins. Last but not least, we have to demonstrate whether and how small heat shock proteins work with such non-protein components as phospholipids of the lipid bilayer (or even the RNA and DNA molecules) to exhibit a protective effect under stress conditions.

Acknowledgements The author would like to thank Mr. Yang Liu (Peking University) for his kind assistance in preparing the figure, and Prof. Lawrence Hightower (University of Connecticut) for his kind editorial assistance. This work was supported by research grants from the National Basic Research Program of China (973 Program) (No. 2012CB917300) and the National Natural Science Foundation of China (No. 31170738) to Professor Zengyi Chang.

References

Allen SP, Polazzi JO, Gierse JK, Easton AM (1992) Two novel heat shock genes encoding proteins produced in response to heterologous protein expression in *Escherichia coli.* J Bacteriol 174:6938–6947

Amzel LM, Poljak RJ (1979) Three-dimensional structure of immunoglobulins. Annu Rev Biochem 48:961–967

Arrigo AP, Welch WJ (1987) Characterization and purification of the small 28,000-dalton mammalian heat shock protein. J Biol Chem 262:15359–15369

Arrigo AP, Suhan JP, Welch WJ (1988) Dynamic changes in the structure and intracellular locale of the mammalian low-molecular-weight heat shock protein. Mol Cell Biol 8:5059–5071

Ashburner M, Bonner JJ (1979) The induction of gene activity in drosophilia by heat shock. Cell 17:241–254

Behlke J, Lutsch G, Gaestel M, Bielka H (1991) Supramolecular structure of the recombinant murine small heat shock protein hsp25. FEBS Lett 288:119–122

Booth RJ, Harris DP, Loe JM, Watson JD (1988) Antigenic proteins of *M. leprae*: complete sequence of the gene for the 18-kDa protein. J Immunol 140:597–601

Bova MP, Huang Q, Ding L, Horwitz J (2002) Subunit exchange, conformational stability, and chaperone-like function of the small heat shock protein 16.5 from *Methanococcus jannaschii.* J Biol Chem 277:38468–38475

Capozzi V, Weidmann S, Fiocco D, Rieu A, Hols P, Guzzo J, Spano G (2011) Inactivation of a small heat shock protein affects cell morphology and membrane fluidity in *Lactobacillus plantarum* WCFS1. Res Microbiol 162:419–425

Chang Z, Primm TP, Jakana J, Lee IH, Chiu W, Gilbert HF, Quiocho FA (1996) *Mycobacterium tuberculosis* 16-kDa antigen (Hsp16.3) functions as an oligomeric structure *in vitro* to suppress thermal aggregation. J Biol Chem 271:7218–7223

Chen Y, Lu YJ, Wang HW, Quan S, Chang Z, Sui SF (2003) Two-dimensional crystallization of a small heat shock protein HSP16.3 on lipid layer. Biochem Biophys Res Commun 310:360–366

Cheney CM, Shearn A (1983) Developmental regulation of Drosophila imaginal disc proteins: synthesis of a heal-shock protein under non-heat-shock conditions. Dev Biol 95:325–330

Collier NC, Heuser J, Levy MA, Schlesinger MJ (1988) Ultrastructural and biochemical analysis of the stress granule in chicken embryo fibroblasts. J Cell Biol 106:1131–1139

Cooper S, Ruettinger T (1975) Temperature dependent alteration in bacterial protein composition. Biochem Biophys Res Commun 62:584–586

Coucheney F, Gal L, Beney L, Lherminier J, Gervais P, Guzzo J (2005) A small HSP, Lo18, interacts with the cell membrane and modulates lipid physical state under heat shock conditions in a lactic acid bacterium. Biochim Biophys Acta 1720:92–98

Craig EA, Lindquist S (1988) The heat-shock proteins. Annu Rev Genet 22:631–677

Cunningham AF, Spreadbury CL (1998) Mycobacterial stationary phase induced by low oxygen tension: cell wall thickening and localization of the 16-kilodalton alpha-crystallin homolog. J Bacteriol 180:801–808

de Jong WW, Leunissen JAM, Voorter CEM (1993) Evolution of the α-crystallin/small heat-shock protein family. Mol Biol Evol 10:103–126

Ellis J (1987) Proteins as molecular chaperones. Nature 328:378–379

Ezemaduka AN, Yu JY, Shi XD, Zhang KM, Yin CC, Fu X, Chang Z (2014) A small heat shock protein enables *Escherichia coli* to grow at a lethal temperature of 50°C conceivably by maintaining cell envelope integrity. J Bacteriol 196:2004–2011

Fu X, Chang Z (2004) Temperature-dependent subunit exchange and chaperone-like activities of Hsp16.3, a small heat shock protein from *Mycobacterium tuberculosis*. Biochem Biophys Res Commun 316:291–299

Fu X, Li W, Mao Q, Chang Z (2003) Disulfide bonds convert small heat shock protein Hsp16.3 from a chaperone to a non-chaperone: implications for the evolution of cysteine in molecular chaperones. Biochem Biophys Res Commun 308:627–635

Fu X, Liu C, Liu Y, Feng X, Gu L, Chen X, Chang Z (2004) Small heat shock protein Hsp16.3 modulates its chaperone activity by adjusting the rate of oligomeric dissociation. Biochem Biophys Res Commun 310:412–420

Fu X, Shi X, Yan L, Zhang H, Chang Z (2013a) *In vivo* substrate diversity and preference of small heat shock protein IbpB as revealed by using a genetically incorporated photo-crosslinker. J Biol Chem 288:31646–31654

Fu X, Shi XD, Yin LX, Liu JF, Joo KH, Lee JY, Chang Z (2013b) Small heat shock protein IbpB acts as a robust chaperone in living cells by hierarchically activating its multi-type substrate-binding residues. J Biol Chem 288:11897–11906

Gu L, Abulimiti A, Li W, Chang Z (2002) Monodisperse Hsp16.3 nonamer exhibits dynamic dissociation and reassociation, with the nonamer dissociation prerequisite for chaperone-like activity. J Mol Biol 319:517–526

Guzzo J, Delmas F, Pierre F, Jobin MP, Samyn B, Van Beeumen J, Cavin JF, Diviès C (1997) A small heat shock protein from *Leuconostoc oenos* induced by multiple stresses and during stationary growth phase. Lett Appl Microbiol 24:393–396

Henriques AO, Beall BW, Mran CP Jr (1997) CotM of *Bacillus subtilis*, a member of the α-crystallin family of stress proteins, is induced during development and participates in spore outer coat formation. J Bacteriol 179:1887–1897

Hockertz MK, Clark-Lewis I, Candido EP (1991) Studies of the small heat shock proteins of *Caenorhabditis elegans* using anti-peptide antibodies. FEBS Lett 280:375–378

Horvath I, Glatz A, Varvasovszki V, Torok Z, Pali T, Balogh G, Kovacs E, Nadasdi L, Benko S, Joo F, Vigh L (1998) Membrane physical state controls the signaling mechanism of the heat shock response in *Synechocystis* PCC 6803: identification of hsp17 as a "fluidity gene". Proc Natl Acad Sci U S A 95:3513–3518

Horváth I, Multhoff G, Sonnleitner A, Vígh L (2008) Membrane-associated stress proteins: more than simply chaperones. Biochim Biophys Acta 1778:1653–1664

Horwitz J (1992) α-crystallin can function as a molecular chaperone. Proc Natl Acad Sci U S A 89:10449–10453

Ingolia TD, Craig EA (1982) Four small *Drosophila* heat shock proteins are related to each other and to mammalian α-crystallin. Proc Natl Acad Sci U S A 79:2360–2364

Jakob U, Gaestel M, Engel K, Buchner J (1993) Small heat shock proteins are molecular chaperones. J Biol Chem 268:1517–1520

Jiao W, Qian M, Li P, Zhao L, Chang Z (2005) The essential role of the flexible termini in the temperature-responsiveness of the oligomeric state and chaperone-like activity for the polydisperse small heat shock protein IbpB from *Escherichia coli*. J Mol Biol 347:871–884

Jiao W, Hong W, Li P, Sun S, Ma J, Qian M, Hu M, Chang Z (2008) The dramatically increased chaperone activity of Small heat shock protein IbpB is retained for an extended period of time after the stress condition is removed. Biochem J 410:63–70

Jobin MP, Delmas F, Garmyn D, Diviès C, Guzzo J (1997) Molecular characterization of the gene encoding an 18-kilodalton small heat shock protein associated with the membrane of *Leuconostoc oenos*. Appl Environ Microbiol 63:609–614

Kampinga HH, Hageman J, Vos MJ, Kubota H, Tanguay RM, Bruford EA, Cheetham ME, Chen B, Hightower LE (2009) Guidelines for the nomenclature of the human heat shock proteins. Cell Stress Chaperones 14:105–111

Kappé G, Boelens WC, de Jong WW (2010) Why proteins without the α-crystallin domain should not be included in the human small heat shock protein family HSPB. Cell Stress Chaperones 15:457–461

Kelley P, Schlesinger MJ (1978) The effect of amino-acid analogues and heat shock on gene expression in chicken embryo fibroblasts. Cell 15:1277–1286

Kennaway CK, Benesch JLP, Gohlke U, Wang L, Robinson CV, Orlova EV, Saibi HR, Keep NH (2005) Dodecameric structure of the small heat shock protein Acr1 from *Mycobacterium tuberculosis*. J Biol Chem 280:33419–33425

Key JL, Lin CY, Chen YM (1981) Heat shock proteins of higher plants. Proc Natl Acad Sci U S A 78:3526–3530

Kim KK, Kim R, Kim S-H (1998a) Crystal structure of a small heat-shock protein. Nature 394:595–599

Kim R, Kim KK, Yokota H, Kim SH (1998b) Small heat shock protein of *Methanococcus jannaschii*, a hyperthermophile. Proc Natl Acad Sci U S A 95:9129–9133

Kitagawa M, Miyakawa M, Matsumara Y, Tsuchido T (2002) *Escherichia coli* small heat shock proteins, IbpA and IbpB, protect enzymes from inactivation by heat and oxidants. Eur J Biochem 269:2907–2917

Koninkx JF (1976) Protein synthesis in salivary glands of *Drosophila hydei* after experimental gene induction. Biochem J 58:623–628

Laksanalamai P, Robb FT (2004) Small heat shock proteins from extremophiles: a review. Extremophiles 8:1–11

Laskowska E, Wawrzynów A, Taylor A (1996) IbpA and IbpB, the new heat-shock proteins, bind to endogenous *Escherichia coli* proteins aggregated intracellularly by heat shock. Biochimie 78:117–122

Lee BY, Hefta SA, Brennan PJ (1992) Characterization of the major membrane protein of virulent *Mycobacterium tuberculosis*. Infect Immun 60:2066–2074

Lentze N, Aquilina JA, Lindbauer M, Robinson CV, Narberhaus F (2004) Temperature and concentration-controlled dynamics of rhizobial small heat shock proteins. Eur J Biochem 271:2494–2503

Lewis K (2007) Persister cells, dormancy and infectious disease. Nat Rev Microbiol 5:48–56

Lewis MJ, Pelham HRB (1985) Involvement of ATP in the nuclear and nucleolar functions of the 70 kd heat shock protein. EMBO J 4:3137–3143

Lewis M, Helmsing PJ, Ashburner M (1975) Parallel changes in puffing activity and patterns of protein synthesis in salivary glands of *Drosophila*. Proc Natl Acad Sci U S A 72:3604–3608

Maitre M, Weidmann S, Rieu A, Fenel D, Schoehn G, Ebel C, Coves J, Guzzo J (2012) The oligomer plasticity of the small heat-shock protein Lo18 from *Oenococcus oeni* influences its role in both membrane stabilization and protein protection. Biochem J 444:97–104

Maitre M, Weidmann S, Dubois-Brissonnet F, David V, Covès J, Guzzo J (2014) Adaptation of the wine bacterium *Oenococcus oeni* to ethanol stress: role of the small heat shock protein Lo18 in membrane integrity. Appl Environ Microbiol 80:2973–2980

Mao Q, Ke D, Feng X, Chang Z (2001) Preheat treatment for *Mycobacterium tuberculosis* Hsp16.3: correlation between a structural Phase change at 60°C and a dramatic increase in chaperone-like activity. Biochem Biophys Res Commun 284:942–947

Matuszewska M, Kuczyńska-Wiśnik D, Laskowska E, Liberek K (2005) The small heat shock protein IbpA of *Escherichia coli* cooperates with IbpB in stabilization of thermally aggregated proteins in a disaggregation competent state. J Biol Chem 280:12292–12298

Mirault M-E, Goldschmidt-Clermont M, Moran L, Arrigo AP, Tissieres A (1978) The effect of heat shock on gene expression in Drosophila melanogaster. Cold Spring Harb Symp Quant Biol 42:819–827

Miyake T, Araki S, Tsuchido T (1993) Synthesis and sedimentation of a subset of 15 kDa heat shock proteins in *Escherichia coli* cells recovering from sublethal heat stress. Biosci Biotechnol Biochem 57:578–583

Mogk A, Deuerling E, Vorderwülbecke S, Vierling E, Bukau B (2003a) Small heat shock proteins, ClpB and the DnaK system form a functional triade in reversing protein aggregation. Mol Microbiol 50:585–595

Mogk A, Schlieker C, Friedrich KL, Schönfeld H-J, Vierling E, Bukau B (2003b) Refolding of substrates bound to small Hsps relies on a disaggregation reaction mediated most efficiently by ClpB/DnaK. J Biol Chem 278:31033–31042

Munro S, Pelham HRB (1986) An Hsp70-like protein in the ER: identity with the 78 kd glucose-regulated protein and immunoglobulin heavy chain binding protein. Cell 46:291–300

Nagao RT, Czarnecka E, Gurley WB, Schoffl F, Key JL (1985) Genes for low-molecular-weight heat shock proteins of soybeans: sequence analysis of a multigene family. Mol Cell Biol 5:3417–3428

Nakamoto H, Vígh L (2007) The small heat shock proteins and their clients. Cell Mol Life Sci 64:294–306

Narberhaus F (2002) α-Crystallin-type heat shock proteins: socializing minichaperones in the context of a multichaperone network. Microbiol Mol Biol Rev 66:64–93

Nerland AH, Mustafa AS, Sweetser D, Godal T, Young RA (1988) A protein antigen of *Mycobacterium leprae* is related to a family of small heat shock proteins. J Bacteriol 170:5919–5921

Pelham HRB (1986) Speculations on the functions of the major heat shock and glucose-regulated proteins. Cell 46:959–961

Ratajczak E, Strózecka J, Matuszewska M, Zietkiewicz S, Kuczyńska-Wiśnik D, Laskowska E, Liberek K (2010) IbpA the small heat shock protein from Escherichia coli forms fibrils in the absence of its cochaperone IbpB. FEBS Lett 584:2253–2257

Ritossa F (1962) A new puffing pattern induced by temperature shock and DNP in drosophila. Experientia 18:571–573

Ritossa F (1996) Discovery of the heat shock response. Cell Stress Chaperones 1:97–98

Roy SK, Nakamoto H (1998) Cloning, characterization, and transcriptional analysis of a gene encoding an α-crystallin-related, small heat shock protein from the thermophilic cyanobacterium *Synechococcus vulcanus*. J Bacteriol 180:3997–4001

Russnak RH, Jones D, Candido EP (1983) Cloning and analysis of cDNA sequences coding for two 16 kilodalton heat shock proteins (hsps) in *Caenorhabditis elegans*: homology with the small hsps of *Drosophila*. Nucleic Acids Res 11:3187–3205

Sauer U, Dürre P (1993) Sequence and molecular characterization of a DNA region encoding a small heat shock protein of *Clostridium acetobutylicum*. J Bacteriol 175:3394–3400

Servant P, Mazodier P (1995) Characterization of *Streptomyces albus* 18-kilodalton heat shock-responsive protein. J Bacteriol 177:2998–3003

Shearstone JR, Baneyx F (1999) Biochemical characterization of the small heat shock protein IbpB from *Escherichia coli*. J Biol Chem 274:9937–9945

Shi X, Wang Z, Yan L, Ezemaduka AN, Fan G, Wang R, Fu X, Yin C, Chang Z (2011) Small heat shock protein AgsA forms dynamic fibrils. FEBS Lett 585:3396–3402

Studer S, Narberhaus F (2000) Chaperone activity and homo- and hetero-oligomer formation of bacterial small heat shock proteins. J Biol Chem 275:37212–37218

Sun Y, MacRae TH (2005) Small heat shock proteins: molecular structure and chaperone function. Cell Mol Life Sci 62:2460–2476

Thomas JG, Baneyx F (1998) Roles of the *Escherichia coli* small heat shock proteins IbpA and IbpB in thermal stress management: comparison with ClpA, ClpB, and HtpG *in vivo*. J Bacteriol 180:5165–5172

Tissieres A, Mitchell HK, Tracy VM (1974) Protein synthesis in salivary glands of *Drosophila melanogaster:* relation to chromosome puffs. J Mol Biol 84:389–398

Török Z, Goloubinoff P, Horváth I, Tsvetkova NM, Glatz A, Balogh G, Varvasovszki V, Los DA, Vierling E, Crowe JH, Vigh L (2001) *Synechocystis* HSP17 is an amphitropic protein that stabilizes heat-stressed membranes and binds denatured proteins for subsequent chaperone-mediated refolding. Proc Natl Acad Sci U S A 98:3098–3103

Tsvetkova NM, Horváth I, Török Z, Wolkers WF, Balogi Z, Shigapova N, Crowe LM, Tablin F, Vierling E, Crowe JH, Vigh L (2002) Small heat-shock proteins regulate membrane lipid polymorphism. Proc Natl Acad Sci U S A 99:13504–13509

van Montfort RLM, Basha E, Friedrich KL, Slingsby C, Vierling E (2001) Crystal structure and assembly of a eukaryotic small heat shock protein. Nat Struct Biol 8:1025–1030

van Noort JM, Bsibsi M, Nacken P, Gerritsen WH, Amor S (2012) The link between small heat shock proteins and the immune system. Int J Biochem Cell Biol 44:1670–1679

Veinger L, Diamant S, Buchner J, Goloubinoff P (1998) The small heat-shock protein IbpB from *Escherichia coli* stabilizes stress-denatured proteins for subsequent refolding by a multichaperone network. J Biol Chem 273:11032–11037

Verbon A, Hartskeerl RA, Schuitema A, Kolk AH, Young DB, Lathigra R (1992) The 14,000-molecular-weight antigen of *Mycobacterium tuberculosis* is related to the alpha-crystallin family of low-molecular-weight heat shock proteins. J Bacteriol 174:1352–1359

Wang YB, Ezemaduka AN, Tang Y, Chang ZY (2009) Understanding the mechanism of the dormant dauer formation of *C. elegans*: from genetics to biochemistry. IUBMB Life 61:607–612

Williams AF, Barclay AN (1988) The immunoglobulin superfamily–domains for cell surface recognition. Annu Rev Immunol 6:381–405

Yamamori T, Ito K, Nakamura Y, Yura T (1978) Transient regulation of protein synthesis in *Escherichia coli* upon shift-up of growth temperature. J Bacteriol 134:1133–1140

Yang H, Huang S, Dai H, Gong Y, Zheng C, Chang Z (1999) The *Mycobacterium tuberculosis* small heat shock protein Hsp16.3 exposes hydrophobic surfaces at mild conditions: conformational flexibility and molecular chaperone activity. Protein Sci 8:174–179

Young DB, Garbe TR (1991) Heat shock proteins and antigens of *Mycobacterium tuberculosis*. Infect Immun 59:3086–3093

Yuan Y, Crane DD, Barry CE 3rd (1996) Stationary phase-associated protein expression in *Mycobacterium tuberculosis*: function of the mycobacterial α-crystallin homolog. J Bacteriol 178:4484–4492

Yuan Y, Crane DD, Simpson RM, Zhu YQ, Hickey M, Sherman DR, Barry CE III (1998) The 16-kDa alpha-crystallin (Acr) protein of *Mycobacterium tuberculosis* is required for growth in macrophages. Proc Natl Acad Sci U S A 95:9578–9583

Zhang H, Fu X, Jiao W, Zhang X, Liu C, Chang Z (2005) The association of small heat shock protein Hsp16.3 with the plasma membrane of *Mycobacterium tuberculosis*: dissociation of oligomers is a prerequisite. Biochem Biophys Res Commun 330:1055–1061

Chapter 23
How to Stabilize Both the Proteins and the Membranes: Diverse Effects of sHsps in Neuroprotection

Melinda E. Tóth, Miklós Sántha, Botond Penke, and László Vígh

Abstract Small heat shock proteins (sHsps) are ubiquitously expressed evolutionarily conserved proteins which are upregulated by different stressors and in various pathological conditions in the brain. The most important function of sHsps is to inhibit the aggregation of incorrectly folded proteins by binding to non-native proteins, and to maintain them in a refolding-competent state. They also exhibit anti-apoptotic, antioxidant activities and can bind to the cytoskeleton and membranes, stabilizing and protecting them against stress. These properties enable sHsps to protect neurons against various brain damaging effects. Here, we summarize our current view on the role of sHsps in neurodegeneration, chaperon function, membrane protection, oxidative stress, apoptosis, protein degradation, insulin resistance and blood-brain barrier function.

Keywords sHsps • Neurodegeneration • Chaperon function • Membrane protection • Oxidative stress • Apoptosis • Protein degradation • Insulin resistance • Blood brain barrier

23.1 sHsps in Neurodegeneration

Small heat shock proteins (sHsps) are ubiquitously expressed evolutionarily conserved proteins which are upregulated by different stressors and in various pathological conditions in the brain. The family of sHsps consists of 11 members, characterized by a conserved crystallin domain flanked by variable N- and C-termini

M.E. Tóth (✉) • M. Sántha • L. Vígh (✉)
Institute of Biochemistry, Biological Research Centre,
Hungarian Academy of Sciences, P.O. Box 521, 6701 Szeged, Hungary
e-mail: toth.erzsebetmelinda@brc.mta.hu;
toth.e.melinda@gmail.com; vigh.laszlo@brc.mta.hu

B. Penke
Department of Medical Chemistry, Faculty of Medicine,
University of Szeged, 11. Dóm square, Szeged, Hungary
e-mail: penke.botond@med.u-szeged.hu

R.M. Tanguay, L.E. Hightower (eds.), *The Big Book on Small Heat Shock Proteins*, Heat Shock Proteins 8, DOI 10.1007/978-3-319-16077-1_23

(Kampinga et al. 2009). Mammalian neurons express 5 of the 11 members, and at least 4 of them are upregulated after certain stress conditions (Kirbach and Golenhofen 2011; Bartelt-Kirbach and Golenhofen 2014). HspB1 (Hsp27) and HspB5 (αB-crystallin), the most extensively studied members, have been found to be upregulated in neurons after stress conditions such as ischemia (Kato et al. 1994; Minami et al. 2003), heat shock (Bechtold and Brown 2000; Krueger-Naug et al. 2000) or kainic acid-induced seizures (Kato et al. 1999). HspB6 (Hsp20) has been demonstrated to be upregulated after ischemia (Niwa et al. 2009) and hypoxia (David et al. 2006), and HspB8 (Hsp22) after proteasome inhibition (Yew et al. 2005, reviewed by Bartelt-Kirbach and Golenhofen 2014). A comprehensive study revealed that rat hippocampal neurons can upregulate more than one sHsp under different types of stress conditions, suggesting that they act together, though their expression profiles differ (Kirbach and Golenhofen 2011; Bartelt-Kirbach and Golenhofen 2014).

Most of the Hsps are molecular chaperones and have essential roles in the biosynthesis, folding, transport and assembly of other proteins (Morimoto et al. 1992). The most important function of sHsps is to inhibit the aggregation of incorrectly folded proteins by binding to non-native proteins, and to maintain them in a refolding-competent state (Haslbeck and Buchner 2002). They also exhibit anti-apoptotic, antioxidant activities and can bind to the cytoskeleton and membranes, stabilizing and protecting them against stress (Wang and Spector 1996; Preville et al. 1999; Arrigo et al. 2002; Sun and MacRae 2005). These properties enable sHsps to protect neurons against various brain-damaging effects (Fig. 23.1).

Several sHsps are found to be upregulated in damaged and diseased brains, and are colocalized with protein aggregates in neurodegenerative diseases such as Alzheimer's disease (AD), Parkinson's disease (PD), Huntington's disease (HD) or amyotrophic lateral sclerosis (ALS) (Iwaki et al. 1992, for reviews see Sun and

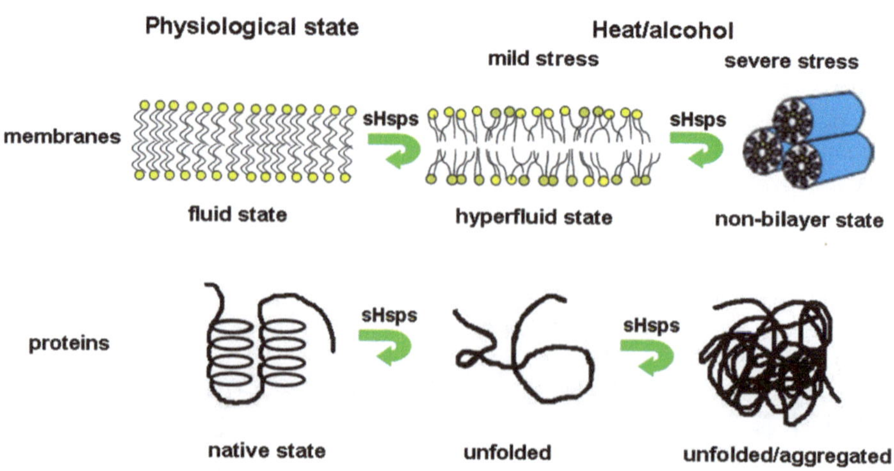

Fig. 23.1 Protective effect of sHsps on membrane and protein structures under different stress conditions (Based on Nakamoto and Vigh 2007; Eyles and Gierasch 2010)

MacRae 2005; Wilhelmus et al. 2007; Brownell et al. 2012). In AD, the HspB1 levels in the cortex of the patients are highly elevated and the expression of HspB1 increases with the duration of the disease and the degree of dementia (Renkawek et al. 1994a, b). The concentration of HspB5 is also elevated in the cerebral cortex of AD patients (Shinohara et al. 1993; Renkawek et al. 1994b; Yoo et al. 2001). Immunohistochemically, HspB5 is localized in astrocytes, microglia and oligodendrocytes, while HspB1 is present in degenerating neurons (Shinohara et al. 1993; Renkawek et al. 1994b). Specific sHsps such as HspB1, HspB2, HspB5 and HspB6 have been found to be associated with senile plaques in the AD brain (Wilhelmus et al. 2006a). Three sHsp family members co-immunoprecipitate with human amyloid-β (Aβ) in transgenic *Caenorhabditis elegans* (Fonte et al. 2002), and human recombinant HspB5 also interacts with Aβ in vitro as revealed by fluorescence energy transfer experiments (Liang 2000). HspB1 has been demonstrated to bind preferentially to hyperphosphorylated tau protein rather than to the non-phosphorylated form (Shimura et al. 2004).

Similarly as in AD, HspB1 and HspB5 are upregulated in the reactive astrocytes in PD with severe dementia, and there is a strong association between the level of sHsps and the number of tangles in the hippocampus (Renkawek et al. 1994a, 1999, reviewed by Brownell et al. 2012; Sun and MacRae 2005). There is a strong expression of HspB5 in the Lewy bodies (Jellinger 2000). Increases in HspB1 and HspB5 expression are likewise observed in the alpha-SynA53T transgenic mouse, a model for PD (Wang et al. 2008).

Strong immunostaining for HspB5 has been reported in the spinal cord of patients with ALS (Iwaki et al. 1992). Similarly, mutant SOD1-overexpressing mouse, a model of ALS, revealed elevated levels of HspB5 and HspB1 in the spinal cord (Vleminckx et al. 2002; Wang et al. 2008, reviewed by Brownell et al. 2012). These studies additionally showed that HspB1 is upregulated in the neurons and glial cells in the affected regions in these mice. Mutant SOD1 proteins bind to HspB5 as demonstrated by co-immunoprecipitation (Shinder et al. 2001), and HspB1 is also coprecipitated and colocalized with mutant SOD1 in the neurons and astrocytes of SOD1 mutant mice (Strey et al. 2004, reviewed by Sun and MacRae 2005).

The expression of sHsps may be increased in prion disease (reviewed by Brownell et al. 2012; Sun and MacRae 2005). HspB5-positive neurons, astrocytes and oligodendrocytes were observed in the brain of Creutzfeldt-Jakob disease (CJD) patients, (Iwaki et al. 1992; Renkawek et al. 1992). An increased level of HspB1 was found in a murine model of bovine spongiform encephalopathy (BSE) (Tortosa et al. 2008) The expression level of HspB1 is not significantly elevated in scrapie, however its expression pattern is altered (Vidal et al. 2009).

The expression of HspB1 is enhanced in the temporal cortex of patients with epilepsy (Bidmon et al. 2004).

Elevated levels of sHsps has been found not only in chronic neurodegenerative diseases, but also after acute brain damage (reviewed by Brownell et al. 2012). The neuronal expression of HspB5 has been observed in the brain of patients with cerebral infarction (Lowe et al. 1992; Minami et al. 2003). Transient middle cerebral artery occlusion in rats dramatically increased the expression of HspB1, but not of

HspB5, while chemical ischemic stress induced both sHsps in cultured glial cells (Imura et al. 1999). These findings clearly indicate that sHsps and neurological diseases are linked. It was initially considered that the sHsps are pathological in these disease states because they are found in the areas of damage, but later studies suggested that sHsps exert a protective function in these diseases (reviewed, for example, by Muchowski and Wacker 2005; Sun and MacRae 2005; Mymrikov et al. 2011; Brownell et al. 2012). The overexpression of HspB1 for instance, ameliorates certain symptoms of AD in APP/Psen transgenic mice, such as Aβ plaque formation, learning dysfunctions and synaptic abnormalities (Tóth et al. 2013). An increased expression of HspB1 has a potent protective anti-apoptotic effect against the damaging effects of α-synuclein, the main protein component of the Lewy bodies, inclusion bodies in PD (Zourlidou et al. 2004). HspB1, and to a smaller degree HspB5, are able to prevent α-synuclein-induced toxicity, and HspB1 can also reduce α-synuclein aggregation in a cell culture model (Outeiro et al. 2006).

Treatment of a polyQ (polyglutamine) disease model with an antiulcer drug, geranylgeranylacetone, which can induce the expression of different Hsps in the central nervous system (CNS), can suppress the accumulation of pathogenic proteins and ameliorate the related phenotype (Katsuno et al. 2005). Oral administration of a geldanamycin derivative markedly suppresses polyQ-induced neurodegeneration and lethality in *Drosophila* models of two polyQ diseases, HD and spinocerebellar ataxia (Fujikake et al. 2008). The overexpression of HspB1 has been shown to decrease polyQ toxicity without suppressing inclusion body formation, by protecting against reactive oxygen species (ROS) (Wyttenbach et al. 2002).

The hydroxylamine derivatives are known to be effective in the treatment of various protein conformational diseases in animal models. These compounds are co-inducers of Hsps; they can increase Hsp expression only under stressed conditions by prolonging the activation of heat shock factor 1 (Hsf1) (Hargitai et al. 2003). NG-094 is markedly effective in ameliorating polyQ-dependent paralysis, and it reduced the number of protein aggregates in a *Caenorhabditis elegans* model expressing polyQ expansions. The drug confers protection against polyQ proteotoxicity, even when administered after the onset of the disease (Haldimann et al. 2011).

Induction of molecular chaperones by treatment with another hydroxylamine derivative, arimoclomol, has been reported to delay the disease progression significantly in a mouse model of ALS. Arimoclomol-treated SOD1 mice display pronounced improvements in hind limb muscle function and motoneuron survival, leading to an increase in lifespan (Kieran et al. 2004). The protective effects of HspB1 in ALS were confirmed by Sharp et al. (2008), who demonstrated a delayed decline in motor strength, a significant improvement in the number of functional motor units and an improved survival of the spinal motor neurons in SOD1/HspB1 double transgenic mice as compared with SOD1 single transgenics during the early phase of the disease.

The overexpression of HspB1 significantly reduced the infarct size in a mouse model of cerebral ischemia (van der Weerd et al. 2010). Virally delivered HspB1 also reduced the lesion volume in a rat middle cerebral artery occlusion model of focal cerebral ischemia (Badin et al. 2006). Administration of HspB5 reduced both the stroke volume and the inflammatory cytokines associated with the pathology of stroke, even when treatment was started 12 h after the stroke onset (Arac et al. 2011, reviewed by Brownell et al. 2012).

HspB1 overexpression is neuroprotective against kainate-induced neuronal loss in a model of epilepsy (Akbar et al. 2003) and can moderate the damaging effects of acute and chronic ethanol administration in the brain (Tóth et al. 2010). sHsps can protect neurons from heat-stroke-associated cell death both in a nematode model and in mammalian cell cultures (Kourtis et al. 2012).

Mutations in the genes of certain sHsps correlate with the development of different diseases. Mutations of HspB1 and HspB2 for instance are associated with hereditary neuropathies (Irobi et al. 2004; Evgrafov et al. 2004; Tang et al. 2005; Houlden et al. 2008, reviewed by Bartelt-Kirbach and Golenhofen 2014), and a mutation in the heat-shock element region of the HspB1 promoter was found in certain ALS patients (Dierick et al. 2007).

These results suggest a neuroprotective function of sHsps, but the mechanism of this protection is not fully understood. As molecular chaperones, one of their most important functions is to inhibit protein aggregation, but they probably function at several levels not linked exclusively to their chaperone function. We present below some possible ways in which sHsps can exert their neuroprotective functions.

23.2 Chaperone Function of sHsps

Probably the most characteristic function of sHsps is their chaperone function. sHsps, together with other members of the chaperone network, such as the Hsp70 protein family, and with cell components responsible for protein degradation, such as proteasome and lysosome, maintain the normal protein homeostasis that is essential for the cellular function. During aging, however, the activity of members of the chaperone network and the inducibility of various chaperones are impaired, concomitantly with a decrease in proteasome activity (Söti and Csermely 2002; Finka et al. 2011). Hsps are therefore thought to play an essential role in longevity and aging (Murshid et al. 2013). It has been shown that the exposure of adult *Caenorhabditis elegans* to amyloid-binding agents such as Thioflavin T (ThT), curcumin and rifampicin resulted in an extended lifespan and slowed aging. ThT also suppressed human Aβ-associated toxicity. These beneficial effects of ThT depend, among others, on the protein homeostasis network regulator Hsf1. These results demonstrate that pharmacological maintenance of the protein homeostatic network has a profound impact on aging (Alavez et al. 2011).

As the capacity of cells to maintain protein homeostasis weakens during aging, it is not surprising that the first symptoms of the neurodegenerative diseases termed protein-misfolding disorders typically appear in elderly persons. To prevent protein aggregation, the levels of several chaperones, including sHsps, are elevated in these neurodegenerative diseases and during aging. On the other hand, the inducibility of Hsps is impaired (reviewed by Söti and Csermely 2002). As an example, cells containing cytoplasmic polyQ protein aggregates exhibited a delayed expression of Hsp72 after heat shock (Cowan et al. 2003). To make matters worse, despite increased intracellular levels, sHsps are not fully able to prevent the accumulation of misfolded proteins in neurodegenerative diseases, possibly because of their decreased chaperone activity. In a mouse model of HD HspB1 overexpression cannot improve the disease phenotype. Native protein gel analysis revealed that overexpressed HspB1, which is otherwise activable upon heat shock, exists predominantly as low molecular weight, inactive species in the brain of the disease model animals (Zourlidou et al. 2007). Moreover, the colocalization of different Hsps with the toxic protein aggregates may reduce the availability of these chaperones, thereby disrupting their normal functions, which leads to the enhanced vulnerability of the cells to stress conditions that occur within the normal lifespan. Indeed, activation of the cellular stress response in cells that express the polyQ-expanded truncated AR protein increased the overall extent of polyQ protein aggregation (Cowan et al. 2003). The overexpression of the polyQ-green fluorescent protein similarly increased stress-induced protein aggregation following stress in neural SH-SY5Y cells (Ding et al. 2002). The impairment of chaperone activity and protein degradation cause the massive accumulation of misfolded proteins and finally can lead to neurodegeneration (Söti and Csermely 2002).

The different neurodegenerative disorders are characterized by the accumulation of intra- or extracellular protein aggregates. Certain aggregates accumulate intracellularly, such as α-synuclein in PD, superoxide dismutase (SOD) in ALS, huntingtin in HD, and hyperphosphorylated tau in AD. Prion protein in prion diseases accumulates extracellularly, while Aβ peptides are accumulated both extra- and intracellularly in AD (reviewed by Penke et al. 2012). These different aggregation-prone proteins exhibit some common structural features, e.g. they can fold into a β-sheet-rich structure (Haass and Selkoe 2007).

In AD, Aβ peptides appear to occupy a central role in the disease pathology. The amyloid cascade hypothesis postulates that the cerebral Aβ balance is disturbed because of an enhanced production or reduced clearance of the peptide. The relative increase in $A\beta_{42}$ enhances oligomer formation, these oligomers adversely affecting the synaptic structure and plasticity (Haass and Selkoe 2007). In parallel, $A\beta_{42}$ polymerizes into insoluble fibrils that form microscopically visible plaques in the brain. Although the Aβ plaques do not appear to be as neurotoxic as the soluble oligomers, they can provoke local inflammatory responses that later give rise to chronic inflammation in the brain (Khandelwal et al. 2011). These events eventually lead to a series of biochemical changes, such as oxidative stress and altered calcium homeostasis, that can cause neuronal damage (Haass and Selkoe 2007). PD is characterized by the loss of dopaminergic neurons in the substantia nigra and with the formation of α-synuclein-enriched Lewy bodies in the neurons (Dauer and Przedborski 2003). Like Aβ, PD associated mutant form of α-synuclein can form pore-like, annular oligomeric structures in vitro (Lashuel et al. 2002). In polyQ diseases such as HD, spinobulbar muscular atrophy and spinocerebellar ataxias, the affected neurons contain accumulations of mutant, polyQ-expanded polypeptides in insoluble aggregates. However, some evidence suggests that the aggregates are in fact not the pathogenic basis, but may rather play a role in sequestration of the pathogenic protein, and the misfolded monomers and low molecular weight oligomeric species are the most cytotoxic forms of the disease-causing proteins (reviewed by Zoghbi and Orr 1999; Haass and Selkoe 2007). In cell-culture studies of HD, for example aggregation of polyQ-rich huntingtin protein prolonged cell survival, probably by reducing intracellular level of soluble huntingtin (Arrasate et al. 2004). Proteolysis-sensitive forms of prion proteins are normally present in the brain, and they cannot form large aggregates. Misfolded prions are aggregation-prone, β-sheet enriched infectious proteins that are responsible for the transmissible spongiform encephalopathies, e.g. BSE, scrapie, and CJD (Prusiner 1998).

These mutant, misfolded, aggregation-prone proteins can bind to a variety of biomolecules, including lipids, proteins and proteoglycans. The intracellular Aβ hypothesis emphasizes the primary role of intracellular Aβ in initiating the disease by interaction with cytoplasmic proteins and membranes of cell organelles such as mitochondria and endoplasmic reticulum (ER), thereby triggering apoptosis (Penke et al. 2012). PolyQ proteins have also been found to interact with other cellular proteins, such as transcription factors, proteasome subunits and cytoskeletal proteins, eventually leading to the repression of transcription, impairment of the protein degradation system and alteration of the neurofilament network (Nagai et al. 1999; Steffan et al. 2000; Bence et al. 2001, reviewed by Fujikake et al. 2008).

The most important function of the sHsps is to bind to incorrectly folded, partially denatured proteins, thereby preventing their irreversible self-aggregation or aggregation with other cellular components. They are unable alone to actively refold the misfolded proteins, but can maintain them in a refolding-competent state (Haslbeck and Buchner 2002; Nakamoto and Vígh 2007). This is an ATP-independent process that is essential during stress conditions, because the ATP level in stressed cells may be decreased considerably. After the cells have recovered and the ATP level has been restored, the bound proteins can be transferred to ATP-dependent chaperones such as Hsp70, which can facilitate the refolding of the sequestered, damaged proteins or can transmit them to the protein degradation machines, e.g. proteasomes or autophagosomes (Garrido et al. 2003; Kappe et al. 2003; Sun and MacRae 2005). In this way, the Hsps and protein degradation systems form a complex network, as many of the members may interact with each other (reviewed, for example, by Mymrikov et al. 2011). HspB1, for instance, was reported to be copurified with HspB5 and it seems that these proteins form large complexes that dissociate during heat shock (Kato et al. 1992; Zantema et al. 1992). HspB1 and Hsp70 are usually co-expressed in response to stress stimuli (Finka et al. 2011). Yeast sHsps, Hsp26 and Hsp42, form distinct, large oligomeric complexes. In response to heat shock, these large complexes dissociate into smaller, active species that bind aggregation-prone proteins, producing large globular assemblies (Haslbeck et al. 1999, 2005). In this way, the sHsps maintain the aggregate in a state that allows efficient disaggregation by Hsp104. Indeed it has been demonstrated that Hsp26 facilitates the protein disaggregation by Hsp104 after a heat shock in vivo and in vitro. sHsps and Hsp104 cooperate to antagonize polyQ aggregation and toxicity. Importantly, Hsp26 can perform its solubilizing function only when it co-aggregates with the protein substrate, and not after substrate aggregation (Cashikar et al. 2005). Haslbeck et al. (2005) reported that the collaboration of Hsp26, Hsp70 and Hsp104 chaperone families is needed for the disintegration of aggregates formed in the yeast cytosol upon heat stress. A recent study led to the finding that Hsp26 and Hsp42 in yeast work together in a distinct and synergistic manner to prevent Sup35 prion formation. sHsps promote the disaggregation of Sup35 prions by Hsp104, Hsp70 and Hsp40. Moreover, sHsps enhance the Hsp104-catalyzed disaggregation of α-synuclein and polyQ in vitro. Human HspB5 stimulates the depolymerization of α-synuclein by human Hsp110, Hsp70 and Hsp40 (Duennwald et al. 2012). Mammalian HspB1 has also been reported to bind non-native proteins under heat shock conditions in vitro. Under permissive folding conditions, misfolded proteins can be released from HspB1, and reactivated in the presence of Hsp70 and ATP (Ehrnsperger et al. 1997). Interestingly, a significant reduction in polyQ toxicity was observed when sHsps were overexpressed alone, without major changes in the number or morphology of polyQ aggregates (Cashikar et al. 2005), and it was suggested that the binding of polyQ-containing proteins by sHsps prevents their harmful interactions with other proteins. These findings are concordant with the results of Wyttenbach et al. (2002), who reported that HspB1 suppressed polyQ death without suppressing polyQ aggregation. Instead, HspB1 seemed to protect against oxidative stress in cells expressing mutant huntingtin. These studies indicate that under stress conditions, sHsps may sequester non-native proteins, preventing their unwanted interactions. In this way, they might protect cells against protein aggregation-mediated toxicity, but for effective disaggregation ATP-dependent chaperones are necessary.

As sHsps can prevent protein aggregation, they might inhibit different steps in the amyloidogenesis of various disease proteins, as suggested by the above-mentioned studies. Indeed, various sHsps have been shown to bind to these proteins in vitro or in vivo. Human HspB1 effected a significant reduction in amyloid formation in vitro. However, when HspB1 was added to pre-formed amyloid there was a more moderate decrease in amyloid formation (Kudva et al. 1997). HspB5, HspB6 and HspB8 can reduce Aβ fibril formation and inhibit Aβ-mediated toxicity toward cerebrovascular cells (Wilhelmus et al. 2006b, c). However it was reported earlier that, by preventing the in vitro fibril formation of $A\beta_{40}$, HspB5 keeps it in a nonfibrillar, but highly toxic form (Stege et al. 1999). HspB5 is also a potent in vitro inhibitor of α-synuclein fibrillization (Rekas et al. 2004). By binding to tau, HspB1 can alter the conformation of pathological hyperphosphorylated tau and reduce its concentration (Shimura et al. 2004). HspB8 can prevent the in vivo aggregation of polyQ-containing proteins (Carra et al. 2005). Interestingly, Duennwald et al. (2012) found that the heat-shock treatment of yeast Hsp26 reduced its ability to inhibit Sup35 prionogenesis, while simultaneously enhancing its ability to prevent the aggregation of a chemically denatured substrate. They speculated that at physiological temperatures Hsp26 inhibit prion formation, whereas at elevated temperatures it loses this ability and switches to inhibiting the aggregation of heat-denatured proteins.

In summary, sHsps can bind aggregation-prone proteins until they become actively unfolded by the ATP-dependent chaperones or degraded by the proteasome. During aging, however the activity of the chaperone network declines, leading to the loss of normal protein homeostasis, which favors the accumulation of various toxic proteins and the development of neurodegenerative diseases.

23.3 Membrane Protection

Under stress conditions and in different diseases, the cellular plasma membrane is a sensitive target for damage. Stress factors can influence membrane fluidity and denature membrane proteins. Heat shock and ethanol stress responses, for example, exert similar membrane-disruptive effects, and they both denature proteins and cause similar changes in the plasma membrane protein composition (Piper 1995). A number of studies have pointed to a potential link between membrane function and neurodegenerative diseases. Certain studies indicate that Aβ can interact with the plasma membrane altering its fluidity, which triggers different events catalyzed by perturbation of the conformation of the membrane proteins, which finally leads to neurotoxicity (Kanfer et al. 1999). This hypothesis suggests that the neurotoxic cascade of Aβ is initiated in the cell membrane. There are several binding sites for Aβ on the cell membranes, including lipids, proteoglycans and different proteins (reviewed in Verdier et al. 2004). Soluble Aβ oligomers can interact with the cell membrane and also with the membranes of subcellular organelles, thereby influencing the membrane fluidity. $A\beta_{40}$ has been found to reduce the fluidity of a brain membrane preparation (Müller et al. 1995). It was shown later that unaggregated Aβ had little or no effect on the membrane fluidity, whereas aggregated Aβ decreased the membrane fluidity in a time- and dose-dependent manner in a model membrane system. The aggregation rate and surface hydrophobicity were strongly correlated with the extent of the decrease in membrane fluidity. These results suggest that the aggregation of Aβ exposes hydrophobic sites inducing a decrease in membrane fluidity (Kremer et al. 2000). However, other studies have revealed that both soluble and aggregated $A\beta_{40}$ increase the fluidity of the isolated synaptic plasma membrane of the cerebral cortex and the hippocampus but not in cerebellum (Mason et al. 1999; Chochina et al. 2001), which suggests that the effect of Aβ on the membrane fluidity might depend on the membrane composition. It has also been demonstrated that Aβ fibrillogenesis is accelerated in the presence of lysosomal, endosomal and plasmamembranes, and the addition of Aβ to these membranes decreases the fluidity. However, the interaction of Aβ with the Golgi membrane did not induce fibril formation, and it increased the Golgi bilayer fluidity (Waschuk et al. 2001, reviewed by Verdier et al. 2004). In another study, it was found that the effect of Aβ on the membrane fluidity depends on the membrane cholesterol content (Yip et al. 2001). Aβ-induced changes in the physical properties of the membrane may have deleterious consequences on the cellular functioning, e.g. it might lead to alterations in the hippocampal synaptic efficacy, which is an early symptom of AD. It is also important that α-, β-, and γ-secretases are associated with the cellular membranes, and their functions, such as amyloid precursor protein (APP) processing depend strongly on the membrane fluidity. For example, oligomeric Aβ can reduce the membrane fluidity, which in turn stimulates the amyloidogenic processing of APP (Peters et al. 2009).

Amyloidogenic APP processing and Aβ fibrillogenesis are thought to take place in cholesterol- and ganglioside-enriched membrane microdomains, termed lipid rafts. Changes induced in the lipid rafts by changes in the ganglioside or cholesterol content or their interaction with mutant proteins result in the development of specific types of dementia (Schengrund 2010). Several studies have identified abnormalities in the ganglioside metabolism in the brain of human AD patients and mouse models of AD (reviewed in Ariga et al. 2008). In vitro studies suggest that Aβ aggregation proceeds in ganglioside-enriched lipid rafts (Kakio et al. 2003). Indeed, it has been reported that Aβ binds selectively to the membranes containing gangliosides, and the presence of ganglioside accelerates the rate of amyloid assembly (Choo-Smith et al. 1997; Yamamoto et al. 2005). It has also been established that there are significant alterations in numerous receptors and cell signaling proteins in the cortical lipid rafts isolated from the cognitively impaired mouse model of AD, as only 17 % of the raft proteins were similar between the AD model and age-matched control mice (Chadwick et al. 2010). There is a growing body of evidence of a crucial role of cholesterol in the pathogenesis of AD (reviewed, for instance, in Verdier et al. 2004). Cholesterol can stimulate the proteolytic activity of β-secretase, for example (Kalvodova et al. 2005), while cholesterol depletion inhibits the activity of both β- and γ-secretase (Grimm et al. 2008) and also the generation of Aβ in the hippocampal neurons (Simons et al. 1998). $A\beta_{40}$ polymers have been reported to bind preferentially to cholesterol rather than to fatty acids and phosphatidylcholine (Avdulov et al. 1997). It seems that cholesterol-induced changes in membrane fluidity can affect Aβ-cell surface interactions and cell toxicity (Yip et al. 2001). It has been proposed that age-related changes in the asymmetric distribution of cholesterol in the neuronal plasma membranes provide a cooperative environment for the accumulation of Aβ, as the exofacial leaflet of the synaptic plasma membranes of aged mice contains twice as much cholesterol as that in to younger mice (Wood et al. 2002).

The amyloid channel hypothesis holds that Aβ can exert its neurotoxic function, at least in part, by forming aberrant ion channels in neuronal membranes and thereby disrupting neuronal ion (especially Ca^{2+}) homeostasis. This was first tested by Arispe et al. (1993), and a number of studies later validated the phenomenon (reviewed by Shirwany et al. 2007). Through use of the patch-clamp technique, $A\beta_{40}$ was demonstrated to form cation-selective channels in membranes derived from hypothalamic neurons, suggesting that the interactions between Aβ protein and neuronal membranes also occur in vivo (Kawahara et al. 1997). It was reported later that, at neurotoxic concentrations, $A\beta_{42}$ can form slightly cation selective, voltage-independent ion channels in planar bilayer membranes (Hirakura et al. 1999). Indeed, atomic force microscopy of $A\beta_{42}$ reconstituted in a planar lipid bilayer revealed multimeric channel-like structures (Lin et al 2001). Moreover, not only Aβ, but also other neurotoxic proteins have been reported to have pore-like activities. PolyQ forms cation-selective channels when incorporated into artificial planar lipid bilayer membranes. The channel-forming activity was found to depend critically on the length of the polyQ chains: ion channels were detected with 40-residue stretches, but not with 29-residue tracts (Monoi et al. 2000). Analysis of protofibrillar α-synuclein by atomic force microscopy and electron microscopy indicated that the oligomers consist of spheres, chains and rings. α-Synuclein protofibrils permeabilize synthetic vesicles and form pore-like assemblies on the surface of brain-derived vesicles (Rochet et al. 2004). The formation of pore-like oligomeric structures may explain the membrane permeabilization activity of α-synuclein protofibrils (Lashuel et al. 2002).

Protein misfolding leads to the exposure of hydrophobic regions of proteins, and hence to their interaction with lipids inducing membrane damage (Kourie and Henry 2002). As molecular chaperones, sHsps can inhibit protein misfolding and the exposure of hydrophobic regions, thereby preventing the interactions between these toxic peptides and cellular membranes. However Hsps have also been shown to give rise to direct alterations in certain attributes of the membrane lipid phase, such as fluidity and permeability, and can thereby maintain the membrane stability under stress conditions. A number of studies have pointed to a potential link between membrane function and Hsps, which can bind to the membranes, primarily to lipid rafts, and increase the physical order (reviewed by Nakamoto and Vígh 2007 and Horváth et al. 2008). After heat shock, most of the newly synthesized Hsp17 in *Synechocystis* was found to be associated with the thylakoid membranes (Horváth et al. 1998; Glatz et al. 1999), binding in particular to certain heat-shock lipid (Balogi et al. 2005). A noteworthy finding was that an increased level of thylakoid association of Hsp17 provided improved resistance to UV-B damage in *Synechocystis* (Balogi et al. 2008). After a short-term ethanol treatment, molecular chaperones DnaK and GroEL are activated and recruited in the membrane in *Bacillus subtilis*, which suggests that these chaperones contribute to an early step in the cell response to ethanol stress, by protecting the membranes (Seydlová et al. 2012). Similar results were observed in *Oenococcus oeni*. In the early stages of ethanol stress, the expression of the sHsp Lo18 is increased, and this is then addressed to the membrane and participates in its rigidification (Maitre et al. 2014). In *Lactobacillus plantarum*, it has been demonstrated that a sHsp can affect the membrane fluidity after ethanol stress. Hsp18.55-overexpressing *L. plantarum* cells exhibited a significant reduction in membrane fluidification after ethanol treatment, whereas the deletion mutant strain displayed an increased membrane fluidification after the same stress (Capozzi et al. 2011). An early study demonstrated that *E. coli* GroEL can increase the molecular order of the lipid bilayers under heat-shock conditions (Török et al. 1997). It was later found that Hsp17 of the blue alga *Synechocystis* is associated with lipid membranes, resulting in an elevated degree of physical order of the membranes and reduced membrane fluidity (Török et al. 2001). HspB5 and *Synechocystis* Hsp17 have been demonstrated to be able to regulate the bilayer-non-bilayer phase equilibrium and to exert a bilayer-stabilizing effect on model membranes in vitro (Tsvetkova et al. 2002). That study clearly indicated that the nature of sHsp-membrane interactions depends strongly on the lipid composition, and especially on the overall extent of lipid unsaturation of the host membranes. Accordingly, it was suggested that the association between sHsps and specific lipids or lipid domains of the membranes may constitute a general mechanism that preserves membrane integrity during thermal fluctuations. We recently provided evidence of the cholesterol-controlled lipid raft interaction of the mammalian Hspb11 (Hsp16.6) (Török et al. 2012). Overall, these data suggest that the membrane-associated pool(s) of Hsps may play a pivotal role in membrane protection under very varied stress conditions (Nakamoto and Vigh 2007; Balogi et al. 2008; Horváth et al. 2008, 2012; Horváth and Vígh 2010).

23.4 Oxidative Stress

Oxidative stress is characteristic of different neurological disorders, such as protein misfolding diseases, stroke and ethanol induced neurodegeneration, and seems to be one of the most important processes in aging. Markers of lipid peroxidation, for example including 4-hydroxinonenal (4-HNE) and malondialdehyde (MDA), and protein nitration, a marker of protein oxidation, have been identified in the brain of patients with AD, PD or HD and in the spinal cord of patients with ALS (reviewed by Andersen 2004 and Reed 2011). Lipid peroxidation is a complex process involving the interactions of ROS with polyunsaturated fatty acids, resulting in a variety of highly reactive electrophilic aldehydes such as MDA and 4-HNE. These products are able to bind to proteins by forming stable adducts with cysteine, lysine and histidine amino acid residues (for a review see Reed 2011). The brain is particularly vulnerable to oxidation because brain lipids are rich in polyunsaturated fatty acids. An elevated level of 4-HNE leads to the disruption of Ca^{2+} homeostasis, the disturbance of mitochondrial functions, alterations in the cytoskeletal structure, membrane damage and cell death (Esterbauer et al. 1991). 4-HNE cross-linked proteins have been reported to inhibit proteasome as a non-competitive inhibitor (Friguet and Szweda 1997), and 4-HNE-modified proteins therefore accumulate during oxidative stress. 4-HNE-proteasome conjugates have also been detected, and could be responsible for the loss of proteasome activities (Okada et al. 1999).

In the course of aging, oxidative stress increases, while the ability of cells to respond to oxidative protein damage decreases, and this can result in enhanced levels of oxidatively modified proteins that might be resistant to proteasomal degradation. Increased oxidative alterations to aggregation-prone proteins might result in a greater degree of protein misfolding and impaired degradation (reviewed by Andersen 2004). In turn, the pathological deposition of abnormal proteins is associated with alterations in redox state homeostasis and a mitochondrial dysfunction, and aggregation-prone proteins might therefore themselves produce oxidative stress. As an example, Aβ causes increased levels of H_2O_2 and lipid peroxides in cultured cells, and Aβ toxicity can be blocked by reagents that inhibit flavin oxidases (Behl et al. 1994). The overexpression of α-synuclein in a neuronal cell line resulted in the formation of inclusion-like structures and concomitant mitochondrial alterations accompanied by increased levels of free radicals (Hsu et al. 2000). The transfection of cells with mutated forms of the parkin gene led to elevated levels of protein carbonyls and lipid peroxidation (Hyun et al. 2002). Mutant huntingtin can induce oxidative stress by increasing intracellular ROS levels in different cell lines, and the degree of ROS generation increased in parallel with the length of polyQ expansions (Wyttenbach et al. 2002). The brain of the mouse model of HD (R6/2 mice) exhibited a significant reduction in mitochondrial complex IV activity and elevated extent of the generation of nitric oxide and superoxide radicals (Tabrizi et al. 2000). Transfection of cell lines with mutant SOD1 coding genes, associated with familial ALS, produced higher levels of lipid peroxidation, 3-nitrotyrosine and protein carbonyl (Lee et al. 2001).

Ethanol-induced neuronal damage is also related to oxidative stress and lipid peroxidation. Both acute and chronic ethanol administration resulted in enhanced formation of ROS, a decreased glutathione level and increased lipid peroxidation in different brain regions of rats (Uysal et al. 1989; Montoliu et al. 1994; Calabrese et al. 1996, 1998). In a study of the association between ROS production and the level of membrane fluidity in cell cultures (Sergent et al. 2005), ethanol treatment resulted in an increase in ROS production and in parallel in an increase in membrane fluidity, which could be prevented by pretreatment with an antioxidant, indicating that the metabolism of ethanol and the formation of ROS are involved in the elevation of membrane fluidity. It also emerged that membrane-stabilizing agents decreased the levels of ROS production, lipid peroxidation and cell death in response to ethanol incubation. In contrast, membrane-fluidizing compounds enhanced the degree of ethanol-induced oxidative stress. These data suggested that ROS formation can influence membrane fluidity, which in turn further amplifies the oxidative stress and lipid peroxidation, eventually leading to cell death. In neurological diseases, oxidative stress can result in damage to different cell components, triggering intracellular pathways that finally lead to cell death, the brain being believed to be particularly susceptible to the damaging effects of ROS.

There are a number of experimental data indicating that sHsps exert a protective effect against oxidative stress thanks to their ability to regulate intracellular redox homeostasis (reviewed, for example, by Arrigo 1998; Arrigo et al. 2005; Mymrikov 2011). Large aggregates of HspB1 are known to be able to modulate ROS and glutathione and to generate cellular protection against oxidative stress, whereas phosphorylation down-regulates these activities through the dissociation of sHsp complexes to tetramers (Mehlen et al. 1997; Rogalla et al. 1999). sHsp expression also buffers the increase in protein oxidation following H_2O_2 treatment, and protects several key enzymes against inactivation. sHsps significantly increased glucose-6-phosphate dehydrogenase activity and to a lesser extent the activities of glutathione reductase and glutathione transferase (Preville et al. 1999). Lipid peroxidation and protein oxidation are also inhibited by the expression of HspB1 or HspB5 (Mehlen et al. 1996), as is oxidative DNA damage (Park et al. 1998). The expression of human HspB1 or HspB5 in mammalian cell lines decreases the intracellular level of iron in parallel with a diminution in the level of oxidized proteins

(Arrigo et al. 2005). HspB1 has been reported to suppress polyQ mediated cell death without decreasing polyQ aggregation, whereas it decreases ROS in cells expressing mutant huntingtin, suggesting that this chaperone protects cells against polyQ-induced oxidative stress (Wyttenbach et al. 2002). The cytoskeleton is a sensitive target of oxidative stress. Overexpression of the wild-type, but not a non-phosphorylatable form of human HspB1, resulted in increased resistance against oxidative stress-induced actin fragmentation and improved the cell survival following H_2O_2 treatment (Huot et al. 1996). However, another study demonstrated that phosphorylation is not essential for the protective activity of HspB1 against H_2O_2-induced actin disruption, and that only a small fraction of HspB1 is colocalized with actin microfilaments, suggesting that the protection against actin network disruption is probably a consequence of the modulation of the redox change rather than a direct effect of HspB1 to actin (Preville et al. 1998). These data suggest that the antioxidant action of sHsps involves different mechanisms, e.g. modulation of the glutathione system, chaperone-like activity to prevent protein aggregation, and protection of the cytoskeleton (Mymrikov et al. 2011). It has been proposed that HspB1 has a role in the presentation of oxidized proteins to the proteasome degradation machinery (Arrigo et al. 2005).

23.5 Apoptosis

Apoptotic pathways have been reported to be engaged in various neurological disorders, including AD, PD, HD, stroke, brain trauma and ALS (for reviews, see Mattson 2000; Friedlander 2003; Wyttenbach and Arrigo 2009). The hallmarks of these diseases, including oxidative stress, perturbed calcium homeostasis, a mitochondrial dysfunction and disturbances of the cytoskeleton, can all activate cell death cascades.

Increased DNA fragmentation, caspase activity, and alterations in the expression of apoptosis-related genes have been detected in neurons associated with Aβ plaques in AD (Su et al. 1994; Masliah et al. 1998, reviewed by Mattson 2000). Aβ can directly induce apoptosis in cultured neurons and transgenic mice (Loo et al. 1993; LaFerla et al. 1995; Forloni et al. 1996). Moreover, the APP is known to be cleaved directly and efficiently by caspases during apoptosis, and neurons undergoing apoptosis thereby generate elevated levels of Aβ (LeBlanc 1995; Gervais et al. 1999). In turn, the caspase-mediated cleavage of APP can release a carboxy-terminal peptide which is a potent inducer of apoptosis (Lu et al. 2000).

Pronounced apoptosis-related DNA damage and gene activation have been observed in the brain of PD patients (Jenner and Olanow 1998). α-Synuclein proteins can induce apoptotic cell death in human neuroblastoma SH-SY5Y cells (El-Agnaf et al. 1998). Transgenic mice expressing a dominant negative inhibitor of caspase-1 are resistant to MPTP-induced apoptotic cell death (Klivenyi et al. 1999).

Activation of caspases and the release of cytochrome c have been demonstrated in the striatal neurons of HD patients and in the R6/2 mouse strain (Kiechle et al.

2002). In this transgenic mouse model of HD, the expression of a dominant-negative caspase-1 mutant prolongs the survival and delays the appearance of the symptoms of the disease (Ona et al. 1999). Inhibition of caspase-1 by minocycline treatment also delays disease progression in the R6/2 mouse model (Chen et al. 2000). Similarly to Aβ, huntingtin is cleaved by caspases. As the disease progresses, enhanced caspase-mediated cleavage of huntingtin increases the generation of huntingtin fragments and depletes wild-type huntingtin (Ona et al. 1999, reviewed by Friedlander 2003).

In ALS, motor neurons most probably die as a result of apoptosis. Caspase activation and cytochrome c release have been detected in the transgenic mouse model of ALS and also in spinal cord samples from patients with ALS (Martin 1999; Guegan et al. 2001, for reviews, see Friedlander 2003; Rowland and Shneider 2001). In mutant SOD1 transgenic mice crossed with a mouse line expressing a dominant negative inhibitor of caspase-1 in neurons, the symptomatic progression of the disease was slowed and death was delayed (Friedlander et al. 1997a). Furthermore, the intracerebroventricular administration of a broad caspase inhibitor delays the disease onset and death in SOD1 transgenic mice (Li et al. 2000).

Necrotic and apoptotic cell death both play roles in mediating tissue injury after ischemia or brain trauma (reviewed by Friedlander. 2003). It has been suggested that apoptosis occur first after acute cerebral ischemia, but in the core of the infarction the process is aborted because of a severe energy depletion, whereas energy-dependent caspase activation observed in the penumbra, in which apoptosis can develop fully because of the residual blood supply (Benchoua et al. 2001). The inhibition of caspases reduces trauma-mediated brain tissue damage or the infarct volume after ischemia (Friedlander et al. 1997b; Fink et al. 1999; Rabuffetti et al. 2000).

The cytoprotective effect of Hsps is related to their ability to inhibit apoptosis. Several studies have demonstrated that sHsps, and especially HspB1 and HspB5, can interfere with different steps of the apoptotic pathway (reviewed in detail by Sreedhar and Csermely 2004; Wyttenbach and Arrigo 2009; Mymrikov et al. 2011). Heat-shocked and HspB1-overexpressing cells exhibit an increased resistance to drug -induced apoptosis (Samali and Cotter 1996). The use of herpes simplex virus-based vectors revealed that HspB1 can protect dorsal root ganglion neurons from the apoptosis induced by nerve growth factor withdrawal, and ND7 neuronal cells from retinoic acid-induced apoptosis (Wagstaff et al. 1999). The blocking of HspB1 by an antisense strategy can trigger the release of the mitochondrial protein Smac, which suggests that HspB1 inhibits the release of this mitochondria-derived activator of caspases (Chauhan et al. 2003). HspB1 can prevent the cytochrome c-dependent activation of procaspase-9 and the subsequent activation of procaspase-3 (Garrido et al. 1999; Bruey et al. 2000), and it inhibits the release of cytochrome c from the mitochondria (Paul et al. 2002). HspB1 also affects the Fas-induced Daxx-mediated apoptotic pathway (Charette et al. 2000). The transgenic overexpression of HspB1 can significantly reduce the severity of kainate-induced seizures and markedly reduce neuronal cell death in the CA3 region of the hippocampus, in parallel with a concomitant attenuation of caspase-3 activation and apoptosis (Akbar et al. 2003).

HspB5 has also been found to process resistance to apoptosis in different cells and tissues. For example, it can protect against H_2O_2-induced apoptosis in murine cells by interacting with p53, a pro-apoptotic protein which might prevent the translocation of p53 from the cytoplasm to the mitochondria (Liu et al. 2007). In the HspB5-overexpressing C6 astroglioma cell line, H_2O_2-induced apoptosis was reduced by 60 % as compared with the parent cells, while the caspase-3 activity was markedly suppressed. Moreover, immunoprecipitation revealed the interaction of HspB5 with partially processed caspase-3 (Shin et al. 2009). The overexpression of HspB5 blocks the activation of RAS to inhibit ERK1/2 activation, and attenuates calcimycin-induced apoptosis (Li et al. 2005). These studies demonstrate that sHsps are potentially important suppressors of death-signaling pathways. However, the full details of the molecular mechanism of modulation of the apoptotic process have not been deciphered. The chaperone function and capacity of the sHsps to protect against changes in the cellular redox homeostasis and to stabilize the cytoskeleton enable them to promote cell survival (Sreedhar and Csermely 2004).

23.6 Protein Degradation

Besides the chaperone network, degradation pathways such as the ubiquitin-proteasome pathway (UPP) and the lysosomal system also have crucial roles in maintaining the protein homeostasis. Unlike acute stress-induced unfolded proteins, the intrinsically misfolded proteins that accumulate in degenerative disorders can usually not be refolded. Thus, chaperones can only prevent their aggregation, and the long-term solution would be their elimination, for instance by degradation. Unfortunately, aging is accompanied by a deterioration in the activity of the proteasomal and lysosomal system, which can favor the accumulation of misfolded proteins (reviewed by Söti and Csermely 2002). Moreover, aggregated proteins cannot be degraded properly, and ubiquitinated proteins are enriched in neurofibrillary tangles and senile plaques and, besides Hsps, components of the UPP are sequestered in these aggregates (Perry et al. 1987; Choi et al. 2004; Ross and Pickart 2004; Wyttenbach and Arrigo 2009). Depletion of these proteins may contribute to the pathogenesis of diseases.

Moreover misfolded, aggregated proteins are effective inhibitors of the UPP. Certain brain regions of patients with AD exhibit a significant decrease in proteasome activity (Keller et al. 2000; Keck et al. 2003). Paired helical filaments (PHF)-tau proved to co-precipitate during proteasome immunoprecipitation, and proteasome subunits could be co-isolated during the isolation of PHFs from the AD brain. Furthermore the incubation of isolated proteasomes with PHF-tau either isolated from the AD brain, or assembled from human recombinant tau protein resulted in the inhibition of proteasome activity (Keck et al. 2003). Aβ oligomers have also been found to inhibit the proteasome in vivo. Young 3xTg-AD model mice displayed an impaired proteasome activity that correlated with the detection of intraneuronal Aβ oligomers. Blockade of the proteasome function in pre-pathological

3xTg-AD mice resulted in increased Aβ and tau accumulation, while Aβ immunotherapy diminished the Aβ oligomers and reversed the deficits in proteasome activity (Tseng et al. 2008). A component of the UPP, ubiquitin C-terminal hydrolase L1 (Uch-L1), that enhances the recycling of ubiquitin, is necessary for normal synaptic and cognitive functions. A reduction in the level of soluble Uch-L1 protein was observed in the hippocampus of APP/PS1 AD model mice, and the inhibition of Uch-L1 activity led to the inhibition of hippocampal long-term potentiation (LTP). However, transduction of the Uch-L1 protein restores the normal synaptic function both in hippocampal slices treated with oligomeric Aβ and in the APP/PS1 mouse model of AD (Gong et al. 2006). These results suggest that the activity of UPP is decreased in AD, which can lead to the further accumulation of misfolded proteins.

A proportion of the APP is internalized into the cell by endocytosis and enters the endosomes, late endosomes and lysosomes. Since the membranes of the endosomal–lysosomal (EL) system are enriched in APP and β- and γ-secretases (Yu et al. 2004), APP internalization by endocytosis increases Aβ generation. Although most of the Aβ formed during autophagy is normally degraded within the lysosomes, in AD it is accumulated in the EL system, resulting in increased lysosomal membrane permeability, and this stimulates the release of the lysosomal enzymes into the cytoplasm, inducing apoptotic cell death (reviewed in Nixon 2006; Penke et al. 2012).

As reviewed earlier (Ross and Pickart 2004; Wyttenbach and Arrigo 2009), impairment of the UPP has a central role in the neuropathology of PD and polyQ diseases. Mutations in the parkin gene, an E3 ubiquitin ligase, have been shown to cause an autosomal recessive form of PD. Most parkin mutations are thought to be loss-of-function mutations, and the failure of parkin to ubiquitinate and remove substrates may lead to their accumulation and subsequent toxicity (Healy et al. 2004). Another gene implicated in PD is Uch-L1 (Healy et al. 2004). Furthermore, aggregated α-synuclein and (with lower affinity) monomeric α-synuclein inhibit 26S ubiquitin-dependent proteasomal activity. Binding studies have demonstrated that both aggregated and monomeric α-synuclein selectively bind to a 19S proteasoma subunit (Snyder et al. 2003). Moreover, the pathogenic mutant forms of α-synuclein bind to lysosomal membrane receptors, and appear to act as uptake blockers, inhibiting protein degradation (Cuervo et al. 2004). Transient expression of a huntingtin fragment containing a pathogenic polyQ repeat also caused nearly complete inhibition of the UPP (Bence et al. 2001).

It is possible that molecular chaperones facilitate neuroprotection through their ability to promote autophagy or proteasome-mediated protein degradation. CHIP (carboxyl terminus of Hsc70-interacting protein) is a cofactor that interacts with Hsc70 and Hsp70, and has intrinsic E3 ubiquitin ligase activity, thereby accelerating the ubiquitin-dependent degradation of chaperone substrates. CHIP interacts with the stress-responsive ubiquitin-conjugating enzyme family UBCH5 (Jiang et al. 2001). Another Hsp70 co-chaperone, BAG-1 possesses an ubiquitin-like domain at its amino terminus and it can associate to the 26S proteasome, providing a physical link between Hsp70 and proteasome (Luders et al. 2000, reviewed by Muchowski and Wacker 2005). These results suggest that Hsp70 contributes to the delivery of protein substrates to the degradation systems. Since sHsps create a reservoir of non-native proteins available to Hsp70, they also participate in this process indirectly. However, certain sHsps, such as HspB1 or HspB8, have been reported to assist in proteasome degradation directly (for a review see Mymrikov et al. 2011). Under stress conditions, HspB1 promotes the ubiquitylation of certain proteins (Parcellier et al. 2003, 2006). HspB1 preferentially binds to pathological hyperphosphorylated tau and PHF-tau, thereby facilitating its proteasomal elimination (Shimura et al. 2004). It has also been demonstrated that HspB1 can prevent the cell death induced by administration of proteasome inhibitors (Goldbaum et al. 2009). Certain results suggest that HspB1 partially restores the proteasome activities impaired by mutant huntingtin (Wyttenbach and Arrigo 2009).

HspB8 appears to be involved in both the proteasome and the macroautophagy-mediated proteolysis of unfolded proteins (reviewed in Mymrikov et al. 2011). HspB8 collaborates with BAG-3, a member of the co-chaperone family, to create a chaperone complex that stimulates the degradation of misfolded protein substrates by macroautophagy. It has been proposed that the role of HspB8 in this complex is to recognize the misfolded proteins, while BAG-3 might recruit and activate the macroautophagy machinery (Carra et al. 2008). Indeed, HspB8 has been shown to increase mutant SOD1 clearance via autophagy, and immunoprecipitation studies have demonstrated that mutant SOD1 interacts with the HspB8/BAG3/Hsc70/CHIP multiheteromeric complex (Crippa et al. 2010, reviewed by Mymrikov et al. 2011). Overexpression of HspB8 in a transgenic mouse leads to the increased expression of both the 19S and 20S proteasome subunits, an increase in 20S catalytic activity, and a redistribution of proteasome from the cytosol to the nuclear periphery. The perinuclear proteasome is co-localized with and interacts with HspB8, which suggests that HspB8 can affect the stability and intracellular localization of proteasomes (Hedhli et al. 2008). HspB7 was recently identified as a potent suppressor of polyQ aggregation and polyQ-induced toxicity. This action of HspB7 is independent of the Hsp70 machinery or the proteasomal activity, but the anti-aggregation activity of HspB7 is substantially reduced in cells that are defective in macroautophagy (Vos et al. 2010). Finally, the chymotrypsin-like activity of the proteasome is significantly suppressed in consequence of tau expression, while the proteasome activity can be enhanced by the Hsf1 activator geldanamycin, resulting in increased tau degradation (Oppatova et al. 2013). Thus, it seems that the Hsps are also involved in mediating the protein degradation systems.

23.7 Insulin Resistance

There is mounting evidence suggesting that the insulin and insulin-like growth factor (IGF) signaling pathway is a major factor in neurodegeneration and aging. Type-2 diabetes mellitus (T2DM) and obesity may contribute to the development of AD, but by themselves are probably not sufficient to cause the disease (reviewed in detail by de la Monte and Wands 2008). As an example, there is an increased risk of the development of a mild cognitive impairment, dementia, or AD in individuals with T2DM (Pasquier et al. 2006). In obese T2DM model rats, it emerged that learning and memory were impaired and the plasma membrane association of the insulin-sensitive glucose transporter GLUT4 was reduced in the hippocampus (Winocur et al. 2005). High-fat-diet feeding for 16 weeks led to obesity and T2DM caused brain atrophy with insulin resistance, oxidative stress and a cytoskeleton degradation (Moroz et al. 2008). An improved cognitive performance was observed in animal models and AD patients after treatment with insulin-sensitizer agents or intranasal insulin (Landreth 2006; Reger et al. 2008, reviewed by de la Monte and Wands 2008).

Reduced glucose utilization and a deficient energy metabolism occur early during AD, suggesting a role for impaired insulin signaling in the pathogenesis of the disease. Extensive abnormalities in the insulin and insulin-like growth factor type I and II (IGF-I and IGF-II) signaling mechanisms have been observed in AD brains. The reduced CNS expression of insulin, IGF-I, and IGF-II, and their receptors, suggests that AD may represent a distinct neuroendocrine disorder that is similar to diabetes mellitus in some features (Steen et al. 2005). Impairments in the insulin and IGF-I and IGF-II signaling become more severe as the disease progresses (Rivera et al. 2005). de la Monte et al. therefore proposed the term type 3 diabetes and suggested that AD might represent a brain-specific form of diabetes (reviewed in de la Monte and Wands 2008). They later demonstrated that many of the characteristic features of AD-type neurodegeneration, including increased glycogen synthase kinase-3β (GSK-3β) activation, increased tau phosphorylation and decreased neuronal survival, can be induced experimentally through the chemical depletion of insulin with a single intracerebral injection of streptozotocin (STZ). The ic-STZ-injected rats did not display elevated blood glucose levels, but their brains were reduced in size and exhibited neurodegeneration, gliosis, and increased immunoreactivity for active GSK-3β, phospho-tau, ubiquitin and Aβ. These abnormalities were associated with the reduced expression of insulin, IGF and their receptors, and reduced ligand binding to the insulin and IGF-II receptors (Lester-Coll et al. 2006). It was proposed that impaired insulin/IGF signaling leads to a greater degree of oxidative stress, the activation of GSK-3β and a mitochondrial dysfunction, which induce intensive APP gene expression and cleavage and tau phosphorylation (reviewed by de la Monte and Wands 2008). It has also been reported that Aβ can interfere with insulin receptor signaling directly by reducing the affinity of insulin binding to the insulin receptor and inhibiting the autophosphorylation of insulin receptors (Ling et al. 2002; Xie et al. 2002). Moreover, evidence is increasing that the neuronal glucose metabolism and its control by the insulin signal transduction cascade play essential roles in memory formation and retrieval. The neurotransmitter acetylcholine is synthesized from acetyl-CoA, a compound involved in the metabolism of glucose, and the synthesis is controlled by insulin, a mediator of acetylcholine transferase. Any impairment of the neuronal glucose metabolism can therefore lead to disturbances in memory function, as observed in sporadic AD (reviewed by Hoyer 2003). Overall, these data suggest that selective impairments in insulin and IGF signaling mechanisms may participate in the development of AD.

Clinical data indicate that some patients with PD are characterized by impaired glucose tolerance and insulin dysregulation, and results from both animal and in vitro studies point to a clear role for insulin in the regulation of the brain dopaminergic activity. Losses of insulin receptor immunoreactivity and mRNA in the substantia nigra have been found in neuropathological studies of patients with PD. Certain studies have concluded that patients with HD have a higher prevalence of diabetes and insulin resistance (reviewed by Craft and Watson 2004). Interestingly, ethanol induced neurotoxicity and fetal alcohol syndrome may also be mediated, at least in part, by the inhibition of insulin and IGF-I signaling. Insulin and IGF signaling are required during early neurological development and in order to maintain neuronal viability. Ethanol can interfere with IGF-I-mediated cell survival in primary cultured cerebellar granule neurons and inhibit tyrosine autophosphorylation of the IGF-I receptor in human neuroblastoma cells (Zhang et al. 1998; Seiler et al. 2001). Cerebellar neurons isolated from ethanol-treated rat pups displayed inhibited insulin-stimulated neuronal viability, a decreased mitochondrial function and an enhanced pro-apoptotic gene expression (de la Monte and Wands 2002).

In the past decade, evidence has been accumulating of a possible connection between diabetes and the heat-shock response. Metabolic disturbances in diabetes induce various alterations in protein homeostasis (Dancso et al. 2010). In obese Zucker rats, for example large aggregates of ubiquitinated proteins were observed in insulin-expressing β-cells (Kaniuk et al. 2007). The elevated levels of Hsf1, Hsp70 and Hsp90 found in the pancreatic cells of diabetic monkeys might reflect a compensatory mechanism for the altered protein homeostasis (Kavanagh et al. 2009, reviewed by Dancso et al. 2010). On the other hand, the levels of inducible stress proteins, including HspB1, and their response to stress are low in tissues that are insulin-sensitive; this is particularly true for skeletal muscle and liver, probably because of the deactivation of Hsf1 (Kurucz et al. 2002; Kavanagh et al. 2009, reviewed by Hooper et al. 2014).

As defects in the stress response occur prior to the development of glucose intolerance, it was proposed that loss of the cellular stress response in insulin-responsive tissues occur early in the pathogenesis of T2DM, that disrupts the metabolic homeostasis, leading to a cascade of pathological outcomes (Hooper et al. 2014). Indeed, several results suggest that measures aimed at improving the cellular stress response via Hsp induction can an effective therapy for patients with T2DM. Acute exercise, for instance, which induces Hsp expression, can improves the whole-body insulin sensitivity and glucose tolerance (reviewed by Hooper et al. 2014). Treatment of T2DM with repeated, partial submersion in a hot tub for 3 weeks resulted in an improvement in fasting glucose, a trend toward weight loss, and the relief of neuropathic symptoms (Hooper 1999). Heat treatment also improved glucose tolerance, restored the insulin-stimulated glucose transport, and increased insulin signaling in skeletal muscles of high-fat-diet rats, while the expression of Hsp72 and HspB1 was upregulated (Gupte et al. 2009). A bimoclomol derivative, BRX-220, was able to reduce insulin resistance in both STZ-treated and Zucker rats, and dose-dependently improved diabetes-related deficits in muscle motor and sensory nerve function in STZ-induced diabetes (Kürthy et al. 2002). Heat shock and treatment with a Hsp co-inducer, BGP-15, successfully prevented diet- or obesity-induced hyperglycemia, hyperinsulinemia, glucose intolerance and insulin resistance in fat-fed mice (Chung et al. 2008). BGP-15 treatment additionally significantly improved the insulin sensitivity in insulin-resistant, nondiabetic human patients and prevented the olanzapine induction of insulin resistance (Literáti-Nagy et al. 2009, 2012).

Such studies suggest that the modulation of Hsps may be a promising therapeutic strategy for improving insulin and IGF-I signaling. Pharmacological induction of the heat-shock response may provide a therapeutic approach for the treatment of several insulin-associated neurodegenerative disorders (Urban et al. 2012).

23.8 Blood-Brain Barrier

The permeability of the blood-brain barrier (BBB) increases with normal aging, and there is a further increase in patients with AD or vascular dementia (Farrall and Wardlaw 2009). In approximately 90 % of AD brains, Aβ has been reported to accumulate in the walls of small arteries, arterioles. However, cerebral amyloid angiopathy is also frequently present in the elderly without AD and it is a well-documented risk factor for cerebral hemorrhage (Vinters 1987). BBB has a special structure in the CNS, its function being to separate the circulating blood components from the neurons and to maintain the brain homeostasis. The BBB consists of the endothelial cells of brain capillaries connected by tight junctions. The BBB is not a completely impermeable barrier, but regulates the passage of ions, glucose and amino acids (Zlokovic 2008). A number of studies have shown the association of a dysfunction of tight junction proteins and the destruction of the BBB with different neurological disorders. The Evans blue assay demonstrated increased BBB permeability in different transgenic mouse models of AD (Paul et al. 2007). In the Tg2576 strain, an increase in BBB permeability was found to precede Aβ plaque formation and the development of cognitive deficits, and it was compromised as early as 4 months of age (Ujiie et al. 2003). However, immunization with Aβ can decrease the permeability of the BBB and Aβ plaque formation in these AD model mice (Dickstein et al. 2006), in the brain of which a significant tight junction dysfunction has been demonstrated (Biron et al. 2011). Relative to controls, AD model mice exhibit a higher incidence of abnormal, punctate staining of the proteins occludin and zonula occludens-1 (ZO-1), and reductions in occludin level in the cortex and hippocampus.

Although the literature on the effects of Hsps on BBB is rather sparse, some data suggest that there is a link between chaperones and the permeability of the BBB. Thus, the expression pattern of Hsp72 was found to be correlated with the regional pattern of BBB breakdown after mild brain injury in rats (Tanno et al. 1993). In an in vitro BBB model of porcine brain capillary endothelial cells, a brief period of hyperthermia caused a transient loss in BBB integrity, but significantly lower loss of BBB integrity was found after a second heat treatment, indicating the development of thermotolerance. In fact, the degree of thermotolerance increased as the pre-conditioning temperature was increased (Klein and Bobilya 2008). Hsp overexpression achieved through a whole-body hyperthermia can prevent the osmotic stress induced dysfunction of the BBB (Lu et al. 2004). The permeability of the BBB was significantly increased and the levels of occludin and ZO proteins were significantly decreased after D-mannitol treatment, whereas they were preserved if the animals were heat-shock-treated 24 h before the D-mannitol infusion. Moreover, a co-immunoprecipitation study revealed that, following heat shock, Hsp72 interacted with the major tight junction proteins. Another study led to the finding that heat stress could increase occludin expression and that this was mediated by the activation and binding of Hsf1 to the occludin promoter (Dokladny et al. 2008). High-dose geldanamycin post-treatment significantly attenuated BBB disruption, brain edema formation and neurological deficits in a mouse model of intracerebral hemorrhage (Manaenko et al. 2010). HspB1 was recently shown to maintain the integrity of the BBB in mice subjected to middle cerebral artery occlusion (Leak et al. 2013). These results are indicative that Hsps might have an important role in stabilizing the tight junction proteins and thereby maintaining the integrity of the BBB under stress conditions.

23.9 Conclusions

There is increasing evidence to suggest, that sHsps may play a protective role in different neurological disorders: not only in protein misfolding diseases, but also in acute injuries such as ischemic stroke. In these pathological conditions, the whole cellular homeostasis is disturbed, which eventually leads to cell death. Although deleterious effects can influence almost all cellular components and pathways, these are not isolated events but are rather mutual consequences of each other. For example, oxidative stress leads to a disruption of the cytoskeleton, which can then induce apoptosis. sHsps affect various processes in cooperation with other chaperones, these not being discrete effects. One of the most important properties of sHsps is that they can bind misfolded proteins, and some of their additional protective effects can be consequences of this function. Thanks to their preventing misfolded proteins from aggregating with other proteins and cell components, they can protect membrane integrity or inhibit apoptosis. It is an interesting observation that specific sets of Hsps can influence the different diseases (reviewed by Kakkar et al. 2014). For example, HspB1 can prevent protein aggregation in AD, but not in HD, in which disease it protects against oxidative stress. It is also important that the whole chaperone network is necessary for the complete removal of the unfolded proteins. sHsps are unable alone to actively refold the misfolded proteins, but can maintain them in a refolding-competent state until members of the Hsp70 facilitate their refolding or transmit them to the protein degradation machinery. The overexpression of a single Hsp might therefore not be sufficient to ameliorate the disease, or it might even be detrimental. HspB5, for instance, can increase the toxic effect of Aβ, probably by keeping it in a nonfibrillar, highly toxic form (Stege et al. 1999). The overexpression of several chaperones might be beneficial in neurodegenerative diseases, prevention of which is likely to be a more efficient than their subsequent treatment. Some in vitro studies have demonstrated that sHsps can perform their solubilizing function only when they are co-aggregated with the protein substrate, whereas when they added to pre-formed aggregates there is a smaller decrease in amyloid formation. Heat-shock response is known to be impaired during aging and in neurodegenerative diseases. Sequestration of different Hsps within protein inclusions may reduce the availability of these chaperones, thereby disrupting their normal functions, and leading to a greater vulnerability of the cells to the stress conditions that occur within the normal lifespan. Indeed, activation of the cellular stress response in cells that express mutant aggregation-prone proteins can lead to an overall increase in protein aggregation, which means that the stress response can exacerbate the disease symptoms. It should also be mentioned that the excessive overexpression of a Hsp can lead to the instability of the stress response and can cause unwanted side-effects. The mild, unforced overexpression of Hsps by natural methods might help in preventing these diseases. The therapeutic use of Hsp co-inducers could also be a solution. Hydroxylamine derivatives, for example increase Hsp expression only under stressed conditions by prolonging the activation of Hsf1, and therefore increase the Hsp level only in diseased cells.

References

Akbar MT, Lundberg AM, Liu K, Vidyadaran S, Wells KE, Dolatshad H, Wynn S, Wells DJ, Latchman DS, de Belleroche J (2003) The neuroprotective effects of heat shock protein 27 overexpression in transgenic animals against kainate-induced seizures and hippocampal cell death. J Biol Chem 278:19956–19965

Alavez S, Vantipalli MC, Zucker DJ, Klang IM, Lithgow GJ (2011) Amyloid-binding compounds maintain protein homeostasis during ageing and extend lifespan. Nature 472:226–229

Andersen JK (2004) Oxidative stress in neurodegeneration: cause or consequence? Nat Med 10(Suppl):S18–S25, Review

Arac A, Brownell SE, Rothbard JB, Chen C, Ko RM, Pereira MP, Albers GW, Steinman L, Steinberg GK (2011) Systematic augmentation of alpha B crystallin, a novel immunomodulatory target, provides therapeutic benefit twelve hours post-stroke onset. Proc Natl Acad Sci U S A 108:13287–13292

Ariga T, McDonald MP, Yu RK (2008) Role of ganglioside metabolism in the pathogenesis of Alzheimer's disease—a review. J Lipid Res 49:1157–1175

Arispe N, Pollard HB, Rojas E (1993) Giant multilevel cation channels formed by Alzheimer disease amyloid beta-protein [A beta P-(1–40)] in bilayer membranes. Proc Natl Acad Sci U S A 90:10573–10577

Arrasate M, Mitra S, Schweitzer ES, Segal MR, Finkbeiner S (2004) Inclusion body formation reduces levels of mutant huntingtin and the risk of neuronal death. Nature 431:805–810

Arrigo AP (1998) Small stress proteins: chaperones that act as regulators of intracellular redox state and programmed cell death. Biol Chem 379:19–26

Arrigo AP, Paul C, Ducasse C, Manero F, Kretz-Remy C, Virot S, Javouhey E, Mounier N, Diaz-Latoud C (2002) Small stress proteins: novel negative modulators of apoptosis induced independently of reactive oxygen species. Prog Mol Subcell Biol 28:185–204

Arrigo AP, Virot S, Chaufour S, Firdaus W, Kretz-Remy C, Diaz-Latoud C (2005) Hsp27 consolidates intracellular redox homeostasis by upholding glutathione in its reduced form and by decreasing iron intracellular levels. Antioxid Redox Signal 7:414–422, Review

Avdulov NA, Chochina SV, Igbavboa U, Warden CS, Vassiliev AV, Wood WG (1997) Lipid binding to amyloid ß-peptide aggregates: preferential binding of cholesterol as compared with phosphatidylcholine and fatty acids. J Neurochem 69:1746–1752

Badin RA, Lythgoe MF, van der Weerd L, Thomas DL, Gadian DG, Latchman DS (2006) Neuroprotective effects of virally delivered HSPs in experimental stroke. J Cereb Blood Flow Metab 26:371–381

Balogi Z, Török Z, Balogh G, Jósvay K, Shigapova N, Vierling E, Vigh L, Horváth I (2005) "Heat shock lipid" in cyanobacteria during heat/light-acclimation. Arch Biochem Biophys 436:346–354

Balogi Z, Cheregi O, Giese KC, Juhász K, Vierling E, Vass I, Vígh L, Horváth I (2008) A mutant small heat shock protein with increased thylakoid association provides an elevated resistance against UV-B damage in *Synechocystis* 6803. J Biol Chem 283:22983–22991

Bartelt-Kirbach B, Golenhofen N (2014) Reaction of small heat-shock proteins to different kinds of cellular stress in cultured rat hippocampal neurons. Cell Stress Chaperones 19:145–153

Bechtold DA, Brown IR (2000) Heat shock proteins Hsp27 and Hsp32 localize to synaptic sites in the rat cerebellum following hyperthermia. Brain Res Mol Brain Res 75:309–320

Behl C, Davis JB, Lesley R, Schubert D (1994) Hydrogen peroxide mediates amyloid- protein toxicity. Cell 77:817–827

Bence NF, Sampat RM, Kopito RR (2001) Impairment of the ubiquitin-proteasome system by protein aggregation. Science 292:1552–1555

Benchoua A, Guegan C, Couriaud C, Hosseini H, Sampaïo N, Morin D, Onténiente B (2001) Specific caspase pathways are activated in the two stages of cerebral infarction. J Neurosci 21:7127–7134

Bidmon HJ, Görg B, Palomero- Gallagher N, Behne F, Lahl R, Pannek HW, Speckmann EJ, Zilles K (2004) Heat shockprotein- 27 is upregulated in the temporal cortex of patients with epilepsy. Epilepsia 45:1549–1559

Biron KE, Dickstein DL, Gopaul R, Jefferies WA (2011) Amyloid triggers extensive cerebral angiogenesis causing blood brain barrier permeability and hypervascularity in Alzheimer's disease. PLoS ONE 6(8):e23789

Brownell SE, Becker RA, Steinman L (2012) The protective and therapeutic function of small heat shock proteins in neurological diseases. Front Immunol 3:74. doi:10.3389/fimmu.2012.00074. eCollection 2012

Bruey JM, Ducasse C, Bonniaud P, Ravagnan L, Susin SA, Diaz-Latoud C, Gurbuxani S, Arrigo AP, Kroemer G, Solary E, Garrido C (2000) Hsp27 negatively regulates cell death by interacting with cytochrome c. Nat Cell Biol 2:645–652

Calabrese V, Renis M, Calderone A, Russo A, Barcellona ML, Rizza V (1996) Stress proteins and SH-groups in oxidant-induced cell damage after acute ethanol administration in rat. Free Radic Biol Med 20:391–398

Calabrese V, Renis M, Calderone A, Russo A, Reale S, Barcellona ML, Rizza V (1998) Stress proteins and SH-groups in oxidant induced cellular injury after chronic ethanol administration in rat. Free Radic Biol Med 24:1159–1167

Capozzi V, Weidmann S, Fiocco D, Rieu A, Hols P, Guzzo J, Spano G (2011) Inactivation of a small heat shock protein affects cell morphology and membrane fluidity in lactobacillus plantarum WCFS1. Res Microbiol 162:419–425

Carra S, Sivilotti M, Chavez Zobel AT, Lambert H, Landry J (2005) HspB8, a small heat shock protein mutated in human neuromuscular disorders, has in vivo chaperone activity in cultured cells. Hum Mol Genet 14:1659–1669

Carra S, Seguin SJ, Landry J (2008) HspB8 and Bag3: a new chaperone complex targeting misfolded proteins to macroautophagy. Autophagy 4:237–239

Cashikar AG, Duennwald M, Lindquist SL (2005) A chaperone pathway in protein disaggregation. Hsp26 alters the nature of protein aggregates to facilitate reactivation by Hsp104. J Biol Chem 280:23869–23875

Chadwick W, Brenneman R, Martin B, Maudsley S (2010) Complex and multidimensional lipid raft alterations in a murine model of Alzheimer's disease. Int J Alzheimers Dis 2010:604792

Charette SJ, Lavoie JN, Lambert H, Landry J (2000) Inhibition of Daxx-mediated apoptosis by heat shock protein 27. Mol Cell Biol 20:7602–7612

Chauhan D, Li G, Hideshima T, Podar K, Mitsiades C, Mitsiades N, Catley L, Tai YT, Hayashi T, Shringarpure R, Burger R, Munshi N, Ohtake Y, Saxena S, Anderson KC (2003) Hsp27 inhibits release of mitochondrial protein Smac in multiple myeloma cells and confers dexamethasone resistance. Blood 102:3379–3386

Chen M, Ona VO, Li M, Ferrante RJ, Fink KB, Zhu S, Bian J, Guo L, Farrell LA, Hersch SM, Hobbs W, Vonsattel JP, Cha JH, Friedlander RM (2000) Minocycline inhibits caspase-1 and caspase-3 expression and delays mortality in a transgenic mouse model of Huntington disease. Nat Med 6:797–801

Chochina SV, Avdulov NA, Igbavboa U, Cleary JP, O'Hare EO, Wood WG (2001) Amyloid beta-peptide 1–40 increases neuronal membrane fluidity: role of cholesterol and brain region. J Lipid Res 42:1292–1297

Choi J, Levey AI, Weintraub ST, Rees HD, Gearing M, Chin LS, Li L (2004) Oxidative modifications and down-regulation of ubiquitin carboxyl-terminal hydrolase L1 associated with idiopathic Parkinson's and Alzheimer's diseases. J Biol Chem 279:13256–13264

Choo-Smith LP, Garzon-Rodriguez W, Glabe CG, Surewicz WK (1997) Acceleration of amyloid fibril formation by specific binding of Abeta-(1–40) peptide to ganglioside-containing membrane vesicles. J Biol Chem 272:22 987–22 990

Chung J, Nguyen AK, Henstridge DC, Holmes AG, Chan MH, Mesa JL, Lancaster GI, Southgate RJ, Bruce CR, Duffy SJ, Horvath I, Mestril R, Watt MJ, Hooper PL, Kingwell BA, Vigh L, Hevener A, Febbraio MA (2008) HSP72 protects against obesity-induced insulin resistance. Proc Natl Acad Sci U S A 105:1739–1744

Cowan KJ, Diamond MI, Welch WJ (2003) Polyglutamine protein aggregation and toxicity are linked to the cellular stress response. Hum Mol Genet 12:1377–1391

Craft S, Watson GS (2004) Insulin and neurodegenerative disease: shared and specific mechanisms. Lancet Neurol 3:169–178

Crippa V, Sau D, Rusmini P, Boncoraglio A, Onesto E, Bolzoni E, Galbiati M, Fontana E, Marino M, Carra S, Bendotti C, De Biasi S, Poletti A (2010) The small heat shock protein B8 (HspB8) promotes autophagic removal of misfolded proteins involved in amyotrophic lateral sclerosis (ALS). Hum Mol Genet 19:3440–3456

Cuervo AM, Stefanis L, Fredenburg R, Lansbury PT, Sulzer D (2004) Impaired degradation of mutant alpha-synuclein by chaperone-mediated autophagy. Science 305(5688):1292–1295

Dancsó B, Spiró Z, Arslan MA, Nguyen MT, Papp D, Csermely P, Soti C (2010) The heat shock connection of metabolic stress and dietary restriction. Curr Pharm Biotechnol 11:139–145

Dauer W, Przedborski S (2003) Parkinson's disease: mechanisms and models. Neuron 39:889–909

David JC, Boelens WC, Grongnet JF (2006) Up-regulation of heat shock protein HSP 20 in the hippocampus as an early response to hypoxia of the newborn. J Neurochem 99:570–581

de la Monte SM, Wands JR (2002) Chronic gestational exposure to ethanol impairs insulin-stimulated survival and mitochondrial function in cerebellar neurons. Cell Mol Life Sci 59:882–893

de la Monte SM, Wands JR (2008) Alzheimer's disease is type 3 diabetes-evidence reviewed. J Diabetes Sci Technol 2:1101–1113

Dickstein DL, Biron KE, Ujiie M, Pfeifer CG, Jeffries AR, Jefferies WA (2006) A{beta} peptide immunization restores blood-brain barrier integrity in Alzheimer disease. Faseb J 20:426–433

Dierick I, Irobi J, Janssens S, Theuns J, Lemmens R, Jacobs A, Corsmit E, Hersmus N, Van Den Bosch L, Robberecht W, De Jonghe P, Van Broeckhoven C, Timmerman V (2007) Genetic variant in the HSPB1 promoter region impairs the HSP27 stress response. Hum Mutat 28:830

Ding Q, Lewis JJ, Strum KM, Dimayuga E, Bruce-Keller AJ, Dunn JC, Keller JN (2002) Polyglutamine expansion, protein aggregation, proteasome activity, and neural survival. J Biol Chem 277:13935–13942

Dokladny K, Ye D, Kennedy JC, Moseley PL, Ma TY (2008) Cellular and molecular mechanisms of heat stress-induced up-regulation of occludin protein expression: regulatory role of heat shock factor-1. Am J Pathol 172:659–670

Duennwald ML, Echeverria A, Shorter J (2012) Small heat shock proteins potentiate amyloid dissolution by protein disaggregases from yeast and humans. PLoS Biol 10(6):e1001346

Ehrnsperger M, Gräber S, Gaestel M, Buchner J (1997) Binding of non-native protein to Hsp25 during heat shock creates a reservoir of folding intermediates for reactivation. EMBO J 16:221–229

El-Agnaf OM, Jakes R, Curran MD, Middleton D, Ingenito R, Bianchi E, Pessi A, Neill D, Wallace A (1998) Aggregates from mutant and wildtype α-synuclein proteins and NAC peptide induce apoptotic cell death in human neuroblastoma cells by formation of β-sheet and amyloid-like filaments. FEBS Lett 440:71–75

Esterbauer H, Schaur RJ, Zollner H (1991) Chemistry and biochemistry of 4- hydroxynonenal, malonaldehyde and related aldehydes. Free Radic Biol Med 11:81–128

Evgrafov OV, Mersiyanova I, Irobi J, Van Den Bosch L, Dierick I, Leung CL, Schagina O, Verpoorten N, Van Impe K, Fedotov V, Dadali E, Auer-Grumbach M, Windpassinger C, Wagner K, Mitrovic Z, Hilton- Jones D, Talbot K, Martin J-J, Vasserman N, Tverskaya S, Polyakov A, Liem RKH, Gettemans J, Robberecht W, De Jonghe P, Timmerman V (2004) Mutant small heat-shock protein 27 causes axonal Charcot-Marie-Tooth disease and distal hereditary motor neuropathy. Nat Genet 36:602–606

Eyles SJ, Gierasch LM (2010) Nature's molecular sponges: small heat shock proteins grow into their chaperone roles. Proc Natl Acad Sci U S A 107:2727–2728

Farrall AJ, Wardlaw JM (2009) Blood-brain barrier: ageing and microvascular disease–systematic review and meta-analysis. Neurobiol Aging 30:337–352

Fink KB, Andrews LJ, Butler WE, Ona VO, Li M, Bogdanov M, Endres M, Khan SQ, Namura S, Stieg PE, Beal MF, Moskowitz MA, Yuan J, Friedlander RM (1999) Reduction of post-traumatic brain injury and free radical production by inhibition of the caspase-1 cascade. Neuroscience 94:1213–1218

Finka A, Mattoo RU, Goloubinoff P (2011) Meta-analysis of heat- and chemically upregulated chaperone genes in plant and human cells. Cell Stress Chaperones 16:15–31

Fonte V, Kapulkin V, Taft A, Fluet A, Friedman D, Link CD (2002) Interaction of intracellular b amyloid peptide with chaperone proteins. Proc Natl Acad Sci U S A 99:9439–9444

Forloni G, Bugiani O, Tagliavini F, Salmona M (1996) Apoptosis-mediated neurotoxicity induced by beta-amyloid and PrP fragments. Mol Chem Neuropathol 28:163–171

Friedlander RM (2003) Apoptosis and caspases in neurodegenerative diseases. N Engl J Med 348:1365–1375

Friedlander RM, Brown RH, Gagliardini V, Wang J, Yuan J (1997a) Inhibition of ICE slows ALS in mice. Nature 388(31) [Erratum, Nature 1998;392:560.]

Friedlander RM, Gagliardini V, Hara H, Fink KB, Li W, MacDonald G, Fishman MC, Greenberg AH, Moskowitz MA, Yuan J (1997b) Expression of a dominant negative mutant of interleukin-1 beta converting enzyme in transgenic mice prevents neuronal cell death induced by trophic factor withdrawal and ischemic brain injury. J Exp Med 185(5):933–940

Friguet B, Szweda LI (1997) Inhibition of the multicatalytic proteinase (proteasome) by 4-hydroxy-2-nonenal cross-linked protein. FEBS Lett 405:21–25

Fujikake N, Nagai Y, Popiel HA, Okamoto Y, Yamaguchi M, Toda T (2008) Heat shock transcription factor1-activating compounds suppress polyglutamine-induced neurodegeneration through induction of multiple molecular chaperones. J Biol Chem 283:26188–26197

Garrido C, Bruey JM, Fromentin A, Hammann A, Arrigo AP, Solary E (1999) HSP27 inhibits cytochrome c-dependent activation of procaspase-9. FASEB J 13:2061–2070
Garrido C, Schmitt E, Candé C, Vahsen N, Parcellier A, Kroemer G (2003) HSP27 and HSP70: potentially oncogenic apoptosis inhibitors. Cell Cycle 2:579–584
Gervais FG, Xu D, Robertson GS, Vaillancourt JP, Zhu Y, Huang J, LeBlanc A, Smith D, Rigby M, Shearman MS, Clarke EE, Zheng H, Van Der Ploeg LH, Ruffolo SC, Thornberry NA, Xanthoudakis S, Zamboni RJ, Roy S, Nicholson DW (1999) Involvement of caspases in proteolytic cleavage of Alzheimer's amyloid-beta precursor protein and amyloidogenic A beta peptide formation. Cell 97:395–406
Glatz A, Vass I, Los DA, Vigh L (1999) The synechocystis model of stress: from molecular chaperones to membranes. Plant Physiol Biochem 37:1–12
Goldbaum O, Riedel M, Stahnke T, Richter-Landsberg C (2009) The small heat shock protein HSP25 protects astrocytes against stress induced by proteasomal inhibition. Glia 57:1566–1577
Gong B, Cao Z, Zheng P, Vitolo OV, Liu S, Staniszewski A, Moolman D, Zhang H, Shelanski M, Arancio O (2006) Ubiquitin hydrolase Uch-L1 rescues β-amyloid-induced decreases in synaptic function and contextual memory. Cell 126:775–788
Grimm MO, Grimm HS, Tomic I, Beyreuther K, Hartmann T, Bergmann C (2008) Independent inhibition of Alzheimer disease β- and γ-secretase cleavage by lowered cholesterol levels. J Biol Chem 283:11302–11311
Guegan C, Vila M, Rosoklija G, Hays AP, Przedborski S (2001) Recruitment of the mitochondrial-dependent apoptotic pathway in amyotrophic lateral sclerosis. J Neurosci 21:6569–6576
Gupte AA, Bomhoff GL, Swerdlow RH, Geiger PC (2009) Heat treatment improves glucose tolerance and prevents skeletal muscle insulin resistance in rats fed a high-fat diet. Diabetes 58:567–578
Haass C, Selkoe DJ (2007) Soluble protein oligomers in neurodegeneration: lessons from the Alzheimer's amyloid beta-peptide. Nat Rev Mol Cell Biol 8:101–112
Haldimann P, Muriset M, Vígh L, Goloubinoff P (2011) The novel hydroxylamine derivative NG-094 suppresses polyglutamine protein toxicity in Caenorhabditis elegans. J Biol Chem 286:18784–18794
Hargitai J, Lewis H, Boros I, Rácz T, Fiser A, Kurucz I, Benjamin I, Vígh L, Pénzes Z, Csermely P, Latchman DS (2003) Bimoclomol, a heat shock protein co-inducer, acts by the prolonged activation of heatshock factor-1. Biochem Biophys Res Commun 307:689–695
Haslbeck M, Buchner J (2002) Chaperone function of sHSPs. Prog Mol Subcell Biol 28:37–59
Haslbeck M, Walke S, Stromer T, Ehrnsperger M, White HE, Chen S, Saibil HR, Buchner J (1999) Hsp26: a temperature-regulated chaperone. EMBO J 18:6744–6751
Haslbeck M, Miess A, Stromer T, Walter S, Buchner J (2005) Disassembling protein aggregates in the yeast cytosol. The cooperation of Hsp26 with Ssa1 and Hsp104. J Biol Chem 280:23861–23868
Healy DG, Abou-Sleiman PM, Wood NW (2004) PINK, PANK, or PARK? A clinicians' guide to familial parkinsonism. Lancet 3:652–662
Hedhli N, Wang L, Wang Q, Rashed E, Tian Y, Sui X, Madura K, Depre C (2008) Proteasome activation during cardiac hypertrophy by the chaperone H11 Kinase/Hsp22. Cardiovasc Res 77:497–505
Hirakura Y, Lin MC, Kagan BL (1999) Alzheimer amyloid abeta1-42 channels: effects of solvent, pH, and Congo Red. J Neurosci Res 57:458–466
Hooper PL (1999) Hot-tub therapy for type 2 diabetes mellitus. N Engl J Med 341:924–925
Hooper PL, Balogh G, Rivas E, Kavanagh K, Vigh L (2014) The importance of the cellular stress response in the pathogenesis and treatment of type 2 diabetes. Cell Stress Chaperones 19(4):447–464 [Epub ahead of print]
Horváth I, Vígh L (2010) Cell biology: stability in times of stress. Nature 463:436–438
Horváth I, Glatz A, Varvasovszki V, Török Z, Páli T, Balogh G, Kovács E, Nádasdi L, Benkö S, Joó F, Vígh L (1998) Membrane physical state controls the signaling mechanism of the heat shock response in Synechocystis PCC6803: identification of hsp17 as a "fluidity gene". Proc Natl Acad Sci U S A 95:3513–3518

Horváth I, Multhoff G, Sonnleitner A, Vígh L (2008) Membrane-associated stress proteins: more than simply chaperones. Biochim Biophys Acta 1778:1653–1664

Horváth I, Glatz A, Nakamoto H, Mishkind ML, Munnik T, Saidi Y, Goloubinoff P, Harwood JL, Vígh L (2012) Heat shock response in photosynthetic organisms: membrane and lipid connections. Prog Lipid Res 51:208–220

Houlden H, Laura M, Wavrant-DeVrieze F, Blake J, Wood N, Reilly MM (2008) Mutations in the HSP27 (HSPB1) gene cause dominant, recessive, and sporadic distal HMN/CMT type 2. Neurology 71:1660–1668

Hoyer S (2003) Memory function and brain glucose metabolism. Pharmacopsychiatry 36:S62–S67

Hsu LJ, Sagara Y, Arroyo A, Rockenstein E, Sisk A, Mallory M, Wong J, Takenouchi T, Hashimoto M, Masliah E (2000) Alpha-synuclein promotes mitochondrial deficit and oxidative stress. Am J Pathol 157:401–410

Huot J, Houle F, Spitz DR, Landry J (1996) HSP27 phosphorylation-mediated resistance against actin fragmentation and cell death induced by oxidative stress. Cancer Res 56:273–279

Hyun DH, Lee M, Hattori N, Kubo S, Mizuno Y, Halliwell B, Jenner P (2002) Effect of wild-type or mutant Parkin on oxidative damage, nitric oxide, antioxidant defenses, and the proteasome. J Biol Chem 277:28572–28577

Imura T, Shimohama S, Sato M, Nishikawa H, Madono K, Akaike A, Kimura J (1999) Differential expression of small heat shock proteins in reactive astrocytes after focal ischemia: possible role of beta- adrenergic receptor. J Neurosci 19:9768–9779

Irobi J, Van Impe K, Seeman P, Jordanova A, Dierick I, Verpoorten N, Michalik A, De Vriendt E, Jacobs A, Van Gerwen V, Vennekens K, Mazanec R, Tournev I, Hilton-Jones D, Talbot K, Kremensky I, Van Den Bosch L, Robberecht W, Van Vandekerckhove J, Van Broeckhoven C, Gettemans J, De Jonghe P, Timmerman V (2004) Hot-spot residue in small heat-shock protein 22 causes distal motor neuropathy. Nat Genet 36:597–601

Iwaki T, Wisniewski T, Iwaki A, Corbin E, Tomokane N, Tateishi J, Goldman JE (1992) Accumulation of alpha B-crystallin in central nervous system glia and neurons in pathologic conditions. Am J Pathol 140:345–356

Jellinger KA (2000) Cell death mechanisms in Parkinson's disease. J Neural Transm 107:1–29

Jenner P, Olanow CW (1998) Understanding cell death in Parkinson's disease. Ann Neurol 44:S72–S84

Jiang J, Ballinger CA, Wu Y, Dai Q, Cyr DM, Höhfeld J, Patterson C (2001) CHIP is a U-box-dependent E3 ubiquitin ligase: identification of Hsc70 as a target for ubiquitylation. J Biol Chem 276:42938–42944

Kakio A, Nishimoto S, Kozutsumi Y, Matsuzaki K (2003) Formation of a membrane-active form of amyloid beta-protein in raft-like model membranes. Biochem Biophys Res Commun 303:514–518

Kakkar V, Meister-Broekema M, Minoia M, Carra S, Kampinga HH (2014) Barcoding heat shock proteins to human diseases: looking beyond the heat shock response. Dis Model Mech 7:421–434

Kalvodova L, Kahya N, Schwille P, Ehehalt R, Verkade P, Drechsel D, Simons K (2005) Lipids as modulators of proteolytic activity of BACE: involvement of cholesterol, glycosphingolipids, and anionic phospholipids in vitro. J Biol Chem 280:36815–36823

Kampinga HH, Hageman J, Vos MJ, Kubota H, Tanguay RM, Bruford EA, Cheetham ME, Chen B, Hightower LE (2009) Guidelines for the nomenclature of the human heat shock proteins. Cell Stress Chaperones 14:105–111

Kanfer JN, Sorrentino G, Sitar D (1999) Amyloid beta peptide membrane perturbation is the basis for its biological effects. Neurochem Res 24:1621–1630

Kaniuk NA, Kiraly M, Bates H, Vranic M, Volchuk A, Brumell JH (2007) Ubiquitinated-protein aggregates form in pancreatic beta-cells during diabetes-induced oxidative stress and are regulated by autophagy. Diabetes 56:930–939

Kappe G, Franck E, Verschuure P, Boelens WC, Leunissen JA, de Jong WW (2003) The human genome encodes 10 a-crystallin-related small heat shock proteins: HspB1-10. Cell Stress Chaperones 8:53–61

Kato K, Shinohara H, Goto S, Inaguma Y, Morishita R, Asano T (1992) Copurification of small heat shock protein with alpha B crystallin from human skeletal muscle. J Biol Chem 267:7718–7725

Kato H, Liu Y, Kogure K, Kato K (1994) Induction of 27-kDa heat shock protein following cerebral ischemia in a rat model of ischemic tolerance. Brain Res 634:235–244

Kato K, Katoh-Semba R, Takeuchi IK, Ito H, Kamei K (1999) Responses of heat shock proteins hsp27, alphaB-crystallin, and hsp70 in rat brain after kainic acid-induced seizure activity. J Neurochem 73:229–236

Katsuno M, Sang C, Adachi H, Minamiyama M, Waza M, Tanaka F, Doyu M, Sobue G (2005) Pharmacological induction of heat shock proteins alleviates polyglutamine-mediated motor neuron disease. Proc Natl Acad Sci U S A 102:16801–16806

Kavanagh K, Zhang L, Wagner JD (2009) Tissue-specific regulation and expression of heat shock proteins in type 2 diabetic monkeys. Cell Stress Chaperones 14:291–299

Kawahara M, ArispeN Kuroda Y, Rojas E (1997) Alzheimer's disease amyloid beta-protein forms Zn(2+)-sensitive, cation-selective channels across excised membrane patches from hypothalamic neurons. Biophys J 73:67–75

Keck S, Nitsch R, Grune T, Ullrich O (2003) Proteasome inhibition by paired helical filament-tau in brains of patients with Alzheimer's disease. J Neurochem 85:115–122

Keller JN, Hanni KB, Markesbery WR (2000) Impaired proteasome function in Alzheimer's disease. J Neurochem 75(1):436–439

Khandelwal PJ, Herman AM, Moussa CE (2011) Inflammation in the early stages of neurodegenerative pathology. J Neuroimmunol 238:1–11

Kiechle T, Dedeoglu A, Kubilus J, Kowall NW, Beal MF, Friedlander RM, Hersch SM, Ferrante RJ (2002) Cytochrome c and caspase-9 expression in Huntington's disease. Neuromol Med 1:183–195

Kieran D, Kalmar B, Dick JR, Riddoch-Contreras J, Burnstock G, Greensmith L (2004) Treatment with arimoclomol, a coinducer of heat shock proteins, delays disease progression in ALS mice. Nat Med 10:402–405

Kirbach BB, Golenhofen N (2011) Differential expression and induction of small heat shock proteins in rat brain and cultured hippocampal neurons. J Neurosci Res 89:162–175

Klein C, Bobilya D (2008) Transient changes in blood-brain barrier integrity, thermotolerance, and heat shock protein expression following brief hyperthermia in an in vitro model. FASEB J 22(734):736

Klivenyi P, Andreassen O, Ferrante RJ, Schleicher JR Jr, Friedlander RM, Beal MF (1999) Transgenic mice expressing a dominant negative mutant interleukin-1β converting enzyme show resistance to MPTP neurotoxicity. Neuroreport 10:635–638

Kourie JI, Henry CL (2002) Ion channel formation and membrane-linked pathologies of misfolded hydrophobic proteins: the role of dangerous unchaperoned molecules. Clin Exp Pharmacol Physiol 29:741–753

Kourtis N, Nikoletopoulou V, Tavernarakis N (2012) Small heat-shock proteins protect from heatstroke-associated neurodegeneration. Nature 490:213–218

Kremer JJ, Pallitto MM, Sklansky DJ, Murphy RM (2000) Correlation of beta-amyloid aggregate size and hydrophobicity with decreased bilayer fluidity of model membranes. Biochemistry 39:10309–10318

Krueger-Naug AM, Hopkins DA, Armstrong JN, Plumier JC, Currie RW (2000) Hyperthermic induction of the 27-kDa heat shock protein (Hsp27) in neuroglia and neurons of the rat central nervous system. J Comp Neurol 428:495–510

Kudva YC, Hiddinga HJ, Butler PC, Mueske CS, Eberhardt NL (1997) Small heat shock proteins inhibit in vitro Ab1)42 amyloidogenesis. FEBS Lett 416:117–121

Kurthy M, Mogyorosi T, Nagy K, Kukorelli T, Jednakovits A, Talosi L, Biro K (2002) Effect of BRX-220 against peripheral neuropathy and insulin resistance in diabetic rat models. Ann N Y Acad Sci 967:482–489

Kurucz I, Morva A, Vaag A, Eriksson KF, Huang X, Groop L, Koranyi L (2002) Decreased expression of heat shock protein 72 in skeletal muscle of patients with type 2 diabetes correlates with insulin resistance. Diabetes 51:1102–1109

LaFerla FM, Tinkle BT, Bieberich CJ, Haudenschild CC, Jay G (1995) The Alzheimer's A beta peptide induces neurodegeneration and apoptotic cell death in transgenic mice. Nat Genet 9:21–30

Landreth G (2006) PPAR gamma agonists as new therapeutic agents for the treatment of Alzheimer's disease. Exp Neurol 199:245–248

Lashuel HA, Petre BM, Wall J, Simon M, Nowak RJ, Walz T, Lansbury PT Jr (2002) α-Synuclein, especially the Parkinson's disease-associated mutants, forms porelike annular and tubular protofibrils. J Mol Biol 322:1089–1102

Leak RK, Zhang L, Stetler RA, Weng Z, Li P, Atkins GB, Gao Y, Chen J (2013) HSP27 protects the blood-brain barrier against ischemia-induced loss of integrity. CNS Neurol Disord Drug Targets 12:325–337

LeBlanc A (1995) Increased production of 4 kDa amyloid beta peptide in serum deprived human primary neuron cultures: possible involvement of apoptosis. J Neurosci 15:7837–7846

Lee M, Hyun D, Jenner P, Halliwell B (2001) Effect of overexpression of wild-type and mutant Cu/Zn-superoxide dismutases on oxidative damage and antioxidant defences: relevance to Down's syndrome and familial amyotrophic lateral sclerosis. J Neurochem 76:957–965

Lester-Coll N, Rivera EJ, Soscia SJ, Doiron K, Wands JR, de la Monte SM (2006) Intracerebral streptozotocin model of type 3 diabetes: relevance to sporadic Alzheimer's disease. J Alzheimers Dis 9:13–33

Li M, Ona VO, Guegan C, Chen M, Jackson-Lewis V, Andrews LJ, Olszewski AJ, Stieg PE, Lee JP, Przedborski S, Friedlander RM (2000) Functional role of caspase-1 and caspase-3 in an ALS transgenic mouse model. Science 288:335–339

Li DW, Liu JP, Mao YW, Xiang H, Wang J, Ma WY, Dong Z, Pike HM, Brown RE, Reed JC (2005) Calcium- activated RAF/MEK/ERK signaling pathway mediates p53-dependent apoptosis and is abrogated by alpha B-crystallin through inhibition of RAS activation. Mol Biol Cell 16:4437–4453

Liang JJ-N (2000) Interaction between b-amyloid and lens aB-crystallin. FEBS Lett 484:98–101

Lin H, Bhatia R, Lal R (2001) Amyloid beta protein forms ion channels: implications for Alzheimer's disease pathophysiology. Faseb J 15(13):2433–2444

Ling X, Martins RN, Racchi M, Craft S, Helmerhorst E (2002) Amyloid beta antagonizes insulin promoted secretion of the amyloid beta protein precursor. J Alzheimers Dis 4:369–374

Literáti-Nagy B, Kulcsár E, Literáti-Nagy Z, Buday B, Péterfai E, Horváth T, Tory K, Kolonics A, Fleming A, Mandl J, Korányi L (2009) Improvement of insulin sensitivity by a novel drug, BGP-15, in insulin-resistant patients: a proof of concept randomized double-blind clinical trial. Horm Metab Res 41:374–380

Literáti-Nagy Z, Tory K, Literáti-Nagy B, Kolonics A, Török Z, Gombos I, Balogh G, Vígh L Jr, Horváth I, Mandl J, Sümegi B, Hooper PL, Vígh L (2012) The HSP co-inducer BGP-15 can prevent the metabolic side effects of the atypical antipsychotics. Cell Stress Chaperones 17:517–521

Liu S, Li J, Tao Y, Xiao X (2007) Small heat shock protein alphaB- crystallin binds to p53 to sequester its translocation to mitochondria during hydrogen peroxide-induced apoptosis. Biochem Biophys Res Commun 354:109–114

Loo D, Copani A, Pike C, Whittemore E, Walencewicz A, Cotman CW (1993) Apoptosis is induced by β-amyloid in cultured central nervous system neurons. Proc Natl Acad Sci U S A 90:7951–7955

Lowe J, Errington DR, Lennox G, Pike I, Spendlove I, Landon M, Mayer RJ (1992) Ballooned neurons in several neurodegenerative diseases and stroke contain alpha B crystallin. Neuropathol Appl Neurobiol 18:341–350

Lu DC, Rabizadeh S, Chandra S, Shayya RF, Ellerby LM, Ye X, Salvesen GS, Koo EH, Bredesen DE (2000) A second cytotoxic proteolytic peptide derived from amyloid precursor protein. Nat Med 6:397–404

Lu TS, Chen HW, Huang MH, Wang SJ, Yang RC (2004) Heat shock treatment protects osmotic stress-induced dysfunction of the blood-brain barrier through preservation of tight junction proteins. Cell Stress Chaperones 9:369–377

Luders J, Demand J, Hohfeld J (2000) The ubiquitin-related BAG-1 provides a link between the molecular chaperones Hsc70/Hsp70 and the proteasome. J Biol Chem 275:4613–4617

Maitre M, Weidmann S, Dubois-Brissonnet F, David V, Covès J, Guzzo J (2014) Adaptation of the wine bacterium oenococcus oeni to ethanol stress: role of the small heat shock protein Lo18 in membrane integrity. Appl Environ Microbiol 80:2973–2980

Manaenko A, Fathali N, Chen H, Suzuki H, Williams S, Zhang JH, Tang J (2010) Heat shock protein70 upregulation by geldanamycin reduces brain injury in a mouse model of intracerebral hemorrhage. Neurochem Int 57:844–850

Martin LJ (1999) Neuronal death in amyotrophic lateral sclerosis is apoptosis: possible contribution of a programmed cell death mechanism. J Neuropathol Exp Neurol 58:459–471

Masliah E, Mallory M, Alford M, Tanaka S, Hansen LA (1998) Caspase dependent DNA fragmentation might be associated with excitotoxicity in Alzheimer disease. J Neuropathol Exp Neurol 57:1041–1052

Mason RP, Jacob RF, Walter MF, Mason PE, Avdulov NA, Chochina SV, Igbavboa U, Wood WG (1999) Distribution and fluidizing action of soluble and aggregated amyloid beta-peptide in rat synaptic plasma membranes. J Biol Chem 274:18801–18807

Mattson MP (2000) Apoptosis in neurodegenerative disorders. Nat Rev Mol Cell Biol 1:120–129

Mehlen P, Kretz-Remy C, Préville X, Arrigo AP (1996) Human hsp27, Drosophila hsp27 and human _B-crystallin expression-mediated increase in glutathione is essential for the protective activity of these protein against TNF-induced cell death. EMBO J 15:2695–2706

Mehlen P, Hickey E, Weber LA, Arrigo AP (1997) Large unphosphorylated aggregates as the active form of hsp27 which controls intracellular reactive oxygen species and glutathione levels and generates a protection against TNFalpha in NIH-3 T3-ras cells. Biochem Biophys Res Commun 241:187–192

Minami M, Mizutani T, Kawanishi R, Suzuki Y, Mori H (2003) Neuronal expression of alpha B crystallin in cerebral infarction. Acta Neuropathol 105:549–554

Monoi H, Futaki S, Kugimiya S, Minakata H, Yoshihara K (2000) Poly-L-glutamine forms cation channels: relevance to the pathogenesis of the polyglutamine diseases. Biophys J 78:2892–2899

Montoliu C, Vallés S, Renau-Piqueras J, Guerri C (1994) Ethanol-induced oxygen radical formation and lipid peroxidation in rat brain: effect of chronic alcohol consumption. J Neurochem 63:1855–1862

Morimoto RI, Sarge KD, Abravaya K (1992) Transcriptional regulation of heat shock genes. A paradigm for inducible genomic responses. J Biol Chem 267:21987–21990

Moroz N, Tong M, Longato L, Xu H, de la Monte SM (2008) Limited Alzheimer-type neurodegeneration in experimental obesity and type 2 diabetes mellitus. J Alzheimers Dis 15:29–44

Muchowski PJ, Wacker JL (2005) Modulation of neurodegeneration by molecular chaperones. Nat Rev Neurosci 6:11–22

Muller WE, Koch S, Eckert A, Hartmann H, Scheuer K (1995) Beta-amyloid peptide decreases membrane fluidity. Brain Res 674:133–136

Murshid A, Eguchi T, Calderwood SK (2013) Stress proteins in aging and life span. Int J Hyperthermia 29:442–447

Mymrikov EV, Seit-Nebi AS, Gusev NB (2011) Large potentials of small heat shock proteins. Physiol Rev 91:1123–1159

Nagai Y, Onodera O, Chun J, Strittmatter WJ, Burke JR (1999) Expanded polyglutamine domain proteins bind neurofilament and alter the neurofilament network. Exp Neurol 155(2):195–203

Nakamoto H, Vígh L (2007) The small heat shock proteins and their clients. Cell Mol Life Sci 64:294–306

Niwa M, Hara A, Taguchi A, Aoki H, Kozawa O, Mori H (2009) Spatiotemporal expression of Hsp20 and its phosphorylation in hippocampal CA1 pyramidal neurons after transient forebrain ischemia. Neurol Res 31:721–727

Nixon RA (2006) Autophagy in neurodegenerative disease: friend, foe or turn coat? Trends Neurosci 29:528–535

Okada K, Wangpoengtrakul C, Osawa T, Toyokuni S, Tanaka K, Uchida K (1999) 4-Hydroxy-2-nonenal-mediated impairment of intracellular proteolysis during oxidative stress. Identification of proteasomes as target molecules. J Biol Chem 274:23787–23793

Ona VO, Li M, Vonsattel JP, Andrews LJ, Khan SQ, Chung WM, Frey AS, Menon AS, Li XJ, Stieg PE, Yuan J, Penney JB, Young AB, Cha JH, Friedlander RM (1999) Inhibition of caspase-1 slows disease progression in a mouse model of Huntington's disease. Nature 399:263–267

Opattova A, Filipcik P, Cente M, Novak M (2013) Intracellular degradation of misfolded tau protein induced by geldanamycin is associated with activation of proteasome. J Alzheimers Dis 33:339–348

Outeiro TF, Klucken J, Strathearn KE, Liu F, Nguyen P, Rochet JC, Hyman BT, McLean PJ (2006) Small heat shock proteins protect against alpha-synuclein-induced toxicity and aggregation. Biochem Biophys Res Commun 351:631–638

Parcellier A, Schmitt E, Gurbuxani S, Seigneurin-Berny D, Pance A, Chantome A, Plenchette S, Khochbin S, Solary E, Garrido C (2003) HSP27 is a ubiquitin-binding protein involved in I-kappa Balpha proteasomal degradation. Mol Cell Biol 23:5790–5802

Parcellier A, Brunet M, Schmitt E, Col E, Didelot C, Hammann A, Nakayama K, Nakayama KI, Khochbin S, Solary E, Garrido C (2006) HSP27 favors ubiquitination and proteasomal degradation of p27Kip1 and helps S-phase re-entry in stressed cells. FASEB J 20:1179–1181

Park YM, Han MY, Blackburn RV, Lee YJ (1998) Overexpression of HSP25 reduces the level of TNF alpha-induced oxidative DNA damage biomarker, 8-hydroxy-2′-deoxyguanosine, in L929 cells. J Cell Physiol 174:27–34

Pasquier F, Boulogne A, Leys D, Fontaine P (2006) Diabetes mellitus and dementia. Diabetes Metab 32:403–414

Paul C, Manero F, Gonin S, Kretz-Remy C, Virot S, Arrigo AP (2002) Hsp27 as a negative regulator of cytochrome C release. Mol Cell Biol 22:816–834

Paul J, Strickland S, Melchor JP (2007) Fibrin deposition accelerates neurovascular damage and neuroinflammation in mouse models of Alzheimer's disease. J Exp Med 204:1999–2008

Penke B, Tóth AM, Földi I, Szűcs M, Janáky T (2012) Intraneuronal β-amyloid and its interactions with proteins and subcellular organelles. Electrophoresis 33:3608–3616

Perry G, Friedman R, Shaw G, Chau V (1987) Ubiquitin is detected in neurofibrillary tangles and senile plaque neurites of Alzheimer disease brains. Proc Natl Acad Sci U S A 84:3033–3036

Peters I, Igbavboa U, Schütt T, Haidari S, Hartig U, Böttner S, Copanaki E, Deller T, Kögel D, Wood WG, Müller WE, Eckert GP (2009) The interaction of beta-amyloid protein with cellular membranes stimulates its own production. Biochim Biophys Acta 1788:964–972

Piper PW (1995) The heat shock and ethanol stress responses of yeast exhibit extensive similarity and functional overlap. FEMS Microbiol Lett 134:121–127

Préville X, Gaestel M, Arrigo AP (1998) Phosphorylation is not essential for protection of L929 cells by Hsp25 against H2O2-mediated disruption of actin cytoskeleton, a protection which appears related to the redox change mediated by Hsp25. Cell Stress Chaperones 3:177–187

Preville X, Salvemini F, Giraud S, Chaufour S, Paul C, Stepien G, Ursini MV, Arrigo AP (1999) Mammalian small stress proteins protect against oxidative stress through their ability to increase glucose-6-phosphate dehydrogenase activity and by maintaining optimal cellular detoxifying machinery. Exp Cell Res 247:61–78

Prusiner SB (1998) Prions. Proc Natl Acad Sci U S A 95:13363–13383

Rabuffetti M, Sciorati C, Tarozzo G, Clementi E, Manfredi AA, Beltramo M (2000) Inhibition of caspase-1-like activity by Ac-Tyr-Val-Ala-Asp-chloromethyl ketone induces long-lasting neuroprotection in cerebral ischemia through apoptosis reduction and decrease of proinflammatory cytokines. J Neurosci 20:4398–4404

Reed TT (2011) Lipid peroxidation and neurodegenerative disease. Free Radic Biol Med 51(7):1302–1319

Reger MA, Watson GS, Green PS, Wilkinson CW, Baker LD, Cholerton B, Fishel MA, Plymate SR, Breitner JC, DeGroodt W, Mehta P, Craft S (2008) Intranasal insulin improves cognition and modulates beta-amyloid in early AD. Neurology 70:440–448

Rekas A, Adda CG, Andrew Aquilina J, Barnham KJ, Sunde M, Galatis D, Williamson NA, Masters CL, Anders RF, Robinson CV, Cappai R, Carver JA (2004) Interaction of the molecular chaperone alphaB-crystallin with alpha-synuclein: effects on amyloid fibril formation and chaperone activity. J Mol Biol 340:1167–1183

Renkawek K, deJong WW, Merck KB, Frenken CW, van Workum FP, Bosman GJ (1992) Alpha B-crystallin is present in reactive glia in Creutzfeldt-Jakob disease. Acta Neuropathol 83:324–327

Renkawek K, Bosman GJ, de Jong WW (1994a) Expression of small heat-shock protein hsp27 in reactive gliosis in Alzheimer disease and other types of dementia. Acta Neuropathol 87:511–519

Renkawek K, Voorter CEM, Bosman GJ, van Workum FPA, de Jong WW (1994b) Expression of αB-crystallin in Alzheimer's disease. Acta Neuropathol 87:155–160

Renkawek K, Stege GJ, Bosman GJ (1999) Dementia, gliosis and expression of the small heat shock proteins hsp27 and alpha B-crystallin in Parkinson's disease. Neuroreport 10:2273–2276

Rivera EJ, Goldin A, Fulmer N, Tavares R, Wands JR, de la Monte SM (2005) Insulin and insulin-like growth factor expression and function deteriorate with progression of Alzheimer's disease: link to brain reductions in acetylcholine. J Alzheimers Dis 8:247–268

Rochet JC, Outeiro TF, Conway KA, Ding TT, Volles MJ, Lashuel HA, Bieganski RM, Lindquist SL, Lansbury PT (2004) Interactions among alpha-synuclein, dopamine, and biomembranes. J Mol Neurosci 23:23–33

Rogalla T, Ehrnsperger M, Preville X, Kotlyarov A, Lutsch G, Ducasse C, Paul C, Wieske M, Arrigo AP, Buchner J, Gaestel M (1999) Regulation of HSP27 oligomerization, chaperone function, and protective activity against oxidative stress/tumor necrosis factor a by phosphorylation. J Biol Chem 274:18947–18956

Ross CA, Pickart CM (2004) The ubiquitin-proteasome pathway in Parkinson's disease and other neurodegenerative diseases. Trends Cell Biol 14:703–711

Rowland LP, Shneider NA (2001) Amyotrophic lateral sclerosis. N Engl J Med 344:1688–1700

Samali A, Cotter TG (1996) Heat shock proteins increase resistance to apoptosis. Exp Cell Res 223:163–170

Schengrund CL (2010) Lipid rafts: keys to neurodegeneration. Brain Res Bull 82:7–17

Seiler AE, Ross BN, Rubin R (2001) Inhibition of insulin-like growth factor-1 receptor and IRS-2 signaling by ethanol in SH-SY5Y neuroblastoma cells. J Neurochem 76:573–581

Sergent O, Pereira M, Belhomme C, Chevanne M, Huc L, Lagadic-Gossmann D (2005) Role for membrane fluidity in ethanol-induced oxidative stress of primary rat hepatocytes. J Pharmacol Exp Ther 313:104–111

Seydlová G, Halada P, Fišer R, Toman O, Ulrych A, Svobodová J (2012) DnaK and GroEL chaperones are recruited to the Bacillus subtilis membrane after short-term ethanol stress. J Appl Microbiol 112:765–774

Sharp PS, Akbar MT, Bouri S, Senda A, Joshi K, Chen HJ, Latchman DS, Wells DJ, de Belleroche J (2008) Protective effects of heat shock protein 27 in a model of ALS occur in the early stages of disease progression. Neurobiol Dis 30:42–55

Shimura H, Miura-Shimura Y, Kosik KS (2004) Binding of tau to heat shock protein 27 leads to decreased concentration of hyperphosphorylated tau and enhanced cell survival. J Biol Chem 279:17957–17962

Shin JH, Kim SW, Lim CM, Jeong JY, Piao CS, Lee JK (2009) alphaB-crystallin suppresses oxidative stress-induced astrocyte apoptosis by inhibiting caspase-3 activation. Neurosci Res 64:355–361

Shinder GA, Lacourse M-C, Minotti S, Durham HD (2001) Mutant Cu/Zn-superoxide dismutase proteins have altered solubility and interact with heat shock/stress proteins in models of amyotrophic lateral sclerosis. J Biol Chem 276:12791–12796

Shinohara H, Inaguma Y, Goto S, Inagaki T, Kato K (1993) Alpha B crystallin and HSP28 are enhanced in the cerebral cortex of patients with Alzheimer's disease. J Neurol Sci 119:203–208

Shirwany NA, Payette D, Xie J, Guo Q (2007) The amyloid beta ion channel hypothesis of Alzheimer's disease. Neuropsychiatr Dis Treat 3:597–612

Simons M, Keller P, De Strooper B, Beyreuther K, Dotti CG, Simons K (1998) Cholesterol depletion inhibits the generation of beta-amyloid in hippocampal neurons. Proc Natl Acad Sci U S A 95:6460–6464

Snyder H, Mensah K, Theisler C, Lee J, Matouschek A, Wolozin B (2003) Aggregated and monomeric alpha-synuclein bind to the S6′ proteasomal protein and inhibit proteasomal function. J Biol Chem 278(14):11753–11759

Söti C, Csermely P (2002) Chaperones and aging: role in neurodegeneration and in other civilizational diseases. Neurochem Int 41:383–389

Sreedhar AS, Csermely P (2004) Heat shock proteins in the regulation of apoptosis: new strategies in tumor therapy: a comprehensive review. Pharmacol Ther 101:227–257

Steen E, Terry BM, Rivera EJ, Cannon JL, Neely TR, Tavares R, Xu XJ, Wands JR, de la Monte SM (2005) Impaired insulin and insulin-like growth factor expression and signaling mechanisms in Alzheimer's disease—is this type 3 diabetes? J Alzheimers Dis 7:63–80

Steffan JS, Kazantsev A, Spasic-Boskovic O, Greenwald M, Zhu YZ, Gohler H, Wanker EE, Bates GP, Housman DE, Thompson LM (2000) The Huntington's disease protein interacts with p53 and CREB-binding protein and represses transcription. Proc Natl Acad Sci U S A 97:6763–6768

Stege GJ, Renkawek K, Overkamp PS, Verschuure P, van Rijk AF, Reijnen-Aalbers A, Boelens WC, Bosman GJ, de Jong WW (1999) The molecular chaperone alphaB-crystallin enhances amyloid beta neurotoxicity. Biochem Biophys Res Commun 262:152–156

Strey CW, Spellman D, Stieber A, Gonatas JO, Wang X, Lambris JD, Gonatas NK (2004) Dysregulation of stathmin, a microtubule-destabilizing protein, and up-regulation of Hsp25, Hsp27, and the antioxidant peroxiredoxin 6 in a mouse model of familial amyotrophic lateral sclerosis. Am J Pathol 165:1701–1718

Su JH, Anderson AJ, Cummings B, Cotman CW (1994) Immunocytochemical evidence for apoptosis in Alzheimer's disease. Neuroreport 5:2529–2533

Sun Y, MacRae TH (2005) The small heat shock proteins and their role in human disease. FEBS J 272:2613–2627

Tabrizi SJ, Workman J, Hart PE, Mangiarini L, Mahal A, Bates G, Cooper JM, Schapira AH (2000) Mitochondrial dysfunction and free radical damage in the Huntington R6/2 transgenic mouse. Ann Neurol 47:80–86

Tang BS, Zhao GH, Luo W, Xia K, Cai F, Pan Q, Zhang RX, Zhang FF, Liu XM, Chen B, Zhang C, Shen L, Jiang H, Long ZG, Dai HP (2005) Small heat-shock protein 22 mutated in autosomal dominant Charcot-Marie-Tooth disease type 2L. Hum Genet 116:222–224

Tanno H, Nockels RP, Pitts LH, Noble LJ (1993) Immunolocalization of heat shock protein after fluid percussive brain injury and relationship to breakdown of the blood-brain barrier. J Cereb Blood Flow Metab 13:116–124

Török Z, Horvath I, Goloubinoff P, Kovács E, Glatz A, Balogh G, Vígh L (1997) Evidence for a lipochaperonin: association of active proteinfolding GroESL oligomers with lipids can stabilize membranes under heat shock conditions. Proc Natl Acad Sci U S A 94:2192–2197

Török Z, Goloubinoff P, Horvath I, Tsvetkova NM, Glatz A, Balogh G, Varvasovszki V, Los DA, Vierling E, Crowe JH, Vigh L (2001) Synechocystis HSP17 is an amphitropic protein that stabilizes heatstressed membranes and binds denatured proteins for subsequent chaperone-mediated refolding. Proc Natl Acad Sci U S A 98:3098–3103

Török Z, Pilbat A-M, Gombos I, Hocsak E, Sümegi B, Horváth I, Vígh L (2012) Evidence on cholesterol-controlled lipid raft interaction of the small heat shock protein Hspb11. In: Handerson B, Pockley AD (eds) Cellular trafficking of cell stress proteins in health and disease. Springer, Dordrecht, pp 75–86. ISBN 978-9400747395

Tortosa R, Vidal E, Costa C, Alamillo E, Torres JM, Ferrer I, Pumarola M (2008) Stress response in the central nervous system of a transgenic mouse model of bovine spongiform encephalopathy. Vet J 178:126–129
Toth ME, Gonda S, Vigh L, Santha M (2010) Neuroprotective effect of small heat shock protein, Hsp27, after acute and chronic alcohol administration. Cell Stress Chaperones 15:807–817
Toth ME, Szegedi V, Varga E, Juhasz G, Horvath J, Borbely E, Csibrány B, Alföldi R, Lenart N, Penke B, Santha M (2013) Overexpression of Hsp27 ameliorates symptoms of Alzheimer's Disease in APP/PS1 mice. Cell Stress Chaperones 18:759–771
Tseng BP, Green KN, Chan JL, Blurton-Jones M, LaFerla FM (2008) Abeta inhibits the proteasome and enhances amyloid and tau accumulation. Neurobiol Aging 29:1607–1618
Tsvetkova NM, Horváth I, Török Z, Wolkers WF, Balogi Z, Shigapova N, Crowe LM, Tablin F, Vierling E, Crowe JH, Vigh L (2002) Small heat-shock proteins regulate membrane lipid polymorphism. Proc Natl Acad Sci U S A 99:13504–13509
Ujiie M, Dickstein D, Carlow D, Jefferies WA (2003) Blood-brain barrier permeability precedes senile plaque formation in an Alzheimer disease model. Microcirculation 10:463–470
Urban MJ, Dobrowsky RT, Blagg BS (2012) Heatshock response and insulin-associated neurodegeneration. Trends Pharmacol Sci 33:129–137
Uysal M, Kutalp G, Ozdemirler G, Aykac G (1989) Ethanol-induced changes in lipid peroxidation and glutathione content in rat brain. Drug Alcohol Depend 23:227–230
van der Weerd L, Tariq Akbar M, Aron Badin R, Valentim LM, Thomas DL, Wells DJ, Latchman DS, Gadian DG, Lythgoe MF, de Belleroche JS (2010) Overexpression of heat shock protein 27 reduces cortical damage after cerebral ischemia. J Cereb Blood Flow Metab 30:849–856
Verdier Y, Zarándi M, Penke B (2004) Amyloid beta-peptide interactions with neuronal and glial cell plasma membrane: binding sites and implications for Alzheimer's disease. J Pept Sci 10:229–248
Vidal E, Acín C, Foradada L, Monzón M, Márquez M, Monleón E, Pumarola M, Badiola JJ, Bolea R (2009) Immunohistochemical characterisation of classical scrapie neuropathology in sheep. J Comp Pathol 141:135–146
Vinters HV (1987) Cerebral amyloid angiopathy. A critical review. Stroke 18:311–324
Vleminckx V, Van Damme P, Goffin K, Delye H, Van Den Bosch L, Robberecht W (2002) Upregulation of HSP27 in a transgenic model of ALS. J Neuropathol Exp Neurol 61:968–974
Vos MJ, Zijlstra MP, Kanon B, van Waarde-Verhagen MA, Brunt ER, Oosterveld-Hut HM, Carra S, Sibon OC, Kampinga HH (2010) HSPB7 is the most potent polyQ aggregation suppressor within the HSPB family of molecular chaperones. Hum Mol Genet 19:4677–4693
Wagstaff MJ, Collaco-Moraes Y, Smith J, de Belleroche JS, Coffin RS, Latchman DS (1999) Protection of neuronal cells from apoptosis by Hsp27 delivered with a herpes simplex virus-based vector. J Biol Chem 274:5061–5069
Wang K, Spector A (1996) α-Crystallin stabilizes actin filaments and prevents cytochalasin-induced depolymerization in a phosphorylation-dependent manner. Eur J Biochem 242:56–66
Wang J, Martin E, Gonzales V, Borchelt DR, Lee MK (2008) Differential regulation of small heat shock proteins in transgenic mouse models of neurodegenerative diseases. Neurobiol Aging 29:586–597
Waschuk SA, Elton EA, Darabie AA, Fraser PE, McLaurin JA (2001) Cellular membrane composition defines a beta–lipid interactions. J Biol Chem 276:33561–33568
Wilhelmus MM, Otte-Höller I, Wesseling P, de Waal RM, Boelens WC, Verbeek MM (2006a) Specific association of small heatshock proteins with the pathological hallmarks of Alzheimer's disease brains. Neuropathol Appl Neurobiol 32:119–130
Wilhelmus MM, Boelens WC, Otte-Holler I, Kamps B, Kusters B, Maat-Schieman ML, de Waal RM, Verbeek MM (2006b) Small heat shock protein HspB8: its distribution in Alzheimer's disease brains and its inhibition of amyloid-beta protein aggregation and cerebrovascular amyloid-beta toxicity. Acta Neuropathol (Berl) 111:139–149

Wilhelmus MM, Boelens WC, Otte-Holler I, Kamps B, de Waal RM, Verbeek MM (2006c) Small heat shock proteins inhibit amyloid-beta protein aggregation and cerebrovascular amyloid beta protein toxicity. Brain Res 1089:67–78

Wilhelmus MM, de Waal RM, Verbeek MM (2007) Heat shock proteins and amateur chaperones in amyloid-Beta accumulation and clearance in Alzheimer's disease. Mol Neurobiol 35:203–216

Winocur G, Greenwood CE, Piroli GG, Grillo CA, Reznikov LR, Reagan LP, McEwen BS (2005) Memory impairment in obese Zucker rats: an investigation of cognitive function in an animal model of insulin resistance and obesity. Behav Neurosci 119:1389–1395

Wood WG, Schroeder F, Igbavboa U, Avdulov NA, Chochina SV (2002) Brain membrane cholesterol domains, aging and amyloid beta-peptides. Neurobiol Aging 23:685–694

Wyttenbach A, Arrigo AP (2009) The role of heat shock proteins during neurodegeneration in Alzheimer's, Parkinson's and Huntington's disease. In: Richter-Landsberg C (ed) Heat shock proteins in neural cells, Neuroscience Intelligence Unit. pp 81–99, Springer New York

Wyttenbach A, Sauvageot O, Carmichael J, Diaz-Latoud C, Arrigo AP, Rubinsztein DC (2002) Heat shock protein 27 prevents cellular polyglutamine toxicity and suppresses the increase of reactive oxygen species caused by huntingtin. Hum Mol Genet 11:1137–1151

Xie L, Helmerhorst E, Taddei K, Plewright B, Van Bronswijk W, Martins R (2002) Alzheimer's beta-amyloid peptides compete for insulin binding to the insulin receptor. J Neurosci 22:RC221

Yamamoto N, Matsuzaki K, Yanagisawa K (2005) Cross seeding of wild-type and hereditary variant-type amyloid β- proteins in the presence of gangliosides. J Neurochem 95:1167–1176

Yew EH, Cheung NS, Choy MS, Qi RZ, Lee AY, Peng ZF, Melendez AJ, Manikandan J, Koay ES, Chiu LL, Ng WL, Whiteman M, Kandiah J, Halliwell B (2005) Proteasome inhibition by lactacystin in primary neuronal cells induces both potentially neuroprotective and proapoptotic transcriptional responses: a microarray analysis. J Neurochem 94:943–956

Yip CM, Elton EA, Darabie AA, Morrison MR, McLaurin J (2001) Cholesterol, a modulator of membrane-associated Aβ-fibrillogenesis and neurotoxicity. J Mol Biol 311:723–734

Yoo BC, Kim SH, Cairns N, Fountoulakis M, Lubec G (2001) Deranged expression of molecular chaperones in brains of patients with Alzheimer's disease. Biochem Biophys Res Commun 280:249–258

Yu WH, Kumar A, Peterhoff C, Shapiro Kulnane L, Uchiyama Y, Lamb BT, Cuervo AM, Nixon RA (2004) Autophagic vacuoles are enriched in amyloid precursor protein-secretase activities: implications for beta-amyloid peptide over-production and localization in Alzheimer's disease. Int J Biochem Cell Biol 36:2531–2540

Zantema A, Verlaan-De Vries M, Maasdam D, Bol S, van der Eb A (1992) Heat shock protein 27 and alpha B-crystallin can form a complex, which dissociates by heat shock. J Biol Chem 267:12936–12941

Zhang FX, Rubin R, Rooney TA (1998) Ethanol induces apoptosis in cerebellar granule neurons by inhibiting insulin-like growth factor 1 signaling. J Neurochem 71:196–204

Zlokovic BV (2008) New therapeutic targets in the neurovascular pathway in Alzheimer's disease. Neurotherapeutics 5:409–414

Zoghbi HY, Orr HT (1999) Polyglutamine diseases: protein cleavage and aggregation. Curr Opin Neurobiol 9:566–570

Zourlidou A, Payne Smith MD, Latchman DS (2004) HSP27 but not HSP70 has a potent protective effect against alpha-synuclein- induced cell death in mammalian neuronal cells. J Neurochem 88:1439–1448

Zourlidou A, Gidalevitz T, Kristiansen M, Landles C, Woodman B, Wells DJ, Latchman DS, de Belleroche J, Tabrizi SJ, Morimoto RI, Bates GP (2007) Hsp27 overexpression in the R6/2 mouse model of Huntington's disease: chronic neurodegeneration does not induce Hsp27 activation. Hum Mol Genet 16:1078–1090

Chapter 24
Small Heat Shock Proteins and Diapause in the Crustacean, *Artemia franciscana*

Thomas H. MacRae

Abstract Diapause is a physiological state of metabolic depression and increased stress tolerance observed in many organisms, and especially prominent in insects. The small heat shock proteins (sHsps) mediate protein storage as well as stress tolerance and their accumulation within cells is controlled during diapause. Partially denatured proteins are bound by sHsps preventing their irreversible denaturation, and these proteins are released to Hsp70 and other molecular chaperones for either ATP-dependent folding or degradation. During diapause, which is characterized by exposure to varying levels of stress, proteins are sequestered by sHsps, thus, by enhancing sHsp synthesis the protection of other proteins is maximized. In the crustacean, *Artemia franciscana*, diapause occurs in encysted gastrula stage embryos (cysts), entailing a profound reduction in metabolism and extreme stress tolerance. Three sHsps, namely p26, ArHsp21 and ArHsp22, accumulate to varying levels in diapause-destined *A. franciscana* embryos. Experiments performed in vivo by the use of RNA interference (RNAi) demonstrate that p26 has an important role in stress tolerance while also influencing embryo development and diapause maintenance. The activities of ArHsp21 and ArHsp22 are less well defined although the former sHsp may have a minor role in stress tolerance. The data described herein reveal that the activities of sHsps during diapause are more diverse than previously thought and they contribute to the general understanding of sHsp function.

Keywords Diapause • Small heat shock protein • Stress tolerance • Protein protection • *Artemia*

24.1 Diapause, an Alternate Developmental Pathway

Diapause is a widely distributed physiological process of developmental delay characterized by metabolic suppression, dormancy and increased stress tolerance (Denlinger 2002; Košťál 2006; MacRae 2010; Hahn and Denlinger 2011;

T.H. MacRae (✉)
Department of Biology, Dalhousie University,
1355 Oxford St., Halifax, NS B3H 4R2, Canada
e-mail: tmacrae@dal.ca

R.M. Tanguay, L.E. Hightower (eds.), *The Big Book on Small Heat Shock Proteins*, Heat Shock Proteins 8, DOI 10.1007/978-3-319-16077-1_24

Ptak et al. 2012). In some organisms diapause is induced by photoperiod, temperature and crowding, while in others diapause occurs independent of environmental conditions. Survival under adverse circumstances is improved during diapause (Rinehart et al. 2007; Denekamp et al. 2009; MacRae 2010; Clark et al. 2012). Diapause can be divided into the overlapping periods of initiation, maintenance and termination (Košťál 2006), with initiation driven by differential gene expression leading to altered feeding regimes, changes in the synthesis and activity of cell proteins, and modification of behavior and morphology. Subsequent to initiation, organisms enter maintenance which coincides with the greatest stress tolerance and the most profound dormancy, the intensity of both varying substantially from one organism to another (Tammariello and Denlinger 1998; Reynolds and Hand 2009). The last phase of diapause is termination where either continuation of development or quiescence ensues, depending respectively on whether conditions for growth are favorable or unfavorable. Stress tolerance persists during quiescence but wanes as metabolism and growth resume.

Differential gene expression during diapause modifies the production of many proteins some of which are essential to the process (MacRae 2010; Hahn and Denlinger 2011; King and MacRae 2012). In particular, molecular chaperone mRNAs and proteins, including the sHsps, either increase or decrease in amount or remain constant throughout diapause, with changes occurring before stress is encountered. Throughout diapause, and for cells exposed to stress such as heat, cold, anoxia and drying, molecular chaperones shield cells from damage that may lead to death, however, the responses to diapause and abiotic stress are not identical. Exposure to abiotic stressors decreases overall protein synthesis in cells, an effect not necessarily seen upon entry into diapause. Moreover, changes in molecular chaperones are more persistent in diapause as opposed to stressed cells, and chaperones may be down-regulated during diapause, a response not seen in response to abiotic stress (Rinehart et al. 2007).

24.2 Diapause in the Extremophile Crustacean, *Artemia franciscana*

Embryos of *A. franciscana* undergo either ovoviviparous or oviparous development, resulting respectively in the production of swimming nauplii and gastrulae enclosed in a chitinous shell termed cysts (Liang and MacRae 1999; MacRae 2003; Qiu and MacRae 2010). Nauplii undergo several molts, developing into adults, whereas upon release from females, cysts undergo profound reduction in metabolism over several days (Clegg et al. 1996; Clegg and Jackson 1998; Patil et al. 2013) and they enter diapause characterized by tolerance to physiological and environmental stressors. As examples of stress tolerance, diapause embryos of *A. franciscana* withstand repeated freezing and thawing, drying, heat and years of anoxia (Clegg 1997, 2005, 2007, 2011; Clegg and Jackson 1998; Hengherr et al. 2011) during which their protein composition undergoes little apparent change (Clegg 2007). Cold and/or dehydration end diapause in *A. franciscana* but the molecular basis of termination is obscure (Robbins et al. 2010).

How proteins are protected from irreversible denaturation during stress associated with diapause in *A. franciscana* has received considerable attention. *A. franciscana* cysts contain high levels of trehalose (Clegg and Jackson 1998; Clegg 2001) and in concert with late embryogenesis abundant (LEA) proteins (Sharon et al. 2009; Menze et al. 2009; Hand et al. 2011; Warner et al. 2010, 2012; Wu et al. 2011; Boswell et al. 2014; Toxopeus et al. 2014) this non-reducing sugar may protect embryos either by the formation of amorphous glasses, known as vitrification, or by replacing H_2O lost during dehydration (Clegg 2001; Hand et al. 2011; Hengherr et al. 2011). Cysts of *A. franciscana* are endowed with large amounts of two cyst-specific molecular chaperones, namely artemin, a heat tolerant ferritin homologue (Tanguay et al. 2004; Chen et al. 2007; Hu et al. 2011; Shahangian et al. 2011; King et al. 2014) and p26, a sHsp (Jackson and Clegg 1996; Liang and MacRae 1999; Sun et al. 2004, 2006; King and MacRae 2012). As determined by the immunoprobing of western blots artemin and p26 each represent about 7 % of the soluble protein in cysts (King et al. 2013, 2014).

24.3 sHsps Are Oligomeric, ATP-Independent Molecular Chaperones

sHsps, a family of ATP-independent molecular chaperones, consist of monomers ranging in molecular mass from 15 to 42 kDa and containing three domains, an amino-terminus, the conserved α-crystallin domain, and a carboxyl-terminal extension (Sun and MacRae 2005a; Mchaourab et al. 2009; Laganowsky et al. 2010; Hilario et al. 2011; Waters 2013). sHsp monomers associate into oligomers either of homogeneous or heterogeneous quaternary structure which tend, respectively, to be representative of non-metazoan and metazoan sHsps (Laganowsky et al. 2010; Jehle et al. 2011; Quinlan et al. 2013). Oligomerization and chaperone activity depend on all three sHsp domains (Hilario et al. 2011; Jehle et al. 2011; Basha et al. 2012; McDonald et al. 2012; Hanazono et al. 2012; Peschek et al. 2013; Quinlan et al. 2013), while the amino- and carboxyl-terminal domains, the latter encompassing the IXI motif, also contribute to quaternary dynamics and protein solubility (Basha et al. 2012; McDonald et al. 2012; Peschek et al. 2013; Hilton et al. 2013; Quinlan et al. 2013). sHsps are thought to interact with the exposed hydrophobic regions of partially denatured proteins, an activity promoted by changes in oligomer structure that depend upon stress exposure and phosphorylation, and entailing either disassembly of oligomers and/or increases in hydrophobicity by structural rearrangement (Hilario et al. 2011; McDonald et al. 2012; Palmieri et al. 2013; Peschek et al. 2013) (Fig. 24.1). sHsps bind many different substrates, the so-called interactome, including proteins involved in metabolism, transport, transcription, translation, the

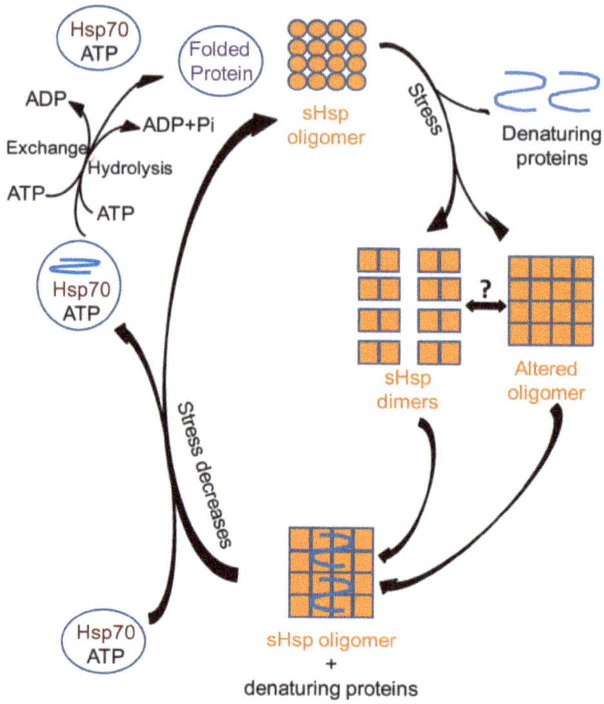

Fig. 24.1 sHsps prevent irreversible protein denaturation. In response to stress sHsp oligomers are structurally altered and/or dissociate into dimers which increases hydrophobicity (*right* side of figure). The modified sHsps bind denaturing proteins and prevent aggregation. A single sHsp oligomer may bind several substrates. When diapause terminates and metabolism increases proteins are removed from sHsps by Hsp70 and either refolded (*left* side of figure) or degraded (not shown). After ATP hydrolysis ADP is removed and replaced by ATP, ensuring Hsp70 is ready for a subsequent round of activity

cell cycle and apoptosis, among other activities (Fu et al. 2013; Arrigo 2013; Arrigo and Gibert 2013; Fan et al. 2014; Andley et al. 2014). The sequestered proteins are protected from irreversible denaturation/aggregation until release, a process contingent upon ATP-dependent chaperones such as Hsp70 (Ehrnsperger et al. 1997; Lee and Vierling 2000; Peschek et al. 2013). By protecting proteins from irreversible denaturation (Fu et al. 2013) and inhibiting apoptosis (Villeneuve et al. 2006; Acunzo et al. 2012; Hamann et al. 2013), sHsps represent the initial line of defense in organisms exposed to stress and undergoing diapause.

24.4 sHsps in *A. franciscana* Cysts

An abundant 26 kDa protein that assembles into oligomers and undergoes reversible translocation into the nucleus during anoxia, heat stress and diapause was first reported in *A. franciscana* cysts by Clegg et al. (1994). It was proposed at the time that the 26 kDa protein, termed p26, is a molecular chaperone, possibly a sHsp, and/

```
ArHsp21   -MSGMRLARSLLLLGRPQSRHLF------WGRRTWDPFEELRMIMR   39
ArHsp22   MTTLVPWTDQW---TDPWEDPFADLPVETFTGRWRDPFAADVYKPY   43
p26       -MALNPWYGGFGGMTDPWSDPFG------FGG-FGGGMDLDIDRPF   38
            :             * .  :          :     . :

ArHsp21   EMENQFQNINQNVFKALPSSFKEETAVPVISSKGDDNMYRLVLDLS   85
ArHsp22   GLPRTHLHRRRRRRRIRTVQRVFSRKGTDVRTREDDKEWEITMQLP   89
p26       --------RRRMMRRGPDTSRALKELATPGSLRDTADEFQVQLDVG   76
                  .:   :     .   .   .     :    . :.: :::

ArHsp21   GFKPEDVKIDLMDRNLRVTGKCEQKTS---DGCRMYHETQREYLLP   128
ArHsp22   GFLPSDITVNSTDKEIIVHGVHKERPDYEGEEGYVSREIRRRFVPP   135
p26       HFLPNEITVKTTDDDILVHGKHDERSD---EYGHVQREFRRRYRLP   119
           * *.::.:.   * :: * *   .::..    :    : :* :*.:   *

ArHsp21   ENVNLNELKSAFTDSGYLTIEAPMPEGMKPKE--IPINRGAQQIES   172
ArHsp22   KTINPGELSSTFSSDGELRIHAPKAIPGEPRQRRIQIMPAPIGSRF   181
p26       EHVKPESVSSTLSSDGVLTIHAPKTALSSPTERIVPITPAPAVGRI   165
          : ::   .:.*::::..* * *.**  .     .* :  :   .     .. :

ArHsp21   ESKE------SKRED-----------   181
ArHsp22   EG--------ENEEEWP----------   190
p26       EGGTTGTTTGSTASSTPARTTRSGGAA   192
                   .. ..
```

Fig. 24.2 Alignment of the sHsps produced in diapause-destined *A. franciscana* embryos. The amino acid sequences of ArHsp21 (ABD19712), ArHsp22 (ABD19713) and p26 (AF031367) were aligned with CLUSTALW. The α-crystallin domains, containing a highly conserved arginine (*shaded*), are *boxed* and followed closely by the IXI motifs (IPI, IQI, VPI) in the carboxyl terminal extensions. Identical residues are indicated by (*), conserved substitutions by (:) and semi-conserved substitutions by (.) (From King et al. 2013)

or a metabolic regulator (Clegg et al. 1994, 1995). Sequencing revealed that p26 purified to apparent homogeneity from cysts is a sHsp featuring a conserved α-crystallin domain flanked by amino- and carboxyl-terminal regions, a result verified by the cloning and sequencing of p26 cDNA (Liang et al. 1997a, b) (Fig. 24.2). Structural/functional studies of p26 demonstrated an amino acid reside (R114) corresponding to a highly conserved arginine found in sHsps from other organisms and appearing, based on site-directed mutagenesis, to be crucial for p26 activity (Sun et al. 2006). Protein modeling and mutagenesis indicated that β-strand 7 within the α-crystallin domain plays a significant role in p26 oligomerization, as shown for other sHsps.

p26 synthesized in *A. franciscana* cysts, transformed *Escherichia coli* and transfected COS-1 cells forms heterogeneous oligomers, and as determined by turbidimetric assays, purified p26 exhibits molecular chaperone activity in vitro (Liang et al. 1997a, b; Crack et al. 2002; Day et al. 2003; Sun et al. 2004, 2006; Sun and MacRae 2005b). The heat tolerance of cyst-derived, p26-containing, first instar nauplii is greater than that of direct developing nauplii which lack p26, supporting the proposal that p26 has chaperone activity. Additionally, transformed *E. coli* synthesizing p26 are more heat resistant than bacteria lacking this protein (Liang and MacRae 1999; Crack et al. 2002; Sun and MacRae 2005b; Sun et al. 2006).

p26 is synthesized only in diapause-destined *A. franciscana* embryos, with mRNA and protein respectively appearing 2 and 3 days after oocyte fertilization (Jackson and Clegg 1996; Liang and MacRae 1999). p26 mRNA decays as activated post-diapause cysts resume development but the protein declines only when nauplii begin to emerge from cysts and, as revealed by immunofluorescent staining, p26 is last seen in salt gland nuclei of instar II nauplii (Liang and MacRae 1999). p26 moves into the nuclei of diapause-destined *A. franciscana* embryos soon after synthesis and cyst nuclei purified in Percoll gradients have p26 in discrete foci perhaps associated with the structural matrix or with protein complexes active in DNA replication and other molecular processes (Liang et al. 1997b; Liang and MacRae 1999; Willsie and Clegg 2002).

The use of suppressive subtractive hybridization to study gene expression in diapause-destined *A. franciscana* embryos showed the up-regulation of several genes 2 days post-fertilization (Qiu et al. 2007). Among these genes were two sHsps in addition to p26, termed ArHsp21 and ArHsp22, both of which contain an α-crystallin domain (Fig. 24.2), form oligomers and behave as molecular chaperones in vitro (Table 24.1) (Qiu et al. 2007; Qiu and MacRae 2008a, b). The use of polyclonal antibodies specific to each of these sHsps demonstrated that like p26, ArHsp21 and ArHsp22 are developmentally regulated, appearing in diapause-destined embryos, but not in embryos developing directly into nauplii (Qiu and MacRae 2008a, b). Although the *A. franciscana* sHsps share many similarities, they differ in that p26 and ArHsp22 localize to nuclei whereas ArHsp21 does not, and only ArHsp22 is induced by heat shock, an event restricted to the adult stage of growth (Table 24.1).

Table 24.1 Diapause-specific *Artemia* sHsps

	Artemia sHsps		
Property	p26	ArHsp21	ArHsp22
Molecular mass (kDa)	20.8	21.1	22.4
Amino acid residues	192	181	190
α-Crystallin domain residue numbers	61–152	70–162	74–168
Oligomerization	+	+	+
Chaperone activity in vitro	+	+	+
Chaperone activity in vivo	+	±	ND
Nuclear localization	+	–	+
Developmentally regulated	+	+	+
Stress induced	–	–	+
			(Adults only)
% of cyst soluble protein	7.0	1.2	0.4

(Adults only), sHsp stress induced only in adults, + observed, – not observed, ± uncertain, *ND* not determined

24.5 In Vivo Studies of *A. franciscana* sHsps

The presence of three similar sHsps in diapause-destined *A. franciscana* embryos, all with molecular chaperone activity in vitro, prompt questions regarding their roles in cysts. To address these issues, *A. franciscana* sHsps were studied in vivo by RNA interference (RNAi) (King and MacRae 2012; King et al. 2013). Initially, *A. franciscana* females were injected in the egg sac with double stranded RNA (dsRNA) for p26 before fertilization (Fig. 24.3), mated, and liberated cysts were recovered for analysis. Injection of females with dsRNA resulted in almost complete knockdown of p26 mRNA and the reduction of p26 to levels undetectable by immunoprobing of western blots containing cyst proteins (King and MacRae 2012) (Fig. 24.4). p26 loss greatly reduced the resistance of cysts to desiccation and freezing, stresses required to terminate diapause, thereby confirming that this protein is a molecular chaperone (Fig. 24.5). Unexpectedly, the loss of p26 slowed embryo development such that the release of cysts from females occurred 7 days post-fertilization as opposed to 5 days when p26 is present (Fig. 24.6). Equally surprising, diapause cysts obtained from females in the laboratory lacking p26 hatched spontaneously after long term incubation in sea water at room temperature, whereas cysts with p26 did not hatch under the same conditions (Fig. 24.7) (King and MacRae 2012). These results indicate that p26 prevents diapause termination, or in other words, is required to maintain diapause.

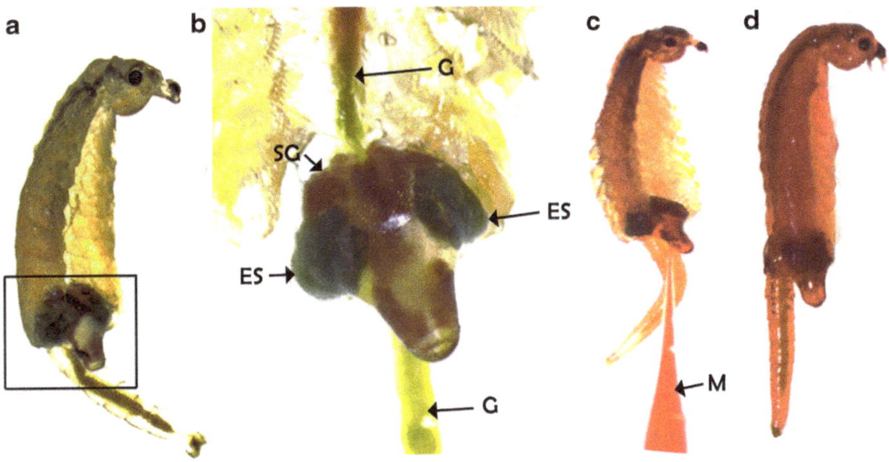

Fig. 24.3 Injection of an *A. franciscana* female with dsRNA. (**a–d**) Light micrographs showing the injection of an *A. franciscana* female primed to produce cysts; all images are of the same female. (**a**) Female prior to fertilization with the shell gland and egg sacs boxed; (**b**) boxed region of (**a**) enlarged; (**c**) injection into the egg sac of dsRNA mixed with phenol *red*; (**d**) an injected female which has retained phenol *red*, and thus dsRNA, for 2 h. Only morphologically normal females retaining phenol *red* for at least 2 h after injection were used in knock down experiments. *ES* egg sac, *G* gut, *M* micropipette, *SG* shell gland (Adapted from King and MacRae 2012)

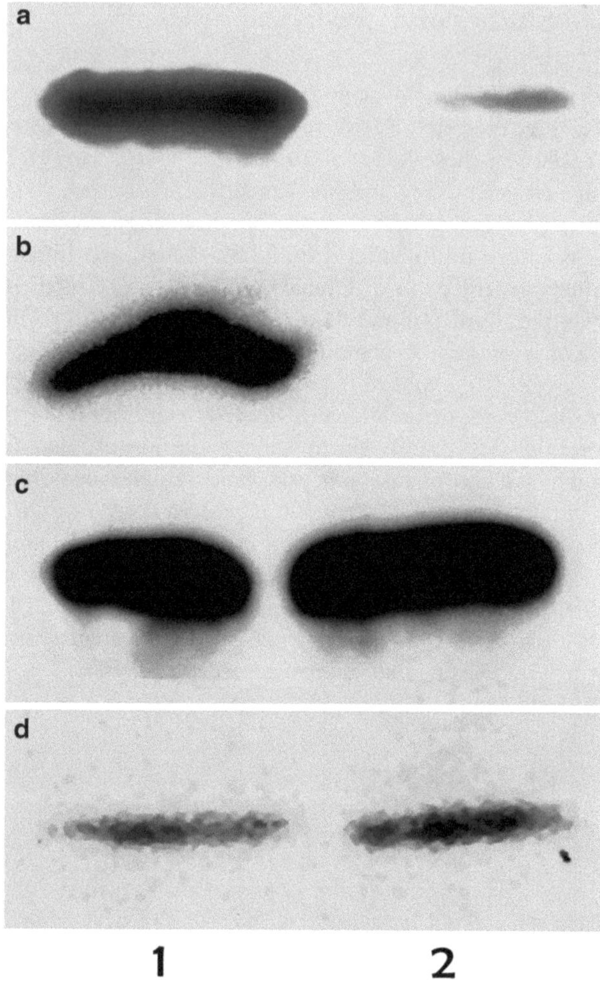

Fig. 24.4 p26 dsRNA specifically knocks down p26 mRNA and protein in *A. franciscana* cysts. (**a**) PCR amplification of p26 mRNA in cysts released by females injected with either control solution (*1*) or p26 dsRNA (*2*). Amplification products were resolved in 1.2 % agarose gels and stained with Sybersafe. Protein from 25 cysts (**b**) and 40 cysts (**c**, **d**) produced by females injected with either control solution (*1*) or p26 dsRNA (*2*) were resolved in 12.5 % SDS polyacrylamide gels and blotted to nitrocellulose. The blots were probed with antibody specific to p26 (**b**), ArHsp21 (**c**) and ArHsp22 (**d**), followed by HRP-conjugated goat anti-rabbit IgG antibody. Antibody reactive proteins were visualized by chemiluminescence (From King and MacRae 2012)

In subsequent experiments *A. franciscana* females were injected with dsRNA for ArHsp21 and ArHsp22 (King et al. 2013). Knock down of ArHsp21 had no effect on ArHsp22 mRNA and protein as anticipated from earlier knock down experiments. Unlike the situation with p26, the loss of ArHsp21 caused only marginal reduction in cyst stress tolerance and had no apparent effect on embryo development

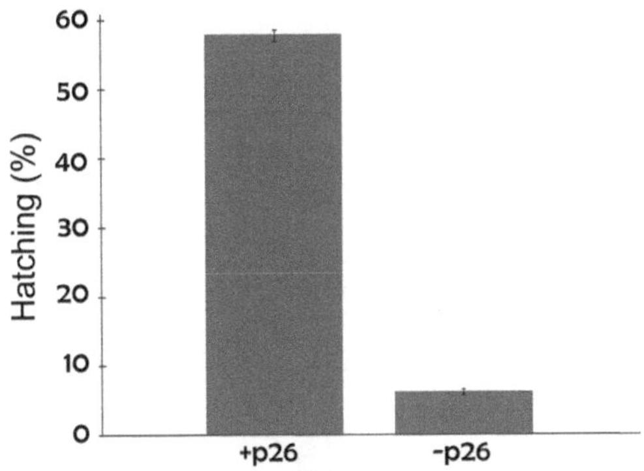

Fig. 24.5 p26 enhances the stress tolerance of *A. franciscana* cysts. Diapause of *A. franciscana* cysts either containing (+p26) or lacking (−p26) p26 was terminated by desiccation and freezing. The cysts were then incubated in sea water at room temperature and hatched nauplii were counted and removed. The results are given as the percentage of cysts that hatched which was equated to the level of diapause termination and cyst viability. The experiment was done in triplicate with separate broods of cysts and the error bars represent standard error (From King and MacRae 2012)

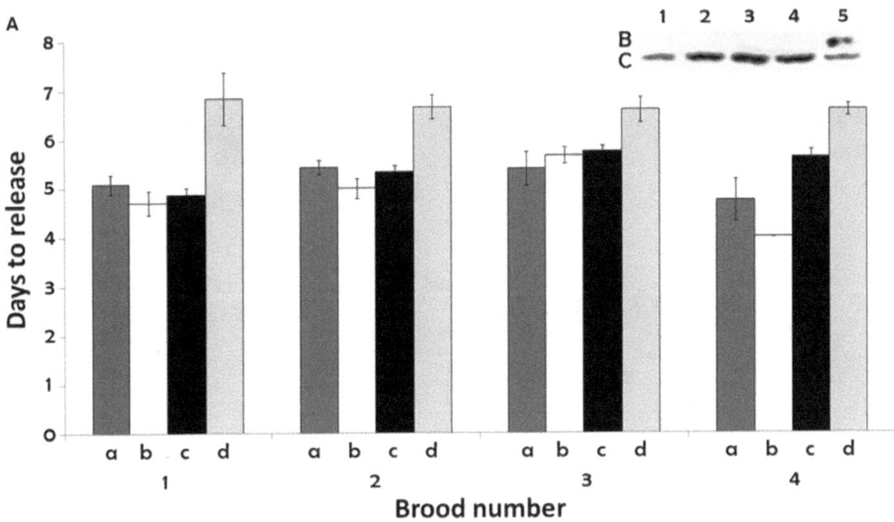

Fig. 24.6 p26 knock down slows the development of diapause-destined *A. franciscana* embryos. (**A**) The time to release in days for cysts and nauplii from females injected with control solution (**a**, **c**) and p26 dsRNA (**b**, **d**) for broods 1–4. *a* nauplii, *b* nauplii, *c* cysts, *d* cysts. N is 3–12 with lower values in later broods due to the death of females with increased culture time. The *bars* indicate standard error for each measurement. *Inset B*, protein extracts from cysts in broods 1–4 were resolved by SDS polyacrylamide gel electrophoresis, blotted to nitrocellulose and probed with antibody to p26 followed by HRP-conjugated goat anti-rabbit IgG. *Inset C*, after stripping the blot was reprobed with antibody to ArHsp21. Lane *1* brood 1, *2* brood 2, *3* brood 3, *4* brood 4, *5* protein extract from commercially obtained cysts containing p26 and ArHsp21. Proteins reacting with antibodies were visualized by chemiluminescence (From King and MacRae 2012)

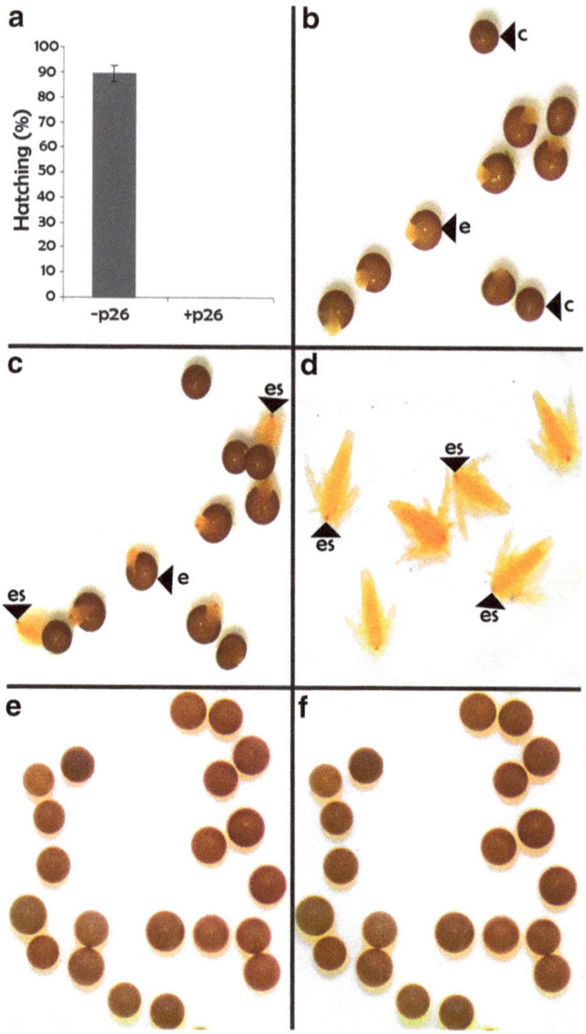

Fig. 24.7 *A. franciscana* cysts lacking p26 terminate diapause spontaneously. (**a**) Cysts lacking (–p26) and containing (+p26) p26 were incubated without agitation in sealed tubes at room temperature for at least 90 days, transferred to weigh boats and observed with a dissecting microscope until 5 days after the last cyst hatched. The experiments were done in duplicate with different broods of cysts and the error bar represents the standard error. (**b–f**) Light micrographs showing the development, or absence thereof, of cysts lacking (**b–d**) and containing (**e, f**) p26. *c* cyst, *e* emerged cyst, *es* eye spot. The nauplii shown in Fig. **d** originated from commercial cysts and they are representative of nauplii obtained from laboratory reared cysts lacking p26 (From King and MacRae 2012)

and the maintenance of diapause (King et al. 2013). The divergent results for p26 and ArHsp21 suggest different functions in *A. franciscana* cysts, a possibility reinforced by the greater abundance of p26 than ArHsp21 (Table 24.1) (King et al. 2013). p26 is therefore likely to interact with a larger amount of compromised protein in diapause cysts than is ArHsp21 and to bind a greater diversity of protein substrates.

Injection of dsRNA for ArHsp22 into *A. franciscana* female and male adults resulted in death, rendering the examination of ArHsp22 activity in embryo development and cyst properties impossible (King et al. 2013). ArHsp22 is the only sHsp induced in adult *A. franciscana* by heat stress (Qiu and MacRae 2008b). With this in mind, it is possible that injection represents a stress for adults which is countered, at least in part, by the synthesis of ArHsp22, a response that is inhibited by dsRNA for ArHsp22.

24.6 sHsps and Diapause in Animals Other than *Artemia*

Hsp synthesis occurs in animals other than *A. franciscana* when they experience diapause and in many cases Hsps are linked to desiccation tolerance. The monogonont rotifer, *Brachionus plicatilis*, undergoes diapause when hydrated and it becomes resistance to desiccation and heat (Denekamp et al. 2009; Clark et al. 2012). Transcription profiles produced by Illumina short read sequencing and the generation of expressed sequence tag (EST) libraries uncovered the up-regulation of several molecular chaperones including Hsp60, Hsp70, Hsp40 co-chaperones, the Hsc70/Hsp90 organizing protein (HOP), and sHsps in mitic or resting eggs of *B. plicatilis* (Denekamp et al. 2009; Clark et al. 2012). These chaperones may play a role in protecting dormant stages of *B. plicatilis*, which survive stressful environments for decades. mRNA for the sHsp, Hsp22, increases during diapause in the marine copepod, *Calanus finmarchicus*, where it has the potential to protect proteins from denaturation (Aruda et al. 2011). However, mRNAs for most Hsps examined in *C. finmarchicus* do not change during diapause even though Hsp genes respond to handling stress, perhaps because diapause in this species does not involve exposure to the extreme stresses endured by other species. *Daphnia magna* produces dormant as opposed to subitaneous eggs when environmental parameters deteriorate and the resting, as opposed to the active stage, exhibits more Hsp60 (Pauwels et al. 2007). The sHsps have not been investigated during diapause in *D. magna*, as is true for the killifish, *Austrofundulus limnaeus*, where an inducible Hsp70 increases during embryonic diapause II (Podrabsky and Somero 2007; Podrabsky et al. 2010). For tardigrades, which are tolerant of stress even in their active stage, only a limited role for Hsps in diapause and cryptobiosis, as shown by their up-regulation during stress, has been demonstrated (Förster et al. 2009; Mali et al. 2010). The possibility that sufficient Hsps are present in tardigrades prior to stress to ensure protein homeostasis has yet to be investigated. That Hsps are more involved in protein repair upon rehydration of tardigrades, rather than as protective molecules during dehydration, is possible (Reuner et al. 2010; Guidetti et al. 2011; Møbjerg et al. 2011).

24.7 Conclusions

Protein association with sHsps, a process that occurs in the absence of ATP, is the first line of defense for cells against stress-induced protein loss. Effective storage of aberrant proteins conserves cell resources by ensuring proteins are not irreversibly denatured nor are they repaired during stress only to be denatured again. The liberation of substrate from sHsps and either their folding or degradation once metabolism and ATP increase is likely to occur under the direction of ATP-dependent molecular chaperones such as Hsp70, Hsp60 and Hsp90. Refolded proteins are then available for use in the cell, promoting post-diapause growth by reducing energy expensive transcription and translation.

The loss of p26 from *A. franciscana* cysts reduces stress tolerance suggesting this sHsp interacts with proteins structurally perturbed upon experiencing stresses associated with diapause. In contrast, ArHsp21, a protein present in lesser quantities than p26, appears to be relatively unimportant in cyst stress tolerance. It is not possible to test the role of ArHsp22 in stress tolerance by RNAi because injection of dsRNA for this protein kills adults. Equally interesting, the actions of p26 extend beyond a role in stress tolerance because its depletion affects the rate of embryo development and diapause termination. How sHsps function during metabolic suppression and protein maintenance during diapause is only partially defined, but clearly the role of sHsps is broader than previously thought. In this context, it will be particularly interesting to identify substrates bound by sHsps during diapause. This approach may identify proteins playing key roles in diapause and contribute to the general understanding of sHsp function in all cells.

Acknowledgements This work was supported by a Natural Sciences and Engineering Research Council of Canada Discovery Grant to THM.

References

Acunzo J, Katsogiannou M, Rocchi P (2012) Small heat shock proteins HSP27 (HspB1), αB-crystallin (HspB5) and HSP22 (HspB8) as regulators of cell death. Int J Biochem Cell Biol 44:1622–1631

Andley UP, Malone JP, Townsend RR (2014) *In vivo* substrates of the lens molecular chaperones αA-crystallin and αB-crystallin. PLoS One 9(4):e95507

Arrigo A-P (2013) Human small heat shock proteins: protein interactomes of homo- and hetero-oligomeric complexes: an update. FEBS Lett 587:1959–1969

Arrigo A-P, Gibert B (2013) Protein interactomes of three stress inducible small heat shock proteins: HspB1, HspB5 and HspB8. Int J Hyperth 29:409–422

Aruda AM, Baumgartner MF, Reitzel AM, Tarrant AM (2011) Heat shock protein expression during stress and diapause in the marine copepod *Calanus finmarchicus*. J Insect Physiol 57:665–675

Basha E, O'Neill H, Vierling E (2012) Small heat shock proteins and α-crystallins: dynamic proteins with flexible functions. Trends Biochem Sci 37:106–117

Boswell LC, Moore DS, Hand SC (2014) Quantification of cellular protein expression and molecular features of group 3 LEA proteins from embryos of *Artemia franciscana*. Cell Stress Chaperones 19:329–341

Chen T, Villeneuve TS, Garant KA, Amons R, MacRae TH (2007) Functional characterization of artemin, a ferritin homolog synthesized in *Artemia* embryos during encystment and diapause. FEBS J 274:1093–1101

Clark MS, Denekamp NY, Throne MAS, Reinhardt R, Drungowski M, Albrecht MW, Klages S, Beck A, Kube M, Lubzens E (2012) Long-term survival of hydrated resting eggs from Brachionus plicatilis. PLoS One 7(1):e29365

Clegg JS (1997) Embryos of *Artemia franciscana* survive four years of continuous anoxia: the case for complete metabolic rate depression. J Exp Biol 200:467–475

Clegg JS (2001) Cryptobiosis – a peculiar state of biological organization. Comp Biochem Physiol B 128:613–624

Clegg JS (2005) Desiccation tolerance in encysted embryos of the animal extremophile, *Artemia*. Integr Comp Biol 45:715–724

Clegg JS (2007) Protein stability in *Artemia* embryos during prolonged anoxia. Biol Bull 212:74–81

Clegg JS (2011) The unusual response of encysted embryos of the animal extremophile, *Artemia franciscana*, to prolonged anoxia. In: Altenbach AV, Bernhard JM, Seckbach J (eds) Anoxia: paleontological strategies and evidence for eukaryotic survival. Springer, Dordrecht/New York, pp 189–203

Clegg JS, Jackson SA (1998) The metabolic status of quiescent and diapause embryos of *Artemia franciscana* (Kellogg). Arch Hydrobiol Spec Issues Adv Limnol 52:425–439

Clegg JS, Jackson SA, Warner AH (1994) Extensive intracellular translocations of a major protein accompany anoxia in embryos of *Artemia franciscana*. Exp Cell Res 212:77–83

Clegg JS, Jackson SA, Liang P, MacRae TH (1995) Nuclear-cytoplasmic translocations of protein p26 during aerobic-anoxic transitions in embryos of *Artemia franciscana*. Exp Cell Res 219:1–7

Clegg JS, Drinkwater LE, Sorgeloos P (1996) The metabolic status of diapause embryos of *Artemia franciscana* (SFB). Physiol Zool 69:49–66

Crack JA, Mansour M, Sun Y, MacRae TH (2002) Functional analysis of a small heat shock/α--crystallin protein from *Artemia franciscana* Oligomerization and thermotolerance. Eur J Biochem 269:933–942

Day RM, Gupta JS, MacRae TH (2003) A small heat shock/α-crystallin protein from encysted *Artemia* embryos suppresses tubulin denaturation. Cell Stress Chaperones 8:183–193

Denekamp NY, Thorne MAS, Clark MS, Kube M, Reinhardt R, Lubzens E (2009) Discovering genes associated with dormancy in the monogonont rotifer *Brachionus plicatilis*. BMC Genomics 10:108

Denlinger DL (2002) Regulation of diapause. Annu Rev Entomol 47:93–122

Ehrnsperger M, Gräber S, Gaestel M, Buchner J (1997) Binding of non-native protein to Hsp25 during heat shock creates a reservoir of folding intermediates for reactivation. EMBO J 16:221–229

Fan Q, Huang L-Z, Zhu X-J, Zhang K-K, Ye H-F, Luo Y, Sun X-H, Zhou P, Lu Y (2014) Identification of proteins that interact with alpha A-crystallin using a human proteome microarray. Mol Vis 20:117–124

Förster F, Liang C, Shkumatov A, Beisser D, Engelmann JC, Schnölzer M, Frohme M, Müller T, Schill RO, Dandekar T (2009) Tardigrade workbench: comparing stress-related proteins, sequence-similar and functional protein clusters as well as RNA elements in tardigrades. BMC Genomics 10:469

Fu X, Shi X, Yan L, Zhang H, Chang Z (2013) *In vivo* substrate diversity and preference of small heat shock protein IbpB as revealed by using a genetically incorporated photo-cross-linker. J Biol Chem 288:31646–31654

Guidetti R, Altiero T, Rebecchi L (2011) On dormancy strategies in tardigrades. J Insect Physiol 57:567–576

Hahn DA, Denlinger DL (2011) Energetics of insect diapause. Annu Rev Entomol 56:103–121

Hamann S, Métrailler S, Schorderet DF, Cottet S (2013) Analysis of the cytoprotective role of α-crystallins in cell survival and implication of the αA-crystallin C-terminal extension domain in preventing bax-induced apoptosis. PLoS One 8(2):e55372

Hanazono Y, Takeda K, Yohda M, Miki K (2012) Structural studies on the oligomeric transition of a small heat shock protein, StHsp14.0. J Mol Biol 422:100–108

Hand SC, Menze MA, Toner M, Boswell L, Moore D (2011) LEA proteins during water stress: not just for plants anymore. Annu Rev Physiol 73:115–134

Hengherr S, Schill RO, Clegg JS (2011) Mechanisms associated with cellular desiccation tolerance of *Artemia* encysted embryos from locations around the world. Comp Biochem Physiol A 160:137–142

Hilario E, Martin FJM, Bertolini MC, Fan L (2011) Crystal structures of *Xanthomonas* small heat shock protein provide a structural basis for an active molecular chaperone oligomer. J Mol Biol 408:74–86

Hilton GR, Hochberg GKA, Laganowsky A, McGinnigle SI, Baldwin AJ, Benesch JLP (2013) C-terminal interactions mediate the quaternary dynamics of αB-crystallin. Philos Trans R Soc B 368:20110405

Hu Y, Bojikova-Fournier S, King AM, MacRae TH (2011) The structural stability and chaperone activity of artemin, a ferritin homologue from diapause-destined *Artemia* embryos, depend on different cysteine residues. Cell Stress Chaperones 16:133–141

Jackson SA, Clegg JS (1996) Ontogeny of low molecular weight stress protein p26 during early development of the brine shrimp, *Artemia franciscana*. Dev Growth Differ 38:153–160

Jehle S, Vollmar BS, Bardiaux B, Dove KK, Rajagopal P, Gonen T, Oschkinat H, Klevit RE (2011) N-terminal domain of αB-crystallin provides a conformational switch for multimerization and structural heterogeneity. Proc Natl Acad Sci U S A 108:6409–6414

King AM, MacRae TH (2012) The small heat shock protein p26 aids development of encysting Artemia embryos, prevents spontaneous diapause termination and protects against stress. PLoS One 7(8):e43723

King AM, Toxopeus J, MacRae TH (2013) Functional differentiation of small heat shock proteins in diapause-destined *Artemia* embryos. FEBS J 280:4761–4772

King AM, Toxopeus J, MacRae TH (2014) Artemin, a diapause-specific chaperone, contributes to the stress tolerance of *Artemia franciscana* cysts and influences their release from females. J Exp Biol 217:1719–1724

Koštál V (2006) Eco-physiological phases of insect diapause. J Insect Physiol 52:113–127

Laganowsky A, Benesch JLP, Landau M, Ding L, Sawaya MR, Cascio D, Huang Q, Robinson CV, Horwitz J, Eisenberg D (2010) Crystal structures of truncated alphaA and alphaB crystallins reveal structural mechanisms of polydispersity important for eye lens function. Protein Sci 19:1031–1043

Lee GJ, Vierling E (2000) A small heat shock protein cooperates with heat shock protein 70 systems to reactivate a heat-denatured protein. Plant Physiol 122:189–197

Liang P, MacRae TH (1999) The synthesis of a small heat shock/α-crystallin protein in *Artemia* and its relationship to stress tolerance during development. Dev Biol 207:445–456

Liang P, Amons R, MacRae TH, Clegg JS (1997a) Purification, structure and *in vitro* molecular-chaperone activity of *Artemia* p26, a small heat-shock/α-crystalin protein. Eur J Biochem 243:225–232

Liang P, Amons R, Clegg JS, MacRae TH (1997b) Molecular characterization of a small heat shock/α-crystallin protein in encysted *Artemia* embryos. J Biol Chem 272:19051–19058

MacRae TH (2003) Molecular chaperones, stress resistance and development in *Artemia franciscana*. Semin Cell Dev Biol 14:251–258

MacRae TH (2010) Gene expression, metabolic regulation and stress tolerance during diapause. Cell Mol Life Sci 67:2405–2424

Mali B, Grohme MA, Förster F, Dandekar T, Schnölzer M, Reuter D, Welnicz W, Schill RO, Frohme M (2010) Transcriptome survey of the anhydrobiotic tardigrade *Milnesium tardigradum* in comparison with *Hypsibius dujardini* and *Richtersius coronifer*. BMC Genomics 11:168

McDonald ET, Bortolus M, Koteiche HA, Mchaourab HS (2012) Sequence, structure, and dynamic determinants of Hsp27 (HspB1) equilibrium dissociation are encoded by the N-terminal domain. Biochemistry 51:1257–1268

Mchaourab HS, Godar JA, Stewart PL (2009) Structure and mechanism of protein stability sensors: the chaperone activity of small heat-shock proteins. Biochemistry 48:3828–3837

Menze MA, Boswell L, Toner M, Hand SC (2009) Occurrence of mitochondria-targeted late embryogenesis abundant (LEA) gene in animals increases organelle resistance to water stress. J Biol Chem 284:10714–10719

Møbjerg N, Halberg KA, Jørgensen A, Persson D, Bjørn M, Ramløv H, Kristensen RM (2011) Survival in extreme environments – on the current knowledge of adaptations in tardigrades. Acta Physiol 202:409–420

Palmieri V, Maulucci G, Maiorana A, Papi M, De Spirito M (2013) α-crystallin modulates its chaperone activity by varying the exposed surface. ChemBioChem 14:2362–2370

Patil YN, Marden B, Brand MD, Hand SC (2013) Metabolic downregulation and inhibition of carbohydrate catabolism during diapause in embryos of *Artemia franciscana*. Physiol Biochem Zool 86:106–118

Pauwels K, Stoks R, Verbiest A, De Meester L (2007) Biochemical adaptation for dormancy in subitaneous and dormant eggs of *Daphnia magna*. Hydrobiologia 594:91–96

Peschek J, Braun N, Rohrberg J, Back KC, Kriehuber T, Kastenmüller A, Weinkauf S, Buchner J (2013) Regulated structural transitions unleash the chaperone activity of αB-crystallin. Proc Natl Acad Sci U S A 110:E3780–E3789

Podrabsky JE, Somero GN (2007) An inducible 70 kDa-class heat shock protein is constitutively expressed during early development and diapause in the annual killifish *Austrofundulus limnaeus*. Cell Stress Chaperones 12:199–204

Podrabsky JE, Tingaud-Sequeira A, Cerdà J (2010) Metabolic dormancy and responses to environmental desiccation in fish embryos. In: Lubzens E, Cerdà J, Clark MS (eds) Dormancy and resistance in harsh environments. Springer, Berlin/Heidelberg, pp 203–226

Ptak GE, Tacconi E, Czernik M, Toschi P, Modlinski JA, Loi P (2012) Embryonic diapause is conserved across mammals. PLoS One 7(3):e33027

Qiu Z, MacRae TH (2008a) ArHsp21, a developmentally regulated small heat-shock protein synthesized in diapausing embryos of *Artemia franciscana*. Biochem J 411:605–611

Qiu Z, MacRae TH (2008b) ArHsp22, a developmentally regulated small heat shock protein produced in diapause-destined *Artemia* embryos, is stress inducible in adults. FEBS J 275:3556–3566

Qiu Z, MacRae TH (2010) A molecular overview of diapause in embryos of the crustacean, Artemia francsicana. In: Lubzens E, Cerdà J, Clark MS (eds) Dormancy and resistance in harsh environments. Springer, Berlin/Heidelberg, pp 165–187

Qiu Z, Tsoi SCM, MacRae TH (2007) Gene expression in diapause-destined embryos of the crustacean, *Artemia franciscana*. Mech Dev 124:856–867

Quinlan RA, Zhang Y, Lansbury A, Williamson I, Pohl E, Sun F (2013) Changes in the quaternary structure and function of MjHSP16.5 attributable to deletion of the IXI motif and introduction of the substitution, R107G, in the α-crystallin domain. Philos Trans R Soc B 368:20120372

Reuner A, Hengherr S, Mali B, Förster F, Arndt D, Reinhardt R, Dandekar T, Frohme M, Brümmer F, Schill RO (2010) Stress response in tardigrades: differential gene expression of molecular chaperones. Cell Stress Chaperones 15:423–430

Reynolds JA, Hand SC (2009) Decoupling development and energy flow during embryonic diapause in the cricket, *Allonemobius socius*. J Exp Biol 212:2065–2074

Rinehart JP, Li A, Yocum GD, Robich RM, Hayward SAL, Denlinger DL (2007) Up-regulation of heat shock proteins is essential for cold survival during insect diapause. Proc Natl Acad Sci U S A 104:11130–11137

Robbins HM, Van Stappen G, Sorgeloos P, Sung YY, MacRae TH, Bossier P (2010) Diapause termination and development of encysted *Artemia* embryos: roles for nitric oxide and hydrogen peroxide. J Exp Biol 213:1464–1470

Shahangian SS, Rasti B, Sajedi RH, Khodarahmi R, Taghdir M, Ranjbar B (2011) Artemin as an efficient molecular chaperone. Protein J 30:549–557

Sharon MA, Kozarova A, Clegg JS, Vacratsis PO, Warner AH (2009) Characterization of a group 1 late embryogenesis abundant protein in encysted embryos of the brine shrimp *Artemia franciscana*. Biochem Cell Biol 87:415–430

Sun Y, MacRae TH (2005a) Small heat shock proteins: molecular structure and chaperone function. Cell Mol Life Sci 62:2460–2476

Sun Y, MacRae TH (2005b) Characterization of novel sequence motifs within the N- and C-terminal extensions of p26, a small heat shock protein from *Artemia franciscana*. FEBS J 272:5230–5243
Sun Y, Mansour M, Crack JA, Gass GL, MacRae TH (2004) Oligomerization, chaperone activity, and nuclear localization of p26, a small heat shock protein from *Artemia franciscana*. J Biol Chem 279:39999–40006
Sun Y, Bojikova-Fournier S, MacRae TH (2006) Structural and functional roles for β-strand 7 in the α-crystallin domain of p26, a polydisperse small heat shock protein from *Artemia franciscana*. FEBS J 273:1020–1034
Tammariello SP, Denlinger DL (1998) G0/G1 cell cycle arrest in the brain of *Sarcophaga crassipalpis* during pupal diapause and the expression pattern of the cell cycle regulator, proliferating cell nuclear antigen. Insect Biochem Mol Biol 28:83–89
Tanguay JA, Reyes RC, Clegg JS (2004) Habitat diversity and adaptation to environmental stress in encysted embryos of the crustacean *Artemia*. J Biosci 29:489–501
Toxopeus J, Warner AH, MacRae TH (2014) Group 1 LEA proteins contribute to the desiccation and freeze tolerance of *Artemia franciscana* embryos during diapause. Cell Stress Chaperones (in Press)
Villeneuve TS, Ma X, Sun Y, Oulton MM, Oliver AE, MacRae TH (2006) Inhibition of apoptosis by p26: implications for small heat shock protein function during *Artemia* development. Cell Stress Chaperones 11:71–80
Warner AH, Miroshnychenko O, Kozarova A, Vacratsis PO, MacRae TH, Kim J, Clegg JS (2010) Evidence for multiple group 1 late embryogenesis abundant proteins in encysted embryos of *Artemia* and their organelles. J Biochem 148:581–592
Warner AH, Chakrabortee S, Tunnacliffe A, Clegg JS (2012) Complexity of the heat-soluble LEA proteome in *Artemia* species. Comp Biochem Physiol D 7:260–267
Waters ER (2013) The evolution, function, structure, and expression of the plant sHSPs. J Exp Bot 64:391–403
Willsie JK, Clegg JS (2002) Small heat shock protein p26 associates with nuclear lamins and Hsp70 in nuclei and nuclear matrix fractions from stressed cells. J Cell Biochem 84:601–614
Wu G, Zhang H, Sun J, Liu F, Ge X, Chen W-H, Yu J, Wang W (2011) Diverse LEA (late embryogenesis abundant) and *LEA-like* genes and their responses to hypersaline stress in post-diapause embryonic development of *Artemia franciscana*. Comp Biochem Physiol B 160:32–39

Chapter 25
Drosophila Small Heat Shock Proteins: An Update on Their Features and Functions

Geneviève Morrow and Robert M. Tanguay

Abstract Small heat shock proteins (sHsps) are present in varying numbers in all organisms. In *Drosophila melanogaster* there are 12 sHsps, which have distinctive developmental expression patterns, intracellular localizations and substrate specificities. Even if most of *Drosophila* sHsps do not have a known mammalian ortholog, they share their involvement in multiple cellular processes such as cytoskeleton modulation, apoptosis and autophagy. New data on *Drosophila* sHsps have arisen from high-throughput genomic and proteomic studies as well as from deletion experiments. In addition to showing the complexity of this family, these experiments suggest the involvement of sHsps in new cellular processes such as the involvement of Hsp27 in piRNA biosynthesis. The goal of this review is to summarize the new findings on each *Drosophila* sHsp and to highlight its similarity to other sHsps as well as its distinctive features.

Keywords Small heat shock protein (sHsp) • *Drosophila melanogaster* • Chaperone • Alpha-crystallin domain (ACD) • Apoptosis • Cytoskeleton • Autophagy • PolyQ • Protein misfolding diseases • Aging • Development

25.1 Introduction

The heat shock protein (Hsp) family is composed of proteins that are involved in multiple processes and that generally have a protective role upon stress. It is composed of six different subfamilies based on sequence homology and molecular weight, namely, Hsp100, Hsp90, Hsp70, Hsp60, Hsp40 and small Hsp (sHsp).

The sHsp family is present in all three domains of life and has even recently been found in cyanophages (Maaroufi and Tanguay 2013). It is composed of proteins ranging in size from 12 to 42 kDa that share a conserved C-terminal domain that

G. Morrow • R.M. Tanguay (✉)
Department of Molecular Biology, Medical Biochemistry & Pathology Medical School Lab Cell & Developmental Genetics IBIS, Pavillon C.E. Marchand 1030 Ave de la médecine Université Laval, Québec, QC, Canada G1V 0A6
e-mail: robert.tanguay@ibis.ulaval.ca

R.M. Tanguay, L.E. Hightower (eds.), *The Big Book on Small Heat Shock Proteins*, Heat Shock Proteins 8, DOI 10.1007/978-3-319-16077-1_25

was originally described in crystallin proteins, hence its designation as the α-crystallin domain (ACD, pfam family: PF00011 (Finn et al. 2014)). The ACD is flanked by a highly variable and often disordered N-terminal sequence and a short flexible C-terminal extension. It is rich in ß-strands and has been shown to be responsible for dimerization. Most of the sHsps assemble into large oligomers from 12 to 40 subunits, using dimers as the building block. These oligomers are dynamic and the rate of subunit exchange is thought to govern their ability to bind substrates (Delbecq and Klevit 2013).

Small Hsps are one of the most up-regulated classes of Hsps following stress. Small Hsps are important players in the maintenance of protein homeostasis due in part to their ability to bind misfolded proteins in an ATP-independent manner. The nature of their substrates is diversified but accumulated evidence indicates that they preferentially protect certain classes of functional proteins, such as metabolic enzymes and proteins involved in translation (Fu 2014). Under non-stressed conditions, some sHsps have been shown to interact with components of the cytoskeleton, apoptotic pathway and autophagy pathway among others, showing their importance at the cellular level. Hence mutations in sHsps have been associated with neurodegenerative diseases, cardiomyopathies and cataracts (Carra et al. 2013; Boncoraglio et al. 2012; Garrido et al. 2012).

The number of sHsps is different from one kingdom to another and even between species, making sHsp a very heterogeneous family. For example, while mammals have ten sHsps (termed HspB1-HspB10), the number of plant sHsps varies between 19 and 36 according to the species (Basha et al. 2012; Waters 2013). Their level and timing of expression, tissue distribution and intracellular localization display considerable differences suggesting that sHsps have adapted to their environment and to specific cell needs. In this review, we will focus on *Drosophila melanogaster* sHsps, the organism in which the heat shock response was first discovered.

25.1.1 Drosophila melanogaster *sHsps*

The genome of *D. melanogaster* encodes for 12 proteins containing the characteristic ACD (pfam family: PF00011, (Finn et al. 2014)). Among those 12 sHsps, 7 are located at position 67B on the left arm of chromosome 3 (Table 25.1). Hsp22, Hsp23, Hsp26 and Hsp27 were the first sHsps reported followed by Hsp67Ba and Hsp67Bc (Ayme and Tissieres 1985; Craig and McCarthy 1980).

CG43851, the latest sHsp annotated (FlyBase annotation: FBrf0217539), is located on the left arm of chromosome 2 like CG13133 but at a different locus (Table 25.1). One transcript and one polypeptide are associated with CG43851 but little information aside from electronic inference is available on them. Nearest neighbor analysis shows the divergence of CG43851 compared to the other sHsps (Fig. 25.1a). Interestingly, CG43851 shares more similarity with plant sHsps than with any other *Drosophila* sHsp. Further analysis of the CG43851 sequence reveals that it lacks the loop containing ß6 as does *Arabidopsis thaliana* Hsp18.5

Table 25.1 *Drosophila melanogaster* sHsps

CG number	Name	Chromosome localization	Locus	Transcripts	Protein size	Protein forms on 2D-PAGE/ post-translational modifications	Intracellular localization
CG14207	–	X	18D	4	21.8/20.8/ 17.1/17.2	1 phosphoserine on 21.8 kDa (Bodenmiller et al. 2007)	Cytosolic (Zhang et al. 2011a; Vos 2009)
CG43851	–	2L	23A	1	16.9		Not determined
CG13133	–	2L	31A	1	24.5		Cytosolic (Vos 2009; Zhang et al. 2011a)
CG4533	l(2)efl	2R	59F	2	21.3/17.2		Cytosolic (Vos 2009; Zhang et al. 2011a)
CG7409	–	3L	66A	1	18.0		Cytosolic (Vos 2009; Zhang et al. 2011a)
CG4461	–	3L	67B	1	23.8		Cytosolic (Vos 2009; Zhang et al. 2011a)
CG4167	Hsp67Ba	3L	67B	1	26.6		Cytosolic (Vos 2009; Zhang et al. 2011a)
CG4183	Hsp26	3L	67B	1	23.0	5 forms (Marin et al. 1993)/3 phosphoserines (Bodenmiller et al. 2007)	Cytosolic (Vos 2009; Zhang et al. 2011a; Michaud et al. 2002)
CG4190	Hsp67Bc	3L	67B	1	22.2		Cytosolic (Vos 2009; Zhang et al. 2011a)
CG4460	Hsp22	3L	67B	1	19.8	1 form/Tanguay, unpublished 1 phosphothreonine (Zhai et al. 2008)	Mitochondrial (Morrow et al. 2000)
CG4463	Hsp23	3L	67B	1	20.6	2 forms (Marin et al. 1996)	Cytosolic (Vos 2009; Zhang et al. 2011a; Michaud et al. 2002)
CG4466	Hsp27	3L	67B	1	23.6	4 forms (Marin et al. 1996)/2 phosphoserines (Bodenmiller et al. 2007; Zhai et al. 2008)	Nuclear (Beaulieu et al. 1989; Michaud et al. 2008)

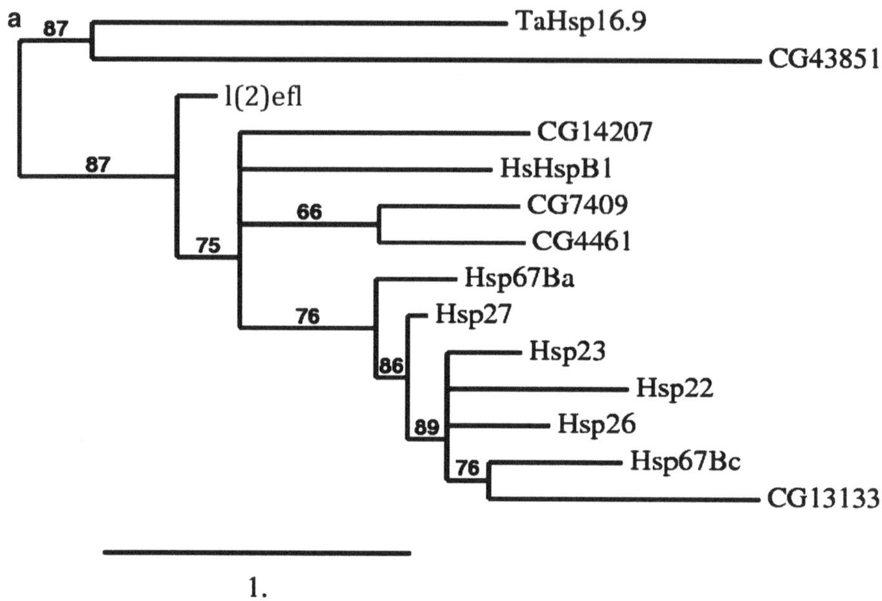

b

```
               ß2       ß3               ß4        ß5              ß6
TaHsp16.9  ARVDWKETPEAHVFKVDLPGVKKEEVKVEVED-GNVLVVSGERSREKEDKNDKWHRVERS -102
AtHsp18.5  -QLNWTETPTAHVFKAYLPGVDQDEVIAFVDEEGYLQICTGD------------------ -101
CG43851    -RVSW---------EFDMRDFPIENICIQLKD-SHIYVRAHYKN----------------- -99
            ::.*         :  : ..  :::   :.: . : : :

                 ß7            ß8      ß9                        ß10
TaHsp16.9  SGKFVRRFRLPEDAKVEEVKAGLEN-GVLTVTVPKAEVKKP----------EVKAIEISG -151
AtHsp18.5  -NKFMSRFKLPNNALTDQVTAWMED-EFLVVFVEKDASSSPPQLPEIEENRNVRVVEITG -159
CG43851    -VNIKQKILIPQNVDTSKLEAVLTASGILAISVPIVSPIS-----------DNQKMRF-  -151
             ::  :: :*::. ..:: * :    .*.: *      .          : : :.:
```

Fig. 25.1 Phylogeny of *Drosophila* sHsps. (**a**) The phylogeny was made with the following sequences from Uniprot; P02515 (Hsp22), P02516 (Hsp23), P02517 (Hsp26), P02518 (Hsp27), P05812 (Hsp67Ba), P22979 (Hsp67Bc), P82147 (l(2)efl), M9NHC5 (CG14207), Q9VSX2 (CG4461), Q9VL41 (CG13133), Q9VSA9 (CG7409) and M9PCC9 (CG43851). *Triticum aestivum* Hsp16.9 (TaHsp16.9, P12810) and *Homo sapiens* HspB1 (HsHspB1, P04792) sequences were used as controls for the alignment. Only the ACD (as defined by pfam (Finn et al. 2014)) and the C-terminal extension were used for multiple alignment with MAFFT. PhyML was used to determine the phylogeny between the sequences and TreeDyn to visualize the phylogenic tree (Chevenet et al. 2006; Dereeper et al. 2008, 2010). The numbers in the tree represent the branch values (similar to bootstraps). Branches with value between 40 % and 50 % were collapsed. (**b**) Alignment of ACD and C-terminal extension from CG43851 (M9PCC9), *Arabidopsis thaliana* Hsp18.5 (AtHsp18.5, O64564) and TaHsp16.9 (P12810). The alignment was generated using ClustalW. Secondary structural elements are based on TaHsp16.9 (van Montfort et al. 2001)

(Fig. 25.1b), suggesting that, like this sHsp, it exists as a dimer that forms large oligomers only when bound to substrate (Basha et al. 2013).

For most *D. melanogaster* sHsps, the human ortholog has not been clearly identified. Hsp67Bc is the functional equivalent of HspB8 (Carra et al. 2010) and l(2)efl would be the ortholog of HspB5 (Wang et al. 2011).

25.1.2 Intracellular Localization

Similar to plant sHsps, *Drosophila* sHsps have different intracellular localizations; Hsp22 is located in the mitochondrial matrix (Morrow et al. 2000), Hsp27 is nuclear (Beaulieu et al. 1989; Michaud et al. 2008) and the other studied sHsps are cytoplasmic (Vos 2009; Zhang et al. 2011a; Michaud et al. 2002). Analysis of the CG43851 sequence predicts nuclear or cytoplasmic localization depending on the prediction tool used (Euk-mPLoc 2.0 (Chou and Shen 2010), SLP-Local (Matsuda et al. 2005), CELLO v2.5 (Yu et al. 2006), Sherloc2 (Briesemeister et al. 2009), MultiLoc2 (Blum et al. 2009)). In mammals, all sHsps are mainly located in the cytoplasm but some shuttle to the nucleus or to the mitochondria (Bryantsev et al. 2007; den Engelsman et al. 2013; Marunouchi et al. 2013; Nakagawa et al. 2001; van den IJssel et al. 2003). Since *Drosophila* sHsps are localized in different compartments, it has long been thought that their shuttling to other cellular compartments was not needed.

Recently, Hsp26 has been found in the nuclear matrix of *Drosophila* Schneider 2 (S2) cells and embryos together with Hsp27 (Kallappagoudar et al. 2010). To our knowledge, this is the first report showing a nuclear localization for this cytoplasmic sHsp. However, only a small amount of Hsp26 may be nuclear since whole cell immunostaining shows mainly a cytoplasmic localization (Tanguay, personal communication; Michaud et al. 2002). A dual localization cytosol/nucleus has also been reported for HspB1 and HspB5 in certain types of unstressed cells (Bryantsev et al. 2007; van den IJssel et al. 2003). While phosphorylation is not a prerequisite for HspB1 entry in the nucleus (Bryantsev et al. 2007), it seems to be involved in the case of HspB5 (den Engelsman et al. 2013). Marin et al. (1993) have reported the existence of five isoforms of Hsp26. The exact nature of the post-translational modifications of those isoforms is not known but three serines can be phosphorylated and ubiquitination of Hsp26 has been observed in fly neurons (Bodenmiller et al. 2007; Franco et al. 2011). Both types of post-translational modifications can influence the intracellular localization of a protein but their involvement in the nuclear localization of Hsp26 has not been assessed.

Hsp27 is present in up to four isoforms according to the tissue and developmental stage (Marin et al. 1996). Like Hsp26, it can be ubiquitinated and phosphorylated on two serines (Bodenmiller et al. 2007; Franco et al. 2011; Zhai et al. 2008). However, Hsp27 is mainly found in the nucleus where it adopts at least two different localizations, one diffuse and one associated to speckles (Michaud et al. 2008). The role of post-translational modifications on the intra-nuclear localization of Hsp27 has not been investigated in detail like human HspB5 (den Engelsman et al. 2013). Nevertheless, phosphorylation of serine 75 has been linked to the presence of Hsp27 in the centrosome of *Drosophila* Kc cells and early embryos (Habermann et al. 2012). Interestingly, Hsp27 has also been identified with Hsp23 and Hsp26 in a microtubule-associated complex suggesting an additional cytoplasmic localization for Hsp27 (Fisher et al. 2008; Hughes et al. 2008). This is not surprising since Hsp27 contains a nuclear export signal (NES) consensus sequence in addition to its nuclear localization sequence (NLS) (Michaud et al. 2008).

25.1.3 Developmental Expression and Function

Hsp22, Hsp23, Hsp26, Hsp27, Hsp67Ba, Hsp67Bc and l(2)efl have tissue- and stage-specific expression patterns during development that have been reviewed previously (Michaud et al. 2002). They are mostly found in germline, nervous system and muscle cells and are induced by the molting hormone ecdysone. The expression of CG14207 is also tightly regulated; indeed it is restricted to muscle founder cells and to a subset of somatic myoblasts in embryogenesis (Artero et al. 2003; Estrada et al. 2006; Tomancak et al. 2007). To our knowledge, the larval and pupal expression patterns of CG14207 have not been reported nor has the developmental expression pattern for CG13133, CG7409, CG4461 and CG43851. However, their mRNA expression levels can be found on FlyAtlas and modENCODE projects websites (Celniker et al. 2009; Chintapalli et al. 2007).

The function of sHsps during development is still not well understood, but their specific regulation suggests that they are involved in specific function(s). By their ability to bind actin and microtubule, Hsp23, Hsp26 and Hsp27 could be involved in embryo morphogenesis (Fisher et al. 2008; Goldstein and Gunawardena 2000; Gong et al. 2004; Hughes et al. 2008). Accordingly, Hsp23 has been identified in a proteomic screen aimed at identifying proteins involved in ventral furrow formation (Gong et al. 2004). Moreover, Hsp26 has been shown to interact with myosin 10A, the *Drosophila* myosin XV homolog, a protein involved in regulating filopodial dynamics during dorsal closure (Liu et al. 2008). Interestingly, the role of myosin 10A is to maintain the balance between actin and microtubule cytoskeleton components (Liu et al. 2008). Therefore, this interaction places Hsp26 in the same category as mammalian HspB1, HspB5 and HspB6, which all contribute to the organization/regulation of the intermediate filament network (Seit-Nebi et al. 2013; Wettstein et al. 2012).

An interaction between Hsp27 and vreteno has been shown by immunoprecipitation in ovaries (Handler et al. 2011; Zamparini et al. 2011). Vreteno is one of the four factors required for primary piwi-interacting RNA (piRNA) biogenesis in the ovarian soma and germline and is therefore involved in the gonad-specific small RNA silencing pathway that protects the fly genome against the deleterious activity of transposable elements. While the functional significance of the interaction between Hsp27 and vreteno has not been pushed further, it is intriguing since Hsp27 has a well-characterized stage-specific distribution during oogenesis (Marin and Tanguay 1996).

Hsp27 is also involved in eye development and abolishing its expression results in rough eye phenotype with fused and enlarged ommatidia (Chen et al. 2012). While expressing the autophagy gene Atg7 restores the eye phenotype (see later section) the pathway by which Hsp27 intervenes in eye development has not been elucidated (Chen et al. 2012). Recently, Hsp27 was shown to interact with XPORT (exit protein of rhodopsin and TRP) together with Hsp83 during TRP and rhodopsin biosynthesis (Rosenbaum et al. 2011).

While the expression of sHsps is tightly regulated during development, eliminating the expression of one or another does not have deleterious effects. This is especially so for CG14207, Hsp22, Hsp23 and Hsp27, suggesting a non-essential role of those sHsps during fly development under laboratory conditions (Hao et al. 2007; Michaud and Tanguay 2003; Morrow et al. 2004a; Zimmermann et al. 2006). While this may be due to a certain redundancy of sHsps function, even the deletion of locus 67B (seven sHsps) has no deleterious effect during development (Geiger-Thornsberry and Mackay 2004).

Rather then having an essential developmental role, Hsp22, Hsp67Ba and Hsp67Bc would be involved in the maintenance of phenotypic stability under different environmental conditions (Takahashi et al. 2010). Indeed the transcriptional knockdown of Hsp22 was shown to increase fluctuation asymmetry of bristle number, while Hsp67Ba absence increases both fluctuation asymmetry of bristle number and wing shape. Therefore, in the absence of those sHsps, an increased number of phenotypic aberrations are observed suggesting that sHsps would act as developmental buffers ensuring developmental stability and canalization. This is reminiscent of the buffering ability of Hsp83, which supports its designation as a capacitor for morphological evolution (Rutherford and Lindquist 1998).

25.1.4 sHsp Expression in Adult Flies

As they do during development, sHsps exhibit tissue-specific expression patterns in flies, most of them being expressed in at least one of the following tissues: gonads, nervous system and muscle/heart (Kapelnikov et al. 2008; McGraw et al. 2008; Michaud et al. 2002). The expression pattern of *CG7409* and *hsp22* in the fly head was even found to vary according to the circadian rhythm (Ceriani et al. 2002).

During aging, the expression of most of the sHsps is up-regulated with the exception of CG7409 and CG4461 (Table 25.2). *Hsp22* is the sHsp preferentially expressed during aging and is mostly up-regulated in the head, while *l(2)efl* and *hsp23* are mostly up-regulated in thoraces and abdomen (Girardot et al. 2006; King and Tower 1999). Interestingly, an earlier onset of *hsp22* and *hsp23* expression was observed in flies genetically selected for increased longevity (Kurapati et al. 2000; Zhao et al. 2005), suggesting a beneficial role of sHsps during aging. Accordingly, the over-expression of l(2)efl, Hsp22, Hsp23, Hsp26 and Hsp27 were all shown to increase lifespan (Table 25.2). However, the timing of over-expression is important as flies over-expressing Hsp22 from day 4 and onward did not benefit from the same protective effect as flies over-expressing Hsp22 from embryogenesis (Bhole et al. 2004; Morrow et al. 2004b). The matter of timing has also been reported in other studies. For example, the degree of the protective effect of Hsp26 over-expression is higher in adults than in larvae where it has only a small thermoprotective effect on larval locomotion and no effect on neural function (Liao et al. 2008; Mileva-Seitz et al. 2008; Wang et al. 2004).

Table 25.2 *Drosophila* sHsps and longevity

sHsps	Aging regulation	Effect on longevity: Over-expression	Effect on longevity: Knockdown	References
CG14207	Up	ND	ND	Landis et al. (2012), Zou et al. (2000)
l(2)efl	Up	Increased	ND	Girardot et al. (2006), Landis et al. (2004, 2012), Wang et al. (2005)
CG7409	Down	ND	ND	Landis et al. (2012)
CG4461	Down	ND	ND	Landis et al. (2012)
Hsp26	Up	Increased	ND	Landis et al. (2012), Liao et al. (2008), Wang et al. (2004)
Hsp67Bc	Up	ND	ND	Landis et al. (2012)
Hsp22	Up	Increased	Decreased	Grover et al. (2009), King and Tower (1999), Morrow et al. (2004a, b), Moskalev et al. (2009), Yang and Tower (2009)
Hsp23	Up	Increased	Decreased	Tanguay (unpublished), King and Tower (1999), Landis et al. (2012), Tanguay and Morrow (2008)
Hsp27	Up	Increased	Decreased	Hao et al. (2007), Landis et al. (2012), Liao et al. (2008), Wang et al. (2004)

No data were available for CG43851, CG13133 and Hsp67Ba
ND not determined

While eliminating sHsp expression does not have deleterious effects on embryo development, it does affect the lifespan of adult flies. Indeed, eliminating Hsp22, Hsp23 or Hsp27 expression resulted in a decreased lifespan (Table 25.2). In the case of Hsp22 and Hsp23, the decreased longevity was also associated with a decreased resistance to thermal and oxidative stress (Tanguay unpublished; Morrow et al. 2004a). In contrast, flies without Hsp27 were as resistant as control flies to thermal and oxidative stress, but were more sensitive to starvation and more susceptible to infection (Chen et al. 2010; Hao et al. 2007). These differences may be due in part to the specific intracellular localization of those sHsps and reflect their involvement in different life promoting pathways.

25.1.5 Hsp22: An Aging Biomarker

Hsp22 is the sHsp preferentially expressed during aging and its level of expression is partially predictive of longevity in individual flies (Grover et al. 2009; King and Tower 1999; Yang and Tower 2009). Indeed, using the *hsp22* promoter to drive the expression of GFP or DsRed, a spike of reporter expression was consistently observed in the hours preceding and overlapping the death of the flies (Grover et al. 2009). This was also true for *hsp70*-reporter (Grover et al. 2009; Yang and Tower 2009), suggesting that both Hsp22 and Hsp70 are aging biomarkers. Moreover, it was shown that flies experiencing stress sooner in life, as shown by *hsp22*-reporter expression, are also the ones that die sooner (Yang and Tower 2009). Therefore, the level of *hsp22*-reporter expression is indicative of an individual susceptibility to stress and even to imminent mortality.

The use of the *hsp22*-GFP reporter construct has also unveiled cell-specific and cell lineage-specific patterns of mitochondrial failure during aging (Tower et al. 2014). Indeed, the expression of *hsp22*-GFP during aging is up-regulated in a subset of oenocytes (liver-like cells) that accumulate fewer age pigments. Interestingly, two proteins were shown to increase the preferential expression of the *hsp22*-GFP reporter in aging oenocytes and to decrease the age pigment accumulation, the mitochondrial MnSOD and Hsp22 itself. While over-expression of MnSOD was already shown to result in *hsp22* up-regulation (Curtis et al. 2007), it is the first time that Hsp22 is shown to have an effect on its own expression. Interestingly, the cells that do express the *hsp22*-GFP reporter have no detectable enrichment for MitoSOX-Red suggesting a decreased accumulation of superoxide (Tower et al. 2014). It was therefore suggested that those cells underwent a change of mitochondrial metabolic activity that could be mediated by Hsp22 and MnSOD, which both affect transcription of such enzymes upon over-expression (Curtis et al. 2007; Kim et al. 2010).

25.1.6 Small Hsps and dFoxo

The up-regulation of the Jun-N-terminal Kinase (JNK) pathway and the down-regulation of the Insulin/IGF pathway signaling (IIS) result in an increased lifespan and resistance to stress (Giannakou and Partridge 2007; Tatar et al. 2001; Wang et al. 2003, 2005). Both pathways converge to the transcription factor Foxo, which integrates nutritional and stress signals to regulate growth, cell proliferation and stress tolerance (Accili and Arden 2004; Wang et al. 2005). In *Caenorhabditis elegans*, Daf-16/Foxo influences life span by inducing expression of Hsp-16.1, Hsp-12.6, Hsp-16.49, and Hsp-16.11 (Hsu et al. 2003; Murphy et al. 2003). In *Drosophila*, dFoxo has been shown to regulate the expression of *CG14207* (Wang et al. 2005), *l(2)efl* (Hull-Thompson et al. 2009; Wang et al. 2005), *hsp22* (Harvey et al. 2008; Hull-Thompson et al. 2009) and *hsp23* (Harvey et al. 2008). Additionally, *hsp67Bc* is up-regulated upon dFoxo gain of function but not *hsp26* nor *hsp67Ba* suggesting that not all of the sHsps are regulated by this transcription factor (Harvey et al. 2008). Interestingly, four of the five sHsps regulated by dFoxo are also the ones showing an increased expression during aging suggesting a role of dFoxo in their aging-induced expression. Accordingly, *dfoxo* null flies have a reduced lifespan and display reduced age-induced expression of Hsp22 (Morrow, unpublished) and *l(2)efl* (Shen and Tower 2010). Since l(2)efl, Hsp22 and Hsp23 increase longevity upon over-expression they could account, at least in part, for the life span extension mediated by dFoxo (Wang et al. 2005).

25.1.7 Thermal Induction

Ten out of the 12 sHsps are heat-inducible; only CG14207 is not and no data is available on CG43851 (Landis et al. 2012; Telonis-Scott et al. 2013; Vos 2009). The set of sHsps induced by the heat shock response is stage-dependent; while heat-shocked embryos express *hsp22*, *hsp26*, *hsp27* and *hsp67Bc* (Leemans et al. 2000), 10 days-old flies additionally express *hsp67Ba*, *CG7409* and *l(2)efl* (Landis et al. 2012). *CG4461*, *hsp23*, *hsp26*, *hsp27* and *hsp67Bc* are part of the early heat-responding genes in flies and are also the more abundant *shsp* transcripts following heat-shock in S2 cells (Telonis-Scott et al. 2013; Vos 2009). The heat-shock response is driven by the heat shock factor (Hsf), which has been shown to bind tightly to *hsp22*, *hsp23*, *hsp26*, *hsp27*, *hsp67Ba* and *hsp67Bc* promoter while the *CG4461* promoter was apparently not bound (Birch-Machin et al. 2005). While Hsf is responsible for the majority of heat-induced expression of *l(2)efl*, *hsp23* and *hsp27*, it is only responsible for a small proportion of *hsp26* and *hsp22* heat-induced expression suggesting the involvement of other transcription factor(s) or mechanisms in the heat shock response (Neal et al. 2006). Accordingly, a possible involvement of RNAi machinery in the heat shock response in conjunction with RNA polymerase II has recently been proposed and will need further investigation (Cernilogar et al. 2011).

Hsp22, *hsp23*, *hsp26* and *hsp27* are also induced by cold stress while *hsp67Ba* is not altered (Colinet et al. 2010a, b, 2013; Zhang et al. 2011b). Eliminating the expression of Hsp22 and Hsp23 affects the recovery from chill coma suggesting that these two proteins contribute to the adaptive response to fluctuating thermal conditions (Colinet et al. 2010a). Since their level of cold-induction is stage-specific it has been proposed that they act through different mechanisms, which are still not elucidated (Colinet et al. 2013). Consistent with a protective role of sHsps, *hsp22*, *hsp23* and *hsp26* have been shown to contribute to cold hardening (Colinet et al. 2013; Qin et al. 2005).

Not surprisingly and in line with the thermal modulation of their expression, sHsps are involved in fly climatic adaptation. This is especially so for *D. pseudoobscura hsp23*, *hsp26*, *hsp27* and *hsp67Ba* (Graham et al. 2012), *D. virilis hsp67Bc* and *CG4461* (Vesala et al. 2012) and *D. subobscura hsp26* (Laayouni et al. 2007). In addition, studies on flies from sub-Saharan Africa and Europe have revealed the consistent over-expression of *CG7409* in African flies and of *hsp23* in European flies (Catalan et al. 2012; Muller et al. 2011).

25.1.8 sHsps and Oxidative Stress

CG14207, *l(2)efl*, *hsp22*, *CG7409* and *hsp23* are all induced by different oxidative stressors while *hsp26* and *CG4461* are up- or down-regulated according to the stressor (Grover et al. 2009; Gruenewald et al. 2009; Hirano et al. 2012; Landis et al. 2004, 2012; Yang and Tower 2009; Zou et al. 2000). Interestingly, the regulation of *hsp67Ba*, *hsp27* and *hsp67Bc* expression is not affected by any kind of oxidative stress suggesting their involvement in different regulatory processes (Landis et al. 2012). In the case of Hsp27, these data are consistent with the fact that eliminating its expression has no effect on oxidative stress resistance (Hao et al. 2007).

The role of oxidative stress in aging is debated, but it was recently shown that aging shares the largest number of gene expression changes with hyperoxia (Landis et al. 2012). Consistently, most of the sHsps up-regulated by oxidative stress are also the ones that are up-regulated during aging (Landis et al. 2012). In line with a beneficial role for sHsp in oxidative stress conditions, over-expression of l(2)efl, Hsp22, Hsp23, Hsp26 or Hsp27 increases resistance to this stress (Gruenewald et al. 2009; Morrow et al. 2004b; Tanguay and Morrow 2008; Wang et al. 2004, 2005).

25.1.9 *l(2)efl and Proteostasis*

In a recent study aimed at comparing the changes caused by heat, ionizing radiation, hyperoxia and hydrogen peroxide to changes observed during normal aging, *l(2)efl* was identified as one of the 18 up-regulated genes in all conditions besides *hsp70*, *hsp83* and *ref(2)P* which is involved in the mitochondrial unfolding protein response (UPR^{mt}) (Landis et al. 2012; Shen and Tower 2013). This places l(2)efl at the center of proteostasis during mild stress treatment and aging (Shen and Tower 2013).

25.1.10 *sHsps and Other Stressors*

Multiple genome-wide studies on the stress response to various compounds are available, each showing the preferential up/down regulation of only some sHsps. The implication of those sHsps in each stress will not be discussed here, but an overview of sHsp regulation by different stressors is presented in Table 25.3.

25.1.11 *sHsps and the Immune System*

Drosophila host defense against pathogenic bacteria, fungi and viruses involves Toll, Imd, JNK, JAK-STAT and p38 MAPK pathways (Eleftherianos and Castillo 2012; Kingsolver et al. 2013). This latter pathway activates Hsf and requires the proper expression of Hsp26, Hsp27, Hsp60D and Hsp70Bc to mediate host defense (Chen et al. 2010). Consistent with the role of Hsp27 in the immune system, *hsp27* mutant flies are more susceptible to pathogenic infection (Chen et al. 2010; Hao et al. 2007).

Table 25.3 Regulation of *Drosophila* sHsps by different stress

Stressors	*shsps*	References
Ethanol	*hsp26* (up)	Awofala et al. (2012), Morozova et al. (2006)
	l(2)efl (up)	
	CG14207 (up)	
Electromagnetic field	*hsp27* (up)	Herranz et al. (2012), Lee et al. (2008), Li et al. (2013)
	hsp22 (up/down)	
	hsp23 (up)	
Methotrexate	*hsp67Bc* (down)	Chahine and O'Donnell (2010)
Salicylate	*hsp67Bc* (down)	Chahine and O'Donnell (2010)
Hypoxia	*hsp23* (up)	Azad et al. (2009), Baird et al. (2006)
Ionizing radiation	*l(2)efl* (up)	Landis et al. (2012)
	hsp67Bc (up)	
	CG14207 (up)	
	CG4461 (up)	
	hsp67Ba (down)	
	CG7409 (down	
	hsp27 (down)	
Açai (anti-oxidant)	*l(2)efl* (up)	Sun et al. (2010)
Nectarine extract (anti-oxidant)	*l(2)efl* (down)	Boyd et al. (2011)
Endosulfane	*hsp23* (up)	Sharma et al. (2011, 2012)
	hsp22 (up)	
Cadmium	*hsp22* (up)	Yepiskoposyan et al. (2006)
Curcumin	*hsp22* (up)	Soh et al. (2013)
	hsp27 (up)	
Benzene, toluene, xylene	*hsp26* (up, individually)	Singh et al. (2009, 2010)
	hsp26 (down, mixture)	
Destruxin A (insecticide)	*hsp27* (down)	Pal et al. (2007)
	hsp23 (up)	
Pyrethrum (insecticide)	*l(2)efl* (up)	Jensen et al. (2006)

Pseudomonas aeruginosa eludes the host defenses of *Drosophila* by suppressing antimicrobial peptide gene expression (Apidianakis et al. 2005). The primary steps of the host response to the bacterial challenge involve the down-regulation of *hsp22*, *hsp26* and *CG7409* (Apidianakis et al. 2005; Chakrabarti et al. 2012). Interestingly, *Wolbachia* bacteria, which are obligate endosymbionts, also use this strategy. Indeed, they down-regulate the expression level of *hsp22*, *hsp27* and *hsp67Bc* (Xi et al. 2008). The roles of mammalian sHsps in the immune system are also being uncovered and include the activation of macrophages (van Noort et al. 2012). In *Drosophila*, a role for Hsp26 in phagosome internalization has been suggested from a proteomic study aimed at detailing the protein-protein interactions of this subcellular compartment (Stuart et al. 2007).

25.1.12 sHsps and Apoptosis

In *Drosophila*, apoptotic stimuli converge to activate a common death program through transcriptional activation of *reaper*, *hid* and *grim* (Steller 2008). The resulting three proteins bind IAPs (inhibitor of apoptosis proteins), resulting in caspase-activation and apoptosis. While reaper and grim are expressed only in cells destined to die, hid is also present during development and its expression is regulated by EGFR/RAS signaling, Hippo tumor suppressor pathway, miRNAs and the hormone ecdysone (Bilak and Su 2009).

Hsp27 reduces hid-induced lethality but has no effect on the lethality induced by reaper and grim (Liao et al. 2008). The exact mechanism by which Hsp27 protects *Drosophila* against hid-induced apoptosis is not known but it could be by regulating the Ras/MAPK pathway in a similar way to Hsp83 i.e. by facilitating signaling through conformational modulation of its substrate (Raf in the case of Hsp83) (van der Straten et al. 1997).

While Hsp26 has no effect on fly survival to hid-, reaper- and grim-induced lethality when over-expressed in neurons (Liao et al. 2008), it was shown to interact with activated Cdc42 kinase (ack) together with Hsp23 and some other proteins in an affinity assay on S2 cells (Schoenherr et al. 2012). Ack is an anti-apoptotic protein, which blocks programmed cell death induced by hid and reaper. While the knockdown of Hsp23 had no effect on ack anti-apoptotic activity, the knockdown of Hsp26 was shown to suppress this activity (Schoenherr et al. 2012). These results demonstrate the distinctive features of *Drosophila* sHsps and suggest that Hsp26 functions directly with ack or in a parallel pathway to regulate survival.

Hsp23 was identified in a study aimed at determining which proteins undergo caspase-dependent changes during apoptosis in a sub-line of S2 cells named D-Mel2 (Creagh et al. 2009). The Hsp23 isoform identified appears due to actinomycin D-induced activation of caspases since the protein was absent when using the caspase inhibitor z-VAD-fmk. The nature of this isoform is not discussed further, but its experimental pI is reminiscent of the two Hsp23 isoforms previously reported and which display a tissue-specific expression pattern in flies (Marin et al. 1996). In mammalian cells, actinomycin D prevents HspB1 degradation depending on the cell type and HspB1 knockdown potentiates actinomycin D-induced apoptosis through caspase activation (Cervantes-Gomez et al. 2009; Ma et al. 2013). Moreover, actinomycin D increases HspB1 phosphorylation, a post-translational modification that would favor small oligomers and cytochrome c binding (Boncoraglio et al. 2012; Ma et al. 2013). In light of these results, the appearance of the Hsp23 isoform in D-Mel2 cells treated with actinomycin D could reflect the attempt of the cells to counteract the apoptotic pathway.

The anti-apoptotic function of sHsps is sometimes taken as an advantage for tumor cells to grow and proliferate. Thus mammalian sHsps are up-regulated in many different cancer cell types and are often linked to bad prognoses (Boncoraglio et al. 2012; Kampinga and Garrido 2012). In *Drosophila*, *hsp27* and *hsp26* are up-regulated in brain tumors (Loop et al. 2004). Moreover, cancer cells over-expressing

Hsp22 were shown to form aggressive tumors and to acquire an anti-cancer drug-resistant phenotype. These effects would be due in part to the interaction of Hsp22 with p53 resulting in a mitochondrial sequestration of the tumor suppressor (Wadhwa et al. 2010).

25.1.13 sHsps and Misfolding Protein Diseases

In vitro, Hsp22, Hsp23, Hsp26 and Hsp27 prevent heat-induced aggregation of citrate synthase and luciferase and maintain heat-denatured luciferase in a refoldable state albeit with different efficiencies (Morrow et al. 2006). The chaperone-like activity of *Drosophila* sHsps to assist substrate refolding was also analyzed in vivo in cultured cells using luciferase, EGFP-HttQ119 and HttQ128 as reporters (Table 25.4). As for mammalian sHsps, *Drosophila* sHsps have different abilities to prevent substrate aggregation and to assist refolding (Carra et al. 2013). Indeed, CG14207 and CG7409 have the strongest ability to assist refolding of heat-denatured luciferase but are only partially able to reduce the level of polyQ insoluble protein aggregates and have no effect on the level of soluble polyQ proteins. On the other hand, Hsp67Bc is not able to assist luciferase refolding but strongly reduces the level of soluble polyQ proteins and insoluble aggregates (Carra et al. 2010; Vos 2009). The only other sHsp that was shown to reduce the level of soluble polyQ protein is l(2)efl (see next section for the mechanism, (Carra et al. 2010)).

Consistent with their different ability to prevent substrate aggregation and assist protein refolding in cultured cells, *Drosophila* sHsps also have different abilities to suppress degeneration in misfolded-protein disease models (Table 25.5). Usually the strongest effect of sHsps on degeneration is observed with mildly toxic proteins;

Table 25.4 Chaperone-like activity of *Drosophila* sHsps in cultured cells

sHsps	Activities	Methods	Effects	References
Hsp67Bc	Preventing aggregation	Htt-128Q and Hsp67Bc/S2 cells	Reduce the soluble level and the amount of high MW aggregates	Carra et al. (2010)
		EGFP-Htt-Q119 and Hsp67Bc/S2 cells	Strong reduction of the level of insoluble proteins/decreased amount of soluble proteins	Vos (2009)
	Assisting refolding	Luciferase and Hsp67Bc/S2 cells	Do not assist refolding	Vos (2009)
CG14207	Preventing aggregation	Htt-128Q and CG14207/S2 cells	No effect on soluble level neither on high MW	Carra et al. (2010)
		EGFP-Htt-Q119 and CG14207/S2 cells	Partial reduction of the level of insoluble proteins/no effect on soluble proteins	Vos (2009)
	Assisting refolding	luciferase and CG14207/S2 cells	Strong ability to assist refolding/requires Hsp70 machinery	Vos (2009)

(continued)

Table 25.4 (continued)

sHsps	Activities	Methods	Effects	References
l(2)efl	Preventing aggregation	Htt-128Q and l(2)efl/S2 cells	Reduce the soluble level of proteins/no effect on high MW	Carra et al. (2010)
		EGFP-Htt-Q119 and l(2)efl/S2 cells	Partial reduction of the level of insoluble proteins/no effect on soluble proteins	Vos (2009)
	Assisting refolding	Luciferase and l(2)efl/S2 cells	Assist refolding	Vos (2009)
Hsp23	Preventing aggregation	EGFP-Htt-Q119 and Hsp23/S2 cells	Partial reduction of the level of insoluble proteins/partial reduction on soluble proteins	Vos (2009)
	Assisting refolding	Luciferase and Hsp23/S2 cells	Assist refolding	Vos (2009)
Hsp26	Preventing aggregation	EGFP-Htt-Q119 and Hsp26/S2 cells	Partial reduction of the level of insoluble proteins/partial reduction on soluble proteins	Vos (2009)
	Assisting refolding	Luciferase and Hsp26/S2 cells	Assist refolding	Vos (2009)
Hsp27	Preventing aggregation	EGFP-Htt-Q119 and Hsp27/S2 cells	Partial reduction of the level of insoluble proteins/no effect on soluble proteins	Vos (2009)
	Assisting refolding	Nuclear luciferase and Hsp27/S2 cells	Assist refolding/requires Hsp70 machinery	Vos (2009)
Hsp67Ba	Preventing aggregation	EGFP-Htt-Q119 and Hsp67Ba/S2 cells	Partial reduction of the level of insoluble proteins/no effect on soluble proteins	Vos (2009)
	Assisting refolding	Luciferase and Hsp67Ba/S2 cells	Do not assist refolding	Vos (2009)
CG4461	Preventing aggregation	EGFP-Htt-Q119 and CG4461/S2 cells	Partial reduction of the level of insoluble proteins/no effect on soluble proteins	Vos (2009)
	Assisting refolding	Luciferase and CG4461/S2 cells	Do not assist refolding	Vos (2009)
CG7409	Preventing aggregation	EGFP-Htt-Q119 and CG7409/S2 cells	Partial reduction of the level of insoluble proteins/no effect on soluble proteins	Vos (2009)
	Assisting refolding	Luciferase and CG7409/S2 cells	Strong ability to assist refolding	Vos (2009)
CG13133	Preventing aggregation	EGFP-Htt-Q119 and CG13133/S2 cells	Partial reduction of the level of insoluble proteins/no effect on soluble proteins	Vos (2009)
	Assisting refolding	luciferase and CG13133/S2 cells	Do not assist refolding	Vos (2009)
Hsp22	Assisting refolding	Mito-luciferase and Hsp22/HeLa cells	Assist refolding	Tanguay (unpublished)

No data were available for CG43851
MW molecular weight

Table 25.5 Effect of *Drosophila* sHsps over-expression on protein misfolding diseases

sHsps	Misfolding disease model	Suppression of degeneration	Effect on protein	References
Hsp67Bc	UAS SCA3-Q78/gmr-Gal4 UAS-Hsp67Bc	Partial	ND	Vos (2009)
	UAS SCA3-78Q/gmr-Gal4 UAS-Hsp67Bc	Partial	ND	Carra et al. (2010)
CG7409	UAS SCA3-Q78/gmr-Gal4 UAS-CG7409	Partial	ND	Vos (2009)
CG4461	UAS SCA3-Q78/gmr-Gal4 UAS-CG4461	No effect	ND	Vos (2009)
CG14207	UAS SCA3-Q78/gmr-Gal4 UAS-CG14207	No effect	ND	Vos (2009)
	UAS SCA3tr-Q78/gmr-Gal4 EP 1348	Partial – mild	ND	Bilen and Bonini (2007)
	UAS SCA3-Q84/gmr-Gal4 EP 1348	Partial – strong	Increased solubility of monomers/no effect on oligomers	Bilen and Bonini (2007)
	UAS TDP-43/gmr-Gal4 EP 1348	Partial	Increased solubility of monomers	Gregory et al. (2012)
	UAS TDP-25/gmr-Gal4 (mild) EP 1348	Complete	Clearance of monomers and oligomers	Gregory et al. (2012)
Hsp27	41Q/gmr-Gal4 (mild) UAS-Hsp27	Complete	ND	Chen et al. (2012)
	63Q/gmr-Gal4 UAS-Hsp27	No effect	ND	Chen et al. (2012)
	41Q/gmr-Gal4 (mild) UAS-Hsp27	Complete	ND	Liao et al. (2008)
	127Q/gmr-Gal4 UAS-Hsp27	No effect	ND	Liao et al. (2008)
	UAS GFAP^{R79H}/repo-Gal4 UAS-Hsp27	Partial	Decreased number of inclusion bodies	Wang et al. (2011)
Hsp26	41Q/gmr-Gal4 (mild) UAS-Hsp26	No effect	ND	Liao et al. (2008)
	127Q/gmr-Gal4 UAS-Hsp26	No effect	ND	Liao et al. (2008)
	UAS GFAP^{R79H}/repo-Gal4 UAS-Hsp26	Partial	Decreased number of inclusion bodies	Wang et al. (2011)
l(2)efl	UAS GFAP^{R79H}/repo-Gal4 UAS-l(2)efl	Partial	Decreased number of inclusion bodies	Wang et al. (2011)

No data were available for Hsp22, Hsp23, CG43851, CG13133 and Hsp67Ba
ND not determined

this is particularly so for CG14207 (TDP25 vs TDP43, (Gregory et al. 2012)) and Hsp27 (41Q vs 63Q or 127Q, (Chen et al. 2012)), suggesting that the balance between toxic proteins and chaperones is very important. However, CG14207 was shown to be a strong modifier of full length Ataxin-3 but only a moderate suppressor of the truncated protein (Sca84Q vs Sca78Q, (Bilen and Bonini 2007)), suggesting that the affinity of the sHsps for the substrate is also an important factor. This could explain why some sHsps, like Hsp26, have an effect only on specific misfolded protein substrates (41Q vs GFAP^{R79H}, (Liao et al. 2008; Wang et al. 2011)). Alternatively, the localization of the sHsp versus the substrate may also be an important factor to bear in mind.

One way by which misfolded proteins cause their toxic effect is by titrating out proteins involved in protein quality control such as sHsps. This is especially true for l(2)efl which is found in inclusions together with GFAP^{R79H} in Alexander disease (Head and Goldman 2000; Wang et al. 2011). Consistently, eliminating the expression of sHsps enhances the toxicity of aggregation prone proteins. Indeed, the knockdown of Hsp67Bc significantly worsens the SCA3 degenerative eye phenotype and the knockdown of Hsp27 enhances the pigmentation phenotype observed in the eye of flies expressing 41Q (Carra et al. 2010; Chen et al. 2012).

A biphasic *hsp* expression profile has been reported in different polyQ disease models; the heat shock response is induced by short-term polyQ protein expression and reduced upon long-term expression (Chan et al. 2011; Chou et al. 2008; Hands et al. 2008; Huen and Chan 2005). A recent study has shown that *hsp22*, *hsp40* and *hsp70* are induced when MJDQ84 is localized in the cytoplasm but not when it is located in the nucleus in the absence of xpo1 (Chan et al. 2011). Xpo1 is an exportin that preferentially exports expanded polyQ protein to the cytosol and which has also been shown to suppress Aß toxicity in a *C. elegans* model of Alzheimer's disease (Chan et al. 2011; Treusch et al. 2011). Interestingly, MDJQ84 was shown to directly bind to *hsp70* promoter, suggesting a mechanism whereby the nuclear accumulation of expanded polyQ proteins could block *hsp* expression (Chan et al. 2011). Alternatively, some toxic polyQ proteins have a higher affinity for transcription factor than their wild-type counterpart, thereby affecting their normal cellular function(s). This is notably the case for hTBP80Q, which has a high affinity for suppressor of hairless (Su(H)) (Ren et al. 2011). While *hsp23*, *hsp26* and *hsp27* are down-regulated in flies expressing the toxic hTBP80Q, the involvement of Su(H) in this regulation has not been assessed (Ren et al. 2011).

25.1.14 Additional Mechanisms of sHsps

As mentioned in the previous section, most sHsps can bind to misfolded proteins to prevent their aggregation and/or assist their refolding by ATP-dependent chaperones. Another way by which l(2)efl and Hsp67Bc (but not CG14207) prevent the deleterious effect of misfolded proteins is by inhibiting protein synthesis. The inhibition of protein synthesis occurs via eIF2α phosphorylation and is responsible for the decreased amount of soluble polyQ proteins observed in their presence (Table 25.4, (Carra et al. 2010)). Among those three sHsps, only Hsp67Bc also promotes autophagy to allow the clearance of high molecular weight aggregates (Table 25.4, (Carra et al. 2010)). The stimulation of autophagy by Hsp67Bc occurs via its binding to starvin in a similar manner to HspB8 and Bag3 (Carra et al. 2010, 2009). Interestingly, in addition to having no interaction with starvin and to having no effect on LC3 lipidation in cultured cells, l(2)efl over-expression was even shown to reduce the activation of autophagy in the Alexander disease model ($GFAP^{R79H}$) (Carra et al. 2010; Wang et al. 2011). Therefore l(2)efl shares the ability to modulate protein synthesis with Hsp67Bc but both proteins have the opposite effect on autophagy. These results illustrate well the functional specificity of sHsps.

A link between Hsp27 and autophagy has recently been unveiled (Chen et al. 2012). Indeed, it was shown that autophagy-related gene 7 (Atg7) acts downstream of Hsp27 to regulate eye morphology, polyQ toxicity and even lifespan. This genetic interaction is specific and does not involve other Atgs nor Hsp22. Further studies will be needed to assess if Hsp27 functions through chaperone-mediated autophagy (CMA) like Hsp70 or through chaperone-assisted selective autophagy (CASA) like HspB8 (Arias and Cuervo 2011; Arndt et al. 2010). Autophagy serves to clear unnecessary or dysfunctional cellular components and Atg7 expression attenuates polyQ toxicity (Chen et al. 2012). The specific interplay between Hsp27 and Atg7 may explain in part why this sHsp was shown to be more effective than Hsp26 in suppressing 41Q induced neurodegeneration (Table 25.5, (Liao et al. 2008)).

Another way to clear misfolded proteins apart from autophagy is by promoting their degradation through the proteasome. While both HspB1 and HspB5 have been shown to bind proteasome subunits (Boelens et al. 2001; Parcellier et al. 2003), a direct interaction between proteasome subunits and *Drosophila* sHsps remains to be demonstrated. However, *hsp23* and *hsp27* were shown to be up-regulated in parallel to both 19S and 20S proteasome subunits, suggesting that these two sHsps are regulated components of the metazoan proteasome network. Moreover, an interaction between Hsp26 and lawc has been demonstrated by two-hybrid screening (Brandt and Corces 2008). While lawc is not a proteasome component, it does interact with the nuclear proteasome regulator dREGγ to ensure proper transcription by RNA polymerase II. The exact role of Hsp26 interaction with lawc has not been discussed, but it could involve its ability, also shared by Hsp23 and Hsp27, to bind to the ubiquitin-conjugating enzyme 9 (DmUbc9, (Joanisse et al. 1998)).

Finally, using a model for atrial fibrillation, a distinctive functional feature of Hsp23 was unveiled. Indeed, it was shown that over-expression of Hsp23 protects against tachycardia remodeling while Hsp27, Hsp67Bc, CG4461, CG7409 and CG14207 had absolutely no effect (Zhang et al. 2011a). While the lack of effect of Hsp67Bc suggests the non-involvement of autophagy and the lack of effect of Hsp27 suggests the necessity for cytoplasmic localization, the lack of effect of the three other sHsps suggests that a general chaperone-like activity is not required. Therefore, the ability of Hsp23 to protect against tachycardia remodeling involves a specific, yet undetermined, client.

25.1.15 sHsps and the Mitochondria

Drosophila Hsp22 is localized in the mitochondrial matrix (Morrow et al. 2000). It is rapidly induced by oxidative stress and preferentially up-regulated during aging (King and Tower 1999; Landis et al. 2012). Interestingly, it is also strongly up-regulated by the disruption of mitochondrial protein synthesis while *hsp23* and *hsp27* are mildly induced (Fernandez-Ayala et al. 2010). Since the disruption of mitochondrial protein synthesis has been shown to create a mito-nuclear imbalance of the electron transport chain subunits and the subsequent activation of the UPR^{mt} in *C. elegans* (Houtkooper et al. 2013), it was suggested that Hsp22 is part of *Drosophila* UPR^{mt} (Shen and Tower 2013). In *Drosophila*, the UPR^{mt} is involved in the aging process as well as in different types of stress, as demonstrated by the up-regulation of *ref(2)P* (Landis et al. 2012). In support of a role in UPR^{mt}, over-expression of Hsp22 results in alterations of many nuclear transcripts and also influences its own level of expression (Kim et al. 2010; Shen and Tower 2013; Tower et al. 2014). Due to the fact that the presence of Hsp22 in oenocyte cells was associated with the absence of MitoSox-Red staining enrichment in an aging model, it was suggested that Hsp22 could act by repressing mitochondrial metabolic activity (Shen and Tower 2013; Tower et al. 2014). Interestingly, CG13133 has also been shown to repress the basic mitochondrial function. Indeed, abolishing the expression of this sHsp results in increased citrate synthase activity (Chen et al. 2008). The way by which CG13133 interferes with mitochondrial function has not been pushed further but since it is a cytosolic protein it most likely acts on the mitochondrial outer membrane as does HspB2, HspB5 and HspB8 (Jin et al. 2008; Nakagawa et al. 2001). Alternatively, CG13133 could have an additional sub-mitochondrial localization such as HspB1 (Marunouchi et al. 2013). Unfortunately, few data are available on CG13133 except that it is expressed in *Drosophila* auditory organ (Senthilan et al. 2012).

25.2 Concluding Remarks

The genome of *Drosophila melanogaster* encodes for 12 sHsps. While Hsp22, Hsp23, Hsp26 and Hsp27 remain the most studied, the ongoing research on other members allows the appreciation of the diversity of *Drosophila* sHsps. Their distinctive developmental expression pattern, intracellular localization and in vivo substrate specificity, suggest that the amino acid sequence outside of the ACD is important for function determination. Since the functions of sHsps are only partially overlapping, much work remains to be done in order to understand their functions.

Acknowledgements We would like to thank Dr. Halim Maaroufi for his advice regarding the sHsp sequence analysis. This work was supported by grants from the Canadian Institutes of Health Research and the National Sciences Engineering Research Council to RMT.

References

Accili D, Arden KC (2004) FoxOs at the crossroads of cellular metabolism, differentiation, and transformation. Cell 117(4):421–426

Apidianakis Y, Mindrinos MN, Xiao W, Lau GW, Baldini RL, Davis RW, Rahme LG (2005) Profiling early infection responses: pseudomonas aeruginosa eludes host defenses by suppressing antimicrobial peptide gene expression. Proc Natl Acad Sci U S A 102(7):2573–2578

Arias E, Cuervo AM (2011) Chaperone-mediated autophagy in protein quality control. Curr Opin Cell Biol 23(2):184–189. doi:10.1016/j.ceb.2010.10.009

Arndt V, Dick N, Tawo R, Dreiseidler M, Wenzel D, Hesse M, Fürst DO, Saftig P, Saint R, Fleischmann BK, Hoch M, Höhfeld J (2010) Chaperone-assisted selective autophagy is essential for muscle maintenance. Curr Biol 20:143–148. doi:10.1016/j.cub.2009.11.022

Artero R, Furlong EE, Beckett K, Scott MP, Baylies M (2003) Notch and Ras signaling pathway effector genes expressed in fusion competent and founder cells during Drosophila myogenesis. Development 130(25):6257–6272

Awofala AA, Davies JA, Jones S (2012) Functional roles for redox genes in ethanol sensitivity in Drosophila. Funct Integr Genomics 12(2):305–315

Ayme A, Tissieres A (1985) Locus 67B of Drosophila melanogaster contains seven, not four, closely related heat shock genes. EMBO J 4(11):2949–2954

Azad P, Zhou D, Russo E, Haddad GG (2009) Distinct mechanisms underlying tolerance to intermittent and constant hypoxia in Drosophila melanogaster. PLoS One 4(4):e5371

Baird NA, Turnbull DW, Johnson EA (2006) Induction of the heat shock pathway during hypoxia requires regulation of heat shock factor by hypoxia-inducible factor-1. J Biol Chem 281(50):38675–38681

Basha E, O'Neill H, Vierling E (2012) Small heat shock proteins and alpha-crystallins: dynamic proteins with flexible functions. Trends Biochem Sci 37(3):106–117

Basha E, Jones C, Blackwell AE, Cheng G, Waters ER, Samsel KA, Siddique M, Pett V, Wysocki V, Vierling E (2013) An unusual dimeric small heat shock protein provides insight into the mechanism of this class of chaperones. J Mol Biol 425(10):1683–1696

Beaulieu JF, Arrigo AP, Tanguay RM (1989) Interaction of Drosophila 27,000 Mr heat-shock protein with the nucleus of heat-shocked and ecdysone-stimulated culture cells. J Cell Sci 92 (Pt 1):29–36

Bhole D, Allikian MJ, Tower J (2004) Doxycycline-regulated over-expression of hsp22 has negative effects on stress resistance and life span in adult Drosophila melanogaster. Mech Ageing Dev 125(9):651–663

Bilak A, Su TT (2009) Regulation of Drosophila melanogaster pro-apoptotic gene hid. Apoptosis 14(8):943–949

Bilen J, Bonini NM (2007) Genome-wide screen for modifiers of ataxin-3 neurodegeneration in Drosophila. PLoS Genet 3(10):1950–1964

Birch-Machin I, Gao S, Huen D, McGirr R, White RA, Russell S (2005) Genomic analysis of heat-shock factor targets in Drosophila. Genome Biol 6(7):R63

Blum T, Briesemeister S, Kohlbacher O (2009) MultiLoc2: integrating phylogeny and gene ontology terms improves subcellular protein localization prediction. BMC Bioinformatics 10:274

Bodenmiller B, Mueller LN, Pedrioli PG, Pflieger D, Junger MA, Eng JK, Aebersold R, Tao WA (2007) An integrated chemical, mass spectrometric and computational strategy for (quantitative) phosphoproteomics: application to Drosophila melanogaster Kc167 cells. Mol Biosyst 3(4):275–286

Boelens WC, Croes Y, de Jong WW (2001) Interaction between alphaB-crystallin and the human 20S proteasomal subunit C8/alpha7. Biochim Biophys Acta 1544(1–2):311–319

Boncoraglio A, Minoia M, Carra S (2012) The family of mammalian small heat shock proteins (HSPBs): implications in protein deposit diseases and motor neuropathies. Int J Biochem Cell Biol 44(10):1657–1669

Boyd O, Weng P, Sun X, Alberico T, Laslo M, Obenland DM, Kern B, Zou S (2011) Nectarine promotes longevity in Drosophila melanogaster. Free Radic Biol Med 50(11):1669–1678
Brandt T, Corces VG (2008) The Lawc protein is required for proper transcription by RNA polymerase II in Drosophila. Mol Genet Genomics 280(5):385–396
Briesemeister S, Blum T, Brady S, Lam Y, Kohlbacher O, Shatkay H (2009) SherLoc2: a high-accuracy hybrid method for predicting subcellular localization of proteins. J Proteome Res 8(11):5363–5366
Bryantsev AL, Kurchashova SY, Golyshev SA, Polyakov VY, Wunderink HF, Kanon B, Budagova KR, Kabakov AE, Kampinga HH (2007) Regulation of stress-induced intracellular sorting and chaperone function of Hsp27 (HspB1) in mammalian cells. Biochem J 407(3):407–417
Carra S, Brunsting JF, Lambert H, Landry J, Kampinga HH (2009) HspB8 participates in protein quality control by a non-chaperone-like mechanism that requires eIF2{alpha} phosphorylation. J Biol Chem 284(9):5523–5532
Carra S, Boncoraglio A, Kanon B, Brunsting JF, Minoia M, Rana A, Vos MJ, Seidel K, Sibon OC, Kampinga HH (2010) Identification of the Drosophila ortholog of HSPB8: implication of HSPB8 loss of function in protein folding diseases. J Biol Chem 285(48):37811–37822
Carra S, Rusmini P, Crippa V, Giorgetti E, Boncoraglio A, Cristofani R, Naujock M, Meister M, Minoia M, Kampinga HH, Poletti A (2013) Different anti-aggregation and pro-degradative functions of the members of the mammalian sHSP family in neurological disorders. Philos Trans R Soc Lond B Biol Sci 368(1617):20110409
Catalan A, Hutter S, Parsch J (2012) Population and sex differences in Drosophila melanogaster brain gene expression. BMC Genomics 13:654
Celniker SE, Dillon LA, Gerstein MB, Gunsalus KC, Henikoff S, Karpen GH, Kellis M, Lai EC, Lieb JD, MacAlpine DM, Micklem G, Piano F, Snyder M, Stein L, White KP, Waterston RH, mod EC (2009) Unlocking the secrets of the genome. Nature 459(7249):927–930
Ceriani MF, Hogenesch JB, Yanovsky M, Panda S, Straume M, Kay SA (2002) Genome-wide expression analysis in Drosophila reveals genes controlling circadian behavior. J Neurosci 22(21):9305–9319
Cernilogar FM, Onorati MC, Kothe GO, Burroughs AM, Parsi KM, Breiling A, Lo Sardo F, Saxena A, Miyoshi K, Siomi H, Siomi MC, Carninci P, Gilmour DS, Corona DF, Orlando V (2011) Chromatin-associated RNA interference components contribute to transcriptional regulation in Drosophila. Nature 480(7377):391–395
Cervantes-Gomez F, Nimmanapalli R, Gandhi V (2009) Transcription inhibition of heat shock proteins: a strategy for combination of 17-allylamino-17-demethoxygeldanamycin and actinomycin d. Cancer Res 69(9):3947–3954
Chahine S, O'Donnell MJ (2010) Effects of acute or chronic exposure to dietary organic anions on secretion of methotrexate and salicylate by Malpighian tubules of Drosophila melanogaster larvae. Arch Insect Biochem Physiol 73(3):128–147
Chakrabarti S, Liehl P, Buchon N, Lemaitre B (2012) Infection-induced host translational blockage inhibits immune responses and epithelial renewal in the Drosophila gut. Cell Host Microbe 12(1):60–70
Chan WM, Tsoi H, Wu CC, Wong CH, Cheng TC, Li HY, Lau KF, Shaw PC, Perrimon N, Chan HY (2011) Expanded polyglutamine domain possesses nuclear export activity which modulates subcellular localization and toxicity of polyQ disease protein via exportin-1. Hum Mol Genet 20(9):1738–1750
Chen J, Shi X, Padmanabhan R, Wang Q, Wu Z, Stevenson SC, Hild M, Garza D, Li H (2008) Identification of novel modulators of mitochondrial function by a genome-wide RNAi screen in Drosophila melanogaster. Genome Res 18(1):123–136
Chen J, Xie C, Tian L, Hong L, Wu X, Han J (2010) Participation of the p38 pathway in Drosophila host defense against pathogenic bacteria and fungi. Proc Natl Acad Sci U S A 107(48):20774–20779
Chen SF, Kang ML, Chen YC, Tang HW, Huang CW, Li WH, Lin CP, Wang CY, Wang PY, Chen GC, Wang HD (2012) Autophagy-related gene 7 is downstream of heat shock protein 27 in the

regulation of eye morphology, polyglutamine toxicity, and lifespan in Drosophila. J Biomed Sci 19:52

Chevenet F, Brun C, Banuls AL, Jacq B, Christen R (2006) TreeDyn: towards dynamic graphics and annotations for analyses of trees. BMC Bioinformatics 7:439

Chintapalli VR, Wang J, Dow JA (2007) Using FlyAtlas to identify better Drosophila melanogaster models of human disease. Nat Genet 39(6):715–720

Chou KC, Shen HB (2010) A new method for predicting the subcellular localization of eukaryotic proteins with both single and multiple sites: Euk-mPLoc 2.0. PLoS One 5(4):e9931

Chou AH, Yeh TH, Ouyang P, Chen YL, Chen SY, Wang HL (2008) Polyglutamine-expanded ataxin-3 causes cerebellar dysfunction of SCA3 transgenic mice by inducing transcriptional dysregulation. Neurobiol Dis 31(1):89–101

Colinet H, Lee SF, Hoffmann A (2010a) Knocking down expression of Hsp22 and Hsp23 by RNA interference affects recovery from chill coma in Drosophila melanogaster. J Exp Biol 213(Pt 24):4146–4150

Colinet H, Lee SF, Hoffmann A (2010b) Temporal expression of heat shock genes during cold stress and recovery from chill coma in adult Drosophila melanogaster. FEBS J 277(1):174–185

Colinet H, Siaussat D, Bozzolan F, Bowler K (2013) Rapid decline of cold tolerance at young age is associated with expression of stress genes in Drosophila melanogaster. J Exp Biol 216(Pt 2):253–259

Craig EA, McCarthy BJ (1980) Four Drosophila heat shock genes at 67B: characterization of recombinant plasmids. Nucleic Acids Res 8(19):4441–4457

Creagh EM, Brumatti G, Sheridan C, Duriez PJ, Taylor RC, Cullen SP, Adrain C, Martin SJ (2009) Bicaudal is a conserved substrate for Drosophila and mammalian caspases and is essential for cell survival. PLoS One 4(3):e5055

Curtis C, Landis GN, Folk D, Wehr NB, Hoe N, Waskar M, Abdueva D, Skvortsov D, Ford D, Luu A, Badrinath A, Levine RL, Bradley TJ, Tavare S, Tower J (2007) Transcriptional profiling of MnSOD-mediated lifespan extension in Drosophila reveals a species-general network of aging and metabolic genes. Genome Biol 8(12):R262

Delbecq SP, Klevit RE (2013) One size does not fit all: the oligomeric states of alphaB crystallin. FEBS Lett 587(8):1073–1080

den Engelsman J, van de Schootbrugge C, Yong J, Pruijn GJ, Boelens WC (2013) Pseudophosphorylated alphaB-crystallin is a nuclear chaperone imported into the nucleus with help of the SMN complex. PLoS One 8(9):e73489

Dereeper A, Guignon V, Blanc G, Audic S, Buffet S, Chevenet F, Dufayard JF, Guindon S, Lefort V, Lescot M, Claverie JM, Gascuel O (2008) Phylogeny.fr: robust phylogenetic analysis for the non-specialist. Nucleic Acids Res 36(Web Server issue):W465–W469

Dereeper A, Audic S, Claverie JM, Blanc G (2010) BLAST-EXPLORER helps you building datasets for phylogenetic analysis. BMC Evol Biol 10:8

Eleftherianos I, Castillo JC (2012) Molecular mechanisms of aging and immune system regulation in Drosophila. Int J Mol Sci 13(8):9826–9844

Estrada B, Choe SE, Gisselbrecht SS, Michaud S, Raj L, Busser BW, Halfon MS, Church GM, Michelson AM (2006) An integrated strategy for analyzing the unique developmental programs of different myoblast subtypes. PLoS Genet 2(2):e16

Fernandez-Ayala DJ, Chen S, Kemppainen E, O'Dell KM, Jacobs HT (2010) Gene expression in a Drosophila model of mitochondrial disease. PLoS One 5(1):e8549

Finn RD, Bateman A, Clements J, Coggill P, Eberhardt RY, Eddy SR, Heger A, Hetherington K, Holm L, Mistry J, Sonnhammer EL, Tate J, Punta M (2014) Pfam: the protein families database. Nucleic Acids Res 42(Database issue):D222–D230

Fisher KH, Deane CM, Wakefield JG (2008) The functional domain grouping of microtubule associated proteins. Commun Integr Biol 1(1):47–50

Franco M, Seyfried NT, Brand AH, Peng J, Mayor U (2011) A novel strategy to isolate ubiquitin conjugates reveals wide role for ubiquitination during neural development. Mol Cell Proteomics 10(5):M110 002188

Fu X (2014) Chaperone function and mechanism of small heat-shock proteins. Acta Biochim Biophys Sin (Shanghai) 46(5):347–356

Garrido C, Paul C, Seigneuric R, Kampinga HH (2012) The small heat shock proteins family: the long forgotten chaperones. Int J Biochem Cell Biol 44(10):1588–1592

Geiger-Thornsberry GL, Mackay TF (2004) Quantitative trait loci affecting natural variation in Drosophila longevity. Mech Ageing Dev 125(3):179–189

Giannakou ME, Partridge L (2007) Role of insulin-like signalling in Drosophila lifespan. Trends Biochem Sci 32(4):180–188

Girardot F, Lasbleiz C, Monnier V, Tricoire H (2006) Specific age-related signatures in Drosophila body parts transcriptome. BMC Genomics 7:69

Goldstein LS, Gunawardena S (2000) Flying through the drosophila cytoskeletal genome. J Cell Biol 150(2):F63–F68

Gong L, Puri M, Unlu M, Young M, Robertson K, Viswanathan S, Krishnaswamy A, Dowd SR, Minden JS (2004) Drosophila ventral furrow morphogenesis: a proteomic analysis. Development 131(3):643–656

Graham AM, Merrill JD, McGaugh SE, Noor MA (2012) Geographic selection in the small heat shock gene complex differentiating populations of Drosophila pseudoobscura. J Hered 103(3):400–407

Gregory JM, Barros TP, Meehan S, Dobson CM, Luheshi LM (2012) The aggregation and neurotoxicity of TDP-43 and its ALS-associated 25 kDa fragment are differentially affected by molecular chaperones in Drosophila. PLoS One 7(2):e31899

Grover D, Yang J, Ford D, Tavare S, Tower J (2009) Simultaneous tracking of movement and gene expression in multiple Drosophila melanogaster flies using GFP and DsRED fluorescent reporter transgenes. BMC Res Notes 2:58

Gruenewald C, Botella JA, Bayersdorfer F, Navarro JA, Schneuwly S (2009) Hyperoxia-induced neurodegeneration as a tool to identify neuroprotective genes in Drosophila melanogaster. Free Radic Biol Med 46(12):1668–1676

Habermann K, Mirgorodskaya E, Gobom J, Lehmann V, Muller H, Blumlein K, Deery MJ, Czogiel I, Erdmann C, Ralser M, von Kries JP, Lange BM (2012) Functional analysis of centrosomal kinase substrates in Drosophila melanogaster reveals a new function of the nuclear envelope component otefin in cell cycle progression. Mol Cell Biol 32(17):3554–3569

Handler D, Olivieri D, Novatchkova M, Gruber FS, Meixner K, Mechtler K, Stark A, Sachidanandam R, Brennecke J (2011) A systematic analysis of Drosophila TUDOR domain-containing proteins identifies Vreteno and the Tdrd12 family as essential primary piRNA pathway factors. EMBO J 30(19):3977–3993

Hands S, Sinadinos C, Wyttenbach A (2008) Polyglutamine gene function and dysfunction in the ageing brain. Biochim Biophys Acta 1779(8):507–521

Hao X, Zhang S, Timakov B, Zhang P (2007) The Hsp27 gene is not required for Drosophila development but its activity is associated with starvation resistance. Cell Stress Chaperones 12(4):364–372

Harvey KF, Mattila J, Sofer A, Bennett FC, Ramsey MR, Ellisen LW, Puig O, Hariharan IK (2008) FOXO-regulated transcription restricts overgrowth of Tsc mutant organs. J Cell Biol 180(4):691–696

Head MW, Goldman JE (2000) Small heat shock proteins, the cytoskeleton, and inclusion body formation. Neuropathol Appl Neurobiol 26(4):304–312

Herranz R, Larkin OJ, Dijkstra CE, Hill RJ, Anthony P, Davey MR, Eaves L, van Loon JJ, Medina FJ, Marco R (2012) Microgravity simulation by diamagnetic levitation: effects of a strong gradient magnetic field on the transcriptional profile of Drosophila melanogaster. BMC Genomics 13:52

Hirano Y, Kuriyama Y, Miyashita T, Horiuchi J, Saitoe M (2012) Reactive oxygen species are not involved in the onset of age-related memory impairment in Drosophila. Genes Brain Behav 11(1):79–86

Houtkooper RH, Mouchiroud L, Ryu D, Moullan N, Katsyuba E, Knott G, Williams RW, Auwerx J (2013) Mitonuclear protein imbalance as a conserved longevity mechanism. Nature 497(7450):451–457

Hsu AL, Murphy CT, Kenyon C (2003) Regulation of aging and age-related disease by DAF-16 and heat-shock factor. Science 300(5622):1142–1145

Huen NY, Chan HY (2005) Dynamic regulation of molecular chaperone gene expression in polyglutamine disease. Biochem Biophys Res Commun 334(4):1074–1084

Hughes JR, Meireles AM, Fisher KH, Garcia A, Antrobus PR, Wainman A, Zitzmann N, Deane C, Ohkura H, Wakefield JG (2008) A microtubule interactome: complexes with roles in cell cycle and mitosis. PLoS Biol 6(4):e98

Hull-Thompson J, Muffat J, Sanchez D, Walker DW, Benzer S, Ganfornina MD, Jasper H (2009) Control of metabolic homeostasis by stress signaling is mediated by the lipocalin NLaz. PLoS Genet 5(4):e1000460

Jensen HR, Scott IM, Sims SR, Trudeau VL, Arnason JT (2006) The effect of a synergistic concentration of a Piper nigrum extract used in conjunction with pyrethrum upon gene expression in Drosophila melanogaster. Insect Mol Biol 15(3):329–339

Jin JK, Whittaker R, Glassy MS, Barlow SB, Gottlieb RA, Glembotski CC (2008) Localization of phosphorylated alphaB-crystallin to heart mitochondria during ischemia-reperfusion. Am J Physiol Heart Circ Physiol 294(1):H337–H344

Joanisse DR, Inaguma Y, Tanguay RM (1998) Cloning and developmental expression of a nuclear ubiquitin-conjugating enzyme (DmUbc9) that interacts with small heat shock proteins in Drosophila melanogaster. Biochem Biophys Res Commun 244(1):102–109

Kallappagoudar S, Varma P, Pathak RU, Senthilkumar R, Mishra RK (2010) Nuclear matrix proteome analysis of Drosophila melanogaster. Mol Cell Proteomics 9(9):2005–2018

Kampinga HH, Garrido C (2012) HSPBs: small proteins with big implications in human disease. Int J Biochem Cell Biol 44(10):1706–1710

Kapelnikov A, Zelinger E, Gottlieb Y, Rhrissorrakrai K, Gunsalus KC, Heifetz Y (2008) Mating induces an immune response and developmental switch in the Drosophila oviduct. Proc Natl Acad Sci U S A 105(37):13912–13917

Kim H, Morrow G, Westwood TJ, Michaud S, Tanguay RM (2010) Gene expression profiling implicates OXPHOS complexes in aging and lifespan extension in flies overexpressing a small mitochondrial chaperone, Hsp22. J Exp Gerontol 45(7–8):611–620

King V, Tower J (1999) Aging-specific expression of Drosophila hsp22. Dev Biol 207(1):107–118

Kingsolver MB, Huang Z, Hardy RW (2013) Insect antiviral innate immunity: pathways, effectors, and connections. J Mol Biol 425(24):4921–4936

Kurapati R, Passananti HB, Rose MR, Tower J (2000) Increased hsp22 RNA levels in Drosophila lines genetically selected for increased longevity. J Gerontol A Biol Sci Med Sci 55(11):B552–B559

Laayouni H, Garcia-Franco F, Chavez-Sandoval BE, Trotta V, Beltran S, Corominas M, Santos M (2007) Thermal evolution of gene expression profiles in Drosophila subobscura. BMC Evol Biol 7:42

Landis GN, Abdueva D, Skvortsov D, Yang J, Rabin BE, Carrick J, Tavare S, Tower J (2004) Similar gene expression patterns characterize aging and oxidative stress in Drosophila melanogaster. Proc Natl Acad Sci U S A 101(20):7663–7668

Landis G, Shen J, Tower J (2012) Gene expression changes in response to aging compared to heat stress, oxidative stress and ionizing radiation in Drosophila melanogaster. Aging (Albany NY) 4(11):768–789

Lee KS, Choi JS, Hong SY, Son TH, Yu K (2008) Mobile phone electromagnetic radiation activates MAPK signaling and regulates viability in Drosophila. Bioelectromagnetics 29(5):371–379

Leemans R, Egger B, Loop T, Kammermeier L, He H, Hartmann B, Certa U, Hirth F, Reichert H (2000) Quantitative transcript imaging in normal and heat-shocked Drosophila embryos by using high-density oligonucleotide arrays. Proc Natl Acad Sci U S A 97(22):12138–12143

Li SS, Zhang ZY, Yang CJ, Lian HY, Cai P (2013) Gene expression and reproductive abilities of male Drosophila melanogaster subjected to ELF-EMF exposure. Mutat Res 758(1–2):95–103

Liao PC, Lin HY, Yuh CH, Yu LK, Wang HD (2008) The effect of neuronal expression of heat shock proteins 26 and 27 on lifespan, neurodegeneration, and apoptosis in Drosophila. Biochem Biophys Res Commun 376(4):637–641

Liu R, Woolner S, Johndrow JE, Metzger D, Flores A, Parkhurst SM (2008) Sisyphus, the Drosophila myosin XV homolog, traffics within filopodia transporting key sensory and adhesion cargos. Development 135(1):53–63

Loop T, Leemans R, Stiefel U, Hermida L, Egger B, Xie F, Primig M, Certa U, Fischbach KF, Reichert H, Hirth F (2004) Transcriptional signature of an adult brain tumor in Drosophila. BMC Genomics 5(1):24

Ma W, Teng Y, Hua H, Hou J, Luo T, Jiang Y (2013) Upregulation of heat shock protein 27 confers resistance to actinomycin D-induced apoptosis in cancer cells. FEBS J 280(18):4612–4624

Maaroufi H, Tanguay RM (2013) Analysis and phylogeny of small heat shock proteins from marine viruses and their cyanobacteria host. PLoS One 8(11):e81207

Marin R, Tanguay RM (1996) Stage-specific localization of the small heat shock protein Hsp27 during oogenesis in Drosophila melanogaster. Chromosoma 105(3):142–149

Marin R, Valet JP, Tanguay RM (1993) hsp23 and hsp26 exhibit distinct spatial and temporal patterns of constitutive expression in Drosophila adults. Dev Genet 14(1):69–77

Marin R, Landry J, Tanguay RM (1996) Tissue-specific posttranslational modification of the small heat shock protein HSP27 in Drosophila. Exp Cell Res 223(1):1–8

Marunouchi T, Abe Y, Murata M, Inomata S, Sanbe A, Takagi N, Tanonaka K (2013) Changes in small heat shock proteins HSPB1, HSPB5 and HSPB8 in mitochondria of the failing heart following myocardial infarction in rats. Biol Pharm Bull 36(4):529–539

Matsuda S, Vert JP, Saigo H, Ueda N, Toh H, Akutsu T (2005) A novel representation of protein sequences for prediction of subcellular location using support vector machines. Protein Sci 14(11):2804–2813

McGraw LA, Clark AG, Wolfner MF (2008) Post-mating gene expression profiles of female Drosophila melanogaster in response to time and to four male accessory gland proteins. Genetics 179(3):1395–1408

Michaud S, Tanguay RM (2003) Expression of the Hsp23 chaperone during Drosophila embryogenesis: association to distinct neural and glial lineages. BMC Dev Biol 3:9

Michaud S, Morrow G, Marchand J, Tanguay RM (2002) Drosophila small heat shock proteins: cell and organelle-specific chaperones? Prog Mol Subcell Biol 28:79–101

Michaud S, Lavoie S, Guimond MO, Tanguay RM (2008) The nuclear localization of Drosophila Hsp27 is dependent on a monopartite arginine-rich NLS and is uncoupled from its association to nuclear speckles. Biochim Biophys Acta 1783(6):1200–1210

Mileva-Seitz V, Xiao C, Seroude L, Robertson RM (2008) Tissue-specific targeting of Hsp26 has no effect on heat resistance of neural function in larval Drosophila. Cell Stress Chaperones 13(1):85–95

Morozova TV, Anholt RR, Mackay TF (2006) Transcriptional response to alcohol exposure in Drosophila melanogaster. Genome Biol 7(10):R95

Morrow G, Inaguma Y, Kato K, Tanguay RM (2000) The small heat shock protein Hsp22 of Drosophila melanogaster is a mitochondrial protein displaying oligomeric organization. J Biol Chem 275(40):31204–31210

Morrow G, Battistini S, Zhang P, Tanguay RM (2004a) Decreased lifespan in the absence of expression of the mitochondrial small heat shock protein Hsp22 in Drosophila. J Biol Chem 279(42):43382–43385

Morrow G, Samson M, Michaud S, Tanguay RM (2004b) Overexpression of the small mitochondrial Hsp22 extends Drosophila life span and increases resistance to oxidative stress. FASEB J 18(3):598–599

Morrow G, Heikkila JJ, Tanguay RM (2006) Differences in the chaperone-like activities of the four main small heat shock proteins of Drosophila melanogaster. Cell Stress Chaperones 11(1):51–60

Moskalev A, Shaposhnikov M, Turysheva E (2009) Life span alteration after irradiation in Drosophila melanogaster strains with mutations of Hsf and Hsps. Biogerontology 10(1):3–11

Muller L, Hutter S, Stamboliyska R, Saminadin-Peter SS, Stephan W, Parsch J (2011) Population transcriptomics of Drosophila melanogaster females. BMC Genomics 12:81

Murphy CT, McCarroll SA, Bargmann CI, Fraser A, Kamath RS, Ahringer J, Li H, Kenyon C (2003) Genes that act downstream of DAF-16 to influence the lifespan of Caenorhabditis elegans. Nature 424(6946):277–283

Nakagawa M, Tsujimoto N, Nakagawa H, Iwaki T, Fukumaki Y, Iwaki A (2001) Association of HSPB2, a member of the small heat shock protein family, with mitochondria. Exp Cell Res 271(1):161–168

Neal SJ, Karunanithi S, Best A, So AK, Tanguay RM, Atwood HL, Westwood JT (2006) Thermoprotection of synaptic transmission in a Drosophila heat shock factor mutant is accompanied by increased expression of Hsp83 and DnaJ-1. Physiol Genomics 25(3):493–501

Pal S, St Leger RJ, Wu LP (2007) Fungal peptide Destruxin A plays a specific role in suppressing the innate immune response in Drosophila melanogaster. J Biol Chem 282(12):8969–8977

Parcellier A, Schmitt E, Gurbuxani S, Seigneurin-Berny D, Pance A, Chantome A, Plenchette S, Khochbin S, Solary E, Garrido C (2003) HSP27 is a ubiquitin-binding protein involved in I-kappaBalpha proteasomal degradation. Mol Cell Biol 23(16):5790–5802

Qin W, Neal SJ, Robertson RM, Westwood JT, Walker VK (2005) Cold hardening and transcriptional change in Drosophila melanogaster. Insect Mol Biol 14(6):607–613

Ren J, Jegga AG, Zhang M, Deng J, Liu J, Gordon CB, Aronow BJ, Lu LJ, Zhang B, Ma J (2011) A Drosophila model of the neurodegenerative disease SCA17 reveals a role of RBP-J/Su(H) in modulating the pathological outcome. Hum Mol Genet 20(17):3424–3436

Rosenbaum EE, Brehm KS, Vasiljevic E, Liu CH, Hardie RC, Colley NJ (2011) XPORT-dependent transport of TRP and rhodopsin. Neuron 72(4):602–615

Rutherford SL, Lindquist S (1998) Hsp90 as a capacitor for morphological evolution. Nature 396(6709):336–342

Schoenherr JA, Drennan JM, Martinez JS, Chikka MR, Hall MC, Chang HC, Clemens JC (2012) Drosophila activated Cdc42 kinase has an anti-apoptotic function. PLoS Genet 8(5):e1002725

Seit-Nebi AS, Datskevich P, Gusev NB (2013) Commentary on paper: small heat shock proteins and the cytoskeleton: an essential interplay for cell integrity? (Wettstein et al.). Int J Biochem Cell Biol 45(2):344–346

Senthilan PR, Piepenbrock D, Ovezmyradov G, Nadrowski B, Bechstedt S, Pauls S, Winkler M, Mobius W, Howard J, Gopfert MC (2012) Drosophila auditory organ genes and genetic hearing defects. Cell 150(5):1042–1054

Sharma A, Mishra M, Ram KR, Kumar R, Abdin MZ, Chowdhuri DK (2011) Transcriptome analysis provides insights for understanding the adverse effects of endosulfan in Drosophila melanogaster. Chemosphere 82(3):370–376

Sharma A, Mishra M, Shukla AK, Kumar R, Abdin MZ, Chowdhuri DK (2012) Organochlorine pesticide, endosulfan induced cellular and organismal response in Drosophila melanogaster. J Hazard Mater 221–222:275–287

Shen J, Tower J (2010) Drosophila foxo acts in males to cause sexual-dimorphism in tissue-specific p53 life span effects. Exp Gerontol 45(2):97–105

Shen J, Tower J (2013) Aging, MnSOD, and hormesis mechanisms converge on liver mUPR. Cell Cycle 12(20):3237–3238
Singh MP, Reddy MM, Mathur N, Saxena DK, Chowdhuri DK (2009) Induction of hsp70, hsp60, hsp83 and hsp26 and oxidative stress markers in benzene, toluene and xylene exposed Drosophila melanogaster: role of ROS generation. Toxicol Appl Pharmacol 235(2):226–243
Singh MP, Ram KR, Mishra M, Shrivastava M, Saxena DK, Chowdhuri DK (2010) Effects of co-exposure of benzene, toluene and xylene to Drosophila melanogaster: alteration in hsp70, hsp60, hsp83, hsp26, ROS generation and oxidative stress markers. Chemosphere 79(5):577–587
Soh JW, Marowsky N, Nichols TJ, Rahman AM, Miah T, Sarao P, Khasawneh R, Unnikrishnan A, Heydari AR, Silver RB, Arking R (2013) Curcumin is an early-acting stage-specific inducer of extended functional longevity in Drosophila. Exp Gerontol 48(2):229–239
Steller H (2008) Regulation of apoptosis in Drosophila. Cell Death Differ 15(7):1132–1138
Stuart LM, Boulais J, Charriere GM, Hennessy EJ, Brunet S, Jutras I, Goyette G, Rondeau C, Letarte S, Huang H, Ye P, Morales F, Kocks C, Bader JS, Desjardins M, Ezekowitz RA (2007) A systems biology analysis of the Drosophila phagosome. Nature 445(7123):95–101
Sun X, Seeberger J, Alberico T, Wang C, Wheeler CT, Schauss AG, Zou S (2010) Acai palm fruit (Euterpe oleracea Mart.) pulp improves survival of flies on a high fat diet. Exp Gerontol 45(3):243–251
Takahashi KH, Rako L, Takano-Shimizu T, Hoffmann AA, Lee SF (2010) Effects of small Hsp genes on developmental stability and microenvironmental canalization. BMC Evol Biol 10:284
Tanguay RM, Morrow G (2008) Neuronal expression of small heat shock proteins influences longevity and resistance to oxidative stress. In: Asea AAA, Brown IR (eds) Heat shock proteins and the brain: implications for neurodegenerative diseases and neuroprotection. Springer, New York, pp 319–336
Tatar M, Kopelman A, Epstein D, Tu MP, Yin CM, Garofalo RS (2001) A mutant Drosophila insulin receptor homolog that extends life-span and impairs neuroendocrine function. Science 292(5514):107–110
Telonis-Scott M, van Heerwaarden B, Johnson TK, Hoffmann AA, Sgro CM (2013) New levels of transcriptome complexity at upper thermal limits in wild Drosophila revealed by exon expression analysis. Genetics 195(3):809–830
Tomancak P, Berman BP, Beaton A, Weiszmann R, Kwan E, Hartenstein V, Celniker SE, Rubin GM (2007) Global analysis of patterns of gene expression during Drosophila embryogenesis. Genome Biol 8(7):R145
Tower J, Landis G, Gao R, Luan A, Lee J, Sun Y (2014) Variegated expression of Hsp22 transgenic reporters indicates cell-specific patterns of aging in Drosophila oenocytes. J Gerontol A Biol Sci Med Sci 69(3):253–259
Treusch S, Hamamichi S, Goodman JL, Matlack KE, Chung CY, Baru V, Shulman JM, Parrado A, Bevis BJ, Valastyan JS, Han H, Lindhagen-Persson M, Reiman EM, Evans DA, Bennett DA, Olofsson A, DeJager PL, Tanzi RE, Caldwell KA, Caldwell GA, Lindquist S (2011) Functional links between Abeta toxicity, endocytic trafficking, and Alzheimer's disease risk factors in yeast. Science 334(6060):1241–1245
van den IJssel P, Wheelock R, Prescott A, Russell P, Quinlan RA (2003) Nuclear speckle localisation of the small heat shock protein alpha B-crystallin and its inhibition by the R120G cardiomyopathy-linked mutation. Exp Cell Res 287(2):249–261
van der Straten A, Rommel C, Dickson B, Hafen E (1997) The heat shock protein 83 (Hsp83) is required for Raf-mediated signalling in Drosophila. EMBO J 16(8):1961–1969
van Montfort RL, Basha E, Friedrich KL, Slingsby C, Vierling E (2001) Crystal structure and assembly of a eukaryotic small heat shock protein. Nat Struct Biol 8(12):1025–1030
van Noort JM, Bsibsi M, Nacken P, Gerritsen WH, Amor S (2012) The link between small heat shock proteins and the immune system. Int J Biochem Cell Biol 44(10):1670–1679

Vesala L, Salminen TS, Laiho A, Hoikkala A, Kankare M (2012) Cold tolerance and cold-induced modulation of gene expression in two Drosophila virilis group species with different distributions. Insect Mol Biol 21(1):107–118

Vos MJ (2009) Small heat shock proteins; implications for neurodegeneration and longevity. Thesis, University of Groningen, The Netherlands.

Wadhwa R, Ryu J, Gao R, Choi IK, Morrow G, Kaur K, Kim I, Kaul SC, Yun CO, Tanguay RM (2010) Proproliferative functions of Drosophila small mitochondrial heat shock protein 22 in human cells. J Biol Chem 285(6):3833–3839

Wang MC, Bohmann D, Jasper H (2003) JNK signaling confers tolerance to oxidative stress and extends lifespan in Drosophila. Dev Cell 5(5):811–816

Wang HD, Kazemi-Esfarjani P, Benzer S (2004) Multiple-stress analysis for isolation of Drosophila longevity genes. Proc Natl Acad Sci U S A 101(34):12610–12615

Wang MC, Bohmann D, Jasper H (2005) JNK extends life span and limits growth by antagonizing cellular and organism-wide responses to insulin signaling. Cell 121(1):115–125

Wang L, Colodner KJ, Feany MB (2011) Protein misfolding and oxidative stress promote glial-mediated neurodegeneration in an Alexander disease model. J Neurosci 31(8):2868–2877

Waters ER (2013) The evolution, function, structure, and expression of the plant sHSPs. J Exp Bot 64(2):391–403

Wettstein G, Bellaye PS, Micheau O, Bonniaud P (2012) Small heat shock proteins and the cytoskeleton: an essential interplay for cell integrity? Int J Biochem Cell Biol 44(10):1680–1686

Xi Z, Gavotte L, Xie Y, Dobson SL (2008) Genome-wide analysis of the interaction between the endosymbiotic bacterium Wolbachia and its Drosophila host. BMC Genomics 9:1

Yang J, Tower J (2009) Expression of hsp22 and hsp70 transgenes is partially predictive of drosophila survival under normal and stress conditions. J Gerontol A Biol Sci Med Sci 64(8):828–838

Yepiskoposyan H, Egli D, Fergestad T, Selvaraj A, Treiber C, Multhaup G, Georgiev O, Schaffner W (2006) Transcriptome response to heavy metal stress in Drosophila reveals a new zinc transporter that confers resistance to zinc. Nucleic Acids Res 34(17):4866–4877

Yu CS, Chen YC, Lu CH, Hwang JK (2006) Prediction of protein subcellular localization. Proteins 64(3):643–651

Zamparini AL, Davis MY, Malone CD, Vieira E, Zavadil J, Sachidanandam R, Hannon GJ, Lehmann R (2011) Vreteno, a gonad-specific protein, is essential for germline development and primary piRNA biogenesis in Drosophila. Development 138(18):4039–4050

Zhai B, Villen J, Beausoleil SA, Mintseris J, Gygi SP (2008) Phosphoproteome analysis of Drosophila melanogaster embryos. J Proteome Res 7(4):1675–1682

Zhang D, Ke L, Mackovicova K, Van Der Want JJ, Sibon OC, Tanguay RM, Morrow G, Henning RH, Kampinga HH, Brundel BJ (2011a) Effects of different small HSPB members on contractile dysfunction and structural changes in a Drosophila melanogaster model for Atrial Fibrillation. J Mol Cell Cardiol 51(3):381–389

Zhang J, Marshall KE, Westwood JT, Clark MS, Sinclair BJ (2011b) Divergent transcriptomic responses to repeated and single cold exposures in Drosophila melanogaster. J Exp Biol 214(Pt 23):4021–4029

Zhao Y, Sun H, Lu J, Li X, Chen X, Tao D, Huang W, Huang B (2005) Lifespan extension and elevated hsp gene expression in Drosophila caused by histone deacetylase inhibitors. J Exp Biol 208(Pt 4):697–705

Zimmermann G, Furlong EE, Suyama K, Scott MP (2006) Mes2, a MADF-containing transcription factor essential for Drosophila development. Dev Dyn 235(12):3387–3395

Zou S, Meadows S, Sharp L, Jan LY, Jan YN (2000) Genome-wide study of aging and oxidative stress response in Drosophila melanogaster. Proc Natl Acad Sci U S A 97(25):13726–13731

Index

R.M. Tanguay, L.E. Hightower (eds.), *The Big Book on Small Heat Shock Proteins*, Heat Shock Proteins 8, DOI 10.1007/978-3-319-16077-1

Kolibri 360 GmbH
Ferdinand-Jühlke-Straße 7
99095 Erfurt, Deutschland
produktsicherheit@kolibri360.de